Springer Texts in Business and Economics

More information about this series at http://www.springer.com/series/10099

Giancarlo Gandolfo

International Finance and Open-Economy Macroeconomics

Second Edition

With Contributions by Daniela Federici

 Springer

Giancarlo Gandolfo
Classe di Scienze Morali
Accademia Nazionale dei Lincei
Rome, Italy

ISSN 2192-4333   ISSN 2192-4341   (electronic)
Springer Texts in Business and Economics
ISBN 978-3-662-49860-6   ISBN 978-3-662-49862-0   (eBook)
DOI 10.1007/978-3-662-49862-0

Library of Congress Control Number: 2016945040

Printed on acid-free paper

This Springer imprint is published by Springer Nature
The registered company is Springer-Verlag GmbH Berlin Heidelberg

*To the memory of my father
Edgardo Gandolfo*

Preface

Following standard practice in international economics, I have always treated the theory of international trade separately from international finance, thus writing my *International Economics* textbook as a two-volume course (Vol. 1 on the theory of international trade and Vol. 2 on international monetary theory). This text had two editions, several reprints, and translations in other languages. However, the different pace of the revisions of the two volumes suggested to make them self-contained and independent from each other. The volume covering trade was published in 1998 (second edition 2014) under the title *International Trade Theory and Policy*. This is the second edition of the volume covering international finance (the first edition was published in 2001). This new edition contains a wealth of additional material that has been introduced thanks to the suggestions of colleagues and students and to the comments contained in book reviews. All has been thoroughly classroom tested in both undergraduate and graduate courses in various universities in Italy and other countries.

In the Preface to the first edition (1986) of *International Economics*, I wrote:

There is no lack of good international economics textbooks ranging from the elementary to the advanced, so that an additional drop in this ocean calls for an explanation. In the present writer's opinion, there seems still to be room for a textbook which can be used in both undergraduate and graduate courses, and which contains a wide range of topics, including those usually omitted from other textbooks. These are the intentions behind the present book, which is an outcrop from undergraduate and graduate courses in international economics that the author has been holding at the University of Rome 'La Sapienza' since 1974, and from his ongoing research work in this field.

Accordingly, the work is organised as two-books-in-one by distributing the material between text and appendices.

The treatment in the text is directed to undergraduate students and is mainly confined to graphic analysis and to some elementary algebra, but it is assumed that the reader will have a basic knowledge of economics. Each chapter has a mathematical appendix, where (i) the topics treated in the text are examined at a level suitable for advanced undergraduate or first-year graduate students, and (ii) generalizations and/or topics not treated in the text (including some at the frontiers of research) are formally examined.

The text is self-contained, and the appendices can be read independently of the text and can, therefore, also be used by students who already know 'graphic' international economics and want to learn something about its mathematical counterpart. Of course the connections between text and appendices are carefully indicated, so that the latter can be used as mathematical appendices by the student who has mastered the text, and the text can be used as graphic and literary exposition of the results derived mathematically in the appendices by the student who has mastered these. [...]

The present book maintains the same approach, in particular the unique two-tier feature and the ample coverage of topics, including many at the frontiers of research, whose often obscure mathematical aspects are fully clarified in the second tier. This new edition has been thoroughly revised and enriched thanks to the numerous contributions by Professor Daniela Federici that bring the book up-to-date.

I am grateful to the students from all over the world who have written me over the years to indicate unclear points and misprints and to Guido Ascari, Marianna Belloc, Andrea Bubula, Nicola Cetorelli, Giuseppe De Arcangelis, Vivek H. Dehejia, Kieran P. Donaghy, Michele Gambera, the late Carlo Giannini, Bernardo Maggi, Giovanna Paladino, Luca Ricci, Francesca Sanna Randaccio, the late Jerome L. Stein, for their advice and comments.

None of the persons mentioned has any responsibility for possible deficiencies that might remain.

Rome, Italy   Giancarlo Gandolfo
January 2016

Contents

List of Figures

List of Tables

Part I

Introduction

International Finance and International Macroeconomics: An Overview

While several specialistic fields of economics have been developed as distinct branches of general economic theory only in relatively recent times, the presence of a specific treatment of the theory of international economic transactions is an old and consolidated tradition in the economic literature. Various reasons can be advanced to explain the need for this specific treatment, but the main ones are the following.

The first is that factors of production are generally less mobile between countries than within a single country. Traditionally, this observation has been taken as a starting point for the development of a theory of international trade based on the extreme assumption of perfect national mobility and perfect international immobility of the factors of production, accompanied by the assumption of perfect mobility (both within and between countries) of the commodities produced, exception being made for possible restrictive measures on the part of governments.

The second is the fact that the mere presence of different countries as distinct political entities each with its own frontiers gives rise to a series of problems which do not occur in general economics, such as the levying of duties and other impediments to trade, the existence of different national currencies whose relative prices (the exchange rates) possibly vary through time, etc.

The specialistic nature of international economics—a discipline of increasing importance given the increasing openness and interdependence of the single national economic systems—does *not* mean that its methods and tools of analysis are different from those of general economic theory: on the contrary, international economics makes ample use of the methods and tools of microeconomics and macroeconomics, as we shall see presently.

As in any other discipline, also in international economics we can distinguish a *theoretical* and a *descriptive* part. The former is further divided into the *theory of international trade* and *international monetary economics*. All these distinctions are of a logical and pedagogical nature, but of course both the descriptive and

© Springer-Verlag Berlin Heidelberg 2016
G. Gandolfo, *International Finance and Open-Economy Macroeconomics*,
Springer Texts in Business and Economics, DOI 10.1007/978-3-662-49862-0_1

the theoretical part, both the trade and the monetary branch, are necessary for an understanding of the international economic relations in the real world.

The descriptive part, as the name clearly shows, is concerned with the description of international economic transactions just as they happen and of the institutional context in which they take place: flows of goods and financial assets, international agreements, international organizations like the World Trade Organization and the European Union, and so forth.

The theoretical part tries to go beyond the phenomena to seek general principles and logical frameworks which can serve as a guide to the understanding of actual events (so as, possibly, to influence them through policy interventions). Like any economic theory, it uses for this purpose abstractions and models, often expressed in mathematical form. The theoretical part can be further divided, as we said above, into trade and monetary theory each containing aspects of both *positive* and *normative* economics; although these aspects are strictly intertwined in our discipline, they are usually presented separately for didactic convenience.

A few words are now in order on the distinction between international trade theory and international monetary economics.

The *theory of international trade* (which has an essentially microeconomic nature) deals with the causes, the structure and the volume of international trade (that is, which goods are exported, which are imported, and why, by each country, and what is their amount); with the gains from international trade and how these gains are distributed; with the determination of the relative prices of goods in the world economy; with international specialization; with the effects of tariffs, quotas and other impediments to trade; with the effects of international trade on the domestic structure of production and consumption; with the effects of domestic economic growth on international trade and vice versa; and so on. The distinctive feature of the theory of international trade is the assumption that trade takes place in the form of *barter* (or that money, if present, is only a veil having no influence on the underlying real variables but serving only as a reference unit, the *numéraire*). A by-no-means secondary consequence of this assumption is that the international accounts of any country vis-à-vis all the others always balance: that is, no balance-of-payments problem exists.

This part of international economics was once also called the *pure* theory of international trade, where the adjective "pure" was meant to distinguish it from *monetary* international economics.

International finance, also called *international monetary economics* (which is essentially of a macroeconomic nature) is often identified with *open-economy macroeconomics* or *international macroeconomics* because it deals with the monetary and macroeconomic relations between countries. Although there are nuances in the meaning of the various labels, we shall ignore them and take it that our field deals with the problems deriving from balance-of-payments disequilibria in a monetary economy, and in particular with the automatic adjustment mechanisms and the adjustment policies concerning the balance of payments; with the relationships between the balance of payments and other macroeconomic variables; with the various exchange-rate regimes and exchange-rate determination; with international financial markets and the problems of the international monetary systems such as currency crises, debt problems, international policy coordination; with international monetary integration such as the European Monetary Union. deals with the problems deriving from balance-of-payments disequilibria in a monetary economy, and in particular with the automatic adjustment mechanisms and the adjustment policies of the balance of payments; with the relationships between the balance of payments and other macroeconomic variables; with the various exchange-rate regimes; with the problems of international liquidity and other problems of the international monetary system; etc.

In this book we shall treat the theory of international finance. A companion volume (Gandolfo 2014) treats international trade theory and policy, thus following the standard practice of international textbooks and courses.

One last word: in this work we shall be concerned mainly with the theoretical part (both positive and normative) of international finance, even if references to the real world will not be lacking. Thanks to the advances in econometrics and computer power, practically all theories of international macroeconomics have been subjected to a great number of empirical tests. As it would not be possible to consider all these tests, it was necessary to make occasionally arbitrary choices, though we feel that the most important empirical studies have been treated. In any case, where no treatment is given, we have referred the reader to the relevant empirical literature.

1.1   Globalization

Nowadays international economics is intimately linked with globalization. "Globalization" is a much used and abused word. According to The American Heritage®️ Dictionary of the English Language (©️ 2009 by Houghton Mifflin Company), to globalize means "To make global or worldwide in scope or application." In the field of international economics, globalization means different things to different people (see, for example, Gupta ed., 1997, Stern 2009). Some authors (see, for example, Dreher 2006, Dreher et al. 2008) have also suggested indexes to measure the degree of globalization of the various countries, taking into account the three main dimensions of globalization (economic, social, and political). These indexes are available at http://globalization.kof.ethz.ch/.

A by no means exhaustive list of the elements that make up globalization is:

(a) the increase in the share of international and transnational transactions, as measured for example by the share of world trade and world direct investment (carried out by multinational corporations) in world GNP;
(b) the integration of world markets, as measured for example by the convergence of prices and the consequent elimination of arbitrage opportunities;
(c) the growth of international transactions and organizations having a non-economic but political, cultural, social nature;

(d) an increasing awareness of the importance of common global problems (the environment, infectious diseases, the presence of international markets which are beyond the control of any single nation, etc.,

(e) the tendency to eliminate national differences and to an increasing uniformity of cultures and institutions.

The debate on globalization usually considers the following aspects:

(1) the actual degree of integration of markets;

(2) globalization as a process that undermines the sovereignty of the single states, reducing their autonomy in policy making and increasing the power of multinational corporations;

(3) the effects of globalization on world income distribution, both within and across countries;

(4) the possible development of an international government to cope with global problems.

1.2   Old and New Approaches to International Finance

In the field of international monetary economics we can distinguish two different views. On the one hand there is the "old" or traditional view, which considers the balance of payments as a phenomenon to be studied as such, by studying the specific determinants of trade and financial flows. On the other hand there is the "new" or modern view, that considers trade and financial flows as the outcome of intertemporally optimal saving-investment decisions by forward-looking agents.

The new international macroeconomic theory, following the steps of the new closed-economy macroeconomics, has reached a high degree of sophistication where new models are built based on microfoundations, and the open-economy macroeconomic behaviour is obtained as the outcome of intertemporal optimization (possibly in a stochastic environment) by a representative agent (Obstfeld and Rogoff 1996; Turnovsky 1997). More precisely, the new international macroeconomic theory is based on models where:

(a) the aggregate outcome of the choices of all the diverse agents in the macroeconomy can be considered as the choices of one "representative" individual, an assumption that may prove seriously wrong (Kirman 1992);

(b) agents can predict, or know the stochastic properties of, economic variables over their lifetime. "A Ramsey maximizer needs correct expectations out to infinity if there is to be a unique optimum path. It seems to us that no macroeconomic insights can be gained with so extravagant an expectational hypothesis" (Hahn and Solow 1995; p. 140);

(c) intertemporal optimizing models have the saddle path property, namely there is a unique trajectory (the so-called stable arm of the saddle) that converges to equilibrium, all other trajectories being unstable. This requires economic agents to possess the perfect knowledge required to put them on the stable arm of the saddle without error at the initial time, because the slightest error in implementing this stable arm will put the system on a trajectory that will diverge from the optimal steady state;

(d) economic data are known with infinite precision, because in the contrary case, given a nonlinear model, a slightly different value in the initial data might give rise to widely different results in certain cases (the chaos problem).

Further examination of the assumptions of the new international macroeconomics is contained in, e.g., Krugman (2000), Phelps (2007), Solow (2000), but this is not our main interest. In fact, one of the aims of this book is to introduce, alongside with a detailed treatment of the traditional models and their policy implications, some aspects of the new international macroeconomics that we think are most useful to cope with real-life problems.

1.3 Structure of the Book

We shall start (Part II) with the basics, which are theory-independent, such as the market for foreign exchange (including currency derivatives and euromarkets) and the various exchange-rate regimes, the international interest-rate parity conditions, the definition and accounting rules of the balance of payments, the main international organizations. In this part there is also a detailed treatment of the relationships between the balance of payments and the other macroeconomic (real and financial) variables from an accounting point of view, which is indispensable to fit the balance of payments in the context of the whole economic system and to illustrate the meaning of several widely used identities (for example, that the excess of national saving over national investment equals the country's current account). The accounting framework that we propose will render us invaluable services in the course of the examination of the various models.

In treating the traditional approaches to balance-of-payments adjustment (which include the elasticity and multiplier approaches, the Mundell-Fleming model, the monetary approach to the balance of payments, the portfolio-balance model, the relations between growth and the balance of payments) we shall distinguish between flow approaches (Part III) and stock and stock-flow approaches (Part IV), because strikingly different results may occur according as we take the macroeconomic variables involved as being pure flows, or as flows only deriving from stock adjustments, or a mixture of both.

To clarify this distinction let us consider, for example, the flow of imports of consumption commodities, which is part of the flow of national consumption. Suppose that the agent decides how much to spend for current consumption by simply looking at the current flow of income, ceteris paribus. This determines imports as a pure flow. Suppose, on the contrary, that the agent first calculates the desired stock of wealth (based on *current* values of interest rates, income, etc.) and then, looking at the existing stock, decides to adjust the latter toward the former, thus determining a flow of saving (or dissaving) and consequently the flow of consumption. This determines imports as a flow deriving from a stock adjustment.

The exchange-rate will be exhaustively examined in Part V, including the various theories for its determination, foreign exchange speculation and currency crises, the debate on fixed versus flexible exchange rates.

The modern intertemporal approach is examined in Part VI, showing the importance of the saving-investment decisions taken by *forward-looking* optimizing agents in the determination of the international flows of commodities and financial capital. The old absorption approach, commonly examined in the context of the traditional theory, is explained here, because it can be seen as the precursor of the modern approach. The intertemporal approach throws new light on the issues previously examined in the context of the traditional approach, such as current account disequilibria, macroeconomic policy, growth, exchange rate determination.

Part VII deals with the theory of international monetary integration, from simple (optimum) currency areas to full-blown monetary unions with a common currency such as the European Monetary Union.

Finally, the key problems of the international monetary system, from debt problems to international policy coordination, will be examined in Part VIII.

1.4   Small and Large Open Economies

We shall use both *one-country* and *two-country* models. With the expression *one-country* or *small-country* model (also called SOE, *small open economy*) we refer to a model in which the rest of the world is taken as exogenous, in the sense that what happens in the country under consideration (call it country 1) is assumed to have a negligible influence (since this country is small relative to the rest of the world) on the rest-of-the-world variables (income, price level, interest rate, etc.). This means that these variables can be taken as *exogenous* in the model.

With the expression *two-country* or *large country* model we refer to a model in which the effect on the rest-of-the-world's variables of country 1's actions cannot be neglected, so that the rest of the world has to be explicitly included in the analysis (as country 2). It follows that, through the channels of exports and imports of goods and services, and capital movements, the economic events taking place in a country have repercussions on the other country, and vice versa.

Two-country models may seen more realistic, as in the real world inter-country repercussions do take place. However, in such models the various countries making up the rest of the world are assumedly aggregated into a single whole (country 2), which is not necessarily more realistic. In fact, if the world is made up of *n* interdependent countries which interact more and more with one another (*globalization* is the fashionable word for this increasing interdependence and interaction), dealing with it as if it were a two-country world is not necessarily better than using the SOE assumption as a first approximation. These problems can be overcome by the construction of *n*-country models, which will be examined in the relevant Appendixes, given their degree of mathematical difficulty. These will often show that results are not qualitatively different from those obtained with the first approximation.

References

Dreher, A. (2006). Does globalization affect growth? Evidence from a new index of globalization. *Applied Economics, 38*(10), 1091–1110.

Dreher, A., Gaston, N., & Martens, P. (2008). *Measuring globalization – gauging its consequences.* New York: Springer.

Gandolfo, G. (2014). *International trade theory and policy* (2nd ed.). Berlin, Heidelberg, New York: Springer.

Gupta, S. D. (Ed.). (1997). *The political economy of globalization.* Boston/Dordrecht/London: Kluwer Academic Publishers.

Hahn, F., & Solow, R. (1995). *A critical essay on modern macroeconomic theory.* Oxford: Blackwell.

Kirman, A. P. (1992). Whom or what does the representative individual represent? *Journal of Economic Perspectives, 6*, 117–136.

Krugman, P. (2000). How complicated does the model have to be? *Oxford Review of Economic Policy, 16*, 33–42.

Obstfeld, M., & Rogoff, K. (1996). *Foundations of international macroeconomics.* Cambridge (Mass): MIT Press.

Phelps, E. S. (2007). Macroeconomics for a modern economy. *American Economic Review, 97*, 543–561.

Solow, R. M. (2000). Towards a macroeconomics of the medium run. *Journal of Economic Perspectives, 14*, 151–158.

Stern, R. M. (2009). *Globalization and international trade policies* (Chap. 2, Sect. II). Singapore: World Scientific.

Turnovsky, S. (1997). *International macroeconomic dynamics.* Cambridge (Mass): MIT Press.

Part II

The Basics

The Foreign Exchange Market

2

2.1 Introduction

As soon as one comes to grips with the actual problems of international monetary economics it becomes indispensable to account for the fact that virtually every country (or group of countries forming a monetary union) has its own monetary unit (currency) and that most international trade is not barter trade but is carried out by exchanging goods for one or another currency. Besides, there are international economic transactions[1] of a purely financial character, which, therefore, involve different currencies.

From all the above the necessity arises of a *foreign exchange market*, that is, of a market where the various national currencies can be exchanged (bought an sold) for one another. The foreign exchange market, like any other concept of market used in economic theory, is not a precise physical place. It is actually formed apart from institutional characteristics which we shall not go into by banks, brokers and other authorized agents (who are linked by telephone, telex, computer, etc.), to whom economic agents apply to buy and sell the various currencies; thus it has an international rather than national dimension, although in the actual quotation of the foreign exchange rates it is customary to refer to typical places (financial centres) such as New York, London, Paris, Zurich, Milan, Tokyo, Frankfurt, etc.

We must now define the concept of (*foreign*) *exchange rate*. It is a price, and to be exact the price of one currency in terms of another. Since two currencies are

---

[1] An economic transaction is the transfer from one economic agent (individual, corporate body, etc.) of title to a good, the rendering of a service, or title to assets, having economic value. It includes both real transfers (i.e., transfer of title to goods or rendering of services) and financial transfers (i.e. of money, credits and financial assets in general). Transactions can either involve a payment, whatever the form (bilateral transfers) or not (unilateral transfers, such as gifts). As we shall see in detail in the next chapter, what qualifies an economic transaction as international is the fact that the parties are residents of different countries.

© Springer-Verlag Berlin Heidelberg 2016
G. Gandolfo, *International Finance and Open-Economy Macroeconomics*,
Springer Texts in Business and Economics, DOI 10.1007/978-3-662-49862-0_2

involved, there are two different ways of giving the quotation of foreign exchange. One is called the *price quotation system*, and defines the exchange rate as the *number of units of domestic currency per unit of foreign currency* (taking the USA as the home country, we have, say, $1.423 per British pound, $0.00844 per Japanese yen, $1.082 per euro, etc.); this amounts to defining the exchange rate as the *price of foreign currency in terms of domestic currency*.

The other one is called the *volume quotation system*, and defines the exchange rate as the *number of units of foreign currency per unit of domestic currency* and is, obviously, the reciprocal of the previous one (again taking the USA as the home country, we would have 0.702 British pounds per US dollar, 118.536 Japanese yens per US dollar, 0.923 euros per US dollar, etc.); with this definition the exchange rate is the *price of domestic currency in terms of foreign currency*.

We shall adopt the price quotation system, and a good piece of advice to the reader is always to ascertain which definition is being (explicitly or implicitly) adopted, to avoid confusion. The same concept, in fact, will be expressed in opposite ways according to the definition used. Let us consider for example the concept of "*depreciation of currency x*" (an equivalent expression from the point of view of country x is an "*exchange rate depreciation*"). This means that currency x is worth *less* in terms of foreign currency, namely that a greater amount of currency x is required to buy one unit of foreign currency, and, conversely, that a lower amount of foreign currency is required to buy one unit of currency x. Therefore the concept of depreciation of currency x is expressed as an *increase* in its exchange rate if we use the price quotation system (to continue with the example of the US dollar, we have, say, $1.45 instead of $1.423 per British pound, etc.) and as a *decrease* in its exchange rate, if we use the volume quotation system (0.689 instead of 0.702 British pounds per US dollar, etc.). By the same token, expressions such as a fall in the exchange rate or currency x is falling are ambiguous if the definition used is not specified. Similar observations hold as regards an *exchange rate appreciation* (currency x is worth *more* in terms of foreign currency).

Since many important books and articles in the field of exchange rates have been written and are being written by authors using the volume quotation system and by other authors using the price quotation system, the danger of confusion is not to be underrated.

It should now be pointed out that, as there are various monetary instruments which can be used to exchange two currencies, the respective exchange rate may be different: the exchange rate for cash, for example, may be different from that for cheques and from that for bank transfer orders.

These differences depend on various elements, such as the costs of transferring funds, the carrying costs in a broad sense (if bank keeps foreign currency in the form of banknotes in its vaults rather than in the form of demand deposits with a foreign bank, it not only loses interest but also has to bear custody costs). These differences are however very slight, so that henceforth we shall argue as if there were only one exchange rate for each foreign currency, thus also neglecting the *bid-offer spread*, that is, the spread which exists at the same moment and for the same monetary instrument, between the buying and selling price of the same currency in the foreign exchange market.

To conclude this introductory section, we must explain another difference: that between the *spot* and the *forward* exchange rate. The former is that applied to the exchange of two currencies on the spot, that is, for immediate delivery. In practice the currencies do not materially pass from hand to hand except in certain cases, such as the exchange of banknotes; what usually takes place is the exchange of drawings on demand deposits[2] denominated in the two currencies.

The *forward* exchange rate is that applied to the agreement for a future exchange of two currencies at an agreed date (for instance, in three months' time). In other words, we are in the presence of a contract which stipulates the exchange of two currencies at a prescribed future date but at a price (the forward exchange rate) which is fixed in advance (as is the amount) at the moment of the stipulation of the contract. When the contract expires, or, to be exact, two days before the expiry date, it becomes automatically a spot contract, but of course the price remains that fixed at the moment of the stipulation.

The forward exchange rate is quoted for various delivery dates (one week; 1-, 3-, 6-months; etc.; rarely for more than 1 year ahead) and for the main currencies: not all currencies, in fact, have a forward market. The spot and the forward market together constitute the *foreign exchange market*.

Since the exchange rate is, as we have seen, a price which is quoted on a market, the problem comes immediately to mind of whether the exchange rate (spot and forward) is determined in accordance with the law of supply and demand, much as the price of a commodity is determined on the relative market. The problem is very complicated, as it involves the whole of international monetary theory and also depends on the institutional setting; therefore we shall deal with it later, after having introduced the necessary notions (for a general treatment see Chap. 15).

2.2   The Spot Exchange Market

Given n currencies, $n-1$ bilateral (spot) exchange rates of each one vis-à-vis all the others will be defined, thus $n(n-1)$ exchange rates in total. The spot exchange market, however, by way of the so-called *arbitrage* on currencies, enables one to determine all the exchange rates by knowing only $(n-1)$ of them. In other words, arbitrage succeeds in causing actual exchange rates to practically coincide with the values which satisfy certain simple mathematical relations which, from the theoretical point of view, exist between them. Arbitrage on foreign currencies can be defined as the simultaneous buying and selling of foreign currencies to profit

---

[2]To avoid confusion, the reader should note that *demand deposit* is here taken to mean a deposit with a bank from which money can be drawn without previous notice and on which cheques can be drawn (synonyms in various countries are: current account deposit, checking deposit, sight deposit).

from discrepancies between exchange rates existing at the same moment in different financial centres.

Let us first consider the mathematical relations and then the arbitrage activity.

To begin with, the exchange rate of currency i for currency j and the exchange rate of currency j for currency i (of course expressed using the same quotation system) are—theoretically—the reciprocal of each other: this enables us to reduce the exchange rates from $n(n-1)$ to $n(n-1)/2$. If we denote by r_{ji} the exchange rate of currency i $(i = 1, 2, \ldots, n)$ with respect to currency j $(j = 1, 2, \ldots, n; j \neq i)$ in the ith financial centre, that is, given the definition adopted, the number of units of currency i exchanged for one unit of currency j[3](price of currency j in terms of currency i), the *consistency condition* requires that

$$r_{sk}r_{ks} = 1, \tag{2.1}$$

where k and s are any two currencies. In fact, the consistency condition (also called *neutrality condition*) means that by starting with any given quantity of currency k, exchanging it for currency s and then exchanging the resulting amount of currency s for currency k, one must end up with the same initial quantity of currency k. More precisely, starting with x units of currency k and selling it in financial centre s for currency s we obtain xr_{ks} units of currency s; if we then sell this amount of currency s in financial centre k for currency k we end up with $(xr_{ks}) r_{sk}$ units of currency k. The consistency condition $x = (xr_{ks}) r_{sk}$ must therefore hold; if we divide through by x and rearrange terms we obtain Eq. (2.1). From this equation it immediately follows that

$$r_{sk} = 1/r_{ks}, \qquad r_{ks} = 1/r_{sk}, \tag{2.2}$$

which is our initial statement.

If, for example, the exchange rate between the yen (¥) and the US dollar ($) is 118.563 in Tokyo (¥118.563 per $1), mathematically the $/¥ exchange rate in New York is 0.00843 ($0.00843 per Japanese yen). What ensures that the exchange rate between the two currencies in New York is 0.00843—given the exchange rate of 118.563 between them in Tokyo—is indeed arbitrage, which in such cases is called two-point arbitrage, as two financial centres are involved. Let us assume, for example, that while the ¥/$ exchange rate in Tokyo is 118.563, the exchange rate in New York is $0.009 for ¥1. Then the arbitrageur can buy ¥ with $ in Tokyo and sell them for $ in New York, thus obtaining a profit of $0.00057 per ¥, which is the difference between the selling ($0.009) and buying ($0.00843) dollar price of one ¥.

---

[3]The reader should note that the order of the subscripts is merely conventional, so that many authors (as here) use r_{ji} to denote the price of currency j in terms of currency i, whereas others follow the reverse order and use r_{ij} to denote the same concept. It is therefore important for the reader to carefully check which convention is adopted.

It should also be noted that, since everything occurs almost instantaneously and simultaneously on the computer, telephone, telex, or other such means of communication, this arbitrage does not tie up capital, so that no cost of financing is involved and, also, no exchange risk is incurred; the cost is the fee for the utilization of the telephone or other lines.

In this way opposite pressures are put on the yen in Tokyo and in New York. The additional demand for yen (supply of dollars) in Tokyo brings about an appreciation of the yen with respect to the dollar there, and the additional supply of yens (demand for dollars) in New York brings about a depreciation of the yen with respect to the dollar there, that is, an appreciation of the dollar with respect to the yen. This continues as long as arbitrage is no longer profitable, that is, when the exchange rates between the two currencies in the two financial centres have been brought to the point where they satisfy the condition of neutrality.

In practice this condition is never exactly satisfied, because of possible friction and time-lags (such as, for example, transaction costs, different business hours, different time zones, etc.), but in normal times the discrepancies are so small as to be negligible for all purposes.

After examining the relations between the bilateral exchanges rates, we must now introduce the notion of *indirect* or *cross (exchange) rate*. The cross rate of currency i with respect to currency j indicates how many units of currency exchange *indirectly* (this is, through the purchase and sale of a third currency, m) for one unit of currency j. More precisely, with one unit of currency j one can purchase r_{jm} units of currency m in financial centre j; by selling this amount of currency m for currency i in financial centre i at the exchange rate r_{mi}, one obtains $r_{jm}r_{mi}$ units of currency i. The indirect rate between currency i and currency j is thus $r_{jm}r_{mi}$. The consistency (or neutrality) condition obviously requires that the indirect and direct rates should be equal, and as the direct rate of currency i with respect to currency j is r_{ji}, the mathematical relation which must hold is

$$r_{ji} = r_{jm}r_{mi}, \tag{2.3}$$

for any triplet of (different) indexes i, j, m. This condition can also be written—recalling that, from (2.2), we have $r_{ji} = 1/r_{ij}, r_{mi} = 1/r_{im}$—as

$$r_{ji}r_{im}r_{mj} = 1 \text{ or } r_{ij}r_{jm}r_{mi} = 1. \tag{2.4}$$

If, for example, the US dollar/euro rate in new York is 1.082 dollars per one euro, and the yen/dollar exchange rate in Tokyo is 118.563 yen per one dollar in Tokyo, then the euro/yen cross rate in Tokyo is $118.563 \times 1.082 = 128.285$ yen per one euro. It is still arbitrage, this time in the form of *three-point* or *triangular arbitrage* (as three currencies are involved), which *equalizes the direct and indirect exchange rate*.

The considerations made above on the almost instantaneousness and negligible cost of the various operations also explain why these will continue until the direct and indirect exchange rates are brought into line, so as to cause the profit to disappear. This will, of course, occur when, and only when, the direct exchange rate between any two currencies coincides with all the possible cross rates between them. In practice this equalization is never perfect, for the same reasons as in the case of two-point arbitrage, but here too we can ignore these discrepancies.

It can readily be checked that the cross rates between any pair of currencies (i, j) are $n - 2$: in fact, as there are n currencies, it is possible to exchange currencies i and j indirectly through any one of the other $(n - 2)$ currencies. And, since all these cross rates must equal the only direct rate between currencies i and j, it can easily be shown that it is sufficient to know the $n - 1$ direct rates of one currency vis-à-vis all the others to be able to determine the full set of (direct) exchange rates among the n currencies. Let us in fact assume that we know the $n - 1$ direct rates of one currency, say currency 1, vis-à-vis all the others: that is, we know the rates $r_{21}, r_{31}, \ldots, r_{n1}$. From Eq. (2.3) we have, letting $i = 1$,

$$r_{j1} = r_{jm}r_{m1},   \tag{2.5}$$

for any pair of different subscripts j, m. From Eq. (2.5) we immediately get

$$r_{jm} = r_{j1}/r_{m1},   \tag{2.6}$$

whence, account being taken of Eq. (2.3),

$$r_{mj} = 1/r_{jm} = r_{m1}/r_{j1}.   \tag{2.7}$$

Now, since the rates r_{j1} and r_{m1} are known by assumption, from Eqs. (2.6) and (2.7) it is possible to determine all the direct exchange rates between all pairs of currencies (m, j) and therefore the full set of bilateral exchange rates. This completes the proof of the statement made at the beginning of this section.

2.3   The Real Exchange Rate

In general, real magnitudes are obtained from the corresponding nominal magnitudes eliminating the changes solely due to price changes, which can be done in a variety of ways. In the case of exchange rates the question is however more complicated, due to the fact that the exchange rate is intrinsically a *nominal* concept, which is *not* obtained (as is instead the case with nominal income or other nominal magnitudes which have a clear-cut price/quantity decomposition) multiplying a physical quantity by its price. The real exchange rate, like the nominal one, is a relative price, but there is no agreement on which relative price should be called "the" real exchange rate, since currently there are several definitions, a few of which are given here (for surveys see, e.g., Lipschitz and McDonald 1992; Marsh and Tokarick 1996; Hinkle and Montiel 1999; Chinn 2006).

The oldest notion of real exchange rate, and the one which is often (incorrectly) identified with "the" real exchange rate, is probably the ratio of the general price

levels at home and abroad expressed in a common monetary unit, or the nominal
exchange rate adjusted for relative prices between the countries under consideration:

$$r_R = \frac{p_h}{r p_f},\qquad(2.8)$$

or,

$$r_R = \frac{\frac{1}{r}p_h}{p_f},\qquad(2.9)$$

where p_h, p_f are the domestic and foreign price levels in the respective currencies.
This definition is obviously linked to PPP (Purchasing Power Parity, see Chap. 15,
Sect. 15.1). From the economic point of view, it is easy to see that in (2.8)
domestic and foreign prices have been made homogeneous by expressing the latter
in domestic currency before taking their ratio, whilst in (2.9) they have been made
homogeneous by expressing the former in foreign currency; the ratio is of course
the same.

According to another definition, the real exchange rate is the (domestic) relative
price of tradable and nontradable goods,

$$r_R = \frac{p^T}{p^{NT}},\qquad(2.10)$$

or, alternatively

$$r_R = \frac{p^{NT}}{p^T}.\qquad(2.11)$$

The rationale of this definition is that, in a two-sector (tradables-nontradables)
model, the balance of trade depends on p_T/p_{NT} because this relative price measures
the opportunity cost of domestically producing tradable goods, and the ex ante
balance of trade depends on the ex ante excess supply of tradables. According to
the Harrod-Balassa-Samuelson hypothesis (see Chap. 15, Sect. 15.1.1), the long run
PPP holds only for traded goods, and the real exchange rate in the long run is a
function of the relative productivity of traded to non-traded goods in the home and
foreign countries. Definition (2.11) means that a rise is an appreciation, a convention
often adopted in the field of real exchange rates (contrary to what happens with the
price quotation system normally used for nominal exchange rates).

A widely held opinion is that the real exchange rate should give a measure of the
external competitiveness of a country's goods (if non-traded goods are also present,
only tradables should be considered), but even if we so restrict the definition, it
is by no means obvious which index should be taken. In the simple exportables-
importables model of trade the real exchange rate reduces to the notion of *terms of*

trade defined in the theory of international trade, namely

$$r_R = \pi = \frac{p_x}{r p_m},$$

(2.12)

where p_x represents export prices (in terms of domestic currency), p_m import prices (in terms of foreign currency), and r the nominal exchange rate of the country under consideration.

From the point of view of the consumer, π represents the relative price of foreign and domestic goods on which (in accordance with standard consumer's theory) demand will depend. From the point of view of the country as a whole, π represents the amount of imports that can be obtained in exchange for one unit of exports (or the amount of exports required to obtain one unit of imports). Therefore an increase in π is also defined as an improvement in the terms of trade, as it means that a greater amount of imports can be obtained per unit of exports (or, equivalently, that a smaller amount of exports is required per unit of imports). It should also be noted that it is irrelevant whether π is defined as above or as

$$\pi = \frac{\frac{1}{r} p_x}{p_m},$$

(2.13)

since the two formulae are mathematically equivalent.

The terms of trade π can serve both the domestic and the foreign consumer (country) for the relevant price-comparison, because in (2.12) the prices of domestic and foreign goods are expressed in domestic currency, while in (2.13) they are expressed in foreign currency.

It is clear that, since exports are part of domestic output, $p_x = p_h$, and similarly $p_m = p_f$, so that (2.12) and (2.8) coincide.

Another definition of real exchange rate takes the ratio of unit labour costs at home (W_h) to unit labour costs abroad (W_f), expressed in a common monetary unit through the nominal exchange rate (r), as the measure we are looking for, hence

$$r_R = r \left(\frac{W_f}{W_h} \right),$$

(2.14)

where r_R is the real exchange rate. Note that the real exchange rate is defined such that an *increase* (*decrease*) in it means an *improvement* (*deterioration*) in the external competitiveness of domestic goods. In fact, ceteris paribus, a decrease in domestic with respect to foreign unit labour costs (W_f/W_h increases) is reflected (in both perfectly and imperfectly competitive markets) in a decrease of the relative price of domestic with respect to foreign goods. The same result is obtained when, at given W_f/W_h, the exchange rate depreciates (i.e., the nominal exchange rate r increases).

2.4   The Effective Exchange Rate

The concept of *effective* exchange rate must not be confused with the real exchange rate, since the effective exchange rate can be nominal or real.

While the (nominal or real) exchange rate involves two currencies only, it may be desirable to have an idea of the *overall* external value of a currency, namely with respect to the rest of the world (or a subset of it, for example the industrialized countries) and not only with respect to another country's currency. The presence of floating exchange rates makes it difficult to ascertain the behaviour of the external value of a currency. In fact, in a floating regime a currency may simultaneously depreciate with respect to one (or more) foreign currency and appreciate with respect to another (or several others).

In such a situation it is necessary to have recourse to an index number, in which the bilateral exchange rates of the currency under consideration with respect to all other currencies enter with suitable weights. This index is called an *effective exchange rate*. Let us begin with the *nominal* effective exchange rate, which is given by the formula

$$r_{ei} = \sum_{j=1, j \neq i}^{n} w_j r_{ji}, \quad \sum_{j=1, j \neq i}^{n} w_j = 1, \tag{2.15}$$

where

$r_{ei} =$ (nominal) effective exchange rate of currency i,
$r_{ji} =$ nominal exchange rate of currency i with respect to currency j,
$w_j =$ weight given to currency j in the construction of the index; by definition, the sum of the weights equals one.

Usually the effective exchange rate is given as an index number with a base of 100 and presented in such a way that an increase (decrease) in it means an appreciation (depreciation) of the currency under consideration with respect to the other currencies as a whole. This implies that r_{ji} is defined using the volume quotation system.

Unfortunately it is not possible to determine the weights unambiguously: this is an ambiguity inherent in the very concept of index number. Many effective exchange rates thus exist in theory; usually, however, the weights are related to the share of the foreign trade of country i with country j in the total foreign trade of country i. Effective exchange rates are computed and published by the IMF, central banks and private institutions.

If we carry out the same operation defined by Eq. (2.15) using real rather than nominal bilateral exchange rates we shall of course obtain a *real* effective exchange rate. This will give a measure of the overall competitiveness of domestic goods on world markets rather than with respect to another country's goods.

2.5   The Forward Exchange Market

2.5.1   Introduction

The main function of the forward exchange market is to allow economic agents engaged in international transactions (whether these are commercial or financial) to cover themselves against the *exchange risk* deriving from possible future variations in the spot exchange rate.[4] If, in fact, the spot exchange rate were permanently and rigidly fixed, the agent who has in the future to make or receive a payment in foreign currency (or, more generally, who has liabilities and/or assets in foreign currency) does not incur any exchange risk, as he already knows how much he will pay or receive (or, more generally, the value of his liabilities and assets) in terms of his own national currency. But when exchange rates are bound to change through time, as is usually the case (see below, Chap. 3, for the various exchange rate systems), an exchange rate risk arises.

From the point of view of the agent who has to make a future payment in foreign currency (for example, an importer who will have to pay in three months' time for the goods imported now), the risk is that the exchange rate will have depreciated at the time of the payment, in which case he will have to pay out a greater amount of domestic currency to purchase the required amount of foreign currency. From the point of view of the agent who is to receive a future payment in foreign currency (for example, an exporter who will be paid in three months time for the goods exported now) the risk is that the exchange rate will have appreciated at the time of the payment, in which case he will get a smaller amount of domestic currency from the sale of the given amount of foreign currency.

Naturally the agent who has to make a future payment in foreign currency will benefit from an appreciation of the domestic currency and, similarly, a depreciation will benefit the agent who is to receive a future payment in foreign currency. But, if we exclude the category of speculators, the average economic agent is usually risk averse, in the sense that, as he is incapable of predicting the future behaviour of the exchange rate and considers future appreciations and depreciations to be equally likely, he will assign a greater weight to the eventuality of a loss than a gain deriving from future variations in the exchange rate.

The average operator, therefore, will seek cover against the exchange risk, that is, he will *hedge*.[5] In general, hedging against an asset is the activity of making sure to have a zero net position (that is, neither a net asset nor a net liability position)

---

[4]Here, as well as subsequently, "spot exchange rate" is used to denote a generic spot exchange rate belonging to the set of all spot exchange rates.

[5]Some writers (see, for example, Einzig 1961, 1966) distinguish between covering and hedging. Covering (by means of forward exchange) is an arrangement to safeguard against the exchange risk on a payment of a definite amount to be made or received on a definite date in connection with a self-liquidating commercial or financial transaction. Hedging (by means of forward exchange) is an arrangement to safeguard against an indefinite and indirect exchange risk arising from the existence of assets or liabilities, whose value is liable to be affected by changes in spot rates. More

in that asset. As we are considering foreign exchange, to hedge means to have an exact balance between liabilities and assets in foreign currency (of course, this exact balance must hold for each foreign currency separately considered), that is, in financial jargon, to have no open position in foreign exchange, neither a *long* position (more assets than liabilities in foreign currency) nor a *short* position (more liabilities than assets in foreign currency). A particular case of a zero net position in foreign exchange is, of course, to have zero assets and zero liabilities. This can be obtained, for example, by stipulating all contracts in domestic currency. But this hardly solves the problem, because for the other party the contract will then be necessarily in foreign currency, and this party will have to hedge.

Now, one way to cover against the exchange risk is through the forward exchange market. The agent who has to make a payment in foreign currency at a known future date can at once purchase the necessary amount of foreign currency forward: since the price (the forward exchange rate) is fixed now, the future behaviour of the spot exchange rate is irrelevant for the agent; the liability position (the obligation to make the future payment) in foreign currency has been exactly balanced by the asset position (the claim to the given amount of foreign exchange at the maturity of the forward contract). Similarly, the agent who is to receive a payment in foreign currency at a known future date can at once sell the given amount of foreign currency forward.

There are, however, other ways of hedging; the main possibilities will be briefly examined and then compared.

2.5.2   Various Covering Alternatives: Forward Premium and Discount

Let us consider the case of an economic agent who has to make a payment at a given future date, for example an importer of commodities (the case of the agent who is to receive a future payment is a mirror-image of this). Let us also list the main opportunities for cover, including the forward cover mentioned above. The possibilities are these:

(a) The agent can buy the foreign exchange forward. In this case he will not have to pay out a single cent now, because the settlement of the forward contract will be made at the prescribed future date.[6]
(b) The agent can pay immediately, that is, purchase the foreign exchange spot and settle his debt in advance. To evaluate this alternative we must examine its costs and benefits. On the side of costs we must count the opportunity cost of

---

often, however, no distinction is made and hedging (in the broad sense) is taken to include all operations to safeguard against the exchange risk, however it arises.

[6]We are abstracting from possible domestic regulations requiring the immediate deposit of a certain proportion of the value (in domestic currency) of the forward contract.

(domestic) funds, that is, the fact that the economic agent forgoes the domestic interest rate on his funds for the delay granted in payment (if he owns the funds) or has to pay the domestic interest rate to borrow the funds now (if he does not own the funds). For the sake of simplicity, we ignore the spread between the lending and borrowing interest rates, so that the costs are the same whether the agent owns the funds or not. On the side of benefits, we have the discount that the foreign creditor (like any other creditor) allows because of the advance payment; this discount will be related to the foreign interest rate (the creditors domestic interest rate). For the sake of simplicity, we assume that the percentage discount is equal to the full amount of the foreign interest rate and that the calculation is made by using the exact formula $x[1/(1 + i_f)]$ instead of the approximate commercial formula $x - i_f x = x(1 - i_f)$, where x is the amount of foreign currency due in the future and i_f is the foreign interest rate (referring to the given period of time).

(c) The agent can immediately buy the foreign exchange spot, invest it in the foreign country from now till the maturity of the debt and pay the debt at maturity (*spot covering*). The costs are the same as in the previous case; on the side of benefits we must count the interest earned by the agent by investing the foreign exchange abroad.

In practice things do not go so smoothly (think, for example, of foreign drafts which are discounted and rediscounted, etc.), but at the cost of some simplification they can be fitted into these three alternatives.

In the case of an agent who is to receive a payment in the future the alternatives are: (a) sell the foreign exchange forward; (b) allow a discount to the foreign debtor so as to obtain an advance payment, and immediately sell the foreign exchange spot; (c) discount the credit with a bank and immediately sell the foreign exchange spot.

In order to compare these three alternatives, besides the domestic and foreign interest rates, we must also know the exact amount of the divergence between the forward exchange rate and the (current) spot exchange rate. For this we need to define the concept of forward *premium* and *discount*. A *forward premium* denotes that the currency under consideration is more expensive (of course in terms of foreign currency) for future delivery than for immediate delivery, that is, it is more expensive forward than spot. A forward discount denotes the opposite situation, i.e. the currency is cheaper forward than spot. The higher or lower value of the currency forward than spot is usually measured in terms of the (absolute or proportional) deviation of the forward exchange rate with respect to the spot exchange rate.

We observe, incidentally, that in the foreign exchange quotations the forward exchange rates are usually quoted implicitly, that is, by quoting the premium or discount, either absolute or proportional. When the forward exchange rate is quoted explicitly as a price, it is sometimes called an *outright forward exchange rate*. We also observe, as a matter of terminology, that when the spot price of an asset exceeds (falls short of) its forward price, a *backwardation* (*contango*, respectively) is said to occur.

This is one of the cases where it is most important to have a clear idea of how exchange rates are quoted (see Sect. 2.1). If the price quotation system is used, the higher value of the currency forward than spot means that the forward exchange rate is *lower* than the spot exchange rate, and the lower value of a currency forward than spot means that the forward exchange rate is *higher* than the spot rate. But if the volume quotation system is used the opposite is true: the higher (lower) value of a currency on the forward than on the spot foreign exchange market means that the forward exchange rate is *higher* (*lower*, respectively) than the spot rate. If, say, the $ in New York is more expensive forward than spot with respect to the euro, this means that fewer dollars are required to buy the same amount of euros (or, to put it the other way round, that more euros can be bought with the same amount of dollars) on the forward than on the spot exchange market, so that if the USA uses the price quotation system, in New York the $/euro forward exchange rate will be lower than the spot rate, whereas if the USA used the other system, the opposite will be true.

Therefore in the case of the price quotation system the forward *premium* will be measured by a *negative* number (the difference forward minus spot exchange rate is, in fact, negative) and the forward discount by a positive number. This apparently counterintuitive numerical definition (intuitively it would seem more natural to associate premium with a positive number and discount with a negative one) is presumably due to the fact that this terminology seems to have originated in England, where the volume quotation system is used, so that by subtracting the spot from the forward exchange rate one obtains a positive (negative) number in the case of a premium (discount). Be this as it may, having adopted the price quotation system and letting r denote the generic spot exchange rate and r^F the corresponding forward rate of a currency, the *proportional* difference between them,

$$\frac{r^F - r}{r}, \tag{2.16}$$

gives a measure of the forward premium (if negative) and discount (if positive). As there are different maturities for forward contracts, in practice the proportional difference (2.16) is given on a per annum basis by multiplying it by a suitable factor (if, for example, we are considering the 3-month forward rate, the approximate factor is 4) and as a percentage by multiplying by 100. The reason why the forward margin (a margin is a premium or a discount) is expressed in this proportional form is that, in this way, *we give it the dimension of an interest rate* and can use it to make comparisons with the (domestic and foreign) interest rates; expression (2.16) is, in fact, sometimes called an *implicit interest rate* in the forward transaction.

So equipped, we can go back to compare the various alternatives of the agent who has to make a future payment (the case of the agent who has to receive a future payment is perfectly symmetric). We first show that alternatives (b) and (c) are equivalent. We have already seen that the costs are equivalent; as regards the benefits, we can assume that the discount made by the foreign creditor for advance payment (case b) is percentually equal to the interest rate that our debtor might earn on foreign currency invested in the creditors country (case c). More precisely, let i_h and i_f be the home and the foreign interest rate respectively, referring to the period considered in the transaction (if, for example, the delay in payment is three months, these rates will refer to a quarter), and x the amount of the debt in foreign currency. With alternative (b), thanks to the discount allowed by the foreign creditor, it is sufficient to purchase an amount $x/(1+i_f)$ of foreign currency now. The same is true with alternative (c), because by purchasing an amount $x/(1+i_f)$ of foreign currency now and investing it in the creditor's country for the given period at the interest rate i_f, the amount $[x/(1+i_f)](1+i_f) = x$ will be obtained at the maturity of the debt. The purchase of this amount of foreign currency spot requires the immediate outlay of an amount $r[x/(1+i_f)]$ of domestic currency.

Therefore, if we consider the opportunity cost of domestic funds (interest foregone on owned funds, or paid on borrowed funds), referring to the period considered, the *total* net cost of the operation in cases (b) and (c), referring to the maturity date of the debt, is obtained by adding this opportunity cost to the sum calculated above. Thus we have

$$\frac{rx}{1+i_f}(1+i_h). \tag{2.17}$$

Let us now consider case (a): the agent under consideration will have to pay out the sum $r^F x$ in domestic currency when the debt falls due. It is then obvious that alternative (a) will be better than, the same as, or worse than the other one [since (b) and (c) are equivalent, there are actually two alternatives] according as

$$r^F x \lessgtr \frac{rx}{1+i_f}(1+i_h). \tag{2.18}$$

If we divide through by rx we have

$$\frac{r^F}{r} \lessgtr \frac{1+i_h}{1+i_f}, \tag{2.19}$$

whence, by subtracting unity from both sides,

$$\frac{r^F - r}{r} \lessgtr \frac{i_h - i_f}{1+i_f}. \tag{2.20}$$

On the left-hand side we meet our old friend, the *forward margin*; the numerator of the fraction on the right-hand side is the *interest (rate) differential* between the domestic and the foreign economy. Formula (2.20) is often simplified by ignoring the denominator, but this is legitimate only when i_f is very small (for a precise determination of the degree of approximation, see Sect. 4.1). The *condition of indifference* between the alternatives then occurs when the forward margin equals the interest rate differential.

It is interesting to observe that an absolutely identical condition holds in the case of *covered interest arbitrage*, that will be treated in Chap. 4, Sect. 4.1.

We conclude the section by observing that in the forward exchange market the same type of arbitrage operations on foreign exchange takes place as described in relation to the spot market (see Sect. 2.2), so that the direct and indirect (or cross) forward rates come to coincide.

2.6   The Transactors in the Foreign Exchange Market

It is as well to point out at the beginning that the classification of the various transactors will be made on a *functional* rather than personal or institutional basis. In fact, the same economic agent can be a different transactor at different times or even simultaneously belong to different functional categories of transactors: for example, importers and exporters who change the timing of their payments and receipts to get the benefit of an expected variation in the exchange rate are simultaneously traders and speculators. If, for example, a depreciation is expected, and traders do not hedge on the forward market but, on the contrary, pay in advance for the goods they are due to receive in the future (as importers) and delay the collection of payment for the goods already delivered (as exporters), then we are in the presence of speculative activity (speculative exploitation by traders of the *leads and lags* of trade).

A possible classification is based on three categories (within which it is possible to perform further subdivisions): non-speculators, speculators, and monetary authorities. To put this classification into proper perspective, a digression on speculative activity is in order.

2.6.1   Speculators

In general, speculation can be defined as the purchase (sale) of goods, assets, etc. with a view to re-sale (re-purchase) them at a later date, where the motive behind such action is the expectation of a gain deriving from a change in the relevant prices relatively to the ruling price and not a gain accruing through their use, transformation, transfer between different markets, etc. (Kaldor 1939).

The first rigorous formulation of the equilibrium of speculative activity in a broad sense (that is, referred to a generic real or financial asset) is attributed to Keynes (1936) and was subsequently elaborated by Kaldor (1939) and Tsiang (1958), who applied it to foreign exchange speculation.

In general, the agent who expects an increase in the price of an asset is called a *bull*, whereas a *bear* is one who expects a decrease in the price of an asset. Therefore, if we denote by $\tilde{r}$ the expected future spot exchange rate, a bull in foreign currency ($\tilde{r} > r$) will normally buy foreign currency (have a *long position*) and a bear ($\tilde{r} < r$) will normally sell foreign currency (have a *short position*). *Both deliberately incur an exchange risk to profit from the expected variation in the exchange rate.* This risk is usually accounted for by introducing a *risk premium,* which will be the greater, the greater the dispersion of expectations and the size of commitments. More precisely, consider for example a bull, whose expected speculative capital gain in percentage terms is given by $(\tilde{r} - r)/r$. Although the interest rate gain (from placing the funds abroad) and loss (forgone domestic interest) are negligible in speculative activity, they have to be taken into account for a precise evaluation, hence the bull will speculate if $(\tilde{r} - r)/r + \delta + i_f > i_h$, where δ is the risk premium. Similar considerations show that the bear will speculate if $(\tilde{r} - r)/r + \delta + i_f < i_h$. No incentive to speculate in either direction will exist when

$$\frac{\tilde{r} - r}{r} + \delta = i_h - i_f. \tag{2.21}$$

This is speculation on *spot* foreign exchange, besides which a *forward* exchange speculation also exists. The latter derives from a divergence between the *current forward* rate and the *expected spot rate* of a currency. If the expected spot rate is higher than the current forward rate, it is advantageous for the speculator to buy foreign currency forward, as he expects that, when the forward contract matures, he will be able to sell the foreign currency spot at a price (the expected spot rate) higher than the price that he knows he will pay for it (the current forward rate). In the opposite case, namely if the expected spot rate is lower than the current forward rate, it is advantageous for the speculator to sell foreign currency forward, in the expectation of being able to buy it, at delivery time, at a price (the expected spot rate) lower than the price that he knows he will be paid for delivering it (the current forward rate).

We have talked of delivery etc. In practice, the parties of a forward exchange speculative transaction settle the *difference* between the forward exchange rate and the spot exchange rate existing at maturity, multiplied by the amount of currency contemplated in the forward contract. It should also be noted that, in principle, forward speculation *does not require the availability of funds* (neither command over cash nor access to credit facilities) at the moment the contract is stipulated, by the very nature of the forward contract (both payment and delivery are to be made at a future date). In practice banks often require the transactor in forward exchange to put down a given percentage of the contract as collateral; this percentage depends, amongst other things, on the efficiency and development of the forward market, and on possible binding instructions of central banks.

The problems deriving from the presence of speculators will be dealt with in subsequent chapters, especially Sects. 16.2 and 16.3.

2.6.2   Non-Speculators

A second functional category is that of non-speculators. This category includes exporters and importers of goods and services, businesses which carry out invest-ment abroad, individual or institutional savers who wish to diversify their portfolios between national and foreign assets on the basis of considerations of risk and yield (excluding speculative gains), arbitrageurs, etc. Non-speculators are more precisely defined by exclusion, i.e., those agents who are neither speculators nor monetary authorities.

2.6.3 Monetary Authorities

Finally we have the monetary authorities. These are the institutions (usually the central bank, but also exchange equalization agencies where they exist as bodies juridically separate from the central bank) to which the management of the international reserves (for a precise definition of which, see Sect. 5.1.5) of the relative country is attributed. Monetary authorities can intervene in the foreign exchange market both by buying and selling foreign currencies in exchange for their own, and by taking various administrative measures (such as exchange controls).

2.7 Derivatives

The enormous growth of what is known as derivative instruments has also involved foreign exchange transactions (Steinherr 1998; Pilbeam 2013). We shall give a brief introduction to the main types of derivative instruments. Standardized derivatives contracts are traded on organized exchanges, such as the CBOE (Chicago Board Options Exchange), the LIFFE (London International Financial Futures Exchange), the MATIF (Marché À Terme International de France), etc. However, trading can also occur outside of the major exchanges in what is known as the OTC (Over-The-Counter) market, an expression which means that banks and other financial institutions design contracts tailor-made to satisfy the specific needs of their clients.

2.7.1 Futures

A currency futures contract is an agreement between two counterparties to exchange a specified amount of two currencies at a given date in the future at an exchange rate which is pre-determined at the moment of the contract. The definition looks the same as that given in previous sections of currency forward contract. What are then the differences?

The main differences are of practical type, as summarized in the following list.

(1) In forward contracts the amount to be exchanged can be any, as determined by the mutual agreement of the two parties, while currency futures contracts are for *standardized* amounts.
(2) Forward contracts are essentially over-the-counter instruments with the exchange taking place directly between the two parties, while currency futures are traded on an Exchange. Hence the next difference (point 3) follows.

(3) Forward contracts involve a counterparty risk, while futures are guaranteed by the Exchange.
(4) Forward contracts are relatively illiquid assets, because forward contract obligations cannot be easily transferred to a third party. On the contrary, the standardized nature of futures means that they can be easily sold at any time prior to maturity to a third party at the prevailing futures price.
(5) Forward contracts cover over 50 currencies, while futures cover only major currencies.

The asset (in this case the currency) to be delivered in fulfilment of the contract is called the *underlying*. In futures contract involving physical assets (gold or other commodities) the physical delivery of the commodity would be cumbersome, hence most parties enter into what is known as *reversing trade*. This means that they will liquidate their position at the clearing house just prior to maturity so that they neither have to actually receive or actually pay the underlying. Reversing trade is also applied in around 99 % of currency futures contracts.

Apart from these practical differences, currency futures can be used for the same purposes of currency forwards for hedging (see above, Sect. 2.5.2) and speculating (see above, Sect. 2.6.1).

2.7.2 Options

A currency option is a contract that gives the purchaser the right (but *not* the obligation) to buy or sell a currency at a predetermined price (exchange rate) sometime in the future. Hence options are a much more complicated instrument than forwards and futures. They also have a precise terminology, which is the following.

The party selling the option is called the *writer,* while the purchaser is the *holder.* A *call option* gives the holder the right to *purchase* the currency involved, while a *put option* gives the right to *sell* the currency. The currency in which the option is granted is called the *underlying currency*, while the currency in which the price will be paid is the *counter currency*. For example, if the contract specifies the right to sell *euro*1, 000, 000 at $1.05/*euro*1, the euro is the underlying currency while the dollar is the counter currency. The price at which the underlying currency can be bought or sold is the *strike* (or *exercise*) *price*. The price that the holder pays to the writer for an option is known as the *option premium*. The date at which the contract expires is called the *expiry date* or *maturity date*. Finally, a distinction is made between the *American option* (the right to buy or sell the currency at the given price can be exercised any time up to the maturity date) and the *European option* (the right can be exercised only on the maturity date).

There are two main differences between options and forward or futures contracts, which both derive from the fact that the option gives the holder a right but not an obligation.

The *first* is that the option provides the agent interested in hedging with a more flexible instrument, because it enables him to fix a maximum payable price (the sum of the option premium plus the exercise price) while leaving him free to take advantage of favourable movements in the exchange rate. With a forward or futures contract the hedger is obliged to respect the contract in any case, also when the spot exchange rate at maturity is more favourable than the forward rate agreed upon when the contract was signed. On the contrary, with an option the holder can decide not to exercise the right if the spot exchange rate at the expiry date is more favourable than that the exercise price, account being taken of the option premium.

Suppose, for example, that a US company has to make a payment of £1 million in sixth months' time, and that the forward/futures exchange rate is $1.50/£1. Alternatively, the company can buy a call option with an exercise price of $1.50/£1 for 8 cents per pound (the option premium). At maturity, if the spot exchange rate is higher than $1.50/£1, the US firm will exercise the option. This is in any case cheaper than buying pounds spot, but of course more expensive than would have been with the forward/futures contract, given the option premium. If the spot exchange rate is lower than $1.50/£1, the firm will *not* exercise the option and buy pounds spot. The cost will again be higher than with a forward/futures contract, but only if the spot exchange rate is higher than $1.42/£1, because adding the option premium (8 cents per pound) the price paid will be higher than $1.50/£1. If the spot exchange rate is lower than $1.42/£1, the option will have provided a cheaper means of hedging than a forward/futures contract.

The *second* difference concerns the *asymmetry* in the risk-return characteristics of the contract. With a forward/futures contract, for every cent the spot exchange rate at the date of expiry is above (below) the exchange rate established in the forward/futures contract, the buyer makes (loses) a cent and the seller loses (makes) a cent. This means a perfect symmetry. On the contrary, with an options contract, the maximum loss of the option holder equals the option premium, which is also the maximum gain for the option writer, but there is *unlimited potential gain* for the *option holder* and, correspondingly, *unlimited potential loss* for the *option writer*. This feature makes options very attractive for speculators, because speculative holders can combine limited losses (the premium paid) with unlimited potential profit.

2.7.3   Swap Transactions

The presence of the forward exchange market beside the spot exchange market, allows hybrid spot and forward transactions such as *swap* contracts. The swap contracts we are dealing with take place between private agents and are different from swap agreements between central banks, in which the latter exchange their respective currencies between themselves (by crediting the respective accounts held with one another: for example, the Bank of England credits the European Central Bank's account with 100 million pounds, and the European Central Bank credits the Bank of England's account with 151.6 million euros), usually with the obligation to make a reverse operation after a certain period of time.

A swap is a transaction in which two currencies are exchanged in the spot market and, simultaneously, they are exchanged in the forward market in the opposite direction. At first sight the swap contract would not seem to have wide potential use: on the contrary, its market is more important than the outright forward exchange market, second only to that for spot exchange. The swap market is currently organized by the ISDA (International Swap Dealers Association).

An obvious example of swap transaction is that deriving from covered interest arbitrage operations (see Sect. 4.1). If we assume, for instance, that the condition for an outward arbitrage occurs, the arbitrageur will buy foreign exchange spot and simultaneously sell it forward. More precisely, since the arbitrageur covers not only the capital but also the accrued interest against the exchange risk, the quantity of foreign currency sold forward will exceed the quantity of it bought spot by an amount equal to the interest on the latter accrued abroad.

Another example is related to the cash management of multinational corporations. Suppose that a parent company in the US has an excess of liquidity in dollars, which is likely to persist for three months, whereas a subsidiary in England has a temporary shortage of liquidity in pounds, which is likely to last for three months. In such a situation the parent company can sell dollars for pounds spot and lend these to the subsidiary, at the same time selling pounds for dollars forward so as to cover the repayment of the debt by the subsidiary. This is a swap transaction in the pound/dollar market.

A swap agreement can also be used by firms to raise finance more cheaply than would otherwise be the case. Suppose that a European company wants to raise yen funds while a Chinese company wants to raise euro funds. Additionally suppose that Japanese investors are not very desirous to invest in European companies but are eager to invest in Chinese companies, while European investors are not very keen to lend to Chinese companies but are willing to invest in a European company. Then it may be advantageous (in term of cheaper conditions, such as a lower interest rate to be offered to investors) that the Chinese company raises funds in yen, while the European company raises funds in euros; the companies then swap the funds raised and the corresponding obligations. The result is that both companies obtain the funds they need at cheaper cost than they had directly raised the funds.

Swap transactions are also carried out by banks themselves, to eliminate possible mismatches in the currency composition of their assets and liabilities. A bank, for example, may have—for a time horizon of three months—a $50 million excess of dollar loans over dollar deposits, and, simultaneously, an excess of deposits in pounds over loans in pounds of equivalent value. In such a situation the bank can sell the excess of pounds for $50 million spot and simultaneously buy the same amount of pounds for dollars three months forward so as to cover against the exchange risk. Alternatively, the bank could have lent the pound equivalent of $50 million, and borrowed $50 million, in the interbank money market.

Swap transactions involve two exchange rates (the spot and the forward rate); in practice a *swap rate* is quoted, which is a price difference, namely, the difference between the spot and forward rates quoted for the two transactions which form the swap transaction (this difference is quoted in absolute rather than percentage terms).

A swap agreement is basically the same as a forward/futures contract, from which it differs for various practical aspects:

(1) most forward/futures contracts are for a year or less, while swap contracts are often for long periods, from 5 to 20 years and possibly longer. This makes them more attractive to firms which have long-run obligations in foreign currency.

(2) Futures have an active secondary market, which is not the case for swaps. Since swap agreements, as all contracts, can only be cancelled with the consent of both parties, a party who wants to get rid of a swap may not be able to do so.

(3) Futures are standardized contracts, while swaps can be tailored to meet the needs of the client.

(4) Futures contracts are guaranteed by the futures Exchange, while swap agreements present the risk that one of the parties will not fulfil its obligations.

2.7.4   Credit Derivatives

A credit derivative is a financial instrument that transfers credit risk related to an underlying entity or a portfolio of underlying entities from one party to another without transferring the underlying(s). The underlyings may or may not be owned by either party in the transaction. There are several types of credit derivatives, the best known being Credit Default Swaps (CDS) and Collateralized Debt Obligations (CDO).

A CDS is similar to an insurance contract. It is a credit derivative contract between two parties where the buyer ("Protection Buyer") pays a periodic premium (over the maturity period of the CDS) to the seller ("Protection Seller") in exchange for a commitment to a payoff if a third party defaults. Generally used as insurance against default on a credit asset but can also be used for speculation: in fact, the buyer does not need to own the underlying security and does not have to suffer a loss from the event in order to receive payment from the seller. The cost of a CDS, namely the premium, incorporates the evaluation of the risk of default, so that the cost rises when the risk increases.

A CDO is an asset-backed security that pools together cash-flow-generating assets and repackages this asset pool into discrete tranches that can be sold to investors. A collateralized debt obligation is so called because the pooled assets—such as mortgages, bonds and loans—are essentially debt obligations that serve as collateral for the CDO. The tranches in a CDO vary substantially in their risk profile. The safest tranches are called senior tranches, and have first priority on the collateral in the event of default. As a result, the senior tranches of a CDO generally have a higher credit rating and offer lower coupon rates than the junior tranches (the less safe tranches), which offer higher coupon rates to compensate for their higher default risk. Intermediate tranches are called mezzanine tranches.

2.8   Eurodollars and Xeno-Currencies

The description of foreign exchange transactions given in the previous sections is the traditional one. The situation has, however, been complicated by the development, since the late 1950s, of an international money market of a completely new type: the so-called *Eurodollar* system, subsequently extended to other currencies.

In the traditional system, economic agents can obtain loans, hold deposits, etc., in a currency, say currency *j*, only with country *j*'s banks so that, for example, a German resident can hold dollar deposits only with the US banking system. *Eurodollars* are, on the contrary, dollar deposits with European banks. The Eurodollar market began in fact with dollar deposits placed with European banks and used by these to grant loans in dollars. By European banks we mean banks "resident" in Europe (in accordance with the definition of resident which will be examined in detail in Sect. 5.1). Thus a European bank can also be a subsidiary of a US bank.

Note that, in general, a European bank can also accept deposits and grant loans denominated in currencies other than the dollar (and, of course, different from the currency of the country where the bank is resident); so that the denomination *Eurocurrencies* was coined (these include the Eurodollar, Eurosterling, Euroyen, etc.). Still more generally, since similar operations can be carried out by banks outside Europe (Asiadollars, etc.), the general denomination *Xeno-currencies* (from the Greek xenos = foreigner) has been suggested by F. Machlup (1972; p. 120) to indicate deposits and loans denominated in currencies other than that of the country in which the bank is located. An equivalent denomination is *cross-border bank assets&liabilities.*

As regards the Eurodollar market, various reasons have been put forward to explain its birth. According to some, the origin lies in an initiative of the Soviet Union which, during the Korean war, fearing that its dollar deposits in the US might be frozen by the US government, found it convenient to shift these dollar accounts to Europe, largely to London. Others believe that the initiative was taken by London banks which, in order to avoid the restrictions on the credit to foreign trade imposed in the UK in 1957, induced the official agencies of the Soviet Union to deposit their dollar holdings in London by granting favourable interest rates. Still another factor is believed to be the US Federal Reserve System's Regulation Q, which fixed the rates of interest paid on time deposits, but which did not apply to time deposits owned by nonresidents. Thus New York banks began to compete for nonresidents deposits, the interest rates on these rose about 0.25 % above the ceiling in 1958–1959, and London banks were induced to bid for dollar deposits which in turn they re-lent to New York banks. A practical factor may also have had its importance: due to the difference in time zones, European and US banks are open simultaneously only for a short time in the day, so that Europeans who had to borrow or lend dollars found it convenient to do this directly in London rather than in New York through a London bank.

Be this as it may, the enormous growth of the Xeno-currency markets has complicated the international financial market: let it suffice to think of the greater complexity of interest arbitrage operations and of the birth of new types of international banking transactions. As regards these, they can be classified in four main types: *onshore-foreign*, *offshore-foreign*, *offshore-internal*, and *offshore-onshore*. The first word of each pair refers to the currency in which the bank is transacting: if it is that of the country in which the bank is resident the transaction is onshore, whilst if it is the currency of another country the transaction is offshore. The second word refers to the residence of the customer (borrower or lender): the customer is internal if resident in the same country as the bank, foreign if resident in a country different from that where the bank is resident and also different from the country which issues the currency being transacted; in fact, the customer is onshore if resident in the country issuing the currency.

Before the birth of Xeno-currencies, international banking transactions were entirely of the *onshore-foreign* type: an example of an onshore-foreign deposit is a deposit in dollars placed with Chase Manhattan, New York, by a non-US resident. The growth of offshore deposits related to Xeno-currencies has given rise to the multiplication of the three other types of international banking transactions.

An example of an *offshore-foreign* deposit is a deposit in euros placed with a Swiss bank by a Japanese resident.

An example of an *offshore-internal* deposit is a deposit in US dollars placed with a Dutch bank by a Dutch resident.

Finally, an example of an *offshore-onshore* deposit is a deposit in US dollars placed by a US resident with a Japanese bank.

When international capital flows where to some extent subject to controls (this was the normal situation during the Bretton Woods system, and also after its collapse several countries maintained capital controls), a specific analysis of Xeno-markets, by their very nature exempt from national controls (a situation that worried central bankers very much), was very important. But in the early 1990s completely free international mobility of capital became the rule rather than the exception, hence this importance no longer exists.

2.9 Appendix

2.9.1 N-Point Arbitrage

In the text we have described 2- and 3-point arbitrage and shown how these keep bilateral exchange rates equal to (or very near to) their theoretical values. But one might well ask whether more complicated forms of arbitrage involving more than three currencies (in general n) are possible. The answer is theoretically affirmative, but negative in practice, as the arbitrage activity involving more than three centres is in reality extremely rare. This does not derive from the complexity of the calculations, which increases as the number of centres involved increases (but this is not a problem in the current computer era), but from an important theorem, according to which *if three-point arbitrage is not profitable, then k-point arbitrage* $(k = 4, 5, \ldots n)$ *will not be profitable either.*

We begin by noting that, as in the case of three currencies treated in the text, also in the n-currency case if one starts with one unit of a currency and exchanges it successively for all the other currencies in turn so as to return to the initial currency, one must end up exactly with the initial unit. Thus the following condition must hold

$$r_{ji}r_{im}r_{ms} \ldots r_{vz}r_{zj} = 1, \tag{2.22}$$

where $j, i, m, s, \ldots, v, z$ run from 1 to n and are different among themselves. Equation (2.22) is a generalization of Eq. (2.4).

The proof of the theorem can be given by induction, showing that if $(k - 1)$-point arbitrage is not profitable, then k-point arbitrage is not profitable either (Chacholiades 1971). Let us then assume that the left side of (2.22) involves k financial centres: if we eliminate one, for example the i-th, we have $(k - 1)$-point arbitrage and, as this is assumed not to be profitable, it must be true that

$$r_{jm}r_{ms} \ldots r_{vz}r_{zj} = 1. \tag{2.23}$$

Now, if k-point arbitrage were profitable, Eq. (2.22) would not hold, that is

$$r_{ji}r_{im}r_{ms} \ldots r_{vz}r_{zj} = \alpha, \quad \alpha \neq 1. \tag{2.24}$$

Dividing (2.24) by (2.23) we have

$$\frac{r_{ji}r_{im}}{r_{jm}} = \alpha, \tag{2.25}$$

that is, as $r_{jm} = 1/r_{mj}$ by Eq. (2.2) (two-point arbitrage),

$$r_{ji}r_{im}r_{mj} = \alpha. \tag{2.26}$$

Now, if three-point arbitrage is not profitable, Eq. (2.4) holds, which is given here for the reader's convenience

$$r_{ji}r_{im}r_{mj} = 1. \tag{2.27}$$

Therefore in Eq. (2.26) α must necessarily equal 1. Thus (2.24) cannot hold and (2.22) is verified, so that k-point arbitrage is not profitable.

It follows, by mathematical induction, that, when 3-point arbitrage is not profitable, 4-,5-,$\ldots$, n-point arbitrage is also non-profitable.

In the text we mentioned the existence of forward arbitrage and it is easy to see, by replacing r with r^F, that the theorem demonstrated above holds for forward exchange as well. Another interesting theorem is that *if three-point spot arbitrage and (two-point) covered interest arbitrage are not profitable, then three-point forward arbitrage is not profitable either*. In fact, three-point arbitrage insures that (2.27) holds, whilst the non-profitability of (two-point) covered interest arbitrage insures the following conditions (which can be immediately derived from Eqs. (4.2) in Chap. 4)

$$\frac{r_{ji}^F}{r_{ji}} = \frac{1+i_i}{1+i_j}, \quad \frac{r_{im}^F}{r_{im}} = \frac{1+i_m}{1+i_i}, \quad \frac{r_{mj}^F}{r_{mj}} = \frac{1+i_j}{1+i_m}. \tag{2.28}$$

From (2.28) we have

$$r_{ji} = \frac{1+i_j}{1+i_i}r_{ji}^F, \quad r_{im} = \frac{1+i_i}{1+i_m}r_{im}^F, \quad r_{mj} = \frac{1+i_m}{1+i_j}r_{mj}^F, \tag{2.29}$$

and, by substituting from (2.29) into (2.27), we get, after obvious simplifications,

$$r_{ji}^F r_{im}^F r_{mj}^F = 1, \tag{2.30}$$

which is the condition for three-point forward arbitrage (concerning the same three financial centres) to be non-profitable.

References

Chacholiades, M. (1971). The sufficiency of three-point arbitrage to insure consistent cross rates of exchange. *Southern Economic Journal, 38*, 86–88.

Chinn, M. D. (2006). A primer on real effective exchange rates: Determinants, overvaluation, trade flows and devaluation. *Open Economies Review, 17*, 15–143.

Einzig, P. (1961). *The dynamic theory of forward exchange.* London: Macmillan.

Einzig, P. (1966). *A textbook on foreign exchange* (2nd ed.). London: Macmillan.

Hinkle, E. L., & Montiel, P. (1999). *Exchange rate misalignment: Concepts and measurement for developing countries.* Oxford: Oxford University press for the World Bank.

Keynes, J. M. (1936). *The general theory of employment, interest and money.* London: Macmillan; reprinted as Vol. VII of The Collected Writings of J.M. Keynes, London: Macmillan for the Royal Economic Society, 1973.

Kaldor, N. (1939). Speculation and economic stability. *Review of Economic Studies, 7*, 1–27; reprinted, with revisions, In N. Kaldor (1960). *Essays on economic stability and growth* (pp. 17–58). London: G. Duckworth.

Lipschitz, L., & McDonald, D. (1992). Real exchange rates and competitiveness: A clarification of concepts, and some measurements for Europe. *Empirica, 19*, 37–69.

Machlup, F. (1972). Euro-dollars, once again. *Banca Nazionale del Lavoro Quarterly Review, 25*, 119–137.

Marsh, I. W., & Tokarick, S. P. (1996). An assessment of three measures of competitiveness. *Weltwirtschaftliches Archiv, 132*, 700–732.

Pilbeam, K. (2013). *International finance* (4th ed.). London: Macmillan.

Steinherr, A. (1998). *Derivatives: the wild beast of finance.* New York: Wiley.

Tsiang, S. C. (1958). A theory of foreign exchange speculation under a floating exchange system. *Journal of Political Economy, 66*, 399–418.

Exchange-Rate Regimes

In theory a large number of exchange-rate regimes are possible, because between the two extremes of perfectly *rigid* (or *fixed*) and perfectly (freely) *flexible* exchange rates there exists a range of intermediate regimes of *limited flexibility*. A detailed treatment is outside the scope of the present work, so that we shall briefly deal with the main regimes, beginning by the two extremes. Our treatment will be purely descriptive, with no discussion of the pros and cons of the various regimes, for which see Chap. 17.

3.1 The Two Extremes, and Intermediate Regimes

One extreme is given by perfectly and freely flexible exchange rates. This system is characterized by the fact that the monetary authorities do not intervene in the foreign exchange market. Therefore the exchange rate (both spot and forward) of the currency with respect to any foreign currency is left completely free to fluctuate in either direction and by any amount on the basis of the demands for and supplies of foreign exchange coming from all the other operators.

The other extreme is given by rigidly fixed exchange rates. Here various cases are to be distinguished. The first and oldest is the *gold standard*, where each national currency has a precisely fixed gold content (for our purposes it is irrelevant whether gold materially circulates in the form of gold coins or whether circulation is made of paper currency which can be immediately converted into gold on demand). In this case the exchange rate between any two currencies is automatically and rigidly fixed by the ratio between the gold content of the two currencies (which is called the *mint parity*): if, in fact, it were different, a profit could be made by shipping gold between the two countries concerned. Let us assume, for example, that the gold content of the pound sterling and the US dollar is 0.04631 and 0.02857 ounces of gold respectively: the exchange rate between the two currencies is $0.04631/0.02857 \cong 1.621$, that is, 1.621 dollars to the pound. If, in fact, the

© Springer-Verlag Berlin Heidelberg 2016
G. Gandolfo, *International Finance and Open-Economy Macroeconomics*,
Springer Texts in Business and Economics, DOI 10.1007/978-3-662-49862-0_3

monetary authorities stated a different rate, for example, 1.7831 dollars to the pound, anyone could sell pounds for dollars, give these to the US Fed in exchange for gold, ship the gold to England and obtain pounds in exchange for it, thus ending up with 10 % more pounds (with one pound one gets 1.7831 dollars and then $1.7831 \times 0.02857 \cong 0.050943$ ounces of gold in the United States, which are worth £1.1 in England). As long as the exchange rate is out of line with the gold content ratio, the outflow of gold from the United States would continue, a situation which cannot be maintained: either the monetary authorities halt it by administrative measures (for example by suspending convertibility, in which case we have gone off the gold standard system) or are compelled to fix the rate at 1.621 dollars to the pound. It is clear that we would arrive at the same result if the exchange rate were lower, for example 1.474 dollars to the pound (gold would flow from England to the United States, etc.).

To be precise, the exchange rate can diverge from mint parity within certain margins—called the *gold points*—which depend on the cost of transport and insurance of the gold shipped from one country to another. It is self-evident, in fact, that the operations described above are not profitable if these costs exceed the gain deriving from the divergence between the exchange rate and mint parity.

Conceptually similar to the gold standard is the *gold exchange standard*, in which, without itself buying and selling gold, a country stands ready to buy or sell a particular foreign currency which is fully convertible into gold. This system enables the international economy to economize gold with respect to the gold standard, because the ultimate requests for conversion into gold of the convertible foreign currency are normally only a fraction of the latter. It must be emphasized that, for this system to be a *true* gold exchange standard, the convertibility of the foreign currency must be free and full, so that it can be demanded and obtained by *any* agent. In this case the system is equivalent to the gold standard.

If, on the contrary, the convertibility is restricted, for example solely to the requests from central banks, we are in the presence of a *limping* gold exchange standard, in which case the automatic mechanisms governing the gold standard no longer operate, and the concept itself of convertibility has to be redefined: now convertibility simply means that private agents have the right to freely exchange the various currencies between each other at fixes rates. When convertibility into gold is completely eliminated, even between central banks, we have a *pure exchange standard*, in which a country buys and sells foreign exchange (or a stipulated foreign currency) at fixed rates.

Classification of countries into exchange rate regimes is in most cases not clear-cut. Traditionally (since 1950), the IMF *Annual Report on Exchange Arrangements and Exchange Restrictions* is the main source of information about the exchange rate policies pursued by member countries. The classification contained therein has been used to document the evolution of exchange rate regimes over time. There has been an expanding literature on the discrepancy between what policy makers announce they do regarding the exchange rate policy (exchange rate regime *de jure*), and what they actually implement (exchange rate regime *de facto*): countries may not always be following the exchange rate policy that they have officially declared to IMF. To address these shortcomings the IMF adopted a new classification scheme based on the *de facto* policies, which has become official since January 1999 (for the evolution of the IMF's Classification Taxonomies, see Habermeier et al. 2009). The IMF provides annual reports that incorporate both the declared exchange rate regime and the subjective evaluation of its staff.

To bring the reader up to date we report the *de facto* exchange rate arrangement taxonomy as in the latest release of the *Annual Report on Exchange Arrangements and Exchange Restrictions* (65th issue, 2014).

The IMF in its classification system of *official de facto* exchange rate regimes ranks them on the basis of the following broad principles: (a) the degree to which the exchange rate is determined by the market rather than by official action, with market-determined rates being on the whole more flexible; (b) the degree of flexibility of the arrangement or a formal or informal commitment to a given exchange rate path.

The methodology classifies the prevailing exchange rate regimes into four major categories: hard pegs (such as exchange arrangements with no separate legal tender and currency board arrangements); soft pegs (including conventional pegged arrangements, pegged exchange rates within horizontal bands, crawling pegs, stabilized arrangements, and crawl-like arrangements); floating regimes (such as floating and free floating); and a residual category, other managed.

(1) **Hard pegs**

    (a) *Exchange arrangements with no separate legal tender.* The currency of another country circulates as the sole legal tender. Adopting such an arrangement implies complete surrender of the monetary authorities' control over domestic monetary policy. There are two types of exchange arrangements with no separate legal tender. One is formal dollarization where the currency of another country circulates as the sole legal tender. The other is shared legal tender where members belonging to a monetary or currency union share the same legal tender. As stated by the IMF, exchange arrangements of the countries that belong to a monetary or currency union in which the same legal tender is shared by the members of the union are classified under the arrangement governing the joint currency.

    (b) *Currency board.* A currency board arrangement is a monetary arrangement based on an explicit legislative commitment to exchange domestic currency for a specified foreign currency at a fixed exchange rate, combined with restrictions on the issuance authority to ensure the fulfillment of its legal obligation. This implies that domestic currency is usually fully backed by foreign assets, eliminating traditional central bank functions.

    In the regimes (a) and (b) countries do not issue an independent currency.

(2) **Soft pegs**

    (a) *Conventional pegged arrangements.* The country formally or *de facto* pegs its currency at a fixed rate to another currency or a basket of currencies, where the basket is formed, for example, by the currencies of major trading or financial partners and the weights reflect the geographic distribution of trade, services, or capital flows, or the SDR. The anchor currency or basket weights are public or notified to the IMF. The country authorities stand ready to maintain the fixed parity through direct intervention.

(b) *Pegged exchange rates within horizontal bands.* The value of the currency is maintained within certain margins of fluctuation of at least ±1 % around a fixed central rate, or a margin between the maximum and minimum value of the exchange rate that exceeds 2 %. An example is the case of the multilateral exchange rate mechanism (ERM) of the European Monetary System (EMS), replaced with ERM II on January 1, 1999. This system was suggested by Halm (1965) under the name of wider band.

(c) *Crawling pegs.* The currency is adjusted in small amounts at a fixed rate or in response to changes in selected quantitative indicators, such as past inflation differentials vis-à-vis major trading partners or differentials between the inflation target and expected inflation in major trading partners. The commitment to maintaining crawling pegs imposes constraints on monetary policy. This system was suggested by Harrod (1933), Meade (1964), Williamson (1965, 1981).

(d) *Stabilized arrangements.* Classification as a stabilized arrangement entails a spot market exchange rate that remains within a margin of 2 % for six months or more (with the exception of a specified number of outliers or step adjustments) and is not floating. The required margin of stability can be met either with respect to a single currency or a basket of currencies.

(e) *Crawl-like arrangements.* The exchange rate must remain within a narrow margin of 2 % relative to a statistically identified trend for six months or more (with the exception of a specified number of outliers), and the exchange rate arrangement cannot be considered as floating.

(3) **Floating regimes**

(a) *Floating.* A floating exchange rate is largely market determined, without an ascertainable or predictable path for the rate.

(b) *Free floating.* A floating exchange rate can be classified as free floating if intervention occurs only exceptionally and aims to address disorderly market conditions and if the authorities have provided information or data confirming that intervention has been limited to at most three instances in the previous six months, each lasting no more than three business days.

(4) **Other managed** Any arrangement that does not fall into any of the categories described above is assigned to this category. Arrangements characterized by frequent shifts in policies may fall into this category.

Table 3.1 Exchange rate arrangements, 2009–2014 (percent of IMF members)

Exchange rate arrangements	2009	2010	2011	2012	2013	2014
Hard peg	12.2	13.2	13.2	13.2	13.1	13.1
No separate Legal Tender	5.3	6.3	6.8	6.8	6.8	6.8
Currency Board	6.9	6.9	6.3	6.3	6.3	6.3
Soft peg	34.6	39.7	43.2	39.5	42.9	43.5
Conventional peg	22.3	23.3	22.6	22.6	23.6	23.0
Stabilized arrangement	6.9	12.7	12.1	8.4	9.9	11.0
Crawling peg	2.7	1.6	1.6	1.6	1.0	1.0
Crawl-like arrangement	0.5	1.1	6.3	6.3	7.9	7.9
Pegged exchange rate	2.1	1.1	0.5	0.5	0.5	0.5
Floating	42.0	36.0	34.7	34.7	34.0	34.0
Floating	24.5	20.1	18.9	18.4	18.3	18.8
Free floating	17.6	15.9	15.8	16.3	15.7	15.2
Residual						
Other managed arrangement	11.2	11.1	8.9	12.6	9.9	9.4

Source: International Monetary Fund (2014)

Table 3.1 presents information regarding the exchange rate arrangements of IMF's member countries. It is based on members' actual, *de facto* regimes, as classified by the IMF as of April 2014, which may differ from their officially announced arrangements.

3.2 The Bretton Woods System

The exchange rate system that was put into being after the end of World War II and which is called the Bretton Woods system (after the name of the New Hampshire town where the negotiations took place and where the final agreement was signed in 1944), belonged to the category of the limping gold exchange standard with important modifications. To synthesize to the utmost, each country declared a *par value* or *parity* of its own currency in terms of gold, from which the bilateral parities automatically derived. However, at that time, the only currency convertible into gold at the fixed price of $35 per ounce of gold was the US dollar, which in this sense became the *key currency*. The convertibility of the other currencies into dollars qualified the system as a gold exchange standard, limping because the convertibility of dollars into gold was restricted to the requests from central banks.

The member countries were required to stand ready to maintain the declared parity in the foreign exchange market by buying and selling foreign exchange (usually dollars, which thus became the main *intervention currency*); more precisely, the actual exchange rate could vary only within the so-called *support* (or *intervention*) *points*, which were initially set at 1 % above or below parity.

The modifications consisted in the fact that parity, notwithstanding the obligation to defend it, was not immutable, but could be changed in the case of "fundamental disequilibrium" in accordance with certain rules: changes up to 10 % could be made at the discretion of the country, whilst for greater changes the country had first to notify the IMF (the International Monetary Fund, which is one of the international organizations set up by the Bretton Woods agreement) and obtain its assent.

The obligation to maintain the declared parity together with the possibility of changing it gave the system the name of *adjustable peg*. The idea behind it was a compromise between rigidly fixed and freely flexible exchange rates, and it is clear that the greater or lesser extent to which it approached either system depended essentially on the interpretation of the rules for changing parity. The prevailing interpretation was restrictive, in the sense that parity was to be defended at all costs and changed only when it was unavoidable, but this point will be taken up again in Sect. 22.5.

3.2.1 The Monetary Authorities' Intervention

In any case, the defence of a given parity requires a continuous intervention of the monetary authorities in the foreign exchange market: the authorities stand ready to meet both the market excess demand for foreign exchange and the market excess supply when these arise. The alternative to this intervention is to act on other macroeconomic variables of the system, so as to eliminate or reduce the excess demand, or to introduce administrative controls on foreign exchange. In the latter case, the foreign exchange is rationed by the monetary authorities and economic agents cannot freely engage in international transactions. Excluding this case, what happens is that if, for example, at the given parity the market demand for foreign exchange is higher than the supply by a certain amount, the monetary authorities must intervene by supplying the market with that amount, because if they did not do so, the pressure of excess demand for foreign exchange would cause a depreciation in the exchange rate. And vice versa in the case of an excess supply of foreign exchange on the market.

To put this in graphic form, let us consider Fig. 3.1, where a simple partial equilibrium analysis of the foreign exchange market has been depicted, on the assumption that the supply of foreign exchange is a well-behaved (increasing) function of its price (the exchange rate) and the demand for foreign exchange is a decreasing function of the exchange rate. In reality these demands and supplies are not necessarily well behaved (see Chap. 7, Sect. 7.3), and depend on a lot of other factors, which will determine shifts in the schedules, but we shall ignore these complications. We further assume that the market behaves as all other markets, i.e., the price (in our case the price of foreign currency is the exchange rate) tends to increase (decrease) if there is an excess demand (excess supply) in the market.

Let us now suppose that the exchange rate has to be pegged at r' whilst the market is in equilibrium at r_e. In the absence of official intervention, the exchange rate would move towards r_e, driven by the excess supply of foreign exchange.

Fig. 3.1 Monetary
authorities' intervention to
peg the exchange rate

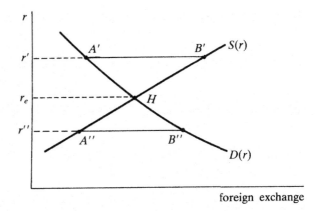

To prevent this from happening, the monetary authorities must absorb, as residual buyers, the excess supply $A'B'$ (providing the market with the corresponding amount of domestic currency). If, on the contrary, the exchange rate were to be pegged at r'', to prevent it from depreciating towards r_e in response to the pressure of excess demand for foreign exchange, the monetary authorities would have to meet (as residual sellers) the excess demand, by supplying an amount $A''B''$ of foreign currency to the market (absorbing from the market the corresponding amount of domestic currency).

It should be pointed out that, as the schedules in question represent *flows*, the monetary authorities must (ceteris paribus) go on absorbing $A'B'$, or supplying $A''B''$, of foreign exchange *per unit of time*. This may well give rise to problems, especially in the case r'', because by continuously giving up foreign exchange the monetary authorities run out of reserves. These problems will be dealt with in Parts III and following.

So much as regards the spot exchange market. As regards the forward market, the Bretton Woods system did not contemplate a similar obligation to intervene. For an examination of the advisability of such an intervention, see Sect. 7.4.3.

The Bretton Woods system collapsed with the declaration of the legal (*de jure*) inconvertibility into gold of the US dollar on August 15, 1971. It was *de jure*, but the dollar had actually been inconvertible (*de facto* inconvertibility) for several years. The amount of dollars officially held by non-US central banks was, in fact, much greater than the official US gold reserve, and the system was able to keep going only because these central banks did not actually demand the conversion of dollars into gold. Therefore a *de facto "dollar standard"* prevailed: see Sects. 17.3.1 and 22.5.

After the collapse, the Bretton Woods system was replaced by a situation in which many countries adopt a regime of *managed* or *dirty float*, where no officially declared parities exist (except for possible agreements among specific countries forming a currency area) and the exchange rates float, albeit with more or less pronounced interventions on the part of the monetary authorities. The member countries of the IMF have agreed (Second Amendment to the Articles of Agreement of the Fund, which came into force in March, 1978) to adhere to certain general principles in their interventions in the exchange markets, amongst which that of not manipulating exchange rates in order to prevent effective balance of payments adjustment or in order to gain an unfair competitive advantage over other members. For a treatment of these problems from both a theoretical point of view and the point of view of actual practice, see, for example, Polak (1999).

The managed float belongs to the category of limited-flexibility exchange-rate systems, of which we have given a classification in Sect. 3.1.

3.3   The Current Nonsystem

After the collapse of the Bretton Woods system, no other replaced it, if by system we mean a coherent set of rules (rights and obligations) and a precise exchange rate regime universally adopted. Williamson (1976) aptly coined the name "nonsystem" to denote such a situation, still in force. In fact, the situation at the moment of going to the press, is that each country can choose the exchange-rate regime that it prefers and notify its choice to the IMF, so that various regimes coexist. Some countries peg their exchange rate to a reference currency (usually the dollar, but also the French franc and other currencies) with zero or very narrow margins; naturally they will follow the reference currency's regime with respect to the other countries. Then there are other countries which peg their currency to a composite currency[1] such as, for example, the IMF's Special Drawing Right (SDR). Groups of countries enter into monetary agreements to form currency areas, by maintaining fixed exchange rates among themselves, or monetary unions with a common currency, such as the European Monetary Union. The situation at the time of writing is given above, Table 3.1, but it is continually changing, hence the reader had better to consult the monthly *International Financial Statistics* published by the International Monetary Fund where a table is present showing the exchange rate arrangements existing at the moment.

We conclude this section by stressing again that we have deliberately abstained from giving a comparative evaluation of the various exchange-rate regimes. The reason is that such an evaluation requires familiarity with notions (adjustment processes of the balance of payments, macroeconomic equilibrium in an open economy, etc.), which will be dealt with in the following chapters. Thus the comparative evaluation of the different exchange-rate regimes, which has led to hot debates in the literature, has to be deferred (see Chap. 17).

---

[1] A composite currency, also called a "basket-currency" is an artificial currency consisting of predetermined amounts of various currencies. Another example of a basket-currency was the ECU (European Currency Unit). On the characteristics of basket currencies in general see Sect. 20.3; on SDRs see Sect. 22.4.

3.4 International Organisations

The international organisations dealing with economic matters are numerous, but here we shall only examine the International Monetary Fund and the World Bank.

The American and British governments, in the hope of avoiding the international economic disorder that had followed World War I, in the early 1940s took the initiative of assembling governments and experts to design rules and institutions for post-World War II monetary and financial relations. The agreements that emerged were adopted by 44 nations at a conference held in Bretton Woods (new Hampshire) in July 1944. The system that was set up took the name of Bretton Woods system and has been described above. Although this system collapsed in 1971, the institutions that were created at the Bretton Woods conference still exist, and are the IMF (International Monetary Fund) and the IBRD (International Bank for Reconstruction and Development, commonly known as the World Bank).

The Fund was set up to deal with monetary matters; the World Bank (as its full name says) to promote a flow of long-term loans for purposes of reconstruction and development. The Fund was of course seen as centre of the international monetary system, and it is unfortunately impossible to examine even cursorily the debate between the plans of White and Keynes, who had completely different ideas as to the role of the future Fund.

3.4.1 The IMF

The International Monetary Fund (http://imf.org) began operations in Washington, DC in May 1946 with 39 members; it now has 188 members. On joining the IMF each member country contributes a certain sum of money called a quota subscription, based broadly on its relative size in the world economy, which determines its maximum contribution to the IMF's financial resources. The IMF uses a quota formula to help assess a member's relative position. The current quota formula is a weighted average of GDP (weight of 50 %), openness (30 %), economic variability (15 %), and international reserves (5 %). Quotas are reviewed every 5 years and can be lowered or raised on the basis of the needs of the IMF and the economic situation of the member. The Fourteenth General Review of Quotas was in December 2012, with a decision to double the IMF's quota resources to SDR 477 billion and a major realignment of quota and voting shares to emerging and developing countries (with a more than 6 % quota shift to dynamic emerging market and developing countries and under-represented countries.

In 1946, the 39 members paid in the equivalent of $7.6 billion; by 2015, the 188 members had paid in the equivalent of about $327 billion, and there is a proposal to raise quotas to a still higher value.

Quotas serve various purposes:

(1) they constitute the resources from which the IMF draws to make loans to members in financial difficulty;
(2) they are the basis for determining how much a member can borrow from the IMF, or receives from the IMF in periodic allocations of special assets known as SDRs (Special Drawing Rights, on which see Chap. 22, Sect. 22.4);
(3) they determine the voting power of the member. Each IMF member's votes are comprised of basic votes plus one additional vote for each SDR 100,000 of quota.

The highest link of the chain of command in the Fund is the Board of Governors and Alternate Governors (one Governor and one Alternate per member). These persons are ministers of finance or heads of central banks. The Board of Governors meets once each year at the IMF-World Bank Annual Meetings. Twenty-four of the Governors sit on the International Monetary and Financial Committee (IMFC) and normally meet twice each year. The day-to-day management of the Fund is delegated to the Executive Board chaired by the Managing Director. The Executive Board consists of 24 Directors. Together, these 24 board members represent all 188 countries. Large economies, such as the United States and China, have their own seat at the table but most countries are grouped in constituencies representing 4 or more countries. The largest constituency includes 24 countries. By tradition, the Managing Director is a non-US national, while the President of the World Bank is a US national.

The IMF performs three main functions in the interest of an orderly functioning of the international monetary system:

(1) *Surveillance.* After the demise of the Bretton Woods system it seemed that the preeminent role of the IMF would disappear. This has not happened, as under the current system the IMF has been entrusted with the examination of all aspects of any member's economy that are relevant for that member's exchange rate and with the evaluation of the economy's performance for the entire membership. This entails more scope for the IMF to monitor members' policies. The activity we are describing is called by the IMF "surveillance" over members' exchange policies, and is carried out through periodic consultations conducted in the member country.

It provides regular assessment of global prospects in its World Economic Outlook, of financial markets in its Global Financial Stability Report, and of public finance developments in its Fiscal Monitor, and publishes a series of regional economic outlooks.
(2) *Financial assistance.* This is perhaps the most visible activity to the general public: as of July 1998, for example, the IMF had committed about $35 billion to Indonesia, Korea and Thailand to help them with their financial crisis (the Asian crisis) and around $21 billion to Russia to support its economic program.

Deep crises in Latin America and Turkey kept demand for IMF resources high in the early 2000s. IMF lending rose again in late 2008 in the wake of the global financial crisis. In response to the global economic crisis, the IMF strengthened its lending capacity and approved a major overhaul of its financial support mechanisms in April 2009, with further reforms adopted in 2010 and 2011. These reforms focused on enhancing crisis prevention, mitigating contagion during systemic crisis, and tailoring instruments based on members' performances and circumstances.

The general rules for obtaining financial assistance from the Fund are the following.

Lending Procedure Upon request by a member country, IMF provides resources under a lending "arrangement," which may entail specific economic policies and measures a country agrees to implement to resolve its balance of payments problem. The economic policy program underlying an arrangement is formulated by the country in consultation with the IMF, and is in most cases presented to the Fund's Executive Board in a "Letter of Intent" and is further detailed in the annexed "Memorandum of Understanding".

Access to Financing The amount of financing a member can obtain from the IMF (its access limit) is based on its quota. For example, under Stand-By and Extended Arrangements, a member can borrow up to 200 % of its quota annually and 600 % cumulatively.

Lending Instruments The IMF has developed various loan instruments tailored to address the specific circumstances of members. Low-income countries may borrow on concessional terms through the Extended Credit Facility (ECF), the Standby Credit Facility (SCF) and the Rapid Credit Facility (RCF). Concessional loans carry zero interest rates until the end of 2016. Non-concessional loans are provided mainly through Stand-By Arrangements (SBA), the Flexible Credit Line (FCL), the Precautionary and Liquidity Line (PLL), and the Extended Fund Facility (which is useful primarily for medium-term needs). The IMF also can provide emergency assistance via the Rapid Financing Instrument (RFI) to all its members facing urgent balance of payments needs. All non-concessional facilities are subject to the IMF's market-related interest rate, known as the "rate of charge," and large loans carry a surcharge. The rate of charge is based on the SDR interest rate, which is revised weekly to take into account changes in short-term interest rates in major international money markets. The maximum amount that a country can borrow from the IMF, its access limit, differs depending on the loan category, but is typically a multiple of the country's IMF quota. This limit may be exceeded in exceptional circumstances.

(3) *Technical Assistance.* Members (for example developing countries, countries moving from planned to market economy such as Russia, Eastern European countries, etc.), may sometimes lack expertise in highly technical areas of

central banking and public finance, and thus turn to the IMF for technical assistance, including advice by Fund's experts, training of the member's officials in Washington or locally. In particular, the IMF provides technical assistance in its areas of core expertise: macroeconomic policy, tax policy and revenue administration, expenditure management, monetary policy, the exchange rate system, financial sector stability, legislative frameworks, and macroeconomic and financial statistics.

3.4.2   The World Bank

The International Bank for Reconstruction and Development, commonly know as World Bank (http://worldbank.org), provides loans and development assistance to creditworthy poor countries as well as to middle-income countries. Its organization is conceptually similar to that of the IMF: it is like a cooperative, made up of 188 member countries which are shareholders with voting power proportional to the members' capital subscriptions, that in turn are based on each country's economic strength. These member countries, or shareholders, are represented by a Board of Governors, who are the ultimate policymakers at the World Bank. Generally, the governors are member countries' ministers of finance or ministers of development. They meet once a year at the Annual Meetings of the Boards of Governors of the World Bank Group and the International Monetary Fund.

Each member appoints a Governor and an Alternate Governor, who meet once a year. The day-to-day management of the World Bank is carried out by a Board consisting of 25 Executive Directors chaired by a President (by tradition a national of the United States). The five largest shareholders appoint an executive director, while other member countries are represented by elected executive directors.

While the task of the IMF is to promote a well functioning and orderly international monetary system, the main task of the World Bank is to promote growth of poorer countries. Contrary to the IMF, whose resources are the members' quotas, the World Bank raises almost all its funds in financial markets by selling bonds and other assets. Its average annual loans are around $45–50 billion.

Over the years the World Bank has become a group, consisting of five institutions: the IBRD proper (lends to governments of middle-income and creditworthy low-income countries), the IDA (International Development Association, that provides interest-free loans, or credits, and grants to governments of the poorest countries), IFC (International Finance Corporation, which provides loans, equity, and advisory services to stimulate private sector investment in developing countries), MIGA (Multilateral Investment Guarantee Agency, which provides political risk insurance or guarantees to foreign investors against losses caused by non-commercial risk to facilitate foreign direct investment in developing countries), and ICSID (International Center for Settlement of Investment Disputes, which arbitrates disputes between foreign investors and the country where they have invested).

References

Halm, G. (1965). *The "band" proposal: The limits of permissible exchange rate variations.* Special Papers in International Economics, No. 6, International Finance Section, Princeton University.

Habermeier, K., Kokenyne, A., Veyrune, R., & Anderson, H. (2009). *Revised system for the classification of exchange rate arrangement.* IMF WP No. 211.

Harrod, R. F. (1933). *International economics.* Cambridge (UK): Cambridge University Press.

International Monetary Fund (2014). *Annual report on exchange arrangements and exchange restrictions.*

Meade, J. E. (1964). The international monetary mechanism. *Three Banks Review*, September, 3–25.

Polak, J. J. (1999). *Streamlining the financial structure of the International Monetary Fund.* Essays in International Finance, No. 216, International Finance Section, Princeton University.

Williamson, J. (1965). *The crawling peg.* Essays in International Finance, No. 90, International Finance Section, Princeton University.

Williamson, J. (1976). The benefits and costs of an international nonsystem. In E. M. Bernstein et al., (Eds.), *Reflections on Jamaica*, Essays in International Finance, No. 115, International Finance Section, Princeton University.

Williamson, J. (Ed.) (1981). *Exchange rate rules: the theory, performance and prospects of the crawling peg.* London: Macmillan.

International Interest-Rate Parity Conditions 4

The relations between interest rates (domestic and foreign) and exchange rates (spot and forward) that were already mentioned in Chap. 2, Sect. 2.5.2 are very important and frequently used in international finance. Hence, we give here a general overview, with additional important considerations on the efficiency of the foreign exchange market and on capital mobility.

4.1 Covered Interest Arbitrage, and Covered Interest Parity (CIP)

In general, *interest arbitrage* is an operation that aims to benefit from the short-term employment of liquid funds in the financial centre where the yield is highest: we are in the presence of economic agents engaged in purely financial operations. As, however, these agents are not speculators, they will cover themselves against exchange risk (by having recourse to the forward exchange market), hence the denomination of *covered* interest arbitrage. Since Keynes (1923) is credited with the first precise treatment of this problem, the theory is also referred to as the Keynesian theory of covered interest arbitrage.

Let us consider, for example, an agent who has to place a certain amount of domestic currency short-term, and assume that the interest rates are independent of the amount of funds placed or that this amount is not so huge as to give its owner the power to influence market interest rates significantly, so that we can reason at the unit level. For each unit of domestic currency placed at home short-term, the agent will obtain, after the stipulated period has elapsed, the amount $(1 + i_h)$, where i_h is referred to this same period. Alternatively, the agent can buy foreign currency spot and place it abroad for the same period of time[1]: as $(1/r)$ of foreign currency is

---

[1] To make the two alternatives comparable, the domestic and foreign assets must have similar characteristics (including similar risk of default).

© Springer-Verlag Berlin Heidelberg 2016
G. Gandolfo, *International Finance and Open-Economy Macroeconomics*,
Springer Texts in Business and Economics, DOI 10.1007/978-3-662-49862-0_4

obtained per unit of domestic currency, the amount $(1/r)(1+i_f)$ of foreign currency will be obtained at the end of the period, where i_f is the foreign interest rate referring to this same period. To eliminate any exchange risk, the agent can now sell that amount of foreign currency forward: thus he will obtain, after the stipulated period has elapsed, the amount $r^F(1/r)(1+i_f)$ of domestic currency with no exchange risk.

Now, if, for the sake of simplicity, we assume that the costs of the operations are equal, it is obvious that the agent will place the funds at home or abroad according as $(1+i_h) \gtreqless r^F(1/r)(1+i_f)$, whilst he will be indifferent in the case of equality. Since as can be easily checked the same conditions hold when the arbitrageur does not own the funds but has to borrow them, or when the funds are in foreign currency, it follows that funds will flow in (inward arbitrage), have no incentive to move, flow out (outward arbitrage) according as

$$(1+i_h) \gtreqless \frac{r^F}{r}(1+i_f). \tag{4.1}$$

If we divide through by $\left(1+i_f\right)$ and exchange sides, this condition can be written as

$$\frac{r^F}{r} \lesseqgtr \frac{1+i_h}{1+i_f}, \tag{4.2}$$

whence, by subtracting one from both sides,

$$\frac{r^F - r}{r} \lesseqgtr \frac{i_h - i_f}{1+i_f}. \tag{4.3}$$

Note that this inequality coincides with the inequality concerning the various covering alternatives of commercial traders, see Eq. (2.20), of course when both refer to the same period of time (as we said above, the computations are usually normalized on a per annum basis).

The condition of equality in (4.2) or in (4.3), that is when funds have no incentive to move from where they are placed, is called the *neutrality condition* and the forward rate is said to be at *interest parity* or simply that *covered interest parity*(CIP) prevails, and the corresponding forward exchange rate is called the *parity* forward rate. When there is a difference between the forward margin and the interest rate differential such that funds tend to flow in (out), we say that there is an *intrinsic* premium (discount) for the domestic currency.

Sometimes the condition of equality is wrongly called equilibrium condition: in fact, as we shall see in Sect. 7.4, the foreign exchange market can be in equilibrium even if the neutrality condition does not occur.

The equations that define CIP

$$\frac{1 + i_h}{1 + i_f} = \frac{r^F}{r}, \tag{4.4}$$

$$\frac{i_h - i_f}{1 + i_f} = \frac{r^F - r}{r}, \tag{4.5}$$

can be written in alternative specifications, that are often used in the literature. From the algebra of logarithms we recall that

$$\ln(1 + x) \simeq x \tag{4.6}$$

for small x. If we let $x = (y - z)/z$ we also have

$$\ln[y/z] \equiv \ln[1 + (y - z)/z] \simeq (y - z)/z. \tag{4.7}$$

Now, if we take the logarithms of both members of Eq. (4.4), we get

$$\ln(1 + i_h) - \ln(1 + i_f) = \ln(r^F/r), \tag{4.8}$$

whence, using the approximation (4.6) and

$$i_h - i_f = \ln r^F - \ln r, \tag{4.9}$$

and, if we also use approximation (4.7),

$$i_h - i_f = \frac{r^F - r}{r}, \quad i_h = i_f + \frac{r^F - r}{r} \tag{4.10}$$

i.e. the interest differential equals the forward margin, or the domestic interest rate equals the foreign interest rate plus the (positive or negative) forward margin. The same relations can be directly obtained from (4.5) by neglecting the denominator of the fraction on the left-hand side.

A precise measure of the approximation error made using (4.10) instead of (4.5) can easily be obtained. Let us consider the identity[2]

$$\frac{i_h - i_f}{1 + i_f} \equiv \left(i_h - i_f\right) + \left(i_h - i_f\right) \frac{-i_f}{1 + i_f}. \tag{4.11}$$

The quantity $-\left(i_h - i_f\right)\left[i_f / \left(1 + i_f\right)\right]$ is thus the measure of the approximation error.

Equations (4.9) and (4.10) are also referred to as the *covered interest parity* conditions.

4.2   Uncovered Interest Parity (UIP)

Let us consider an agent who holds deterministic (or certain) exchange-rate expectations, namely is sure of the exactness of his expectations about the future value of the spot exchange rate. Alternatively we can assume that the agent is *risk neutral*, namely is indifferent to seeking forward cover because, unlike arbitrageurs, only cares about the yield of his funds and not about the risk. Suppose that such an agent has to place a certain amount of funds short-term, that we assume to be denominated in domestic currency (if they are denominated in foreign currency the result will not change). He will consider the alternative between:

(a) investing his funds at home (earning the interest rate i_h), or
(b) converting them into foreign currency at the current spot exchange rate r, placing them abroad (earning the interest rate i_f), and converting them (principal plus interest accrued) back into domestic currency at the end of the period considered, using the expected spot exchange rate ($\tilde{r}$) to carry out this conversion.

---

[2]To obtain this expression consider the power series

$$\frac{1}{1 + i_f} = 1 - i_f + (i_f)^2 - (i_f)^3 + (i_f)^4 - (i_f)^5 + \ldots = \sum_{j=0}^{\infty} (-1)^j (i_f)^j,$$

which converges, since $i_f < 1$. Therefore we get

$$\left(i_h - i_f\right) \frac{1}{1 + i_f} = \left(i_h - i_f\right)\left[1 - i_f + (i_f)^2 - (i_f)^3 + (i_f)^4 - (i_f)^5 + \ldots\right]$$

$$= \left(i_h - i_f\right) + \left(i_h - i_f\right) \sum_{j=1}^{\infty} (-1)^j (i_f)^j$$

$$= \left(i_h - i_f\right) + \left(i_h - i_f\right) \frac{-i_f}{1 + i_f}.$$

The agent will be indifferent between the two alternatives when

$$(1 + i_h) = \left[\frac{1}{r}(1 + i_f)\right]\tilde{r}, \tag{4.12}$$

where the interest rates and expectations are referred to the same time horizon. If the two sides of Eq. (4.12) are not equal, the agent can earn a profit by shifting funds in or out of the country according as the left-hand side of Eq. (4.12) is greater or smaller than the right-hand side.

If we divide both members of (4.12) by $(1 + i_f)$, we get

$$\frac{1 + i_h}{1 + i_f} = \frac{\tilde{r}}{r},$$

whence, subtracting one from both members,

$$\frac{i_h - i_f}{1 + i_f} = \frac{\tilde{r} - r}{r}. \tag{4.13}$$

From these relations we obtain, using the logarithmic approximations (4.6) and (4.7),

$$i_h - i_f = \ln \tilde{r} - \ln r = \frac{\tilde{r} - r}{r}, \tag{4.14}$$

or

$$i_h = i_f + \frac{\tilde{r} - r}{r}. \tag{4.15}$$

This condition, according to which the interest differential is equal to the expected variation in the spot exchange rate or, equivalently, the domestic interest rate equals the foreign interest rate plus the expected variation in the exchange rate, is called the *uncovered interest parity* (UIP) condition.

4.3 Uncovered Interest Parity with Risk Premium

Both deterministic expectations and risk neutrality are rather strong assumptions, so that in the normal case of agents who are uncertain about the future value of the exchange rate and/or are risk averse, a risk coefficient or *risk premium* has to be introduced. The reasoning is the same as that used in relation to foreign-exchange speculators—see Sect. 2.6.1, in particular Eq. (2.21). Thus we have

$$i_h = i_f + \frac{\tilde{r} - r}{r} + \delta, \tag{4.16}$$

where δ is the risk coefficient or risk premium, expressed in proportional or percentage terms like the other variables appearing in the equation. It is not surprising that (4.16) is equal to (2.21): the motivations are different, but the underlying economic calculations are the same. In fact, for speculators the main element of profit is the expected change in the exchange rate, and the interest rates enter the picture rationally to compare the alternatives. For financial investors the main element of profit is the interest differential, and the expected variation in the exchange rate enters the picture rationally to compare the alternatives. The final result is in any case the same.

4.4   Real Interest Parity

The interest rates so far considered are nominal rates. It may however be interesting to reason in terms of *real* interest rates. According to the well-known Fisher definition, real and nominal interest rates are related by the *expected* inflation rate (naturally referred to the same time horizon as the interest rates). More precisely,

$$i_{Rh} = i_h - \frac{\tilde{p}_h - p_h}{p_h}, \qquad (4.17)$$

where i_{Rh} is the real interest rate and $(\tilde{p}_h - p_h)/p_h$ is the expected inflation rate. A similar definition holds for the rest-of-the-world real interest rate, namely

$$i_{Rf} = i_f - \frac{\tilde{p}_f - p_f}{p_f}. \qquad (4.18)$$

Let us now consider the uncovered interest parity condition (4.15), that we report here for the reader's convenience,

$$i_h = i_f + \frac{\tilde{r} - r}{r},$$

and assume that purchasing power parity (PPP) holds. PPP will be dealt with at some length in Sect. 15.1, and if we apply it to expected changes, one of its implications is that the expected variation in the exchange rate equals the difference between the expected inflation rates in the two countries, namely

$$\frac{\tilde{r} - r}{r} = \frac{\tilde{p}_h - p_h}{p_h} - \frac{\tilde{p}_f - p_f}{p_f}. \qquad (4.19)$$

This relation is also called *ex ante* PPP, since it is obtained applying the expectation operator to (the relative version of) PPP. If we now substitute the expected variation in the exchange rate, as given by Eq. (4.19), into the UIP condition, and use the definitions (4.17) and (4.18), we get the relation

$$i_{Rh} = i_{Rf}, \qquad (4.20)$$

which is called *real interest parity*. It should be noted that, if we accept the orthodox neoclassical theory, according to which the real interest rate equals the marginal productivity of capital, Eq. (4.20) is equivalent to the equalization of the price of the factor capital (the factor-price equalization theorem, well known in the theory of international trade: see Gandolfo 2014). It is then interesting to observe that, if we assume from the beginning that factor-price equalization holds (i.e., Eq. (4.20) becomes our initial assumption), then—by reasoning backwards—we *prove* PPP as a *result* of the analysis (see Sect. 15.1).

4.5 Efficiency of the Foreign Exchange Market

According to the generally accepted definition of Fama (1970, 1976), a market is efficient when it fully uses all available information or, equivalently, when current prices fully reflect all available information and so there are no unexploited profit opportunities (there are various degrees of efficiency, but they need not concern us here). Then, by definition, in an efficient foreign exchange market both covered and uncovered interest parity must hold.

We have in fact seen (Sect. 4.1) that, if there is a discrepancy between the two sides of the CIP equation, it will be possible to make profits by shifting funds at home or abroad, according to the sign of the discrepancy. But, if assume perfect capital mobility (so that no impediments exist to capital flows), such a profit opportunity cannot exist if the market is efficient.

Similarly, a risk-neutral agent will be able to make profits by shifting funds at home or abroad, as the case may be, if the UIP condition (4.15) does not hold. If the agent is risk averse or has uncertain expectations, we shall have to consider the possible discrepancy between the two sides of Eq. (4.16). This last observation also holds for speculators, as we have shown in Sects. 2.6.1 and 4.3.

If we assume the interest rates as given, in the foreign exchange market the variables that must reflect the available information are the spot (both current and expected) and forward exchange rate. Hence, since both CIP and UIP hold if the market is efficient, we have

$$\frac{r^F - r}{r} = \frac{\tilde{r} - r}{r} \qquad (4.21)$$

from which

$$r^F = \tilde{r}, \qquad (4.22)$$

namely the forward exchange rate and the expected spot exchange rate (both referred to the same time horizon) coincide. In a stochastic context (see the Appendix) it turns out that the forward exchange rate is an unbiased and efficient predictor of the future spot exchange rate.

In the case of risk premium, using (4.10) and (4.16) we get

$$\frac{r^F - r}{r} = \frac{\tilde{r} - r}{r} + \delta, \tag{4.23}$$

namely the forward margin is equal to the sum of the expected variation in the spot exchange rate and the risk coefficient. In terms of levels we have

$$r^F = \tilde{r} + RP, \tag{4.24}$$

where $RP = \delta r$ is the risk premium in levels.

The assumption of efficiency of the foreign exchange market, together with the various parity conditions, has been subjected to innumerable empirical studies. These have generally not given favourable results for either the market efficiency hypothesis or the parity conditions (see, for example, the surveys by Froot and Thaler 1990, MacDonald and Taylor 1992a; Sect. III, 1992b, Taylor 1995, Chinn 2006, Olmo and Pilbeam 2011, Engel 2014).

4.6   Perfect Capital Mobility, Perfect Asset Substitutability, and Interest Parity Conditions

From the theoretical point of view the verification of the interest parity conditions requires *perfect capital mobility*, by which we mean that asset holders are completely free instantaneously to move their funds abroad or to repatriate them. Imperfect capital mobility could derive, e.g., from administrative obstacles such as controls on capital movements, from high transaction costs, and so on. Asset holders can thus instantaneously realize the desired portfolio composition by moving their funds. An equivalent way of saying this is that the speed of adjustment of the actual to the desired portfolio is infinite, so that the adjustment of the actual to the desired portfolio is instantaneous. It follows that the current portfolio is always equal to the desired one.

Perfect capital mobility thus implies covered interest parity: the domestic interest rate is equal to the foreign interest rate plus the forward margin. It does not, however, necessarily imply uncovered interest parity, unless a further assumption is introduced: the assumption of *perfect substitutability* between domestic and foreign assets. This means that assets holders are indifferent as to the composition (in terms of domestic and foreign bonds) of their portfolios so long as the expected yield of domestic and foreign bonds (of equivalent characteristics such as maturity, safety, etc.) is identical when expressed in a common numéraire. Taking for example the domestic currency as numéraire, the expected yield of domestic bonds is the domestic interest rate, while the expected yield of foreign bonds is the foreign interest rate, to which the expected variation in the exchange rate must be added algebraically (to account for the expected capital gains or losses when the foreign currency is transformed into domestic currency). But this is exactly the uncovered interest parity condition.

When domestic and foreign assets are not perfect substitutes, UIP cannot hold, but uncovered interest parity with risk premium will hold. The factors that, besides exchange risk, make domestic and foreign bonds imperfect substitutes are, amongst others, political risk, the risk of default, the risk of the introduction of controls on capital movements, liquidity consideration, and so on.

It is now as well to clarify a point. Perfect capital mobility plus perfect asset substitutability imply, as we have just seen, that both CIP and UIP hold, namely $i_h = i_f + \left(r^F - r\right)/r = i_f + (\tilde{r} - r)/r$. It is however often stated that perfect capital mobility plus perfect asset substitutability imply that the domestic interest rate cannot deviate from the foreign interest rate, i.e. $i_h = i_f$. This is not automatically true, unless further assumptions are made. More precisely, the additional property of (nominal) interest rate equalization requires that the agents involved expect the future spot exchange rate to remain equal to its current value (*static expectations*). This amounts to assuming $\tilde{r} = r$ and so, since there is no perceived exchange risk, the forward exchange rate is also equal to the current spot exchange rate, $r^F = r$.

In the case of fixed exchange rates for this to hold it is sufficient that the various agents are convinced of the fixity of the exchange rate: the current spot exchange rate, in other words, is *credible*. In the case of flexible exchange rates, static expectations are a bit less plausible. However, they remain valid in the case of agents who do not hold any particular expectation about the future spot exchange rate, i.e. hold appreciations and depreciations in the exchange rate as equally likely. If so, the best forecast is exactly $\tilde{r} = r$. More technically, this means that agents implicitly assume a (driftless) random walk, namely a relation of the type $\tilde{r} = r + \varepsilon$, where ε is a random variable with zero mean, constant finite variance, and zero autocovariance.

It should be stressed, in conclusion, that the distinction between perfect capital mobility and perfect asset substitutability (advocated by Dornbusch and Krugman 1976) is not generally accepted. Other authors, in fact (for example, Mundell 1963; p. 475; MacDonald 1988; Sect. 2.5) include in the notion of perfect capital mobility not only instantaneous portfolio adjustment, but also perfect asset substitutability. We agree with Dornbusch and Krugman in considering the distinction between the two notions important and useful, so that we accept it (see, for example, Sect. 15.3).

4.7   Appendix

4.7.1   Rational Expectations and Efficiency of the Foreign Exchange Market

Let us consider an agent who forms his exchange rate expectations according to the rational expectations hypothesis (REH). As is well known, rational expectations, introduced by Muth (1960, 1961), mean that the forecasting agent uses all available information, including the knowledge of the true model that determines the evolution of the variable(s) concerned. We recall that expectations formed according to REH have the following properties:

(a) there are no *systematic* forecast errors, as these errors are stochastically distributed with mean zero (*unbiasedness*). If $_{t-1}\tilde{r}_t$ denotes the expected exchange rate (expectations are formed at time $t - 1$ with reference to period t) then we have $_{t-1}\tilde{r}_t - r_t = \epsilon_t$, where $E(\epsilon_t) = 0$ for all t.

(b) Forecast errors are *uncorrelated*, i.e. $E(\epsilon_t \epsilon_{t-i}) = 0$ for any integer i.

(c) Forecast errors are not correlated with the past history of the variable being forecasted nor with other variable contained in the information set, I_{t-1}, available at the time of forecast (*orthogonality*, i.e. $E(\epsilon_t \mid I_{t-i}) = 0$). This property implies that RE are statistically *efficient*, namely the variance of the forecast error is lower than that of forecasts obtained with any other method (minimum variance property).

Rational expectations are a particularly convenient way of describing expectation formation in an efficient market. This is because REH, like EMH (efficient market hypothesis: see Fama 1970), presumes that economic agents do not make systematic errors when making their predictions. If we consider expectations formed at time t for period $t + 1$ and apply property (a), we have

$$r_{t+1} =_t \tilde{r}_{t+1} + \epsilon_{t+1}, \tag{4.25}$$

which states that the actual future spot exchange rate corresponds to that which was anticipated by rational economic agents plus a (positive or negative) "well behaved" random error. Here "rational" is taken to mean "using REH", and "well behaved" is taken to mean "independently distributed with mean zero" (properties (b) and (c) above).

Let us next assume that economic agents are risk neutral, so that there is no risk premium (see Sects. 4.2 and 4.3). As we know (see Eq. (4.22) in the text), in this case EMH implies

$$r_t^F =_t \tilde{r}_{t+1}, \tag{4.26}$$

where r_t^F is the forward exchange rate quoted at time t (for delivery at time $t + 1$). If we substitute Eq. (4.26) into Eq. (4.25) we have

$$r_{t+1} = r_t^F + \epsilon_{t+1}. \tag{4.27}$$

This equation states that, so long as agents have RE and there is no risk premium, the forward exchange rate is the best (i.e., minimum variance) unbiased predictor of the future spot exchange rate.

This conclusion does no longer hold if there is a risk premium. Instead of Eq. (4.22) we must now use Eq. (4.24), so that instead of Eq. (4.27) we get

$$r_{t+1} = r_t^F + RP_t + \epsilon_{t+1}. \tag{4.28}$$

Equation (4.28) shows that the use of the forward rate to predict the future spot rate may give rise to systematic forecast errors because of the presence of a risk premium.

The efficiency of the foreign exchange markets has been subject to extensive empirical testing. The result of these tests reject the EMH hypothesis (for a survey see MacDonald and Taylor 1992a; Sect. III). The fact is that what is usually tested is the joint hypothesis of rational expectations and risk neutrality. Hence it is not possible to say which leg of the joint hypothesis is responsible for the failure of EMH (both may be). Dropping the assumption of risk neutrality and introducing time-varying risk premia gives rise to mixed and inconclusive results. Hence, the failure of the joint efficiency hypothesis should perhaps be traced to the expectations leg of the hypothesis (MacDonald and Taylor 1992a,b). In fact, tests on the validity of the rational expectations hypothesis in foreign exchange markets have invariably led to reject this hypothesis (Ito 1990; Harvey 1999).

4.7.2 The Peso Problem

The consequences, on the relationships expressing REH and EMH (see the previous section), of the existence of a positive probability of a drastic event (for example the abandonment of fixed exchange rate) constitute the so called "peso problem". The name derives from the fact that these consequences were first observed in the foreign exchange market for the Mexican peso. In the period from April 1954 to August 1976, the peso/US$ spot exchange rate remained fixed at 0.080 dollars per peso. However, during all this period the forward exchange rate of the peso vis-à-vis the dollar was always smaller than the spot rate prevailing on the day of delivery (which was of course the given fixed spot rate). One possible interpretation of this evidence is to take it as rejecting REH and/or EMH. Another possibility is to interpret this evidence as reflecting the existence of a risk premium.

On 31 August 1976 the Mexican authorities abandoned the fixed parity, and the peso dropped to around 0.050 dollars per peso. Forward contracts stipulated before

31 August 1976 with a maturity after that date, used a forward exchange rate that
turned out to be higher than the post-devaluation spot rate, thus denoting that agents
had underestimated the extent of the devaluation. This, again, could be interpreted
as rejecting REH and EMH.

It was however shown by Lizondo (1983) that the peso-market evidence is
perfectly consistent with both REH and EMH if economic agents hypothesize the
existence, also in the fixed-parity period, of a positive probability of a devaluation in
forming their expectations. For further studies of the peso paradox see Lewis (1992,
2011).

4.7.3 The Siegel Paradox

We have seen in the text that the spot, forward, and expected exchange rates can be
replaced by their logarithms in defining CIP [see Eq. (4.9)] and UIP [see Eq. (4.14)].
In empirical analyses the use of logarithms is customary. This also serves to avoid
the *Siegel paradox*.

Siegel (1972) noted that, if the equality between the forward exchange rate and
the expected spot rate holds, it must hold on both sides of the foreign exchange.
This means that it must also hold for the foreign country. If we use the expectation
operator E and recall the consistency conditions (see Sect. 2.2), we have

$$r_t^F = E_t \left(r_{t+1} \right) \tag{4.29}$$

for the home country, and

$$\frac{1}{r_t^F} = E_t \left(\frac{1}{r_{t+1}} \right) \tag{4.30}$$

for the foreign country. However, a well-known theorem in statistics (Jensen's
inequality: see, for example, Mood et al. 1974; Sect. 4.5) states that $E(1/x) >
1/E(x)$ for any stochastic variable x. It follows that (4.29) and (4.30) cannot
simultaneously be true. Although it has been demonstrated that Siegel's paradox
is irrelevant in empirical work (McCulloch 1975; see, however, Sinn 1989, for an
opposite view), the conceptual problem remains. This can be avoided if one defines
the variables in logarithms. Then

$$\ln r_{t+1}^F = E_t \left(\ln r_{t+1} \right) \tag{4.31}$$

and

$$\ln \left(\frac{1}{r_{t+1}^F} \right) = E_t \left(\frac{1}{\ln r_{t+1}} \right). \tag{4.32}$$

Since $\ln\left(1/r^F_{t+1}\right) = -\ln r^F_{t+1}$, and $E_t(1/\ln r_{t+1}) = -E_t(\ln r_{t+1})$, conditions (4.31) and (4.32) are simultaneously true.

Another way to avoid the Siegel paradox has been suggested by Chu (2005), who observes that Jensen's inequality is actually a weak inequality [i.e., $E(1/x) \geq 1/E(x)$], which holds as an equality when x is a degenerate distribution. Chu shows that this is indeed the case when exchange rates are concerned, hence the paradox does not exist.

References

Chinn, M. D. (2006). The (partial) rehabilitation of interest rate parity in the floating rate era: longer horizons, alternative expectations, and emerging markets. *Journal of International Money and Finance, 25*, 7–21.

Chu, K. H. (2005). Solution to the Siegel paradox. *Open Economies Review, 16*, 399–405.

Dornbusch, R. & Krugman, P. (1976). Flexible exchange rates in the short run. *Brookings Papers on Economic Activity, 3*, 537–575.

Engel, C. (2014). Exchange rates and interest parity. In G. Gopinath, E. Helpmand & K. Rogoff (Eds.). *Handbook of international economics* (Vol. 4, Ch. 8, pp. 453–522). Amsterdam: Elsevier.

Fama, E. F. (1970). Efficient capital markets: A review of theory and empirical work. *Journal of Finance, 22*, 383–417.

Fama, E. F.(1976). *Foundations of finance*. New York: Basic Books.

Froot, K. A. & Thaler, R. H. (1990). Foreign exchang. *Journal of Economic Perspectives, 4*, 179–192.

Gandolfo, G. (2014). *International trade theory and policy* (Chap. 5, Sect. 5.2, 2nd ed.). Berlin, Heidelberg, New York: Springer.

Harvey, J. T. (1999). The nature of expectations in the foreign exchange market: A test of competing theories. *Journal of Post Keynesian Economics, 21*, 181–200.

Ito, T. (1990). Foreign exchange rate expectations: Micro survey data. *American Economic Review, 80*, 434–449.

Keynes, J. M. (1923). *A tract on monetary reform*. London: Macmillan; reprinted as Vol. IV of the Collected Writings of J. M. Keynes. London: Macmillan for the Royal Economic Society, 1971.

Lewis, K. K. (1992). Peso problem. In P. Newman, M. Milgate, & J. Eatwell (Eds.), *The new Palgrave dictionary of money and finance* (Vol. 3). London: Macmillan.

Lewis, K. K. (2011). Global asset pricing. *Annual Review of Financial Economics, 3*, 435–467.

Lizondo, J. S. (1983). Foreign exchange futures prices under fixed exchange rates. *Journal of International Economics, 14*, 69–84.

MacDonald, R. (1988). *Floating exchange rates: theories and evidence*. London: Unwin Hyman.

MacDonald, R., & Taylor, M. P. (1992a). Exchange rate economics: A survey. *IMF Staff Papers, 39*, 1–57.

MacDonald, R., & Taylor, M. P. (1992b). *Exchange rate economics*. Aldershot: Elgar.

McCulloch, J. H. (1975). Operational aspects of the Siegel paradox. *Quarterly Journal of Economics, 89*, 170–172.

Mood, A. M., Graybill, F. A., & Boes, D. C. (1974). *Introduction to the theory of statistics* (3rd ed.). New York: McGraw-Hill.

Mundell, R. A. (1963). Capital mobility and stabilization policy under fixed and flexible exchange rates. *Canadian Journal of Economics and Political Science, 29*, 475–485. Reprinted In R. E. Caves & H. G. Johnson (Eds.) (1968). *Readings in internationaleEconomics* (pp. 487–499). London: Allen&Unwin, and In R. A. Mundell (1968). *International economics* (Chap. 18). New York: Macmillan.

Muth, J. F. (1960). Optimal properties of exponentially weighted forecasts. *Journal of the American Statistical Association, 55*, 299–306.

Muth, J. F. (1961). Rational expectations and the theory of price movements. *Econometrica, 29*, 315–335.

Olmo, J, & Pilbeam, K. (2011). Uncovered interst parity and the efficiency of the foreign exchange market: A re-examination of the evidence. *International Journal of Ecomics & Finance, 16*, 189–204.

Siegel, J. J. (1972). Risk, interest, and forward exchange. *Quarterly Journal of Economics, 86*, 303–309.

Sinn, H. W. (1989). Expected utility and the Siegel paradox. *Zeitschrift für Nationalökonomie, 50*, 257–268.

Taylor, M. P. (1995). The economics of exchange rates. *Journal of Economic Literature, 33*, 13–47.

The Balance of Payments

5

5.1 Balance-of-Payments Accounting and Presentation

5.1.1 Introduction

Before coming to grips with the adjustment processes of the balance of payments, we must have a clear idea of what a balance of payments is and be able to understand the content of the statistical data presented therein. Although the various national presentations look different, all obey a common set of accounting rules and definitions, which can be given a general treatment. To enable the international comparison of these statistics, the International Monetary Fund (IMF) has, in cooperation with experts from national and international institutions, developed a framework for compiling the balance of payments and the international investment position, the Balance of Payments and International Investment Position *Manual*. This contains the recommended concepts, rules, definitions, etc., to guide member countries in making their regular reports on the balance of payments, as stipulated in the Fund's Articles of Agreement; at the time of writing the latest edition is the sixth (the first was IMF, 1948), released by the IMF in 2009 (henceforth referred to as the *Manual*). This last edition takes into account important developments occurred in the global economy since the release of the fifth edition, such as: the increased use of cross-border production processes, the complex international company structures, the international labour mobility and financial innovation.

Furthermore, where possible, the IMF publishes the balances of payments of all member countries in a standardized presentation (see the IMF's publications Balance of Payments Statistics Yearbook (IMF, 2016) and International Financial Statistics) in accordance with the classification scheme of the Fund's Manual.

Therefore we shall first explain the general principles (in this and the following section) then treat the standard classification scheme. This well enable the student, independently of his or her country of residence, to understand what a balance of payments is and to obtain information on any country's balance of payments

© Springer-Verlag Berlin Heidelberg 2016
G. Gandolfo, *International Finance and Open-Economy Macroeconomics*,
Springer Texts in Business and Economics, DOI 10.1007/978-3-662-49862-0_5

through the Fund's publications (these, it should be noted, contain references to the original national sources, to which all those interested in a particular country can turn). For uniformity of presentation the Yearbook data are expressed in U.S. dollars. The national sources usually publish the country's balance of payments in national currency. For countries that do not report in U.S. dollars, balance of payments data are converted using normally the average domestic exchange rates for the relevant period taken from the International Financial Statistics (IFS).

In synthesis, the balance of payments of a country is a systematic record of all economic transactions which have taken place during a given period of time between the residents of the reporting country and residents of foreign countries (also called, for brevity, "nonresidents", "foreigners", or "rest of the world"). This record is normally kept in terms of the domestic currency of the compiling country.

This definition needs some clarification, especially as regards the concepts of economic transaction and resident. As regards the period of time, it can be a year, a quarter, or a month, though other periods are in principle possible; note also that, as the balance of payments refers to a given time period, it is a *flow* concept.

(a) The term *economic transaction* means the transfer from one economic agent (individual, business, etc.) to another of an economic value. It includes both real transfers [i.e., transfer of (title to) an economic good or rendering of an economic service] and financial transfers [i.e.. transfer of (title to) a financial asset, including the creation of a new one or the cancellation of an existing one]. Furthermore, an economic transaction may involve either a *quid pro quo* (that is, the transferee gives an economic value in return to the transferrer: a two-way or *bilateral* transfer) or it may not (a one-way or *unilateral* or *unrequited* transfer). Thus we have five basic types of economic transactions:

  (1) purchase or sale of goods and services with a financial *quid pro quo* (the latter can be, e.g., a payment in cash, the granting of a credit, etc.): one real and one financial transfer;

  (2) exchange of goods and services for goods and services (barter): two real transfers;

  (3) exchange of financial items for financial items (for example purchase of bonds with payment in cash; cancellation of an outstanding debt against the creation of a new one, etc.): two financial transfers;

  (4) transfer of goods and services without a *quid pro quo* (for example, a gift in kind): one real transfer;

  (5) transfer of financial items without a *quid pro quo* (for example, a gift of money): one financial transfer.

This classification of economic transactions is obviously valid in general and not only in international economics; what qualifies a transaction as international is that it takes place between a resident and a nonresident (for exceptions to this rule see Sects. 5.1.3.1 and 5.1.4). Illegal transactions are treated the same way as legal actions. Illegal transactions are those that are forbidden by law (e.g. smuggled goods). As illegal transactions generally affect other legal transactions (e.g., certain legal external financial claims may be created through illegal exports of goods) their exclusion could lead to an imbalance in the international accounts

(b) The second concept to be clarified is that of *resident*. To begin with, the concept of resident does *not* coincide with that of citizen, though a considerable degree of overlapping normally exists. In fact, as regards individuals, residents are the persons whose general centre of interest is considered to rest in the given economy, that is, who consume goods and services, participate in production, or engage in other economic activities in the territory of an economy on other than a temporary basis, even if they have a foreign citizenship. On the basis of this definition, the Fund's *Manual* indicates a set of rules to solve possible doubtful cases: for example, migrants are to be considered as residents of the country in which they work, even if they maintain the citizenship of the country of origin; students, tourists, commercial travellers are considered residents of the country of origin provided that their stay abroad is for less than 1 year; official diplomatic and consular representatives, members of the armed forces abroad are in any case residents of the country of origin; etc.

As regards non-individuals, the general government (central, state, and local governments, etc.), and private non-profit bodies serving individuals, are residents of the relative country; enterprises (either private or public) have more complicated rules. In fact, the international character of many enterprises often makes it necessary to divide a single legal entity (for example a parent company operating in one economy and its unincorporated branches operating in other economies) into two or more separate enterprises, each to be considered as resident of the country where it operates, instead of being attributed to the economy of its head office. according to the rules of the Fund.

International organisations, i.e. political, administrative, economic etc., international bodies in which the members are governments (the United Nations, the International Monetary Fund, the World Bank, etc.) are not considered residents of any national economy, not excluding that in which they are located or conduct their affairs. It follows that all economic transactions (for example, sale of goods and services) of a country with the international organizations located in its territory are to be included in that country's balance of payments.

The balance of payments statement divides international transactions into three accounts, according to the nature of the economic resources provided and received: the current account, the capital account and the financial account. The current account deals with international trade in goods and services and with earnings on labour and investments. The capital account consists of capital transfers and the acquisition and disposal of non-produced, non-financial assets. The financial account records transfers of financial capital and non-financial capital (including changes in the country's international reserves). A classification of the main components of these three sections will be given in Sect. 5.1.3; before that, however, we must examine the basic accounting principles on which the balance of payments rests.

5.1.2 Accounting Principles

(a) The first basic principle is that the balance of payments for an economy is to be compiled under standard double-entry bookkeeping from the perspective of the residents of that economy. Therefore each international transaction of the residents of a country will result in two entries that have exactly equal values but opposite signs: a credit and a debit entry in that country's balance of payments. The result of this accounting principle is that the total value of debit entries necessarily equals the total value of credit entries (so that the net balance of all the entries is necessarily zero), that is, the balance of payments always balances (the relation of this with the problem of balance-of-payments equilibrium and disequilibrium will be examined in Sect. 5.2).

Naturally, the credit and debit entries are not arbitrary but must follow precise rules. As stated in the *Manual*, under the conventions of the system a compiling economy records credit entries (i) for real resources denoting exports and (ii) for financial items reflecting reductions in an economy's foreign assets or increases in an economy's foreign liabilities. Conversely, a compiling economy records debit entries (i) for real resources denoting imports and (ii) for financial items reflecting increases in assets or decreases in liabilities. In other words, for assets—whether real or financial—a positive figure (credit) represents a decrease in holdings, and a negative figure (debit) represents an increase. In contrast, for liabilities, a positive figure shows an increase, and a negative figure shows a decrease. Transfers are shown as credits when the entries to which they provide the offsets are debits and as debits when those entries are credits.

Experience shows that those who do not have a background in accounting get confused by the convention concerning financial items, as it seems odd that an *increase* in foreign *assets* owned by the country should give rise to a *debit* entry, etc. Various ways have been devised to prevent this possible confusion; we list the main ones, in the hope that at least one will be to the student's taste.

A first rule of thumb is that all transactions giving rise to a (actual or prospective) payment *from* the rest of the world are to be recorded as credit entries, whilst all transactions giving rise to a (actual or prospective) payment to the rest of the world are to be recorded as *debit* entries. The (actual or prospective) payment is to be recorded as an offsetting entry (and so as a debit or credit entry, respectively) to the transactions which are its cause, so as to respect double-entry bookkeeping. The obvious corollary of this principle is that an increase in the country's foreign assets (or decrease in liabilities) is a debit entry and that a decrease in the country's assets (or increase in liabilities) is a credit entry.

In fact, suppose that Italy (the compiling country) exports goods to the United States; since the US importer will have to pay for these in some way, this transaction gives rise to a (actual or prospective) payment from US to Italy and so it must be recorded as a credit (precisely, as an item of merchandise in the current account). The payment can take various forms: for example, a bank draft in dollars, or the granting of a credit by the Italian exporter to the US importer: in both cases there is an increase in Italy's foreign assets, that must be recorded as a debit to offset the credit entry.

The convention concerning the recording of financial items implies that a *capital outflow* is a *debit* in the financial account, as it gives rise to an increase in foreign assets, or to a decrease in foreign liabilities, of the country; similarly, *a capital inflow* is a *credit* in the financial account, as it gives rise to an increase in foreign liabilities or a decrease in foreign assets.

Finally, as regards unrequited transfers, which have no economic *quid pro quo*, the offsetting entry is created through the *unrequited (or unilateral) transfers* account: thus, for example, a merchandise gift is recorded as an export of goods (credit entry) by the donor country and an offsetting debit entry is made in the unrequited transfers account.

A second rule of thumb is that all economic transactions giving rise to a demand for foreign exchange result in debit entries in original accounts and, conversely, all transactions giving rise to a supply of foreign exchange result in credit entries. Thus an export of goods and services gives rise to a supply of foreign exchange, and is thus a credit entry, whilst an export of capital (capital outflow) gives rise to a demand for foreign exchange (for example by residents to pay for the foreign assets purchased) and so is a debit entry. Similarly, an import of goods and services gives rise to a demand for foreign exchange and is thus a debit entry, whilst an import of capital (capital inflow) gives rise to a supply of foreign exchange (for example by nonresidents to pay for the domestic assets, i.e. the assets of the compiling country, that they purchase). By the same token, the payment of foreign debt (decrease in liabilities) gives rise to a demand for foreign exchange and so is a debit entry, whilst the granting by the rest of the world of a loan (increase in liabilities) gives rise to a supply of foreign currency (the proceeds of the loan) and is thus a credit entry.

This rule is not applicable to gifts in kind made by private agents and governments, for which the same considerations hold as were made above in relation to unrequited transfers.

Third rule of thumb: learn the accounting convention by heart and forget about explanations, as the explanation may be more confusing than the convention itself.

As we said above, since balance-of-payments accounting is kept according to the principle of double-entry bookkeeping, the balance of payments *always* balances, that is, the total value of credit entries equals the total value of debit entries or, in other words, the net balance of all the entries is zero. In practice this never occurs, either because sometimes the double-entry principle cannot be complied with (for some transactions it is possible to record only one side of it and the other has to be estimated[1]) or because of the inevitable material errors, inconsistencies in the estimates, omissions from the statement, etc. Therefore an item for (*net*) *errors and omissions* is introduced as an offset to the over or understatement of the recorded components, so that the net balance of all the entries, including the net errors and omissions, is zero.

(b) The second basic principle concerns the *timing of recording*, that is, the time at which transactions are deemed to have taken place. In general, various rules are possible, such as the payments basis (transactions are recorded at the time of the payment), the contract or commitment basis (transactions are recorded at the time of contract), the movement basis (transactions are recorded when the economic value changes ownership). The principle adopted is that suggested by the Fund's *Manual*, namely the *change of ownership* principle. The term "economic ownership" is introduced in the sixth edition of the *Manual*, which is central in determining the time of recording on an accrual basis for transactions in goods, non-produced non-financial assets and financial assets. By convention, the time of change of ownership is normally taken to be the time that the parties concerned record the transaction in their books. Rules of thumb have to be applied in the case of transactions that do not actually involve a change of ownership (for example goods made available under financial lease arrangements), for which we refer the reader to the *Manual*.

It should be noted that under this principle an import of goods with deferred payment gives rise to a debit entry in the current account at the moment of change of ownership of the goods, with a simultaneous credit entry in the capital account for the increase in liabilities (the importer's debt). When the importer settles the debt (possibly in a different period from that covered by the balance of payments in which the import was recorded), there will be two offsetting entries in the capital account: a debit for the decrease in foreign liabilities (the extinction of the debt), and a credit for the decrease in foreign assets or increase in foreign liabilities involved in the payment of the debt.

(c) The third basic principle is that of the *uniformity of valuation* of exports and imports. Commodities must be valued on a consistent basis, and this may give rise to problems if, for example, the exporting country values exports on a f.o.b. (*free on board*) basis, whilst the importing country values the same commodities as imports on a c.i.f. (*cost, insurance, and freight*) basis. The Fund suggests that all exports and imports should be valued f.o.b. to achieve uniformity of valuation. This matter will be taken up again in Sect. 5.1.3.

---

[1]Think, for example, of a gift of money from a resident of country 1 (the reporting country) to a resident of country 2 which is deposited in an account that country 2's resident holds with a bank in country 1. The foreign liabilities of country 1 have increased, and this is recorded as a credit. But, as there is no independent information on the gift, it is necessary to make an estimate of these gifts to record as a debit in the unrequited transfers account. A similar problem arises for country 2.

5.1.3 Standard Components

We shall examine here the main standard components of the three sections (current account, capital account, and financial account) into which the typical balance of payments is divided.

5.1.3.1 Current Account

The current account includes flows of goods, services, primary income, and secondary income between residents and nonresidents. The term primary income is introduced in the sixth edition of the *Manual* and corresponds to the fifth edition concept of "income" plus some current transfers items (taxes on production and imports, subsidies and rents). The term secondary income, also introduced in the last edition of the *Manual*, broadly corresponds to "current transfers" of the fifth edition of the *Manual*. These changes bring the *Manual* in line with national accounts.

(A) Goods and Services

 (A.1) *Exports and Imports of Goods*. These are also called the "visible" items of trade (the invisible items being the services). The main recording problem that arises in this connection is that of uniformity of valuation, mentioned in section 5.1.2 .

  Once it was customary to record the value of traded goods at the country's border. This meant that imports were valued *c.i.f.* (that is, including cost, insurance, and freight), whilst exports were valued *f.o.b.* (free on board, that is, including, in addition to the cost of the goods, only the cost of stevedoring or loading the merchandise into the carrier) or *f.a.s.* (free alongside: the difference between *f.a.s.* and *f.o.b.* being the cost of stevedoring or loading).

  From this two difficulties arose. The first was that the value of world exports was different from the value of world imports (naturally, to calculate these totals a single currency is used), since the latter exceeded the former by an amount determined by insurance and freight. This statistical discrepancy is illogical, as we are dealing with the *same goods*, first as exports, then as imports.

  The second and perhaps more serious difficulty is that there was a confusion between visible and invisible items in the balance of payments: insurance and freight, in fact, are services and must be properly reported as such rather than being included in the merchandise item.

  The principle of uniformity of valuation of course requires that all goods should be valued in the same way; the rule adopted by the Fund's *Manual* is that they should be valued at the place of uniform valuation, i.e. the customs frontier of the economy from which they are exported, up to and including any loading of the goods on board the carrier at that frontier. That is, merchandise exports *and* imports are to be valued *f.o.b.* The f.o.b. valuation point means that export taxes are treated as payable by the exporter and that import duties and other taxes of the importing economy are payable by the importer.

In this last edition of the *Manual* the exclusive criterion to record international merchandise trade is the transfer of ownership and it removes some exceptions to this principle included in the previous edition. This gives rise to the following innovations: (a) a change to the treatment of goods that cross the frontier for processing without change of ownership; these goods are no longer included in the gross flows of imported and exported goods. The performance of just the service of processing is included among services; (b) net exports of goods through merchanting transactions, for example the purchase by a European operator of goods from a non-resident operator and the subsequent resale of the same goods (usually with the achievement of a profit) to another non-resident operator without the goods physically crossing the European frontier, are recorded as trade in goods and no longer included among services.

(A.2) *Exports and Imports of Services*, also called the "invisible" items of trade. They include both services proper and items classified under this heading for convenience. Examples are: maintenance and repair services, freight services, personal travel (tourism expenditure, including goods purchased by tourists), passenger transportation, insurance and pension services, financial services, charges for the use of proprietary rights (such as patents, trademarks, copyrights, industrial processes and designs including trade secrets, franchises).

Conceptual doubts may arise in relation to other items. For example, at the beginning of the process of standardization of balance of payments accounting, there was much discussion as to whether payments for interest on investment and for profits should be considered unrequited transfers or payments for services. The second solution was adopted, which is in agreement with the neoclassical tradition for which interest is the return for the service rendered by the capital loaned and profits are the return for the efforts of entrepreneurs.[2] Differently from the previous editions, the sixth edition of the *Manual* suggests that factor incomes (compensation of employees and investment income) be included in a separate sub-account of the current account. Thus the current account should have three subsections: *A.* Goods and services; *B.* Primary Income; *C.* Secondary Income.

Another example of a doubtful case arises in relation to tourist expenditures, which, as we have said above, are all included in the item travel (which also contains similar expenditure by commercial travellers, students abroad for less than a year, etc.); note that the international carriage of travellers is not included here, as it is covered in passenger services.

---

[2]We may note, incidentally, that this a case in which accounting conventions reflect the dominant economic theory.

The examination of the services account allows us to illustrate a possible *exception* to the general rule that the balance of payments of a country records solely transactions between residents and nonresidents. This possible exception concerns shipment (a comprehensive term used for brevity to denote freight, insurance, and other distributive services related to merchandise trade) and depends on the fact that there are various possible methods for estimating these services. One method proposes that the compiling country should enter as credit all shipping services performed by residents, both as to exports and to imports, and as debits all shipping services performed as to imports, both by residents and by nonresidents. It follows that, when an importer of the compiling country purchases a shipment directly from a resident in the same country, this transaction should be entered both as a credit (as a shipment performed by a resident) and as a debit (as a shipment performed on imports). These two entries offset each other, as they should, but the fact remains that a transaction involving two residents has been recorded in the balance of payments.

Although this method is used by some countries and is analogous to the one employed by the UN's system of national accounts in its rest-of-the-world account, the Fund's *Manual* recommends a different method (there is also a third possible method, not examined here), according to which the compiling country enters as credit all shipping services performed by residents on its exports and as debits all shipping services performed by foreigners on its imports. The main reason for suggesting this method is that it eliminates any offsetting flows of services between residents and thus complies with the general rule on balance-of payments accounting.

(B) Primary Income

Primary income represents the return that accrues directly from work, financial assets and natural resources. The *Manual* distinguishes the following main types of primary income: compensation of employees; dividends; reinvested earnings; interest; investment income attributable to policyholders in insurance, pension funds; rent; and taxes and subsidies on products and production.

(C) Secondary Income

The secondary income account reports current transfers between residents and nonresidents. Various types of current transfers are recorded in this account to show their role in the process of income distribution between the economies.

Transfers may be made in cash or in kind. Capital transfers are shown in the capital account (see below). The *Manual* classifies the following types of current transfers: Personal transfers and other current transfers.

The sixth edition of the *Manual* introduces the concept of "personal transfers", which includes all current transfers in cash or in kind between resident households and non-resident households, independent of the source of income and the relationship between the households. The concept of "workers' remittances", present in the fifth edition of the *Manual*, is part of personal transfers. The "other current transfers" sub-account includes current taxes on income, wealth, etc.; social contributions; social benefits; net nonlife insurance premiums; net nonlife insurance premiums;

nonlife insurance claims; current international cooperation, and miscellaneous current transfers.

Current transfers can be further classified by institutional sectors receiving or providing the transfers.

5.1.4 Capital Account

The capital account shows (a) capital transfers receivable and payable between residents and nonresidents and (b) the acquisition and disposal of nonproduced, nonfinancial assets between residents and nonresidents. In economic literature, "capital account" is often used to refer to what is called the financial account in the IMF *Manual* and in the National Accounts. The use of the term "capital account" in the *Manual* is designed to be consistent with the National Accounts, which distinguishes between capital transactions and financial transactions. Nonproduced, nonfinancial assets consist of natural resources (mineral rights, forestry rights, water, fishing rights, air space, etc.); contracts, leases, and licenses (marketable operating leases, permissions to use natural resources, etc.); and marketing assets (and goodwill) such as brand names, mastheads, trademarks, logos, and domain names. In the sixth edition of the *Manual*, personal property and other assets of people changing residence are no longer recorded as transactions, i.e. a change in ownership is no longer imputed. Since the residence of the owner changes, but not the ownership of the assets, the change in cross-border assets (such as bank balances and real estate ownership) and liabilities between economies are recorded as reclassifications in "other changes in volume". In the fifth edition of the *Manual*, transactions were imputed when persons changed residence.

5.1.5 Financial Account

The financial account records transactions that involve financial assets and liabilities. In general several classification criteria can be used:

(1) the nature of the operation, or type of capital: direct investment, portfolio investment, financial derivatives (other than reserves) and employee stock options, other investment, reserves.

  The nature of operations (or functional categories) are adopted to facilitate analysis by distinguishing operations that exhibit different economic motivations and patterns of behavior. The main feature of direct investment is taken to be the fact that the direct investor seeks to have, on a lasting basis, an effective voice in the management of a nonresident enterprise; this gives rise to accounting conventions which will be examined in Sect. 16.1. Portfolio investment covers investment in financial assets (bonds, corporate equities other than direct investment, etc.). On reserves see below. "Other investment" is a residual category.

(2) The length of the operation: long-term and short-term capital. The convention adopted is based on the *original* contractual maturity of more than 1 year (long term) and 1 year or less (short-term). Assets with no stated maturity (e.g. corporate equities, property rights) are also considered as long-term capital. The initial maturity convention may give rise to problems: for example the purchase of, say, a foreign bond with an original maturity of 3 years, but only 6 months to maturity when the purchase is made, is nonetheless recorded as a long-term capital movement. The inevitable convention derives from the fact that usually no data are available on the time to maturity of securities when international transactions occur.

(3) The nature of the operator: private and official, the latter possibly divided in general government, central monetary institutions (central banks etc.) and other official institutions.

The *Manual* adopts the first criterion, and categorizes between: 1. Direct investment, 2. Portfolio investment, 3. Financial derivatives and employee stock options, 4. Other investment, 5. Reserve assets. Within these categories further subdivisions are present, which use the other criteria. In the last edition of the *Manual*, several assets and liabilities are added to the financial account. Special drawing rights (SDRs) are now a liability of the recipient (central banks). Pension entitlements are recognised as a financial instrument, including the accrued obligations of unfunded pension schemes. Provisions for calls under standardised guarantees are also recognised in the system and are treated similarly to insurance technical reserves. Employee stock options are recognised as an instrument together with financial derivatives.

"Reserves" is a *functional category* comprising all those assets available for use by the central authorities of a country in meeting balance of payments financing needs, for intervention in exchange rate markets to affect the currency exchange rate. This availability is not linked in principle to criteria of ownership or nature of the assets. In practice, however, the term reserves is taken to include monetary gold, special drawing rights in the Fund, reserve position in the Fund, foreign exchange and other claims available to the central authorities for the use described above. For a more detailed treatment of these items we refer the reader to Chap. 23. These are also called *gross official reserves*; if we deduct the central banks short-term foreign liabilities we obtain *net official reserves*. A notion of gross official reserves in the wide sense is also defined, i.e. gross official reserves plus the central banks medium and long term foreign assets. Incidentally, note that the flow of interest earned by the central bank on these assets (which increases the stock of reserves) is entered in the services account of the current account. Finally, by deducting all foreign liabilities of the central bank from gross official reserves in the wide sense, we obtain net official reserves in the wide sense.

As regards ownership, the concept of reserves, which requires these assets to be strictly under the central authorities' direct and effective control, usually limits the consideration to foreign claims actually owned by the central authorities. But this is not a necessary condition, as private deposit money banks may sometimes be allowed to hold foreign assets, but may have permission to deal in them only on the terms specified by the authorities and with their express approval. These assets can thus be considered as under the authorities direct and effective control, so that the (net) foreign asset position of banks can, in such a case, be considered as part of the country's reserves.

A problem related to reserves is "reclassification" or "revaluation", a term which includes both the changes in the value of reserve assets resulting from fluctuations in their price ("valuation changes") and the changes due to reserve creation, i.e. to allocation of SDRs (on which see Sect. 22.4) and to monetization or demonetization of gold (i.e., when the authorities increase their holdings of monetary gold by acquiring commodity gold, or decrease their holdings of monetary gold for nonmonetary purposes, for example by selling it to private agents). Under the Bretton Woods system valuation changes were a rare event, but under the present system of widespread floating and no official price for gold (this has been abolished: see Sect. 22.7) the various reserve components can show frequent valuation changes relative to each other. Therefore the *Manual* suggests that for each type of reserve asset the main entry should refer to changes other than revaluation changes, with supplementary information on the total change (i.e., including revaluation changes), the valuation change and the reserve creation change, if applicable. This information, since the fifth edition of the *Manual* should no longer be shown in the balance of payments (as was previously), but in a completely separate account, the international investment position (on which see below).

The examination of the financial and capital accounts allows us to mention other exceptions to the general rule that the balance of payments includes all and only the transactions between residents and nonresidents.

These derive from the (practical) fact that, when a transaction in a foreign asset or liability takes place, it is sometimes impossible for the compiler to know the identities of both parties and thus to ascertain whether the resident-nonresident principle is met. Thus the compiler may not be able to find out whether a nonresident who is reported as having acquired or relinquished a financial claim on a resident did so by transacting with another nonresident or with a resident and, similarly, whether a resident who is reported as having acquired or relinquished a financial claim on a nonresident dealt with another resident or with a nonresident. Since is important that the balance of payments covers all transactions in foreign liabilities and assets, the *Manual*'s recommendation is that the transactions exemplified above should be included in the balance of payments. Thus a transaction between two nonresidents as well as a transaction between two residents may happen to be included in the compiling country's balance of payments.

To conclude this section we point out that, in addition to the balance of payments, there also exists the *balance of indebtedness* (also called by the *Manual* the International investment position). The sixth edition of the *Manual* is characterized by an increasing role of balance sheets (reflected in the international investment position). This is due to the growing recognition of balance sheet analysis in understanding sustainability and vulnerability, including currency mismatches, sector and interest rate composition of debt, and the effect of the maturity structure on liquidity.

In general, the balance of indebtedness records the outstanding claims of residents on nonresidents and the outstanding claims of nonresidents on residents at a given point in time, such as the end of the year. Therefore this balance is concerned with *stocks*, unlike the balance of payments, which refers to *flows*. The classification of the International investment position components is consistent with that of the financial account of the balance of payments.

Various balances of indebtedness can be reported according to the claims considered. One is that which considers only the stock of foreign assets and liabilities of the central authorities; another considers the direct investment position, etc.

5.1.5.1 Clandestine Capital Movements

By clandestine capital movements we mean capital flows that take place outside legal channels for various purposes. For example if in a country there are exchange controls and controls on capital movements, clandestine capital movements provide a way to evade these controls. But clandestine capital movements can take place even if there is perfect capital mobility, for example to create slush funds to be used for a range of illegal activities such as money laundering, bribery, terrorism etc. These activities are not limited to "ships in the night" operations (inflows and outflows of goods through illegal places of entry, or shipment of banknotes) but also include illegal trade through legal checkpoints or, more generally, illegal transactions through legal channels (such as misinvoicing, direct compensation, etc.). The former type does not appear in the balance of payments, but the latter does, insofar as it gives rise to inexact entries in the balance of payments and/or to entries in the wrong accounts.

Merchandise trade can serve as a channel for clandestine capital movements, which are usually outward movements. To achieve this, imports will be overinvoiced, or exports underinvoiced. In the former case the foreign exporter, who receives a greater amount than the true value of the goods, will credit the difference to, say, a secret account that the domestic importer holds abroad. In the latter case the foreign importer, who has to pay a lower amount than the true value of the goods, will credit the difference to, say, the account that the domestic exporter secretly holds abroad. These differences are (clandestine) capital movements, which of course are not recorded as such in the capital account, since they are hidden in merchandise trade.

Clandestine capital movements can also be hidden in other current account items, such as travel, labour income and workers' remittances. As regards travel, foreign tourists coming to visit the country may purchase the domestic currency on the black market. This will give rise to a decrease in the credit entries recorded in the travel item in the current account.

As regards labour income and workers' remittances, the best known device is that of *direct compensation* through an illegal organisation. Let us consider Mr. Y, a worker who has (temporarily or definitively) come to work in country 2 from country 1, and who wishes to send (part of) his earnings to his relatives in country 1. Mr. Y can either use the official banking, postal and other authorised channels (in which case there will be a record in the balance of payments), or give the money to the organisation's representative in country 2, who offers him perhaps more favorable conditions, or immediate delivery, or hidden delivery (Mr. Y may be an illegal migrant and doesn't want to be discovered). The money remains in country 2 to be used by the organisation, and the organisation's representative in country 2 only has to instruct his counterpart in country 1 to pay the agreed amount to Mr. Y's relatives out of the organisation's funds held in country 1.

In conclusion, no money has moved, but a clandestine capital outflow from country 1 will have taken place; this will be reflected in lower credit entries in the labour income and migrants' remittances items.

5.2   The Meaning of Surplus, Deficit, and Equilibrium in the Balance of Payments

As we saw in Chap. 1, the examination of balance-of-payments disequilibria and of their adjustments forms the starting point of international monetary theory. It is therefore indispensable to clarify the meaning of equilibrium and disequilibrium in the balance of payments (Machlup 1950, 1958).

We recall from Sect. 5.1.2 that, as the economic transactions between residents and nonresidents are reported under double-entry bookkeeping, the balance of payments always balances. It is therefore a concept of (economic) equilibrium to which one refers when one talks of equilibrium and disequilibrium of the balance of payments. In order to avoid terminological confusion, we shall use the term *equilibrium* to denote economic equilibrium, and *balance* to denote accounting identities. We shall also use the terminology *surplus* and *deficit* to qualify a disequilibrium of the balance of payments, avoiding the terminology "favourable" and "unfavourable" balance of payments (which derives from the Mercantilists), as this terminology implies the identities surplus = good on the one hand, and deficit = evil on the other, which is not necessarily true.

An appropriate definition of deficit and surplus is important for economic analysis and policy, and requires the classification of all the standard components of the balance of payments into two categories. The first includes all those items whose net sum constitutes the surplus (if positive), the deficit (if negative) or indicates equilibrium (if zero); the second includes all the remaining items, whose net sum is necessarily the opposite of the former (since the grand total must, as we know, be zero) and is sometimes said to "finance" or "settle" the imbalance. If we imagine all the standard components being arranged in a column in such a way that all the components included in the first category are listed first, then the balance of payments may be visualized as being separated into the two categories by drawing a horizontal line between the last component of the first category and the first component of the second category (hence the accountant's terminology *to draw the line*). Therefore the transactions whose net sum gives rise to a surplus or deficit (or indicates equilibrium, if zero) are said to be *above the line*, whilst the remaining ones are said to be *below the line*.

However, in most treatments the term "balance of payments" is still used both in the sense of the complete accounting statement (treated in the previous section) and in the sense of the net sum of the items included in the first category, hence the usual terminology "balance-of-payments surplus" and "balance-of-payments deficit".

To recapitulate: we shall say that the balance of payments is in *equilibrium* when the net sum of the items *above the line* is zero, in *disequilibrium* when this net sum is different from zero, showing a *surplus* if positive, a *deficit* if negative.

All this is very fine, but how are we to divide the items into the two categories, i.e. how shall we draw the line? This is the crucial point to give an operational content to the concepts we have been talking about.

The typical balances generally considered are few in number (although individual countries may build other types for the examination of some particular aspect of international economic relationships). We give a list in increasing order of coverage.

(1) *Trade Balance*. As the name indicates, this is the balance between exports and imports of goods or commercial balance. The trade balance in the strict sense or in the broad sense (i.e. taken to include also invisible trade, in which case it is now more usual to speak of balance on goods and services) was that considered by mercantilists and by the traditional theory of the processes of adjustment of the balance of payments.

(2) *Balance on Goods and Services*. In addition to visible trade, also invisibles are put above the line. This balance measures the net transfer of real resources between an economy and the rest of the world.

(3) *Current (Account) Balance*. This is obtained by adding net unilateral transfers to the balance on goods and services, and represents the transactions that give rise to changes in the economy's stock of foreign financial items.

(4) *Balance on Current Account and Long-Term Capital*, also called *Basic Balance*. This includes the flow of long-term capital above the line, besides the items of the current account balance. It is intended to be a rough indicator of long-term trends in the balance of payments. The idea behind it is that short-term capital flows are temporary and reversible, whilst long-term ones are less volatile and more permanent. This may be true in principle, but in practice serious difficulties arise. One is due to the accounting convention according to which the distinction between long-term and short-term capital is made on the basis of the initial maturity of the claim involved, so that, if an investor buys a foreign bond which had an initial maturity of 5 years, but now has only a few months to maturity, what is clearly a short-term investment is recorded as a long-term one. Another is related to the fact that all equities are considered long-term items, but many of them are easily marketable and transactions in stocks (e.g. for speculative purposes) can behave very much like short-term flows.

(5) *Overall Balance*. This considers all components except changes in reserve assets to be above the line. The idea behind it is to provide a measure of the residual imbalance that is financed through the use and acquisition of reserves. Conventionally, the net errors and omissions item is included above the line. The obvious reason is that the use and acquisition of reserves is usually known

and measured with a high degree of accuracy, so that errors and omissions must pertain to items above the line.

To conclude this section we must examine whether an answer can be given to the question that is said to have been asked by one writer on financial matters: "All I want is one number, with no ifs, buts or maybes". Unfortunately no answer to such a question can be given. The definition of a surplus or deficit is an analytic rather than an accounting problem, and the use of one concept of payments balance or the other cannot be separated from the nature of the problem that one wants to analyse. Therefore *the* payments balance does not exist, and different types may have to be examined on different occasions or even simultaneously on the same occasion.

Similar considerations hold for the notion of balance-of-payments equilibrium. An equilibrium in the overall balance, for example, can occur in various ways: with a simultaneous equilibrium within the main items separately considered (i.e., zero balance in merchandise trade, in invisible trade, in unrequited transfers, in long-term capital movements, in short-term capital movements) or with an imbalance in some items offset by an opposite imbalance in other items (for instance, a deficit in the current account balance offset by a surplus in the capital account balance). The former case, which we may define as *full* equilibrium, is altogether hypothetical, since in reality even the more equilibrated situations belong to the latter. One should also take the time period into account, since a situation may be one of equilibrium, but only in the *short-run* and not in the *long-run*: for example, a situation of continuing deficits in the balance on goods and services offset by continuing short-term capital inflows cannot be maintained in the long-run (see, for example, Sect. 11.2.2). The situation might be different if the said deficits were offset by long-term capital inflows, in which case the basic balance would be in equilibrium; the problem of the increase in the stock of long-term foreign liabilities would however remain (see, for example, Sect. 16.1).

It is therefore important always to clearly state which type of balance is being considered when one examines the adjustment processes of the balance of payments.

We note, in conclusion, that when two-country (or n-country) models are used, it is often expedient to use the constraint that the sum of the balances of payments (measured in terms of a common unit) of all countries is zero; this is also called an international consistency condition. It should however be added that this condition holds true insofar as no reserve creation (on which see Sects. 5.1.4 and 22.4) occurs. Therefore, the reader should bear in mind that, when we use this constraint in the following chapters, we shall implicitly be assuming that no reserve creation is taking place.

In any case, whichever type of balance one uses, a necessary constraint or *international consistency condition* (Mundell, 1968) is

$$\sum_{i=1}^{n} B_i = 0, \tag{5.1}$$

namely the balance of payments of all countries (of course the same type of balance, and expressed in a common monetary unit) must algebraically sum up to zero. This follows from the fact that the surplus of one country must necessarily be matched by a corresponding deficit of one or more other countries, provided that no reserve creation takes place (more generally, the sum of the balances of payments of all countries, i.e., the world balance of payments, equals the increase in net world reserves: see Mundell 1968).

From Eq. (5.1) it follows that only $n - 1$ balances of payments can be independently determined. For example, a maximum of $n - 1$ countries can independently set balance-of-payments targets: the nth country has to accept the outcome or the system will be inconsistent.

References

IMF (International Monetary Fund) (1948). *Balance of payments manual*, 1st edn.; 2nd edn. 1950; 3rd edn. 1961; 4th edn. 1977; 5th edn. 1993; 6th edn. 2009.

IMF (International Monetary Fund) (2016) *Balance of payments yearbook* (yearly); *International financial statistics* (monthly).

Machlup, F. (1950). Three concepts of the balance of payments and the so-called dollar shortage. *Economic Journal, 60*, 46–68.

Machlup, F. (1958). Equilibrium and disequilibrium: Misplaced concreteness and disguised politics. *Economic Journal, 68*, 1–24.

Mundell, R. A. (1968). *International economics* (Chap. 10). New York: Macmillan.

The data contained in the publications of the IMF are also available on line on the IMF's site (http://imf.org) against payment of a fee. Other sources of online data at the international level are:

- the World Bank (http://worldbank.org), free
- the Organization for Economic Cooperation and Development (http://oecd.org), against payment of a fee
- the Bank for International Settlements (http://bis.org), free
- the statistical office of the European Union (http://europa.eu.int/eurostat), partly free
- the European Central Bank (http://ecb.int), free

Real and Financial Flows in an Open Economy

6

6.1   Introduction

In this chapter we draw from national economic accounting and flow-of-funds analysis some elementary relations among the main macroeconomic (real and financial) aggregates in an open economy in order to fit the balance of payments into the context of the whole economic system, always keeping within an accounting framework. The reason for giving this treatment in an international economics textbook is not that we are particularly fond of accounting, but that the accounting framework that we propose will render us invaluable services in the course of the examination of the adjustment processes of the balance of payments.

We have built a framework—which, though very simplified, is sufficient to include the basic elements—in which there are only five sectors and six markets or transaction categories (real resources and financial assets). This framework is *exhaustive*, in the sense that it includes all transactors and transactions. In other words, all transactors are included in one of the sectors and all the transactions they carry out are included in one of the categories. It should also be pointed out that this accounting framework records *flows*, that is movements, or changes, that have occurred during a given time period.

The framework under consideration can be represented schematically in a table (see Table 6.1), where the columns refer to sectors and the rows to markets. A tilde (∼) means that the entry is assumed absent for simplicity's sake, while a dash (−) means that there can be no entry for logical reasons; a capital delta (Δ) means a variation.

Let us begin by clarifying the meaning of the *sectors*.

The *private sector* includes all transactors who do not belong to any other sector. The adjective "private" must not, therefore, be interpreted as the opposite of "public" in a juridical sense. For example, public enterprises which sell to the public most of the goods or services they produce are included here (unless they are banks). The private sector therefore includes the producing and household sectors.

© Springer-Verlag Berlin Heidelberg 2016
G. Gandolfo, *International Finance and Open-Economy Macroeconomics*,
Springer Texts in Business and Economics, DOI 10.1007/978-3-662-49862-0_6

Table 6.1 An accounting matrix for real and financial flows

Market	Sector Private	Govern-ment	Banking	Central Bank	Rest of the World	Row Totals
Goods and services	$I - S$	$G - T$	$\sim$	$\sim$	CA	0
Domestic monetary base	ΔH_p	$\sim$	ΔH_b	ΔH^c	$\sim$	0
Domestic bank deposits	ΔD_p	$\sim$	ΔD^b	$-$	ΔD_f	0
Domestic securities	ΔN_p	ΔN^g	ΔN_b	ΔN_c	ΔN_f	0
Foreign money	ΔR_p	$\sim$	ΔR_b	ΔR_c	ΔR^f	0
Foreign securities	ΔF_p	$\sim$	ΔF_b	ΔF_c	ΔF^f	0
Column Totals	0	0	0	0	0	

The subscript/superscript indicates the holding/issuing sector respectively

The *government sector* refers to the *general* government and includes all departments, establishments, and agencies of the country's central, regional, and local governments (excluding the central bank if this is institutionally part of the government and not an independent body). The *banking sector* includes commercial banks, savings banks and all financial institutions other than the central bank.

The *central bank* includes, besides the central bank, the exchange stabilization fund, if this exists as an institutionally separate body.

The *rest-of-the-world* sector includes all nonresidents in the sense explained in Sect 5.1.1.

Let us now turn to an examination of the categories of transactions or *markets*.

Goods and services includes all transactions on goods and services (production, exchange and transfers) and gives rise to the real market. Then there are the markets concerning national or domestic money, distinguished into *monetary base* (also called high-powered money, primary money, etc.) and *bank deposits*. Generally the *monetary base* is the liability of the central bank and consists of coin, banknotes and the balances which the banks keep with the central bank. As regards *bank deposits*, it is outside the scope of the present work to discuss whether only demand deposits or also time deposits (and other types of deposits or financial assets) are to be considered as money. In any case what is *not* included as money here, comes under the heading of the *national* (or *domestic*) *securities* item, which includes, besides securities proper, any form of marketable debt instrument. As regards *foreign money*, we do not make the distinction between monetary base and bank deposits, since all the foreign sectors (including the central bank and the banking sector) have been consolidated into a single sector (the rest-of-the-world or foreign sector), and for residents foreign exchange in either of its forms (cash or deposits) is foreign money. There are, finally, *foreign securities*, for which similar considerations hold as for domestic ones.

The table can be read along either the rows or the columns: in any case, as this is an accounting presentation, the algebraic sum of the magnitudes in any row is zero, as is the algebraic sum of the magnitudes in each column. It should be noted that the items have an intrinsic sign: see below, page 91.

More precisely, the fact that the row totals are zero reflects the circumstance that, as the magnitudes considered represent excess demands (positive or negative: a negative excess demand is, of course, an excess supply) by the single sectors for the item which gives the name to the row, the total quantity of it actually exchanged is necessarily the same both from the point of view of demand and from the point of view of supply. In other words, the *ex post total amount* demanded and the *ex post total amount* supplied are necessarily equal, as they are one and the same thing. It should be stressed that this is a mere fact of accounting, which must *not* be confused with the equilibrium of the market where the item is being transacted, or equality between *ex ante* magnitudes.

The equality to zero of the column totals reflects the *budget constraint* of each transactor, that is, the fact that total receipts and total outlays must necessarily coincide, where *receipts* and *outlays* are of course taken to include the change in financial liabilities and assets. Thus an accounting link is established between real and financial flows, according to which for any transactor the excess of investment over saving coincides with the change in his net liabilities (i.e. the change in liabilities net of the change in assets). Since the budget constraint must hold for each transactor it will hold for their aggregate, i.e. for the sector.

This said, we can go on to a detailed examination of the accounting relationships present in the table, beginning with the rows.

6.2   The Row Identities

The first row gives the relation

$$(I - S) + (G - T) + CA = 0. \tag{6.1}$$

Note that the tildes ($\sim$) in correspondence to the third and fourth columns are due to the simplifying assumptions that the real transactions of the banking sector and the central bank are negligible, i.e. that these sectors neither consume nor produce goods and (real) services, and thus do not transact in the real market, but only in the financial markets.

In Eq. (6.1), I and S denote the private sector's investment and saving respectively; G and T denote government expenditure and revenue (fiscal receipts net of transfer payments); CA indicates the current account balance, namely exports minus imports of goods and services plus net unilateral transfers of goods and services.[1] If we recall that $C + S = Y_d$ by definition, where C, Y_d are the private sector's consumption and disposable income, it follows that $I - S = (I + C) - Y_d$. Therefore, $(I - S)$ is the private sector's excess demand for goods and services, while $(G - T)$ is the excess demand of the government (budget deficit), and CA the excess demand of the rest of the world.

If we remember that the private sector's disposable income is given by national income minus taxes net of transfer payments, i.e. $Y_d = Y - T$, whence $S = Y - T - C$, and if we define a new aggregate called national absorption (or expenditure) A as the sum $C + I + G$, we can rewrite (6.1) as

$$CA = Y - A, \qquad (6.2)$$

namely the current account equals the difference between national income and absorption. Alternatively, if we aggregate the private and public sector and define *national* excess demand for goods and services as $I_N - S_N = (I - S) + (G - T)$ we have

$$CA = S_N - I_N, \qquad (6.3)$$

which states that the current account equals the difference between national saving and investment.

In the subsequent rows, the symbol Δ denotes the changes (which may be positive or negative). Recalling that the subscripts and superscripts indicate the holding or issuing sector respectively, and considering the second row, we have

$$\Delta H_p + \Delta H_b + \Delta H^c = 0, \qquad (6.4)$$

where H denotes the domestic monetary base, issued by the central bank (ΔH^c) and held by the private sector (ΔH_p) and the banking sector (ΔH_b). The tildes in correspondence to the second and fifth columns reflect the simplifying assumptions that the government does not hold or issue monetary base and that the rest of the world does not hold domestic monetary base.

---

[1]We are using the term Current Account Balance in a sense slightly different from the Fund's definition previously illustrated, since we exclude net unilateral transfers of financial assets. This difference is, however, negligible.

The third row gives

$$\Delta D_p + \Delta D_f + \Delta D^b = 0, \tag{6.5}$$

which incorporates the simplifying assumption that only the private and foreign sectors hold deposits with the banking sector which, of course, issues them. It should be noted the dash (instead of the tilde) under the fourth column indicates the fact the central bank does not hold deposits with the banking sector and that any deposits with the central bank to be considered as monetary base.

From the fourth row we obtain

$$\Delta N_p + \Delta N^g + \Delta N_b + \Delta N_c + \Delta N_f = 0, \tag{6.6}$$

where the simplifying assumption is that domestic securities are issued solely by the government.[2]

The fifth row yields

$$\Delta R_p + \Delta R_b + \Delta R_c + \Delta R^f = 0, \tag{6.7}$$

where the government is assumed to hold no foreign money.

Finally from the sixth row we get

$$\Delta F_p + \Delta F_p + \Delta F_c + \Delta F^f = 0, \tag{6.8}$$

where the simplifying assumption is that the government does not hold foreign securities, which of course are issued by the rest of the world.

6.3 The Column Identities

Let us now consider the *budget constraints* (column totals). The first column gives

$$(I - S) + \Delta H_p + \Delta D_p + \Delta N_p + \Delta R_p + \Delta F_p = 0, \tag{6.9}$$

which is the private sector's budget constraint. An alternative way of writing it is

$$S - I = \Delta H_p + \Delta D_p + \Delta N_p + \Delta R_p + \Delta F_p, \tag{6.10}$$

where the excess of saving over investment (if we consider it to be positive) of the private sector is employed by this sector to accumulate monetary base (ΔH_p),

---

[2]One might relax this assumption slightly with no change in the results by hypothesizing that any domestic securities issued by other sectors are held solely by the issuing sector, so that they cancel out within the sector.

deposits (ΔD_p), domestic securities (ΔN_p), foreign money (ΔR_p), and foreign securities (ΔF_p). A short-hand for writing this is

$$S - I = \Delta W_p, \tag{6.11}$$

where $W_p = H_p + D_p + N_p + R_p + F_p$ is the private sector's stock of financial wealth.

If we remember (see above) that the private sector's saving equals disposable income (given by national income minus taxes net of transfer payments) minus consumption, i.e. $S = Y_d - C = Y - T - C$, and if we define a new aggregate called "absorption" of the private sector A_p as the sum $C + I$, so that $S - I = Y_d - A_p$, we can rewrite (6.11) as

$$\Delta W_p = Y_d - A_p, \tag{6.12}$$

namely the change in the private sector's stock of wealth equals the difference between disposable income and absorption. This also called the *net acquisition of financial assets (NAFA)* or the *financial surplus* (that of course can be either positive or negative) of the private sector.

Another way of writing the private sector's budget constraint is

$$I - S = -\Delta H_p - \Delta D_p - \Delta N_p - \Delta R_p - \Delta F_p, \tag{6.13}$$

that is, the excess of investment over saving (if we now assume this excess to be positive) of the private sector is financed by this sector through decumulation (i.e., decrease in the stocks owned) of monetary base ($-\Delta H_p$), deposits ($-\Delta D_p$), domestic securities $\left(-\Delta N_p\right)$, foreign money ($-\Delta R_p$), and foreign securities ($-\Delta F_p$).

It should be pointed out that it is not necessary for all the stocks to increase in the case of (6.10) or to decrease in the case of (6.13). It is in fact perfectly possible for a divergence between S and I to give rise to changes in only one of the financial stocks, and more complicated intermediate cases are also possible in which some stocks vary in one direction and the others in the opposite direction (though always respecting the budget constraint).

The second column gives the government budget constraint

$$G - T + \Delta N^g = 0, \tag{6.14}$$

that is

$$G - T = -\Delta N^g, \tag{6.15}$$

which states that the government budget deficit (excess of expenditure over receipts) is financed by issuing securities (a negative excess demand, that is an excess supply, equal to $-\Delta N^g$). Remember that in such a constraint there are the simplifying assumptions that the government does not issue domestic monetary base and does not hold foreign money or securities. If we dropped these assumptions, it can easily be seen that the government can also meet a budget deficit by issuing monetary base and reducing its stocks of foreign money and securities.

The budget constraint of the banking sector is

$$\Delta H_b + \Delta D^b + \Delta N_b + \Delta R_b + \Delta F_b = 0, \tag{6.16}$$

which expresses the fact that banks—given the simplifying assumption that they do not transact in the real market—record on the credit side a change in the domestic monetary base owned (ΔH_b), a change in foreign money (ΔR_b), and a change in the loans extended by buying domestic (ΔN_b) and foreign (ΔF_b) securities; on the debit side the change in deposits (ΔD^b). To avoid confusion it should be remembered that in a balance sheet the credit items have an *intrinsic* positive sign whilst the debit items have an *intrinsic* negative sign, so that: change in assets+change in liabilities = 0, since the two totals are numerically equal but with opposite sign. In other words, an increase in a credit item, say H^b, means $\Delta H^b > 0$ and a decrease in H^b means $\Delta H^b < 0$. Conversely, D^b is a debit item and carries with it an intrinsic negative sign, so that an increase in deposits (D^b numerically greater) means a negative ΔD^b and a decrease in deposits (D^b numerically smaller) means a positive ΔD^b. These observations on the intrinsic negative sign of the debit entries hold, of course, for all the accounting identities considered here.

The fourth column yields

$$\Delta H^c + \Delta N_c + \Delta R_c + \Delta F_c = 0, \tag{6.17}$$

which is the budget constraint—always referring to changes—of the central bank, in which the issuing of monetary base (ΔH^c) is a debit item, whilst the acquisition of national (ΔN_c) and foreign (ΔF_c) securities and of foreign exchange (ΔR_c) is a credit item.

Finally, the last column gives the rest-of-the-world's budget constraint:

$$CA + \Delta D_f + \Delta N_f + \Delta F^f + \Delta R^f = 0. \tag{6.18}$$

6.4 Derived Identities

The various row and column constraints can be combined to derive other meaningful constraints. Taking for example Eq. (6.18), using Eqs. (6.7) and (6.8), and rearranging terms, we get

$$CA + \{(\Delta D_f + \Delta N_f) - [(\Delta R_p + \Delta F_p) + (\Delta R_b + \Delta F_b)]\} = \Delta R_c + \Delta F_c. \tag{6.19}$$

Equation (6.19) is simply the expression of the *overall balance of payments (B)* already examined in Sect. 5.2.[3] In fact, *CA* is the current account balance[4] and the expression in braces is the autonomous-capital balance, consisting of the change in domestic assets (deposits, ΔD_f, and securities, ΔN_f) owned by non residents plus the change in foreign assets (money, ΔR, and securities, ΔF) owned by residents, who are subdivided into private sector (hence $\Delta R_p + \Delta F_p$) and banking sector (hence $\Delta R_b + \Delta F_b$); note that the minus sign before the square bracket reflects the accounting convention illustrated in Sect. 5.1.2. The offsetting item is given by the change in official international reserves (ΔR) in the wide sense (for this notion see Sect. 5.2), subdivided into liquid assets (foreign money, ΔR_c) and medium/long term assets (foreign securities, ΔF_c) owned by the central bank. Hence the identity

$$B = \Delta R, \tag{6.20}$$

which states that the overall balance of payments coincides with the change in international reserves.

Another derived identity is obtained aggregating the columns "Banking" and "Central Bank", which gives the budget constraint of the aggregate banking sector

$$(\Delta H_b + \Delta H^c) + \Delta D^b + (\Delta N_b + \Delta N_c) + (\Delta R_b + \Delta R_c) + (\Delta F_b + \Delta F_c) = 0,$$

from which, rearranging terms,

$$(\Delta R_c + \Delta F_c) + (\Delta R_b + \Delta N_b + \Delta F_b + \Delta N_c) = -\Delta D^b - (\Delta H_b + \Delta H^c).$$

Let us now observe that from the second row of Table 6.1 we have

$$-(\Delta H_b + \Delta H^c) = \Delta H_p,$$

and from the third we get

$$-\Delta D^b = \Delta D_p + \Delta D_f.$$

---

[3]It should be noted that the balance of payments does not conform exactly with the rest-of-the-world account in SNA (the United Nation's *System of National Accounts*, which forms the basis of national economic accounting in most countries). For our purposes, however, we can ignore these differences, as they are of secondary importance with respect to the fact that the agreement on underlying principles makes the balance of payments consistent with the overall framework of SNA. For a detailed comparison of balance-of-payments classification with external transactions in SNA, see IMF, 2009, *Balance of Payments Manual*, 6th edition.

[4]For simplicity's sake, we have neglected unilateral transfers of financial assets, which however could be included in the capital account.

Therefore, if we define

$$\Delta R = \Delta R_c + \Delta F_c$$

$$\Delta Q = \Delta R_b + \Delta N_b + \Delta F_b + \Delta N_c,$$

$$\Delta M = \Delta D_p + \Delta D_f + \Delta H_p,$$

we obtain the identity

$$\Delta M - \Delta Q = \Delta R = B, \tag{6.21}$$

which states that the balance of payments (change in international reserves) equals the change in the money stock (ΔM : M2 definition) minus the change in all other financial assets (ΔQ) held by the aggregate banking sector.

6.5   Identities Are Only Identities

After completing the examination of the table, we must again emphasize that we are in the presence of a *mere accounting framework, from which it would be logically invalid to draw causal relations automatically*. The better to drive this important point home we shall offer a few examples. If we consider, for instance, Eq. (6.1) and rewrite it in the form:

$$- CA = (I - S) + (G - T), \tag{6.22}$$

we might be induced to believe that the government budget deficit ($G - T > 0$) "determines" the current account deficit ($-CA > 0$, i.e., $CA < 0$), so that the "cause" of an increase in this external deficit is to be seen in an increase in the budget deficit; conversely, the current account improves if the budget deficit is reduced. All this might well be true, but it is logically illegitimate to derive it from Eq. (6.22), which merely states that, *ex post*, we observe from the accounts that the excess of imports over exports equals the algebraic sum of the private sector's excess of investment over saving and the government sector's excess of expenditure over receipts. To continue with the example, we might rewrite (6.1) in yet another form:

$$G - T = -CA + (S - I) \tag{6.23}$$

and be induced to claim that it is the current account deficit ($-CA > 0$, i.e., $CA < 0$) which "determines" the budget deficit ($G - T > 0$)! Again, all this might well be true, but it is not correct to derive it from identities: since the generating relation of (6.22) and (6.23) is Eq. (6.1), that is an accounting identity, also the derived relations maintain the nature of mere accounting identities with no causal content (this should be found elsewhere, if any). Similarly, by considering the identity (6.21) and writing it as

$$B = \Delta M - \Delta Q, \tag{6.24}$$

we might be induced to claim that the "cause" of a balance-of-payments deficit ($B < 0$) is a decrease in domestic monetary holdings of the private sector ($\Delta D_p + \Delta H_p < 0$, hence $\Delta M < 0$) or in domestic financial assets held by the aggregate banking sector (for example $\Delta N_b > 0$, hence $\Delta Q > 0$).

In conclusion, given an *accounting identity*, it is logically inadmissible to draw *causal relations* from it simply by shifting terms from one side to the other of the equality sign.

Part III

Flow Approaches

The Elasticity Approach

7

7.1 Introduction

The balance of payments considered here and in the following two chapters is the balance on goods and services. The neglect of capital movements might seem unrealistic, since nowadays most transactions in the balance of payments are of financial type. We first note that, when the theories we are going to treat were elaborated, capital movements were relatively unimportant, which warranted their neglect. Apart from this, the study of the current account on its own remains important: "One can hardly exaggerate the role played by trade elasticities in translating economic analysis into policy recommendations" (Hooper et al. 2000; page 1). Finally, the study of the current account can be seen as a first step towards a more general theory in which capital movements are also considered.

The adjustment through exchange-rate changes relies upon the effect of the relative price of domestic and foreign goods (considered as not perfectly homogeneous) on the trade flows with the rest of the world. This relative price, or (international) terms of trade, is defined by

$$\pi = \frac{p_x}{r p_m}, \tag{7.1}$$

where p_x represents export prices (in terms of domestic currency), p_m import prices (in terms of foreign currency), and r the nominal exchange rate of the country under consideration. The meaning of π has already been examined in Eq. (2.12).

The idea behind the adjustment process under consideration is that a change in the relative price of goods, ceteris paribus, brings about a change in the demands for the various goods by both domestic and foreign consumers, thus inducing changes in the flows of exports and imports which will hopefully adjust a disequilibrium in the payments balance considered.

The terms of trade may vary both because of a change in the prices p_x and p_m expressed in the respective national currencies, and because r changes. The analysis

© Springer-Verlag Berlin Heidelberg 2016
G. Gandolfo, *International Finance and Open-Economy Macroeconomics*,
Springer Texts in Business and Economics, DOI 10.1007/978-3-662-49862-0_7

with which we are concerned in this chapter focuses on the changes in r and so assumes that p_x and p_m as well as all other variables that might influence the balance of payments, are constant. It is, therefore, a partial equilibrium analysis, in which the ceteris paribus clause is imposed when the exchange rate varies (see the Appendix, Sect. 7.5.1.2 for the case in which prices are allowed to vary). We shall look more closely into this interpretation of the analysis later on. It is important to observe at this point that the problem of the effects of exchange-rate changes does not vary whether we consider a free movement of the exchange rate in a flexible exchange rate regime or a discretionary or managed movement in an adjustable peg or other limited-flexibility regime (see Sect. 3.3). In the latter case (i.e. the case of a policy-determined change), we are in the presence, in Johnson's (1958) terminology, of an *expenditure switching* policy, that is of a policy aiming at restoring balance-of-payments equilibrium by effecting a switch of expenditure (by residents and foreigners) between domestic and foreign goods. By contrast, if we consider a deficit, an *expenditure reducing* policy involves measures inducing a decrease in residents' total expenditure (and thus in that part of it which is directed to foreign goods, i.e. imports) by monetary or fiscal restriction.

Another preliminary observation is that the problem of the effects on the balance of payments of a variation in the exchange rate was identified in the traditional literature with that of the effects on the foreign exchange market of the same variation. This is of course only true under the assumption that the demand for, and supply of, foreign exchange as functions of the exchange rate exclusively derive from trade in goods and services, which is warranted given the neglect of capital flows (see above).

7.2   Critical Elasticities and the So-Called Marshall-Lerner Condition

To begin with, we observe that the ceteris paribus clause enables us to consider exchange-rate variations as the sole cause of changes in export and import flows. A depreciation in the exchange rate (i.e., a depreciation of the domestic currency) at unchanged domestic and foreign prices in the respective currencies, in fact, makes domestic goods cheaper in foreign markets and foreign goods more expensive in the domestic market. The opposite is true for an appreciation. Thus we can say, on the basis of conventional demand theory, that exports vary in the same direction as the exchange rate (an increase in the exchange rate, that is a depreciation, stimulates exports and a decrease, that is an appreciation, lowers them) whilst imports vary in the opposite direction to the exchange rate. Strictly speaking, this is true as regards the foreign demand for domestic goods (demand for exports) and the domestic demand for foreign goods (demand for imports). To be able to identify the demand for exports with exports and the demand for imports with imports, we need the further assumption that the relevant supplies (supply of domestic goods by domestic producers to meet foreign demand, and of foreign goods by foreign producers to meet our demand) are perfectly elastic. The consequences of dropping this assumption will be examined in the Appendix.

However, the fact that exports vary in the same direction as the exchange rate whilst imports vary in the opposite direction to it is not sufficient to allow us to state that suitable exchange-rate variations (a depreciation in the case of a deficit, an appreciation in the case of a surplus) will equilibrate the balance of payments. The balance of payments is, in fact, expressed in monetary terms, and it is not certain that a movement of the *quantities* of exports and imports in the right direction ensures that their *value* also changes in the right direction. The change in receipts and outlays depends on the elasticities, as the student knows from microeconomics.

We define the *exchange-rate elasticity* of exports, η_x, and of imports, η_m, as any price-elasticity, that is as the ratio between the proportional change in quantity and the proportional change in price (here represented by the exchange rate). Thus, letting x and m denote the quantities of exports and imports respectively, we have

$$\eta_x \equiv \frac{\Delta x/x}{\Delta r/r} = \frac{\Delta x}{\Delta r}\frac{r}{x}, \quad \eta_m \equiv -\frac{\Delta m/m}{\Delta r/r} = -\frac{\Delta m}{\Delta r}\frac{r}{m}, \tag{7.2}$$

where Δ as usual denotes a change, and the minus sign before the second fraction serves to make it a positive number (Δm and Δr have, in fact, opposite signs because of what we said at the beginning, so that the fraction by itself is negative).

7.2.1 The Balance of Payments in Domestic Currency

Since each country normally records its balance of payments in terms of domestic currency, we begin by considering the payments balance in domestic currency

$$B = p_x x - r p_m m, \tag{7.3}$$

where the value of imports in terms of foreign currency (p_m, we remember, is expressed in foreign currency) has to be multiplied by the exchange rate to transform it into domestic currency units; as p_x is expressed in terms of domestic currency, the value of exports, $p_x x$, is already in domestic currency units.

To examine the effects of a variation in the exchange rate, let us consider a depreciation by a small amount, say Δr. Correspondingly, exports and imports change by $\Delta x > 0$ and by $\Delta m < 0$. The new value of the balance of payments is then

$$p_x(x + \Delta x) - (r + \Delta r)p_m(m + \Delta m), \tag{7.4}$$

and by subtracting the previous value B as given by Eq. (7.3) we obtain the change in the balance of payments

$$\Delta B = p_x \Delta x - p_m m \Delta r - p_m r \Delta m - p_m \Delta r \Delta m.$$

Since Δr, Δm are small magnitudes, their product is of the second order of smalls and can be neglected. By simple manipulations we obtain

$$\Delta B = p_x \Delta x - p_m m \Delta r - p_m r \Delta m = p_m m \Delta r \left[\frac{\Delta x}{\Delta r} \frac{p_x}{p_m m} - 1 - \frac{\Delta m}{\Delta r} \frac{r}{m} \right]$$

$$= p_m m \Delta r \left[\left(\frac{\Delta x}{\Delta r} \frac{r}{x} \right) \frac{x}{r} \frac{p_x}{p_m m} - 1 - \frac{\Delta m}{\Delta r} \frac{r}{m} \right]$$

$$= p_m m \Delta r \left[\eta_x \frac{p_x x}{r p_m m} - 1 + \eta_m \right], \tag{7.5}$$

where in the last passage we have used the definitions (7.2). Since $\Delta r > 0$, the sign of ΔB will depend on the sign of the expression in square brackets, hence an exchange-rate depreciation will improve the balance of payments ($\Delta B > 0$) when

$$\frac{p_x x}{r p_m m} \eta_x + \eta_m > 1. \tag{7.6}$$

If we consider a situation near equilibrium, such that $p_x x \simeq r p_m m$, the condition becomes

$$\eta_x + \eta_m > 1, \tag{7.7}$$

namely a depreciation will improve the balance of payments if the sum of the export and import elasticities is greater than unity. Inequality (7.7) is generally called the Marshall-Lerner condition, but should be more correctly called the Bickerdicke-Robinson condition.[1] If we wish to eschew problems of historical attribution, we may call it the critical elasticities condition.

Let us note that it is unlikely that a depreciation will occur when the balance of payments is near equilibrium. On the contrary, it will normally occur when the balance of payments shows a deficit (which is the case in which an exchange-rate depreciation normally comes about, either spontaneously or through discretionary

---

[1] Lerner (1944) showed that the critical point lies where the sum of the export and import elasticities equals one, and observed that when this sum is greater (lower) than one, an exchange-rate depreciation will improve (worsen) the balance of payments, which is indeed the condition we are examining. Many doubts, on the contrary, exist as regards Marshall. His contributions, in fact, refer to the pure theory of international trade and the conditions that he developed concern the stability of barter international equilibrium analyzed in terms of offer curves, which are not suitable instruments for analyzing the problem of how an exchange-rate variation influences the balance of payments in a monetary economy. Therefore the attribution of the condition under examination to Marshall is not convincing. Besides, if one wishes to make a question of chronological priority, there exists a treatment by Joan Robinson (1937) prior to Lerner's and, if we go further back still in time, we find the contribution of Bickerdicke (1920), who seems to have been the first to give the full and correct formal conditions for an exchange-rate depreciation to improve the balance of payments.

intervention by the monetary authorities). A deficit means $rp_m m > p_x x$, and we should apply the more general condition (7.6), which is more stringent than (7.7). In fact, since $p_x x / rp_m m < 1$, the value of η_x will be multiplied by a factor smaller than one, hence it may well happen, if the sum of the elasticities is just above unity and the deficit is huge, that condition (7.6) is not satisfied.

7.2.2 The Balance of Payments in Foreign Currency

Conditions (7.7) and (7.6) have been assumedly obtained with reference to the balance of payments expressed in terms of domestic currency. For certain purposes (for example, to ascertain what happens to foreign-currency reserves) it may be useful to consider the foreign-currency balance. Now, if $B = p_x x - rp_m m$ is the domestic-currency balance, then

$$B' = \frac{1}{r}B = \frac{1}{r}p_x x - p_m m \tag{7.8}$$

is the foreign-currency balance. In the case of an initial equilibrium, the condition for a depreciation to improve the balance is the same as before (the mathematical derivations are left to the Appendix), i.e.

$$\eta_x + \eta_m > 1, \tag{7.9}$$

while in the case of an initial disequilibrium the condition becomes

$$\eta_x + \frac{rp_m m}{p_x x}\eta_m > 1. \tag{7.10}$$

It is quite clear that, in the presence of an initial deficit, condition (7.10) can more easily occur than condition (7.6), the elasticities being the same. In the extreme case, it is possible, when there is an initial deficit, that a depreciation will improve the foreign-currency balance and at the same time worsen the domestic-currency balance. It is in fact possible that following a depreciation, there will be a decrease in the outlay of foreign exchange (payment for imports) in absolute value greater than the decrease in foreign exchange receipts (from exports), whilst there will be an increase in the outlay of domestic currency (payment for imports) in absolute value greater than the increase in the domestic-currency value of exports. This means that the balance of payments expressed in foreign currency improves whilst that expressed in domestic currency deteriorates.

It is therefore important, when one deals with depreciation and balance of payments, to specify whether this is expressed in domestic or foreign currency; it goes without saying that when an initial equilibrium situation is considered, the Marshall-Lerner condition ensures that the payments balance will improve in terms of both domestic and foreign currency.

7.2.3 Elasticity Optimism vs Pessimism

A debate went on after the second world war and in the 1950s (for a survey see, for example, Sohmen 1969; Chap. I, Sect. 3), and occasionally crops up in modern times, between elasticity pessimism and elasticity optimism. In the pessimists' opinion, the elasticities were too low to satisfy the critical condition, so that an exchange rate depreciation, instead of improving the payments balance, would have worsened it. The opposite view was, of course, held by the optimists. And the copious empirical analyses were not able to settle the controversy: for a complete survey of the older empirical studies aimed at estimating these elasticities, see Stern et al. 1976; for more recent analyses see, e.g., Himarios 1989, Upadhyaya and Dhakal 1997, Bahmani-Oskooee 1998, Bussière et al. 2013, Imbs and Mejean 2015. But this is not surprising, because these analyses were based on single equations relating export and import demands to relative prices and possibly to other variables (such as national income) in a partial equilibrium context, thus neglecting the interrelationship among the different variables, which can be evaluated only in a more general setting. A study by Hooper et al. (2000) recognizes that there is simultaneity among income, prices, and trade, and consequently uses cointegration methods to estimate long-run elasticities for the G-7 countries. Their finding is that, with the exception of France and Germany, price elasticities for exports and imports satisfy the Marshall-Lerner condition.

7.3 Foreign Exchange Market Equilibrium and Stability

Like any other market, the foreign exchange market will be in equilibrium when demand equals supply. We know that demands and supplies of foreign exchange come from a variety of real and financial operators and depend, in addition to the exchange rate, on a number of other variables (see Chap. 2, Sect. 2.6). This gives rise to a very complicated set of real and financial interrelations.

Here we study the exchange rate that clears the market under the restrictive assumptions made in Sect. 7.1, namely that the demand for foreign exchange (we denote it by D) only comes from importers and equals the foreign-exchange value of imports, whilst the supply (S) of foreign exchange is only due to exporters and equals the foreign-exchange value of exports. This is certainly unrealistic, since nowadays most foreign exchange transactions are of financial type. The introduction of financial assets in the determination of the exchange rate will be done in Chap. 15.

Given the international nature of the foreign exchange market, it should be noted that it makes no difference whether some or all of the transactions are settled in domestic currency. If, for example, domestic importers pay out domestic currency to foreign suppliers, these will sell it in exchange for their own currency, which is a demand for foreign currency from the point of view of the importing country. Similarly, if domestic exporters are paid in domestic currency, this means that foreign importers sell their own currency to purchase the exporters currency, which is a supply of foreign currency from the point of view of the home country.

Thus to examine the second problem mentioned in Sect. 7.1 (effects of an exchange-rate variation in the foreign exchange market) we start from

$$D(r) = p_m m(r), \quad S(r) = \frac{1}{r} p_x x(r), \tag{7.11}$$

where, owing to the ceteris paribus clause, imports and exports are functions solely of the exchange rate. The condition of equilibrium in the foreign exchange market is that typical of all markets, i.e. the equality between demand and supply:

$$D(r) = S(r), \quad E(r) \equiv D(r) - S(r) = 0, \tag{7.12}$$

which determines the equilibrium exchange rate; it goes without saying that this is a *partial equilibrium*, for the reasons explained in Sect. 7.2. The analogy with any other kind of market whatsoever is obvious: here the "commodity" exchanged is foreign exchange, and the equality between demand and supply determines the equilibrium price, i.e., the equilibrium exchange rate (remember that the exchange rate is quoted as the price of foreign currency: see Sect. 2.1). Let us now observe that by substituting from Eq. (7.11) into Eq. (7.12) and rearranging terms we get

$$p_x x(r) - r p_m m(r) = 0, \quad \frac{1}{r} p_x x(r) - p_m m(r) = 0, \tag{7.13}$$

so that the equilibrium in the foreign exchange market coincides with the equilibrium in the balance of payments (no matter whether the latter is expressed in domestic or foreign currency). It can also easily be seen that the presence of an *excess demand* for foreign exchange is equivalent to a *disequilibrium* situation in the balance of payments. To be precise, from the previous equations we get

$$E(r) \gtreqless 0 \quad \text{is equivalent to } B \lesseqgtr 0, B' \lesseqgtr 0, \tag{7.14}$$

that is, a positive (negative) excess demand for foreign exchange is equivalent to a deficit (surplus) in the balance of payments however expressed. This is of course intuitive given the assumption made at the beginning on the sources of the demand for and supply of foreign exchange. However, these demand and supply schedules present some peculiarities which need clarification.

7.3.1 Derivation of the Demand and Supply Schedules: Multiple Equilibria and Stability

The main peculiarity of demand and supply schedules for foreign exchange is the fact that they are *derived* or *indirect* schedules in the sense that they come from the underlying demand schedules for goods (demand for domestic goods by nonresidents and demand for foreign goods by residents). In other words, in the context we are considering, transactors do not directly demand and supply foreign exchange as such, but demand and supply it as a consequence of the underlying demands for goods. From these elementary and apparently irrelevant considerations important consequences follow, and precisely that even we assume that the underlying demand schedules for goods are perfectly normal, the resulting supply schedules for foreign exchange may show an abnormal behaviour and even give rise to multiple equilibria.

Let us then take up Eq. (7.11) again and assume that import and export demands are normal, that is, the demand for imports monotonically decreases as r increases (since the foreign price of imports is assumed constant, an increase in r means an increase in the domestic-currency price of imports), and the demand for exports monotonically increases as r increases (since the domestic price of exports is assumed constant, an increase in r means a decrease in the foreign-currency price of exports). Is this enough for us to be able to state that the demand and supply schedules for foreign exchange are normal, i.e. that the demand for foreign exchange is a monotonically decreasing function of r and the supply of foreign exchange is a monotonically increasing function of r?

The answer is partly in the negative. In fact, as regards the demand for foreign exchange, since it is obtained simply by multiplying the demand for imports by the foreign price of imports p_m, a constant by assumption, the behaviour of the former demand coincides with that of the latter. Hence we can conclude that, on the assumption that imports decrease when the exchange rate depreciates (i.e., r increases), the demand for foreign exchange $D(r)$ is a monotonically decreasing function of r.

On the contrary, the supply of foreign exchange is rather complicated, as it is obtained by multiplying the demand for exports by the factor p_x/r, which varies inversely with r since p_x is a constant. Thus when r increases x increases but p_x/r decreases and their product, namely $S(r)$, can move either way.

Since $S(r)$ is total revenue of foreign exchange from exports (determined by export demand), it is sufficient to recall from elementary microeconomics the relations between total revenue and elasticity of demand, which tell us that total revenue depends on the elasticity of demand. If the elasticity of exports is greater than one, an exchange-rate depreciation of, say, 1 %, causes an increase in the volume of exports greater than 1 %, which thus more than offsets the decrease in the foreign currency price of exports: total receipts of foreign exchange therefore increase. The opposite is true when the elasticity is lower than one.

We must now distinguish two cases and the respective consequences on the foreign exchange market.

(1) The demand for exports has an elasticity everywhere greater than one or everywhere smaller than one. In this case the receipts (and so the supply) of

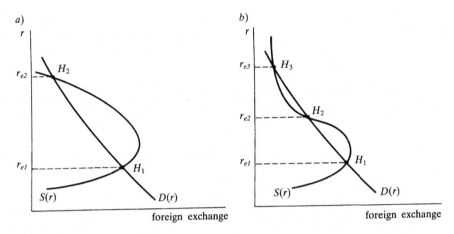

Fig. 7.1 Multiple equilibria and stability

foreign exchange will be either a monotonically increasing function of r (if $\eta_x > 1$) or a monotonically decreasing function of r (if $\eta_x < 1$).

(2) The demand for exports has an elasticity greater than one in some stretch(es), and smaller than one in other stretch(es). This is a perfectly normal: for example, we remember from elementary microeconomics that a simple linear demand curve has an elasticity greater than one in the upper part, equal to one at the intermediate point, lower than one in the lower part. In this case, foreign exchange receipts (and thus the supply) will increase in some stretch(es) and decrease in other stretch(es).

It is important to note that in case (2) multiple equilibria (i.e., multiple intersections between the demand and supply curves) may occur, a quite normal occurrence with varying-elasticity export demand. This is shown in Fig. 7.1.

To examine the consequences on the foreign exchange market of the two cases (and relative subcases) we must first introduce behaviour hypotheses. The hypothesis made here is that the exchange rate tends to depreciate when there is a positive excess demand for foreign exchange and to appreciate in the opposite case. This is an extension to the foreign exchange market (where, as we said above, the good transacted is foreign exchange) of the usual hypothesis concerning the change in the price of a good determined by the forces of demand and supply. We also assume that we are in a regime of freely flexible exchange rates, so that there is no intervention on the part of the monetary authorities to peg the exchange rate.

Under the price-adjustment assumption, the condition for the market to be stable is that a price increase tends to reduce excess demand (and, symmetrically, a price decrease tends to reduce excess supply). In other words, an exchange-rate depreciation should reduce excess demand for foreign exchange. The condition for this to happen is quite easily found: if we remember Eq. (7.14), we immediately see that the condition for a depreciation to reduce excess demand coincides with the condition for a depreciation to improve the balance of payments in foreign currency, namely condition (7.10).

This proves that in the circumstances hypothesized the conditions concerning the twin problems of balance-of-payments adjustment and of foreign-exchange market stability coincide.

Since condition (7.10) is certainly satisfied when $\eta_x > 1$, but may be violated when $\eta_x < 1$, it is easy to see that in the varying-elasticity case the possible multiple equilibria may be either stable or unstable (in general they will be alternatively stable and unstable).

In Fig. 7.1 we have illustrated just two of the several possible occurrences. In Fig. 7.1a there is a case of two equilibrium points, due to the fact that the export demand elasticity is initially higher than one (so that the supply of foreign exchange is increasing, as explained in relation to case (1) above) and then falls below one. It is possible to immediately verify that H_1 is a stable equilibrium point, since for r below (above) r_{e1}, demand is higher (lower) than supply, and so r increases (decreases) towards r_{e1}. On the contrary, H_2 is an unstable equilibrium point, because for r below (above) r_{e2}, demand is lower (higher) than supply, and so r decreases (increases) away from r_{e2}.

In Fig. 7.1b a case of three equilibrium points is illustrated: by the usual reasoning, the reader can check that H_1 and H_3 are stable equilibrium points whilst H_2 is an unstable one.

Since the considerations on foreign-exchange receipts and outlays previously explained can be applied to each equilibrium point, it follows that stable equilibrium points will be characterized by condition (7.10) [or (7.9), if we consider a neighbourhood of the point], whilst unstable points will be characterized by these conditions not being fulfilled.

It has already been stated above that the nature of the supply schedule of foreign exchange makes the presence of multiple equilibria a normal occurrence; the present graphic analysis has shown that stable and unstable equilibria usually alternate. We say usually because the extreme case of $S(r)$ being tangent to $D(r)$ at one point cannot be excluded: this equilibrium point will be stable on one side, unstable on the other.

7.4   Interrelations Between the Spot and Forward Exchange Rate

To examine the interrelations between the spot and the forward exchange rate, we must first examine the determination of the equilibrium in the forward exchange market. For this purpose we have to consider the supplies of and demands for forward exchange by the various operators.

7.4.1   The Various Excess Demand Schedules

(a) *Covered Interest Arbitrage*

We have seen in Sect. 4.1 that short-term funds will tend to flow in, remain where they are, or tend to flow out according to inequality (4.2), that we rewrite in the form

$$r^F \lesseqgtr r\frac{1+i_h}{1+i_f}. \tag{7.15}$$

It is clear that a *supply* (negative excess demand) of forward exchange corresponds to an *outflow* of funds (demand for spot exchange) and vice versa. Therefore, if we denote by E_{AF} the (positive or negative) excess demand for forward exchange and by r_N^F the interest-parity forward rate, i.e. that which satisfies the neutrality condition

$$r_N^F = r\frac{1+i_h}{1+i_f}, \tag{7.16}$$

we have

$$E_{AF} \gtreqless 0 \quad \text{according as } r^F \lesseqgtr r_N^F. \tag{7.17}$$

We must now examine the features of the E_{AF} schedule. Since it is profitable to move funds (inwards or outwards as the case may be) as long as $r^F \neq r_N^F$ because there is an interest gain with no exchange risk, it might seem obvious to assume an infinite elasticity of these funds in the neighbourhood of the forward exchange rate which satisfies the neutrality condition. If it were so, the traditional theory would be right to identify the neutrality condition with an equilibrium condition. In fact, any deviation[2] of the forward exchange rate from r_N^F would set into motion unlimited flows of arbitrage funds, which would prevail over the demands and supplies of other operators (commercial hedgers and speculators) and so would bring the forward exchange rate immediately back to r_N^F. In other words, if the elasticity of arbitrage funds (that is, of the arbitrageurs excess demand for forward exchange) were indeed infinite, the other components of the total demand and supply in the forward market would be altogether irrelevant for the determination of the forward exchange rate. In this case the neutrality condition would also be the equilibrium condition and interest parity would prevail.

---

[2]More precisely, the deviation should be such as to yield a profit rate above a certain threshold, that is the value below which arbitrageurs do not find it worth their while to move funds from one centre to another. This minimum, also called *arbitrage incentive*, may vary according to the historical period and institutional context. Keynes (1923; p. 28) stated the value of 1/2 % per annum; Stern (1973; p. 50) believes that in the postwar period arbitrage was undertaken for as little as 1/10 or 1/32 % per annum. At any given moment, however, this threshold is given and so we need not worry about its determination.

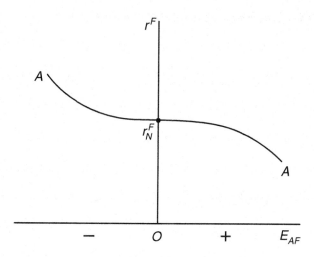

Fig. 7.2 The arbitrageurs' excess demand for forward exchange

However, various considerations can be made to support the argument that arbitrage funds are not infinitely elastic, with the consequent possibility of the forward rate deviating from r_N^F. Apart from the obvious fact that arbitrage funds are in any case a finite amount, so that the forward rate moves towards r_N^F until arbitrageurs run out of funds (or cannot borrow any more), the theory of portfolio selection tells us that when no foreign assets exist in the agent's portfolio, a small differential return is sufficient to induce heavy purchases, thus making the demand schedule highly elastic at the initial point. However, as the stock of foreign assets in the agent's portfolio increases, higher and higher covered interest differentials will be necessary to induce further additions to that stock, which explains why the schedule becomes more and more rigid (and there is a possibility that it will become perfectly rigid at a certain level of forward commitments). In graphic terms, all these considerations point to an arbitrageurs' excess demand schedule for forward exchange like that drawn in Fig. 7.2.

(b) *Commercial Hedging*

We refer to those commercial operators whose export and import contracts are stipulated in foreign exchange and do not contemplate immediate payment. If these operators do not cover the exchange risk, they become (functionally) speculators. We shall consider here only the operators who wish to hedge against the exchange risk. These, as we saw in Sect. 2.5.2, can hedge either through a spot transaction associated with suitable financing (in which case they also act as arbitrageurs[3]) or by having recourse to the forward market. That is, the

---

[3]For example, an importer who hedges a future payment by purchasing foreign exchange spot acts as an arbitrageur, buying the foreign exchange spot and selling it forward, and then as a commercial

Fig. 7.3 Commercial
traders' excess demand for
forward exchange

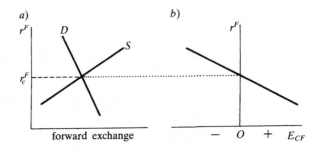

importer purchases the necessary foreign exchange forward, and the exporter
sells the future foreign exchange receipts forward. It is presumable that, in
general, the demand for forward exchange (coming from importers) will be
ceteris paribus a decreasing function of the price of forward exchange, i.e. of
the forward exchange rate (as this rate increases the profitability of importing
will be affected; besides, some importers will reduce the hedging or will not
hedge at all, etc., see Grubel 1966; Chap. 5), and the supply of forward exchange
(coming from exporters) will be, for analogous reasons, an increasing function
of the forward exchange rate. Thus we get Fig. 7.3, where r_c^F denotes that value
of the forward exchange rate which equates the demand for and supply of
forward exchange. In Fig. 7.3a we have drawn the demand and supply schedules
(assumed to be linear for simplicity's sake) and in Fig. 7.3b the resulting excess
demand schedule.

(c) *Speculation*

Forward speculation derives from a divergence between the *current forward*
rate and the *expected spot* rate (namely, the future spot exchange rate expected
to exist at the maturity of the forward contract) of a currency.

The extent of the speculative position, i.e. of the speculative excess demand
for forward exchange, is in absolute value an increasing function of the gap
between the current forward rate and the expected future spot rate, and a
decreasing function of risk. Given a certain evaluation of risk we get a certain
functional relation between the speculative position and the said gap; it is
possible, in other words, to subsume the evaluation of risk in this functional
form. In Fig. 7.4 we have drawn two alternative schedules of speculators' excess
demand for forward exchange, where $\tilde{r}$ is the spot exchange rate expected to
exist at the maturity of the forward contract.

Both schedules (assumed to be linear for simplicity's sake) cross the vertical
axis at $r^F = \tilde{r}$ (no forward speculative position will be opened in this case);
schedule SS corresponds to the case in which the evaluation of risk (which includes

---

hedger, by purchasing from himself (in the former capacity) the forward exchange. As we saw in
Sect. 2.5.2, the indifference line between covering on the spot or on the forward market is given by
the neutrality condition.

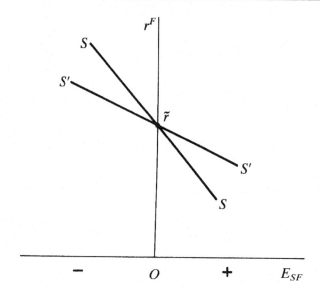

Fig. 7.4 Speculators' excess demand for forward exchange

the speculators' attitude towards risk) is such as to give a higher marginal risk coefficient in the speculators' calculation than schedule $S'S'$. It is clear that if we take $\tilde{r}$ as exogenous, we can write the speculators' excess demand for forward exchange as a function of r^F only. This procedure, which may be legitimate in the context of a static or uniperiodal equilibrium, is no longer so in a dynamic context (see the Appendix).

7.4.2   Forward Market Equilibrium and the Spot Rate

We now have all the elements to determine the equilibrium in the forward exchange market. Equilibrium means, as usual, the equality between (total) demand and (total) supply, or the equality to zero of the total excess demand, which in turn is the sum of the excess demands examined above. In graphic terms, we bring together into a single diagram (see Fig. 7.5) the various excess demand schedules from Figs. 7.2, 7.3b and 7.4. Note that for graphic convenience we have also drawn the AA schedule linear; furthermore we have assumed that the excess demand by commercial hedgers is zero at the forward rate which satisfies the neutrality condition, r_N^F. The equilibrium forward exchange rate is r_E^F, since the algebraic sum of the various excess demands becomes zero at this value.

In fact, $OB'' = OB' + OB$ in absolute value; we must remember that OB'' has to be given a positive sign and OB, OB' a negative one.

Now, this representation is perfectly satisfactory in the context of a uniperiodal static equilibrium and it might also be in the context of any model where exchange

Fig. 7.5 Equilibrium in the
forward exchange market

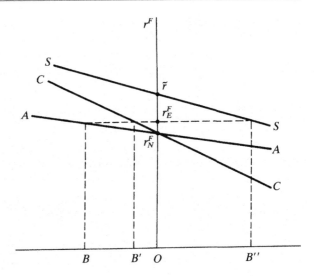

rates are rigid and expected to remain so, whence $r = \tilde{r} =$ given constant not only within a single period but also over any relevant time interval. Unfortunately the representation in question is less satisfactory in a different exchange-rate regime, where r and/or $\tilde{r}$ vary in time. When r changes, r^F_N changes as well and so the AA (and possibly the CC) schedule will shift; when $\tilde{r}$ changes the SS schedule will shift. Thus a different value of r^F_N will be determined at any particular moment; the time path followed by r^F_N can be obtained only by way of a dynamic analysis which should also consider the interrelations between the spot and forward exchange rates, that simultaneously determine these rates.

It should be pointed out that, if no such interrelations existed, or, more precisely, if only one-way relations did (from the spot to the forward market but not vice versa), the problem would admit of an easy and immediate solution. In fact, the spot exchange market would determine r (and possibly $\tilde{r}$) independently of r^F; by substituting these values for r and $\tilde{r}$ in the forward-market excess demands and by solving the equilibrium equation, one would then determine the forward exchange rate and its time path. However, in a complete model the influence of r^F on r cannot be neglected so that the causation runs both ways and the problem can be solved only by a simultaneous determination of the spot and forward exchange rate in a dynamic context. This will be dealt with in the Appendix.

7.4.3 The Monetary Authorities' Intervention

We conclude by briefly examining the problem of the monetary authorities' intervention on the forward market. This has been discussed in the literature both as a policy tool alternative to other tools (for example to an intervention on the domestic interest rate) and in relation to the effectiveness of the forward support of a currency undergoing speculative pressures. As regards the former aspect, it should be recalled from Sect. 2.5.2 that in order to stimulate an inflow of (arbitrage) funds the authorities should act so as to make the inequality

$$r^F < r\frac{1 + i_h}{1 + i_f} \tag{7.18}$$

hold. It follows that, if we consider r and i_f as exogenous, the alternative to an increase in i_h is a reduction in r^F. Therefore, if the monetary authorities want to attract capital, for example to offset a current account deficit, instead of managing i_h (this manoeuvre will be dealt with in Sect. 11.2) they can try to cause a decrease in r^F by selling foreign exchange forward.

As regards the latter aspect, it is a moot question whether official support of the forward exchange rate of a currency undergoing speculative (spot) attack in a devaluative direction is appropriate or not. This official support consists in the monetary authorities selling foreign exchange forward so as to prevent r^F from increasing too much.

On the one hand, to begin with, we observe that the official support tends to favour the inflow (or put a brake on the outflow) of covered interest arbitrage funds and thus brings an immediate benefit to the authorities' reserves subject to the speculative drain. Secondly, the official support, especially if very strong and resolute, helps to restore confidence in the currency under attack and so may ease the speculative pressure. Finally, in the absence of the support, the cost of forward cover to commercial traders might become prohibitive, with the consequence of inducing these operators to omit hedging and to become speculators.

On the other hand, we observe in the first place that, if the monetary authorities do not succeed in overcoming the speculative attack and the spot exchange rate depreciates, the forward support will cause them heavy additional losses, since when the forward contracts mature they have to procure the foreign exchange spot at a higher price. Secondly, the support decreases the speculators risk: in fact, if the forward discount is very high, though smaller than the expected depreciation (i.e., $\tilde{r} > r^F > r$), the speculators, by opening a forward speculative position, run the risk of heavy losses if the spot exchange rate does not depreciate when the forward contract matures (they will in fact have to pay out the difference $r^F - r$ per unit of foreign currency). By supporting the forward rate, the difference $r^F - r$ is reduced and so is the risk of speculators, who will intensify their activity.

It is not easy to strike a balance between the opposite views, partly because the appropriateness of official support may depend on circumstances. Broadly speaking, if the speculative attack is caused by a temporary loss of confidence in a currency which is basically sound (in the sense that in the long run the current exchange rate could be maintained without depreciation) then the official support of the forward exchange is advisable and effective. If, on the contrary, the monetary authorities themselves believe that the current spot rate cannot be maintained in the long run because of an irreversible fundamental disequilibrium in the balance of payments, then the support in question is a costly way of putting off the inevitable.

On the immediate and delayed effects of the official support of the forward exchange see Baillie and Osterberg (1997), Levin (1988), Tseng (1993a,b).

7.5 Appendix

7.5.1 The Critical Elasticities Condition

We shall first examine the simple case in which the supplies are perfectly elastic and then the general case.

7.5.1.1 The Simple Case

The condition for a variation in the exchange rate to make the payments balance move in the same direction is obtained by differentiating B with respect to r and ascertaining the conditions for $dB/dr > 0$. Given the definition of B we have

$$\frac{dB}{dr} = \frac{d\left[p_x x - r p_m m\right]}{dr} = p_x \frac{dx}{dr} - p_m m - r p_m \frac{dm}{dr}$$

$$= p_m m \left(\frac{p_x}{p_m m} \frac{dx}{dr} - 1 - \frac{r}{m} \frac{dm}{dr} \right). \tag{7.19}$$

If we multiply and divide the first term in parentheses by rx we obtain

$$\frac{dB}{dr} = p_m m \left(\frac{p_x x}{r p_m m} \frac{r}{x} \frac{dx}{dr} - 1 - \frac{r}{m} \frac{dm}{dr} \right) \tag{7.20}$$

$$= p_m m \left(\frac{p_x x}{r p_m m} \eta_x - 1 + \eta_m \right),$$

where

$$\eta_x \equiv \frac{dx}{x} / \frac{dr}{r} \equiv \frac{r}{x} \frac{dx}{dr}, \eta_m \equiv -\frac{dm}{m} / \frac{dr}{r} \equiv -\frac{r}{m} \frac{dm}{dr}. \tag{7.21}$$

Since $p_m m > 0$, dB/dr will be positive if and only if

$$\frac{p_x x}{r p_m m} \eta_x - 1 + \eta_m > 0. \tag{7.22}$$

Let us now consider the balance of payments expressed in terms of foreign currency, $B' = (1/r)B$. We have

$$\frac{dB'}{dr} = \frac{d\left[(1/r)\,p_x x - p_m m\right]}{dr}$$

$$= -\frac{1}{r^2}p_x x + \frac{1}{r}p_x\frac{dx}{dr} - p_m\frac{dm}{dr}$$

$$= \frac{p_x x}{r^2}\left(-1 + \frac{r}{x}\frac{dx}{dr} - \frac{r^2 p_m}{p_x x}\frac{dm}{dr}\right). \tag{7.23}$$

If we multiply and divide the last term in parentheses by m, we get

$$\frac{dB'}{dr} = \frac{p_x x}{r^2}\left(-1 + \frac{r}{x}\frac{dx}{dr} - \frac{rp_m m}{p_x x}\frac{r}{m}\frac{dm}{dr}\right)$$

$$= \frac{p_x x}{r^2}\left(-1 + \eta_x + \frac{rp_m m}{p_x x}\eta_m\right), \tag{7.24}$$

where the elasticities are defined as above. Since $\left(p_x x/r^2\right) > 0$, dB'/dr will be positive if and only if

$$-1 + \eta_x + \frac{rp_m m}{p_x x}\eta_m > 0, \tag{7.25}$$

whence condition (7.10) immediately follows.

Condition (7.22) can be rewritten as

$$\frac{rp_m m}{p_x x}\eta_m + \eta_x > \frac{rp_m m}{p_x x}. \tag{7.26}$$

The necessary and sufficient condition for the foreign-currency balance to move in the right direction and the domestic-currency balance to move in the wrong direction at the same time, is that (7.25) occurs whilst (7.22) does *not*, that is

$$\frac{rp_m m}{p_x x} > \frac{rp_m m}{p_x x}\eta_m + \eta_x > 1, \tag{7.27}$$

where the left-hand side of this inequality means that (7.26) is *not* satisfied and the right-hand side that (7.25) *is* satisfied. It can be seen immediately that the double inequality (7.27) can occur only when there is a balance-of-payments deficit.

Likewise we can find the necessary and sufficient condition for the domestic-currency balance to move in the right direction and the foreign-currency balance to move in the wrong direction at the same time. It turns out to be

$$\frac{p_x x}{r p_m m} > \eta_m + \frac{p_x x}{r p_m m} \eta_x > 1, \tag{7.28}$$

which can be verified only when there is a balance-of-payments surplus.

The simple analysis carried out in the text and in the present section *considers only the demand elasticities*: demand for foreign goods by the home country (demand for imports) and demand for domestic goods by the rest of the world (demand for exports). This analysis is therefore based on the assumption that the respective supply elasticities are infinite, so that supply adjusts to demand with no price adjustment. In other words, producers supply any quantity of goods demanded without changing the supply price (expressed in terms of their own currency), as is shown by the fact that p_m and p_x are assumedly constant. It is not, of course, necessary that this should be true everywhere, as it is sufficient that it holds in the range within which the demand changes (triggered by the exchange-rate changes) fall. From this point of view the assumption seems plausible in the context of economies with less than full employment, where it is not infrequent that increases in demand—especially if coming from abroad-are met at the going price; besides, the downward rigidity of prices justifies the fact that decreases in demand do not usually cause a fall in selling prices.

This explains why the problem of the effects of an exchange-rate variation is usually dealt with by way of the simple analysis. However, both for theoretical completeness and because supply effects cannot be neglected, we deal with the general case in the next section.

7.5.1.2 The General Case

In general, supply will be an increasing function of the price (expressed in the supplier's currency) of the commodity exchanged; the equilibrium between demand and supply determines both the quantity exchanged and the price, which is thus no longer a datum. Let us begin by considering the exports of the home country, and let $S_x = S_x(p_x)$ be the supply of exports, an increasing function of their domestic-currency price p_x; the demand for exports by the foreign buyers will be $D_x = D_x(\frac{1}{r}p_x)$, a decreasing function of the price expressed in foreign currency ($\frac{1}{r}p_x$, neglecting transport costs, etc.). Thus we have the system

$$
\begin{aligned}
S_x &= S_x(p_x), \\
D_x &= D_x(\tfrac{1}{r}p_x), \\
S_x &= D_x,
\end{aligned}
\tag{7.29}
$$

whose solution—in correspondence to any given exchange rate r—determines the equilibrium price p_x and the equilibrium quantity, which we denote by $x = S_x = D_x$. In equilibrium we thus have the system

$$
\begin{aligned}
x - S_x(p_x) &= 0, \\
x - D_x(\tfrac{1}{r}p_x) &= 0,
\end{aligned}
\tag{7.30}
$$

which is a system of two implicit functions in the three variable x, p_x, r. By using the implicit function theorem (we assume that the required condition on the Jacobian occurs) we can express x and p_x as differentiable functions of r and then compute the derivatives dx/dr and dp_x/dr by the method of comparative statics (see, for example, Gandolfo 2009; Chap. 20). Thus, by differentiating (7.30) with respect to r, we have

$$
\begin{aligned}
\frac{dx}{dr} - \frac{dS_x}{dp_x}\frac{dp_x}{dr} &= 0, \\
\frac{dx}{dr} - \frac{dD_x}{d[(1/r)p_x]}\frac{d[(1/r)p_x]}{dr} &= 0.
\end{aligned}
\tag{7.31}
$$

Since $d(\tfrac{1}{r}p_x)/dr = -\left(1/r^2\right)p_x + (1/r)dp_x/dr$, after rearranging terms we have

$$
\begin{aligned}
\frac{dx}{dr} - \frac{dS_x}{dp_x}\frac{dp_x}{dr} &= 0, \\
\frac{dx}{dr} - \frac{1}{r}\frac{dD_x}{d[(1/r)p_x]}\frac{dp_x}{dr} &= -\frac{1}{r^2}\frac{dD_x}{d[(1/r)p_x]}p_x,
\end{aligned}
\tag{7.32}
$$

whose solution yields the required derivatives dx/dr, dp_x/dr. These turn out to be

$$
\frac{dx}{dr} = \frac{-\dfrac{dD_x}{d[(1/r)p_x]}p_x\dfrac{dS_x}{dp_x}}{\dfrac{dS_x}{dp_x} - \dfrac{1}{r}\dfrac{dD_x}{d[(1/r)p_x]}}, \quad
\frac{dp_x}{dr} = \frac{-\dfrac{dD_x}{d[(1/r)p_x]}p_x}{\dfrac{dS_x}{dp_x} - \dfrac{1}{r}\dfrac{dD_x}{d[(1/r)p_x]}}.
\tag{7.33}
$$

We now define the elasticities of the demand for and supply of exports

$$
\eta_x \equiv -\frac{dD_x}{d[(1/r)p_x]}\frac{(1/r)p_x}{D_x}, \quad \varepsilon_x \equiv \frac{dS_x}{dp_x}\frac{p_x}{S_x},
\tag{7.34}
$$

and manipulate Eq. (7.33) so as to express the derivatives in terms of elasticities. Beginning with dx/dr and multiplying numerator and denominator by the same quantity $p_x/x = p_x/S_x = p_x/D_x$ (remember that we are considering the equilibrium point) we get

$$\frac{dx}{dr} = \frac{-\dfrac{dD_x}{d\left[(1/r)\,p_x\right]}p_x \dfrac{dS_x}{dp_x}\dfrac{p_x}{S_x}}{\dfrac{dS_x}{dp_x}\dfrac{p_x}{S_x} - \dfrac{dD_x}{d\left[(1/r)\,p_x\right]}\dfrac{(1/r)\,p_x}{D_x}} = \frac{-\dfrac{dD_x}{d\left[(1/r)\,p_x\right]}p_x\varepsilon_x}{\varepsilon_x + \eta_x}. \tag{7.35}$$

If we multiply and divide the numerator by the same quantity $D_x/(1/r)$ we get

$$\frac{dx}{dr} = \frac{-\dfrac{D_x}{1/r}\dfrac{dD_x}{d\left[(1/r)\,p_x\right]}\dfrac{(1/r)\,p_x}{D_x}\varepsilon_x}{\varepsilon_x + \eta_x} = \frac{x\eta_x\varepsilon_x}{\varepsilon_x + \eta_x}, \tag{7.36}$$

where in the last passage we have used the fact that $D_x = x$ and introduced the assumption that the exchange rate equals one in the initial situation (this does not involve any loss of generality as it simply implies a suitable definition of the units of measurement).

As regards dp_x/dr, by a similar procedure (multiply numerator and denominator by $x/p_x = S_x/p_x = D_x/p_x$, then multiply and divide the numerator by $1/r$) and assuming $r = 1$ initially, we obtain

$$\frac{dp_x}{dr} = \frac{p_x\eta_x}{\varepsilon_x + \eta_x}. \tag{7.37}$$

Let us now consider the imported commodity and let $S_m = S_m(p_m)$ be its supply as a function of its price p_m in foreign currency (the currency of the producing country), and $D_m = D_m(rp_m)$ its demand by the importing country as a function of its domestic-currency price (rp_m). Thus we have the system

$$\begin{aligned} S_m &= S_m(p_m), \\ D_m &= D_m(rp_m), \\ S_m &= D_m, \end{aligned} \tag{7.38}$$

whose solution determines—at any given value of r—the equilibrium price (p_m) and the equilibrium quantity which we denote by $m = S_m = D_m$. If we consider the system of implicit functions

$$\begin{aligned} m - S_m(p_m) &= 0, \\ m - D_m(rp_m) &= 0, \end{aligned} \tag{7.39}$$

and use the comparative static method as explained in relation to (7.30) above, we get

$$
\begin{aligned}
\frac{dm}{dr} - \frac{dS_m}{dp_m}\frac{dp_m}{dr} &= 0, \\
\frac{dm}{dr} - \frac{dD_m}{d\,(rp_m)}\left(p_m + r\frac{dp_m}{dr}\right) &= 0,
\end{aligned}
\tag{7.40}
$$

that is

$$
\begin{aligned}
\frac{dm}{dr} - \frac{dS_m}{dp_m}\frac{dp_m}{dr} &= 0, \\
\frac{dm}{dr} - \frac{dD_m}{d\,(rp_m)}r\frac{dp_m}{dr} &= \frac{dD_m}{d\,(rp_m)}p_m,
\end{aligned}
\tag{7.41}
$$

whose solution yields

$$
\frac{dm}{dr} = \frac{\dfrac{dS_m}{dp_m}\dfrac{dD_m}{d\,(rp_m)}p_m}{\dfrac{dS_m}{dp_m} - \dfrac{dD_m}{d\,(rp_m)}r}, \quad \frac{dp_m}{dr} = \frac{\dfrac{dD_m}{d\,(rp_m)}p_m}{\dfrac{dS_m}{dp_m} - \dfrac{dD_m}{d\,(rp_m)}r}.
\tag{7.42}
$$

We now define the import demand and supply elasticities

$$
\eta_m \equiv -\frac{dD_m}{d\,(rp_m)}\frac{rp_m}{D_m}, \varepsilon_m \equiv \frac{dS_m}{dp_m}\frac{p_m}{S_m},
\tag{7.43}
$$

and manipulate Eq. (7.42) to express them in terms of elasticities. Beginning with dm/dr we multiply numerator and denominator by the same quantity p_m/m, then multiply and divide the numerator by r/D_m; thus we arrive at

$$
\frac{dm}{dr} = \frac{\dfrac{p_m}{S_m}\dfrac{dS_m}{dp_m}\dfrac{dD_m}{d\,(rp_m)}\dfrac{rp_m}{D_m}\dfrac{D_m}{r}}{\dfrac{p_m}{S_m}\dfrac{dS_m}{dp_m} - \dfrac{dD_m}{d\,(rp_m)}\dfrac{rp_m}{D_m}} = \frac{-m\varepsilon_m\eta_m}{\varepsilon_m + \eta_m},
\tag{7.44}
$$

where in the last passage we have used Eq. (7.43) and set $r = 1$ in the initial situation.

By a similar procedure we get

$$
\frac{dp_m}{dr} = \frac{\dfrac{1}{r}\dfrac{rp_m}{D_m}\dfrac{dD_m}{d\,(rp_m)}p_m}{\dfrac{p_m}{S_m}\dfrac{dS_m}{dp_m}\dfrac{dD_m}{d\,(rp_m)}\dfrac{rp_m}{D_m}} = \frac{-p_m\eta_m}{\varepsilon_m + \eta_m},
\tag{7.45}
$$

where in the last passage we have again used Eq. (7.43) and set $r = 1$.

We thus have all the elements to examine the effects on the balance of payments of an exchange-rate variation in the general case. To begin with, we consider the balance of payments in domestic currency $B = p_x x - r p_m m$ and differentiate it totally with respect to r, remembering that also p_x and p_m are functions of r as shown above. Thus we have

$$\frac{dB}{dr} = \frac{dp_x}{dr} x + p_x \frac{dx}{dr} - p_m m - r \frac{dp_m}{dr} m - r p_m \frac{dm}{dr}. \tag{7.46}$$

If we now substitute expression (7.36), (7.37), (7.44) and (7.45) into (7.46), we get

$$\frac{dB}{dr} = p_x x \frac{\eta_x}{\varepsilon_x + \eta_x} + p_x x \frac{\eta_x \varepsilon_x}{\varepsilon_x + \eta_x} - p_m m + r p_m m \frac{\eta_m}{\varepsilon_m + \eta_m} + r p_m m \frac{\varepsilon_m \eta_m}{\varepsilon_m + \eta_m}, \tag{7.47}$$

whence, by collecting terms (as we set $r = 1$, we have $r p_m m = p_m m$)

$$\frac{dB}{dr} = r p_m m \left[\frac{p_x x}{r p_m m} \frac{\eta_x (1 + \varepsilon_x)}{\varepsilon_x + \eta_x} + \frac{\eta_m (1 + \varepsilon_m)}{\varepsilon_m + \eta_m} - 1 \right]. \tag{7.48}$$

The condition for $dB/dr > 0$ is thus

$$\frac{p_x x}{r p_m m} \frac{\eta_x (1 + \varepsilon_x)}{\varepsilon_x + \eta_x} + \frac{\eta_m (1 + \varepsilon_m)}{\varepsilon_m + \eta_m} - 1 > 0. \tag{7.49}$$

When we assume $B = 0$ initially, then condition (7.49) can be rewritten by simple manipulations as

$$\frac{\eta_x \eta_m (\varepsilon_x + \varepsilon_m + 1) + \varepsilon_m \varepsilon_x (\eta_x + \eta_m - 1)}{(\varepsilon_x + \eta_x)(\varepsilon_m + \eta_m)} > 0. \tag{7.50}$$

Conditions (7.49) and (7.50) are the two forms usually found in the literature (apart from notational differences), and are called the Bickerdicke-Robinson conditions (Bickerdicke 1920; Robinson 1937) after the names of the two authors who independently derived them. These conditions were popularised by Metzler (1949; p. 226), hence they are also called the Bickerdicke-Robinson-Metzler conditions.

When ε_m and ε_x tend to infinity, by evaluating the limit of (7.49) we get

$$\frac{p_x x}{r p_m m} \eta_x + \eta_m - 1 > 0, \tag{7.51}$$

which is the condition holding in the simple case as shown in (7.22).

It is important to stress that the consideration of the supply elasticities makes the situation more favourable, in the sense that the balance of payments may move in the right direction even if the sum of the demand elasticities is smaller than one. In fact, if we consider (7.50), we see that the fraction may be positive even if $\eta_x + \eta_m < 1$, provided that the supplies are sufficiently rigid: in the extreme case, for $\varepsilon_x = \varepsilon_m = 0$ (absolutely rigid supplies) the condition is *always* satisfied no matter how small the demand elasticities are (but neither can be zero). In this case, in fact, the quantities exported and imported cannot deviate from the given supplies and all the adjustment falls on prices. If we consider, for example, a depreciation, we see that, on the one hand, the increase in the demand for exports is checked by an increase in p_x: as the supply is perfectly rigid, the excess demand causes an increase in p_x up to the point where the demand falls back to its initial amount. This means that $(1/r)\,p_x$ must go back to its initial value for $D_x = D_x(\frac{1}{r}p_x)$ to remain the same, hence the percentage increase in p_x is exactly equal to the percentage increase in r and the domestic-currency receipts from exports increase by the same percentage as the depreciation.

On the other hand, the decrease in the demand for imports induced by the depreciation will—since the foreign supply is perfectly rigid—cause a decrease in the foreign price p_m such as to bring the demand back to its initial level. This decrease must be such that rp_m goes back to its initial value for $D_m = D_m(rp_m)$ to remain the same. This means that the outlay for imports will return to its initial value, and so—as the receipts from exports have increased—the balance of payments must necessarily improve.

Similar observations hold for the more general condition (7.49), which for $\varepsilon_x = \varepsilon_m = 0$ and η_x, η_m however small but positive, certainly occurs.

We now turn to the foreign-currency balance, $B' = (1/r)p_x x - p_m m$; total differentiation with respect to r gives

$$\frac{dB'}{dr} = -\frac{1}{r^2}p_x x + \frac{1}{r}\frac{dp_x}{dr}x + \frac{1}{r}p_x\frac{dx}{dr} - \frac{dp_m}{dr}m - p_m\frac{dm}{dr}, \tag{7.52}$$

which, after substitution of (7.36), (7.37), (7.44) and (7.45) into it, becomes

$$\frac{dB'}{dr} = \frac{1}{r}p_x x\left[-1 + \frac{\eta_x(1 + \varepsilon_x)}{\varepsilon_x + \eta_x} + \frac{rp_m m}{p_x x}\frac{\eta_m(1 + \varepsilon_m)}{\varepsilon_m + \eta_m}\right]. \tag{7.53}$$

The condition for $dB'/dr > 0$ thus is

$$\frac{\eta_x(1 + \varepsilon_x)}{\varepsilon_x + \eta_x} + \frac{rp_m m}{p_x x}\frac{\eta_m(1 + \varepsilon_m)}{\varepsilon_m + \eta_m} - 1 > 0. \tag{7.54}$$

When an initial situation of equilibrium ($B' = 0$) is assumed, condition (7.54) can be reduced to (7.50) as well.

Finally, if we evaluate the limit (for $\varepsilon_x, \varepsilon_m$ tending to infinity) of the expression on the left-hand side of (7.54), we get

$$\eta_x + \frac{rp_m m}{p_x x}\eta_m > 1, \tag{7.55}$$

which coincides with (7.25).

7.5.1.3 Effects on the Terms of Trade

We conclude this section by examining the effects of an exchange-rate variation on the terms of trade $\pi = p_x/rp_m$. It is obvious that when p_x and p_m are constant (this corresponds to the simple case examined in Sect. 7.5.1.1), π varies in the opposite direction to r. In the general case, by totally differentiating π with respect to r, using Eqs. (7.37) and (7.45), and (where convenient) the fact that $r = 1$ in the initial situation, we have

$$
\frac{d\pi}{dr} = \frac{\frac{dp_x}{dr}rp_m - \left(p_m + r\frac{dp_m}{dr}\right)p_x}{r^2 p_m^2} = \frac{1}{rp_m}\frac{p_x\eta_x}{\varepsilon_x + \eta_x} - \frac{1}{rp_m}\left(p_x - \frac{rp_x}{rp_m}\frac{p_m\eta_m}{\varepsilon_m + \eta_m}\right)
$$

$$
= \frac{p_x}{rp_m}\left(\frac{\eta_x}{\varepsilon_x + \eta_x} - 1 + \frac{\eta_m}{\varepsilon_m + \eta_m}\right), \tag{7.56}
$$

whence, by simple manipulations,

$$
\frac{d\pi}{dr} = \pi\frac{\eta_x\eta_m - \varepsilon_x\varepsilon_m}{(\varepsilon_x + \eta_x)(\varepsilon_m + \eta_m)}, \tag{7.57}
$$

so that $d\pi/dr \gtrless 0$ according as $\eta_x\eta_m - \varepsilon_x\varepsilon_m \gtrless 0$. we remember that an increase in π is usually classified as an improvement: an increase in π, in fact, means that with a given amount of exports we obtain a greater amount of imports, or that less exports are required to obtain the same amount of imports. Thus, we can conclude that an exchange-rate depreciation will improve the terms of trade if the product of the demand elasticities is greater than the product of the supply elasticities. This will always be the case if the least one of the supply elasticities is zero, provided that both demand elasticities, no matter how small, are positive. On the contrary, when $\varepsilon_x, \varepsilon_m$ go to infinity, the terms of trade will necessarily worsen.

7.5.2 The Stability of the Foreign Exchange Market

The coincidence—in the assumed situation (i.e. that the supply and demand for foreign exchange come exclusively from transactions in goods and services)— between the conditions for an exchange-rate variation to make the balance of payments move in the same direction, and the stability conditions in the foreign exchange market can be shown as follows.

The dynamic assumption according to which the exchange rate depreciates (appreciates) when there is a positive (negative) excess demand for foreign exchange gives rise to the differential equation

$$
\dot{r} = \Phi[E(r)], \tag{7.58}
$$

where Φ is a sign-preserving function and $\Phi'[0] \equiv k > 0$. If we linearise the Φ function at the equilibrium point we get

$$\dot{r} = kE(r),   \tag{7.59}$$

where k can be interpreted as an adjustment speed. Since, in the assumed situation, $E(r) = -B'(r)$, we have

$$\dot{r} = -kB'(r).   \tag{7.60}$$

To examine the local stability it is sufficient to linearise $B'(r)$ at the equilibrium point,[4] thus obtaining

$$\dot{\bar{r}} = -k\frac{dB'}{dr}\bar{r},   \tag{7.61}$$

where $\bar{r} = r - r_e$ denotes the deviations from equilibrium and dB'/dr is evaluated at r_e. The solution to this simple differential equation is

$$\bar{r}(t) = Ae^{-k(dB'/dr)t},   \tag{7.62}$$

where A depends on the initial deviation from equilibrium. The necessary and sufficient stability condition is

$$-k\frac{dB'}{dr} < 0,   \tag{7.63}$$

that is, as $k > 0$,

$$\frac{dB'}{dr} > 0,   \tag{7.64}$$

which proves the stated coincidence.

7.5.3   A Model for the Simultaneous Determination of the Spot and Forward Exchange Rate

In any market, equilibrium is reached when the algebraic sum of all operators' excess demands is zero. We must, then, formally express these excess demands as described in Sect. 2.6; we begin by considering the *spot market*.

---

[4]In the case of multiple equilibria the linearization will have to be performed at each equilibrium separately; it will then be possible to ascertain the stable or unstable nature of each point.

A first category of transactors consists in non-speculators. By definition, their supplies of and demands for foreign exchange are influenced by the current and not by the expected exchange rate. Thus, by assuming a linear relation for simplicity, we can write

$$E_{n_t} = a_1 r_t + A \cos \omega t, \quad a_1 < 0, \quad A > 0, \tag{7.65}$$

where E_{n_t} is non-speculators' excess demand at time t and $A \cos \omega t$ represents exogenous factors, for example seasonal influences, acting on both the demand and the supply. More complicated functions could be used, but we wish to simplify to the utmost to convey the basic idea. In (7.65) the stability condition examined in the previous section is assumed to be satisfied: in fact, $dE_{n_t}/dr = a_1 < 0$.

A second category consists in (covered interest) arbitrageurs. From what was explained in Sects. 2.5 and 2.6 it follows that an excess demand for spot exchange corresponds to an excess demand for forward exchange, but with is sign reversed.

Thus we have, in general[5]

$$E_{A_t} = -jE_{AF_t}, \tag{7.67}$$

where E_{A_t} denotes the arbitrageurs' excess demand for forward exchange and j is a coefficient depending on the interest rate. Let us assume, for example, that condition (4.2) holds with the $>$ sign and let X be the amount of foreign exchange that arbitrageurs wish to place in the foreign centre. Thus we have a demand, i.e. a positive excess demand (or a non-supply of spot foreign exchange, if the funds are already abroad; the non-supply can be considered as a negative excess supply i.e. a positive excess demand) for spot exchange equal to X. At the same time the arbitrageurs sell forward not only the capital but also the interest accrued on it (the non-consideration of the interest accrued induced some authors erroneously to set E_{A_t} and E_{AF_t} equal in absolute value), that is to say there is a supply (negative

---

[5]Strictly speaking we should also add the excess demand for spot exchange coming from the liquidation of forward contracts stipulated at time $t - \tau$ and maturing at time t. It is however easy to show that, if the forward market is in equilibrium at each instant, that is, if the following relation holds

$$E_{AF_t} + E_{CF_t} + E_{SF_t} = 0, \tag{7.66}$$

which, as we shall see, expresses forward market equilibrium, then the excess demand under consideration (denoted by E_{L_t}) is zero. In fact

$$E_{L_t} = E_{AF_{t-\tau}} + E_{CF_{t-\tau}} + E_{SF_{t-\tau}};$$

now, since (7.66) holds at each instant, and so also at $t - \tau$, we have

$$E_{AF_{t-\tau}} + E_{CF_{t-\tau}} + E_{SF_{t-\tau}} = 0, \text{ that is, } E_{L_t} = 0.$$

excess demand) of forward exchange equal to $(1 + i_f) X$. Therefore, as $E_{A_t} = X$, $E_{AF_t} = -(1 + i_f) X$, it follows that $E_{AF_t} = -(1 + i_f) E_{A_t}$, i.e.

$$E_{A_t} = -\frac{1}{1 + i_f} E_{AF_t}.$$

Speculators make up the third category. Their excess demand is an increasing function of the discrepancy between the expected and the current spot exchange rate: the greater this discrepancy (and so the expected profit) the greater the speculative position (for a detailed examination of this, see, for example, Cutilli and Gandolfo 1973; pp. 35–40). Thus we have (a linear relation is used for simplicity's sake)

$$E_{St} = k(\tilde{r}_t - r_t), \quad k > 0, \tag{7.68}$$

where $\tilde{r}_t$ is the spot exchange rate expected to hold in the future (expectations are, of course, formed at time t).

As regards the *forward market,* we recall that covered interest arbitrageurs demand (supply) forward exchange when the conditions exist for profitably placing short-term liquid funds at home (abroad), so that their excess demand for forward exchange can be taken as an increasing function of the discrepancy between the forward rate satisfying the neutrality condition and the current forward rate:

$$E_{AF_t} = j_1(r^F_{N_t} - r^F_t) = j_1 \left(\frac{1 + i_h}{1 + i_f} r_t - r^F_t \right), \quad j_1 > 0. \tag{7.69}$$

The excess demand of commercial traders hedging in the forward market is, as clarified in the text, a decreasing function of the forward exchange rate, that is

$$E_{CF_t} = j_2 + j_3 \cos \omega t + j_4 r^F_t, \quad j_2 > 0, j_3 > 0, j_4 < 0, \tag{7.70}$$

where the introduction of the term $j_3 \cos \omega t$ is due to the fact that it seems legitimate to assume that the exogenous factors exert their influence not only on the part of trade settled spot—see Eq. (7.65)—but also on the part of it settled forward.

The forward speculators' excess demand is to be considered as an increasing function of the discrepancy between the expected spot rate and the current forward rate, that is, to a linear approximation,

$$E_{SF_t} = j_5(\tilde{r}_t - r^F_t), \quad j_5 > 0. \tag{7.71}$$

The simultaneous determination of the spot and forward exchange rate is obtained by solving the system

$$\begin{aligned} E_{n_t} + E_{A_t} + E_{S_t} &= 0, \\ E_{AF_t} + E_{CF_t} + E_{SF_t} &= 0. \end{aligned} \tag{7.72}$$

It can be readily verified that the solution of this system does not present any difficulty if $\tilde{r}_t$ is assumed to be an exogenous datum. But this assumption cannot be seriously maintained. Expectations can be formed in various ways but will certainly include elements based on the behaviour of endogenous variables (in our simplified model, current and past values of both the spot and the forward rate), so that it is not possible to solve the system if $\tilde{r}_t$ is not specified.

We know that, in the context of an efficient foreign exchange market with rational expectations, the forward exchange rate is an unbiased predictor of the future spot rate; thus the expected spot rate can be fairly well approximated by the current forward rate. However, the empirical evidence does not seem to support this opinion (see the Appendix to Chap. 4).

The conclusion is that, since there is no universally accepted way of specifying $\tilde{r}_t$, different results will be obtained with different specifications.

References

Bahmani-Oskooee, M. (1998). Cointegration approach to estimate the long-run trade elasticities in LDCs. *International Economic Journal, 12*, 89–96.

Baillie, R. T., & Osterberg, W. P. (1997). Central bank intervention and risk in the forward market. *Journal of International Economics, 43*, 483–497.

Bickerdicke, C. F. (1920). The instability of foreign exchange. *Economic Journal, 30*, 118–122.

Bussière, M., Callegari, G., Ghironi, F., Sestieri, G., & Yamano, N. (2013). Estimating trade elasticities: Demand composition and the trade collapse of 2008–2009. *American Economic Journal: Macroeconomics, 5*, 118–151.

Cutilli, B., & Gandolfo, G. (1973). Un contributo alla teoria della speculazione in regime di cambi oscillanti. Roma: Ente per gli studi monetari bancari e finanziari Luigi Einaudi. *Quaderno di Ricerche, 10*.

Gandolfo, G. (2009). *Economic dynamics* (4th ed.). Berlin, Heidelberg, New York: Springer.

Grubel, H. G. (1966). *Forward exchange, speculation, and the international flow of capital.* Stanford: Stanford University Press.

Himarios, D. (1989). Do devaluations improve the balance of payments? The evidence revisited. *Economic Inquiry, 27*, 143–168.

Hooper, P., Johnson, K., & Marquez, J. (2000). *Trade elasticities for the G-7 countries.* Princeton University, International Economics Section: Princeton Studies in International Economics No. 87.

Imbs, J., & Mejean, I. (2015). Elasticity optimism. *American Economic Journal: Macroeconomics, 7*, 43–83.

Johnson, H. G. (1958). Towards a general theory of the balance of payments, Chap. VI in H. G. Johnson, *International trade and economic growth.* London: Allen&Unwin. Reprinted In R. E. Caves & H. G. Johnson (Eds.) (1968). *Readings in international economics.* London: Allen&Unwin, and In R. N. Cooper (Ed.) (1969). *International finance - Selected readings.* Harmondsworth: Penguin.

Keynes, J. M. (1923). *A tract on monetary reform.* London: Macmillan; reprinted as Vol. IV of *The collected writings of J.M. Keynes.* London: Macmillan for the Royal Economic Society, 1971.

Lerner, A. P. (1944). *The economics of control* (pp. 377–379). London: Macmillan.

Levin, J. H. (1988). The effects of government intervention in a dynamic model of the spot and forward exchange markets. *International Economic Journal 2*, 1–20.

Metzler, L. A. (1949). The theory of international trade. In H. S. Ellis (Ed.), *A survey of contemporary economics* (Chap. 6, Sect. III). Philadelphia: Blakiston; reprinted In L. A. Metzler (1973). *Collected papers.* Cambridge (Mass.): Harvard University Press.

Robinson, J. (1937). The foreign exchanges. In J. Robinson, *Essays in the theory of employment.* Oxford: Blackwell; reprinted In H. S. Ellis & L. A. Metzler (Eds.) (1949). *Readings in the theory of international trade* (pp. 83–103). Philadelphia: Blakiston.

Sohmen, E. (1969). *Flexible exchange rates* (2nd ed.). Chicago: University of Chicago Press.

Stern, R. M. (1973). *The balance of payments: theory and economic policy.* Chicago: Aldine.

Stern, R. M., Francis, J., & Schumacher, B. (1976). *Price elasticities in international trade: An annotated bibliograph.* London: Macmillan.

Tseng, H. -K. (1993a). Forward market intervention, endogenous speculation, and exchange rate variability. *Atlantic Economic Journal, 21,* 37–55.

Tseng, H. -K. (1993b). Forward intervention, risk premium and the fluctuations of international financial markets. *Economia Internazionale, 46,* 276–287.

Upadhyaya, K., & Dhakal, D. (1997). Devaluation and the trade balance: estimating the long run effect. *Applied Economics Letters, 4,* 343–345.

The Multiplier Approach

8

The multiplier analysis of the balance of payments (also called the foreign trade multiplier), that was introduced by Harrod (1933) before the Keynesian theory of the multiplier, has a twofold relevance. On the one hand, it can be seen as the counterpart, always in a partial equilibrium context, to the analysis based solely on exchange-rate variations, and thus as a step towards the integration of the two mechanisms. On the other hand, it has an importance of its own insofar as it is applicable to an institutional setting in which the exchange rate and prices are rigid.

We examine the small country case, that is the multiplier with no foreign repercussions, which implies that exports are entirely exogenous. In fact, the small country assumption means that what happens in the country under consideration has no appreciable effect on the rest-of-the-world variables (and, in particular, that changes in the country's imports, which are the rest-of-the-world exports, have no appreciable effect on the rest-of-the-world income and therefore on its imports); consequently, in the model, these variables can be considered as exogenous.

Several important feedbacks are neglected by assuming no foreign repercussions. In fact, the imports of the country under consideration (henceforth referred to as country 1) are the exports of one or more other countries and so enter into their income determination; similarly the exports of country 1 are the imports of one or more other countries. Thus, for example, an increase in income in country 1 causes (through the increase in this country's imports, which means an increase in the exports of one or more other countries) an increase in the income of these other countries and therefore in their imports. All or part of these additional imports will be directed to country 1, which will experience an increase in exports and so in income, a consequent increase in imports, and so on, with a chain of repercussions whose final result (assuming that the process converges) will certainly be different from that obtaining if no such repercussions occurred. A complete analysis of the problem of repercussions requires some heavy algebra, and will be treated in the Appendix.

© Springer-Verlag Berlin Heidelberg 2016
G. Gandolfo, *International Finance and Open-Economy Macroeconomics*,
Springer Texts in Business and Economics, DOI 10.1007/978-3-662-49862-0_8

The restrictive assumptions common to all such models are the usual ones: underemployed resources, rigidity of all prices (including the exchange rate and the rate of interest), absence of capital movements (so that balance of payments is synonymous with balance on goods and services), and all exports are made out of current output.

For simplicity of exposition, we assume linear functions; the case of general functions will be treated in the Appendix.

8.1   The Basic Model

The model used is the standard Keynesian textbook model with the inclusion of the foreign sector; the equations are as follows:

$$C = C_0 + by, \quad 0 < b < 1, \tag{8.1}$$

$$I = I_0 + hy, \quad 0 < h < 1, \tag{8.2}$$

$$m = m_0 + \mu y, \quad 0 < \mu < 1, \tag{8.3}$$

$$x = x_0, \tag{8.4}$$

$$y = C + I + x - m. \tag{8.5}$$

The equations represent, in this order: the consumption function (C_0 is the autonomous component, b is the marginal propensity to consume, and y is national income), the investment function (the autonomous component is I_0, and h is the marginal propensity to invest), the import function (m_0 is the autonomous component, and μ is the marginal propensity to import), the export function (the absence of foreign repercussions, as we said above, is reflected in the fact that exports are entirely exogenous), the determination of national income.

The meaning of Eq. (8.5) is simple: in an open economy, total demand for domestic output is no longer $C + I$, but $C + I - m + x$ which is composed of $C + I - m$ (aggregate demand for domestic output by residents) and x (demand for domestic output by nonresidents). In fact, in $C + I$ both home and foreign goods and services are now included, and the demand for foreign goods and services by residents in our simplified model is m: therefore, by subtracting m from $C + I$ we obtain the demand for domestic output by residents. Government expenditure is not explicitly included in Eq. (8.5) both because it can be considered as present in the autonomous components of the appropriate expenditure functions and because its inclusion as an additive term G in the r.h.s. of Eq. (8.5) as is usually done may be a source of potential error. In fact, the often used equation

$$y = C + I + x - m + G, \tag{8.6}$$

may convey the impression that any increase in government expenditure is income generating. This is not true, because government expenditure on foreign goods and services is *not* income generating: in this case, the increase in G is matched by an (exogenous) increase in m_0. The use of (8.6) tends to obscure this fact; the use of (8.5), on the contrary, draws our attention to the fact that government expenditure, if present, should be appropriately included in C_0, I_0, m_0 as the case may be (taxation T is not considered for simplicity's sake, but could easily be introduced).

It goes without saying that the use of Eq. (8.6) is not incorrect, provided that one bears in mind that the part (if any) of G which is used to purchase foreign goods and services must also be counted in m_0.

Equation (8.5) can be written in several alternative forms. For example, if we shift C and m to the left-hand side and remember that $y - C$ is, by definition, saving (S), we have

$$S + m = I + x, \tag{8.7}$$

which is the extension to an open economy of the well-known $S = I$ closed-economy condition. From (8.7) we obtain

$$S - I = x - m, \tag{8.8}$$

$$I - S = m - x, \tag{8.9}$$

that is, the excess of exports over imports is equal to the excess of saving over investment, viz the excess of imports over exports is equal to the excess of investment over saving. See Chap. 6, Eq. (6.22), where G and T are explicitly present.

Equations (8.1)–(8.5) form a complete system by means of which the foreign multiplier can be analyzed. Since, however, we are interested in balance of payments adjustment, we add the equation which defines the balance of payments B (since prices and the exchange rate are rigid, they can be normalized to one):

$$B = x - m. \tag{8.10}$$

The problem we are concerned with is to ascertain whether, and to what extent, balance-of-payments disequilibria can be corrected by income changes. Suppose, for example, that a situation of equilibrium is altered by an increase in exports, so that B shifts to a surplus situation. What are the (automatic) corrective forces that tend to re-equilibrate the balance of payments? The answer is simple: via the multiplier the increase in exports brings about an increase in income, which in turn determines an induced increase in imports via the marginal propensity to import. This increase in imports tends to offset the initial increase in exports, and we must ascertain whether the former increase exactly matches the latter (so that the balance of payments returns to an equilibrium situation) or not. In the second case the situation usually depicted is that the balance of payments will show a surplus, although smaller than the initial increase in exports: in other words, the induced change in imports will not be sufficient to re-equilibrate the balance of payments.

However, we shall see presently that the contrary case (that is, when induced imports increase more than the initial increase in exports) as well as the borderline case cannot be excluded on *a priori* grounds.

The case of an exogenous increase in imports is harder to examine. It might seem that this increase would cause, via the multiplier, a decrease in income and thus an induced decrease in imports, which tends to offset the initial autonomous increase. Things are not so simple as that however. In fact, we must check whether or not the autonomous increase in imports is accompanied by a simultaneous autonomous decrease in the demand for home output by residents. If the answer is yes, this will indeed cause a decrease in income etc. If not (which means that the autonomous increase in imports is matched by an autonomous decrease in saving), no depressive effect on income takes place and no adjustment via induced imports occurs. Intermediate cases are of course possible.

To rigorously analyse these and similar problems the first step is to find the formula for the multiplier. If we substitute from Eqs. (8.1)–(8.4) into Eq. (8.5) and solve for y, we obtain

$$y = \frac{1}{1 - b - h + \mu}(C_0 + I_0 - m_0 + x_0), \tag{8.11}$$

where of course

$$1 - b - h + \mu > 0 \tag{8.12}$$

for the solution to be economically meaningful. If we then consider the variations (denoted by Δ), we get

$$\Delta y = \frac{1}{1 - b - h + \mu}(\Delta C_0 + \Delta I_0 - \Delta m_0 + \Delta x_0). \tag{8.13}$$

Note that if we assume no induced investment, the multiplier is reduced to the familiar formula $1/(s + \mu)$, where $s = 1 - b$ is the marginal propensity to save. Also observe that, as in all multiplier analyses, the autonomous components are included in the multiplicand (where Δm_0 appears) whereas the coefficients concerning the induced components enter into the multiplier (where μ is included).

The open-economy multiplier is smaller than that for the closed economy $[1/(b + h)]$—of course if we assume that b and h are the same in both the closed and the open economy—because of the additional leakage due to imports.

Let us now introduce the usual dynamic behaviour assumption: producers react to excess demand by making adjustments in output: if aggregate demand exceeds (falls short of) current output, the latter will be increased (decreased); this mechanism operates independently of the origin of excess demand. It can be shown (see the Appendix) that the necessary and sufficient condition for dynamic stability is the same as (8.12): therefore, results for dynamic stability and meaningful comparative statics go hand in hand, as is often the case (Samuelson's correspondence principle).

From (8.12) we obtain

$$b + h < 1 + \mu, \tag{8.14}$$

or

$$b + h - \mu < 1. \tag{8.15}$$

Condition (8.14) means that the marginal propensity to spend, $(b + h)$, must be smaller than one plus the marginal propensity to import. Condition (8.15) means that the marginal propensity to spend on domestic output by residents, $(b + h - \mu)$, must be smaller than one.

To clarify the meaning of $b + h - \mu$, consider a unit increment in income, which causes an increment in the induced components of the various expenditure functions. The increment in consumption and investment is $b + h$ and contains both national and foreign goods. The part pertaining to the latter is thus a part of the total increment $b + h$, and coincides, in our pure flow model, with μ.

Algebraically, let the subscripts d and f denote domestic and foreign goods (and services) respectively; then

$$b = b_d + b_f, \tag{8.16}$$

$$h = h_d + h_f, \tag{8.17}$$

and in our simplified model

$$b_f + h_f = \mu. \tag{8.18}$$

Therefore

$$b + h - \mu = b_d + h_d \tag{8.19}$$

measures the marginal propensity to spend on domestic output by residents.

8.2   Balance-of-Payments Adjustment in the Case of an Exogenous Increase in Exports

Let us now consider the balance of payments. By substituting from Eqs. (8.3) and (8.4) into Eq. (8.10) and considering the variations we have

$$\Delta B = \Delta x_0 - \Delta m_0 - \mu \Delta y, \tag{8.20}$$

which states that the change in the balance of payments is equal to the exogenous change in exports minus the change in imports, the latter being partly exogenous (Δm_0) and partly induced ($\mu \Delta y$, where Δy is given by the multiplier formula (8.13) found above).

Here we have all that is needed to analyse the balance-of-payments adjustment problem. Consider first the case of an exogenous increase in exports. By assumption, no other exogenous change occurs, so that $\Delta C_0 = \Delta I_0 = \Delta m_0 = 0$, and the equations of change become

$$\Delta B = \Delta x_0 - \mu \Delta y, \quad \Delta y = \frac{1}{1-b-h+\mu} \Delta x_0,$$

whence

$$\Delta B = \Delta x_0 - \mu \frac{1}{1-b-h+\mu} \Delta x_0$$

$$= \frac{1-b-h}{1-b-h+\mu} \Delta x_0, \qquad (8.21)$$

which expresses the final change in the balance of payments. The reader will note that the simple mathematical procedure followed is nothing more than the algebraic transposition of the verbal reasoning made above; but it enables us to find the precise conditions under which the adjustment is incomplete, complete, or more than complete. These conditions are easily derived from (8.21).

If the marginal propensity to spend is smaller than one, $b+h < 1$, then $1-b-h > 0$ and so $\Delta B > 0$; furthermore, since $1-b-h < 1-b-h+\mu$, the fraction in the right-hand-side of (8.21) is smaller than one, whence $\Delta B < \Delta x_0$. The conclusion is that adjustment is *incomplete*: the induced increase in imports is not great enough to match the initial exogenous increase in exports, so that the balance of payments will show a surplus ($\Delta B > 0$), although smaller than the initial one ($\Delta B < \Delta x_0$).

Figure 8.1 gives a graphic representation of the situation. If we consider Eq. (8.8) we can draw the $(x - m)$ schedule—that is, the balance-of-payments schedule—and

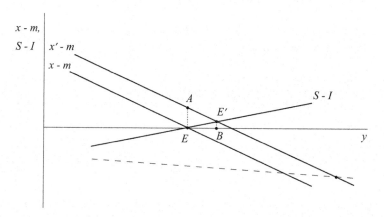

Fig. 8.1 Exogenous increase in exports, the multiplier, and the balance of payments

the $(S-I)$ schedule, both as functions of y; equilibrium will obtain at the intersection of these schedules.

In Fig. 8.1, the $(x-m)$ schedule is downward sloping because we are subtracting an ever greater amount of imports from an exogenously given amount of exports $(x - m = x_0 - m_0 - \mu y)$. The positive intercept reflects the assumption that the autonomous component of imports is smaller than exports; this assumption is necessary to ensure that it is in principle possible to reach balance-of-payments equilibrium at a positive level of income. The $(S - I)$ schedule is increasing, on the assumption that the marginal propensity to spend is smaller than one $[S - I = (1 - b - h)y - (C_0 + I_0)]$. The fact that the two schedules intersect at a point lying on the y axis reflects the assumption, already made above, that in the initial situation the balance of payments is in equilibrium.

An increase in exports shifts the $(x-m)$ schedule to $(x'-m)$; the new intersection occurs at E' where the balance of payments shows a surplus BE'. This is smaller than the initial increase in exports, measured by the vertical distance between $(x' - m)$ and $(x - m)$, for example AE.

As we said above, the case of underadjustment examined so far is not the only one possible. From Eq. (8.21) we see that adjustment is complete ($\Delta B = 0$) when $1 - b - h = 0$, that is when the marginal propensity to spend equals one. In this borderline case the induced increase in imports exactly offsets the initial exogenous increase in exports. But the case of overadjustment is also possible: when the marginal propensity to spend is greater than one, then $1 - b - h < 0$, and $\Delta B < 0$, that is, the induced increase in imports is *greater* than the initial exogenous increase in exports. From the economic point of view it is easy to understand why this is so: the greater the marginal propensity to spend, the greater *ceteris paribus* the multiplier; this means a higher income increase given the exogenous increase in exports, and finally, a greater increase in induced imports.

In terms of Fig. 8.1, the case of overadjustment implies that the $(S - I)$ schedule is downward sloping (as shown by the broken line); the slope, however, must be smaller in absolute value than the slope of the $(x - m)$ schedule for stability to obtain: in fact, from Eq. (8.14) we get $(b + h - 1) < \mu$. Therefore, overadjustment cannot be ruled out on the basis of considerations of stability.

It is true that if the country is stable in isolation, $b + h < 1$ and only underadjustment can occur. But since we are dealing with an open economy, what matters is that it is stable *qua* open economy, and to impose the condition that it should also be stable in isolation seems unwarranted. Thus, on theoretical grounds we must accept the possibility of overadjustment (as well as the borderline case of exact adjustment), and the assertion that the multiplier is incapable of restoring equilibrium in the balance of payments is incorrect.

8.3   Balance-of-Payments Adjustment in the Case of an Exogenous Increase in Imports

Let us now consider the case of an exogenous increase in imports. As we said in Sect. 8.1, the problem is complicated by the fact that we must check what happens to the autonomous component of residents' expenditure on domestic output, which is included—together with their autonomous expenditure on foreign output—in C_0 and I_0.

The commonly followed procedure of considering a Δm_0 while keeping C_0 and I_0 constant, implicitly assumes that the increase in the exogenous expenditure on foreign output is accompanied by a simultaneous decrease of the same amount in the exogenous expenditure by residents on domestic output. This is a very restrictive assumption, because it implies that domestic and foreign output are perfect substitutes. At the opposite extreme there is the assumption that Δm_0 leaves the exogenous expenditure on domestic output unaffected (i.e., Δm_0 entirely derives from an exogenous decrease in savings), which means that $C_0 + I_0$ increases by the same amount as m_0. Intermediate cases are of course possible, and they will be examined in the Appendix; here we limit ourselves to an examination of the two extremes.

(i) When only m_0 varies, Eqs. (8.13) and (8.20) become

$$\Delta y = -\frac{1}{1-b-h+\mu}\Delta m_0,$$

$$\Delta B = -\Delta m_0 - \mu\Delta y,$$

whence

$$\Delta B = \frac{b+h-1}{1-b-h+\mu}\Delta m_0. \tag{8.22}$$

Since $1 - b - h + \mu > 0$ by the stability condition, underadjustment, exact adjustment, overadjustment will take place according as $b + h \lessgtr 1$.

Therefore, if the marginal propensity to spend is smaller than one, the induced decrease in imports following the decrease in income caused by the initial exogenous increase in imports is not enough to completely restore balance-of-payments equilibrium. In the opposite case the balance of payments will go into surplus.

These shifts are illustrated in Fig. 8.2, where the exogenous increase in imports is represented by a shift of the $(x - m)$ schedule to $(x - m')$. The same remarks made above concerning the marginal propensity to spend hold here too.

(ii) When the exogenous increase in imports is *not* accompanied by any reduction in exogenous expenditure on domestic output by residents, we have $\Delta m_0 = \Delta C_0 + \Delta I_0$. From Eq. (8.13) we see that there is *no effect on income*. Therefore,

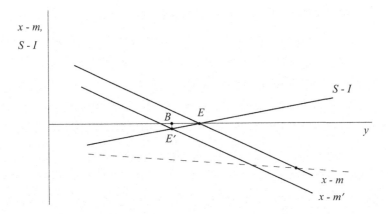

Fig. 8.2 Exogenous increase in imports, the multiplier, and the balance of payments: case (i)

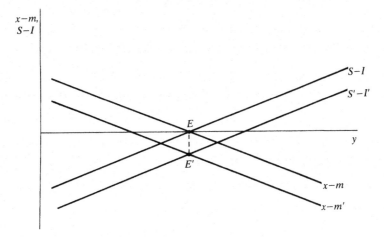

Fig. 8.3 Exogenous increase in imports, the multiplier, and the balance of payments: case (ii)

no adjustment is possible through induced changes in imports, and the balance of payments deteriorates by the full amount of the exogenous increase in imports, $\Delta B = -\Delta m_0$.

In terms of Fig. 8.3, the downward shift of the $(x - m)$ schedule is accompanied by an identical downward shift of the $(S - I)$ schedule, so that the value of y does not change and the balance of payments shows a deficit $EE' = \Delta m_0$.

By means of the same procedure so far illustrated we can examine the effects on the balance of payments of all other kinds of shifts in the exogenous components. But this can be left as an exercise for the reader.

8.4   Intermediate Goods and the Multiplier

In the equation for determining income in an open economy, which for convenience
is rewritten here

$$y = C + I + x - m, \tag{8.23}$$

the symbol y represents national income (product), the calculation of which is
carried out on the side of value added, where (as is well known from national
economic accounting) intermediate goods are not included. But the total imports
of goods and services m also include intermediate goods, and account must be taken
of this fact when the various multipliers are calculated (Miyazawa 1960; Milana
1984). The formulae examined in the previous sections are obviously still valid if it
is assumed that domestic production does not require imported intermediate goods.
On the other hand, in the case where imports of intermediate goods are present,
these formulae remain valid *only* if the content of intermediate goods in the various
categories of final goods (consumer, investment and export) which make up the
national product is the same. Otherwise they must be modified. The proof by Metzler
(1973; but written in 1963 independently of Miyazawa's work) that imported raw
materials do not change the results in any essential respect was based on the implicit
assumption of identical requirements of intermediate goods. The same assumption
is implicit in Meade's analysis of the role of raw materials in multiplier analysis
(Meade 1948; pp. 497–498).

    In order to clarify this point we must distinguish total imports m into imports of
final goods m^F and imports of intermediate goods m^R, where obviously $m^F + m^R =
m$. Imports of final goods can be related directly to income by way of the import
function

$$m^F = m_0^F + \mu^F y, \tag{8.24}$$

while, as far as imports of intermediate goods are concerned, we have to prelim-
inarly establish the requirements of intermediate goods per unit of the national
product. Let us assume that this requirement is a constant independent of the type
of good. Let us also assume that a constant fraction of this requirement consists of
imported intermediate goods. We can then write

$$m^R = \mu^R y, \quad 0 < \mu^R < 1, \tag{8.25}$$

where μ^R is the marginal propensity (assumed to coincide with the average
propensity) to import intermediate goods. The total import function thus becomes

$$m = m^F + m^R = m_0^F + \left(\mu^F + \mu^R\right) y = m_0^F + \mu y, \tag{8.26}$$

where μ is the total marginal propensity to import. It can be seen at once that (8.26) coincides with (8.3) in Sect. 8.1 (except for the qualification that the autonomous component entirely consists of final goods), so that the rest of the treatment and in particular the formulae for multipliers remain unchanged.

The case of different requirements of imported intermediate inputs according to the kind of good produced will be examined in the Appendix.

8.5   The Empirical Relevance of the Multiplier

It might seem that the foreign trade multiplier which, together with the elasticity approach, forms the core of the traditional theory, must nowadays be considered not only theoretically obsolete, but also of little help in analyzing actual problems, such as fiscal-policy international transmission. In fact, the effects of fiscal policy on incomes in a multiple-country world are analysed by using linked econometric models of the countries concerned and simulating the change in fiscal policy.

The complexity of these models might lead one to think that no hope exists for the poor old foreign multiplier, so why bother studying it. Twenty-five years ago Deardoff and Stern (1979) set forth the opposite view, namely that (p. 416) "the linked econometric models, as a group, do not appreciably add to our knowledge about fiscal-policy transmission beyond what is suggest by our calculations using a simple and relatively naive model": meaning that based on the foreign multiplier! In fact, these authors compared the results obtained from simulations of several linked econometric models (a linked econometric model is a set of econometric models of different countries linked together via the respective foreign sectors; for a description see Sect. II of their paper) with those calculated using the naive multiplier. The surprising outcome was that most results obtained by these naive calculations fell between the simulation extremes.

This exercise was repeated by Ferrara (1984) and Rotondi (1989) for different periods, who obtained similarly good results. Of course, as Deardoff and Stern note, the comparison of fiscal-policy multipliers leaves open the question whether the linked models can provide useful information on other issues. In our opinion the results of these exercises are not to be seen from a negative view-point (i.e., as a symptom of the limited contributions of the linked multi-country models to our understanding of the problem at hand) but rather from a positive one, that is as an indication of the usefulness of the foreign multiplier at least to obtain a first, rough idea of fiscal-policy transmission by simple, "back-of-the-envelope" calculations.

8.6   The Transfer Problem

Let us consider a two-country world, and suppose that country 1 makes a transfer of funds to country 2. The reason why we treat it here is that it has many connections with multiplier analysis.

The transfer traditionally considered consists of war reparations and, in fact, although the transfer problem had already been studied in remote times, the culmination of the debate was in relation to the war reparations that Germany had to pay after World War I to the victors. Noteworthy is the debate that was carried out between Keynes (1929) and Ohlin (1929) over the effects of such payments.

Keynes argued that the transferor would undergo a balance-of-payments deficit and, consequently, a deterioration in the terms of trade to eliminate it. Ohlin, on the contrary, pointed out that the payer's terms of trade would not need to deteriorate if the recipient spent the transfer on the payer's goods.

Before going on, it is important to point out that all kinds of transfers fall under this topic: not only war reparations, but also donations, long-term capital movements (direct investment, portfolio investment, international loans, etc.). Although donations and war reparations are *unilateral* transfers whilst capital movements are *bilateral* ones (see Sect. 5.1), in the current period both types of transaction give rise to the transfer of purchasing power from one country to another. It has therefore been correctly noted that an abrupt increase (by a cartel of producers) in the price of an input not substitutable in the short term (and so having a rigid short-term demand), such as for example oil, also gives rise to a transfer, from the importing to the producing countries.

Thus, in general, a transfer takes place whenever there is an international movement of funds temporarily (for example, the case of a loan, which will subsequently be repaid) or definitively (for example, the case of war reparations) without any *quid pro quo*. In case of a loan, of course, the *quid pro quo* exists in the form of an increase in the stock of indebtedness etc. written in the books, but we are looking at the *actual flow* of purchasing power. It goes without saying that, when a loan is granted, there is a transfer of funds in one direction, and when the loan is repaid, a transfer in the opposite direction will take place; in the case of a gradual repayment there will be a series of transfers. One can use a similar argument in the case of the subsequent disinvestment of funds transferred for direct or portfolio investment purposes.

Besides the *financial* aspect of the transfer, we must also consider its *real* effects, that is the movement of goods between country 1 and 2 induced by the movement of funds. The *transfer problem* consists in ascertaining whether, account being taken of all the secondary effects induced by the transfer, the balance of payments of the transferor (country 1) improves by a sufficient amount to "effect" the transfer. To be precise, three cases are possible in theory:

(a) Country 1s balance of trade improves by less than (the absolute amount of) the transfer. In this case the transfer is said to be *undereffected*, and country 1s current account balance worsens.

(b) Country 1s trade balance improves by an amount exactly equal to the transfer. In this case the transfer is said to be *effected*, and country 1s current account does not change.

(c) Country 1s trade balance improves by more than the transfer, which is then said to be *overeffected*; country 1s current account improves.

We recall that the transfer problem consists in determining whether the transferor (country 1) will achieve the *trade surplus* necessary to effect the transfer. This is equivalent to saying that after the financial transfer, country 1 "transfers" goods to country 2 (a real transfer) through the trade surplus (which means that country 1 has released to country 2 goods having a value greater than that of the goods acquired) so as to obtain the funds required to effect the initial financial transfer.

From this point of view it has often been stated that *the transfer problem can be considered as the "inversion" of the balance-of-payments problem,* since any actual balance-of-payments disequilibrium involves a real transfer from the surplus to the deficit country. Therefore, the problem of rectifying balance-of-payments disequilibria can be framed as the problem of creating either a real transfer of equal amount in the opposite direction (i.e. from the deficit to the surplus country) or a financial transfer from the surplus to the deficit country.

All this is undoubtedly true in the traditional context, in which the focus of the analysis is the current account. In this context the only way of effecting the initial transfer of funds is to achieve a trade surplus (i.e. to make a real transfer) and, inversely, a deficit country, by receiving (in value) more goods than it releases, receives a real transfer. But as soon as one considers the overall balance of payments which includes capital movements, none of the above statements is necessarily true. The initial transfer of funds can also be effected by way of a capital flow determined by an increase in the interest rate and, inversely, a deficit country can be in deficit because, though it enjoys a trade surplus (so that it makes, instead of receiving, a real transfer), it suffers from a higher deficit in the capital account. It is therefore advisable, when one reads the literature on the transfer problem, to pay a great deal of attention to the context in which the writer develops the argument.

Going back to the Keynes-Ohlin debate, it boils down to whether the transfer is undereffected (Keynes's opinion) or effected (Ohlin's opinion). If the transfer is undereffected, under the classical theory the terms of trade of country 1 will have to worsen so as to bring about the residual adjustment in the trade balance (the elasticity approach).

The theoretical problem, is, then, to find the conditions under which each of the three possible cases listed above occurs. In the context of the traditional theory one can distinguish the classical and the multiplier transfer theory. Paradoxically, Keynes was reasoning on the basis of the classical theory: in 1929 he had not yet written his famous book, *The General Theory.*

8.6.1 The Classical Theory

The basic proposition of the classical theory is that the transfer will be undereffected, effected, overeffected, according as the sum of those proportions of the expenditure changes in the two countries which fall on imports (in our simple model they can be identified with the marginal propensities to spend on imports) is less than, equal to, or greater than unity.

The underlying assumptions are that the two countries are in continuous *full employment* and were in external equilibrium before the transfer. In both, the entire income is spent in the purchase of goods; prices and exchange rate are constant. It is also assumed that the transfer is financed (by the transferor country 1) and disposed of (by the transferee country 2) in such a way as to reduce the aggregate expenditure of country 1 and increase the aggregate expenditure of country 2, by the *exact* amount of the transfer.

It follows that country 1s imports will decrease by an amount equal to its marginal propensity to spend on imports (μ_1) applied to the expenditure reduction, i.e. to the value of the transfer, and country 2s imports (which are country 1s exports) will increase by an amount equal to its marginal propensity to spend on imports (μ_2) applied to the expenditure increase, i.e. to the value of the transfer.

Thus we shall have three effects on country 1s balance of payments:

(i) initial deterioration by an amount equal to the transfer, T,
(ii) improvement due to lower imports, by an amount $\mu_1 T$,
(iii) improvement due to higher exports, by an amount $\mu_2 T$.

The sum of these gives the overall change in the balance of payments of country 1:

$$\Delta B_1 = -T + \mu_1 T + \mu_2 T = (\mu_1 + \mu_2 - 1)T. \tag{8.27}$$

It is then clear that the balance of payments will improve ($\Delta B_1 > 0$: overeffected transfer), remain unchanged (effected transfer), deteriorate (undereffected transfer) according as the sum $\mu_1 + \mu_2$ is greater than, equal to, less than unity.

In the case of an overeffected or undereffected transfer, the balance of payments which was initially in equilibrium shows a surplus or deficit respectively, and since in the classical theory no multiplier effects can take place, the only way to restore external equilibrium is by a modification in the terms of trade (which can come about either through a change in the absolute price levels under a gold standard regime, or through an exchange-rate variation), which will bring about adjustment provided that the suitable elasticity conditions are fulfilled.

The traditional opinion was that the sum of the marginal propensities to spend on imports was likely to be smaller than one, so that the transfer would be undereffected and the terms of trade would have to move against country 1, to allow the further adjustment via relative prices to take place. This was Keynes's position.

8.6.2 The Multiplier Theory

The multiplier theory is also called the Keynesian theory, but since Keynes reasoned on the basis of the classical theory, it is as well to point out that it was developed by other authors, notably Metzler (1942), Machlup (1943; Chap. IX), and Johnson (1956), who gave it its name.

The multiplier theory differs from the classical theory in two respects. Firstly, it drops the assumption that the transferor's aggregate expenditure decreases, and the transferee's increases, by the amount of the transfer. It is, in fact, quite possible that the transfer is financed in such a way that aggregate expenditure in country 1 decreases by less than the transfer, the difference coming out of saving, and that it is disposed of in such a way that aggregate expenditure in country 2 increases by less, the difference going to increase saving. Secondly, any change in aggregate expenditure due to the transfer is to be taken as an exogenous change which gives rise to multiplicative effects on income in both countries (assumed to be *underemployed*), which have to be accounted for in calculating the induced changes in imports.

The chain of effects is thus much more complicated than that present in the classical theory, and can be summed up as follows:

(1) initial deterioration in country ls balance of payments by an amount equal to the transfer, T;
(2) changes, concomitant to the transfer, in the autonomous components of expenditure in both countries (including the autonomous components of imports, $\Delta m_{01}, \Delta m_{02}$);
(3) multiplier effects of these changes on both countries income (via the multiplier with foreign repercussions);
(4) induced changes in imports, $\mu_1 \Delta y_1$ and $\mu_2 \Delta y_2$.

Let us then introduce the following relations:

$$\begin{aligned}
\Delta C_{01} &= -b_1' T, & \Delta C_{02} &= b_2' T, \\
\Delta I_{01} &= -h_1' T, & \Delta I_{02} &= h_2' T, \\
\Delta m_{01} &= -\mu_1' T, & \Delta m_{02} &= \mu_2' T,
\end{aligned} \tag{8.28}$$

where $\Delta C_{01}, \Delta I_{01}, \Delta m_{01}$ denote the changes, concomitant to the transfer, in the exogenous components of country ls expenditure functions, and b_1', h_1', μ_1' are coefficients that relate these exogenous changes to the transfer (the minus sign indicates that these changes are negative, since country 1 is the transferor). The primes are used to distinguish them from the "normal" marginal propensities referring to the induced changes depending on "ordinary" changes in income. Similar considerations hold for country 2, the transferee.

If we examine country ls balance of payments we see that

$$\Delta B_1 = -T + (\Delta m_{02} - \Delta m_{01}) - \mu_1 \Delta y_1 + \mu_2 \Delta y_2, \tag{8.29}$$

which are the effects mentioned in points (1), (2), (4), in the order. Note that $\Delta m_{01}, \Delta m_{02}$ come from (8.28) while $\Delta y_1, \Delta y_2$ come from the multiplier with foreign repercussions (the effect under (3) above).

The net result of all these effects can be determined only through a formal mathematical analysis (see the Appendix), but it is easy to understand that all will depend on the various propensities (to spend on domestic goods and on imports) as well as on the size of the transfer-induced changes in the autonomous components of expenditure (determined by the primed coefficients).

Hence all cases are possible, though the case of undereffectuation is more likely. Paradoxically again, the opinion of Keynes in his debate with Ohlin is validated through an application by other authors of his later multiplier theory.

If the transfer is undereffected, then—in the context of the traditional theory—a deterioration of the terms of trade of the transferor will have to take place so as to effect the residual adjustment. Hence the conclusions are qualitatively similar to those of the classical theory.

8.6.3   Observations and Qualifications

The traditional opinion (whether classical or multiplier) represents a simplification because it amounts to saying that the effectuation of the transfer can be separated into two phases. The first consists of ascertaining whether the transfer is effected, by applying the chosen model (either the classical or the Keynesian). If it is not, there will remain a balance-of-payments disequilibrium, which must be corrected by some adjustment process (second phase). This process is a change in the terms of trade (according to the classical theory), or some other adjustment process (excluding, of course, the multiplier mechanism which has already worked itself out in the first phase of the Keynesian approach).

But, more rigorously, the transfer problem ought to be tackled from the very beginning in the context of the adjustment process chosen for the second phase in the two-stage traditional approach (note that this phase is not only possible but very likely, since the conditions for the transfer to be effected are unlikely to occur).

This approach does not present any special difficulty, because the *transfer problem can be examined as any comparative-static problem in the context of the model chosen*. It excludes, however, the possibility of enunciating a *general theory* of the transfer, as the traditional approach (whether classical or Keynesian) proposed to do. In fact, different comparative-static results and so different conditions for the transfer to be effected will prevail according as one or the other model is used. We shall give examples in the following chapters (see Sects. 9.1.3 and 10.2.2.1).

8.7 Appendix

8.7.1 The Multiplier Without Foreign Repercussions

8.7.1.1 Basic Results

Consider the following model

$$
\begin{aligned}
C &= C(y, \alpha_C), & 0 &< \partial C/\partial y < 1, \ \partial C/\partial \alpha_C > 0, \\
I &= I(y, \alpha_I), & 0 &< \partial I/\partial y < 1, \ \partial C/\partial \alpha_I > 0, \\
m &= m(y, \alpha_m), & 0 &< \partial m/\partial y < 1, \ \partial m/\alpha_m > 0, \\
x &= \alpha_x, \\
y - C & - I - x + m = 0,
\end{aligned}
\tag{8.30}
$$

where the αs are shift parameters which can be interpreted as the exogenous components of the various expenditure functions. No loss of generality is involved if we assume that $\partial C/\partial \alpha_C$, $\partial C/\partial \alpha_I$, $\partial m/\alpha_m$ are all equal to one. The model written in the text can be considered as a suitable linear approximation to system (8.30), with $b \equiv \partial C/\partial y$ etc.

By means of the implicit function theorem, y can be expressed as a differentiable function of the αs (provided that the condition $1 - \partial C/\partial y - \partial I/\partial y \neq 0$ is satisfied), and consequently, exercises of comparative statics can be carried out. If we differentiate the last equation in (8.30) with respect to the relevant parameters, bearing in mind that y is a function of the αs, we obtain the multiplier formula

$$
\frac{\partial y}{\partial \alpha_C} = \frac{\partial y}{\partial \alpha_I} = \frac{\partial y}{\partial \alpha_x} = -\frac{\partial y}{\partial \alpha_m} = \frac{1}{1 - b - h + \mu},
\tag{8.31}
$$

and so

$$
dy = \frac{1}{1 - b - h + \mu} \left(d\alpha_C + d\alpha_I + d\alpha_x - d\alpha_m \right),
\tag{8.32}
$$

where b, h, μ denote $\partial C/\partial y, \partial I/\partial y, \partial m/\partial y$ evaluated at the equilibrium point.

The correspondence principle enables us to determine the sign of the denominator of the fraction and so to obtain determinate comparative statics results. The usual dynamic behaviour assumption is that producers react to excess demand by making adjustments in output: if aggregate demand exceeds (falls short of) current output, the latter will be increased (decreased); this mechanism operates independently of the origin of excess demand. Formally,

$$
\dot{y} = f \left[(C + I + x - m) - y \right],
\tag{8.33}
$$

where f is a sign-preserving function and $f'[0] > 0$. In order to examine local stability, we take a linear approximation at the equilibrium point and obtain

$$\dot{\bar{y}} = k\,(b + h - \mu - 1)\,\bar{y}, \tag{8.34}$$

where $\bar{y}$ denotes the deviations from equilibrium, k is $f'[0]$ and can be interpreted as a speed of adjustment, and b, h, μ are $\partial C/\partial y, \partial I/\partial y, \partial m/\partial y$ evaluated at the equilibrium point. The stability condition derived from the solution of the differential equation (8.34) is

$$k(b + h - \mu - 1) < 0,$$

that is

$$1 - b - h + \mu > 0, \tag{8.35}$$

which ensures that the multiplier is positive, and has been discussed in Sect. 8.1.

Going back to the multiplier (8.32), we introduce a coefficient q, $0 \le q \le 1$, which measures the relationship between the change in the autonomous expenditure by residents on domestic output and the change in their autonomous expenditure on foreign output (imports). Letting h and f denote home and foreign output respectively, we have

$$d\alpha_C = d\alpha_{Ch} + d\alpha_{Cf}, d\alpha_I = d\alpha_{Ih} + d\alpha_{If}, d\alpha_{Cf} + d\alpha_{If} = d\alpha_m, \tag{8.36}$$

and assuming for the sake of simplicity that q is the same for both consumption and investment expenditure, we have

$$d\alpha_{Ch} = -q d\alpha_{Cf}, d\alpha_{Ih} = -q d\alpha_{If}. \tag{8.37}$$

From Eqs. (8.36) and (8.37) we obtain

$$d\alpha_C + d\alpha_I - d\alpha_m = -q d\alpha_m, \tag{8.38}$$

and so the multiplier formula (8.32) becomes

$$dy = \frac{1}{1 - b - h + \mu}\,(d\alpha_x - q d\alpha_m). \tag{8.39}$$

According to some authors, it is not sufficient simply to subtract the proportion of imports in autonomous expenditure from the multiplicand, for the multiplier also should be changed. This can be carried out by making imports a function of aggregate demand $(C + I)$ or of C and I separately (with different import coefficients) instead of y, namely

$$m = m(C, I, \alpha_m), 0 < \frac{\partial m}{\partial C} < 1, 0 < \frac{\partial m}{\partial I} < 1, \frac{\partial m}{\partial \alpha_m} > 0, \frac{\partial m}{\partial C} \neq \frac{\partial m}{\partial I}.$$

If we substitute this import function in the place of the third equation in system (8.30) we get a new model, that the reader can easily examine as an exercise. On this point, see Meade (1948), Miyazawa (1960), Kennedy and Thirwall (1979, 1980), Thirwall (1980; pp. 57 ff.), and below, Sect. 8.7.3.

8.7.1.2 The Balance of Payments

Consider now the balance of payments, $B = x - m$, and differentiate it totally, obtaining

$$dB = d\alpha_x - d\alpha_m - \mu dy, \tag{8.40}$$

where dy is given by (8.32) or (8.39) as the case may be. When exports change, we have

$$dB = \frac{1 - b - h}{1 - b - h + \mu} d\alpha_x, \tag{8.41}$$

and when imports exogenously change the result, account being taken of (8.39), is

$$dB = \frac{b + h - 1 - \mu(1 - q)}{1 - b - h + \mu} d\alpha_m. \tag{8.42}$$

The cases discussed in the text correspond to the extreme cases $q = 1$ (domestic and foreign output are perfect substitutes) and $q = 0$ (autonomous expenditure by residents on domestic output is not influenced at all by their autonomous expenditure on foreign output). Formula (8.42) makes it possible to examine all intermediate cases.

We can add a further refinements to our analysis, not discussed in the text, if we relax the assumption that all expenditures on exports are income-creating. Following Holzman and Zellner (1958), we suppose that only a proportion β of (the change in) exports is income-creating, $0 < \beta \leq 1$. The effect on national income and on the balance of payments of an autonomous change in exports is given by the formulae

$$dy = \frac{\beta}{1 - b - h + \mu} d\alpha_x, dB = \frac{1 - b - h + \mu(1 - \beta)}{1 - b - h + \mu} d\alpha_x, \tag{8.43}$$

which of course reduce to the previous ones when $\beta = 1$. Note that, when $\beta < 1$, the effect on the balance of payments is greater than when $\beta = 1$, for $1 - b - h + \mu(1 - \beta) > 1 - b - h$. In fact, when $\beta < 1$ the multiplier is smaller, the induced increase in income is smaller, and the induced increase in imports is smaller.

Suppose now that the imports and exports exogenously change by the same amount, say $d\alpha$, so that $d\alpha_x = d\alpha_m = d\alpha$. We have

$$dy = \frac{\beta - q}{1 - b - h + \mu}d\alpha, \quad dB = \frac{\mu(q - \beta)}{1 - b - h + \mu}d\alpha. \tag{8.44}$$

If one considers as normal the case in which both β and q are equal to one (this, in fact, is the case often implicitly presented in textbooks), the conclusion is that the *balanced-trade multiplier* is zero and that the balance of payments does not change. The general expressions (8.44) give the outcome in all possible cases.

Finally, let us consider the problem of the *import content of exports*. For many nations it is true that exports contain a significant amount of imports; in this case, any change in exports will be accompanied by a change in imports, and if we call γ the parameter which measures the import content of exports, to any $d\alpha_x$ a $d\alpha_m = \gamma d\alpha_x$ will correspond. The consequences are easily found, for it is enough to put $(1 - \gamma)d\alpha_x$ in the place of $d\alpha_x$ wherever the later appears in the relevant formulae. This will not change the results qualitatively so long as $0 < \gamma < 1$, a reasonable assumption. On the specific problem of imported intermediate goods see below, Sect. 8.7.3.

8.7.2 Foreign Repercussions in a n-Country Model

8.7.2.1 The General Model

The general model is composed of n sets of equations of the form

$$
\begin{aligned}
&C_i = C_i(y_i, \alpha_{iC}), && 0 < \partial C_i/\partial y_i < 1, \quad \partial C_i/\partial \alpha_{iC} > 0, \\
&I_i = I_i(y_i, \alpha_{iI}), && 0 < \partial I_i/\partial y_i < 1, \quad \partial I_i/\partial \alpha_{iI} > 0, \\
&m_i = m_i(y_i, \alpha_{im}), && 0 < \partial m_i/\partial y_i < 1, \quad \partial m_i/\partial \alpha_{im} > 0, \\
&m_i \equiv \sum_{j=1/j\neq i}^{n} m_{ji}(y_i, \alpha_{jim}), && 0 \leq \partial m_{ji}/\partial y_i < 1, \quad \sum_{j=1/j\neq i}^{n}\frac{\partial m_{ji}}{\partial y_i} \equiv \frac{\partial m_i}{\partial y_i}, \\
&&& \partial m_{ji}/\partial \alpha_{jim} \geq 0, \quad \sum_{j=1/j\neq i}^{n}\frac{\partial m_{ji}}{\partial \alpha_{jim}} \equiv \frac{\partial m_i}{\partial \alpha_{im}}, \\
&x_i = \sum_{k=1/k\neq i}^{n} m_{ik}(y_k, \alpha_{ikm}), \\
&y_i - C_i - I_i - x_i + m_i = 0.
\end{aligned}
\tag{8.45}
$$

The fourth and fifth equations in (8.45) need some clarification. In a n-country world, the $m_i(y_i, \alpha_{im})$ function shows how *total* imports of country i are related to national income, y_i, of the importing country, but we must also know the countries of origin of these imports. As Metzler (1950) made clear, we can think of the total import function of the ith country as composed of a number of subfunctions each showing how much the ith country imports from the jth country at any given level of y_i. By $m_{ji}\left(y_i, \alpha_{jim}\right)$ we denote the imports of the ith country from the jth country, expressed as a function of income in the ith country so that the fourth equation in (8.45) follows by definition. In the real world, some of the m_{ji} functions may be zero, because each country does not necessarily import from all other countries, but this does not present any difficulty. The relationship between the partial derivatives of the subfunctions and the partial derivatives of the total import function follow from the fact that we have an identity.

As regards the exports of country i, they are the sum of what all the other countries import from it, and can be obtained by summing over all countries the subfunctions which express the imports of the kth country from the ith country. This gives the fifth equation in (8.45).

Straightforward substitutions yield the following set of n implicit functions:

$$y_i - C_i(y_i, \alpha_{iC}) - I_i\left(y_i, \alpha_{iI}\right) - \sum_{k=1/k\neq i}^{n} m_{ik}(y_k, \alpha_{ikm}) + m_i\left(y_i, \alpha_{im}\right) = 0,$$
$$i = 1, 2, \ldots, m.$$
(8.46)

The Jacobian matrix of these functions with respect to the y_i is

$$\mathbf{J} \equiv \begin{bmatrix} 1 - b_1 - h_1 + \mu_1 & -\mu_{12} & \cdots & -\mu_{1n} \\ -\mu_{21} & 1 - b_2 - h_2 + \mu_2 & \cdots & -\mu_{2n} \\ \cdots & \cdots & \cdots & \cdots \\ -\mu_{n1} & -\mu_{n2} & \cdots & 1 - b_n - h_n + \mu_n \end{bmatrix}$$
(8.47)

where μ_{ji} is $\partial m_{ji}/\partial y_i$ evaluated at the equilibrium point, and the other symbols have the usual meanings. By using the implicit function theorem, if $\mathbf{J}$ is non-singular at the equilibrium point we can express the y_i as differentiable functions of all the parameters (the αs). Without loss of generality we can assume as before that the various partial derivatives of the expenditure functions with respect to the parameters ($\partial C_i/\partial \alpha_{iC}$ etc.) are equal to one, so that total differentiation of (8.46) yields

$$\mathbf{J}d\mathbf{y} = d\boldsymbol{\alpha},$$
(8.48)

where $\mathbf{J}$ is the matrix (8.47) and $d\mathbf{y}$, $d\boldsymbol{\alpha}$ are the columns vectors

$$d\mathbf{y} \equiv \begin{bmatrix} dy_1 \\ dy_2 \\ \cdots \\ dy_n \end{bmatrix}, \quad d\boldsymbol{\alpha} \equiv \begin{bmatrix} d\alpha_{1C} + d\alpha_{1I} - d\alpha_{1m} + \sum_{k=2}^{n} \frac{\partial m_{1k}}{\partial \alpha_{1km}} d\alpha_{1km} \\ d\alpha_{2C} + d\alpha_{2I} - d\alpha_{2m} + \sum_{\substack{k=1 \\ k\neq 2}}^{n} \frac{\partial m_{2k}}{\partial \alpha_{2km}} d\alpha_{2km} \\ \\ d\alpha_{nC} + d\alpha_{nI} - d\alpha_{nm} + \sum_{k=1}^{n-1} \frac{\partial m_{nk}}{\partial \alpha_{nkm}} d\alpha_{nkm} \end{bmatrix}.$$
(8.49)

Therefore, the solution of (8.48), which is

$$\mathbf{dy} = \mathbf{J}^{-1}\mathbf{d\alpha},\tag{8.50}$$

summarizes all possible multipliers in a n-country framework. Since the non-singularity of the Jacobian lies at the basis of our analysis, $\mathbf{J}^{-1}$ exists; but to obtain determinate comparative static results we must use the correspondence principle as before.

8.7.2.2 Stability Analysis

The usual dynamic behaviour assumption (in each country varies according to excess demand) gives rise to the following system of differential equations

$$\dot{y}_i = f_i\left[(C_i + I_i + x_i - m_i) - y_i\right], \quad i = 1, 2, \ldots, n,\tag{8.51}$$

where f_i are sign-preserving functions, and $f_i'[0] \equiv k_i > 0$. To study local stability we perform a linear approximation at the equilibrium point, which reduces (8.51) to the linear differential system with constant coefficients

$$\dot{\bar{\mathbf{y}}} = \mathbf{k}\left[-\mathbf{J}\right]\bar{\mathbf{y}},\tag{8.52}$$

where $\bar{\mathbf{y}}$ is the vector of the deviations and $\mathbf{k}$ is the diagonal matrix of the speeds of adjustment k_i. We assume for the moment that $k_i = 1$ for all i; we shall show afterwards that this does not involve any loss of generality. The matrix $[-\mathbf{J}]$ is a Metzlerian matrix, that is a matrix with non-negative off-diagonal elements. In this case, necessary ad sufficient stability conditions (Gandolfo 2009; Chap. 18) are that the leading principal minors of the matrix $[-\mathbf{J}]$ alternate in sign, beginning with minus:

$$b_1 + h_1 - \mu_1 - 1 < 0, \quad \begin{vmatrix} b_1 + h_1 - \mu_1 - 1 & \mu_{12} \\ \mu_{21} & b_2 + h_2 - \mu_2 - 1 \end{vmatrix} > 0, \ldots,$$

$$\mathrm{sgn}\ \det\ (-\mathbf{J}) = \mathrm{sgn}\ (-1)^n.\tag{8.53}$$

This implies that each country's foreign multiplier without repercussions must be stable $(1 - b_i - h_i + \mu_i > 0)$ and that all subsets of $2, 3, \ldots, n-1$ countries must give rise to stable foreign multipliers with repercussions. The approach to equilibrium will not be necessarily monotonic (whereas in the two-country model it was); a very special case of monotonic movement occurs when the partial marginal propensity of country i to import from country j is equal to the partial marginal propensity of country j to import from country i. In this case $\mu_{ij} = \mu_{ji}$ and the characteristic roots of $[-\mathbf{J}]$ are all real because it is a symmetric matrix.

Other interesting results concerning stability can be obtained if we consider only sufficient or only necessary stability conditions. A sufficient stability condition is that the matrix $[-\mathbf{J}]$ has a dominant negative diagonal (Gandolfo 2009; Chap. 18), namely

$$b_i + h_i - \mu_i - 1 < 0,$$

$$|b_i + h_i - \mu_i - 1| > \sum_{j=1/j\neq i}^{n} \mu_{ji}, \text{ from which } 1 - b_i - h_i > 0. \qquad (8.54)$$

Since $1 - b_i - h_i > 0$ is the stability condition for the closed economy multiplier, we conclude that the n-country model is stable *if* each country is stable in isolation, namely if the marginal propensity to spend is smaller than one in all countries. Conversely, it can be shown that the n-country model is unstable if each country is unstable in isolation: in other words, $b_i + h_i > 1$ for all i is a sufficient *instability* condition. We must distinguish two cases. The first is when, for at least one i, $b_i + h_i$ is not only greater than one but also greater than $1 + \mu_i$. In this case at least one of the necessary and sufficient condition stated above is violated and therefore the model is unstable. The second case occurs when $1 < b_i + h_i < 1 + \mu_i$ for all i. Noting that $[-\mathbf{J}]$ is a non-negative matrix, it follows from a theorem on such matrices that min $S_i \leq \lambda_M \leq$ max S_i (Gantmacher 1959; p. 82), where the real number λ_M is the dominant root of the matrix and S_i are the column sums of the matrix. Now,
$$S_i = b_i + h_i - \mu_i - 1 + \sum_{j=1/j\neq i}^{n} \mu_{ji} = b_i + h_i - 1; \text{ therefore, } \lambda_M \text{ is positive if } b_i + h_i - 1 > 0$$
for all i, which proves instability. Of course, if $b_i + h_i = 1 > 0$ for some i, and < 0 for some other i, the model may be stable or instable. In other words, if only some countries are unstable in isolation, the world may still be stable.

We have assumed above that all the speeds of adjustment are equal to one, stating that this does not involve any loss of generality. In fact, the results on D-stability (Gandolfo 2009; Chap. 18) enable us to conclude that if the matrix $[-\mathbf{J}]$ satisfies conditions (8.53) or (8.54), also the matrix $\mathbf{k}[-\mathbf{J}]$ is stable for any positive diagonal matrix $\mathbf{k}$.

8.7.2.3 Comparative Statics: A Comparison Between the Various Multipliers

Let us now come to the problem of comparative statics. If $[-\mathbf{J}]$ satisfies conditions (8.53)—note that if it satisfies conditions (8.54) it will also satisfy conditions (8.53), although the converse is not true (Gandolfo 2009; Chap. 18)—it follows from elementary rules on determinants that the matrix $\mathbf{J}$ will have leading principal minors which are all positive, namely

$$1 - b_1 - h_1 + \mu_1 > 0, \begin{vmatrix} 1 - b_1 - h_1 + \mu_1 & -\mu_{12} \\ -\mu_{21} & 1 - b_2 - h_2 + \mu_2 \end{vmatrix} > 0, \dots, |\mathbf{J}| > 0. \qquad (8.55)$$

Conditions (8.55) are nothing more nor less than the well-known Hawkins-Simon conditions, which ensure that system (8.48) has a non-negative solution corresponding to any non-negative $d\alpha$, namely $\mathbf{J}^{-1}$ is a non-negative matrix. This result can be strengthened if we assume that each country imports directly or indirectly from all other countries (we say that a country i imports indirectly from another country j when country i imports directly from country j_1 which in turn imports directly from country j_2 etc. which in turn imports directly from country j). Under this assumption, which seems reasonable, and remembering that $b_i + h_i - \mu_i > 0$ because μ_i is part of $b_i + h_i$, the matrix

$$
\mathbf{A} = \begin{bmatrix} b_1 + h_1 - \mu_1 & \mu_{12} & & \mu_{1n} \\ \mu_{21} & b_2 + h_2 - \mu_2 & \cdots & \mu_{2n} \\ \cdots & \cdots & \cdots & \cdots \\ \mu_{n1} & \mu_{n2} & & b_n + h_n - \mu_n \end{bmatrix}
$$

is a non-negative *indecomposable* matrix; noting that $\mathbf{J} \equiv [\mathbf{I} - \mathbf{A}]$, it follows from the properties of such matrices (Gandolfo 2009; p. 129) that conditions (8.55) are necessary and sufficient for $\mathbf{J}^{-1}$ to have only positive elements, $\mathbf{J}^{-1} > 0$. In this case system (8.48) has a positive solution corresponding to any non-negative $d\boldsymbol{\alpha}$. Let us note for future reference that, since $\mathbf{J}^{-1} = [\text{adj } \mathbf{J}] / |\mathbf{J}|$, and $|\mathbf{J}| > 0$ from (8.55), $\mathbf{J}^{-1} > 0$ implies $[\text{adj } \mathbf{J}] > 0$, that is

$$
|\mathbf{J}_{rs}| > 0, r, s = 1, 2, \ldots, n, \tag{8.56}
$$

where $|\mathbf{J}_{rs}|$ is the cofactor of the element (r, s) in $\mathbf{J}$.

The relationships between the closed-economy multiplier and open-economy multipliers with and without repercussions in the general model under consideration were studied by Metzler (1950). Consider for example an exogenous increase in investment expenditure on domestic goods in country 1, $d\alpha_{1I} > 0$. Let us examine the inequality

$$
\frac{1}{1 - b_1 - h_1 + \mu_1} < \frac{dy_1}{d\alpha_{1I}} < \frac{1}{1 - b_1 - h_1}, \tag{8.57}
$$

where $dy_1/d\alpha_{1I}$ is the multiplier with foreign repercussions derived from the solution of system (8.48), which turns out to be

$$
\frac{dy_1}{d\alpha_{1I}} = \frac{|\mathbf{J}_{11}|}{|\mathbf{J}|}. \tag{8.58}
$$

Now add all other rows to the first row of $|\mathbf{J}|$ and expand according to the elements of this new row. The result is

$$
|\mathbf{J}| = (1 - b_1 - h_1) |\mathbf{J}_{11}| + (1 - b_2 - h_2) |\mathbf{J}_{12}| + \ldots + (1 - b_n - h_n) |\mathbf{J}_{1n}|, \tag{8.59}
$$

where $|\mathbf{J}_{1k}|, k = 1, 2, 3, \ldots, n$, is the cofactor of the elements $(1, k)$ in $|\mathbf{J}|$. If we substitute (8.59) in (8.58) and divide both numerator and denominator by $|\mathbf{J}_{11}|$, we obtain

$$\frac{dy_1}{d\alpha_{1I}} = \frac{1}{(1 - b_1 - h_1) + \{[(1 - b_2 - h_2)|\mathbf{J}_{12}| + \ldots + (1 - b_n - h_n)|\mathbf{J}_{1n}|] / |\mathbf{J}_{11}|\}}. \tag{8.60}$$

Now, $|\mathbf{J}_{1k}| > 0$ according to (8.56); therefore, if the marginal propensity to spend is smaller than one in all countries, the denominator in (8.60) will be greater than $(1 - b_1 - h_1)$. This proves the right-hand part of inequality (8.57) under the assumption that all countries are stable in isolation.

The left-hand part of inequality (8.57) can be proved by expanding $|\mathbf{J}|$ according to the elements of its first column and then dividing both numerator and denominator of (8.58) by $|\mathbf{J}_{11}|$, which gives

$$\frac{dy_1}{d\alpha_{1I}} = \frac{1}{(1 - b_1 - h_1 + \mu_1) - \{[\mu_{21}|\mathbf{J}_{21}| + \ldots + \mu_{n1}|\mathbf{J}_{n1}|] / |\mathbf{J}_{11}|\}}. \tag{8.61}$$

Since $|\mathbf{J}_{1k}| > 0$ from (8.56), the denominator in (8.61) is smaller than $(1 - b_1 - h_1 + \mu_1)$, and this is the proof of inequality in the left-hand part of (8.57). In conclusion, provided that the necessary and sufficient stability conditions are satisfied, the multiplier with foreign repercussions is always greater than the corresponding multiplier without foreign repercussions, and it is certainly smaller than the closed economy multiplier under the additional (sufficient) assumption that the marginal propensity to spend is smaller than one in all countries.

It can also be shown that, if the marginal propensity to spend is smaller than one in all countries, the i-th country multiplier is greater when the increase in exogenous investment occurs in country i than when it occurs in any other country j, namely $dy_i/d\alpha_{iI} > dy_i/d\alpha_{jI}, j \neq i$. Consider for example country 1: the expression for $dy_1/d\alpha_{1I}$ is given by (8.58), and the expression for, say, $dy_1/d\alpha_{2I}$ is

$$\frac{dy_1}{d\alpha_{2I}} = \frac{|\mathbf{J}_{21}|}{|\mathbf{J}|}, \tag{8.62}$$

so that $dy_1/d\alpha_{1I} > dy_1/d\alpha_{2I}$ is equivalent to $|\mathbf{J}_{11}| - |\mathbf{J}_{21}| > 0$.

We have

$$|\mathbf{J}_{11}| - |\mathbf{J}_{21}|$$

$$= \begin{vmatrix} 1 - b_2 - h_2 + \mu_2 - \mu_{12} & -\mu_{23} - \mu_{13} & \cdots & -\mu_{2n} - \mu_{1n} \\ -\mu_{32} & 1 - b_3 - h_3 + \mu_3 & \cdots & -\mu_{3n} \\ \cdots & \cdots & \cdots & \cdots \\ -\mu_{n2} & -\mu_{n3} & \cdots & 1 - b_n - h_n + \mu_n \end{vmatrix}$$

$$= \det(\mathbf{I} - \mathbf{B}),$$

where

$$
\mathbf{B} \equiv
\begin{bmatrix}
b_2 + h_2 - \mu_2 + \mu_{12} & \mu_{23} + \mu_{13} & \dots & \mu_{2n} + \mu_{1n} \\
\mu_{32} & b_3 + h_3 - \mu_3 & \dots & \mu_{3n} \\
\dots & \dots & \dots & \dots \\
\mu_{n2} & \mu_{n3} & \dots & b_n + h_n - \mu_n
\end{bmatrix}.
$$

Now, $\mathbf{B}$ is a non-negative matrix; its columns sums are equal to $b_j + h_j, j = 2, \dots, n$ and so are all less than one *if* $b_j + h_j < 1$. Therefore it follows from the theorem already used above that the dominant root of $\mathbf{B}$ is smaller than 1. Consequently (Gantmacher 1959; p. 85), $\det(\mathbf{I} - \mathbf{B})$ is positive. The same method of proof can be used for any difference $|\mathbf{J}_{11}| - |\mathbf{J}_{k1}|, k = 3, \dots, n$.

8.7.2.4 The Balance of Payments

We now turn to balance-of-payments adjustment. Since $B_i = x_i - m_i$, and $x_i - m_i = y_i - C_i - I_i$ from the national income equation, it is convenient to use the relation

$$
B_i = y_i - C_i - I_i, \tag{8.63}
$$

which relates each country's balance of payments to the income of that country alone. Total differentiation of (8.63) yields

$$
dB_i = (1 - b_i - h_i)\, dy_i - d\alpha_{iC} - d\alpha_{iI}, \tag{8.64}
$$

from which the effects of an exogenous increase in a country's expenditure can be determined. Take for example an increase in α_{1I}, *ceteris paribus*. We have

$$
\frac{dB_1}{d\alpha_{1I}} = (1 - b_1 - h_1)\frac{dy_1}{d\alpha_{1I}} - 1,
$$

$$
\frac{dB_k}{d\alpha_{1I}} = (1 - b_k - h_k)\frac{dy_k}{d\alpha_{1I}}, \quad k = 2, \dots, n, \tag{8.65}
$$

where the $dy_i/d\alpha_{1I}$, $i = 1, 2, \dots, n$, are given by our previous analysis, so that we know that they are all positive. The effect on the balance of payments depends as usual on the magnitude of the marginal propensity to spend. If we assume that this propensity is smaller than one in all countries, then the right-hand part of inequality (8.57) holds, and so (8.65) gives

$$
-1 < \frac{dB_1}{d\alpha_{1I}} < 0, \quad \frac{dB_k}{d\alpha_{1I}} > 0. \tag{8.66}
$$

Therefore, an increase in exogenous investment in a country moves the balance of payments against that country and in favour of all the other countries. This needs not be true any longer if we drop the assumption concerning the marginal propensity to spend. We know from our dynamic analysis that stability is compatible with one or more (but not all) marginal propensities to spend being greater than one. An extreme situation is therefore conceivable in which $(1 - b_1 - h_1) > 0$, $(1 - b_k - h_k) < 0, k = 2, \ldots, n$. In this case all the other countries will suffer a worsening of their balance of payments and, consequently, country 1s balance of payments will improve. Intermediate situations in which $(1 - b_1 - h_1) > 0$ and $(1 - b_k - h_k) \lessgtr 0$ for different ks are possible. Then some of these countries will suffer a deterioration in their balances of payments, while others will find an improvement; the result on country 1s balance of payments will depend on whether the sum of the deteriorations is greater or smaller in absolute value than the sum of the improvements.[1] Note also that if $(1 - b_1 - h_1) < 0$ the balance of payments of country 1 necessarily deteriorates.

8.7.3 Intermediate Goods and the Multiplier

8.7.3.1 Different Requirements of Intermediate Goods

We examine first the case in which the requirement of intermediate goods differs according to the type of good produced. Imports of intermediate goods are now related to the various components of final demand which are satisfied by domestic production, on the basis of coefficients which express the requirements of imported intermediate goods. In what follows we simplify by assuming that investment is entirely exogenous and we also assume that all exports consist of domestically produced goods (in other words, we exclude the case of simple re-export—that is without any transformation—of imported final goods). We thus have

$$m^R = \lambda_c C_d + \lambda_I I_d + \lambda_x x, \quad 0 < \lambda_c < 1, \quad 0 < \lambda_I < 1, \quad 0 < \lambda_x < 1, \quad (8.67)$$

where the λs are the coefficients mentioned above and C_d, I_d, indicate consumer and investment goods produced domestically. As we know, C and I contain both domestic and foreign goods, that is

$$C = C_d + C_f, \quad I = I_d + I_f, \quad (8.68)$$

and, in our simplified model, the part of consumption and investment goods coming from abroad coincides with the imports of final goods, so that the equation for the determination of income can be re-written as

$$y = C + I + x - m = C_d + C_f + I_d + I_f + x - m^F - m^R \quad (8.69)$$
$$= C_d + I_d + x - m^R.$$

---

[1] It goes without saying that the analysis is based on the usual assumption that the balance of payments of the various countries are expressed in terms of a common unit of measurement, so that the condition $\sum_{i=1}^{n} B_i = 0$ holds.

Given Eq. (8.67), we have

$$y = (1 - \lambda_c) C_d + (1 - \lambda_I) I_d + (1 - \lambda_x) x, \tag{8.70}$$

and given the assumptions made on the various functions, it follows that

$$y = (1 - \lambda_c) (C_{0d} + b_d y) + (1 - \lambda_I) I_{0d} + (1 - \lambda_x) x_0, \tag{8.71}$$

where

$$b_d = b - b_f \tag{8.72}$$

is, as we know, the marginal propensity to consume domestically produced goods, equal to the difference between the total marginal propensity to consume (b), and the marginal propensity to consume foreign goods (b_f, which we can identify, in our simplified model, with the marginal propensity to import final goods, μ^F).

From (8.71) we immediately obtain the multiplier by solving for y and considering the variations

$$\Delta y = \frac{1}{1 - b_d (1 - \lambda_c)} [(1 - \lambda_c) \Delta C_{0d} + (1 - \lambda_I) \Delta I_{0d} + (1 - \lambda_x) \Delta x_0] . \tag{8.73}$$

First, note that the multiplier is now different according to the exogenous variation which occurs, given the presence of various coefficients $[(1 - \lambda_c)$ etc.] applied to the different variations. The traditional multiplier, instead, is the same whatever the exogenous variation may be and is[2]

$$\frac{1}{1 - b + \mu}. \tag{8.74}$$

8.7.3.2 Identical Requirements of Intermediate Goods

It is now possible to demonstrate that the multiplier (8.73) coincides with the traditional multiplier (8.74) *if and only if* the content of imported intermediate goods in the various categories of final goods produced domestically is the same.

As this content is expressed by $\lambda_c, \lambda_I, \lambda_x$, we now assume that these coefficients are equal, that is

$$\lambda_c = \lambda_I = \lambda_x = \lambda, \tag{8.75}$$

---

[2] Equation (8.74) is obtained immediately from (8.13) by putting $h = 0$ given the assumption of entirely autonomous investment.

where λ is their common value.[3] Then (8.73) is reduced to

$$\Delta y = \frac{1-\lambda}{1-b_d\left(1-\lambda\right)}\left(\Delta C_{0d} + \Delta I_{0d} + \Delta x_0\right),\qquad(8.76)$$

and therefore the multiplier (equal for all the exogenous variations), which we denote by k, is

$$k = \frac{1-\lambda}{1-b_d\left(1-\lambda\right)}.\qquad(8.77)$$

It remains to be demonstrated that the multiplier (8.77) coincides with the traditional multiplier (8.74). If we divide the numerator and the denominator of (8.77) by $(1-\lambda)$, we have

$$k = \frac{1}{\frac{1}{1-\lambda} - b_d}.\qquad(8.78)$$

Since, as can be ascertained by direct substitution,

$$\frac{1}{1-\lambda} = 1 + \frac{\lambda}{1-\lambda},$$

Eq. (8.78) becomes

$$k = \frac{1}{1 - b_d + \frac{\lambda}{1-\lambda}}.\qquad(8.79)$$

By adding and subtracting $b_f = \mu^F$ [see (8.72)] we have

$$k = \frac{1}{1 - b + \mu^F + \frac{\lambda}{1-\lambda}}.\qquad(8.80)$$

The last passage consists in demonstrating that $\lambda/\left(1-\lambda\right)$ is nothing other than the marginal propensity to import intermediate goods, μ^R. In fact, if we take (8.67) into consideration, then, given (8.75), we have

$$m^R = \lambda\left(C_d + I_d + x\right),$$

---

[3]It will be clear to the reader that we are demonstrating the sufficiency of this condition (the "if" part of the proposition); the necessity (the "only if" part) is implicit in the fact that (8.73) differs from (8.74) when the various λs are different.

and as it follows from (8.69), given (8.75), that $C_d + I_d + x = y + m^R$, we get,

$$m^R = \lambda \left(y + m^R \right),$$

from which

$$m^R = \frac{\lambda}{(1 - \lambda)} y, \tag{8.81}$$

that expresses the imports of intermediate goods as a function of income. Therefore $\lambda/(1 - \lambda) = \mu^R$, as also can be seen from (8.25), which had been obtained by introducing at the very beginning the assumption that the requirement of imported intermediate goods should be independent of the type of good produced.

Thus, by substituting in (8.80) and remembering that $\mu^F + \mu^R = \mu$, we finally have

$$k = \frac{1}{1 - b + \mu}, \tag{8.82}$$

which completes the demonstration.

8.7.4 The Transfer Problem

As we have seen in the text, the transfer problem can be considered as a comparative statics exercise in the context of a two-country multiplier model with repercussions.

As shown in the text, the changes in the exogenous components of expenditure (the notation has been changed to conform with that used in this Appendix) are given by

$$\begin{aligned}
d\alpha_{1C} &= -b_1'T, \ d\alpha_{2C} = b_2'T, \\
d\alpha_{1I} &= -h_1'T, \ d\alpha_{2I} = h_2'T, \\
d\alpha_{1m} &= -\mu_1'T, \ d\alpha_{2m} = \mu_2'T,
\end{aligned} \tag{8.83}$$

while country 1s balance of payments changes by

$$dB_1 = -T + (d\alpha_{2m} - d\alpha_{1m}) - \mu_1 dy_1 + \mu_2 dy_2 = (\mu_2' + \mu_1' - 1)T - \mu_1 dy_1 + \mu_2 dy_2, \tag{8.84}$$

where dy_1, dy_2 are given by the multiplier with foreign repercussions. Thus the first step is to calculate this multiplier, which is simply a particular case of the general n-country multiplier treated in Sect. 8.7.2.

In the two-country case, the Jacobian matrix (8.47) becomes

$$\mathbf{J} \equiv \begin{bmatrix} 1 - b_1 - h_1 + \mu_1 & -\mu_2 \\ -\mu_1 & 1 - b_2 - h_2 + \mu_2 \end{bmatrix}, \tag{8.85}$$

and Eq. (8.48) is

$$\mathbf{J}d\mathbf{y} = d\boldsymbol{\alpha}, \tag{8.86}$$

where

$$d\mathbf{y} \equiv \begin{bmatrix} dy_1 \\ dy_2 \end{bmatrix}, d\boldsymbol{\alpha} \equiv \begin{bmatrix} -b_1'T - h_1'T + \mu_1'T \\ b_2'T + h_2'T - \mu_2'T \end{bmatrix}. \tag{8.87}$$

The solution $d\mathbf{y} = \mathbf{J}^{-1}d\boldsymbol{\alpha}$ gives

$$dy_1 = \frac{(1 - b_2 - h_2)(-b_1' - h_1' + \mu_2' + \mu_1') + \mu_2(-b_1' - h_1' + b_2' + h_2')}{|\mathbf{J}|}T,$$

$$dy_2 = \frac{(1 - b_1 - h_1)(b_2' + h_2' - \mu_1' - \mu_2') + \mu_1(-b_1' - h_1' + b_2' + h_2')}{|\mathbf{J}|}T,$$

$$\tag{8.88}$$

where

$$|\mathbf{J}| = (1 - b_1 - h_1 + \mu_1)(1 - b_2 - h_2 + \mu_2) - \mu_1\mu_2 \tag{8.89}$$

is positive owing to the stability conditions (8.53).

Substituting (8.88) into (8.84) we obtain

$$dB_1 = [\mu_2' + \mu_1' - \frac{\mu_1(1 - b_1' - h_1')}{1 - b_1 - h_1} - \frac{\mu_2(1 - b_2' - h_2')}{1 - b_2 - h_2}$$

$$-1] \frac{(1 - b_1 - h_1)(1 - b_2 - h_2)}{|\mathbf{J}|}. \tag{8.90}$$

If we assume, as is usually done in transfer theory, that $(1 - b_1 - h_1)$ and $(1 - b_2 - h_2)$ are both positive, the sign of dB_1 will be given by the sign of the expression in square brackets. Thus we conclude that $dB_1 \gtrless 0$, namely the transfer will be undereffected, effected, or overeffected, according as

$$\mu_1' + \mu_2' \lessgtr 1 + \mu_1 \frac{(1 - b_1' - h_1')}{1 - b_1 - h_1} + \mu_2 \frac{(1 - b_2' - h_2')}{1 - b_2 - h_2}. \tag{8.91}$$

In the particular case in which the primed and unprimed propensities coincide (i.e., the transfer affects the various demands exactly as any income change), then (8.91) is obviously verified with the $<$ sign. Another particular case is when the transfer does not affect the demand for imports (μ_1', μ_2' are both zero) and changes aggregate demand either by the amount of the transfer ($b_1' + h_1' = b_2' + h_2' = 1$) or not at all ($b_1' + h_1' = b_2' + h_2' = 0$). It is easy to see that also in this case (8.91) is satisfied with the $<$ sign.

Thus in these particular cases (which were those considered by Machlup 1943, and Metzler 1942) the transfer is undereffected. In the general case (which was derived by Johnson 1956) the inequality can in principle be satisfied with either sign. It is however possible to conjecture that also in the general case the inequality is likely to be satisfied with the $<$ sign. In fact, even if the primed and unprimed propensities do not coincide, it is unlikely that they are very much different. Hence it seems safe to assume that $b'_1 + h'_1 < 1, b'_2 + h'_2 < 1$, and that $\mu'_1 + \mu'_2$ does not exceed unity. With these assumptions, (8.91) is satisfied with the $<$ sign.

References

Deardoff, A. V., & Stern, R. M. (1979). What have we learned from linked econometric models? A comparison of fiscal-policy simulations. *Banca Nazionale del Lavoro Quarterly Review, 32,* 415–432.

Ferrara, L. (1984). Il moltiplicatore in mercato aperto nelle analisi dell'interdipendenza internazionale: teoria e evidenza empirica, unpublished thesis, University of Rome La Sapienza, Faculty of Economics.

Gandolfo, G. (2009). *Economic dynamics* (4th ed.). Berlin, Heidelberg, New York: Springer.

Gantmacher, F. R. (1959). *Applications of the theory of matrices.* New York: Interscience (also as Vol. Two of F. R. Gantmacher, 1959, *The theory of matrices*, New York: Chelsea). Pages refer to the Interscience edition.

Harrod, R. (1933). *International economics.* Cambridge (UK): Cambridge University Press.

Holzman, F. D., & Zellner, A. (1958). The foreign-trade and balanced budget multipliers. *American Economic Review, 48,* 73–91.

Johnson, H. G. (1956). The transfer problem and exchange stability. *Journal of Political Economy, 44,* 212–225; reprinted (with additions) In H. G. Johnson (1958). *International trade and economic growth.* London: Allen&Unwin; In R. E. Caves & H. G. Johnson (Eds.) (1968). *Readings in international economics* (pp. 148–171). London: Allen&Unwin; and In R. N. Cooper (Ed.) (1969). *International finance - selected readings* (pp. 62–86). Harmondsworth: Penguin.

Kennedy, C. & Thirwall, A. P. (1979). The input-output formulation of the foreign trade multiplier. *Australian Economic Papers, 18,* 173–180.

Kennedy, C. & Thirwall, A. P. (1980). The foreign trade multiplier revisited. In D. A. Currie & W. Peters (Eds.). *Contemporary economic analysis* (Vol. 2, pp. 79–100). London: Croom Helm.

Keynes, J. M. (1929). The German transfer problem. *Economic Journal, 39,* 1–7; reprinted In H. S. Ellis & L. A. Metzler (Eds.) (1949). *Readings in the theory of international trade* (pp. 161–169). Philadephia: Blakiston. And In *The collected writings of J.M. Keynes* (Vol. XI, pp. 451–459). London: Macmillan for the Royal Economic Society, 1983.

Machlup, F. (1943). *International trade and the national income multiplier.* Philadelphia: Blakiston; reprinted 1965 by Kelley, New York.

Meade, J. E. (1948). National income, national expenditure and the balance of payments, Part I. *Economic Journal, 58,* 483–505 (continued in 1949, *Economic Journal, 59,* 17–39).

Metzler, L. A. (1942). The transfer problem reconsidered. *Journal of Political Economy, 50,* 397–414; reprinted in: H. S. Ellis & L. A. Metzler (Eds.) (1949). *Readings in the theory of international trade* (pp. 179–187). Philadephia: Blakiston. In L. A. Metzler (1973). *Collected papers.* Cambridge (Mass.): Harvard University Press.

Metzler, L. A. (1950). A multiple region theory of income and trade. *Econometrica, 18,* 329–354.

Metzler, L. A. (1973). Imported raw materials, the transfer problem and the concept of income. In L. A. Metzler, *Collected papers.* Cambridge (Mass.): Harvard University Press.

Milana, C. (1984). Le importazioni di beni intermedi nel moltiplicatore del reddito di un'economia aperta. ISPE Working Papers, Roma, Istituto di Studi per la Programmazione Economica.

Miyazawa, K. (1960). Foreign trade Multiplier, input-output analysis and the consumption function. *Quarterly Journal of Economics,* **7**(4), 53–64; reprinted In K. Miyazawa (1976). *Input-output analysis and the structure of income distribution* (pp. 43–58). Berlin, Heidelberg: Springer-Verlag.

Ohlin, B. (1929). The reparation problem: A discussion. I. Transfer difficulties, real and imagined. *Economic Journal, 39,* 172–178; reprinted In H. S. Ellis & L. A. Metzler (Eds.) (1949). *Readings in the theory of international trade* (pp. 170–178). Philadephia: Blakiston.

Rotondi, Z. (1989). La trasmissione internazionale delle perturbazioni in cambi fissi e flessibili, unpublished thesis, University of Rome La Sapienza, Faculty of Economics.

Thirwall, A. P. (1980). *Balance-of-payments theory and the United Kingdom experience.* London: Macmillan.

An Integrated Approach

9

In the previous chapters we have separately examined the role of exchange-rate variations (Chap. 7) and income changes (Chap. 8) in the adjustment process of the balance of payments. An obvious step forward to be taken while remaining in the context of the traditional theory, is to attempt an integration between the two mechanisms in a broader framework in which the adjustment can simultaneously come from both the exchange rate and income.

Among these attempts, which contain the germs of the more general theory of internal and external macroeconomic equilibrium (treated in the next chapters), we shall examine that of Laursen and Metzler (1950); two other attempts worthy of mention are those of Harberger (1950) and Stolper (1950).

9.1 Interaction Between Exchange Rate and Income in the Adjustment Process

The Laursen and Metzler model will be examined here in a simplified version. The simplification consists in considering a "small country" model, which allows us to neglect repercussions and treat foreign output as a constant. For the original two-country version see the Appendix.

Let Y be national money income; assuming that the prices of domestic goods and services are constant (we can then, without loss of generality, put the price level equal to 1), variations in Y measure variations in the physical output.

Imports and exports depend on the terms of trade (see Chap. 7) hence, given the assumption that not only domestic but also foreign prices are constant, on the exchange rate: exports vary in the same direction as the exchange rate (an increase in the exchange rate, that is a depreciation, stimulates exports and a decrease, that is an appreciation, lowers them) whilst imports vary in the opposite direction to the exchange rate.

© Springer-Verlag Berlin Heidelberg 2016
G. Gandolfo, *International Finance and Open-Economy Macroeconomics*,
Springer Texts in Business and Economics, DOI 10.1007/978-3-662-49862-0_9

Imports are assumed additionally to depend on national income (see Chap. 8).

For brevity, aggregate national expenditure (consumption plus investment) or absorption is indicated as A. The marginal propensity to spend (the marginal propensity to consume plus the marginal propensity to invest) is, of course, positive, and is assumed to be less than 1.

Another variable which influences A is the relative price r of imports. This is an important point and must be further discussed.

Total expenditure A includes expenditure on both domestic and foreign goods. Now, if we did not take account of r as an argument on which A depends, we would implicitly assume that when the price of imports changes, the consequent change in the expenditure on imports is exactly offset by a change of equal absolute amount, and in the opposite direction, in the expenditure on domestic goods, so that total expenditure A remains the same. This sounds rather unrealistic, so that an effect of r on A must be introduced.

It remains to determine the nature of this effect. Let us consider a fall in the relative price of imports. Since the prices of domestic goods have been assumed constant, this fall means a fall in the absolute price of imports. This, of course, increases the real income corresponding to any given level of money income. Now, the short-run consumption function is non-proportional, so that the average propensity to consume decreases as real income increases, and vice versa. From this it follows that "as import prices fall and the real income corresponding to a given money income increases, the amount spent on goods and services out of a given money income will fall. The argument is applicable in reverse, of course, to a rise of import prices. In short, our basic premise is that, other things being the same, the expenditure schedule of any given country rises when import prices rise and falls when import prices fall" (Laursen and Metzler 1950; p. 286). A similar argument was made by Harberger (1950), hence the effect of r on A is called in the literature the Harberger-Laursen-Metzler effect. This effect has been the subject of much debate, but here we shall take it for granted (further discussion is contained in Chap. 18, Sect. 18.2.2).

The model can be reduced to two equations, one expressing the determination of national income in an open economy and the other expressing the balance of payments:

$$y = A(y, r) + B, \tag{9.1}$$

$$B = x(r) - r p_m m(y, r), \tag{9.2}$$

where the price of domestic goods has been normalized to unity, and p_m is a constant.

This model is indeterminate, since there are two equations and three unknowns (y, B, r). Thus we can use the degree of freedom to impose the condition that the balance of payments is in equilibrium, obtaining the model

$$y = A(y, r) + B, \tag{9.3}$$

$$B = x(r) - r p_m m(y, r) = 0, \tag{9.4}$$

which determines the equilibrium point (y_e, r_e).

9.1.1   A Graphic Representation

Diagrammatically, system (9.3–9.4) can be represented as two curves in the (r, y) plane. One gives all (r, y) combinations that satisfy Eq. (9.3), namely that ensure real-market equilibrium. The other gives all points that satisfy Eq. (9.4), namely that ensure balance-of-payments equilibrium. The intersection of these curves will yield the equilibrium point (y_e, r_e).

Let us begin by considering the locus of all the combinations of income and the exchange rate which ensure balance-of-payments equilibrium. This locus is a curve (the *BB* schedule in Fig. 9.1, represented as a straight line for simplicity) in the (r, y) plane, which may be positively or negatively sloped according as the critical elasticities condition is satisfied or not.

As we know from Chap. 7, Sect. 7.2, if the sum of the export and import elasticities is greater than one, the balance of payments changes in the same direction as the exchange rate, that is, ceteris paribus, an increase in r (a depreciation) improves the balance of payments and vice versa. If we examine Fig. 9.1a, we see that a higher income means higher imports and so a deterioration in the balance of payments, which can be offset by a higher exchange rate since the critical elasticities condition occurs. If, on the contrary, the critical elasticities condition does *not* occur, the balance of payments varies in the opposite direction to the exchange rate, so that (see Fig. 9.1b) if income increases (hence also imports), a lower exchange rate is called for to offset the deterioration in the payments balance. In this case we get a decreasing *BB* schedule.

If we now go back to Fig. 9.1a, we see that the higher the sum of the elasticities, the higher the slope of the balance-of-payments equilibrium line: in $B'B'$, for example, this sum is greater than that underlying the *BB* schedule (both sums are of course higher than one for what was said above). In fact, the higher the sum of the elasticities, the higher the effect on the balance of payments of an exchange-rate variation, so that this variation will have to be smaller to offset the same balance-of-payments change due to an income change. If we take H as example of an initial

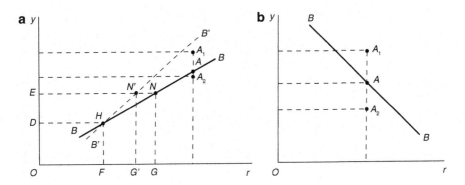

Fig. 9.1 Flexible exchange rates and the level of income: the balance-of-payments schedule

payments equilibrium point, and if we consider income OE instead of OD, we see that along the BB schedule the corresponding exchange rate will have to be OG so as to maintain balance-of-payments equilibrium (point N). We also see that along the $B'B'$ schedule where the sum of the elasticities is assumed higher the exchange rate corresponding to income OE will have to be OG' so as to maintain balance-of-payments equilibrium (point N'). With respect to the initial value of the exchange rate, OG' denotes a smaller depreciation than OG.

Finally, observe that at any point above (below) the BB schedule, independently of its slope, the balance of payments shows a deficit (surplus). In fact, if we consider a point A_1 above the BB schedule (note that what we are saying holds for both Fig. 9.1a and b) we see that this point, at unchanged exchange rate, corresponds to a higher income than the equilibrium point A does. Since higher income means higher imports, and the exchange rate is the same, at A_1 the balance of payments will be in deficit if it was in equilibrium at A. It can be similarly shown that at A_2 the balance of payments shows a surplus.

The real equilibrium schedule is determined by relation (9.3), and it is possible to show in a diagram the locus of all the combinations of income and the exchange rate which ensure the equality expressed by Eq. (9.3). We thus get a positively sloped curve in the (r, y) plane under certain assumptions. Let us first observe that greater values of aggregate demand correspond to greater values of income but, as we assumed that the marginal propensity to spend is lower than one, the increase in demand does not entirely absorb the increase in income. How will the exchange rate have to change for this excess supply to be absorbed? If the critical elasticities condition is satisfied, an increase in r will increase both A and B, so that the excess supply will be eliminated. In this case the real-equilibrium schedule RR is positively sloped.

But if the critical elasticities condition is *not* satisfied, an increase in r will cause on the one hand an increase in A, on the other a decrease in B. Our assumption is that in this second case the increase in A prevails, so that the RR schedule preserves its positive slope.

Besides being increasing, the RR schedule has the property that at all points below (above) it, there is a positive (negative) excess demand for goods, where this excess demand is measured by the difference between aggregate demand A and the national output Y, given balance of payments equilibrium. With reference to Fig. 9.2, consider any point *above* the RR schedule, for example A'. At A' income is higher than at A (which is a point of real equilibrium) and therefore, as the exchange rate is the same at both A and A', at A' aggregate demand will be higher than at A. But since the marginal propensity to spend is assumed lower than one, the higher demand will not entirely absorb the higher output, at unchanged exchange rate. Therefore at A' there will be an excess supply (negative excess demand). It can be similarly shown that at any point *below* the RR schedule (for example, A'') there is a positive excess demand.

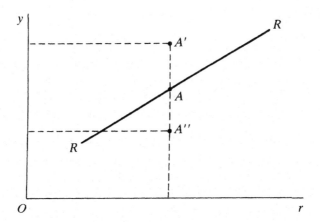

Fig. 9.2 Flexible exchange rates and the level of income: the real-equilibrium schedule

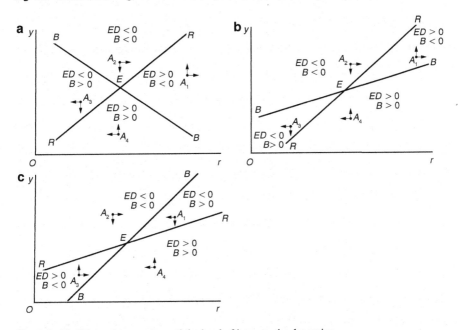

Fig. 9.3 Flexible exchange rates and the level of income: the dynamics

If we now draw schedules *RR* and *BB* in the same diagram, their intersection will determine the point at which real equilibrium and balance-of-payments equilibrium will simultaneously obtain; the coordinates of this point will give us the equilibrium values of income and the exchange rate, namely the values (y_e, r_e), which correspond to the solution of system (9.3)–(9.4). This is done in Fig. 9.3, which also serves to examine the problem of the stability of equilibrium, to which we now turn.

9.1.2   Stability

To examine the dynamic stability of the equilibrium point we must first examine the forces acting on income and the exchange rate when either of these variables is outside equilibrium, and then investigate whether, and under what conditions, these forces cause the deviating variable(s) to converge to the equilibrium point. For this purpose, the dynamic behaviour assumptions usually made are the following:

(a) national income (output) varies in relation to the excess demand for goods and, precisely, it increases if aggregate demand exceeds output (positive excess demand), decreases in the opposite case. This is the usual assumption made in the context of a model of the Keynesian type with underemployment and rigid prices and, in fact, it is the same as that already adopted in the course of the dynamic analysis of the multiplier (see Sect. 8.1).
(b) the exchange rate, that is the price of foreign exchange, increases (decreases) if there is a positive (negative) excess demand for foreign exchange in the foreign exchange market. This is the usual assumption already examined in Sect. 7.3, where it was also shown that—since in our simplified context the supply of foreign exchange comes from exporters and the demand for it from importers— this assumption is equivalent to saying that the exchange rate depreciates (appreciates) if there is a balance-of-payments deficit (surplus).

This said, it is possible to give a simplified graphic representation of the dynamic behaviour of the system and of the stability conditions. It should be emphasized that the intuitive study of stability by way of arrows (see below) is to be taken as a simple expository device and not as giving a rigorous proof (for which see the Appendix).

The procedure is fairly simple. The intersection between the *BB* and *RR* schedules subdivides the space of the first quadrant into four sub-spaces or zones in each of which, on the basis of what was shown in the previous section, it will be possible to give a precise sign to the excess demand for goods and to the balance of payments; these will then indicate the direction of movement of income and the exchange rate in accordance with assumptions (a) and (b) above.

Consider for example Fig. 9.3a, which depicts an unstable equilibrium point. We begin with point A_1. As this point lies below the *RR* and above the *BB* schedule, it implies a positive excess demand for goods (denoted by $ED > 0$ in the diagram) and a balance-of-payments deficit ($B < 0$ in the diagram). Therefore, as a consequence of the dynamic behaviour assumptions made above, income tends to increase, as shown by the vertical arrow (income is measured on the vertical axis) drawn from A_1 and pointing upwards, and the exchange rate tends to depreciate (i.e. to increase), as shown by the horizontal arrow (the exchange rate is measured on the horizontal axis) drawn from A_1 and pointing to the right. Point A_1 will therefore tend to move in a direction included between the two arrows and so will move farther and farther away from the equilibrium point E.

This reasoning holds for any point lying in the same zone as A_1, i.e. below the *RR* and above the *BB* schedule, so that any initial point lying there will diverge from equilibrium. With similar reasoning applied to the other three zones of Fig. 9.3a we can find the suitable arrows (see for example points A_2, A_3, A_4) and conclude that the system will move away from the equilibrium point, which is therefore unstable.

We can now ask the reason for this instability: in graphic terms, it is due to the position of the *BB* relative to the *RR* schedule, namely if we remember what has been said above on the decreasing *BB* schedule the cause of instability is to be seen in the fact that the critical elasticities condition does not occur. So far, there is nothing new with respect to the traditional elasticity approach, which tells us that equilibrium is unstable when the critical elasticities condition does not occur (see Sects. 7.2 and 7.3). But there is more to it than that: the novelty of the present analysis lies in the fact that the equilibrium point may be unstable *even if* the critical elasticities condition is fulfilled. In precise terms, this condition is only necessary but not sufficient for the stability of equilibrium.

To understand this important fact let us have a look at panels *b* and *c* of Fig. 9.3. In Fig. 9.3b the *BB* is an increasing schedule, but as can be seen from an inspection of the arrows (which have been drawn following the same procedure explained in relation to panel *a* of the figure) the points outside equilibrium tend to move away from it rather than converge to it: the equilibrium point is again unstable. In Fig. 9.3c, on the contrary, the equilibrium point is stable, as can be seen from the arrows: compare, for example, point A_1 in panel *c* with point A_1 in panel *b*. It can readily be seen from Fig. 9.3 that in panel *b* the *BB* schedule, though increasing, cuts the *RR* schedule from above, i.e. it is above (below) it to the left (right) of the intersection point. In panel *c*, on the contrary, the increasing *BB* schedule cuts the *RR* schedule from below. This different position of the *BB* schedule depends on the fact that the *BB* has a higher (lower) slope than the *RR* schedule in Figs. 9.3c and b respectively: we also see that *BB* in panel *c* has a higher slope than *BB* in panel *b*. If we remember what has been said in Sect. 9.1.1 on the slope of the *BB* schedule, we see that the sum of the elasticities is greater in *c* than in *b*: in other words, for stability to obtain, it does not suffice that the sum of the elasticities is greater than one, but this sum must be greater than a critical value greater than one.

Leaving aside diagrams and formulae, the economic reason for this result is intuitive. Let us assume, for example, that the balance of payments is in deficit. The exchange rate depreciates and, assuming that the traditional critical elasticities condition occurs, the deficit is reduced. However, the depreciation increases total demand for domestic output; this causes an increase in income and so imports increase, thus opposing the initial favourable effect of the depreciation on the balance of payments: this effect must therefore be more intense than it had to be in the absence of income effects, i.e. the sum of the elasticities must be higher than in the traditional case.

9.1.3   Comparative Statics and the Transfer Problem

As we said in Chap. 8, Sect. 8.6, the transfer problem can be examined as a comparative statics problem in any model where exogenous components of expenditure are present.

Let us then look at the analysis of the transfer problem in the present model, starting from a position of initial equilibrium and assuming that stability prevails. The reason why we consider a stable equilibrium is self-evident: the passage from the old to the new equilibrium point, in fact, can take place only if stability prevails; the study of the comparative static properties of unstable equilibria is, therefore, uninteresting in the context of the analysis of the system, as it is useless to find out the position of the new equilibrium if the system cannot approach it.

In the symbology introduced in Chap. 8, Eq. (8.28), the initial impact (at unchanged exchange rate) of the transfer on the balance of payments is given by

$$\Delta B = -T - \Delta m_{01} = (\mu_1' - 1)T, \tag{9.5}$$

where the term Δm_{02} has to be neglected given the small-country assumption. This expression can be either positive or negative. We shall assume that it is negative, as it is implausible that $\mu_1' > 1$. The impact effect on the real-equilibrium schedule will be given by

$$\Delta RR = \Delta A_{01} + \Delta B_0 = \Delta C_{01} + \Delta I_{01} + \Delta B = (\mu_1' - 1 - b_1' - h_1')T, \tag{9.6}$$

which is clearly negative.

Thus the impact effect is a downward shift of both the BB and RR schedules, as shown in Fig. 9.4. The new equilibrium point is E', where output is lower ($y_E' < y_E$) and the exchange rate higher ($r_E' > r_E$) than initially. The transfer has been effected thanks to a reduction in income of the transferor and an exchange-rate depreciation.

Fig. 9.4 The transfer problem under flexible exchange rates

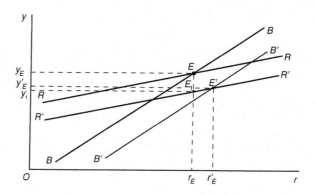

The movement from E to E' can be decomposed into two movements:

(i) the movement from E to E_1, namely the multiplier effect at unchanged exchange rate. Since E_1 is above the new balance-of-payments schedule $B'B'$, in E_1 the balance of payments is still in deficit.

(ii) the movement from E_1 to E', that effects the transfer. Note that the exchange-rate depreciation has a favourable effect on income, which therefore decreases by less than it would have when only multiplier effects are present ($y_E > y'_E > y_1$).

The case represented in the diagram is not the only one; other cases are also possible (see the Appendix). What is important to note is that, if stability of equilibrium obtains, *the transfer will be in any case effected.*

9.2   The J-Curve

It is important to point out that stability of equilibrium also ensures that the balance of payments will improve when the exchange rate is officially devalued in a regime of fixed exchange rates (such as the adjustable peg: see Chap. 3, Sect. 3.1) or is manoeuvred in the direction of a depreciation in a managed float regime. It should however be stressed that the improvement manifests itself in the new equilibrium position that the system will reach *after* all the adjustments have worked themselves out. This does not exclude the eventuality that, in the context of less restrictive assumptions than those adopted here, the balance of payments may deteriorate in the course of the adjustment process because of effects of the exchange-rate devaluation on domestic prices and income and other effects. All these effects may in fact immediately after the devaluation cause a temporary balance-of-payments deterioration (which is often referred to as the *perverse effect of the devaluation*) before the final improvement.

This phenomenon is also called in the literature the *J-curve*, to denote the time path of the payments balance, which initially decreases (deteriorates) and subsequently increases (improves) to a level higher than the one prior to devaluation, thus resembling a *J* slanted to the right in a diagram in which time is measured on the horizontal axis and the balance of payments on the vertical one. This terminology was introduced after the 1967 devaluation of the pound sterling (see NIESR 1968; p. 11), which was followed by a trade deficit which lasted until 1970.

Several studies attempted to explain this phenomenon by introducing adjustment lags, and, more precisely, by distinguishing various periods following the devaluation in which the effects of the devaluation itself take place. These are, in the terminology of Magee (1973), the *currency-contract* period, the *pass-through* period, and the *quantity-adjustment* period.

The currency-contract period is defined as that short period of time immediately following the exchange-rate variation in which the contracts stipulated before the variation mature. It is clear that during this period both prices and quantities are predetermined, so that if we consider a devaluation undertaken to correct a deficit what happens to the payments balance in domestic currency depends on the currency in which the import and export contracts were stipulated. Although from the purely taxonomic viewpoint a case in which import contracts were stipulated in domestic currency and export contracts in foreign currency cannot be excluded (in which case the domestic-currency payments balance), normally both the export and import contracts stipulated before the devaluation are expressed in foreign currency. In fact, in the expectation of a possible devaluation, both domestic and foreign exporters will try to avoid an exchange-rate loss by stipulating contracts in foreign currency. Now, as a consequence of the devaluation, the domestic-currency value of both imports and exports will increase by the same percentage as the devaluation, so that as the pre-devaluation value of the former was assumed higher than that of the latter the deficit will increase (a perverse effect).

The pass-through period is defined as that short period of time following the exchange-rate variation in which prices can change (as they refer to contracts agreed upon after the exchange rate has varied), but quantities remain unchanged due to rigidities in the demand for and/or supply of imports and exports. Consider, for example, the case of a devaluation with a demand for imports by residents of the devaluing country and a demand for the devaluing country's exports by the rest of the world which are both inelastic in the short-run. The domestic-currency price of imports increases as a consequence of the devaluation but the demand does not change, so that the outlay for imports increases. The foreign-currency price of exports decreases as a consequence of the devaluation (by the same percentage as the devaluation, if we assume that domestic exporters, given the domestic currency price, change the foreign-currency price in accordance with the devaluation), but the demand does not change, so that the foreign-currency receipts will decrease and their domestic-currency value will not change. Therefore the domestic-currency balance deteriorates (again a perverse effect). Of course these perverse effects are not inevitable only possible: for a complete classification of the possible effects of a devaluation during the currency-contract period and the pass-through period see Magee (1973; Tables 1 and 2).

Finally, we come to the quantity-adjustment period, in which both quantities and prices can change. Now, if the suitable conditions on the elasticities are fulfilled, the balance of payments ought to improve. This is undoubtedly true from the viewpoint of comparative statics, but from the dynamic point of view it may happen that quantities do not adjust as quickly as prices, owing to reaction lags, frictions etc., so that even if the stability conditions occur the balance of payments may again deteriorate before improving towards the new equilibrium point.

The analysis of the dynamic process of transition from the old to the new equilibrium with different speeds of adjustment of the variables concerned is by no means an easy matter, and we refer the reader to the Appendix.

On the empirical evidence concerning the J curve see Rose and Yellen (1989), Wood (1991), Wilson (1993), Marwah and Klein (1996), Jung and Doroodian (1998), Bahmani-Oskooee and Brooks (1999), Bahmani-Oskooee and Hegerty (2010).

9.3   The S-Curve

The J-curve describes the lag between an exchange-rate depreciation and the improvement in the trade balance. Backus et al. (1994) have found that the trade balance in 11 developed countries "is uniformly countercyclical and is negatively correlated, in general, with current and future movements in the terms of trade, but positively correlated with past movements. We call this asymmetric shape of the cross-correlation function for net exports and the terms of trade the *S-curve*, since it looks like a horizontal S. This finding is reminiscent of earlier work on the J-curve" (Backus et al. 1994; p. 84). The terms-of-trade movements considered may be either way.

Starting from their factual observation, the authors develop a two-country general-equilibrium stochastic growth model in which trade fluctuations reflect, in large part, the dynamics of capital formation.

Considering for example a favourable domestic productivity shock, the authors give an explanation of the empirical regularity in terms of their general equilibrium model of persistent productivity shocks (in the tradition of the real business cycle literature). First observe that such a shock causes an increase in domestic output, a decrease in its relative price vis-à-vis foreign goods, and a deterioration in the terms of trade. "Because the productivity shock is persistent, we also see a rise in consumption and a temporary boom in investment, as capital is shifted to its most productive location. The increases in consumption and investment together are greater than the gain in output, and the economy experiences a trade deficit during this period of high output. This dynamic response pattern gives rise to countercyclical movements in the balance of trade and an asymmetric cross-correlation function much like the ones seen in the data" (Backus et al. 1994; p. 85).

The dynamics of investment is essential in explaining such a result. If there were no capital (hence no investment), the trade balance would improve. In fact, given a favourable domestic productivity shock, the preference for smooth consumption gives rise to an increase in consumption smaller than the increase in output, hence the improvement in the trade balance. Thus the trade balance shows a procyclical rather than a countercyclical behaviour when investment is absent. At the same time, with greater domestic output, the relative price of domestic goods falls, hence an increase in the terms of trade. Thus we would observe positive contemporaneous correlation between the trade balance and the terms of trade, exactly the opposite result as in the case of presence of investment.

It is important to note that the S-curve results have been subsequently found in LDCs. Senhadji (1998) has examined 30 less developed countries asking two questions: first, does a similar empirical regularity in the dynamic behaviour of the trade balance and the terms of trade also exist in less developed countries? Second, if the answer to the first question is in the affirmative, can such a dynamic behaviour be explained in terms of a model similar to that built by Backus et al. (1994)? The answer given by the author is yes to both questions; he also finds a remarkable robustness of the S-curve to significant variations in the key parameters of the model.

Hence we must conclude that the S-curve is a fairly general phenomenon, that does not appear to be typical of developed countries, since it applies to developing countries as well.

9.4   The Alleged Insulating Power of Flexible Exchange Rates, and the International Propagation of Disturbances

If the suitable stability conditions are verified, a regime of freely flexible exchange rates maintains equilibrium in the balance of payments. Therefore, since balance-of-payments disequilibria are the way through which foreign economic perturbations influence the domestic economy and vice versa, it might at first sight seem that the regime under consideration has the power of insulating the domestic economy completely from what happens in the rest of the world as well as of insulating the rest of the world from what happens in the domestic economy (for the origins of this thesis see Laursen and Metzler 1950, who were among the first to effectively criticize it). A simple version of this claim can be given by considering the national income equation

$$y = C + I + B,$$

and assuming that flexible exchange rates maintain the balance on goods and services (B) in continuous equilibrium, so that

$$B = 0, \tag{9.7}$$

whence

$$y = C + I, \tag{9.8}$$

which coincides with the equation valid for a closed economy. Hence the conclusion of the complete insulating power of flexible exchange rates.

This conclusion, however, is incorrect for three reasons.

The first is related to the J curve examined in the previous section: given for example a disturbance which causes a payments imbalance, the adjustment is not instantaneous, i.e. a certain period of time may have to elapse before the exchange-rate variations restore balance-of-payments equilibrium. And this period of time may be sufficiently long for balance-of-payments disequilibria to significantly influence the economy.

But even if one were prepared to accept that no lags exist so that adjustment is instantaneous and the balance of payments is in continuous equilibrium, there is a second reason, related to the Laursen and Metzler model. The variations in the exchange rate have an effect on the *composition* of aggregate demand ($C + I$), insofar as they bring about variations in the relative price of domestic and foreign (imported) goods and so phenomena of substitution between the two categories of goods. The alleged complete insulating power is based on the implicit assumption that these phenomena only alter the composition of aggregate demand, but leave the total unchanged, namely that any change in the expenditure on imported goods is exactly offset by a change in the opposite direction in the residents' expenditure on domestic goods. Only in this case does national income remain unaffected. In fact, if we consider Eq. (9.8) and bear in mind that in C and I *both domestic and foreign (imported) goods are present*, the necessity arises of the offsetting changes mentioned above for $(C + I)$, and so y, to remain the same. But the presence of an influence of r on the level of $A = C + I$ is exactly the essence of the Laursen and Metzler effect (by the way, the main purpose of the Laursen and Metzler 1950 paper was exactly that of disproving the insulating power of flexible exchange rates).

The third reason is related to the fact that in Eq. (9.7) the balance on goods and services is considered, with the exclusion of capital movements. But external equilibrium must be seen in relation to the *overall balance of payments*. In other words, in a world in which capital flows are present in an essential way (these were assumed away for simplicity's sake by Laursen and Metzler 1950), a freely flexible exchange rate will maintain equilibrium in the overall balance of payments. Models for the study of this more general case will be treated in the following chapters, but for our purposes it suffices to point out that the overall equilibrium—leaving aside long-run considerations concerning the increase in the stock of foreign assets or liabilities—may well take place with a non-zero balance on goods and services, offset by a non-zero balance on capital account in the opposite direction. It seems likely that this would be the normal situation, as it is not presumable that a full equilibrium in the balance of payments could be maintained even with freely flexible exchange rates. Therefore, if the normal situation is one in which the balance on goods and services is not zero, the basis itself for the alleged insulating power of freely flexible exchange rates disappears.

As regards the *international propagation of disturbances*, one might believe that the propagation of the *same* disturbance is lower under flexible exchange rates than under fixed exchange rates, which amounts to believing that flexible exchange rates, though not having a *full* insulating power (as shown above), possess this power to a limited extent, in any case greater than that possessed by fixed exchange rates. Even this belief is however incorrect in general terms.

In fact, to examine the international propagation of disturbances foreign repercussions must be taken into account, which requires a two-country model. The necessary mathematical treatment is given in the Appendix; here some very simple considerations are sufficient to give an idea of the ambiguity of the results.

Let us consider, for example, an exogenous decrease in investment in country 1. Country's income decreases and so do its imports. However, it is not certain that country 1s balance of payments improves: in fact, country 2s income has decreased (through the foreign multiplier with repercussions) as a consequence of the exogenous decrease in country 1s investment, so that country 2s imports (which are country 1s exports) decrease as well.

Let us now have a look at country 1s balance of payments: since both imports and exports of country 1 decrease, the initial impact on this country's balance of payments can be in either direction. In any case, no further income movements can take place under fixed exchange rates.

Under flexible exchange rates, the exchange rate will adjust so as to restore balance-of-payments equilibrium, thus causing further income movements. However, given the ambiguity of the initial impact on country 1s balance of payments, we do not know whether the exchange rate will have to depreciate or appreciate (from the point of view of country 1) in order to restore balance-of-payments equilibrium. Suppose that it has to depreciate, which means an appreciation from the point of view of country 2. This appreciation has a negative effect on country 2s income. It follows that country 2 will undergo a further income reduction in the final situation, which means that the transmission has been greater under flexible than under fixed exchange rates.

It is important to point out that this analysis does not consider possible adjustments coming from interest-induced capital flows. For a more general analysis of the shock-insulating properties of fixed versus flexible exchange rates see Sect. 17.2.

9.5   Appendix

9.5.1   A Simplified Version of the Laursen and Metzler Model

The model that we present here is simplified because we consider a small country, and is the mathematical counterpart of the model treated in the text. The original two-country model will be examined in the next section.

The equations of the model are (9.3)–(9.4), that we rewrite as

$$y - \{A(y, r) + [x(r) - r p_m m(y, r)]\} = 0, \qquad (9.9)$$

$$\text{with } 0 < A_y < 1, A_r > 0,$$

$$x(r) - r p_m m(y, r) = 0, \qquad (9.10)$$

$$\text{with } 0 < m_y < 1, x_r > 0, m_r < 0,$$

where a subscript denotes a (partial) derivative ($A_y \equiv \partial A / \partial y$, etc.).

9.5.2   The *BB* and *RR* Schedules

By using the implicit-function differentiation rule we can calculate the derivative of y with respect to r with reference to both equations (9.9) and (9.10), thus obtaining the slopes of the BB and RR schedules in the (r, y) plane. We obtain

$$\left(\frac{dy}{dr}\right)_{BB} = \frac{x_r - p_m m - r p_m m_r}{r p_m m_y} = \frac{p_m m \left(\frac{x}{r p_m m} \eta_x + \eta_m - 1\right)}{r p_m m_y}, \qquad (9.11)$$

$$\left(\frac{dy}{dr}\right)_{RR} = \frac{A_r + p_m m \left(\frac{x}{r p_m m} \eta_x + \eta_m - 1\right)}{1 - A_y + r p_m m_y}, \qquad (9.12)$$

where $\eta_x \equiv (r/x) x_r$, $\eta_m \equiv -(r/m) m_r$ are the elasticities of exports and imports with respect to the exchange rate. It follows from Eq. (9.11) that $(dy/dr)_{BB} \gtrless 0$ according as $[(x/r p_m m) \eta_x + \eta_m - 1] \gtrless 0$; besides, the greater the sum of the two elasticities ceteris paribus, the greater will be this derivative.

As regards $(dy/dr)_{RR}$, we begin by observing that the denominator is certainly positive given the assumption $0 < A_y < 1$. The numerator, given that $A_r > 0$, will certainly be positive if the critical elasticities condition is satisfied, but will remain positive even when $[(x/r p_m m) \eta_x + \eta_m - 1] < 0$ provided that A_r prevails in absolute value. We assume that it is so, not because this is essential for our results, but in order to simplify the graphic treatment in the text, which would give rise to a maze of cases according as the slope of the RR is positive or negative.

If we consider a neighbourhood of the equilibrium point, where the balance of payments is in equilibrium, and so $x = r p_m m$, Eq. (9.11) becomes

$$\left(\frac{dy}{dr}\right)_{BB} = \frac{p_m m (\eta_x + \eta_m - 1)}{r p_m m_y}. \qquad (9.13)$$

As regards the points off the RR and BB schedules, if we define the functions

$$ED = \{A(y, r) + [x(r) - r p_m m(y, r)]\} - y, \ B = x(r) - r p_m m(y, r), \qquad (9.14)$$

we see that the RR and BB schedules are determined by $ED = 0, B = 0$ respectively. We now compute

$$ED_y = A_y - r p_m m_y - 1 < 0, \ B_y = -r p_m m_y < 0, \qquad (9.15)$$

from which it follows that $ED \lessgtr 0$ according as y is higher (lower) than the value for which $ED = 0$ at any given r, and that $B \gtrless 0$ according as y is lower (higher) than the value for which $B = 0$ at any given r.

9.5.2.1 The Dynamics of the System

To examine the stability of the equilibrium point, let us consider the following differential equation system, which is the formal expression of the dynamic behaviour assumed in Sect. 9.1.2

$$\dot{y} = f_1(ED), \quad \dot{r} = f_2(-B), \tag{9.16}$$

where f_1, f_2 are sign-preserving functions, and $f_1'[0] \equiv k_1 > 0, f_2'[0] \equiv k_2 > 0$. The study of the local stability involves the analysis of the linear approximation to system (9.16) at the equilibrium point, that is, of system

$$\dot{\bar{y}} = k_1\{(A_y - rp_m m_y - 1)\bar{y} + [A_r + p_m m (\eta_x + \eta_m - 1)]\bar{r}\},$$
$$\dot{\bar{r}} = k_2[rp_m m_y \bar{y} - p_m m (\eta_x + \eta_m - 1) \bar{r}], \tag{9.17}$$

where a bar over a symbol denotes as usual the deviations from equilibrium. The dynamic path of the system depends on the roots of the characteristic equation

$$\lambda^2 + [k_1(1 - A_y + rp_m m_y) + k_2 p_m m (\eta_x + \eta_m - 1)]\lambda$$
$$+ k_1 k_2\{(1 - A_y + rp_m m_y)p_m m (\eta_x + \eta_m - 1) - rp_m m_y[A_r$$
$$+ p_m m (\eta_x + \eta_m - 1)]\} = 0. \tag{9.18}$$

Necessary and sufficient stability conditions are

$$k_1(1 - A_y + rp_m m_y) + k_2 p_m m (\eta_x + \eta_m - 1) > 0,$$
$$(1 - A_y + rp_m m_y)p_m m (\eta_x + \eta_m - 1) - rp_m m_y[A_r + p_m m (\eta_x + \eta_m - 1)] > 0. \tag{9.19}$$

The second inequality can be written, after simple manipulations, as

$$\eta_x + \eta_m > 1 + \frac{rm_y A_r}{m(1 - A_y)}, \tag{9.20}$$

which shows that the standard critical elasticities (Marshall-Lerner) condition, though necessary, is not sufficient, because the r.h.s. is greater than one, since the fraction is positive. Note that this result crucially depends on the existence of the Harberger-Laursen-Metzler effect, i.e., on $A_r > 0$. When $A_r = 0$ we are back to the traditional condition.

Inequality (9.20) is the crucial one, because when it is satisfied, the first inequality in (9.19) is certainly satisfied.

To relate the math with the diagram shown in Fig. 9.3, we must ask ourselves what happens when the stability conditions are not satisfied. Now, when the elasticity sum is smaller than one, the second inequality in (9.19) is not satisfied, while the first one may or may not be satisfied. This is case (a) in Fig. 9.3. When the elasticity sum is greater than one but smaller than the critical value given in (9.20), the second inequality in (9.19) is not satisfied, while the first one will be satisfied. This is case (b) in Fig. 9.3.

In both cases the constant term in the characteristic equation (9.18) will be negative, which implies one positive and one negative real root. Hence the equilibrium point is a saddle point (see Gandolfo 2009; Table 21.1), which means that all motions will move away from equilibrium, except those that by chance happen to start from a point lying on the stable asymptote.

9.5.2.2 Comparative Statics: The Transfer Problem

To examine the transfer problem, we introduce in Eqs. (9.9)–(9.10) appropriate shift parameters

$$
\begin{aligned}
\Delta B &= -\Delta m_{01} - T = (\mu_1' - 1)T, \\
\Delta RR &= -\Delta m_{01} + \Delta C_{01} + \Delta I_{01} = (\mu_1' - 1 - b_1' - h_1')T,
\end{aligned}
\tag{9.21}
$$

representing the impact effect of the transfer. We assumed in the text that $\Delta B < 0$, $\Delta RR < 0$.

The system becomes

$$
\begin{aligned}
Y - \{A(Y,r) + [x(r) - rp_m m(y,r)] + (\mu_1' - 1 - b_1' - h_1')T\} &= 0, \\
x(r) - rp_m m(y,r) + (\mu_1' - 1)T &= 0.
\end{aligned}
\tag{9.22}
$$

Using the second equation we can simplify the first, so that we have

$$
\begin{aligned}
Y - \{A(Y,r) - (b_1' + h_1')T\} &= 0, \\
x(r) - rp_m m(y,r) + (\mu_1' - 1)T &= 0.
\end{aligned}
\tag{9.23}
$$

If the Jacobian of Eq. (9.22) with respect to y, r is different from zero in the neighbourhood of the equilibrium point, that is if

$$
\begin{aligned}
|\mathbf{J}| &= \begin{vmatrix} 1 - A_y & -A_r \\ -rp_m m_y & p_m m (\eta_x + \eta_m - 1) \end{vmatrix} \\
&= (1 - A_y) p_m m (\eta_x + \eta_m - 1) - rp_m m_y A_r \neq 0,
\end{aligned}
\tag{9.24}
$$

then we can express y and r as continuously differentiable functions of T in the neighbourhood of the equilibrium point, i.e., $y = y(T), r = r(T)$. It is easy to see that, if the equilibrium is stable, then $|\mathbf{J}| > 0$ by (9.20), and so we can calculate the derivatives $dy/dT, dr/dT$ by the method of comparative statics. From Eq. (9.23) we get

$$
\begin{aligned}
(1 - A_y)\frac{dy}{dT} - A_r\frac{dr}{dT} &= -(b_1' + h_1'), \\
-rp_m m_y\frac{dy}{dT} + p_m m (\eta_x + \eta_m - 1)\frac{dr}{dT} &= -(\mu_1' - 1),
\end{aligned}
\tag{9.25}
$$

whose solution is

$$dy/dT = \left[-(b_1' + h_1')p_m m \left(\eta_x + \eta_m - 1\right) - A_r(\mu_1' - 1)\right] / |\mathbf{J}|, \quad (9.26)$$

$$dr/dT = \left[-\left(1 - A_y\right)\left(\mu_1' - 1\right) - rp_m m_y(b_1' + h_1')\right] / |\mathbf{J}|. \quad (9.27)$$

The sign of $dy/dT, dr/dT$ is ambiguous, since the numerators in (9.26), (9.27) contain both positive and negative terms. It is however possible to rule out one of the four possible cases, which is $dy/dT > 0, dr/dT < 0$. In fact, considering for example the first equation in system (9.25), the coefficient of dy/dT is positive while the coefficient of dr/dT is negative, hence with $dy/dT > 0, dr/dT < 0$ the left-hand side would be positive, a contradiction since the right-hand side is negative. Hence it is not possible for the exchange rate to appreciate and income to increase, a quite obvious result from the economic point of view.

In general, income will increase or decrease according to the magnitude of the Laursen-Metzler effect: in fact, from Eq. (9.26) we see that

$$dy/dT \gtrless 0 \text{ according as } A_r \gtrless p_m m \left(\eta_x + \eta_m - 1\right) \frac{b_1' + h_1'}{1 - \mu_1'}. \quad (9.28)$$

A strong Laursen-Metzler effect will, in fact, sustain aggregate demand and hence output in the face of an exchange-rate depreciation required to adjust the balance of payments.

As regards the exchange rate, it will depreciate or appreciate according to the magnitude of the marginal propensity to import: from Eq. (9.27) we get

$$dr/dT \gtrless 0 \text{ according as } rp_m m_y \lessgtr \frac{\left(1 - A_y\right)\left(1 - \mu_1'\right)}{(b_1' + h_1')}. \quad (9.29)$$

A small propensity to import will, in fact, cause only a small adjustment in the balance of payments on the side of income-induced import reduction, with the consequent need for an exchange-rate depreciation.

The case described in the text corresponds to a relatively small Laursen-Metzler effect coupled with a relatively small marginal propensity to import.

9.5.3 The J-Curve

The perverse effects of a devaluation, which may occur in the short period, notwithstanding the occurrence of the critical elasticities condition or of the more general condition (9.20) discussed above, can ultimately be seen as a consequence of *adjustment lags* of various types acting on both quantities and prices. The purpose of this section is to give a truly dynamic treatment of the problem in the context of the model explained in Sect. 9.5.1, duly modified and integrated to account for adjustment lags. Let us rewrite the model with an explicit consideration of import and export prices, i.e.

$$y = A(y, r) + p_x x(r) - p_{hm} m(y, r),$$
$$B = p_x x(r) - p_{hm} m(y, r), \tag{9.30}$$

where $p_{hm} \equiv r p_m$ is the domestic-currency price of imports (their foreign-currency price being p_m). Since we are, by assumption, in the context of a fixed but adjustable exchange-rate regime (adjustable peg), r has to be considered as a parameter and not as an endogenous variable. Given r, system (9.30) determines the values of y and B as differentiable functions of r. By using the method of comparative statics we can calculate dB/dr, which turns out to be

$$\frac{dB}{dr} = x_r - m - rm_r - \frac{rm_y(A_r + x_r - rm_r - m)}{1 - A_y + rm_y}$$
$$= \frac{(x_r - m - rm_r)(1 - A_y) - rm_y A_r}{1 - A_y + rm_y}, \tag{9.31}$$

where we have assumed that not only the international price of imports but also the domestic-currency price of exports remain unchanged as a consequence of the devaluation, hence they can be normalized at unity. Since the denominator in (9.31) is positive, dB/dr will be positive if the numerator is positive, namely

$$x_r - m - rm_r > \frac{rm_y A_r}{1 - A_y},$$

whence

$$\eta_x + \eta_m > 1 + \frac{rm_y A_r}{m(1 - A_y)}.$$

If we assume that the stability conditions (9.20) occurs, it follows that $dB/dr > 0$. Therefore the balance of payments will certainly improve in the final position.

Let us now introduce the adjustment lags by way of suitable *partial adjustment equations* (see Gandolfo 2009; Sect. 12.4):

$$\dot{p}_{hm} = \alpha_1 \left[(r_0 + dr)p_m - p_{hm} \right], \qquad\qquad \alpha_1 > 0,$$
$$\dot{x} = \alpha_2 \left[(x_0 + x_r dr) - x \right], \qquad\qquad \alpha_2 > 0,$$
$$\dot{m} = \alpha_3 \left[\left(m_0 + m_y \frac{A_r + x_r - rm_r - m}{1 - A_y + rm_y} dr + m_r dr \right) - m \right], \; \alpha_3 > 0, \tag{9.32}$$

where, given the assumptions, α_2 and α_3 are much lower than α_1, because prices adjust much more rapidly than quantities, so that the mean time-lag of the latter is much higher than that of the former.

The first equation in (9.32) expresses the fact that—given a devaluation dr—the domestic-currency price of imports, p_{hm}, adjust with a mean time-lag $1/\alpha_1$ to the value corresponding to the new exchange rate (which is r_0+dr, r_0 being the given initial exchange rate) applied to the given international price p_m. The second equation expresses the fact that the quantity of exports adjusts with a mean time-lag $1/\alpha_2$ to the value corresponding to the new exchange rate: this value is $x_0 + dx = x_0 + x_r dr$, where x_0 is the initial quantity. The third equation expresses the fact that the quantity of imports adjust with a mean time-lag $1/\alpha_3$ to the value corresponding to the new exchange rate: this value is $m_0 + dm = m_0 + m_y dy + m_r dr$ where m_0 is the initial quantity; from the first equation in (9.30) we then have[1] $dy = \left[(A_r + x_r - rm_r - m)/(1 - A_y + rm_y)\right]dr$.

System (9.32) is diagonal and has the explicit solution

$$
\begin{aligned}
p_{hm}(t) &= A_1 e^{-\alpha_1 t} + (r_0 + dr)p_m, \\
x(t) &= A_2 e^{-\alpha_2 t} + (x_0 + x_r dr), \\
m(t) &= A_3 e^{-\alpha_3 t} + \left(m_0 + m_y \frac{A_r + x_r - rm_r - m}{1 - A_y + rm_y} dr + m_r dr\right),
\end{aligned}
\tag{9.33}
$$

where the arbitrary constants A_1, A_2, A_3 can be determined by means of the initial conditions. Thus we get

$$
A_1 = -p_m dr, A_2 = -x_r dr, A_3 = -m_y \frac{A_r + x_r - rm_r - m}{1 - A_y + rm_y} dr - m_r dr. \tag{9.34}
$$

For notational convenience we define

$$
r_0 + dr \equiv r_n, x_0 + x_r dr \equiv x_n, m_0 + m_y \frac{A_r + x_r - rm_r - m}{1 - A_y + rm_y} dr + m_r dr \equiv m_n. \tag{9.35}
$$

If we now substitute Eq. (9.33) in the second equation of (9.30) we immediately obtain the time path of the balance of payments

$$
B(t) = A_2 e^{-\alpha_2 t} + x_n - (A_1 e^{-\alpha_1 t} + r_n p_m)(A_3 e^{-\alpha_3 t} + m_n). \tag{9.36}
$$

From this equation we see that, as t tends to infinity, $B(t)$ converges to the value $B_n = x_n - r_n p_m m_n$ that is the value $B_0 + dB$, where $dB = (dB/dr)dr$ is equal, as it must be, to the value which can be derived from Eq. (9.31). Since we have already assumed above that $dB > 0$, it follow that $B_n > B_0$. Thus, when the adjustment process has worked itself out, the balance of payments will certainly improve, because the limit to which $B(t)$ converges is $B_n > B_0$. But what happens in the

---

[1]This implies the simplifying assumption that the income determination equation holds instantaneously. Greater generality would be achieved by introducing an adjustment lag on income as well, but this would complicate the analysis without substantially altering the results.

meantime? To answer this question we can use Eq. (9.36) to calculate the derivative

$$\dot{B} = -\alpha_2 A_2 e^{-\alpha_2 t} + \alpha_1 A_1 e^{-\alpha_1 t} (A_3 e^{-\alpha_3 t} + m_n)$$

$$+ \alpha_3 A_3 e^{-\alpha_3 t} (A_1 e^{-\alpha_1 t} + r_n p_m), \tag{9.37}$$

and see whether $\dot{B} \lessgtr 0$. If $\dot{B}$ is always positive, the balance of payments will improve from the beginning, whilst if $\dot{B} < 0$ for small t, the balance of payments will initially deteriorate before improving (even if initially $\dot{B} < 0$, it will have to be positive subsequently, given the assumption that $B_n > B_0$). Now, for sufficiently small t, the exponential functions in (9.37) are approximately equal to one, so that

$$\left(\frac{dB}{dt}\right)_{t \cong 0} \cong -\alpha_2 A_2 + \alpha_1 A_1 (A_3 + m_n) + \alpha_3 A_3 (A_1 + r_n p_m). \tag{9.38}$$

By using the definitions of A_i, m_n, r_n given in Eqs. (9.34) and (9.35), we get

$$\left(\frac{dB}{dt}\right)_{t \cong 0} \cong \left[\alpha_2 x_r - \alpha_1 p_m m_0 - \alpha_3 \left(m_y \frac{A_r + x_r - r m_r - m}{1 - A_y + r m_y} + m_r\right) r_0 p_m\right] dr, \tag{9.39}$$

whose sign is determined by the expression in square brackets. If the adjustment speeds α_i were approximately equal, the expression under consideration would be positive owing to the assumed positivity of (9.31).[2] But since we have assumed that α_1 is much greater than α_2 and α_3, it is quite possible that the negative term $-\alpha_1 p_m m_0$ prevails (note that the term containing α_3 has an ambiguous sign), so that expression (9.39) may well turn out to be negative.

In conclusion, we can say that the result depends on the adjustment lags of prices and quantities. If these lags are similar, then the balance of payments will presumably improve from the beginning provided that the conditions for its final improvement are fulfilled. If, on the contrary, prices adjust much more rapidly than quantities, then—even if the conditions for a final improvement are fulfilled—the balance of payments will initially deteriorate before improving. Therefore, if one believes that the mean time-lag of quantities is much higher than that of prices, then the J-curve will have to be considered, as it were, a physiological event in the context of the model under consideration.

---

[2] We must bear in mind that the magnitudes appearing on the right-hand side of Eq. (9.31) are evaluated at the initial point, so that $m = m_0$ etc.

9.5.4 The Original Two-Country Version of the Laursen and Metzler Model

9.5.4.1 The Basic Model

Let Y be national money income; assuming that the prices of domestic goods and services are constant (we can then, without loss of generality, put the price level equal to 1), variations in Y measure variations in the physical output. The subscripts 1 and 2 refer to country 1 (e.g., the home country) and country 2 (e.g., the rest of the world), in a two-country world.

Imports M are measured in local currency (i.e. the currency of the importing country), and are assumed to depend on national income and on the relative price of imports with respect to home goods; given the assumption that in each country the price level of home goods is constant, this relative price is measured by the exchange rate. The exchange rate r is defined as the number of units of the currency of country 1 for one unit of the currency of country 2; of course, from the point of view of the other country, the exchange rate is $1/r$. Now, an increase in national income has a positive effect on imports, whereas the effect of a variation in the exchange rate is uncertain, since it depends on the elasticity of the demand for imports (remember that we are dealing not with physical quantities, but with expenditure on imported goods)

Following the original notation (Laursen and Metzler 1950), aggregate national expenditure is indicated as ω. The marginal propensity to spend (the marginal propensity to consume plus the marginal propensity to invest) is, of course, positive, and is assumed to be less than 1.

Thus we have the relations

$$\omega_1 = \omega_1(Y_1, r), \quad 0 < \frac{\partial \omega_1}{\partial Y_1} < 1, \quad \frac{\partial \omega_1}{\partial r} > 0,$$

$$\omega_2 = \omega_2\left(Y_2, \frac{1}{r}\right), \quad 0 < \frac{\partial \omega_2}{\partial Y_2} < 1, \quad \frac{\partial \omega_2}{\partial (1/r)} > 0,$$

$$M_1 = M_1(Y_1, r), \quad 0 < \frac{\partial M_1}{\partial Y_1} < 1, \quad \frac{\partial M_1}{\partial r}?$$

$$M_2 = M_2(Y_2, \frac{1}{r}), \quad 0 < \frac{\partial M_2}{\partial Y_2} < 1, \quad \frac{\partial M_2}{\partial (1/r)}?$$

The static model consists in the following equations

$$Y_1 = \omega_1(Y_1, r) + rM_2(Y_2, \frac{1}{r}) - M_1(Y_1, r),$$

$$Y_2 = \omega_2\left(Y_2, \frac{1}{r}\right) + \frac{1}{r}M_1(Y_1, r) - M_2(Y_2, \frac{1}{r}), \tag{9.40}$$

$$B_2 = \frac{1}{r}M_1(Y_1, r) - M_2(Y_2, \frac{1}{r}).$$

The first and second equation express national income determination in the two countries (remember that exports of country 1 are the imports of country 2 and vice versa; exports must be multiplied by the exchange rate to convert them into local currency). The third equation defines the balance of payments of country 2, expressed in country 2s currency; there is no need of a separate equation for the balance of payments of country 1, since in our two-country world the balance of payments of country 1 equals minus the balance of payments of country 2 multiplied by r.

The above model is indeterminate, since there are three equations and four unknowns (Y_1, Y_2, r, B_2). Thus we can use the degree of freedom to impose the condition that the balance of payments is in equilibrium, i.e. $B_2 = 0$, so that we have

$$
\begin{aligned}
Y_1 &= \omega_1(Y_1, r), \\
Y_2 &= \omega_2(Y_2, \frac{1}{r}), \\
\frac{1}{r} M_1(Y_1, r) &- M_2(Y_2, \frac{1}{r}) = 0,
\end{aligned}
\tag{9.41}
$$

which determine the equilibrium point Y_1^e, Y_2^e, r^e.

9.5.4.2 Stability
Let us now consider the adjustment process. The behaviour assumptions are the usual ones, i.e.

(1) The level of national income (output) varies in relation to excess demand, and, more precisely, it tends to increase if aggregate demand exceeds aggregate supply (i.e. the level of current output), and to decrease in the opposite case. It must be noted that, since we are examining disequilibrium situations, the excess demands must be computed using (9.40) and not (9.41) as the latter equations refer only to the equilibrium situation. Thus excess demand in country 1 is

$$
[\omega_1(Y_1, r) + r M_2(Y_2, 1/r) - M_1(Y_1, r)] - Y_1,
$$

and similarly for country 2.
(2) The rate of exchange, which is the price of the foreign currency, tends to increase (decrease) if in the foreign exchange market there is a positive (negative) excess demand for the foreign currency. We assume that the dominant currency, which is used in international transactions, is the currency of country 2. Thus the relevant price is r; the demand for such currency emanates from the import demand of country 1 and so is $(1/r)M_1$, the supply emanates from export revenue and so is M_2.

The above assumptions can be formally expressed by the following system of differential equations

$$\dot{Y}_1 = k_1[\omega_1(Y_1, r) + rM_2(Y_2, \frac{1}{r}) - M_1(Y_1, r) - Y_1],$$
$$\dot{Y}_2 = k_2[\omega_2(Y_2, \frac{1}{r}) + \frac{1}{r}M_1(Y_1, r) - M_2(Y_2, \frac{1}{r}) - Y_2], \tag{9.42}$$
$$\dot{r} = k_3[\frac{1}{r}M_1(Y_1, r) - M_2(Y_2, \frac{1}{r})],$$

where k_1, k_2, k_3 are positive constants. The linearisation of the system and the study of the stability conditions involves long and tedious, though straightforward, manipulations, for which we refer the reader to Gandolfo (1980; pp. 330–3). For the reader who wants to do them, we just note that at the equilibrium point the following relation holds:

$$M_1 - \frac{\partial M_1}{\partial r} - \frac{\partial M_2}{\partial (1/r)} = M_1(\eta_1 + \eta_2 - 1), \tag{9.43}$$

where η_1, η_2 are the price elasticities of the import demands of the two countries. In fact, denoting by x_1, x_2 the physical quantities of imports (which are measured in units of goods of the exporting country), we have

$$M_1 = rx_1, M_2 = \frac{1}{r}x_2$$

and consequently

$$\frac{\partial M_1}{\partial r} = x_1 + r\frac{\partial x_1}{\partial r} = x_1\left(1 + \frac{r}{x_1}\frac{\partial x_1}{\partial r}\right) = x_1(1 - \eta_1),$$

$$\frac{\partial M_2}{\partial (1/r)} = x_2 + (1/r)\frac{\partial x_2}{\partial (1/r)} = x_2\left(1 + \frac{1/r}{x_2}\frac{\partial x_2}{\partial (1/r)}\right) = x_2(1 - \eta_2),$$

where

$$\eta_1 \equiv -\frac{r}{x_1}\frac{\partial x_1}{\partial r} \quad \eta_2 \equiv -\frac{1/r}{x_2}\frac{\partial x_2}{\partial (1/r)}.$$

We can now choose the units of measurement in such a way that *at the equilibrium point* the exchange rate is equal to 1, so that $M_1 = x_1, M_2 = x_2$, and—given the third equation in (9.42)—$M_1 = M_2$. From all this Eq. (9.43) follows.

The interesting thing is that all the stability conditions boil down to the crucial inequality

$$\eta_1 + \eta_2 > 1 + \frac{1}{M_1}\left(\frac{m_1 s_1}{1 - w_1} + \frac{m_2 s_2}{1 - w_2}\right), \tag{9.44}$$

where $m_i \equiv \partial M_i / \partial Y_i$, $s_1 \equiv \partial \omega_1 / \partial r$, $s_2 \equiv \partial \omega_2 / \partial (1/r)$, $w_i \equiv \partial w_i / \partial Y_i$, and η_1, η_2 are the price elasticities of the import demands of the two countries. The result found in the simple one-country model is therefore confirmed in the context of this more general two-country model.

9.5.4.3 Comparative Statics

The International Propagation of Disturbances
Let us introduce a parameter α_1 representing an exogenous increase in country 1s aggregate demand. Thus we have

$$
\begin{aligned}
Y_1 &= \omega_1 \left(Y_1, r \right) + \alpha_1, \\
Y_2 &= \omega_2 \left(Y_2, \frac{1}{r} \right), \\
\frac{1}{r} M_1 \left(Y_1, r \right) &= M_2 \left(Y_2, \frac{1}{r} \right).
\end{aligned}
\tag{9.45}
$$

On the assumption that the suitable conditions concerning the Jacobian are verified, there exist the differentiable functions

$$
Y_1 = Y_1(\alpha_1), \ Y_2 = Y_2(\alpha_1), \ r = r(\alpha_1).
\tag{9.46}
$$

The multiplier we are interested in is given by the derivative $dY_1/d\alpha_1$; as a by-product we shall also obtain the derivatives $dY_2/d\alpha_1$ and $dr/d\alpha_1$. Differentiating totally Eq. (9.45) with respect to α_1—account being taken of Eq. (9.46)—setting $r = 1$ in the initial equilibrium situation (this involves only a change in units of measurement) and rearranging terms, we obtain

$$
\begin{aligned}
(1 - w_1) \frac{dY_1}{d\alpha_1} - s_1 \frac{dr}{d\alpha_1} &= 1, \\
(1 - w_2) \frac{dY_2}{d\alpha_1} + s_2 \frac{dr}{d\alpha_1} &= 0, \\
-m_1 \frac{dY_1}{d\alpha_1} + m_2 \frac{dY_2}{d\alpha_1} + M_1 (\eta_1 + \eta_2 - 1) \frac{dr}{d\alpha_1} &= 0,
\end{aligned}
\tag{9.47}
$$

where the symbols have been defined above. The solution of this system is

$$
\begin{aligned}
\frac{dY_1}{d\alpha_1} &= \frac{(1 - w_2) M_1 (\eta_1 + \eta_2 - 1) - m_2 s_2}{\Delta}, \\
\frac{dY_2}{d\alpha_1} &= \frac{-m_1 s_2}{\Delta}, \\
\frac{dr}{d\alpha_1} &= \frac{-m_1 (1 - w_2)}{\Delta},
\end{aligned}
\tag{9.48}
$$

where

$$\Delta \equiv \begin{vmatrix} 1 - w_1 & 0 & -s_1 \\ 0 & 1 - w_2 & s_2 \\ -m_1 & m_2 & M_1 (\eta_1 + \eta_2 - 1) \end{vmatrix}$$
$$= (1 - w_1)(1 - w_2) M_1 (\eta_1 + \eta_2 - 1) - s_1 m_1 (1 - w_2) - s_2 m_2 (1 - w_1).$$

$$(9.49)$$

The sign of Δ is indeterminate, since expression (9.49) contains both positive and negative terms and also a term whose sign is not known *a priori*, i.e. $(\eta_1 + \eta_2 - 1)$. However, from the dynamic analysis we know that a necessary (and in this case also sufficient) stability condition is that $\Delta > 0$: see condition (9.44). Moreover, $\Delta > 0$ implies that

$$(1 - w_2) M_1 (\eta_1 + \eta_2 - 1) - m_2 s_2 > 0,$$

as can easily be seen from the fact that $\Delta > 0$ can be rewritten as

$$(1 - w_2) M_1 (\eta_1 + \eta_2 - 1) - m_2 s_2 > s_1 m_1 \frac{1 - w_2}{1 - w_1},$$

where the right-hand side is a positive quantity as $0 < w_i < 1$. Thus we may conclude that $dY_1/d\alpha_1 > 0$, $dr/d\alpha_1 > 0$, $dY_2/d\alpha_1 < 0$.

The fact that, notwithstanding the full flexibility of the exchange rate, a change in income in country 1 brings about a change in income in country 2, is sufficient to show the falsity of the opinion which attributes an insulating property to flexible exchange rates. The fact is that, although the exchange rate adjusts so as to eliminate any effect of the trade balance, keeping it at a zero level, the variations in the exchange rate, however, have a direct effect on aggregate expenditure $C + I$. It follows that $Y = C + I + x - M$ cannot remain constant, although $x - M = 0$. This is the essence of Laursen's and Metzler's argument. It should be stressed that this result crucially depends on the presence of the Laursen-Metzler effect: in fact, if $s_1 = s_2 = 0$, then $dY_2/d\alpha_1 = 0$.

The conclusion reached above on the signs of $dY_1/d\alpha_1$ and of $dr/d\alpha_1$ are no surprise: an increase in autonomous expenditure in country 1, account being taken of all the repercussions, brings about an increase in national income and thus an increase in imports; the consequent deficit in country 1s balance of payments is corrected by a devaluation, that is an increase in r. But the result on the sign of $dY_2/d\alpha_1$ may be somewhat surprising: a boom in one country causes a recession in the other country. However, this result will become intuitively clear if we remember that an increase in r—that is, an *appreciation* in country 2's currency—brings about a *fall* in the aggregate expenditure schedule of country 2, whose national income must therefore decrease, as the balance of trade is in equilibrium.

The problem of the insulating power of flexible exchange rates will be taken up again in Sect. 17.2.

To compare the international transmission of disturbances under fixed and flexible exchange rates we only have to derive the analogous of (9.48) under fixed exchange rates. For this, in addition to eliminating r (since it is fixed, we can normalize it to unity and omit it from the various functions), we must recall that under fixed exchange rates the balance of payments is no longer necessarily in equilibrium, so that we shall have to start from Eq. (9.40) instead of Eq. (9.45). Thus we have

$$
\begin{aligned}
Y_1 &= \omega_1(Y_1) + M_2(Y_2) - M_1(Y_1) + \alpha_1, \\
Y_2 &= \omega_2(Y_2) + M_1(Y_1) - M_2(Y_2), \\
B_2 &= M_1(Y_2) - M_2(Y_1).
\end{aligned}
\tag{9.50}
$$

The Jacobian of this system with respect to Y_1, Y_2, B_2 is

$$
\begin{aligned}
|\mathbf{J}| &=
\begin{vmatrix}
1 - w_1 + m_1 & -m_2 & 0 \\
-m_1 & 1 - w_2 + m_2 & 0 \\
-m_1 & m_2 & 1
\end{vmatrix} \\
&= (1 - w_1 + m_1)(1 - w_2 + m_2) - m_1 m_2 \\
&= (1 - w_1)(1 - w_2) + m_1(1 - w_2) + m_2(1 - w_1) > 0,
\end{aligned}
\tag{9.51}
$$

since $0 < w_i < 1$ by the model's assumptions. Thus we can express Y_1, Y_2, B_2 as continuously differentiable functions of α_1 and compute the derivatives $dY_1/d\alpha_1$, $dY_2/d\alpha_1, dB_2/d\alpha_1$. Thus we have

$$
\begin{aligned}
(1 - w_1 + m_1)(dY_1/d\alpha_1) - m_2(dY_2/d\alpha_1) &= 1, \\
-m_1(dY_1/d\alpha_1) + (1 - w_2 + m_2)(dY_2/d\alpha_1) &= 0, \\
-m_1(dY_1/d\alpha_1) + m_2(dY_2/d\alpha_1) + dB_2/d\alpha_1 &= 0,
\end{aligned}
\tag{9.52}
$$

from which

$$
\begin{aligned}
dY_1/d\alpha_1 &= (1 - w_2 + m_2)/|\mathbf{J}|, \\
dY_2/d\alpha_1 &= m_1/\mathbf{J}, \\
dB_2/d\alpha_1 &= [m_1(1 - w_2)]/|\mathbf{J}|.
\end{aligned}
\tag{9.53}
$$

These results, apart from the different symbology, are obviously the same as those obtained by the multiplier with foreign repercussions in the Appendix to Chap. 8, Eq. (8.51).

We now have all what we need to examine the question. Unfortunately, due to the presence of different terms in (9.48) and (9.53), in particular of the elasticities, it is only possible to compare the results on qualitative grounds.

A negative disturbance in country 1 ($d\alpha_1 < 0$) will cause a decrease in country 1s income under both fixed and flexible exchange rates [$dY_1 = (dY_1/d\alpha_1)d\alpha_1 < 0$ since $(dY_1/d\alpha_1) > 0$]. However, while under fixed exchange rates country 2s income will decrease as well, under flexible exchange rates it will *increase* due to the Laursen-Metzler effect. This reversal in the direction of the effect can hardly be considered as evidence for a lower transmission of the disturbance under flexible exchange rates than under fixed ones.

The negative disturbance under consideration will of course have an effect on the balance of payments under fixed exchange rates: country 2s balance will deteriorate. This is the cause of the result under flexible exchange rates, where the exchange rate will depreciate (from the point of view of country 2; it will appreciate from the point of view of country 1), hence the aggregate expenditure schedule of country 2 will rise owing to the Laursen-Metzler effect, and country 2s income will increase.

The Transfer Problem

Let us suppose that country 2 is the transferor. Introducing the appropriate shift parameters in Eq. (9.41) we have

$$Y_1 = \omega_1(Y_1, r) + (b_1' + h_1')T,$$
$$Y_2 = \omega_2(Y_2, \frac{1}{r}) - (b_2' + h_2')T, \qquad (9.54)$$
$$\frac{1}{r}M_1(Y_1, r) - M_2(Y_2, \frac{1}{r}) + (\mu_1' + \mu_2' - 1)T = 0,$$

or

$$Y_1 - \omega_1(Y_1, r) - (b_1' + h_1')T = 0,$$
$$Y_2 - \omega_2(Y_2, \frac{1}{r}) + (b_2' + h_2')T = 0, \qquad (9.55)$$
$$M_2(Y_2, \frac{1}{r}) - \frac{1}{r}M_1(Y_1, r) - (\mu_1' + \mu_2' - 1)T = 0.$$

The Jacobian determinant of Eq. (9.55) with respect to Y_1, Y_2, r [evaluated at the equilibrium point, where we can set $r = 1$ and use Eq. (9.43)] is

$$|\mathbf{J}| \equiv \begin{vmatrix} 1 - w_1 & 0 & -s_1 \\ 0 & 1 - w_2 & s_2 \\ -m_1 & m_2 & M_1(\eta_1 + \eta_2 - 1) \end{vmatrix}$$
$$= (1 - w_1)(1 - w_2)M_1(\eta_1 + \eta_2 - 1) - s_1 m_1 (1 - w_2)$$
$$- s_2 m_1 (1 - w_1). \qquad (9.56)$$

Since $|\mathbf{J}| > 0$ by the stability conditions—see Eq. (9.44)—we can express Y_1, Y_2, r as continuously differentiable functions of the parameter T and compute the derivatives dY_1/dT, dY_2/dT, dr/dT. Differentiating system (9.56) we have

$$
(1 - w_1)(dY_1/dT) - s_1(dr/dT) = (b_1' + h_1'),
$$
$$
(1 - w_2)(dY_2/dT) + s_2(dr/dT) = -(b_2' + h_2'),
$$
$$
-m_1(dY_1/dT) + m_2(dY_2/dT) + M_1(\eta_1 + \eta_2 - 1)(dr/dT) = (\mu_1' + \mu_2' - 1).
$$
$$(9.57)$$

The solution is

$$
dY_1/dT = N_{Y_1}/|\mathbf{J}|, \quad dY_2/dT = N_{Y_2}/|\mathbf{J}|, \quad dr/dT = N_r/|\mathbf{J}|,
\qquad (9.58)
$$

where

$$
N_{Y_1} \equiv
\begin{vmatrix}
(b_1' + h_1') & 0 & -s_1 \\
-(b_2' + h_2') & 1 - w_2 & s_2 \\
(\mu_1' + \mu_2' - 1) & m_2 & M_1(\eta_1 + \eta_2 - 1)
\end{vmatrix}
$$
$$
= (1 - w_2)[s_1(\mu_1' + \mu_2' - 1) + (b_1' + h_1')M_1(\eta_1 + \eta_2 - 1)]
$$
$$
+ m_2[s_1(b_2' + h_2') - s_2(b_1' + h_1')],
\qquad (9.59)
$$

$$
N_{Y_2} \equiv
\begin{vmatrix}
1 - w_1 & (b_1' + h_1') & -s_1 \\
0 & -(b_2' + h_2') & s_2 \\
-m_1 & (\mu_1' + \mu_2' - 1) & M_1(\eta_1 + \eta_2 - 1)
\end{vmatrix}
$$
$$
= (1 - w_1)[-(b_2' + h_2')M_1(\eta_1 + \eta_2 - 1) + s_2(\mu_1' + \mu_2' - 1)]
$$
$$
+ m_1[s_1(b_2' + h_2') - s_2(b_1' + h_1')],
\qquad (9.60)
$$

$$
N_r \equiv
\begin{vmatrix}
1 - w_1 & 0 & (b_1' + h_1') \\
0 & 1 - w_2 & -(b_2' + h_2') \\
-m_1 & m_2 & (\mu_1' + \mu_2' - 1)
\end{vmatrix}
$$
$$
= (1 - w_1)[(1 - w_2)(\mu_1' + \mu_2' - 1) + m_2(b_2' + h_2')]
$$
$$
- m_1[(1 - w_2)(b_1' + h_1')].
\qquad (9.61)
$$

Since $|\mathbf{J}| > 0$ by the stability conditions, the signs of the derivatives will depend on the signs of the numerators, which are ambiguous since both positive and negative terms. If however we are prepared to assume that the two countries are similar, we can take the various behavioural parameters as similar, namely $b_1' + h_1' \simeq b_2' + h_2'$,

$s_1 \simeq s_2$, etc. It follows that

$$N_{Y_1} \simeq (1 - w_2)[s_1(\mu'_1 + \mu'_2 - 1) + (b'_1 + h'_1)M_1(\eta_1 + \eta_2 - 1)] > 0,$$
$$N_{Y_2} \simeq (1 - w_1)[-(b'_2 + h'_2)M_1(\eta_1 + \eta_2 - 1) + s_2(\mu'_1 + \mu'_2 - 1)] < 0,$$
$$N_r \simeq (1 - w_1)(1 - w_2)(\mu'_1 + \mu'_2 - 1) < 0.$$

This means that the transfer will be undereffected by income changes (a reduction in the transferor's income and an increase in the transferee's income), so that the transferor's exchange rate $(1/r)$ will have to depreciate to effect the transfer.

References

Backus, D. K., Kehoe, P. J., & Kydland, F. E. (1994). Dynamics of trade balance and the terms of trade: The J-curve? *American Economic Review, 84*, 84–103.

Bahmani-Oskooee, M., & Brooks, T. J. (1999). Bilateral J-curve between U.S. and her trading partners. *Weltwirtschaftliches Archiv/ Review of World Economics, 135*, 156–165.

Bahmani-Oskooee, M., & Hegerty, S. W. (2010). The J- and S- curves: A survey of the recent literature. *Journal of Economic Studies, 37*, 580–596.

Gandolfo, G. (1980). *Economic dynamics: methods and models*. Amsterdam: North-Holland.

Gandolfo, G. (2009). *Economic dynamics* (4th ed.). Berlin, Heidelberg, New York: Springer.

Harberger, A. C. (1950). Currency depreciation, income and the balance of trade. *Journal of Political Economy, 58*, 47–60; reprinted In R. E. Caves & H. G. Johnson (Eds.) (1968). *Readings in international economics* (pp. 341–358). London: Allen & Unwin.

Jung, C., & Doroodian, K. (1998). The persistence of Japan's trade surplus. *International Economic Journal, 12*, 25–38.

Laursen S., & Metzler, L. A. (1950). Flexible exchange rates and the theory of employment. *Review of Economics and Statistics, 32*, 281–299; reprinted In L. A. Metzler (1973). *Collected papers* (pp. 275–307). Cambridge (Mass.): Harvard University Press.

Magee, S. P. (1973). Currency contracts, pass-through, and devaluation. *Brookings Papers on Economic Activity, 1*, 303–323.

Marwah, K., & Klein, L. R. (1996). Estimation of J-curves: United States and Canada. *Canadian Journal of Economics, 29*, 523–39.

NIESR (1968). The economic situation. The home economy. *National Institute Economic Review, 44*, 4–17.

Rose, A. K., & Yellen, Y. L. (1989). Is there a J-curve. *Journal of Monetary Economics, 24*, 53–68.

Senhadji, A. S. (1998). Dynamics of the trade balance and the terms of trade in LDCs: The S-curve. *Journal of International Economics, 46*, 105–131.

Stolper, W. F. (1950). The multiplier, flexible exchanges and international equilibrium. *Quarterly Journal of Economics, 64*, 559–582.

Wilson, W. T. (1993). J-curve effect and exchange rate pass-through: An empirical investigation of the United States. *International Trade Journal, 7*, 463–483.

Wood, G. (1991). Valuation effects, currency contract impacts and the J-curve: empirical estimate. *Australian Economic Papers, 30*, 148–163.

The Mundell-Fleming Model

10

10.1 Introductory Remarks

The analysis carried out in the previous chapters was concerned with the "real" side of the economy and the balance of payments, as only the market for goods and services and the relative international flows (the current account) were considered. The introduction of monetary equilibrium, interest rates, and international flows of capital was first carried out through the extension to an open economy of the closed-economy Keynesian model as synthesized by Hicks (1937) in the *IS–LM* analysis. This extension was accomplished by Mundell (1960, 1961a,b, 1963, 1968) and Fleming (1962) in the early 1960s. The two authors wrote independently, and although Mundell's contributions are far more numerous and important, the model is called the Mundell-Fleming model (Frenkel and Razin 1987; on "who was the first" see Boughton 2003 and Mundell 2001). We shall first consider the model under fixed and then under flexible rates; in both cases prices are assumed to be rigid.

The Mundell-Fleming model has important policy implications, which were treated together with the description of the model in the original writings of the two authors. For didactic purposes we shall separate the descriptive and normative aspects. The policy implications of the model will be considered in the next chapter.

It should be pointed out that the Mundell-Fleming model is really not a pure flow model, since adjustments in the money stock play a role, indirectly affecting the balance of payments through their effects on the interest rate and hence on capital movements and real output (see below, Sect. 10.2.2). However, the view of capital movements, which play a crucial role in the adjustment process, is still a pure flow view.

© Springer-Verlag Berlin Heidelberg 2016
G. Gandolfo, *International Finance and Open-Economy Macroeconomics*,
Springer Texts in Business and Economics, DOI 10.1007/978-3-662-49862-0_10

10.2   Fixed Exchange Rates

The model can be reduced to three equations, one which expresses the determination of national income in an open economy (equilibrium in the goods market or real equilibrium), one which expresses the equilibrium in the money market (monetary equilibrium) and the third which expresses balance-of-payments equilibrium (external balance). Since prices and exchange rate are fixed, both can be normalized to unity. This simply means that a suitable choice can be made of the units of measurement so that real and nominal magnitudes coincide. We also consider a one-country model.

Let us begin with real equilibrium. Exports are now exogenous by hypothesis (as the rate of exchange is fixed and repercussions are ignored), while aggregate demand depends on the interest rate and income. With regard to the dependence on income, there are no particular observations to be made with respect to what we already know from previous chapters. As far as the interest rate is concerned, a variation in it causes aggregate demand to vary in the opposite direction: an increase in the interest rate has a depressive influence on total demand (via a reduction in investment and presumably also consumption), while the opposite applies for a drop in the interest rate itself. We thus have the equation for real equilibrium

$$y = A(y, i) + x_0 - m(y, i), \tag{10.1}$$

where i denotes the interest rate. Since national expenditure or absorption ($A = C + I$) includes both domestic and foreign commodities, the introduction of the interest rate as an explanatory variable in A logically implies its introduction as an explanatory variable into the import function with the same qualitative effects (even if, obviously, these effects are quantitatively different). More precisely, if we take account of the fact that a higher interest rate involves a decrease in the demand for both domestic and foreign (imported) commodities, we conclude that $A_i - m_i < 0$, since $A_i - m_i$ is the effect of an interest rate increase on domestic demand for domestic commodities.

We then have the usual equation for monetary equilibrium

$$M = L(y, i), \tag{10.2}$$

where M indicates the money stock and L the demand for money.

The balance of payments includes not only imports and exports of goods, but also capital movements. Net capital flows (inflows less outflows) are expressed as an increasing function of the interest differential. It is in fact clear that the greater the domestic interest rate with respect to the foreign rate, the greater, *ceteris paribus*, will be the incentive to capital inflows and the less to outflows. The capital movements under examination are substantially short-term movements for covered interest arbitrage (see. Sect. 4.1; note that, as this is a one-country model, the foreign interest rate is exogenous, so that the movement of capital is ultimately a function of the domestic interest rate). We can therefore write the following equation for the equilibrium in the balance of payments

$$x_0 - m(y, i) + K(i) = 0, \qquad (10.3)$$

where $K(i) \gtrless 0$ indicates the net private capital inflow (outflow). Let us note, in passing, that the condition that the overall balance of payments is in equilibrium is equivalent to the condition that the stock of international reserves is stationary, as shown in Sect. 6.4.

The system, composed of the three equations studied so far, is determined if the unknowns are also three in number: it is therefore necessary to consider M also as an unknown, in addition to y and i. We shall discuss this fact at length in Sect. 10.2.2.

It is convenient at this point to pass to a graph of the equilibrium conditions, which will be of considerable help in the subsequent analysis.

10.2.1 Graphic Representation of the Equilibrium Conditions

If we plot in the (y, i) plane all the combinations of the interest rate and income which ensure real equilibrium, we obtain a curve (which as usual for simplicity we shall assume to be linear: this is also true for curves, which we shall come across later) which corresponds, in an open economy, to the IS schedule for a closed economy. Mundell called it the XX schedule, but it is now standard practice to call it IS as in the closed economy (Fig. 10.1).

The characteristic of this curve, which as we have just seen, is a locus of equilibrium points, is that it is downward sloping. In fact, if income is higher, aggregate expenditure and imports are likewise higher, but by a smaller amount. For this to be true, it is not necessary that the marginal propensity to aggregate spending $(\Delta A / \Delta y)$ is smaller than one, as it should be in the closed economy: the net change is, in fact, $(\Delta A / \Delta y - \Delta m / \Delta y)$, which can be smaller than one even

Fig. 10.1 Mundell-Fleming under fixed exchange rates: the real equilibrium schedule

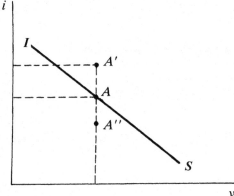

if $\Delta A/\Delta y > 1$. Better to clarify this important point we recall from Chap. 8 that $C + I - m$ is *domestic expenditure* (or domestic demand), namely expenditure on (demand for) domestic output (i.e., domestically produced goods) by residents, so that—see Eq. (8.19)—the difference between the marginal propensity to aggregate spending and the marginal propensity to import is simply the *marginal propensity to domestic expenditure (demand)*, namely the marginal propensity to spend on domestic output by residents, which can safely be considered as smaller than one.

Given this, it is necessary to have a lower value for the rate of interest (so that there will be a further increase in domestic expenditure) to maintain real equilibrium.

Furthermore, the *IS* curve has the property that at all points above it there will be a negative excess demand, while at all points below it the excess demand for goods will be positive.

Consider in fact any point A' above the *IS* curve. Here the rate of interest is greater than at point A, while income is the same. Point A is a point of real equilibrium, being on the *IS* schedule. If, income being equal, the rate of interest is greater, domestic demand will be less, so that at A' there will be *negative* excess demand (excess supply). In the same way, it can be demonstrated that at A'' there is *positive* excess demand.

Let us now examine the monetary equilibrium schedule. Given that the demand for money is an (increasing) function of income and a (decreasing) function of the interest rate, there will be a locus of the combinations of these two variables which make the total demand for money equal to the supply, which is represented by the familiar schedule, *LM* in Fig. 10.2. Mundell called it *LL*, but we shall follow standard practice in calling it *LM*.

The *LM* curve is increasing; in fact, given a certain supply of money, if income is higher the demand for money will also be higher: in consequence, it is necessary to have a higher value for the interest rate, so as to reduce the demand for money itself, to maintain the equality between demand and supply. Furthermore the *LM* schedule has the property that, at all points above it, there is negative excess demand for money; while at all points below it, there is positive excess demand. Consider for example any point above *LM*, for example, A'; there the rate of interest is higher, income being equal, than at point A on *LM*. At A', therefore, there is a lower demand

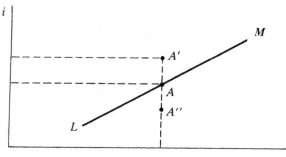

Fig. 10.2 Mundell-Fleming under fixed exchange rates: the monetary equilibrium schedule

Fig. 10.3 Shifts in the monetary equilibrium schedule

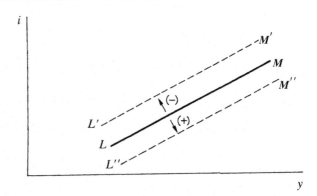

for money than at A. Since at A the demand for money equals the supply and given that at A' the demand for money is lower than at A, it follows that at A' the demand for money is less than the supply. In the same way, it is possible to show that at any point below the LM schedule (for example at A'') the demand for money is greater than the supply.

The LM schedule undergoes shifts as the supply of money varies and, to be precise, it moves to the left (for example to $L'M'$ in Fig. 10.3) if the supply of money is reduced, and to the right (for example to $L''M''$) if the supply of money increases.

In fact, to each given value of income there must correspond a lower rate of interest, if the money supply is higher, so that the greater demand for money will absorb the greater supply so as to maintain equilibrium between demand and supply of money; in consequence, LM must shift downwards and to the right. Similarly, it is possible to demonstrate that LM shifts upwards and to the left if the supply of money is reduced. Thus there is a very precise position for the LM schedule for each level of money supply in the diagram.

The new schedule to be considered in the open-economy extension of the IS-LM model is the external equilibrium schedule. Equilibrium of the balance of payments occurs when the algebraic sum of the current account balance and the capital movements balance is nil. As exports have been assumed exogenous and imports a function of income and the interest rate, and the capital movements balance as a function of the interest rate, it is possible to show in a diagram all the combinations of the interest rate and income which ensure balance-of-payments equilibrium, thus obtaining the BB schedule (Mundell called it the FF schedule, but there is no standardized denomination: BB, BP, FX, NX are all used in the literature).

This schedule is upward sloping: in fact, as exports are given, greater imports correspond to greater income and therefore it is necessary to have a higher interest rate (which tends on the one hand to put a brake on the increase in imports and on the other to improve the capital account) in order to maintain balance-of-payments equilibrium.

We observe that the slope of BB depends, *ceteris paribus*, on the responsiveness of capital flows to the interest rate, i.e. on the degree of international capital mobility.

Fig. 10.4 Mundell-Fleming
under fixed exchange rates:
the external-equilibrium
schedule

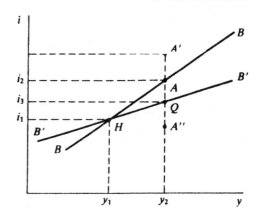

The *higher the mobility of capital, the flatter the BB schedule*. In Fig. 10.4, the
degree of capital mobility underlying schedule $B'B'$ is higher than that underlying
schedule BB. In fact, if we consider for example the external equilibrium point H,
a higher income (for example, y_2 instead of y_1) will mean a balance-of-payments
deficit. This requires—as we have just seen—a higher value of the interest rate to
maintain external equilibrium. Now, the more reactive capital flows to the interest
rate, the lower the required interest rate increase. Given y_2 the interest rate will have
to be i_2 in the case of BB, and only i_3 in the case of $B'B'$. In the limit, namely in the
case of perfect capital mobility, the BB schedule will become a *horizontal* straight
line parallel to the y axis. This case has important policy implications and will be
examined in the next chapter. On the contrary, in the case of insensitivity of capital
movements to the interest rate the BB schedule would become much steeper, in that
the sole effect of the interest rate on the balance of payments would be the direct one
of this rate on imports. Then, if this effect were negligible, the BB schedule would
become a *vertical* line parallel to the i axis with an intercept on the y axis equal to
the only value of income which determines a value for imports equal to the sum
of the value of (exogenous) exports plus the capital movement balance (now also
exogenous). It is clear that in this case we are back at the simple multiplier analysis
dealt with in Chap. 8.

Furthermore, the BB schedule has the property that at all points above it there is
a surplus in the balance of payments, while at all points below it there is a deficit. In
fact, consider any point A' above the line BB. At A', while income is the same, the
interest rate is greater than at A, where the balance of payments is in equilibrium.
Thus, as imports are a decreasing function of the interest rate and as the capital
account balance is an increasing function of the rate itself, at A' imports will be
lower and the capital account balance will be higher—*ceteris paribus*—than at A,
so that if at A the balance of payments is in equilibrium, at A' there must be a surplus.

In the same way, it is possible to show that at all points below BB (see, for
example, point A'') there will be a balance-of-payments deficit.

Finally, it is important to take note of another property of the *BB* schedule. As $K(i)$, the capital account balance, is an increasing function of the interest rate, it will be in surplus for "high" values of i (generally for $i > i_f$), in deficit for "low" values of i ($i < i_f$) and zero for $i = i_f$. On the other hand, the current account will show a surplus for low values of y, for those values, that is, which give rise to lower imports at the given exogenous exports, a deficit for "high" values of y and zero for the value of y at which imports equal exports.[1] There will exist, therefore, only one combination of i and y that gives rise to a *full* equilibrium in the balance of payments, taking "full" equilibrium to mean a situation in which *both* the current account balance *and* the capital account balance are nil. Let us assume that this combination is the one which occurs at point H in Fig. 10.4. It follows from what we have seen above that at all points along *BB* to the right of H there will be a deficit in the current account, exactly matched by net capital inflows (so that the overall balance of payments will be in equilibrium), and that, on the other hand, at all points along *BB* to the left of H, equilibrium in the balance of payments will come about with a surplus in the current account matched by an equal net outflow of capital. These considerations will be useful when one wishes to take into account not only the overall balance of payments, but also its structure.

10.2.2 Simultaneous Real, Monetary and External Equilibrium: Stability

The three schedules *IS*, *BB*, and *LM* so far separately examined, can now be brought together in a single diagram. Given three straight lines, they will not intersect at the same point except by chance. Consider first of all the *IS* and *BB* schedules (Fig. 10.5). They intersect at a point E, where real equilibrium and balance-of-payments equilibrium are simultaneously established with values y_E for income and i_E for the interest rate. Now consider schedule *LM*.

There are two possibilities:

(i) if the quantity of money is considered as given, it is altogether exceptional for the corresponding *LM* schedule to pass through point E. And if the *LM* schedule does not pass through E, it follows that monetary equilibrium does not correspond to real and balance-of-payments equilibrium;

(ii) if, on the other hand, the quantity of money is considered to be variable, then, in principle, it is always possible to find a value for the quantity of money itself such that the corresponding *LM* schedule will also pass through point E. In this case—see Fig. 10.5b—we have the simultaneous occurrence of real, balance of payments, and monetary equilibrium.

---

[1] As imports also depend on i, this value will be calculated by holding i at the value i_f.

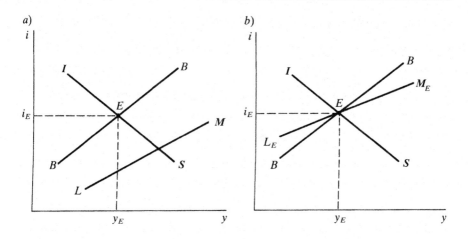

Fig. 10.5 Mundell-Fleming under fixed exchange rates: macroeconomic equilibrium

But, one might ask, do any forces exist which tend to cause the necessary shifts in the *LM* schedule? The answer to this question cannot be given in isolation, but requires a general analysis of the dynamics of the disequilibria in the system, that is, of the behaviour of the system itself when one or more of the equilibrium conditions are not satisfied. For precisely this purpose it is necessary to make certain assumptions regarding the dynamic behaviour of the relevant variables: money supply, income and interest rate. The assumptions are as follows:

(a) the money supply varies in relation to the surplus or deficit in the balance of payments and, precisely, increases (decreases) if there is a surplus (deficit). This assumption—which is identical to the one made in the analysis of the classical theory of the adjustment mechanism of the balance of payments and in the monetary approach to the balance of payments (see Chap. 12)—implies that the monetary authorities do not intervene to sterilize (i.e., to offset) the variations in the quantity of money determined by disequilibria in the balance of payments: in fact, given Eq. (6.21), we have $\Delta M = B$ when the other items are set to zero.

This might give rise to some doubts about the plausibility of the hypothesis itself. In fact, if it is acceptable in a national and international monetary system that is based on the pure gold standard, it appears to be far less acceptable in the present-day institutional context, where the monetary authorities of a country are unlikely to allow the quantity of money to freely vary in relation to the state of the balance of payments. It is as well to warn the reader here that the hypothesis in question will be dropped in the next chapter, which is dedicated to the analysis of monetary and fiscal policy in an open economy. However, the validity of the analysis carried out in the present section lies in the opportunity it offers to study what might be termed the spontaneous behaviour of the system before coming to grips with the problems connected with the use of monetary and fiscal policies to produce external and internal equilibrium in the system itself .

(b) Income varies in relation to the excess demand for goods and, to be exact, it increases (decreases) according to whether this excess demand is positive (negative). This is the same assumption made on more than one occasion (see, for example, Sect. 9.1.2).

(c) The rate of interest varies in relation to the excess demand for money and, more precisely, it increases (decreases) if this excess demand is positive (negative). This is a plausible hypothesis within the context of an analysis of the spontaneous behaviour of the system. In fact, if the interest rate is, in a broad sense, the price of liquidity, an excess demand for liquidity causes an increase on the market in this price and vice versa. The mechanism, commonly described in textbooks, is the following: an excess demand for money—that is, a scarcity of liquidity—induces holders of bonds to offer them in exchange for money: this causes a fall in the price of bonds, and thus an increase in the interest rate (which is inversely related to the price of bonds).

Having made these behavioural hypotheses, it will be seen that the system is stable and will therefore tend to eliminate disequilibria, that is to say, it will tend to reach the point of simultaneous real, monetary and balance-of-payments equilibrium, *if the marginal propensity to domestic expenditure is less than one* (as already previously assumed) and *if, in addition, the marginal propensity to import is below a certain critical value.* It is as well to point out that the condition concerning the marginal propensity to domestic expenditure is only a sufficient condition and not a necessary one, so that the simultaneous equilibrium may still be stable even if the above-mentioned condition is not satisfied. Note the difference from the result obtained in the analysis based only on the international multiplier (see Chap. 8), where the condition under examination besides being sufficient, is also necessary, so that if it does not occur there will inevitably be instability. The reason for this difference lies substantially in the fact that the introduction of the variability in the interest rate and the related effects confers a greater flexibility on the system. The general, necessary and sufficient conditions for stability are analysed in the Appendix.

A simple case of disequilibrium and the related adjustment process can be analysed intuitively on the basis of Fig. 10.6. Assume, for example, that the system is initially at point A. At that point there is real and monetary equilibrium, but not equilibrium of the balance of payments: more exactly, as point A lies below the BB schedule, there is a deficit. Point A is thus a partial or temporary point of equilibrium.

It is necessary here to distinguish two possible cases: if the monetary authorities were to intervene in order to *sterilize* the payments imbalances, *the money supply would remain constant* and the economic system would remain at A; naturally, *a reduction in the stock of international reserves* would correspond to the continued balance-of-payments deficit (except in the case of a country with a reserve currency). This situation could not be sustained indefinitely, since the authorities

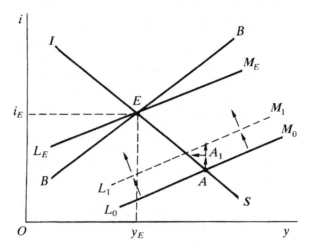

Fig. 10.6 Mundell-Fleming under fixed exchange rates: dynamic analysis of the adjustment process

would eventually run out of international reserves. However, we have assumed—hypothesis (a)—that an intervention of this kind would not take place, so that the supply of money diminishes and the *LM* schedule moves upwards and to the left, for example from L_0M_0 to L_1M_1. Then at A there is a (positive) excess demand for money, so that—hypothesis (c)—the interest rate increases. The increase in the interest rate causes a fall in the demand for goods and thus a (negative) excess demand in the real market, which is confirmed by the fact that point A_1, which is reached from A following the increase in i (vertical arrow from A towards A_1), is above *IS*. Given, as we said, that at A_1 there is negative excess demand on the real market, by hypothesis (b) income falls (horizontal arrow from A_1 towards the left). At A_1, on the other hand, we are still below *BB* and therefore there is still a deficit in the balance of payments; consequently the money supply falls still further and *LM* continues to shift upwards and to the left, so that at A_1 a situation of positive excess demand for money remains, with a further tendency for the interest rate to increase (vertical arrow rising above A_1). We thus have a situation in which y and i approach their respective equilibrium values and also a shift of *LM* toward the position L_EM_E.

Note that a graphical analysis of stability cannot be made simply with a scheme with two arrows of the kind adopted in Sect. 9.1.2 (see Fig. 9.3). In that diagram, in fact, the schedules which represent the various equilibria *do not move*; in our case, on the contrary, the disequilibrium also causes one of the schedules (that of monetary equilibrium) to shift. In effect, in our model, there are *three* variables which adjust themselves: income, interest rate and the quantity of money, so that the two-dimensional graph of this adjustment is necessarily inadequate, differently from the case in which there are only two variables to be adjusted.

We can now ask ourselves what the economic meaning of the conditions of stability might be. As far as concerns the condition that the marginal propensity to domestic expenditure is less than one, the meaning is the usual one: an increase in income, due to a positive excess demand, causes a further increase in domestic demand, but the process is certainly convergent, if the increases in demand are below the increases in income, that is to say, if the marginal propensity to domestic expenditure is less than one. In the opposite case, the process could be divergent, unless other conditions of stability intervene.

With regard to the condition that concerns the marginal propensity to import, the meaning is as follows: if this propensity is too great, it may happen, in the course of the adjustment process, that the reduction in imports (induced by the reduction in y and the increase in i) is such as to bring the balance of payments into surplus (that is to say, point E is passed). An adjustment in the opposite direction is then set into motion: the money supply increases, the rate of interest drops, domestic demand, income and imports all increase (both because y has increased and i decreased). And if the marginal propensity to import is too high, then it may be that the increase in imports is such as to produce a new deficit in the balance of payments. At this point, a new process comes into being, working in the opposite direction and so on, with continual fluctuations around the point of equilibrium which each time may take the system further away from equilibrium itself. In effect, as is shown in the Appendix, the condition that the marginal propensity to import should be less than a critical value serves precisely to exclude the possibility that divergent fluctuations take place. Note that, while it is possible to discover from the graph whether the marginal propensity to domestic expenditure is less than one (a negative slope in *IS*), it is not, on the other hand, possible to ascertain whether the condition related to the marginal propensity to import is satisfied or not, as this condition has no counterpart in the graph in terms of slope and positions of the various schedules (all that one can say is that the slope of the *BB* schedule must not rise above a certain maximum, but this cannot be determined unless the magnitudes of all parameters that appear in the model are known).

10.2.2.1   Observations and Qualifications

It is important to note at this point that the adjustment process described in the previous section is an *automatic* process, ultimately set off by movements in the money stock generated by the balance-of-payments disequilibria. This is a result similar, even if for completely different reasons, to that found in the monetary approach to the balance of payments (see Chap. 12).

Another important point to note is the substantially short-term nature of the analysis. In fact, it is very unlikely that the point of equilibrium E in Fig. 10.6 will coincide with the point of *full* equilibrium of the balance of payments (point H, for which see Fig. 10.4). Normally point E will be to the right or to the left of point H. Assuming, by way of example, that it is to the right, then this means that at E there will be a deficit in the current account matched by an inflow of capital. This is perfectly normal and acceptable in the short term, but in the long one it will be necessary to take into account certain considerations which till now have been ignored. As we are here dealing with *flows*, the continuance of the situation of equilibrium means that, other things being equal, there will be a constant inflow of capital *per unit of time*. This in turn means that the stock of debt to foreign countries is continually growing and so, given the interest rate, the burden of interest payments to the rest of the world will also continue to grow. We know (see Sect. 5.1.3) that these payments go into the services account (under the item "investment income") and therefore feed the deficit in the current account beyond that accounted for in Fig. 10.6. The only way to avoid this problem is to suppose that the owners of the foreign capital that has flowed in should not repatriate the interest due to them, but should capitalize it, leaving it to increment their capital, given the advantage of the favourable interest differential. In the opposite case, in fact, the payment abroad for investment income would cause the form and the position of the *BB* schedule to change. As far as the *form* is concerned, let us consider a given moment in time, in which therefore the stock of debt to foreign countries is given. A variation in *i* thus causes the burden of interest payments on this debt to vary in the same direction. This implies the hypothesis that the new rate of interest will be applied to old debts as well as new ones. This is an entirely plausible assumption, in that the movements of capital considered are short-term ones and so are not incorporated in fixed-income securities (in which case the coupon remains unchanged and the variations in *i* have repercussions on the market value of the securities), but in other financial instruments, for example, bank deposits (in which case the new interest rate starts to run immediately on the entire amount of the capital deposited) or in bonds with variable coupons. Intermediate cases are of course possible, but these have been ignored for the sake of simplicity.

From this the possibility follows that by increasing *i* beyond a certain limit, the *burden* of interest payments will exceed the inflow of *new* capital, in which case the *BB* schedule will bend backwards: in fact, increases in *i* beyond a certain critical limit (i_c) will on the whole cause a *worsening* in the balance of payments, hence the need for lower values of *y* to reduce *m*.

The *BB* schedule *may* therefore assume the form described in Fig. 10.7. The stress is put on *may* to emphasize yet again that we are talking about a possibility: in fact, if the movements of capital are very elastic with respect to *i*, the burden of interest payments will always remain below the inflow of new capital and *BB* will have its normal upward slope. In the case considered in Fig. 10.7 there will be two points of intersection[2] with the *IS* schedule: points *E* and *E'*. With the reasoning followed above it can be ascertained that point *E* is a stable equilibrium (it has in fact the same properties as *E* in Fig. 10.5), while point *E'* is an unstable equilibrium: any shift from *E'* sets into motion forces which cause further movement away. In fact, a downward shift from *E'* sets off the forces of attraction towards point *E* already described. An upward shift, for example to *A*, generates further upward movements. At *A* in fact there is a balance-of-payments deficit: note that in the upper branch of the *BB* curve, the opposite of the rule encountered in Fig. 10.4 is valid (though the rule remains valid for the lower branch), because now greater values of *i* have an

---

[2]We leave aside any particular cases, such as the absence of intersection, the tangency, etc., of the *IS* with the *BB* schedule.

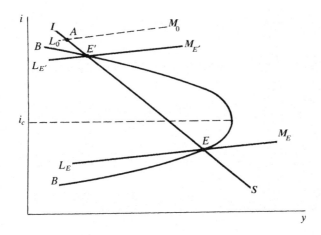

Fig. 10.7 The burden of interest payments and the BB schedule

unfavourable effect on the balance of payments. The deficit causes a reduction in M and therefore a shift of LM from L_0M_0 upwards and to the left, and so on; so that point A tends to move further and further away from E'.

But there is more to it even than this. In fact, let us consider the *position* of the BB curve and, so as to simplify to the maximum, let us assume that the eventuality described in Fig. 10.7 does not come about, so that the BB curve has a normal upward slope. Now, over time, the stock of debt increases and thus the interest payments abroad also increase. This means—with reference to Fig. 10.6—that BB shifts upwards and to the left, as at each given interest rate the growing deficit due to these interest payments can only be offset by lower imports and therefore smaller y. It follows that point E in Fig. 10.6 would gradually shift to the left along the IS schedule and the entire economic system would follow a trajectory of ever-increasing values of i accompanied by ever-decreasing values of y, which would clearly produce a situation that would be intolerable in the long term. But the situation in which the interest is not repatriated, discussed above, cannot be calmly accepted in the long term either. Growing foreign debt, in fact, exposes the country to the risk of insolvency if foreign investors decide to withdraw their funds (interest and capital) or even if they merely cease to allow new funds to flow in. This last eventuality is only too likely if it is believed, on the basis of the theory of portfolio choice, that capital flows induced by a *given* differential in interest are *limited*.

This is the same as considering that the flows under discussion are not pure flows, but ones deriving from stock adjustments. If they were pure flows, in fact, they would remain constant in time as long as a given interest differential held, other things being equal. In the opposite case, as at each given constellation of yields there corresponds an optimum distribution of the stock of funds between various assets in the various countries, once the holders of funds have realized this optimum distribution, the movements of capital cease, even if the interest differential remains. The reason is that the optimum distribution never consists—apart from exceptional cases—in placing the entire stock of funds in a single country even if the yield there is higher (the principle of diversification). Thus the capital inflows could be reactivated only as a result of an increment in the interest differential, that is, an increase in i, as, by hypothesis, i_f is given. In this case the situation already described previously (increasing values of i and decreasing ones for y) again presents itself in a different form.

From all that has been said above, the nature of *short-term equilibrium* of point E is clarified (in accordance with the model of a closed economy from which the basic scheme derives).

Two additional extensions deserve mentioning. The first concerns prices, that have so far been assumed constant. But flexible prices can be easily introduced. As the price level changes over time, the real money supply changes, giving rise to shifts in the *LM* curve. At the same time, the real exchange rate ($r_R = \frac{p_h}{r p_f}$) changes, pushing up or reducing net exports and affecting income. Let us remember that the real exchange rate can move even in a fixed exchange rate regime—the central bank fixes the nominal exchange rate, not the real one. Even if the nominal exchange rate remains unchanged, if prices change, the real exchange rate will change too. The analysis is very similar to that in the basic *IS–LM–BB* model.

The second concerns the country risk. If we allow for *imperfect substitutability* between domestic and foreign assets, an interest-rate differential should be taken into account, $i_h = i_f + \delta$, where δ is the risk premium. A reason for this is the presence of a country risk (for example, political or default risk, risk of depreciation), so borrowers in high-risk countries have to pay higher interest rates to compensate for the greater risk.

What is the effect of an increase in δ?. The *IS* curve shifts back because a higher interest rate reduces investment activity and planned expenditures. The *LM* curve shifts outwards because a higher interest rate reduces demand for money balances, which means that the level of income that satisfies equilibrium in the money market is higher. When δ rises because of a higher country risk, we would also expect this country's currency to depreciate. The model predicts that income should increase as the risk premium rises (since the currency depreciation pushes up net exports). This result is not supported very well in the data as, typically, countries experiencing increases in δ experience periods of recession, not booms.

10.2.3 Comparative Statics

The shifting of the point of equilibrium—and hence the variations in income and interest rate—in consequence of exogenous variations can be analysed graphically by examining the shifts of the various schedules *IS*, *BB*, *LM*. As an example we shall examine the transfer problem and the effects of a devaluation of the exchange rate in a regime of fixed but adjustable exchange rates.

Fig. 10.8 The transfer problem under fixed exchange rates

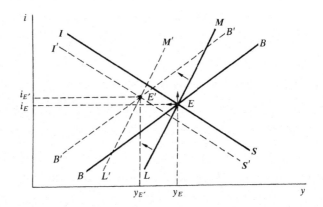

10.2.3.1 The Transfer Problem

As we said in Chap. 8, Sect. 8.6.3, the transfer problem can be examined as a comparative statics problem in any model where exogenous components of expenditure are present. Let us then examine the transfer problem in the present model, starting from a position of initial equilibrium and assuming that stability prevails. It goes without saying that, as this is a flow model, the implicit assumption is that the transfer is itself a flow; besides, we are in the context of a small country model.

The point of departure consists in finding out the initial impact of the transfer on the balance of payments and income of the transferor, so as to determine the shifts in the various schedules and the new equilibrium point. For this purpose we can accept what was said in relation to the Keynesian transfer theory at points (1) and (2) in Sect. 8.6.2: indeed, the idea that the transfer is associated with changes in the autonomous components of the expenditure functions of the two countries has a general validity which goes beyond the Keynesian transfer theory. In the symbology introduced in Chap. 8, Eq. (14.3), the initial impact of the transfer on the balance of payments is given by

$$\Delta B = -T - \Delta m_{01} = (\mu'_1 - 1)T, \qquad (10.4)$$

where the term Δm_{02} has been neglected given the small-country assumption. This expression can be either positive or negative. We shall assume that it is negative, as it is implausible that $\mu'_1 > 1$. Thus the BB schedule will shift to the left. The impact effect on the real-equilibrium schedule will be given by

$$\Delta RR = \Delta A_{01} + \Delta B = \Delta C_{01} + \Delta I_{01} + \Delta B = (\mu'_1 - 1 - b'_1 - h'_1)T, \qquad (10.5)$$

which is clearly negative.

The situation can be easily analysed by a diagram.Consider Fig. 10.8, where E is the initial equilibrium point. Both BB and IS shift to the left, for example to $B'B'$ and $R'R'$. At E we have a balance-of-payments deficit and so a decrease in the supply of money (the *LM* schedule shifts to the left), so that the resulting excess demand for money causes the interest rate to increase (upward-pointing arrow from *E*). At *E* there is also an excess supply of goods and so income tends to decrease (arrow pointing leftwards from *E*).

The reader will have recognized the dynamic adjustment process described in Sect. 10.2.2. We assume that the stability conditions mentioned there occur, so that the system will converge to its new equilibrium point *E'*; the process of transition from the old to the new equilibrium is ensured by the stability conditions.

At the new equilibrium point *E'* income will be lower and the interest rate higher than at the initial equilibrium point *E*.[3] But we are not so much interested in this as in the fact that at *E'* balance-of-payments equilibrium prevails: the transfer is *effected*. This will always be the outcome in the context of the present model, contrary to the Keynesian transfer theory (see Chap. 8, Sect. 8.6.2).

Thus we find the confirmation of what we said in Chap. 8, Sect. 8.6.3, that it is not possible to state *general* rules or formulae as regards the effectuation of the transfer since the result for the same initial situation will depend on the type model being utilized to analyse the transfer problem.

10.2.3.2 Exchange-Rate Devaluation

Consider now a (*once-and-for-all*) *devaluation* of the exchange rate in an adjustable peg regime. The devaluation causes an increase in exports, so that the *IS* schedule shifts to the right. The new equilibrium would *ceteris paribus* occur at *E'*. However, the devaluation also causes an improvement in the balance of payments (assuming that the condition of the critical elasticities is satisfied) and thus a rightward shift of the *BB* curve: in fact, at each given level of the interest rate income can be greater, given that the consequent increase in imports is compensated by the exogenous increase in exports.

At the new point of equilibrium *E''* income is greater; in Fig. 10.9, it can also be seen that at *E''* the increase in income is greater than it would be (with a similar shift in the *IS* schedule) in the case of an exogenous increase in domestic expenditure (point *E'*): this is due to the shift in *BB*.

The variation in the interest rate can be in any direction. In fact, in the face of an increase in imports (as a consequence of the increase in income), there is now an increase in exports, so that the effect on the balance of payments can be of any kind, according to whether the increase in imports is greater or less than the initial exogenous increase in exports: consequently, the variations in the interest rate required to put the balance of payments once more into equilibrium can be either in one direction or the other. Figure 10.9 represents the case in which the interest rate increases; in the opposite case the two schedules *R'R'* and *BB* would be such as to intersect at a point which, though still to the right of *E*, would be below rather than

---

[3]While income is necessarily lower, the interest rate is not necessarily higher, as it may be lower or higher depending on the magnitude of the marginal propensity to aggregate spending. The general case will be examined in the Appendix.

Fig. 10.9 Effects of a once-and-for-all devaluation

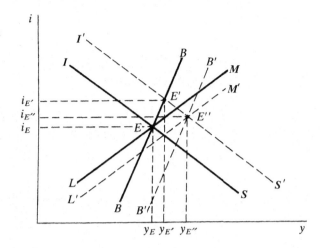

above E. However, note that, even in the most unfavourable case, in which there is a worsening in the balance of payments, this will certainly be less than would have occurred in the case of an exogenous increase in domestic expenditure (see above), so that any increase in the interest rate required to bring the balance of payments back into equilibrium is less than in the above-mentioned case; consequently, the braking effect on domestic expenditure is less and the final increase in income is more, as stated above.

As far as the variation in the quantity of money is concerned, there are two possibilities. If, at the new point of equilibrium, the interest rate is higher, then—following the same reasoning as before—there can be either an increase or a decrease in the quantity of money (Fig. 10.9 shows, by way of example, the case in which LM must shift to the right). If, on the other hand, at the new point of equilibrium, the interest rate is lower, then there is no doubt that the quantity of money must increase. In fact, the greater income and the lower interest rate involve a greater demand for money, and therefore the supply of money must also increase to maintain monetary equilibrium.

In the case examined, the shift in income from y_E to $y_{E'}$ or $y_{E''}$ is the result of the product of the exogenous increase in exports by the appropriate international multiplier. The explicit formula for this multiplier (which differs from the common international multiplier deduced from an analysis that takes account exclusively of real equilibrium) will be presented in the Appendix. What interests us here is to note that the restoration of real equilibrium corresponds likewise to a restoration of balance-of-payments equilibrium (besides monetary equilibrium). Therefore, a *once-and-for-all devaluation has only transitory effects on the balance of payments*: a result analogous even if for different reasons to that which we shall find in the monetary approach to the balance of payments (see Chap. 12).

Note that the process of transition from the old equilibrium E to the new one E' (or E'') is assured by the same dynamic conditions examined in Sect. 10.2.2. In fact, once the various schedules have shifted, the new point of equilibrium is no longer E

(which now becomes a disequilibrium point) but E' or E'', towards which the system is driven by the dynamic forces previously described.

10.3 Flexible Exchange Rates

The model (10.1)–(10.3) can easily be extended to flexible exchange rates (prices are, however, still assumed to be rigid and normalized to unity). We have

$$
\begin{aligned}
y &= A(y, i) + x(r) - rm(y, i, r), \\
M^* &= L(y, i), \\
B &= x(r) - rm(y, i, r) + K(i) = 0.
\end{aligned}
\tag{10.6}
$$

The money supply is indicated with an asterisk because, unlike under fixed exchange rates, it must now be considered as given *in a static context*. In fact, whilst under fixed exchange rates the basic three-equation system would be overdetermined if M were considered given, as there would be only two unknowns (y and i: this is case (a) of Sect. 10.2.2, represented in Fig. 10.5a), now—under flexible exchange rates—there are three basic unknowns (y, i, r), so that the system would be underdetermined if M were considered as an unknown. On the contrary, *in a dynamic context* it is possible to consider M as an endogenous variable as well (see below).

Unfortunately it is not possible to give a simple graphic representation like that used in the case of fixed exchange rates. As a matter of fact, if we take up the *IS* and *BB* schedules again, we see that a different position of these schedules in the (y, i) plane corresponds to each different exchange rate. This position will in turn depend on the critical elasticities condition. Take in fact the *BB* schedule: if the critical elasticities condition is satisfied, we find that higher (lower) values of r imply a balance-of-payments surplus (deficit); hence at any given i, higher (lower) values of y are required, so that the higher (lower) value of m exactly offsets the surplus (deficit). This means that the position of the *BB* schedule will be found more to the right (left) as the exchange rate is higher (lower). The opposite will be true if the critical elasticities condition is not satisfied.

As regards the *IS* schedule, if the critical elasticities condition is satisfied, greater (lower) values of the aggregate demand $[A + (x - rm)]$ correspond to higher (lower) values of r, so that at any given y the interest rate i will have to be higher (lower) to keep total demand at the same value as before. This means that the position of the *IS* schedule will be found further to the right (left) as the exchange rate is higher (lower).

Fortunately the *LM* schedule does not shift as it does not directly depend on the exchange rate. In Fig. 10.10 we have represented the case in which the critical elasticities condition is satisfied.

From the above description it follows that the *BB* and *IS* schedules will shift continuously as r moves. Furthermore, insofar as balance-of-payments disequilibria

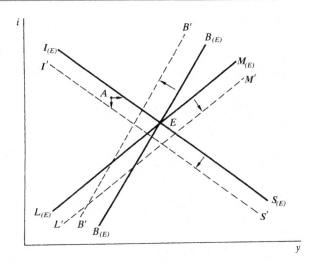

Fig. 10.10 The Mundell-Fleming model under flexible exchange rates

occur, the money supply will change and the *LM* schedule will shift as well, as shown in Sect. 10.2.1.

To visualize the complications of the problem, let us assume we know the equilibrium value of r from the mathematical solution of system (10.6), let it be r_E. We then draw the *IS* and *BB* schedules corresponding to r_E (denoted by $I_{(E)}S_{(E)}$ and $B_{(E)}B_{(E)}$ in Fig. 10.10); the intersection of these determines the equilibrium values of y and i, which are of course the same as those obtained from the mathematical solution of the system. To examine the dynamics of the adjustment process let us assume that the system is initially in equilibrium (point E) and that an accidental disturbance brings it to point A. We must now introduce suitable dynamic behaviour assumptions, which are:

(a) income varies in relation to the excess demand for goods and, to be exact, it increases (decreases) according to whether this excess demand is positive (negative);
(b) the rate of interest varies in relation to the excess demand for money and, more precisely, it increases (decreases) if this excess demand is positive (negative);
(c) the exchange rate varies in relation to the payments imbalance and, to be precise, it increases (decreases) if there is a deficit (surplus);
(d) as regards the money supply, we must distinguish two cases. In the first, the exchange-rate variations described in (c) cannot *instantaneously* maintain the balance of payments in equilibrium. This means that there will be balance-of-payments disequilibria, which will cause changes in the money supply, with consequent shifts in the *LM* schedule. This is what we meant when we said that in a dynamic context it is possible to consider M also as a variable. In the second case, the exchange-rate variations do instead *instantaneously* maintain the balance of payments in equilibrium. This means that there is no effect of the

balance of payments on the money supply, which remains constant (insofar as there are no other causes of variation); hence, the *LM* schedule does not shift.

Assumptions (a), (b), (d) (first case) are the same as those adopted in Sect. 10.2.2, to which we refer the reader. As regards assumption (c), it coincides with that adopted in Sect. 9.1.2, which the reader is referred to. We only add that this assumption can be considered valid not only in the context of freely flexible exchange rates (in which case the cause of exchange-rate variations is to be seen in market forces set into motion by the excess demand for foreign exchange), but also in the context of a managed float if we assume that the monetary authorities manage the exchange rate in relation to balance-of-payments disequilibria. The difference will consist in the speed of adjustment of the exchange rate: very high in the case of a free float, slower in the case of a managed float (as the monetary authorities may deem it advisable to prevent drastic jumps in the exchange rate and so intervene to moderate its movements).

Let us now go back to point *A* in the diagram and examine what happens in relation to the original schedules. Since the exchange rate is for the moment given at r_E, we can apply to the $I_{(E)}S_{(E)}$ and $B_{(E)}B_{(E)}$ schedules the same reasoning shown in Sect. 10.2.1; it follows that, as *A* is below the real-equilibrium schedule and above the external-equilibrium schedule, there will be a positive excess demand for goods and a balance-of-payments surplus. The following effects will then come about at the same time:

(1) the money stock (assumption (d), first case) increases due to the external surplus, the *LM* schedule shifts downwards, and the excess supply of money causes a decrease in the interest rate (downwards-pointing arrow from *A*);

(2) the excess demand for goods induces increases in *y* (rightwards-pointing arrow from *A*);

(3) the external surplus brings about an exchange-rate appreciation, so that the *IS* and *BB* schedules shift from the initial equilibrium position to a new position, for example to $I'S'$ and $B'B'$.

Thus we have, besides the movement of point *A*, movements of the *IS*, *BB*, *LM* schedules, which may give rise to changes in the signs of the excess demands, etc., so that it is not possible to ascertain the final result of all these movements by way of a graphic analysis. The situation is less complex if we adopt the second case of assumption (d), so that the *LM* schedule does not shift. In any case it is indispensable to rely on a mathematical analysis (see the Appendix).

We must now observe that the model employed is subject to the same type of criticism as its fixed exchange rates analogue (see Sect. 10.2.2.1) as regards capital flows, the burden of interest payments, etc. In the meantime we shall continue using it to explain the traditional analysis of the problem of economic policy under flexible exchange rates (see the next chapter).

10.4 Appendix

10.4.1 The Mundell-Fleming Model Under Fixed Exchange Rates

Without loss of generality we can normalize the exchange rate and prices to unity, so that we can achieve a considerable notational simplification by introducing the variable $d(y, i) \equiv A(y, i) - m(y, i)$. Thus the real-equilibrium condition $y = A(y, i) + x_0 - m(y, i)$ can be written as $y = d(y, i) + x_0$, with $d_y \equiv A_y - m_y, d_i \equiv A_i - m_i$. Note that $d(y, i)$ can be interpreted as demand for domestic output by residents. In fact, given that national expenditure A includes expenditure on both domestic and foreign goods, the latter being equal to imports in our pure flow model, the difference $A - m$ is expenditure by residents on domestic output. This operation is of course possible thanks to the normalization of prices and exchange rate to unity.

10.4.1.1 The Slopes of the Various Schedules

Given the three equilibrium equations

$$
\begin{aligned}
d(y, i) + x_0 - y &= 0, & 0 < d_y < 1, & \quad d_i < 0, \\
x_0 - m(y, i) + K(i) &= 0, \; 0 < m_y < 1, & m_i < 0, & \quad K_i > 0, \\
L(y, i) - M &= 0, & L_y > 0, & \quad L_i < 0,
\end{aligned}
\tag{10.7}
$$

if we apply the rule for the differentiation of implicit functions to each one of them, we can calculate the derivatives of the schedules IS, BB and LM, which are

$$
\begin{aligned}
\left(\frac{di}{dy} \right)_{IS} &= \frac{1 - d_y}{d_i} < 0, \\
\left(\frac{di}{dy} \right)_{BB} &= \frac{m_y}{K_i - m_i} > 0, \\
\left(\frac{di}{dy} \right)_{LM} &= -\frac{L_y}{L_i} > 0.
\end{aligned}
\tag{10.8}
$$

As far as points outside the schedules are concerned, consider, for example, the real equilibrium. Once the excess demand for goods function is defined

$$
ED_G = d(y, i) + x_0 - y,
\tag{10.9}
$$

we can calculate the partial derivative

$$
\frac{\partial ED_G}{\partial i} = d_i < 0,
\tag{10.10}
$$

from which it can be seen that for every given y, values of i above or below those which give rise to $ED_G = 0$ (points along the IS schedule) imply $ED_G < 0$. One can proceed in the same way for the other two schedules.

Finally, we shall examine the shifts of *LM*. Taking *M* as a parameter, the third of Eq. (10.7) defines an implicit function in three variables of the type

$$h(M, y, i) = 0. \tag{10.11}$$

On the basis of the implicit-function rule, we can calculate

$$\frac{\partial i}{\partial M} = -\frac{h_M}{h_i} = \frac{1}{L_i} < 0, \tag{10.12}$$

from which it can be seen that for every given *y*, to greater (smaller) values of *M* there correspond smaller (greater) values of *i*, from which the downward (upward) shift of *LM* follows.

10.4.1.2 The Study of Dynamic Stability

The dynamic behaviour assumptions of the text give rise to the following system of differential equations

$$\begin{aligned}
\dot{M} &= a\left[x_0 - m\left(y, i\right) + K(i)\right], \quad a > 0, \\
\dot{y} &= k_1\left[d(y, i) + x_0 - y)\right], \qquad k_1 > 0, \\
\dot{i} &= k_2\left[L(y, i) - M\right], \qquad\quad k_2 > 0,
\end{aligned} \tag{10.13}$$

from which, by linearising at the point of equilibrium, by indicating, as usual, the deviations from equilibrium with a dash above the variable and by putting, for simplicity, $a = k_1 = k_2 = 1$, we have

$$\begin{aligned}
\dot{\overline{M}} &= -m_y \overline{y} + (K_i - m_i)\dot{\overline{i}}, \\
\dot{\overline{y}} &= (d_y - 1)\overline{y} + d_i \dot{\overline{i}}, \\
\dot{\overline{i}} &= -\overline{M} + L_y \overline{y} + L_i \dot{\overline{i}}.
\end{aligned} \tag{10.14}$$

Stability depends on the roots of the characteristic equation

$$\begin{vmatrix}
0 - \lambda & -m_y & (K_i - m_i) \\
0 & (d_y - 1) - \lambda & d_i \\
-1 & L_y & L_i - \lambda
\end{vmatrix}$$

$$\begin{aligned}
&= \lambda^3 + (1 - d_y - L_i)\lambda^2 + \left[K_i - m_i - L_y d_i - L_i(1 - d_y)\right]\lambda \\
&\quad + \left[(1 - d_y)(K_i - m_i) - m_y d_i\right] = 0.
\end{aligned} \tag{10.15}$$

Necessary and sufficient stability conditions are (see Gandolfo 2009)

$$
\begin{aligned}
&d_y < 1 - L_i, \\
&d_y < 1 + \frac{K_i - m_i - d_i L_y}{-L_i}, \\
&d_y < 1 - \frac{m_y d_i}{K_i - m_i}, \\
&m_y < \frac{L_i(K_i - m_i) + (1 - d_y - L_i)\left[-L_y d_i - L_i(1 - d_y)\right]}{-d_i}.
\end{aligned}
\tag{10.16}
$$

Taking account of the signs of the various derivatives, it can at once be observed that $d_y < 1$ is a sufficient (even if not a necessary) condition for the first three inequalities to be satisfied. Furthermore, if $d_y < 1$, the right-hand side of the last inequality is certainly positive, so that this inequality admits positive values of m_y.

Note that the first three inequalities are necessary and sufficient for each real root of the characteristic equation to be negative and therefore exclude monotonic (but not oscillatory) instability; the fourth inequality, on the other hand, is necessary and sufficient, together with the previous ones, for the complex roots to have a negative real part, and therefore, as stated in the text, it excludes oscillatory instability.

We now come to the form and position of BB when interest payments on the stock of foreign debt are considered. The equation for the equilibrium of the balance of payments becomes

$$
x_0 - m(y, i) - i \int_0^t K(i) d\tau + K(i) = 0.
\tag{10.17}
$$

In the current period (t is given) the integral $\int_0^t K(i) d\tau$ is a given constant, which we indicate with γ. By applying the implicit-function rule to (10.17) we have

$$
\left(\frac{di}{dy}\right)_{BB} = \frac{m_y}{K_i - m_i - \gamma}.
\tag{10.18}
$$

If the sensitivity of capital movements to the rate of interest (represented by K_i) is sufficiently great, then the denominator of (10.18) remains positive and BB has the normal positive slope. In the opposite case, BB will have a negative slope. There is however a third possibility: it is in fact to be presumed that K_i will decrease as i increases, because successive equal increases in i lead to progressively smaller inflows of capital (it is in fact kind of "decreasing responsiveness principle", due, for example, to increasing risks). What can then occur is the presence of some critical value of i—say i_c—such that

$$
K_i - m_i - \gamma \gtrless 0 \text{ for } i \lessgtr i_c.
\tag{10.19}
$$

In this case it can be seen at once from (10.18) that

$$\left(\frac{di}{dy}\right)_{BB} \gtrless 0 \text{ for } i \lessgtr i_c \text{ and } \lim_{i \to i_c} \left(\frac{di}{dy}\right)_{BB} = \infty, \tag{10.20}$$

so that BB, initially increasing, bends backwards when i reaches $i = i_c$, as described in Fig. 10.7.

Let us now consider the shifts of BB with the passing of time. Equation (10.17) can be considered as an implicit function in the *three* variables, i, y, t, say

$$\Phi(i, y, t) = 0, \tag{10.21}$$

so that the shifts of BB as time goes on can be determined by calculating, for example, the partial derivative

$$\frac{\partial y}{\partial t} = -\frac{\Phi_t}{\Phi_y}, \tag{10.22}$$

which tells us how y must move in correspondence to each given i, so as to maintain the balance of payments in equilibrium as t varies. By applying formula (10.22) to (10.17), account being taken of the rules for the differentiation of an integral with respect to a parameter (see, for example, Courant (1962)), we have

$$\frac{\partial y}{\partial t} = \frac{iK(i)}{-m_y} < 0, \tag{10.23}$$

from which it follows the shift to the left of BB.

10.4.1.3 Comparative Statics

The Transfer Problem
If we introduce the transfer T as parameter and assume that it shifts the exogenous components of expenditure as assumed in the text, we have

$$\begin{aligned}
y &= d(y, i) + x_0 + a_1 T, \\
x_0 &- m(y, i) + K(i) + a_2 T = 0, \\
L(y, i) &- M = 0,
\end{aligned} \tag{10.24}$$

where

$$a_2 \equiv \mu_1' - 1 < a_1 \equiv \mu_1' - b_1' - b_2' < 0, \tag{10.25}$$

and $T = 0$ initially. If we differentiate the system with respect to T (the relevant Jacobian is different from zero) we obtain

$$
\begin{aligned}
(1 - d_y)(\partial y/\partial T) - d_i(\partial i/\partial T) &= a_1, \\
m_y(\partial y/\partial T) + (m_i - K_i)(\partial i/\partial T) &= a_2, \\
L_y(\partial y/\partial T) + L_i(\partial i/\partial T) - &= 0,
\end{aligned}
\tag{10.26}
$$

from which

$$
\begin{aligned}
\partial y/\partial T &= [a_1(K_i - m_i) - a_2 d_i]/D, \\
\partial i/\partial T &= [-a_2(1 - d_y) + a_1 m_y/D, \\
\partial M/\partial T &= \{a_1[L_i m_y + L_y(K_i - m_i)] - a_2[L_i(1 - d_y) + L_y d_i]\}/D,
\end{aligned}
\tag{10.27}
$$

where

$$
D \equiv (1 - d_y)(K_i - m_i) - m_y d_i.
\tag{10.28}
$$

Since $D < 0$ by the stability conditions (10.16), given the assumed signs of the partial derivatives and (10.25), it follows that $\partial y/\partial T < 0$, namely the transfer unambiguously decreases income. As regards the interest rate, it will certainly increase $(\partial i/\partial T > 0)$ if

$$
d_y + m_y < 1,
\tag{10.29}
$$

namely if the marginal propensity to aggregate spending A_y is smaller than one. Finally, the stock of money will decrease $(\partial M/\partial T < 0)$ if

$$
\frac{m_y}{K_i - m_i} < -\frac{L_y}{L_i},
\tag{10.30}
$$

namely—recalling (10.8)—if the slope of BB is smaller than the slope of LM.

An Exchange-Rate Devaluation

Finally, let us consider an *exogenous variation in the exchange rate* in a regime of adjustable peg. For this purpose, it is necessary to introduce the exchange rate into the various equations *considering it as a parameter*:

$$
\begin{aligned}
y &= x(r) + d(y, i, r), \quad dx/dr > 0, \partial d/\partial r > 0, \\
x(r) - rm(y, i, r) + K(i) &= 0, \quad \partial m/\partial r < 0, \\
M &= L(y, i).
\end{aligned}
\tag{10.31}
$$

Implicit in the second of Eq. (10.31) is the simplifying assumption that a once-and-for-all variation in the exchange rate does not influence the balance on capital account. Differentiating Eq. (10.31) with respect to the parameter r, we have

$$
\begin{aligned}
(1 - d_y)(\partial y/\partial r) - d_i(\partial i/\partial r) &= dx/dr + \partial d/\partial r, \\
m_y(\partial y/\partial r) + (m_i - K_i)(\partial i/\partial r) &= dx/dr - m - r(\partial m/\partial r), \\
L_y(\partial y/\partial r) + L_i(\partial i/\partial r) - \partial M/\partial r &= 0.
\end{aligned}
\tag{10.32}
$$

Note that the sign of the right-hand term of the second equation depends on the elasticities, because

$$
\frac{dx}{dr} - m - r\frac{\partial m}{\partial r} = m\left(\frac{x}{rm}\frac{r}{x}\frac{dx}{dr} - 1 - \frac{r}{m}\frac{\partial m}{\partial r}\right) = m\left(\frac{x}{rm}\eta_x + \eta_m - 1\right).
\tag{10.33}
$$

This expression will be positive if the condition of critical elasticities (7.9) is satisfied. We shall assume that this is the case and for brevity we shall call η the entire expression examined.

Thus, solving system (10.32), we have

$$
\begin{aligned}
\partial y/\partial r &= [(dx/dr + \partial d/\partial r)(K_i - m_i) - \eta d_i]/D, \\
\partial i/\partial r &= [(dx/dr + \partial d/\partial r)m_y - \eta(1 - d_y)]/D, \\
\partial M/\partial r &= L_i(\partial i/\partial r) + L_y(\partial y/\partial r),
\end{aligned}
\tag{10.34}
$$

where D is the usual determinant. It can be seen at once that $\partial y/\partial r > 0$, while the sign of $\partial i/\partial r$ (and therefore of $\partial M/\partial r$) is uncertain. The value of $\partial y/\partial r$ also represents an international multiplier, insofar as it gives the effect on equilibrium income of an exogenous variation in exports and imports: the variation of the parameter r, in fact, causes variations in exports and imports which can to all intents and purposes be considered exogenous variations in the context of the model examined.

10.4.2 The Mundell-Fleming Model Under Flexible Exchange Rates

We consider the model

$$
\begin{aligned}
y &= A(y, i) + x(r) - rm(y, i, r), \quad \text{where } 0 < A_y < 1, A_i < 0, \\
& x_r > 0, 0 < m_y < 1, m_i < 0, m_r < 0, A_i - m_i < 0, \\
M^* &= L(y, i), \quad L_y > 0, L_i < 0, \\
B &= x(r) - rm(y, i, r) + K(i), \quad K_i > 0.
\end{aligned}
\tag{10.35}
$$

The signs of the various derivatives are self-explanatory; we only note that $A_i - m_i < 0$ derives from the fact that, as already noted at the beginning of this appendix, $A_i - m_i$

is the effect of the interest rate on the expenditure on *domestic* output by residents, which is clearly negative.

Let us first examine the case in which—apart from possible policy measures—M^* is constant because the exchange rate instantaneously keeps the balance of payments in equilibrium: this is the second case in point (d) in the text (Sect. 10.3). This means that, for any set of values of y, i, the exchange rate r is always such that $B = 0$. Formally, from the implicit function

$$B(y, i, r) = x(r) - rm(y, i, r) + K(i) = 0 \tag{10.36}$$

we can express r as a function of y, i. This is of course possible provided that the Jacobian of $B(y, i, r)$ with respect to r is different from zero. The condition is

$$B_r = x_r - m - rm_r = m \left(\frac{x}{rm} \eta_x + \eta_m - 1 \right) \neq 0. \tag{10.37}$$

Thus we obtain the differentiable function

$$r = r(y, i), \tag{10.38}$$

whose derivatives are, by the implicit function theorem,

$$\frac{\partial r}{\partial y} = -\frac{B_y}{B_r} = \frac{rm_y}{B_r}, \quad \frac{\partial r}{\partial i} = -\frac{B_i}{B_r} = \frac{rm_i - K_i}{B_r}. \tag{10.39}$$

It is easy to see that $\partial r / \partial y \gtrless 0$ according as $B_r \gtrless 0$, namely according as the critical elasticities condition is satisfied or not. Similarly, $\partial r / \partial i \gtrless 0$ according as $B_r \lessgtr 0$.

Given (10.38), we can write the dynamics of the basic system (10.35) as

$$\begin{aligned} \dot{y} &= k_1 \{ A(y, i) + x[r(y, i)] - r(y, i)m[y, i, r(y.i)] - y \}, \\ \dot{i} &= k_2 [L(y, i) - M^*], \end{aligned} \tag{10.40}$$

namely, by linearising at the equilibrium point, using (10.39), normalizing $r = 1$ at the equilibrium point, and setting $k_1 = k_2 = 1$,

$$\begin{aligned} \dot{\bar{y}} &= \left[A_y - m_y + (x_r - m - m_r) \frac{m_y}{B_r} - 1 \right] \bar{y} + \left[A_i - m_i + (x_r - m - m_r) \frac{m_i - K_i}{B_r} \right] \bar{i}, \\ \dot{\bar{i}} &= L_y \bar{y} + L_i \bar{i}. \end{aligned} \tag{10.41}$$

Using (10.37) and simplifying the system becomes

$$\begin{aligned} \dot{\bar{y}} &= \left(A_y - 1 \right) \bar{y} + (A_i - K_i) \bar{i}, \\ \dot{\bar{i}} &= L_y \bar{y} + L_i \bar{i}. \end{aligned} \tag{10.42}$$

Stability depends on the roots of the characteristic equation

$$\begin{vmatrix} (A_y - 1) - \lambda & A_i - K_i \\ L_y & L_i - \lambda \end{vmatrix} = 0, \tag{10.43}$$

namely

$$\begin{aligned} \lambda^2 + \left(1 - A_y - L_i\right) \lambda \\ + \left[(A_y - 1) L_i - L_y (A_i - K_i)\right] = 0. \end{aligned} \tag{10.44}$$

The necessary and sufficient stability conditions are

$$\begin{aligned} 1 - A_y - L_i &> 0, \\ (A_y - 1) L_i - L_y (A_i - K_i) &> 0. \end{aligned} \tag{10.45}$$

Given the assumed signs of the various derivatives, the stability conditions are certainly satisfied. Thus the critical elasticities condition, that was found to be necessary but not sufficient in the Laursen and Metzler model (see the previous chapter) is now sufficient but not necessary. The reason is that now adjustments can also occur through the rate of interest, which eases the burden on the exchange rate.

The general case, in which the exchange rate cannot instantaneously keep the balance of payments in equilibrium, namely the first case in point (d) in the text (Sect. 10.3), gives rise to the following system of differential equations

$$\begin{aligned} \dot{M} &= a \left[x(r) - rm(y, i, r) + K(i)\right], \\ \dot{y} &= k_1 \left\{[A(y, i) + x(r) - rm(y, i, r)] - y\right\}, \\ \dot{i} &= k_1 \left[L(y, i) - M\right], \\ \dot{r} &= k_3 \left[-x(r) + rm(y, i, r) - K(i)\right], \end{aligned} \tag{10.46}$$

where a, k_1, k_2, k_3 are positive constants which for simplicity will be taken as equal to one. From the first and fourth equation we then observe that

$$\dot{M} + \dot{r} = 0,$$

hence

$$M(t) + r = \alpha,$$

where α is an arbitrary constant. Thus it is possible to set $\bar{M} + \bar{r} = 0$ or $\bar{M} = -\bar{r}$

$$\bar{M} + \bar{r} = 0 \Rightarrow \bar{M} = -\bar{r} \tag{10.47}$$

in the neighbourhood of the equilibrium point, where an overbar as usual denotes the deviations from equilibrium. Thus we can reduce the linearised system to a three-equation system,

$$\dot{\bar{y}} = (A_y - 1)\bar{y} + A_i\bar{i} + B_r\bar{r},$$
$$\dot{\bar{i}} = L_y\bar{y} + L_i\bar{i} + \bar{r}, \qquad (10.48)$$
$$\dot{\bar{r}} = m_y\bar{y} - (K_i - m_i)\bar{i} - B_r\bar{r},$$

where B_r has been defined in (10.37).

The characteristic equation of the system is

$$\begin{vmatrix} (A_y - 1) - \lambda & A_i & B_r \\ L_y & L_i - \lambda & 1 \\ m_y & K_i - m_i & -B_r - \lambda \end{vmatrix} = 0, \qquad (10.49)$$

whence

$$\lambda^3 + \left[B_r + (1 - A_y - L_i)\right]\lambda^2$$
$$+ \left[-m_y B_r + K_i - m_i + B_r(1 - A_y - L_i) + (A_y - 1)L_i - A_iL_y\right]\lambda$$
$$+ \left\{B_r\left[(A_y - 1)L_i - A_iL_y\right] + \left[L_y(K_i - m_i) + L_im_y\right]B_r\right.$$
$$\left. + (1 - A_y)(K_i - m_i) - A_im_y\right\}$$
$$= 0. \qquad (10.50)$$

A set of necessary and sufficient stability conditions (Gandolfo 2009; p. 238) is that (1) the coefficient of λ, (2) the constant term, and (3) the product of the coefficient of λ^2 by the coefficient of λ, minus the constant term, should be all positive. Rather than examining all these inequalities in detail, we concentrate on a crucial issue, namely the relevance of the critical elasticities condition. This condition was found to be necessary, but not sufficient, in previous studies on flexible exchange rates (see Chap. 9). Here we can easily show that it is neither necessary nor sufficient. Consider for example the first of the stability conditions, namely

$$B_r(1 - A_y - m_y - L_i) + \left[(A_y - 1)L_i - A_iL_y + K_i - m_i\right] > 0.$$

The expression $\left[(A_y - 1)L_i - A_iL_y + K_i - m_i\right]$ is certainly positive given the assumed signs of the partial derivatives. On the contrary, the expression $(1 - A_y - m_y - L_i)$ may have either sign. Suppose that it is negative. Then $B_r > 0$ (the critical elasticities condition) is not necessary, since with $B_r < 0$ the stability condition is satisfied. It is not sufficient either, because if $(1 - A_y - m_y - L_i)$ is sufficiently negative, $B_r > 0$ may give rise to instability. Similar results hold for the other stability conditions.

The reason for this result is again to be seen in the role of the interest rate, and in its direct (through capital flows) and indirect (through expenditure) effects on the adjustment process.

References

Boughton, J. M. (2003). On the origins of the Fleming-Mundell model. *IMF Staff Papers, 50*, 1–9.

Courant, R. (1962). *Differential and integral calculus* (Vol. II). Glasgow: Blackie&Sons.

Fleming, J. M. (1962). Domestic financial policy under fixed and under floating exchange rate. *IMF Staff Papers, 9*, 369–379. Reprinted In R. N. Cooper (Ed.) (1969). *International finance - selected readings* (pp. 291–303). Harmondsworth: Penguin; and In J. M. Fleming (1971). *Essays in international economics* (pp. 237–248). London: Allen & Unwin.

Frenkel, J. A., & Razin, A. (1987). The Mundell-Fleming model a quarter century later. *IMF Staff Papers, 34*, 567–620.

Gandolfo, G. (2009). *Economic dynamics*, (4th ed.). Berlin, Heidelberg, New York: Springer.

Hicks, J. R. (1937). Mr. Keynes and the 'classics'. *Econometrica, 5*, 147–159; reprinted In J. R. Hicks (1967). *Critical essays in monetary theory* (Chap. 7). Oxford: Oxford University Press.

Mundell, R. A. (1960). The monetary dynamics of international adjustment under fixed and flexible exchange rates. *Quarterly Journal of Economics, 74*, 227–257.

Mundell, R. A. (1961a). The international disequilibrium system. *Kyklos, 14*, 152–170. Reprinted in R. A. Mundell (1968).

Mundell, R. A. (1961b). Flexible exchange rates and employment policy. *Canadian Journal of Economics and Political Science, 27*, 509–517. Reprinted in R. A. Mundell (1968).

Mundell, R. A. (1963). Capital mobility and stabilization policy under fixed and flexible exchange rates. *Canadian Journal of Economics and Political Science, 29*, 475–485. Reprinted in R. A. Mundell (1968).

Mundell, R. A. (1968). *International economics*. New York: Macmillan.

Mundell, R. A. (2001). On the history of the Mundell-Fleming model. *IMF Staff Papers, 47* (Special Issue), 215–227.

Policy Implications of the Mundell-Fleming Model, and the Assignment Problem

11

11.1 Introduction

In a regime of fixed exchange rates, the problem of achieving and maintaining simultaneous external and internal balance—where *internal balance* (or *equilibrium*) is taken to mean *real equilibrium with full employment*, and *external balance* (or equilibrium) to mean balance-of-payments equilibrium—may seem in certain cases to be insoluble. These are the so-called *dilemma cases*, as for example a balance-of-payments deficit accompanied by a situation of underemployment; we shall go more fully into these questions shortly. From the point of view of the traditional theory of economic policy, however, the conflict between internal and external equilibrium might seem strange, because even if exchange controls and other restrictions are excluded, two instruments of economic policy—fiscal and monetary policy—are still available to achieve the two targets, so that Tinbergen's principle is satisfied.

According to this principle (at the heart of the traditional theory of economic policy[1]), in order to achieve a plurality of independent targets, the number of independent instruments should not be less than the number of targets themselves. It is necessary to emphasize the fact that the instruments must be *independent*, in that, if two or more instruments act in the same way on the same variables, they constitute

---

[1] It must be emphasized that this is a principle which is valid within the ambit of the static theory of economic policy, which still constitutes a point of reference for scholars of this discipline. It is however as well to inform the reader that according to developments of the mathematical "control theory" applied to the instruments-targets problem, in a dynamic context, the Tinbergen principle is no longer generally valid. For further information on the matter, see Petit (1985). It is also important to point out that, after the traditional instruments-(fixed) targets approach, an optimizing approach to economic policy has been developed. This approach involves the maximization of a social welfare function (or of a preference function of the policy-maker) subject to the constraint of the model representing the structure of the economy. In order to avoid further burdening of the present chapter, we postpone the presentation of some considerations on this approach until the end of Sect. 23.8.1.

© Springer-Verlag Berlin Heidelberg 2016
G. Gandolfo, *International Finance and Open-Economy Macroeconomics*,
Springer Texts in Business and Economics, DOI 10.1007/978-3-662-49862-0_11

to all intents and purposes a single instrument. Now, in the theory developed during the 1950s, monetary and fiscal policies were considered *equivalent* means of influencing only aggregate demand (and thus the level of income and imports). If imports and income move in the same direction, it follows from this equivalence that only in certain cases there will not be conflict between the internal and external targets.

Take first of all the case in which there is underemployment and a balance-of-payments surplus; in this case the line to follow is an expansionary (fiscal and monetary) policy, which will cause income to increase and the balance of payments to worsen (seeing that there will be an increase in imports induced by the increase in income).

Now consider the case in which there is excess aggregate demand with respect to full employment income accompanied by a balance-of-payments deficit. The line to follow, in this case, is also clear: a restrictive policy must be adopted, which will reduce the excess aggregate demand, and at the same time improve the balance of payments by way of a reduction in imports connected to the restriction in total demand.

On the other hand, if there is a situation in which there is underemployment and a deficit in the balance of payments, or a situation in which there is excess demand with respect to full employment income and a balance-of-payments surplus, then there appears to be a conflict between internal and external equilibrium. In the first situation, in fact, internal equilibrium requires an expansionary policy, while external equilibrium requires a restrictive one; in the second situation internal equilibrium requires a restrictive policy, while external equilibrium requires an expansionary one. These are the dilemma cases already mentioned.

The first situation is undoubtedly the more difficult. In fact, a surplus in the balance of payments may, at worst, be tolerated indefinitely, insofar as it translates itself into an accumulation of international reserves as a counterpart to national currency (which can be issued in unlimited amounts; naturally to avoid undesirable domestic effects of increases in the quantity of money, the monetary authorities will, if necessary, have to offset the increase in the quantity of money due to the foreign channel by a reduction through other domestic channels). Conversely, a deficit in the balance of payments cannot be tolerated for long and it is necessary to find a remedy before international reserves drop to zero or below the minimum considered acceptable. In the context studied here, therefore, all that remains to be done is to give up the internal target and adopt a restrictive policy, thus eliminating the balance-of-payments deficit at the expense of employment.

The argument has been carried on up to this point on the assumption made at the beginning, that monetary and fiscal policies have the same effect on the same variables. In reality they do not, because monetary policy also has an effect on capital movements by way of variation in the interest rate, while fiscal policy does not have this effect. Therefore, apart from the exceptional case in which capital movements are completely insensitive to variations in the interest rate, or the equally exceptional case of perfect capital mobility (which will be treated later in this chapter, see Sect. 11.4), the two instruments *cannot* be considered equivalent:

the possibility therefore once more arises of obtaining both internal and external equilibrium by means of an appropriate combination of fiscal and monetary policies.

11.2 Internal and External Balance, and the Assignment Problem

Once it has been ascertained that fiscal and monetary policies are distinct instruments, the possibility of obtaining internal and external equilibrium by way of an appropriate use of these two instruments can be studied with a simple graphic analysis, by using the *IS* and *BB* schedules, on which see the previous chapter. The only thing to note is that the position of *IS* now depends on government expenditure (for simplicity, fiscal policy is identified with the management of government expenditure rather than of the budget deficit), so that the schedule shifts to the right if government expenditure increases and to the left if it decreases. In fact, if aggregate demand increases, production must increase at each given level of the interest rate.

Take for example the case in which the system is initially at point E, at which there is external, but not internal, equilibrium (Fig. 11.1), since full-employment income is higher, for example $y_{E'}$. The objectives are to boost income from y_E to $y_{E'}$, while maintaining balance-of-payments equilibrium; that is, to take the system from to E to E'. This requires an expansionary fiscal policy such that the *IS* schedule moves to the position $I'S'$, and an increase in the interest rate from i_E to $i_{E'}$, to attract enough capital to offset the deterioration in the current account caused by the additional imports determined by the increase in income.

In other words, the increase in income is due to the multiplier process set into motion by the additional government expenditure; but, as this increase causes an increase in imports, the balance of payments, initially in equilibrium, goes into deficit, hence the need for an increase in the interest rate, in order to get the balance

Fig. 11.1 Fiscal and monetary policy for internal and external balance under fixed exchange rates: first case

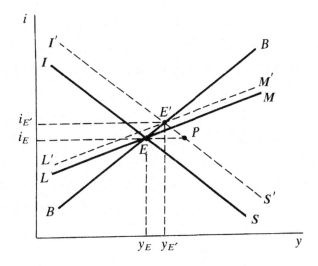

Fig. 11.2 Fiscal and monetary policy for internal and external balance under fixed exchange rates: second case

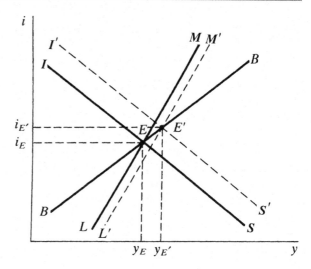

of payments back into equilibrium. It goes without saying that this increase in the interest rate tends to put a brake on the increase in income. In fact, as can be seen from the figure, the increase in income from y_E to $y_{E'}$ is less than it would have been if the interest rate had remained unchanged (abscissa of point P); in other words, the multiplier is less than that met with in the analysis carried out in Chap. 8.

The interest-rate increase requires an appropriate shift in the LM schedule to the position $L'M'$, which entails a *decrease* in the quantity of money (see Fig. 11.1). Note that the consequences can also be the opposite; in fact if the LM schedule had initially been in the position indicated in Fig. 11.2, it would have had to shift to the right, which indicates an *increase* in the quantity of money.

It might seem queer that a restrictive monetary policy (in the sense of a policy that causes an increase in the interest rate) may imply an increase rather than a decrease in the money supply, but the reason for all this will become intuitively clear if we observe that at E' *not only the interest rate but also income* are greater than at E. Hence the demand for money on the one hand tends to decrease and on the other to increase, so that the total demand for money (and therefore, in equilibrium, the supply) can either decrease or increase according to the relative intensity of the two variations.

11.2.1 The Assignment Problem

To realize the possibility of simultaneously achieving external and internal equilibrium, it may be necessary to solve the so-called *assignment problem*, that is of assigning instruments to targets or, more precisely, of assigning each given instrument to the achievement of one single target. This is a problem which arises when

(i) there is decentralization of economic policy decisions and/or
(ii) the policy-maker has incomplete information.

It is essentially a *dynamic* problem: from the *static* point of view, in fact, by solving the system of targets-instruments equations, one obtains those sets of values to be simultaneously attributed to each of the various instruments so as to achieve the various targets. In statics there is thus no sense in talking of "assigning" each instrument to one or the other target.

Now in the case in which all instruments and economic policy decisions are centralized in a single authority, this authority could, in theory, immediately make the instruments assume the appropriate values for the achievement of the targets. But whenever the instruments are in the hands of distinct authorities which are more or less independent of each other, the problem arises of how (that is, in relation to which target) should each of them manage the respective instrument. This gives rise to the problem of pairing off instruments with objectives.

It is however important to note that this problem also arises in the previous case (with centralized decision-making) whenever the numerical values of the various parameters (propensities, etc.) *are not known with sufficient accuracy* (imperfect information: this is the more realistic case) so that the policy-maker must try to approach the targets by way of successive approximations and therefore must know on the basis of which indicators it must move the available instrumental variables. At the very least, the theoretical solution to the assignment problem puts these authorities in a position to cause the system to move towards the targets without any precise numerical information as to the magnitude of the various parameters, but only knowing their signs.

The solution to the assignment problem can be found by applying the Mundell principle (that he called the *principle of effective market classification*) according to which each instrument must be used in relation to the objective upon which it has relatively more influence (Mundell 1962, 1968). The reason is that, otherwise, it may happen that it will not be possible to achieve equilibrium, as we shall see shortly.

Now, it can be argued that monetary policy has a relatively greater effect on external equilibrium. In fact, let us examine a monetary and a fiscal intervention with the same quantitative effects on aggregate demand and therefore on income. They have likewise an effect on the balance of payments, which is greater in the case of monetary policy, because it acts not only on current transactions (by way of variations induced on imports by variations in income, which is an identical effect to that which can be obtained by way of fiscal policy), but it also acts upon capital movements by way of variations in the interest rate.

It follows from all this that monetary policy must be associated with the external target, while fiscal policy must be associated with the internal target; otherwise, as we said, it may be impossible to reach the situation of external and internal balance. A much simplified example may serve to clarify these statements.

Consider a situation in which income is 100 units below the full employment level and there is a balance-of-payments deficit of ten. An increase of 100 in income, obtained by way of fiscal policy, causes, for example, an increase of 20 in imports (assuming the marginal propensity to import to be 0.20) so that the balance-of-payments deficit rises to 30. A restrictive monetary policy is now set into motion, which causes, let us say, a decrease in income of 50, with a consequent reduction in imports of 10; this policy also causes an improvement in the capital movements balance of, say, 15. The net result of the two policies is that an increase in income of 50 is obtained, and a reduction in the balance-of-payments deficit from 10 to 5. If we continue in this way, it is possible, through successive adjustments, to reach a point at which the two targets are simultaneously achieved.

Now consider the same initial situation and pair the instruments and targets in the opposite way to that just examined. An expansionary monetary policy which causes an increase in income of 100 also causes, alongside it, an increase of 20 in imports, a worsening in the capital balance of, lets say, 10; consequently, the balance-of-payments deficit passes from 10 to 40. Now, in order to reduce this deficit, or even simply to bring it down to the initial level of 10, by means of fiscal policy, it is necessary for the fiscal policy adopted to cause a restriction in income of 150 (hence a reduction in imports of 30) which is greater than the initial expansion. The net result is that income is reduced while the balance of payments remains unaltered. The process clearly moves further away from equilibrium.

It is also possible to provide an intuitive graphic analysis of the problem of pairing off instruments and targets by the use of the same diagram used previously.[2] From either Figs. 11.1 or 11.2 we have seen that in the new (full employment) equilibrium (point E') the increase in income must be accompanied by an increase in the interest rate to attract sufficient capital to offset the deterioration in the current account due to the income-induced increase in imports. Now, if we assigned monetary policy to internal equilibrium, the required increase in income could be achieved only thanks to an expansionary monetary policy namely to a *decrease* in the interest rate. This means that the interest rate would move in a direction exactly the opposite of where the new equilibrium point lies. On the contrary, by assigning the interest rate to external equilibrium and fiscal policy to internal equilibrium, we would move the interest rate in an upward direction to counteract the balance-of-payments deficit caused by the income-induced import increase. This is the correct direction of movement.

11.2.2 Observations and Qualifications

What has been said illustrates in simplified form the problem of achieving simultaneous internal and external equilibrium and its solution; for a more rigorous analysis see the Appendix. Here we wish instead to review the various objections which have been raised to the theory so far described, that is, the possibility of achieving internal

---

[2]For an alternative graphic representation, based on a diagram on the axes of which the two instruments are measured, see Mundell (1962).

and external balance by way of an appropriate combination of monetary and fiscal policy. It should be noted that these objections are *internal*, namely they remain within the context of the Keynesian assumptions of the model.

(1) The first observation that one can make is that capital movements induced by manoeuvring the interest rate are short-term movements, which take the place of movements of reserves and of compensatory official financing. They are thus appropriate for correcting temporary and reversible disequilibria in the balance of payments, but not fundamental disequilibria. The manoeuvre in question can therefore cope with conjunctural disequilibria, but not structural ones, which must be cured by different means.

(2) When there is a serious and chronic deficit in the balance of payments, it is probable that even a drastic increase in the rate of interest will have no favourable effect on capital movements, as operators expect a devaluation in the exchange rate and shift their capital to other currencies, notwithstanding the increase in the rate of interest. However, this criticism is valid only in the case when one tries, as it were, to close the stable door after the horse has bolted, that is, in the case in which the policy-maker intervenes by adopting a combination of monetary and fiscal policies when the situation has already become serious. In fact, this kind of combination should be an *habitual* policy, used constantly, precisely with the aim of *preventing* the situation from coming about in which the policy becomes ineffective. A necessary condition for prevention in this way, even if not sufficient in itself, is that the policy-maker has a wealth of information and acts with great speed and timeliness of intervention.

(3) As we have already observed in Sect. 10.2.2.1, it is to be presumed that capital flows induced by a *given* difference between domestic and foreign interest rates will be *limited*. This is a consequence of the general principle of capital stock adjustment. In fact, to each given difference between the rates of interest, there corresponds a certain stock of financial capital which investors wish to place; if the existing stock (that is, the stock they have already placed) is different, there will be a capital flow—spread out over a certain period of time—to bring the stock already in existence up to the level desired. Once the adjustment process is completed, the flows cease. In order to start them off again it is necessary for the difference between the rates of interest to vary, so that the desired stock will be made to diverge once again from that in existence, and so forth.

From all this the following conclusions can be drawn: if the disequilibrium in the current account persists over time, then it is necessary to continue to broaden the difference between the interest rates to keep the equilibrating flows of capital in existence; but it may not turn out to be possible to broaden the difference beyond a certain limit.

(4) According to some writers (for example, Johnson 1965) the capital flows are also sensitive to the level of income, on account of the profitability expected from investment in shares, which would be positively connected to the level of income. If one admits this sensitivity, the comparative advantage of the monetary over the fiscal policy with regard to the balance of payments is no

longer certain. Note that this objection is not directed so much at the theory in general, but rather at the instrument-target assignment mentioned above, which could be turned upside down if this sensitivity is taken into consideration.

(5) As we have already said in Sect. 10.2.2.1, increasing interest rates put an increasing load on interest payments (these payments have not been considered in the previous treatment). At a certain point, these payments could more than offset the advantage of increased inflows of foreign capital. In other words, the sum of the capital movements balance plus the interest payments could react *negatively* to an increase in the domestic interest rate. In this case, of course, the correct assignment is uncertain *a priori* (this objection, like the previous one, is directed at Mundell's assignment and not at the general theory).

(6) In order to achieve the desired simultaneous equilibrium, it may happen that the instruments have to assume values which are politically and institutionally unacceptable (for example, unacceptable levels of fiscal pressure or unacceptably high values for the interest rate). If, let us suppose, the domestic demand is very sensitive to monetary restrictions while capital movements are relatively insensitive to the rate of interest, then in a situation of underemployment and balance-of-payments deficit, it would be necessary to introduce a very considerable increase in the interest rate to attract sufficient capital and a no less considerable increase in public expenditure to compensate for the depressive effect of the increase in the interest rate and to bring income up to full employment. These exceptional increases could turn out to be politically and institutionally unacceptable, as we said, consequently making it impossible to achieve simultaneous internal and external equilibrium. Naturally these are considerations which may or may not be relevant according to the values assumed by the various parameters in each given concrete case.

(7) Last but not least, no account is taken of the financing of the additional government expenditure required to achieve an increase in output and of the consequences on the public debt and on the burden of future additional interest payments.

It is possible to try to take account of some of these observations by appropriately amending the Mundell-Fleming model (see, for example, Frenkel and Razin 1987), but rather than doing this it is preferable to tackle them using more general stock-flow models (see Chap. 13). Remaining in the context of the original Mundell-Fleming model, the critical observations listed above cast serious doubts on the claim that the appropriate combination of monetary and fiscal policies can solve the dilemma cases, even though it cannot be denied that it has some validity as a short-term expedient. If this is the case, the obvious suggestion is to introduce another instrument, such as flexibility of the exchange rate, to which we now turn.

11.3 Flexible Exchange Rates

While under fixed exchange rates both fiscal and monetary policy are engaged in achieving and maintaining internal and external equilibrium, under freely flexible exchange rates external equilibrium is taken care of by the exchange rate (provided that the suitable elasticity conditions occur). If we are under a regime of freely flexible exchange rates, we know that market forces cause the exchange rate to depreciate (appreciate) when there is a deficit (surplus) in the balance of payments. This is a dynamic relation that will adjust the balance of payments if the appropriate elasticity conditions are met. This of course suggests the obvious assignment of the exchange rate (in a regime of adjustable or managed exchange rates) to the external target.

In any case a single instrument is in principle sufficient to achieve internal equilibrium. The other instrument is thus left free, to be used, if desired, to achieve another target. The problem that now has to be solved is one of choice, namely, which of the two traditional instruments (fiscal and monetary policy) is it preferable to use to achieve internal equilibrium? A host of studies were carried out in the 1960s and early 1970s with the aim of examining and comparing the effectiveness of these two instruments under fixed and flexible exchange rates (for surveys see, e.g., Von Neumann Whitman 1970; Shinkai 1975). From these studies it turned out that in certain cases both policies are more effective under flexible than under fixed exchange rates, whilst in other cases solely monetary policy is definitely more effective, the comparative effectiveness of fiscal policy being uncertain. A crucial role is played by capital mobility, as we shall see in Sect. 11.4.

Let us consider what happens under a regime of flexible exchange rates when we apply *fiscal policy* to internal equilibrium. For simplicity's sake we assume that the critical elasticities condition holds, and that the variations in the exchange rate keep the balance of payments in continuous equilibrium, so that the money supply does not change.

In the case in which capital mobility is relatively high, we have the graphic representation of Fig. 11.3.

The relatively high capital mobility gives rise to a position of the *BB* schedule that is flatter than the *LM* schedule.[3] An expansionary fiscal policy causes the *IS* schedule to shift rightwards, to $I'S'$, which intersects the *LM* schedule in E', where both the interest rate and income are higher. As a consequence of the relatively high capital mobility, the capital inflow caused by the increase in the interest rate will more than offset the current account deterioration induced by the income increase. This can be seen from the fact that E' lies above the *BB* schedule, thus denoting a surplus. The surplus causes an immediate appreciation in the exchange rate, which causes the *BB* schedule to shift upwards (to $B'B'$) and the $I'S'$ schedule leftwards, to $I''S''$.

---

[3]On the relationship between the slope of the *BB* schedule and capital mobility, see Sect. 10.2.2.

Fig. 11.3 Flexible exchange rates and policies for internal and external balance: high capital mobility

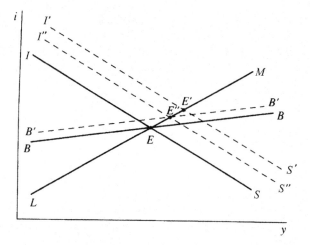

Fig. 11.4 Flexible exchange rates and policies for internal and external balance: low capital mobility

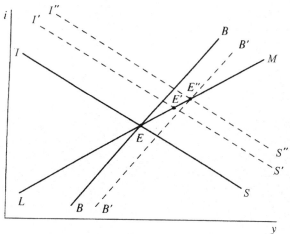

Due to this shift in the *BB* schedule, the restrictive effect of the exchange-rate appreciation on the real-equilibrium schedule will not be such as to completely offset the initial expansion: the new equilibrium E'', though being to the left of E', denotes an income increase all the same. Thus, while it is true that the exchange-rate appreciation contrasts the effectiveness of fiscal policy, this appreciation does not nullify the effects of fiscal policy.

Let us now examine the case of a relatively low capital mobility, in which the *BB* schedule is steeper than the *LM* schedule.

In Fig. 11.4 we see that a fiscal expansion causes a deficit in the balance of payments, as the relatively low capital inflow induced by the interest-rate increase does not offset the current-account deficit (point E; is now below the *BB* schedule). As a consequence the exchange rate depreciates. This reinforces the initial expansion, causing a further rightward shift of the *IS* schedule, to $I''S''$. At the same time the exchange-rate depreciation makes the *BB* schedule shift to the right, to *B'B'*. In the final situation *E''*, there will therefore be an income increase accompanied by an exchange-rate depreciation. It is clear that, the initial fiscal expansion being equal, the final effect on income will be higher than in the previous case.

The case of *monetary policy* is simpler, since its effects are in any case unequivocal. The reader may wish to draw the diagram, in which it will be easy to see that a monetary expansion (the *LM* schedule shifts to the right) initially causes an interest-rate decrease and an income increase. Both changes cause a balance-of-payments deficit and hence an exchange-rate depreciation, which will make the *IS* schedule shift to the right. The *BB* schedule will also shift to the right. In the new equilibrium we shall in any case observe an exchange-rate depreciation and an income increase. These results are valid *independently of the degree of capital mobility*.

11.4 Perfect Capital Mobility

In the case of perfect capital mobility it was shown by Mundell (1968; see also Shinkai 1975) that fiscal policy becomes completely ineffective under flexible exchange rates. This result is symmetric to that of the complete ineffectiveness of monetary policy under fixed exchange rates (and, of course, perfect capital mobility). It should at this point be recalled from Sect. 4.6 that Mundell implicitly assumed perfect capital mobility to imply perfect asset substitutability. Furthermore, in the case of flexible exchange rates static expectations are held by the representative agent, while in the case of fixed exchange rates the given exchange rate is supposed to be credible.

Let us examine Fig. 11.5, where we have drawn the *IS, BB, LM* schedules corresponding to the equilibrium value (r_E) of the exchange rate. This diagram is similar to Figs. 11.2 and 11.3, except for the fact that the *BB* schedule has been drawn parallel to the *y* axis with an ordinate equal to i_f, to denote that the domestic interest rate cannot deviate from the given foreign interest rate (i_f) owing to the assumption of perfect capital mobility. In the initial equilibrium net capital flows are assumed to be zero.

We first consider an expansionary monetary policy: the *LM* schedule shifts rightwards to $L'M'$ and the excess supply of money puts a downward pressure on the domestic interest rate and hence an expansionary effect on output. However, the real-monetary equilibrium point E_H is hypothetical: in fact, as soon as the interest rate tends to decrease below i_f, a capital outflow takes place which brings about a deficit in the balance of payments.

At this point we must distinguish between fixed and flexible exchange rates. Under fixed exchange rates the balance-of-payments deficit causes a decrease in the money stock which pushes the monetary-equilibrium schedule back to the initial position, i.e. from $L'M'$ to *LM*. Any attempt at an expansionary monetary policy gives rise to a loss of international reserves with no effect on national

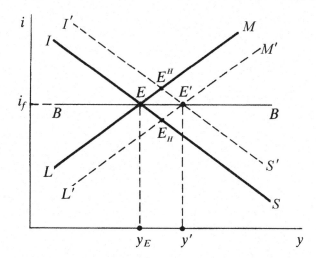

Fig. 11.5 Perfect capital mobility and fiscal and monetary policy under fixed and flexible exchange rates

income: monetary policy is completely ineffective. If the policy maker insists on increasing the money supply the system becomes unstable: when the stock of international reserves is down to zero, the fixed exchange rate regime will have to be abandoned (a currency crisis: see Sect. 16.3).

On the contrary, under flexible exchange rates the exchange rate depreciates as a consequence of the balance-of-payments deficit (we assume that the appropriate elasticity conditions are satisfied). The exchange-rate depreciation also causes a rightward shift in the IS schedule (see Sect. 10.2.3.2) to the position $I'S'$. A new equilibrium is thus established at E' where income is higher: monetary policy has achieved its aim.

We now consider the effects of fiscal policy. An expansionary fiscal policy shifts the IS schedule to $I'S'$ in Fig. 11.5. The pressure of excess demand stimulates output; the increase in income causes an increase in the demand for money which, as the supply is given, exerts an upward pressure on the domestic interest rate. However, the real-monetary equilibrium point E^H is hypothetical: in fact, as soon as the interest rate tends to increase above i_f, a capital inflow takes place which—if the exchange rate is fixed—brings about an increase in the money supply (LM shifts to $L'M'$). The new equilibrium is established at E', where income is higher: fiscal policy has achieved its aim.

On the contrary, under flexible exchange rates the money supply remains constant (the implicit assumption is that we are in the second case of assumption (d) in Sect. 10.3), and the upward pressure on the domestic interest rate is greater: there is, in fact, no increase in the money supply since the exchange rate moves instantaneously to maintain external equilibrium. The increased capital inflow causes an exchange-rate appreciation which nullifies the effects of the expansionary fiscal policy (the real-equilibrium schedule in the upper panel shifts back from $I'S'$ to IS). Income falls back to y_E: fiscal policy is completely ineffective.

The same kind of diagram can be used to examine the effects of other policies under perfect capital mobility. Consider for example *trade policy*, and suppose that the government forces net exports $(x_0 - m)$ to increase (for example by enacting a tariff or quota on imports, or by subsidizing exporting firms). In Fig. 11.5, this would shift the IS schedule outwards (to the position $I'S'$), since for a given equilibrium value (r_E) of the exchange rate, net exports are greater. As we have seen, under flexible exchange rate this only causes an exchange-rate appreciation which completely offsets the direct effects of the trade restriction on net exports (net exports fall back to their original level), so income doesn't change either. However, the total amount of trade is lower: the import restriction reduces imports, and the accompanying currency appreciation reduces exports (assuming that the elasticity conditions hold), so that, even though the balance of payments is the same, total trade flows have dropped.

In contrast, under fixed exchange rates, a rightward shift in the IS curve due to a restrictive trade policy, brings about (as in the fiscal policy case) an outward shift in the LM schedule (to $L'M'$). The new equilibrium has the same exchange rate but a higher level of income. In this case, trade policy induces a change in the money supply, and is effective in raising income.

11.5 Appendix

11.5.1 Monetary and Fiscal Policy Under Fixed Exchange Rates

11.5.1.1 The Static Model
In what follows, monetary policy is identified with interest-rate management (this implies that the money supply is always set at the appropriate level) and fiscal policy with government-expenditure management. Let us consider the system[4] (for the notation see the Appendix to Chap. 10):

$$y - [d(y, i) + x_0 + G] = 0,$$
$$B - [x_0 - m(y, i) + K(i)] = 0,$$
(11.1)

which includes four variables, two of which are objectives (y and B) and two instruments (G and i). On the basis of the implicit-function theorem, it is possible to express two of the variables in terms of the other two, provided the appropriate Jacobian is different from zero; if we wish to express G and i as functions of y and

---

[4]As we have expressed the equation for the determination of income in terms of the demand for domestic goods (on the part of residents and non-residents), G is to be taken as (that part of) public expenditure directed at domestic goods.

B, it is necessary to consider the Jacobian of Eq. (11.1) with respect to G and i:

$$J = \begin{vmatrix} -1 & -d_i \\ 0 & m_i - K_i \end{vmatrix} = K_i - m_i. \tag{11.2}$$

The fact that $J \neq 0$ ensures the absence of functional dependence, and is the mathematical equivalent of the possibility of solving the initial system so as to be able to determine the values of the instruments corresponding to prefixed values of the targets. In fact, once condition $J \neq 0$ is satisfied single-valued functions will exist

$$G = G(y, B), i = i(y, B), \tag{11.3}$$

by which, given that $y = y_F, B = 0$, it is possible to determine the corresponding values of G and i.

From the economic point of view, condition $J \neq 0$ amounts to saying that a direct effect exists of the rate of interest on the balance of payments, given by $(K_i - m_i)$.

Let us now examine the relative effectiveness of the various instruments on the various targets. For this purpose the opposite operation to that just described is carried out, namely y and B are expressed as functions of G and i. This requires the Jacobian of Eq. (11.1) with respect to y and B to be different from zero; this condition certainly occurs because this Jacobian turns out to be

$$\begin{vmatrix} 1 - d_y & 0 \\ m_y & 1 \end{vmatrix} = 1 - d_y.$$

By differentiating Eq. (11.1) we get

$$\frac{\partial y}{\partial G} = \frac{1}{1 - d_y}, \quad \frac{\partial y}{\partial i} = \frac{d_i}{1 - d_y},$$

$$\frac{\partial B}{\partial G} = -m_y \frac{\partial y}{\partial G}, \quad \frac{\partial B}{\partial i} = -m_y \frac{\partial y}{\partial i} + (K_i - m_i), \tag{11.4}$$

from which

$$\frac{\partial B/\partial i}{\partial y/\partial i} = -m_y + \frac{(K_i - m_i)(1 - d_y)}{d_i},$$

$$\frac{\partial B/\partial G}{\partial y/\partial G} = -m_y. \tag{11.5}$$

As the expression $(K_i - m_i)(1 - d_y)/d_i$ is negative, given the signs of the various derivatives, and considering the absolute values of the expressions (11.5), we have

$$\frac{|\partial B/\partial i|}{|\partial y/\partial i|} > \frac{|\partial B/\partial G|}{|\partial y/\partial G|}; \quad \frac{|\partial y/\partial G|}{|\partial B/\partial G|} > \frac{|\partial y/\partial i|}{|\partial B/\partial i|}. \tag{11.6}$$

This means that monetary policy has a relatively greater influence on the balance of payments than fiscal policy and that fiscal policy has a relatively greater influence on income than monetary policy.

In the traditional case in which the instrument of monetary policy is taken to be the money supply, system (11.1) has to be integrated by the monetary-equilibrium equation, $L(y, i) - M = 0$. Thus we get a system of three equations in five unknowns, of which two are exogenous (the instruments G and M) and three (y, B, i) are endogenous (two of these are the targets y and B). The reader can check that both the Jacobian of these equations with respect to G, M, i and the Jacobian with respect to y, B, i are different from zero. It is then possible to calculate the partial derivatives $\partial y/\partial M, \partial B/\partial M; \partial y/\partial G, \partial B/\partial G$. From these expressions it turns out that inequalities (11.6) continue to hold.

11.5.1.2 The Assignment Problem

We now come to the problem of the assignment of instruments to targets. The pairing-off of fiscal policy-internal equilibrium and monetary policy-external equilibrium gives rise to the following system of differential equations

$$\begin{aligned}
\dot{G} &= v_1(y_F - y), & v_1 > 0, \\
\dot{i} &= v_2[m(y, i) - x_0 - K(i)], & v_2 > 0, \\
\dot{y} &= v_3[d(y, i) + x_0 + G - y], & v_3 > 0.
\end{aligned} \tag{11.7}$$

The first two equations are the formal counterpart of the adjustment rules described in the text: government expenditure increases (decreases) if income is lower (higher) than the full employment level and the interest rate increases (decreases) if there is a deficit (surplus) in the balance of payments. The third equation describes the usual process of adjustment of national income in response to excess demand in a context of rigid prices; the constants v_1, v_2, v_3 indicate the adjustment velocities.

Expanding in Taylor's series around the point of full-employment equilibrium with external equilibrium and neglecting non-linear terms, we have

$$\dot{z} = Az, \tag{11.8}$$

where z is the column vector of the deviation $\{\overline{G}, \overline{i}, \overline{y}, \}$ and

$$A \equiv \begin{bmatrix} 0 & 0 & -v_1 \\ 0 & v_2(m_i - K_i) & v_2 m_y \\ v_3 & v_3 d_i & v_3(d_y - 1) \end{bmatrix} \tag{11.9}$$

is the matrix of the system of differential equations. Indicating the characteristic roots by λ and expanding the characteristic equation, we have

$$\lambda^3 + \left[v_2(K_i - m_i) + v_3(1 - d_y)\right]\lambda^2 + \left[v_2 v_3(K_i - m_i)(1 - d_y)\right.$$

$$\left. - v_2 v_3 m_y d_i + v_1 v_3\right]\lambda + v_1 v_2 v_3(K_i - m_i) = 0, \tag{11.10}$$

and the necessary and sufficient conditions of stability are (Gandolfo 2009; Sect. 16.4)

$$v_2(K_i - m_i) + v_3(1 - d_y) > 0,$$
$$v_2 v_3(K_i - m_i)(1 - d_y) - v_2 v_3 m_y d_i + v_1 v_3 > 0,$$
$$v_1 v_2 v_3(K_i - m_i) > 0,$$
$$\left[v_2(K_i - m_i) + v_3(1 - d_y)\right]\left[v_2 v_3(K_i - m_i)(1 - d_y)\right. \tag{11.11}$$
$$\left. - v_2 v_3 m_y d_i + v_1 v_3\right] - v_1 v_2 v_3 (K_i - m_i)$$
$$= \left[v_2(K_i - m_i) + v_3(1 - d_y)\right]\left[v_2 v_3(K_i - m_i)(1 - d_y)\right.$$
$$\left. - v_2 v_3 m_y d_i\right] + v_1 v_3^2(1 - d_y) > 0.$$

It is easy to see that, given the signs of the various derivatives and as, by assumption, $d_y < 1$, all these conditions are satisfied. Starting from an initial situation below full employment, the system will converge to the full-employment equilibrium with external equilibrium.

Let us now consider the assignment of monetary policy to internal equilibrium and fiscal policy to external equilibrium, which gives rise to the following system of differential equations:

$$\dot{G} = k_1 \left[x_0 - m(y, i) + K(i)\right], \quad k_1 > 0,$$
$$\dot{i} = k_2 \left[y - y_F\right], \qquad\qquad k_2 > 0, \tag{11.12}$$
$$\dot{y} = k_3 \left[d(y, i) + x_0 + G - y\right], \ k_3 > 0,$$

from which, by linearising around the full-employment equilibrium point,

$$\dot{z} = A_1 z, \tag{11.13}$$

where, as before, z indicates the vector of the deviations and

$$A_1 \equiv \begin{bmatrix} 0 & k_1 (K_i - m_i) & -k_1 m_y \\ 0 & 0 & k_2 \\ k_3 & k_3 d_i & k_3(d_y - 1) \end{bmatrix} \tag{11.14}$$

is the matrix of the system of differential equations. If we expand the characteristic equation, we have

$$\lambda^3 + k_3 (1 - dy) \lambda^2 + \left(k_1 k_3 m_y + k_2 k_3 d_i\right) \lambda - k_1 k_2 k_3 (K_i - m_i) = 0. \tag{11.15}$$

One can immediately see that, as the constant term is negative, one of the stability conditions is violated and thus the assignment in question gives rise to a movement that diverges from full-employment equilibrium.

11.5.1.3 A Generalization of the Assignment Problem

Consider the following typical target-instrument policy problem

$$\mathbf{x} = \mathbf{Ay}, \tag{11.16}$$

where $\mathbf{x}$ is the vector of the deviations of the n targets from equilibrium, $\mathbf{y}$ is the vector of the deviations of the n instruments from equilibrium, and $\mathbf{A}$ is the $n \times n$ matrix representing the reduced form of the underlying model. The number of independent instruments has been assumed equal to the number of targets. Assume now that the instruments are changed in response to the disequilibria in the targets according to the following general dynamic scheme

$$\dot{\mathbf{y}} = \mathbf{Kx}, \tag{11.17}$$

where the elements of the matrix $\mathbf{K} = [k_{ij}]$ represent the weight of the j-th target in the determination of the adjustment of the i-th instrument. The assignment scheme (or decentralized policy making) is a particular case, in which each instrument is managed in relation to only one target. In this case it is always possible, by appropriately renumbering the variables and rearranging the equations if the case, to write the dynamic system (11.18) as

$$\mathbf{y}' = \mathbf{kx}, \tag{11.18}$$

where $\mathbf{k} = \text{diag}\{k_{11}, k_{22}, \ldots, k_{mm}\}$ is a diagonal matrix. Substituting from Eq. (11.16) into (11.18) we have

$$\mathbf{y}' = \mathbf{kAy}. \tag{11.19}$$

The problem is then to choose the policy parameters k_{ii} in an economically meaningful way so that system (11.19) is stable.

The stability of the model depends on the roots of the matrix $\mathbf{kA}$. If the matrix $\mathbf{A}$ is stable on its own, by using the results on D-stability (see Gandolfo 2009; Sect. 18.2.2.1) it will be possible to choose economically meaningful (i.e., positive) policy parameters such that the matrix $\mathbf{kA}$ remains stable. More generally, it is possible to apply the Fisher and Fuller theorem; this describes the conditions that a matrix $\mathbf{A}$ must fulfil for a diagonal matrix $\mathbf{k}$ to exist capable of making the product matrix $\mathbf{kA}$ stable (see Gandolfo 2009; Sect. 18.2.2.1). The diagonal matrix $\mathbf{k}$ may further be chosen so that the matrix $\mathbf{kA}$ has strictly negative real eigenvalues. It follows that, if $\mathbf{A}$ is unstable but satisfies the required conditions, it is possible to find a matrix $\mathbf{k}$ that not only stabilizes the system but also eliminates possible oscillations.

In our case, we want $k_{ii} > 0$, hence the matrix $\mathbf{A}$ must satisfy conditions (18.94) in Gandolfo (2009).

11.5.2 Monetary and Fiscal Policy Under Flexible Exchange Rates

We consider the model

$$
\begin{aligned}
y &= A(y, i) + x(r) - rm(y, i, r) + G, \quad \text{where } 0 < A_y < 1, A_i < 0, \\
&\quad x_r > 0, 0 < m_y < 1, m_i < 0, m_r < 0, A_i - m_i < 0, \\
M &= L(y, i), \quad L_y > 0, L_i < 0, \\
B &= x(r) - rm(y, i, r) + K(i), \quad K_i > 0.
\end{aligned} \tag{11.20}
$$

The signs of the various derivatives are self-explanatory; we only note that $A_i - m_i < 0$ derives from the fact that $A_i - m_i$ is the effect of the interest rate on the expenditure on *domestic* output by residents, which is clearly negative.

We assume that the exchange rate instantaneously keeps the balance of payments in equilibrium: this means that, for any set of values of y, i, the exchange rate r is always such that $B = 0$. As we have seen in the Appendix to Chap. 10, from the implicit function

$$
B(y, i, r) = x(r) - rm(y, i, r) + K(i) = 0 \tag{11.21}
$$

we can express r as a function of y, i. This is of course possible provided that the Jacobian of $B(y, i, r)$ with respect to r is different from zero. The condition is

$$
B_r = x_r - m - rm_r = m\left(\frac{x}{rm}\eta_x + \eta_m - 1\right) \neq 0. \tag{11.22}
$$

Thus we obtain the differentiable function

$$
r = r(y, i), \tag{11.23}
$$

whose derivatives are, by the implicit function theorem,

$$
\frac{\partial r}{\partial y} = -\frac{B_y}{B_r} = \frac{rm_y}{B_r}, \frac{\partial r}{\partial i} = -\frac{B_i}{B_r} = \frac{rm_i - K_i}{B_r}. \tag{11.24}
$$

It is easy to see that $\partial r / \partial y \gtrless 0$ according as $B_r \gtrless 0$, namely according as the critical elasticities condition is satisfied or not. Similarly, $\partial r / \partial i \gtrless 0$ according as $B_r \lessgtr 0$.

Given (11.21) and (11.23), we can write the basic system (11.20) as

$$
\begin{aligned}
\{A(y, i) + x[r(y, i)] - r(y, i)m[y, i, r(y.i)] + G\} - y &= 0, \\
L(y, i) - M &= 0.
\end{aligned} \tag{11.25}
$$

It is possible to express y and i as differentiable functions of G and M if the Jacobian of (11.25) with respect to y, i is different from zero. We calculate:

$$|\mathbf{J}| = \begin{vmatrix} A_y - m_y + (x_r - m - m_r)\frac{m_y}{B_r} - 1 & A_i - m_i + (x_r - m - m_r)\frac{m_i - K_i}{B_r} \\ L_y & L_i \end{vmatrix}$$

$$= \begin{vmatrix} A_y - 1 & A_i - K_i \\ L_y & L_i \end{vmatrix} = L_i(A_y - 1) - L_y(A_i - K_i) > 0, \tag{11.26}$$

where in writing the second determinant we have used (11.22). Given the assumed signs of the various derivatives, the Jacobian is positive.

Differentiating (11.25) with respect to G we obtain

$$(A_y - 1)(\partial y/\partial G) + (A_i - K_i)(\partial i/\partial G) = -1, \\ L_y(\partial y/\partial G) + L_i(\partial i/\partial G) = 0, \tag{11.27}$$

from which

$$\frac{\partial y}{\partial G} = \frac{-L_i}{|\mathbf{J}|} > 0, \qquad \frac{\partial i}{\partial G} = \frac{L_y}{|\mathbf{J}|} > 0. \tag{11.28}$$

We also have

$$\frac{\partial r}{\partial G} = \frac{\partial r}{\partial y}\frac{\partial y}{\partial G} + \frac{\partial r}{\partial i}\frac{\partial i}{\partial G} = \frac{-\frac{rm_y}{B_r}L_i + \frac{rm_i - K_i}{B_r}L_y}{|\mathbf{J}|} \gtrless 0. \tag{11.29}$$

An expansionary fiscal policy raises income and the interest rate (the interest rate has to rise so as to keep money demand constant in the face of increased income). As regards the exchange rate, it will depreciate or appreciate according to the degree of capital mobility, namely

$$\frac{\partial r}{\partial G} \gtrless 0 \text{ according as } -\frac{rm_y}{B_r}L_i + \frac{rm_i - K_i}{B_r}L_y \gtrless 0. \tag{11.30}$$

Rearranging terms and setting $r = 1$ in the neighbourhood of the equilibrium point we obtain

$$\frac{\partial r}{\partial G} \gtrless 0 \text{ according as } \frac{m_y}{K_i - m_i} \gtrless -\frac{L_y}{L_i}. \tag{11.31}$$

A high capital mobility (K_i is high) will make this inequality satisfied with the $<$ sign, which means that the exchange rate will appreciate when capital is highly mobile. If we recall from the appendix to the previous chapter that

$$
\begin{aligned}
\left(\frac{di}{dy}\right)_{BB} &= \frac{m_y}{K_i - m_i} > 0, \\
\left(\frac{di}{dy}\right)_{LM} &= -\frac{L_y}{L_i} > 0,
\end{aligned}
\tag{11.32}
$$

we can also write inequality (11.31) as

$$
\frac{\partial r}{\partial G} \gtreqless 0 \text{ according as } \left(\frac{di}{dy}\right)_{BB} \gtreqless \left(\frac{di}{dy}\right)_{LM},
\tag{11.33}
$$

which is the relation used in the text.

We next differentiate (11.25) with respect to M and obtain

$$
\frac{\partial y}{\partial M} = \frac{K_i - A_i}{|\mathbf{J}|} > 0, \qquad \frac{\partial i}{\partial M} = \frac{A_y - 1}{|\mathbf{J}|} < 0.
\tag{11.34}
$$

We also have

$$
\frac{\partial r}{\partial M} = \frac{\partial r}{\partial y}\frac{\partial y}{\partial M} + \frac{\partial r}{\partial i}\frac{\partial i}{\partial M} = \frac{\dfrac{rm_y}{B_r}(K_i - A_i) + \dfrac{rm_i - K_i}{B_r}(A_y - 1)}{|\mathbf{J}|} > 0.
\tag{11.35}
$$

An increase in the money supply raises income but the interest rate has to decrease so that money demand increases sufficiently to maintain monetary equilibrium. The exchange rate will in any case depreciate, the more so the higher the capital mobility.

11.5.3 Perfect Capital Mobility

Let us again consider the basic system

$$
\begin{aligned}
&y = A(y, i) + x(r) - rm(y, i, r) + G, \text{ where } 0 < A_y < 1, A_i < 0, \\
&\quad x_r > 0, 0 < m_y < 1, m_i < 0, m_r < 0, A_i - m_i < 0, \\
&M = L(y, i), \quad L_y > 0, L_i < 0, \\
&B = x(r) - rm(y, i, r) + K(i) = 0, \quad K_i > 0.
\end{aligned}
\tag{11.36}
$$

In the case of perfect capital mobility ($K_i \to \infty$) and static expectations the third equation collapses to

$$
i = i^*,
\tag{11.37}
$$

where i^* is the given foreign interest rate. Correspondingly, the slope of the BB schedule in the (y, i) plane goes to zero. In fact, from $B(y, i, r) = 0$, the implicit function theorem gives us

$$\left(\frac{\partial i}{\partial y}\right)_{BB} = -\frac{B_y}{B_i} = \frac{rm_y}{K_i - m_i},$$

(11.38)

from which

$$\lim_{K_i \to \infty} \left(\frac{\partial i}{\partial y}\right)_{BB} = 0.$$

(11.39)

To examine the effects of fiscal and monetary policy under fixed and flexible exchange rates we can then consider the model

$$y - [A(y, i^*) + x(r) - rm(y, i^*, r) + G] = 0,$$
$$M - L(y, i^*) = 0,$$

(11.40)

where we assume that $B_r > 0$.

Under *fixed exchange rates* M ceases to be a policy tool, because with perfect capital mobility any attempt at modifying the money supply from its equilibrium level (namely, the level corresponding to money demand) is immediately offset by the capital flows induced by the pressure of monetary disequilibrium on the interest rate. Hence M has to be treated as an endogenous variable in (11.40). We can express y, M as differentiable functions of G if the Jacobian

$$\begin{vmatrix} 1 - A_y + m_y & 0 \\ -L_y & 1 \end{vmatrix} = \left(1 - A_y + m_y\right)$$

(11.41)

is different from zero, which is indeed the case. Differentiating (11.40) with respect to G we immediately obtain

$$\frac{\partial y}{\partial G} = \frac{1}{1 - A_y + m_y} > 0, \qquad \frac{\partial M}{\partial G} = L_y \frac{\partial y}{\partial G} > 0,$$

(11.42)

which shows that fiscal policy is effective.

Under *flexible exchange rates* from (11.40) we can express y and r as differentiable functions of G, M if the Jacobian

$$\begin{vmatrix} 1 - A_y + m_y & -B_r \\ -L_y & 0 \end{vmatrix} = -B_r L_y$$

(11.43)

is different from zero, which is indeed the case. Differentiating (11.40) with respect to M we obtain

$$(1 - A_y + m_y)(\partial y / \partial M) - B_r(\partial r / \partial M) = 0,$$
$$-L_y(\partial y / \partial M) = -1,$$

(11.44)

from which

$$\frac{\partial y}{\partial M} = \frac{1}{L_y} > 0, \quad \frac{\partial r}{\partial M} = \frac{1 - A_y + m_y}{B_r L} > 0. \tag{11.45}$$

Monetary policy is effective: in the new equilibrium income will be higher and the exchange rate will depreciate.

If we now differentiate (11.40) with respect to G we obtain

$$(1 - A_y + m_y)(\partial y/\partial G) - B_r(\partial r/\partial G) = 1,$$
$$-L_y(\partial y/\partial G) = 0, \tag{11.46}$$

from which

$$\frac{\partial y}{\partial G} = 0, \quad \frac{\partial r}{\partial G} = -\frac{1}{B_r} < 0. \tag{11.47}$$

Fiscal policy is ineffective: its only result is an appreciation of the exchange rate with unchanged income.

References

Frenkel, J. A., & Razin, A. (1987). The Mundell-Fleming model a quarter century later. *International Monetary Fund Staff Papers, 34*, 567–620.

Gandolfo, G. (2009). *Economic dynamics* (4th ed.). Berlin, Heidelberg, New York: Springer.

Johnson, H. G. (1965). Some aspects of the theory of economic policy in a world of capital mobility. *Rivista Internazionale di Scienze Economiche e Commerciali, 12*, 545–559.

Mundell, R. A. (1962). The appropriate use of monetary and fiscal policy under fixed rates. *International Monetary Fund Staff Papers, 9*, 70–79.

Mundell, R. A. (1968). The nature of policy choices, Chap. 14 in R. A. Mundell. *International Economics*. London: Macmillan.

Petit, M. L. (1985). Path controllability of dynamic economic systems. *Economic Notes, 1*, 26–42.

Shinkai, Y. (1975). Stabilization policies in an open economy: A taxonomic discussion. *International Economic Review, 16*, 662–681.

Von Neumann Whitman, M. (1970). *Policies for internal and external balance*. Special Papers in International Economics No. 9, International Finance Section, Princeton University.

Part IV

Stock and Stock-Flow Approaches

The Monetary Approach to the Balance of Payments and Related Approaches

12

12.1 Introduction

The first attack to the dominant paradigm treated in the previous chapters came from the monetary approach to the balance of payments (MABP), which was born during the 1960s and widely diffused during the following decade (see the contributions contained in Frenkel and Johnson eds., 1976).

With a leap of over two hundred years, the supporters of the MABP claimed their allegiance to David Hume (1752), considered the author of the first complete formulation of the classical theory of the mechanism for the adjustment of the balance of payments based on the flows of money (gold).

To put the question into proper perspective we shall briefly examine the classical price-specie-flow mechanism before turning to the MABP.

12.2 The Classical (Humean) Price-Specie-Flow Mechanism

This theory can be summed up as follows: under the gold standard (hence with fixed exchange rates) a surplus in the balance of payments causes an inflow of gold into a country, that is to say—as there is a strict connection between gold reserves and the amount of money—it causes an increase in prices (the quantity theory of money is assumed to be valid). This increase tends on the one hand to reduce exports, as the goods of the country in question become relatively more expensive on the international market, and on the other, it stimulates imports, as foreign goods become relatively cheaper. There is thus a gradual reduction in the balance-of-payments surplus. An analogous piece of reasoning is used to explain the adjustment in the case of a deficit: there is an outflow of gold which causes a reduction in the stock of money and a reduction in domestic prices, with a consequent stimulation of exports and a reduction of imports, which thus lead to a gradual elimination of the deficit itself.

© Springer-Verlag Berlin Heidelberg 2016
G. Gandolfo, *International Finance and Open-Economy Macroeconomics*,
Springer Texts in Business and Economics, DOI 10.1007/978-3-662-49862-0_12

A different way of looking at the same thing is based on the notion of the *optimum distribution of specie* (gold). As the level of prices in each country, under the gold standard, and under the assumption that the quantity theory of money is valid, depends on the quantity of gold that exists in the country and as the balance of payments of a country can be in equilibrium only when the levels of domestic and foreign prices, on which both imports and exports depend, are in fact such as to ensure this equilibrium, then it follows that the equilibrium under consideration corresponds to a very precise distribution (in fact known as the *optimum* distribution), among all countries, of the total amount of gold available in the world. If this balanced distribution is altered in any way, then upsets in the levels of prices will follow, with payments imbalances and therefore flows of gold which, by way of the mechanism described above, will automatically re-establish the balance.

In the reasoning of Hume there is an obvious flaw, that readers of this book will immediately spot. His starting point is that the inflow of gold and the consequent price increase due to a surplus does on the one hand reduce exports, and on the other stimulate imports. This is perfectly correct under his assumptions. However, his conclusion that there will consequently be a gradual reduction in the balance-of-payments surplus is not warranted. For it to be valid it is in fact necessary that the sum of the price-elasticities of exports and imports should be greater than one. This will formally be shown in the Appendix, but is intuitively clear. This condition is actually similar to the one we have already met in Sect. 7.2 when dealing with the effects of a variation in the exchange rate. And in fact, in the context examined, the case of a variation in the exchange rate at constant price levels is equivalent to the case of a variation in price levels with a fixed exchange rate. This confirms the observation already made in Sect. 7.1, with regard to the effects of a variation in the terms of trade $\pi = p_x/r p_m$. In fact, once we accept that the key variable which acts upon the demands for imports and exports is the relative price of the goods, π, it becomes irrelevant whether π varies because the prices (expressed in the respective national currencies) vary while the exchange rate r is given or because r varies while the prices themselves are given. It is therefore not surprising that the conditions concerning the elasticities are formally the same. However, it is necessary at this point to introduce some important qualifications.

In the first place, the analysis of the effects of a variation in the exchange rate on the balance of payments is a partial equilibrium analysis, as explained in Sect. 7.1, whereas the one examined in the present section is a general equilibrium analysis, in which money, prices and quantities interact. The assumption that y is constant (see above) should not be deceptive: national income is in fact not treated as a constant because of a *ceteris paribus* clause, but is given at the level of full employment, thanks to the operation of those spontaneous forces which, according to the classical economists, keep the "real" system in full employment. These forces are of an exclusively real nature, because money is a veil, which only serves to determine the absolute level of prices. It is therefore this dichotomy which makes it possible for us to examine the monetary part of the model with a given y, as we have done.

In the second place, bearing in mind the distinction between flow and stock disequilibria (see Chap. 1), we must note that the analysis of the effects of a variation in the exchange rate on the balance of payments refers to *flow disequilibria*, while the analysis in the present section the present section refers to *stock disequilibria*. It is likewise true that in this case the balance-of-payments disequilibrium also shows itself—and it could not be otherwise—in a divergence between the value of imports and that of exports, this divergence between the flows is ultimately due to a *stock disequilibrium*: as we have seen, the balance-of-payments disequilibrium is determined, ultimately, by an inexact distribution of the stock of money (gold) existing in the world, that is to say, a divergence, in each country, between the existing and the optimum stock. It is in this sense that we have talked, in the present context, about a stock disequilibrium. This in substance is the meaning of the idea of "optimum distribution of specie (gold)".

Hume's theory was accepted by later writers. Schumpeter wrote (1954, p. 367): "In fact it is not far from the truth to say that Hume's theory, including his overemphasis on price movements as the vehicle of adjustments, remained substantially unchallenged until the twenties of this century."

It should however be pointed out that certain classical writers admitted the existence of a direct influence of "purchasing power" (identified with the amount of specie in circulation) on imports, by means of which, they believed, it would be possible to re-establish the balance-of-payments equilibrium *without* movements in prices necessarily intervening. Painstaking research into what the classical economists had to say concerning this influence was carried out by Viner (1937), to whom we refer the reader. See also Gandolfo (1966), where it is demonstrated that Hume was indeed aware of the possible existence of a *direct* influence of purchasing power on imports.

Be it as it may, the main point of Hume's analysis is in that the ultimate cause of the balance-of-payments disequilibria is to be found in monetary disequilibria, namely in a divergence between the quantity of money in existence and the optimum or desired quantity. It is from these considerations that the supporters of the monetary approach to the balance of payments took their cue. We now turn to an examination of this approach in the next section.

12.3   The Monetary Approach to the Balance of Payments

The idea that changes in the money-stock are important in explaining balance-of-payments adjustment was already present in Mundell's model (see Mundell 1968; see also Chap. 10 above, p. 201), but it was developed by the MABP to give these changes an *exclusive* importance in the context of a different model based on a pure stock-adjustment behaviour.

12.3.1 The Basic Propositions and Implications

The MABP can be summed up in a few basic propositions, from which certain implications for economic policy can be derived. The principal propositions are the following.

Proposition I

The balance of payments is essentially a monetary phenomenon and must therefore be analysed in terms of adjustment of money stocks. More precisely, the balance-of-payments disequilibria reflect stock disequilibria in the money market (excess demand or supply) and must therefore be analysed in terms of adjustment of these stocks toward their respective desired levels (it is in fact the disequilibria in stocks which generate adjustment flows). It follows that the demand for money (assumed to be a stable function of few macroeconomic variables) and the supply of money represent the theoretical relations on which the analysis of the balance of payments must be concentrated.

In order to fully understand the meaning of this proposition, we shall begin by examining the initial statement.

That the balance of payments is essentially a monetary phenomenon is obvious (if we take this statement in the trivial sense that the balance of payments has, by its very nature, to do with monetary magnitudes) and a necessary consequence of the accounting relationships examined in Chap. 6. In particular, from the accounting matrix of real and financial flows we obtained

$$\triangle R = \triangle M - \triangle Q. \tag{12.1}$$

Now, as the variation in international reserves is nothing but the overall balance of payments, which coincides on the basis of (12.1) with the difference between the variation in the stock of money and the variation in the other financial assets, it is obvious that the balance of payments is a monetary phenomenon. But we could also say, still on the basis of the accounting identities, that the balance of payments (or more precisely the current account) is a real phenomenon, because, on the basis of (6.2), we have

$$CA = Y - A.$$

Thus, in order to derive operational consequences from the statement that the balance of payments is a monetary phenomenon, it is necessary to go beyond the identity (12.1) and introduce behaviour hypotheses, and this in fact constitutes the aim of the second part of the proposition under examination.

More precisely, the basic idea is that any monetary disequilibria produce an effect on the aggregate expenditure for goods and services (absorption) in the sense that an excess supply of money causes—*ceteris paribus*—absorption to be greater than income and, conversely, an excess demand for money causes absorption to be smaller than income itself. It is necessary at this point to warn the reader that so far we have not made any distinction between money in nominal terms (M) and money in real terms (M/p), given the hypothesis of a constant price level. It is however important to stress that, in a context of variable prices, what is important is money in real terms.

The connection between the demand for money and the other macroeconomic variables is in turn immediate, given the hypothesis that it is a stable function of few of these variables; it is therefore easy, if the existing money stock is known, to determine the excess demand or supply of money.

On the other hand, the divergence between income and absorption which is created in this way, is necessarily translated into an increase or decrease in the stock of assets owned by the public: in fact, this divergence is the equivalent of one between saving and investment and therefore, given the balance constraint of the private sector (see Sect. 6.2), it is translated into a variation in the stock of assets held by this sector. If, for the sake of simplicity, we introduce the hypothesis that the only asset is money, a variation in the money stock comes about, which in turn, given (12.1), coincides with the overall balance of payments.

Ultimately what has happened, through this sequence of effects (for which we shall give a model below) is that an excess demand or supply of money, by causing an excess or deficiency of absorption with respect to national income (product), has been unloaded onto the balance of payments: an excess of absorption means a balance-of-payments deficit (the only way of absorbing more than one produces is to receive from foreign countries more than one supplies to them) and a deficiency in absorption means a balance-of-payments surplus. In other words, if the public has an excess supply of money it gets rid of it by increasing absorption and, ultimately, by passing this excess on to foreign countries in exchange for goods and services (balance-of-payments deficit). If on the other hand the public desires more money than it has in stock, it procures it by reducing absorption and, ultimately, it passes goods and services on to foreign countries in exchange for money (balance-of-payments surplus).

What is implicit in our reasoning is the hypothesis that the level of prices and the level of income are a datum (otherwise the variations in absorption with respect to income could generate variations in prices and/or income) and in fact this is what, among other things, the next propositions refer to.

Proposition II

There exists an efficient world market for goods, services and assets and that implies, as far as goods are concerned (as usual, goods refers to both goods and services), that the goods themselves must have the same price everywhere (law of one price), naturally account being taken of the rate of exchange (which, it must be remembered, is, by hypothesis, fixed) and therefore that the levels of prices must be connected—if we ignore the cost of transport—by the relationship

$$p = rp_f, \tag{12.2}$$

where p_f is the foreign price level expressed in foreign currency, p is the domestic price level in domestic currency, and r is the exchange rate. Equation (12.2), which is also called the equation of *purchasing power parity*,[1] has the effect that, given r and p_f, p will be fixed. This is valid under the hypothesis of a one-country model. It is also possible to extend the analysis to two countries and to determine both p and p_f, but the results of the analysis do not change substantially (see the Appendix).

Proposition III

Production is given at the level of full employment. This proposition implies that the MABP is a long-term theory, in which it is assumed that production tends toward the level of full employment thanks to price and wage adjustments. From this point of view the strict relationship with the classical theory examined in the previous section is clear.

Given these assumptions, important policy implications follow, and precisely:

Implication I. In a regime of fixed exchange rates, monetary policy does not control the country's money supply. This implication comes directly from the propositions given above and particularly Proposition I. In fact, if the monetary authorities try to create a different money supply from that desired by the public, the sole result will be to generate a balance-of-payments disequilibrium, onto which, as we have seen, the divergence between the existing money stock and the demand for money will be unloaded.

Implication II. The process of adjustment of the balance of payments is auto-matic, and the best thing that the monetary authorities can do is to abstain from all intervention. This implication also immediately derives from the propositions given above and particularly from Proposition I. The balance-of-payments disequilibria are in fact monetary symptoms of money-stock disequilibria which correct themselves in time, if the automatic mechanism of variation in the money stock is allowed to work. If, let us say, there is a balance-of-payments deficit (which is a symptom of an excess supply of money), this deficit will automatically cause a reduction in the stock of money—see (12.1), where (as we assumed above) ΔQ is zero—and therefore a movement of this stock towards its desired value. When the desired stock is reached, the deficit in the balance of payments will disappear. It may happen, however, in a fixed exchange regime other than the gold standard, that in the course of this adjustment the stock of international reserves of the country will show signs of exhaustion before equilibrium is reached. In this and only this case, a policy intervention is advisable, which in any case should consist of a monetary restriction (so as to reduce the money supply more rapidly towards its desired value) and not of a devaluation in the exchange rate or any other measure, which are inadvisable palliatives, with purely transitory effects.

---

[1]We shall deal more fully with the purchasing power parity later, among other things so as to examine the reasons for deviations from it, in Sect. 15.1.

12.3.2 A Simple Model

It is possible to give a simple model for the MABP in the basic version so far illustrated. We must remember in the following discussion that the price level is considered to be constant (see the Appendix for a more general model in which prices are allowed to vary). The exchange rate is also constant; for a model in which the exchange rate is allowed to vary (the monetary approach then serves as a theory of exchange-rate determination) see Sect. 15.3.1.

The first behavioural equation expresses the fact that excess or deficiency of absorption with respect to income varies in relation to the divergence between money supply (M) and demand (L) :

$$A - py = \alpha\,(M - L), \quad 0 < \alpha < 1,$$

or   (12.3)

$$A = py + \alpha\,(M - L), \quad 0 < \alpha < 1.$$

The parameter α is a coefficient which denotes the intensity of the effect on absorption of a divergence between M and L. It is assumed positive but smaller than unity because the divergence is not entirely eliminated within one period, due to lags, frictions and other elements.

Equation (12.3) is the crucial assumption of the MABP, and represents expenditure as coming out of a stock adjustment rather than deriving from a relation between flows as in the functions in the Keynesian tradition encountered in previous chapters. Here the representative agent does not directly decide the flow of expenditure in relation to the current flow of income, but does instead decide the flow of (positive or negative) saving out of current income in order to bring the current stock of wealth (here represented by money) to its desired stock (here represented by money demand).

The second behavioural equation expresses the demand for money as a stable function of income:

$$L = kpy.$$ (12.4)

Equation (12.4) can be interpreted as a simple version of the quantity theory of money; alternatively, one can take parameter k (equal to the reciprocal of the velocity of circulation of money) as a function of the interest rate (in which case the demand for money becomes a function of this variable). However, given Proposition II, i is a datum. In fact, the hypothesis of efficient markets (Proposition II) implies that the national interest rate cannot diverge from the value which ensures the condition of interest parity (see Sect. 4.5). Therefore, as the rate of exchange (spot and forward) and the foreign interest rate are given, the national interest rate is fixed, and therefore L practically becomes a function of income only.

We then have the accounting equation

$$\Delta M = py - A, \tag{12.5}$$

which expresses the fact that—if we assume that money is the sole asset in existence—the excess of income over expenditure coincides with an increase in the money stock held by the public (monetary base in the hands of the public plus bank deposits) and vice versa, as can be seen from the matrix of real and financial flows illustrated in Chap. 6 and precisely from Eq. (6.12), where the only asset is assumed to be money (monetary base in the hands of the public plus bank deposits).

Given the hypothesis that the only existing asset is money, the accounting identity (12.1) becomes

$$\Delta R = \Delta M. \tag{12.6}$$

Finally, we have the equation

$$y = \bar{y}, \tag{12.7}$$

which expresses Proposition III ($\bar{y}$ indicates the level of full employment income expressed in real terms).

By substituting (12.3) into (12.5) and then into (12.6), we have

$$\Delta R = \alpha \left(L - M \right), \tag{12.8}$$

which says that the variation in reserves (coinciding with the balance of payments) depends, through coefficient α, on the divergence between the demand and supply of money. Given (12.4) and (12.7), we can also write

$$\Delta R = \alpha \left(kp\bar{y} - M \right), \tag{12.9}$$

where $kp\bar{y}$ is the desired stock of money.

The self-correcting nature of the disequilibria can be clearly seen from Eqs. (12.9) and (12.6): if the existing money stock is excessive, that is if $M > kp\bar{y}$, then ΔR is negative and therefore so is ΔM: thus M decreases and moves toward its equilibrium value $kp\bar{y}$. If the existing stock of money is inadequate, that is if $M < kp\bar{y}$, ΔR is positive and therefore so is ΔM, so that M increases toward its equilibrium value.

12.3.3 Does a Devaluation Help?

One of the striking consequences of the MABP is the ineffectiveness of an exchange-rate devaluation to create an ongoing surplus in the balance of payments. In fact, a devaluation can bring about only a *transitory improvement* in the balance of payments. On the contrary, according to the traditional approach, if the suitable critical elasticities condition occurs, then a devaluation will cause a lasting improvement in the balance of payments. In fact, in the traditional approach, based on flows, the improvement due to the devaluation is *itself* a flow which persists through time, naturally other things being equal.

The fact is that the logical frame of reference is entirely different. In the traditional approach, a devaluation causes variations in relative prices and therefore induces substitution effects and also (where contemplated) effects on income. Nothing of this can happen in the MABP, given the assumption of purchasing power parity (Proposition II) and full employment (Proposition III). What happens in the MABP frame of reference is that a devaluation, by causing the price level to rise, generates an *increase in the desired money stock.* As the existing stock is what it is, and therefore insufficient, the mechanism already described (reduction in absorption, concomitant balance-of-payments surplus i.e. increases in international reserves and the money supply) is set into motion; when the desired money supply is reached, the adjustment comes to an end and the balance-of-payments surplus thus disappears.

On the other hand, the fact that this must be the case according to the MABP is obvious: the flows deriving from stock disequilibria are (unlike pure flows) necessarily transitory, insofar as they are bound to disappear once the stocks have reached their desired levels.

It is important to note that the argument could be conducted in terms of real magnitudes without thereby causing a change in the results. The demand for money in real terms, $L/p = k\bar{y}$, is given. The supply of money in real terms, M/p, varies inversely with the price-level. The increase in p, by causing the real value of the money stock to diminish, causes an excess demand for money and thus the mechanism for the reduction in absorption is set into motion [absorption in real terms is $A/p = \bar{y} + \alpha(M/p - L/p)$], etc., which produces an increase in the nominal stock of money. This increase continues to the point where the real value of the money stock has returned to the initial level of equilibrium (that is, when M has increased in the same proportion as p). It can be seen that, in the long term, the real variables do not change, insofar as the variations in one of the nominal variables (e.g. in the level of prices just examined) set automatic adjustment mechanisms of other nominal variables into motion and these re-establish the equilibrium value of the real variables (the equilibrium value of absorption in real terms is given and equals $\bar{y}$).

12.3.4 Concluding Remarks

The demolishing effect of the MABP on the traditional theory of the adjustment processes is obvious: not only are standard measures, like devaluation, ineffective, but even the more sophisticated macroeconomic policies derived from the Mundell-Fleming model (which have been deal with in Chap. 11) have to be discarded. It is enough to leave the system to its own devices (*Implication II*) for everything to be automatically adjusted.

It is not surprising therefore that the MABP gave rise to a large number of criticisms and rejoinders. Most of this debate, however, did not get the gist of the matter, which is the validity of the behavioural assumption (12.3). If it is valid, the MABP is perfectly consistent, since other controversial points (such as the neglect of financial assets other than money, the constancy of price levels, etc.) are really not essential for its conclusions. If, on the contrary, the representative agent behaves according to a pure flow expenditure function $A = A(y, \ldots)$, then the MABP conclusions are clearly unwarranted.

Be it as it may, the MABP should be given due acknowledgement for one fundamental merit: that of having directed attention to the fact that in the case of balance-of-payments disequilibria and the related adjustment processes, *stock* equilibria and disequilibria must be taken into account. Naturally, this must not be taken in the sense—typical of the cruder versions of the MABP—that *only* stock equilibria and disequilibria matter, but that they *also* matter in addition to pure *flow* equilibria and disequilibria. But if this is the case, then nothing new under the sun. As long ago as 1954 Clower and Bushaw (1954) had drawn attention, with rigorous scholarship, to the fact that there was a need for economic theory in general to consider stocks and flows simultaneously, and had shown that a theory based only on flows or one based solely on stocks was inherently incomplete and could lead to inexact conclusions.

12.4   The New Cambridge School of Economic Policy

This approach was dubbed "new" when it was proposed in the first half of the 1970s (for a historical survey see Vines 1976, and McCallum and Vines 1981) to contrast it with the "old" school (the traditional Keynesian model). Its novelty consisted in turning upside down the traditional assignment of tools to targets in a context in which the exchange rate could be used as a tool. In this context, the traditional assignment was to use fiscal policy for internal equilibrium and the exchange rate for external equilibrium (see Chap. 11). This followed from the generally accepted observation that the exchange rate has a greater relative effectiveness on the balance of payments, hence it should be assigned to the balance-of-payments target on the basis of the assignment principle. On the contrary, the new school claimed that it is fiscal policy that has a greater effectiveness on the balance of payments, hence the opposite assignment.

Although its proponents did not show any awareness of this, from the theoretical point of view the approach under examination is actually an extension of the MABP to the case of (fixed but) adjustable exchange rates. Our claim follows from the fact that the assumed expenditure function of the private sector is based on a stock-adjustment behaviour exactly like the MABP.

To understand the position of the new school it is convenient to start from the accounting framework explained in Chap. 6. The first is Eq. (6.12), namely

$$Y_p - A_p = \triangle W_p, \tag{12.10}$$

which states that the excess of disposable income over private current expenditure equals the change in the private sector's stock of financial assets (net acquisition of financial assets).

The second is Eq. (6.1), namely

$$CA = (A_p - Y_d) + (G - T). \tag{12.11}$$

Up to now we are in the context of accounting identities, from which, as we know (see Chap. 6) it would be incorrect to deduce causal relations. Of this fact the new school—which starts from Eqs. (12.10)–(12.11)—is well aware, so that at this point it introduces behaviour assumptions which, together with the accounting identities, enable it to give a precise vision of the functioning of the economy and to arrive at the policy suggestions stated above.

The first behaviour assumption concerns the functional dependence of private expenditure on disposable income. This relationship is *not*, as in traditional Keynesian models, a relationship directly and originally existing between flows (the flow of expenditure and the flow of income). On the contrary, it is a relationship deriving from an adjustment of stocks. We know that, in the orthodox Keynesian framework, the private sector decides the flow of expenditure corresponding to any given flow of income on the basis of the propensity to consume. Saving has a residual nature, as it is the part of income not spent. In the approach under consideration, on the contrary, the private sector *first* decides how much to save, and consumption has a residual nature: it is the part of income not saved. The saving decision, in turn, is due to a stock adjustment.

More precisely, the private sector decides $S - I$, and hence $Y_d - A_p$, on the basis of the *desired* value of $\triangle W_p$, let us call it $\triangle W_p^*$. This last does, in turn, depend on the desire of the private sector to bring its stock of financial assets from the current to the desired value (W_p^*). For the sake of simplicity it is assumed that the private sector is always in equilibrium,[2] namely that it is able to fully and instantaneously realize its plans. This means that

$$\triangle W_p = \triangle W_p^* = \left(W_p^* - W_p \right). \tag{12.12}$$

---

[2] The consideration of situations of disequilibrium, and of the relative adjustment mechanisms of actual to desired stocks, does not change the substance of the argument, though making it more complicated formally. See McCallum and Vines (1981).

If we take the budget constraint (12.10) and the behaviour assumption (12.12) into account, we can write the private sector's expenditure function as

$$A_p = Y_p + \left(W_p - W_p^*\right). \tag{12.13}$$

In brief, the behaviour assumption is that the private sector, wishing to hold a certain stock of net financial assets, regulates its expenditure, given its disposable income, so as to achieve the suitable (positive or negative) financial surplus to keep the actual stock of net financial assets equal to the desired one.

The vision of the functioning of the economy warrants the statement that the new school is much nearer to the MABP [Eq. (12.13) is very similar to Eq. (12.3) with $\alpha = 1$] than to the old school (the traditional Keynesian model), because, like the MABP, it stresses stock adjustments. The difference would consist only in the fact that, whilst the MABP emphasizes the money market, the new school emphasizes the commodity market, as expressed by Eq. (12.11) and as we shall further show presently.

The new school further assumes that the private sector has a fairly stable desired stock of financial assets, which it is not willing to appreciably modify in response to changes in disposable income. This means that, once W_p, has been brought to its desired level W_p^*, there will be no incentive to accumulate or decumulate financial assets, hence $A_p = Y_p$, as we get from (12.13) when $W_p^* = W_p$. Thus we have, considering Eq. (12.11),

$$CA = G - T. \tag{12.14}$$

Equation (12.14), it should be noted, is *no longer* an identity. It is true that it has been derived from identity (12.11), but by using the private sector's behaviour assumption explained above. Thus it can legitimately be used to find out causal relationships. Of course we cannot yet clam that $(G - T)$ determines CA or the other way round without further behaviour assumptions, but these do not present any particularly novel feature. Let us assume that we start from a situation in which there is a budget deficit and a deficit in the balance on goods and services, which are equal on the basis of Eq. (12.14). Now, suppose there is an increase in G which causes an increase in income (via the multiplier) and hence in imports (IMP). Let us also assume that T is a lump-sum tax (the results do not change with T variable as income varies: see McCallum and Vines 1981), hence $\Delta T = 0$. By applying the formula of the simple multiplier—see Chap. 8, Eq. (8.13)—we get $\Delta y = [1/(1 - b - h + \mu)]\Delta G$, and so $\Delta IMP = \mu \Delta y = [\mu/(1 - b - h + \mu)]\Delta G$. Since exports are assumed to be exogenous, the variation in CA will coincide with ΔIMP.

Now, $A_p = Y_p$ (see above) means that the marginal propensity to spend equals one (i.e., in the symbology of Chap. 8, $b + h = 1$), hence $\Delta IMP = \mu \Delta y = [\mu/\mu]\Delta G = \Delta G$. This proves that the (additional) deficit in the balance on goods and services is exactly equal to the increase in the budget deficit. The opposite will, of course, be true in the case of a decrease in G. Thus we have obtained a relationship that clearly goes from the budget deficit to the deficit in the current account, and in this sense we can say that $(G - T)$ "determines" CA.

Thus we have shown that fiscal policy influences the current account in the ratio of one to one. To complete the policy assignment under examination we must show that exchange-rate variations have a negligible effect on the balance on the current account and an appreciable effect on income. The latter effect is known: an exchange-rate devaluation, for example, stimulates exports (and the demand for domestic goods by residents as well), so that national income increases. As regards the current account, the assumption of the new school is that the values of the elasticities and the marginal propensity to import are such that the improvement (via elasticities) due to the direct effect of the devaluation on EXP and IMP is exactly offset by the deterioration (via marginal propensity to import) indirectly due to the devaluation itself by way of its effect on income.

This said, the application of the assignment principle immediately leads one to assign fiscal policy to external balance and the exchange rate to internal equilibrium.

The fatal criticism to this approach concerns the practical advisability of using the exchange-rate devaluation to promote domestic expansion. This is a macroeconomic measure equivalent to tariffs, export subsidies, and other prohibited forms of trade intervention. Actually it is a beggar-my-neighbour policy, as the stimulus to exports and the brake on imports in the devaluing country implies the stimulus to imports and the brake on exports in the rest of the world, which thus undergoes a depressionary disturbance. The policy under consideration, then, might possibly be successful only if the rest of the world is willing to accept this negative effect. In the contrary case, the rest of the world may well decide to follow the same policy and devalue in its turn, thus giving rise to a chain of competitive devaluations of the type experienced in the 1930s, and international trade would run the risk of again approaching what it was at the time, namely "a desperate expedient to maintain employment at home by forcing sales on foreign markets and restricting purchases" (Keynes 1936; Chap. 24, pp. 382–3).

12.5 Appendix

12.5.1 The Classical Theory

Given the definition of the balance of payments,

$$B = px(p) - p_m m(p), \tag{12.15}$$

where p_m is considered given, the condition by which a variation of p will cause the balance of payments to vary in the opposite direction is obtained by differentiating

B with respect to p and putting $dB/dp < 0$. We have

$$\frac{dB}{dp} = \frac{d\left[px - p_m m\right]}{dp} = p\frac{dx}{dp} + x - p_m\frac{dm}{dp} = x\left(\frac{p}{x}\frac{dx}{dp} + 1 - \frac{p_m}{x}\frac{dm}{dp}\right)$$

$$= x\left(-\eta_x + 1 - \frac{p_m m}{px}\eta_m\right), \qquad (12.16)$$

where

$$\eta_x \equiv -\frac{dx/x}{dp/p} \equiv -\frac{dx}{dp}\frac{p}{x}, \quad \eta_m \equiv \frac{dm/m}{dp/p} \equiv \frac{dm}{dp}\frac{p}{m}. \qquad (12.17)$$

As $x > 0$, then $dB/dp < 0$ if and only if

$$\eta_x + \frac{p_m m}{px}\eta_m > 1. \qquad (12.18)$$

We now come to dynamic analysis. The assumption that the variation in the money supply coincides with the international payments imbalance, gives rise to the differential equation

$$\dot{M} = px(p) - p_m m(p), \qquad (12.19)$$

which, since $MV = yp$ by the quantity theory, and hence $\dot{M} = (y/V)\dot{p}$, becomes

$$\dot{p} = \frac{V}{y}\left[px(p) - p_m m(p)\right]. \qquad (12.20)$$

In order to examine local stability we linearize and consider the deviations from equilibrium, thus obtaining

$$\dot{\bar{p}} = \frac{V}{y}\left(x + p\frac{dx}{dp} - p_m\frac{dm}{dp}\right)\bar{p}, \qquad (12.21)$$

where all the derivatives are calculated at the point of equilibrium. Indicating the expression in brackets with γ, the solution is

$$\bar{p}(t) = \bar{p}_0 e^{(V/y)\gamma t}, \qquad (12.22)$$

where $\bar{p}_0$ indicates the initial deviation. A necessary and sufficient stability condition is that $\gamma < 0$. We note at once that γ coincides with dB/dp obtained above; from this it follows that the condition for dynamic stability is also $\eta_m + \eta_x > 1$ (seeing that we linearize at the point of equilibrium we have $p_m m = px$) and that there is dynamic stability if $dB/dp < 0$.

This model is based on the small country hypothesis by which p_m can be considered as given. However, the results do not substantially change if a two-country model is considered: see Gandolfo (1966). For further studies in the formalization of the classical theory of the balance of payments, see Gandolfo (2009; Sect. 26.8.2) and Samuelson (1980).

12.5.2 The Monetary Approach to the Balance of Payments

We shall examine the more general two-country model, in which price levels are also endogenously determined. This model has the same equations for each country as described in the text.[3] Thus, using differential equations we have

$$\dot{M} = \alpha \left(kp\bar{y} - M \right), \tag{12.23}$$

$$\dot{M}_f = \alpha_f \left(k_f p_f \bar{y}_f - M_f \right). \tag{12.24}$$

As p and p_f are now unknowns, it is necessary to introduce two other equations. One is the already mentioned purchasing power parity relation, which is given here for convenience

$$p = rp_f. \tag{12.25}$$

The other is obtained from the assumption that the world money stock is given constant (in a gold standard regime this given stock is obviously fixed by the amount of gold in the world):

$$M + rM_f = \overline{M}_w, \tag{12.26}$$

where the magnitudes are made homogeneous by expressing them in terms of the currency of the first country.

When both countries are in equilibrium, $M = kp\bar{y}$ and $M_f = k_f p_f \bar{y}_f$ and therefore, by substituting in (12.26) and taking account of (12.25), we have

$$kp\bar{y} + rk_f \frac{1}{r} p\bar{y}_f = \overline{M}_w. \tag{12.27}$$

By solving we obtain the equilibrium price levels

$$p^e = \overline{M}_w / \left(k\bar{y} + k_f \bar{y}_f \right), \quad p_f^e = \frac{1}{r} \overline{M}_w / \left(k\bar{y} + k_f \bar{y}_f \right), \tag{12.28}$$

---

[3]To avoid confusion, it should be noted that in this section—unlike the other ones in this Appendix—a dash over a variable denotes that it is given (and not a deviation from the equilibrium). This is for the sake of conformity with the notation used in the text.

given which, we can determine the optimum or equilibrium distribution of the world money supply:

$$M^e = \bar{y}k\overline{M}_w / \left(k\bar{y} + k_f\bar{y}_f\right), \quad M_f^e = \bar{y}_f \frac{1}{r}k_f\overline{M}_w / \left(k\bar{y} + k_f\bar{y}_f\right). \tag{12.29}$$

We now come to the adjustment process. This is complicated in that, when the system is in disequilibrium, prices also vary. To find the law of price variation we begin by observing that, as world money supply is given, a variation in the money supply of one of the two countries in one direction will correspond to a variation in the money supply of the other country in the opposite direction. Formally, if we differentiate both sides of (12.26), we have

$$\dot{M} + r\dot{M}_f = 0. \tag{12.30}$$

Substituting the values of $\dot{M}$ and $\dot{M}_f$ from Eqs. (12.23) and (12.24) we obtain

$$\alpha \left(kp\bar{y} - M\right) + \alpha_f \left(k_f r p_f \bar{y}_f - rM_f\right) = 0, \tag{12.31}$$

from which, taking into account (12.25) and solving, we have

$$p = \left(\alpha M + \alpha_f rM_f\right) / \left(\alpha k\bar{y} + \alpha_f k_f\bar{y}_f\right), \quad p_f = \left(\alpha \frac{1}{r}M + \alpha_f M_f\right) / \left(\alpha k\bar{y} + \alpha_f k_f\bar{y}_f\right), \tag{12.32}$$

which determine price levels out of equilibrium. It can at once be seen that in equilibrium, as $M = kp\bar{y}$ and $M_f = k_f p_f \bar{y}_f$, Eqs. (12.32) are reduced to identities. The price dynamics out of equilibrium is therefore determined by the dynamics of the division of the world money stock between the two countries. In fact, differentiating (12.32) with respect to time, we have the movement of prices in time

$$\dot{p} = \frac{\alpha}{D}\dot{M} + \frac{\alpha_f r}{D}\dot{M}_f, \quad \dot{p}_f = \frac{\alpha \frac{1}{r}}{D}\dot{M} + \frac{\alpha_f}{D}\dot{M}_f, \tag{12.33}$$

where D denotes the expression in the denominator of Eqs. (12.32). From now on, as $p = rp_f$ and $\dot{p} = r\dot{p}_f$, we shall limit ourselves to the consideration of the price level of the home country. The first of Eqs. (12.33) can be written as

$$\dot{p} = \frac{\alpha_f}{D}\left(\frac{\alpha}{\alpha_f}\dot{M} + r\dot{M}_f\right), \tag{12.34}$$

from which, by using (12.30) and collecting the terms, we have

$$\dot{p} = \frac{\alpha - \alpha_f}{D}\dot{M}. \tag{12.35}$$

The price level, then, varies in the same direction or in the opposite direction to that in which the national money supply moves, according to $\alpha \gtrless \alpha_f$, that is, according to whether the national speed of adjustment of absorption to monetary disequilibria is greater or less than the foreign one. We shall return to this point later; for the moment we shall examine the behaviour of the money stock. Given (12.26) and (12.30), it is sufficient to examine the differential equation (12.23), which— taking (12.32) into account—can be written as

$$\dot{M} = \alpha \left(\frac{k\bar{y}\alpha}{D} M + \frac{k\bar{y}\alpha_f}{D} rM_f - M \right). \tag{12.36}$$

If we substitute $rM_f = \overline{M}_w - M$ from Eq. (12.26) and rearrange terms—account being taken of Eq. (12.29)—we get

$$\dot{M} = \alpha \left(\frac{\alpha_f k\bar{y}}{D} \overline{M}_w + \frac{k\bar{y}\alpha - k\bar{y}\alpha_f - D}{D} M \right)$$

$$= \alpha \left(\frac{\alpha_f k\bar{y}}{D} \overline{M}_w - \frac{\alpha_f \left(k\bar{y} + k_f \bar{y}_f \right)}{D} M \right)$$

$$= \alpha \frac{\alpha_f \left(k\bar{y} + k_f \bar{y}_f \right)}{D} (M^e - M). \tag{12.37}$$

From this differential equation, in which the price movements are also included, it can be seen that when M is below (above) its equilibrium value M^e given by (12.29), then $\dot{M} \gtrless 0$ and thus M increases (decreases), converging towards its equilibrium value. In fact the general solution of (12.37) is

$$M(t) = (M_0 - M^e)e^{-\alpha Ht} + M^e, \tag{12.38}$$

where H indicates the multiplying fraction of $(M^e - M)$ in (12.37). As $H > 0$, the equilibrium is stable. The movements of the balance of payments can easily be obtained from the definitional relationship $B = R = \dot{M}$, by which, taking account of (12.38), we have

$$B(t) = -\alpha H(M_0 - M^e)e^{-\alpha Ht}, \tag{12.39}$$

from which it can be seen that $B(t)$ necessarily tends towards zero.

Given (12.26), if M tends towards its equilibrium value, M_f also tends towards equilibrium; consequently—given (12.32) and (12.28)—p and p_f also tend towards their respective equilibrium levels.

It is interesting to observe that when $\alpha = \alpha_f$, that is to say, when the adjustment speeds are equal, we have—on the basis of (12.35)—$\dot{p} = 0$, that is, prices do no vary. This is due to the fact that, in this case, the *world* demand does not vary, insofar as the greater absorption of one country is exactly compensated by the lesser absorption of the other. In fact, with equal adjustment velocities, the redistribution of money through the balance of payment would leave *world* expenditure unchanged and therefore, at world level, there would be no excess demand and thus no effect on prices. Instead, when the national adjustment velocity exceeds that of the rest of the world ($\alpha > \alpha_f$), then to a surplus in the home country balance of payments (and therefore a deficit in the rest of the world) there will correspond, instant by instant, an increase in world expenditure and therefore—as the world product is given— an excess demand, which in turn will cause the increase in prices. Conversely, if $\alpha_f > \alpha$, then the contraction in expenditure of the rest of the world would more than compensate for the increase in the home country and there would thus be a world excess supply and a reduction in prices.

The fact that the adjustment equation arrived at above can be interpreted in terms of world excess demand can be demonstrated as follows. Remembering that, as explained in the text,

$$A = p\bar{y} + \alpha \left(M - kp\bar{y} \right),$$
$$A_f = p_f\bar{y}_f + \alpha_f \left(M_f - k_f p_f \bar{y}_f \right),$$

(12.40)

we get

$$\dot{M} = \alpha \left(M - kp\bar{y} \right) = A - p\bar{y},$$
$$\dot{M}_f = \alpha_f \left(M_f - k_f p_f \bar{y}_f \right) = A_f - p_f \bar{y}_f.$$

(12.41)

Substituting in (12.34), we have

$$p = \frac{\alpha_f}{D} \left[\frac{\alpha}{\alpha_f} (A - p\bar{y}) + r \left(A_f - p_f \bar{y}_f \right) \right]$$

$$= \frac{1}{D} \left[\alpha (A - p\bar{y}) + \alpha_f r \left(A_f - p_f \bar{y}_f \right) \right].$$

(12.42)

The variation in prices depends on the excess demand in the two countries. It is necessary to bear in mind that, on the basis of Walras' law, the sum of the nominal excess demand of both countries is zero[4]

$$(A - p\bar{y}) + r \left(A_f - p_f \bar{y}_f \right) = 0,$$

(12.43)

where naturally the nominal excess demand of each country is expressed in a common unit of measurement, in this case the currency of the home country. This at once explains that, when $\alpha = \alpha_f$, the level of prices does not vary: in that case, in fact, the expression in square brackets in (12.42) is cancelled out. On the other

---

[4]The fact that Walras' law holds can easily be seen thanks to the constraint (12.30). In fact, by substituting from (12.41) in (12.30) we obtain (12.43).

hand, when $\alpha > \alpha_f$, any positive excess demand in the home country is less than compensated for by the negative excess demand in the rest of the world,[5] so that the expression in square brackets in (12.42) remains positive and $\dot{p} > 0$. The opposite happens when $\alpha < \alpha_f$.

12.5.2.1 The Effects of a Devaluation

We conclude by examining the effects of a devaluation, distinguishing the impact effects from the long-run effects. The impact effect is an improvement in the balance of payments of the country which devalues (and, obviously, a worsening of the balance of payments of the other country) as already illustrated in the text. Formally this can be demonstrated in the following way. Let us assume that we are starting from an equilibrium situation, in which therefore there is no money supply movement and the balance of payments is in equilibrium. Now let us consider a devaluation on the part of the home country. The equilibrium is altered and the impact effect on price levels is obtained by differentiating (12.32) with respect to r:

$$\frac{dp}{dr} = \frac{\alpha_f M_f}{D} > 0, \qquad \frac{dp_f}{dr} = -\frac{\frac{1}{r^2}\alpha M}{D} < 0, \qquad (12.44)$$

from which it can be seen that p increases and p_f decreases. Therefore the demand for money increases in the home country and decreases in the rest of the world, so that in correspondence to existing money supply there is excess demand for money in the home country and excess supply in the other. It follows that, given (12.23) and (12.24), we get $\dot{M} > 0$ (and therefore $B > 0$) and $\dot{M}_f < 0$ (and thus $B_f < 0$).

The long-run effect is the disappearance of every surplus and deficit in the balance of payments: in fact, the dynamic mechanism described above causes M and M_f to coincide eventually with their new equilibrium values. Once these are reached, the money flows cease so that the surplus of the home country and the deficit of the rest of the world are eliminated.

We have thus demonstrated the transitory nature of the effects of a devaluation. It is however interesting to observe that there is an improvement (though of a transitory nature) in the balance of payments without any condition being placed on the elasticities. Indeed, these do not even appear in the analysis. This is due to the fact that, in the model we are examining, there is no substitution effect between domestic and foreign goods, a fact which, on the contrary, as we saw in Sect. 12.2, was at the basis of Hume's theory. It is possible to introduce an exchange rate effect into the MABP (in which case the elasticities come into play once again, on which the stability of the adjustment mechanism comes to depend); the reader who is interested in this exercise can consult, for example, Dornbusch (1980; Chap. 7,

---

[5]The magnitudes $(A - p\bar{y})$ and $r\left(A_f - p_f\bar{y}_f\right)$ are of opposite signs and equal in absolute value on the basis of (12.43), but as the former is multiplied by a greater positive coefficient than that which multiplies the latter, it is the former which prevails.

Sect. III). See also Niehans (1984; Sect. 4.4) for an interesting parallelism between the monetary approach and the elasticity approach.

References

Clower, R. W., & Bushaw, D. W. (1954). Price determination in a stock-flow economy. *Econometrica, 22*, 238–243.

Dornbusch, R. (1980). *Open economy macroeconomics*. New York: Basic Books Inc.

Frenkel, J. A., & Johnson H. G. (Eds.) (1976). *The monetary approach to the balance of payments*. London: Allen&Unwin.

Gandolfo, G. (1966). La teoria classica del meccanismo di aggiustamento della bilancia dei pagamenti. *Economia Internazionale, 19*, 397–432.

Gandolfo, G. (2009). *Economic Dynamics*, 4th edition. Berlin Heidelberg New York: Springer.

Hume, D. (1752). Of the balance of trade, in D. Hume, *Political discourses*. Edinburgh. Reprinted in D. Hume (1955). *Writings on economics*, edited by E. Rotwein. London: Nelson, 60ff, and (partially) in R.N. Cooper (Ed.) (1969). *International finance: Selected readings*. Harmondsworth: Penguin, 25–37.

Keynes J.M. (1936). *The general theory of employment, interest and money*. London: Macmillan.

McCallum, J., & Vines, D. (1981). Cambridge and Chicago on the balance of payments. *Economic Journal, 91*, 439–453.

Mundell, R.A., 1968, *International Economics*, New York: Macmillan.

Niehans, J. (1984). *International monetary economics*. Oxford: Philip Allan.

Samuelson, P.A. (1980). A corrected version of Hume's equilibrating mechanisms for international trade, in J.S. Chipman & C.P. Kindleberger (Eds.), *Flexible exchange rates and the balance of payments-Essays in memory of egon sohmen*. Amsterdam: North-Holland, 141–158.

Schumpeter, J.A. (1954). *History of economic analysis*. Oxford: Oxford University Press.

Viner, J. (1937). *Studies in the theory of international trade*. London: Allen&Unwin.

Vines, D. (1976). Economic policy for an open economy: Resolution of the new school's elegant paradoxes. *Australian Economic Papers, 15*, 207–229.

Portfolio and Macroeconomic Equilibrium in an Open Economy

13.1 Introduction

We have already mentioned (Sects. 10.2.2 and 11.2.2) the problems which arise when capital movements are no longer considered to be pure flows, but flows that derive from stock adjustments. The formulation of these adjustments comes within the framework of the Tobin-Markowitz theory of portfolio equilibrium, extended to an open economy by McKinnon and Oates (1966), McKinnon (1969), Branson (1974, 1979) and numerous other scholars.

The extension of the theory of portfolio equilibrium to international capital movements can be effected in two principal ways. The first is to see the problem as one of partial equilibrium, that is, by examining how the holders of wealth divide up their wealth among the various national and international assets in a context in which national income, current transactions in the balance of payments, etc., are given by hypothesis, that is to say, are exogenous variables. This problem will be dealt with in Sect. 13.2.

The second way is to insert the portfolio analysis of capital movements into the framework of macroeconomic equilibrium, in which therefore national income, current transactions, etc., are endogenous variables. One thus obtains a more general and satisfactory model (which can be considered an evolution of the traditional Mundell-Fleming model), but which is also more complicated, because alongside flow equilibria and disequilibria, it is necessary to consider stock equilibria and disequilibria and also the interrelations between stocks and flows over the short and long terms. We shall deal with these matters in Sects. 13.3 and 13.4.

© Springer-Verlag Berlin Heidelberg 2016
G. Gandolfo, *International Finance and Open-Economy Macroeconomics*,
Springer Texts in Business and Economics, DOI 10.1007/978-3-662-49862-0_13

13.2   Asset Stock Adjustment in a Partial Equilibrium Framework

The central idea of the Tobin-Markowitz (Tobin 1958; Markowitz 1952) theory of portfolio equilibrium is that holders of financial wealth (which is a magnitude with the nature of a stock) divide their wealth among the various assets[1] on the basis of the yield and risk of the assets themselves. Let us suppose therefore that the holders of wealth have a choice between national money, and national and foreign bonds; the exchange rate is assumed to be fixed.[2] If we indicate total wealth by W, considered exogenous, and the three components just mentioned by L, N and F, we have first of all the balance constraint:

$$L + N + F = W, \qquad \frac{L}{W} + \frac{N}{W} + \frac{F}{W} = 1. \tag{13.1}$$

The three fractions are determined, as we said, on the basis of the yield and risk, account also being taken of income (among other things, this accounts for the transactions demand for money). Supposing for the sake of simplicity that the risk element does not undergo any variations, we have the functions

$$\frac{L}{W} = h(i_h, i_f, y), \qquad \frac{N}{W} = g(i_h, i_f, y), \qquad \frac{F}{W} = f(i_h, i_f, y), \tag{13.2}$$

where i_h and i_f indicate as usual the home and foreign interest rates respectively. The three functions (13.2) are not independent of each other insofar as once two of them are known the third is also determined given the balance constraint (13.1).

It is then assumed that these functions have certain plausible properties. First of all, the fraction of wealth held in the form of money is a decreasing function of the yields of both national and foreign bonds: an increase in the interest rates i_h and i_f has, other things being equal, a depressive effect on the demand for money and, obviously, an expansionary effect on the demand for bonds (see below). Also, h is an increasing function of y and this means that, on the whole, the demand for bonds is a decreasing function of y.

The fraction of wealth held in the form of domestic bonds, on account of what has just been said, is an increasing function of i_h; it is, furthermore, a decreasing function of i_f insofar as an increase in the foreign interest rate will induce the holders of

---

[1] In general the wealth holder's portfolio can also contain real goods (for example, fixed capital) or securities that represent them (shares). For simplicity we ignore this component and only take account of the main purely financial assets; another simplifying hypothesis is that private residents cannot hold foreign currencies (for the "currency substitution" models, in which, in theory, private residents also can freely hold foreign currencies, see Sect. 15.8.3.4).

[2] For brevity we omit the study of portfolio equilibrium in a partial equilibrium context under flexible exchange rates, for which we refer the reader, for example, to Branson (1979), Branson and Henderson (1985). In Sect. 13.4 we shall however examine portfolio equilibrium under flexible exchange rates in a general macroeconomic equilibrium context.

wealth to prefer foreign bonds, *ceteris paribus*. Similarly the fraction of wealth held in the form of foreign bonds is an increasing function of i_f and a decreasing function of i_h. Finally the fraction of wealth held totally in the form of bonds, $(N + F)/W$, is—for the reasons given above—a decreasing function of y.

We could at this point introduce similar equations for the rest of the world, but so as to simplify the analysis we shall make the assumption of a small country and thus use a one-country model.[3] This implies that the foreign interest rate is exogenous and that the variations in the demand for foreign bonds on the part of residents do not influence the foreign market for these bonds, so that the (foreign) supply of foreign bonds to residents is perfectly elastic. Another implication of the small-country hypothesis is that non-residents have no interest in holding bonds from this country, so that capital flows are due to the fact that residents buy foreign bonds (capital outflow) or sell them (capital inflow).

Having made this assumption, we now pass to the description of the asset market equilibrium and introduce, alongside the demand functions for the various assets, the respective supply functions, which we shall indicate by M for money and N^S for domestic bonds; for foreign bonds no symbol is needed as the hypothesis that their supply is perfectly elastic has the effect that the supply is always equal to the demand on the part of residents. The equilibrium under consideration is described as usual by the condition that supply and demand are equal, that is[4]

$$M = h\left(i_h, i_f, y\right)W, \quad N^S = g(i_h, i_f, y)W, \quad F = f(i_h, i_f, y)W. \tag{13.3}$$

It is easy to demonstrate that only two of (13.3) are independent and therefore that, when any two of these three equations are satisfied, the third is necessarily also satisfied. This is a reflection of the general rule (also called Walras law) according to which, when n markets are connected by a balance constraint, if any $n - 1$ of them are in equilibrium, then the nth is necessarily in equilibrium. In the case under consideration, let us begin by observing that the given stock of wealth W is the same seen from both the demand and the supply sides, namely

$$M + N^S + F = W. \tag{13.4}$$

From Eqs. (13.1) and (13.2), we obtain

$$h\left(i_h, i_f, y\right)W + g(i_h, i_f, y)W + f(i_h, i_f, y)W = W, \tag{13.5}$$

---

[3]It has been demonstrated that, within the framework under consideration, the results of a two-country model would not be substantially different: see, e.g., De Grauwe (1983; Sect. 10.13).

[4]The demand functions expressed as levels are immediately obtained from (13.2), if we multiply through by W: for example, $L = h(i_h, i_f, y)W$ etc.

and so, if we subtract (13.5) from (13.4), we obtain

$$\left[M - h\left(i_h, i_f, y\right) W\right] + \left[N^S - g(i_h, i_f, y)W\right] + \left[F - f(i_h, i_f, y)W\right] = 0, \qquad (13.6)$$

which is the formal statement of Walras' law. From Eq. (13.6) we see that, if any two of the expressions in square brackets are zero (namely, if any two of Eq. (13.3) are satisfied), the third is also.

Equation (13.3) therefore provide us with two independent equations which, together with (13.4) make it possible to determine the three unknowns, which are the home interest rate (i_h), the stock of foreign bonds held by residents (F), and the stock of domestic money (M): the equilibrium values of these three variables will thus result from the solution of the problem of portfolio equilibrium, while the stock of domestic bonds (N^S) is given as are i_h, y and W.

The system being examined can be represented graphically. In Fig. 13.1, taken from De Grauwe (1983; Chap. 10), we have shown three schedules, LL, NN and FF, derived from Eq. (13.3). The LL schedule represents the combinations of i_h and F which keep the money market in equilibrium, given, of course, the exogenous variables. It is upward sloping on account of the following considerations. An increase in the stock of foreign securities held by residents implies—*ceteris paribus* (and so given N)—a decrease in the money stock (because residents give up domestic money to the central bank in exchange for foreign money to buy the foreign securities). Thus in order to maintain monetary equilibrium, domestic money demand must be correspondingly reduced, which requires an increase in i_h.

The NN schedule represents the combinations of i_h and F which ensure equilibrium in the domestic bond market. It is a horizontal line because, whatever the amount of foreign bonds held by residents, variations in this amount give rise

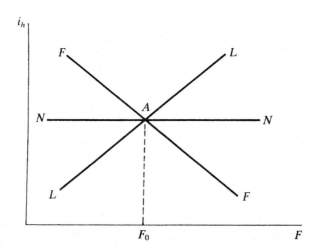

Fig. 13.1 Determination of portfolio equilibrium in an open economy

to variations of equal absolute value in the stock of money, but in the opposite direction, so that W does not change. The mechanism is the same as that described above with regard to LL: an increase (decrease) in F means that residents give up to (acquire from) the central bank national money in exchange for foreign currency. Note that F is already expressed in terms of national currency at the given and fixed rate of exchange. Hence the demand for domestic bonds does not vary and consequently, as its supply is given, i_h cannot vary.

Finally, the FF schedule represents the combinations of i_h and F which keep the demand for foreign bonds on the part of residents equal to the supply (which, remember, is perfectly elastic). It has a negative slope because an increase in i_h by generating, as we have said, a reduction in the residents' demand for foreign bonds, generates an equal reduction in the stock of these bonds held by residents themselves.

The three schedules necessarily intersect at the same point A, thanks to Walras' law, mentioned above. In economic terms, given the stock of domestic bonds N^S and the other exogenous variables, the equilibrium in the market for these bonds determines i_h. Consequently the demand for money is determined and thus the stock of foreign bonds to which it corresponds, given the balance constraint, a stock of money exactly equal to the demand for money itself.

What interests us particularly in all this analysis is to examine what happens to capital movements as a consequence of monetary policy, which, by acting on the national interest rate, generates a portfolio reallocation. In this context, monetary policy influences the interest rate indirectly, by acting on the stock of money. This action can come about in various ways, for example by way of open market operations, in which the central bank trades national bonds for money.[5] Let us suppose then that the monetary authorities increase the supply of domestic bonds, N^S. We can now begin an examination of the shifts in the various schedules (see Fig. 13.2). NN shifts parallely upwards to position $N'N'$: in fact, a greater value of i_h is needed in order to have a greater value of the demand for bonds so as to absorb the greater supply. In concomitance, LL shifts upwards to the left, because, as a consequence of the acquisition of new bonds, the stock of money is reduced. The FF schedule remains where it was, because none of the exogenous variables[6] present in it has changed.

The new point of equilibrium is obviously A' to which there corresponds a stock of foreign bonds (F_1) *smaller* than that (F_0) which occurred in correspondence to the previous point of equilibrium A. The reduction in the stock of foreign bonds from F_0 to F_1 obviously involves an *inflow* of capital, but when this stock has

---

[5]These operations obviously leave the total wealth W unaltered. Other operations on the other hand—for example the financing of the government's deficit by the printing of money—cause variations in W. However, the results do not change qualitatively. See, for example, De Grauwe (1983; Sect. 10.5).

[6]From (13.3) it can be seen that the exogenous variables in FF are W, i_f and y; in NN they are W, i_f, y, N^S (the last of which has increased); in LL they are W, i_f, y and M (the last of which has decreased).

Fig. 13.2 Monetary policy, portfolio equilibrium and capital flows

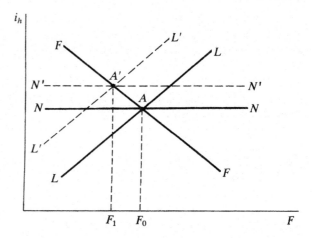

reached the new position of equilibrium, these capital movements will cease, and can begin again only in the case of a further increase in i_h. This provides a rigorous demonstration of what was already said in Sects. 10.2.2.1 and 11.2.2.

We shall conclude with a mention of the dynamic process which takes the system from A to A'. The supply of new domestic bonds on the part of the monetary authorities creates an excess supply of bonds with respect to the previous situation of equilibrium, so that the price falls: thus i_h, which is inversely related to the price of bonds, increases until the demand for bonds increases to a sufficient extent to absorb the greater supply. As the bonds are sold by the monetary authorities in exchange for money, the stock of money is reduced; besides, as i_h has increased, the demand for money falls so that monetary equilibrium is maintained. Finally, as the demand for foreign bonds is in inverse relationship to i_h, the increase in the latter leads to a reduction in F of which we have already spoken. As the quantities being demanded and supplied have the nature of *stocks* and as the total in existence (wealth W) is given,[7] once the new equilibrium stocks have been reached, the adjustment flows will cease, including the capital movements, as already stated.

13.3 Portfolio and Macroeconomic Equilibrium Under Fixed Exchange Rates

13.3.1 Introductory Remarks

When, within a macroeconomic equilibrium framework, one proceeds to the integration of stocks and flows, it is necessary to take a series of problems into

---

[7]It is clear that if W went on increasing in time, then there could be a continuous flow of capital.

account (usually neglected in analyses based solely on flows), amongst which:

(a) the way the budget deficit is financed. It is in fact clear that, to the extent in which it is financed by issuing bonds, the stock of these increases, possibly causing modifications in financial wealth and in portfolio equilibrium. Similar observations hold as regards the financing by printing money.
(b) The presence in private disposable income (on which private expenditure depends) of the flow of interest on bonds (both domestic and foreign) owned by residents.
(c) The payment of interest on government bonds constitutes a component of government expenditure, which must be explicitly taken into account. Besides, the interest on the part of government bonds owned by nonresidents is counted in the balance of payments, as is the interest received by residents on the foreign bonds that they own.

13.3.2 A Simple Model

In the following treatment, we shall adopt certain simplifying hypotheses, which will allow us to make the exposition less complex, without losing sight of the central points. First of all there is the assumption of fixed exchange rates and rigid prices, by which nominal and real magnitudes coincide. We then assume then that there are only two assets, money and securities. We also make the small country assumption and the assumption of *perfect capital mobility* with perfect asset substitutability: this means not only that capital is freely mobile, but also that domestic and foreign securities are homogeneous,[8] so that the domestic interest rate is exogenous (equal to the given foreign interest rate). Finally—with reference to the matrix of real and financial flows treated in Chap. 6—we shall consolidate the sectors of public administration and banking (which in turn includes the central bank and the commercial banks) into a single sector which for brevity we shall call the public sector. This public sector, then, can finance any excess of expenditure over receipts both by issuing bonds and printing money.

The consideration of the payment of interest by the public sector on securities which it has issued makes it necessary to divide total public expenditure, G, into two components: one of course consisting of the payment of this interest, iN^g, and the other of the purchase of goods and services which we shall call G_R. According to a first thesis, the discretionary variable of current public expenditure is therefore G_R, the variable which the government fixes exogenously. According to another thesis, the policy-maker fixes the total expenditure G exogenously, and so the

---

[8]This homogeneity does not mean that, in the portfolios of residents and nonresidents alike, there cannot be both domestic and foreign securities: as there can be different subjective evaluations on the part of both residents and nonresidents which, as we already know, concur, together with objective elements, in determining portfolio equilibrium.

amount of expenditure for goods and services G_R becomes an endogenous variable, as it is determined by difference once interest payments have been accounted for. These two extreme theses are already present in Christ (1979). Finally—as others observe—there are all the intermediate possibilities, which occur when the policy-maker in determining G_R takes account with a certain weight (represented by a parameter between zero and one) of the payment of interest.

Following O'Connell (1984), we shall introduce an auxiliary variable Z, which represents the discretionary variable of economic policy[9] and is made up of the sum of the expenditure for goods and services and of the fraction k of interest payments:

$$Z = G_R + kiN^g, \quad 0 \le k \le 1, \tag{13.7}$$

while the actual total public expenditure is, as stated above,

$$G = G_R + iN^g. \tag{13.8}$$

We see at once that, for $k = 0$, we have $Z = G_R$ (the first of the theses illustrated above), while for $k = 1$ we have $Z = G$ (the second thesis); for $0 < k < 1$ there are all the intermediate cases. By substituting from (13.7) into (13.8), we have

$$G = Z + (1 - k)iN^g. \tag{13.9}$$

Fiscal revenue is assumed, for simplicity, to be proportional to personal income in accordance with a constant rate, u. Personal income is given by the sum of the domestic product Y and residents' income from interest on both domestic and foreign securities held by them. The stock of domestic securities[10] N^g is held in part by residents (N_p) and in part by non-residents (N_f, where[11] $N_p + N_f = N^g$); residents also hold a stock of foreign securities F_p. Personal income is therefore $Y + i(N_p + F_p)$. From the above it follows that the public sector's budget deficit, which we indicate by g, is

$$g = G - u(Y + iN_p + iF_p) = Z + (1 - k)iN^g - u(Y + iN_p + iF_p), \tag{13.10}$$

where in the second passage we have used (13.9).

We now move on to the balance of payments, which we divide into current account (CA) and capital movements (K). The current account includes, besides imports and exports of goods and services, interest payments on domestic securities

---

[9]The two extreme theses are already present in Christ (1979); the introduction of the parameter k makes a simple generalization possible.

[10]These are assumed to be perpetuities, as are foreign bonds.

[11]As we have consolidated the sector of the public administration, the central and commercial banks, any public securities held by the latter two of these sectors to meet issues by the first sector cancel out.

held by nonresidents and income from interest on foreign securities held by residents (item: investment income in the balance of payments: see Sect. 5.1.3[12]). In symbols

$$CA = x_0 - m\left[(1-u)\left(Y + iN_p + iF_p\right), W\right] - i\left(N^g - N_p\right) + iF_p, \qquad (13.11)$$

where exports are as usual assumed to be exogenous, while imports are a function not only of disposable income (with a positive marginal propensity to import, but less than one), but also of the stock of private wealth W, so as to account for a possible wealth effect on total private expenditure (thus on the part of it which is directed to foreign goods). Wealth is defined as the sum of the stock of money M and the stock of domestic[13] and foreign securities held by residents:

$$W = M + N_p + F_p. \qquad (13.12)$$

Capital movements are given by the *flow* of securities, that is by the *variation* in the stock of domestic securities held by non-residents net of the *variation* in the stock of foreign securities held by residents; remembering the recording rules for capital movements, we have

$$K = \Delta\left(N^g - N_p\right) - \Delta F_p = \Delta N^g - \Delta N_p - \Delta F_p. \qquad (13.13)$$

The overall balance is, obviously,

$$B = CA + K. \qquad (13.14)$$

The equation for the determination of income in an open economy is, as usual,

$$Y = d\left[(1-u)\left(Y + iN_p + iF_p\right), W\right] + x_0 + \left(Z - kiN^g\right), \qquad (13.15)$$

where domestic demand d is defined as the difference between absorption and imports, a function of disposable income and wealth. Finally, $Z - kiN^g$ is G_R [see (13.7)] that is the government expenditure for goods and services. The marginal propensity to domestic expenditure is as usual assumed to be positive, but less than one.

We then have the equation for monetary equilibrium

$$M = L(Y, W), \qquad (13.16)$$

---

[12]As we know, unilateral transfers are also present in the current account. On account of their exogenous nature we can ignore them (for example by assuming that they have a nil balance) without substantially altering the results of the analysis.

[13]We do not wish to enter here into the controversy on whether domestic (government) bonds can be considered net wealth and the related "Ricardian equivalence theorem" (see, for example, Tobin 1980, Chap. III, and references therein); we shall simply assume—here and in the following chapters—that they are to be included in W.

where the demand for money is an increasing function of income and of the stock of wealth (this derives from the portfolio equilibrium discussed in Sect. 13.2). As only two assets exist (remember the assumption of homogeneity between domestic and foreign securities), the balance constraint makes it possible to omit the function of demand for securities. The interest rate, being given and constant, has been omitted for brevity.

It is now necessary to stress an important point. While in a static context the stocks of assets are considered given, in a dynamic context they vary and this variation is due to the balance of payments and the budget deficit. This can be seen from the matrix of real and financial flows explained in Chap. 6 (duly consolidated according to the assumptions made at the beginning of the present section), but it can also be simply explained within the framework of the present model.

The public sector finances its own deficit either by issuing securities or by printing money; a part of the variation in the money supply is therefore due to the financing of the public deficit, to which a part of the new issue of securities is also due (variations in the stock of securities). The money stock also varies, as we know, in consequence of disequilibria in the balance of payments (variations in the stock of international reserves), which the monetary authorities can however sterilize in part or in their entirety; in the model before us this sterilization occurs through open market operations, that is, by a sale of government bonds to residents (in order to sterilize a balance-of-payments surplus) and a purchase from them (to sterilize a deficit). Another part of the variation in the stock of domestic securities is therefore due to these sterilization operations.

We can therefore state the relationships

$$\Delta M = sB + hg,$$
$$0 \leq s \leq 1, 0 \leq h \leq 1, \qquad (13.17)$$
$$\Delta N^g = (1 - s)B + (1 - h)g,$$

where s represents the unsterilized fraction of the international payments imbalance and h the fraction of the public deficit financed by the printing of money. The meaning of (13.17) corresponds to what we have just said. In fact, given a payments imbalance B, a fraction $(1 - s)B$ is sterilized by means of open market operations, while the remaining fraction sB gives rise to a change in the money supply. The latter also varies to finance a fraction h of the public deficit g, the remaining part of which, $(1 - h)g$, is financed by the issue of bonds.

13.3.3 Momentary and Long-Run Equilibrium

The model is now fully described and its solution and analysis can be carried out in successive phases. By following a now well-tried methodology for the analysis of stock-flow models, we shall distinguish between a *momentary* or *short-run* and a *long-run equilibrium*. This terminology was introduced with regard to capital accumulation and growth models, but the concepts expressed by this terminology have general validity in the presence of stock-flow models where the stocks vary in time as a result of flows.

The meaning is the following. At any given moment in time, in addition to intrinsically exogenous variables, there are certain stock variables that result from past flows, which can therefore also be dealt with as exogenous variables in finding out the solution to the model; this solution therefore determines a *momentary equilibrium*. With the passing of time, however, the stocks just referred to vary as a result of flows and thus become endogenous variables, whose equilibrium values must therefore be determined together with those of the other endogenous variables in the solution for the *long-run equilibrium*.

With reference to the model we are examining, we shall begin with the observation that at a given moment in time, in which the money supply and the stock of domestic securities are given, Eqs. (13.15), (13.16) and (13.12) constitute a *short-term sub-system*, which determines the momentary-equilibrium values of the domestic product Y, of the total stock of securities (domestic and foreign) held by residents $(N_p + F_p)$, and wealth W, as a function of the other magnitudes $(x_0, Z, i, M, N^g$: the first two are in the nature of pure flows, i is a point variable, i.e. a variable measured at a point in time, and the last two are stocks). Note that, while x_0, Z and i are *intrinsically exogenous* variables in the model, M and N^g are exogenous simply because we are considering an instant in time: in fact, with the passing of time these magnitudes will endogenously vary on the basis of the relationships (13.17).

We now turn to the examination of *long-run equilibrium*, in which all the stocks must be in equilibrium. Stock equilibrium of the private sector means that wealth W must be constant in time. Thus, given definition (13.12), it is necessary for both the money supply M and the total stock of securities held by the private sector $(N_p + F_p)$ to be stationary. Given (13.16), the fact that M and W are constant implies that Y must also be constant.

Since wealth, income, and the interest rates are constant, the composition of the (constant) total stock of securities does not change, so that

$$\Delta N_p = \Delta F_p = 0, \tag{13.18}$$

hence, by Eq. (13.13),

$$K = \Delta N^g. \tag{13.19}$$

The implication of this is that, as the (domestic and foreign) stock of bonds held by residents is determined, the entire amount of the new domestic bonds issued is absorbed by nonresidents.

Furthermore, in the private sector's stock equilibrium, private "saving" is zero, using saving to mean the *unspent* part of disposable income (the spent part including both consumer and investment goods, both domestic and foreign). Alternatively, if saving is defined as that part of disposable income that is *not consumed*, then, in the equilibrium under study, there is equality between saving and investment. In the opposite case, in fact, the stock of assets of the private sector would vary—see (6.10)—contrary to the condition that stocks should be stationary. Note that, if the constancy of the stock of physical capital is also considered among the conditions of stationarity, the investment in question will be that of replacement only.

This in turn implies—as a result of Eq. (6.1)—that the sum of the budget deficit and current account balance is zero:

$$g + CA = 0. \tag{13.20}$$

We now introduce a further long-run equilibrium condition into the analysis, namely that the stock of international reserves is constant (a country cannot go on indefinitely accumulating or losing reserves) and see what it implies. If R=constant, then

$$\Delta R = B = 0. \tag{13.21}$$

Letting $B = 0$ in the first equation of (13.17), where $\Delta M = 0$ as we have seen above, we have

$$hg = 0, \tag{13.22}$$

which requires $g = 0$, unless $h = 0$.

If $g = 0$, from the second equation in (13.17) we have $\Delta N^g = 0$. Putting this together with Eqs. (13.19), (13.20), (13.21), we get $CA = K = 0$, namely a *full equilibrium* in the balance of payments, because not only the overall balance, but also the current account and the capital account balances, taken separately, must be zero.

If $h = 0$, we are in the institutional setting (adopted by several countries, for example the European countries belonging to the European Union) in which there is the prohibition of financing the budget deficit by issuing money. Considering the second equation in (13.17), given $B = 0$, we have

$$g = \Delta N^g. \tag{13.23}$$

Since $\Delta N_p = 0$ by Eq. (13.18), Eq. (13.23) implies that all government bonds issued to finance the budget deficit are absorbed by the rest of the world. Now, the rest of the world, although not explicitly considered thanks to the small-country assumption, cannot be considered in portfolio equilibrium while indefinitely accumulating our domestic securities. Hence a world long-run equilibrium situation will require

$$\Delta N^g = 0, \tag{13.24}$$

and given Eq. (13.23), we have $g = 0$ as above, and full equilibrium in the balance of payments.

Thus we have reached the important result that full stock-flow equilibrium in the long run requires a balanced budget ($g = 0$) and full equilibrium in the balance of payments ($CA = K = 0$).

The dynamic analysis of the convergence of the variables to the long-run equilibrium will be carried out in the Appendix.

We conclude with the observation that we have not yet talked about the problem of *full employment*: it is however easy to see that, as in any case the solution of the model expresses the equilibrium values of the endogenous variables in terms of the exogenous ones, including the policy variable Z, it is in principle possible to fix Z so that the resulting equilibrium value of Y coincides with that which corresponds to full employment Y_F. What happens to the balance of payments can then be determined by applying the analysis carried out above.

13.4   Portfolio and Macroeconomic Equilibrium under Flexible Exchange Rates

13.4.1 Introductory Remarks

The introduction of a plurality of financial assets and the integration between stocks and flows under flexible exchange rates present an additional problem with respect to those mentioned in Sect. 13.3. This problem was usually neglected in the traditional analyses of macroeconomic equilibrium under flexible exchange rates. In fact, these analyses are based on models where the assumption of rigid prices is maintained, being carried over from the fixed exchange rates models. Amongst other consequences this assumption implies the absence of the exchange rate from the traditional monetary-equilibrium equation: the exchange rate, in fact, would be present in that equation insofar as it influenced the price level which, then, could no longer be assumed constant and ought to be explicitly introduced in the equation under consideration (as well as in the other equations of the model, as we shall see presently). In fact, the price level enters in the monetary equilibrium equation in the form $M = pL(y, i)$ or $M/p = L(y, i)$; given the assumption of constancy of p, it is set at one for brevity.

Now, there are good reasons for believing that exchange-rate changes influence the price level. If we consider, for example, a depreciation, this will cause an increase in the domestic-currency price of final foreign imported goods and so, even if the price of domestic goods is assumed constant, an increase in the general price-index. If, in addition, we consider imported intermediate inputs, a depreciation can also cause an increase in the price of domestic goods, insofar as these are set by applying a markup on production costs. Flexible wages that adjust to the price level can also influence the price of domestic goods via markup pricing. To avoid complicating an already difficult treatment we shall neglect the second and third effect (for which see Sect. 17.4.1) and concentrate on the first. Thus we have

$$p = p_h^\alpha \left(rp_f\right)^{1-\alpha}, \quad 0 < \alpha < 1, \tag{13.25}$$

where α and $(1 - \alpha)$ are the given weights of commodities in the general price index p. These commodities are domestically produced (with a price p_h) and foreign imported (with a foreign-currency price p_f). Equation (13.25) implicitly assumes downward (and not only upward) price flexibility. Since p_f is assumed constant, it can be set at one, so that

$$p = p_h^\alpha r^{1-\alpha}. \tag{13.26}$$

By the same token the terms of trade turn out to be

$$\pi = \frac{p_h}{r}. \tag{13.27}$$

As soon as one introduces the variability of the price level, the *expectations* on its future value come into play, for example to determine the real interest rate (given by the difference between the nominal interest rate and the expected proportional rate of change in the price level), which, together with other variables, is an argument in the aggregate demand function. The topic of expectations is difficult and enormously far-reaching; here we shall only consider two extreme cases: *static expectations* and *rational expectations* or *perfect foresight*. The former consist in the belief that no change will occur in the variable under consideration; the latter in forecasting the variable exactly (apart from a stochastic term with mean zero and the properties of unbiasedness, uncorrelation, orthogonality: see the Appendix to Chap. 4). It follows that—*in a deterministic context*—rational expectations imply perfect foresight, so that the expected change coincides with the one that actually occurs. If we denote the change of a variable by putting a dot over it and the expectation by putting a tilde, we have

$$\frac{\tilde{\dot{p}}}{p} = \begin{cases} 0, \\ \dfrac{\dot{p}}{p}, \end{cases} \tag{13.28}$$

in the case of static and rational expectations respectively; the expected changes are expressed as a proportion of the current actual value because, as mentioned above, we are interested in the expected *proportional* change. It should be noted that the case of static expectations is practically equivalent to eliminating expectations wherever they appear and so coincides with the traditional analysis where expectations are not taken into account.

The exchange-rate flexibility suggests the presence of expectations concerning the changes in this variable, which we shall again limit to the two extreme cases:

$$\frac{\tilde{r}}{r} = \begin{cases} 0, \\ \frac{\dot{r}}{r}. \end{cases} \tag{13.29}$$

Expectations on the exchange rate require a new form of the relation between the domestic and the foreign interest rate in the case of a small country with perfect capital mobility and perfect substitutability between domestic-currency and foreign-currency financial assets (these are the same assumptions already employed in the fixed exchange rate model treated in Sect. 13.3). In fact, the domestic interest rate i is no longer equal to the given foreign interest rate i_f, but is related to it by equation

$$i = i_f + \tilde{r}/r, \tag{13.30}$$

where i and i_f refer to the same period of time as $\tilde{r}$. This equation is the UIP interest rate parity condition (4.15) explained in Chap. 4.

13.4.2 The Basic Model

After clarifying these preliminary problems, we go on to the exposition of a model (due to Branson and Buiter 1983) which enables us, under flexible exchange rates, to deal with the same problems as in Sect. 13.3 under fixed exchange rates. We shall however adopt further simplifying assumptions with respect to the framework used there, namely:

(a) we do not introduce the auxiliary variable Z, but consider only the actual overall government expenditure G, given by government expenditure on goods and services[14] plus interest payments on public debt.
(b) we assume that domestic assets are held solely by residents, so that $N^g = N_p$ (for brevity we shall omit both the superscript and the subscript and use the symbol N only).
(c) we do not introduce a taxation function linking fiscal receipts to income, but we assume them to be a discretionary policy variable like G_R; if we call these receipts (in real terms) T, they will be pT in nominal terms.

Thus the budget deficit is

$$g = G - pT - ri_f R = p_h G_R + iN - pT - ri_f R. \tag{13.31}$$

---

[14]As usual we assume that government expenditure in real terms (G_R) is entirely directed to domestically produced output.

With respect to the definition given in Sect. 13.3—see Eq. (13.10)—besides the explicit presence of the price level p, we note the additional term (on the side of receipts) $ri_f R$, where R denotes the stock of international reserves expressed in terms of foreign currency. The assumption behind this term is that the central bank (remember that it is consolidated together with the government in a single sector, the public sector) holds all the international reserves in the form of interest-bearing foreign assets (at the given interest rate i_f).

Let us now consider the balance of payments, divided as usual into current account (CA) and capital movements (K). As we know, the former includes the flow of interest on net foreign assets (denominated in foreign currency) held by both private and public residents.[15] All the other items of the current account are lumped together under "exports" and "imports". Exports (in real terms) are a function of the terms of trade π; imports (in real terms) are a function of the terms of trade and of private aggregate expenditure or absorption (instead of income) in real terms. The idea is that imports, as part of aggregate expenditure, are related directly to it and so indirectly to income, which is one of the arguments in the absorption function.

Thus, if we denote *net exports* (that is exports less imports) by x and absorption by a, we have $x = x(\pi, a)$, where it is assumed:

(i) that the effect of an increase in π on x is negative and vice versa [this implies the assumption that the critical elasticities condition—see Chap. 7—occurs: we must remember that an exchange-rate appreciation causes an increase in π and vice versa, on the basis of Eq. (13.27)], and

(ii) that the effect of an increase in a on x is negative (and vice versa), included between zero and minus one (this amounts to assuming a positive and smaller-than-one marginal propensity to import with respect to absorption).

Since x is expressed in real terms and, more precisely, in terms of domestic output, if we multiply it by p_h we obtain its monetary value. Inherently exports consist of domestic output. Imports m consist of foreign goods, which have a monetary value rm in domestic currency; if we divide this by p_h we express them in terms of domestic output. See the Appendix.

As regards the flow of interest income, since foreigners do not hold domestic government bonds (all foreign lending or borrowing is done in foreign currency-denominated bonds), it will equal $ri_f (F_p + R)$, where F_p is the stock of privately held net foreign assets (note that, since we are talking of the stock of net foreign assets—assets minus liabilities—in general this stock can be either positive or negative); for brevity we shall henceforth omit the subscript p. Thus we have

$$CA = p_h x(\pi, a) + ri_f(F + R). \tag{13.32}$$

---

[15]It should be recalled out that, in general, interest on net foreign assets held by the official authorities is also reported in the investment income item of the current account.

Private capital movements (K) are a flow equal to the change in the stock of net foreign assets held by the private sector, that is, if we express these flows in domestic-currency terms and remember the recording rules explained in Chap. 5,

$$K = -r\dot{F} \tag{13.33}$$

where, according to the convention introduced above, the dot indicates the change. Therefore the overall domestic-currency balance of payments, equal in turn to the change in international reserves (now expressed in terms of domestic currency) is

$$r\dot{R} = p_h x(\pi, a) + r i_f (F + R) - r\dot{F}. \tag{13.34}$$

The equation for the determination of domestic production q is, as usual,

$$q = a(y_d, i - \frac{\widetilde{p}}{p}, \frac{W}{p}) + x(\pi, a) + G_R, \tag{13.35}$$

where a is absorption in real terms, a function of real disposable income (with a marginal propensity to absorb that is smaller than one), the *real* interest rate (with an inverse relationship), and real private financial wealth (with a positive relationship).

We must now define real disposable income. The variability in the price level discussed above makes it necessary first to define *nominal* disposable income, which can then be deflated by p to get *real* disposable income. Nominal disposable income equals the value of domestic output $(p_h q)$ plus interest received by residents on both national (iN) and (net) foreign assets $(r i_f F)$ minus taxes (pT). In real terms we have

$$y_d = \frac{p_h}{p} q + \frac{iN}{p} + \frac{r i_f F}{p} - T. \tag{13.36}$$

We then have the monetary equilibrium equation

$$\frac{M}{p} = L\left(i, q, \frac{W}{p}\right), \tag{13.37}$$

where money demand in real terms is a decreasing function of the rate of interest, and an increasing function of both domestic production (taken as an index of transactions) and real financial wealth (this derives from the portfolio equilibrium discussed in Sect. 13.2). Since there are only two financial assets (money and bonds: remember the assumption of perfect substitutability between national and foreign bonds), the budget constraint allows us to omit the demand-for-bonds function and so to avoid the problems related to the specification of this function.

Wealth is defined as the sum of the stock of money M and the stock of domestic and (net) foreign bonds held by residents[16]; the latter must, of course, be multiplied by the exchange rate to express it in domestic-currency terms:

$$W = M + N + rF. \tag{13.38}$$

If we divide W by the general price-level we get real financial wealth, W/p.

It is now necessary to introduce—as we did under fixed exchange rates, in particular Eq. (13.17)—the ways of financing the budget deficit, and the sources of money creation. The further simplifying assumptions introduced by Branson and Buiter (1983) are the following. Firstly, the government always balances its budget by *endogenous* changes in taxes. In other words, given the other variables entering into g, the level of T is always set by the government so as to make $g = 0$. In terms of comparative-static analysis this means that, when we consider fiscal policy, we shall obtain balanced-budget multipliers. Thus, from Eq. (13.31), we have

$$p_h G_R + iN - pT - ri_f R = 0. \tag{13.39}$$

The second simplifying assumption is that the government does not carry out *continuative* open market operations (it can carry out these operations but only occasionally) and does not sterilize payments imbalances. Together with the previous assumption, this implies that the stock of national bonds does not change (except occasionally) and that the change in the stock of money coincides with the balance of payments, namely

$$\dot{N} = 0, \tag{13.40}$$

$$\dot{M} = r\dot{R}. \tag{13.41}$$

It goes without saying that in the case of perfectly flexible exchange rates and instantaneous adjustment, the balance of payments is always zero, so that

$$\dot{R} = 0, \text{ hence } \dot{M} = 0. \tag{13.42}$$

Having described the model, we now pass to the examination of its solution and properties according to the type of expectations.

---

[16]This implies that the stock of physical capital is ignored. This is a simplifying assumption, common to this type of model, which also allows us to ignore the subdivision of absorption between consumption and investment.

13.4.3 Static Expectations

In the case of static expectations we have $\tilde{r} = 0$ and $\tilde{p} = 0$. As usual (see Sect. 13.3.2), we distinguish between *short-run* or *momentary equilibrium* and *long-run equilibrium* and begin by examining the former.

At a given point in time, the stocks (money, national and foreign bonds, wealth, international reserves) are given; note that the stocks of money, national bonds and international reserves are in any case given owing to the assumptions included in Eqs. (13.42) and (13.40). The stock of foreign bonds can, on the contrary, vary through time, so that it is an exogenous variable only because we are considering a given point in time. The domestic interest rate is given [by Eq. (13.30) and the assumption of static expectations, it equals the given foreign interest rate] and government expenditure is also given, as it is an exogenous variable.

Since the price level depends on the exchange rate, it is possible to reduce the short run model to two equations only, which determine the short run endogenous variables r[17] and q, i.e. the exchange rate and domestic production. A simple interpretation in terms of *IS* and *LM* schedules can then be given (see Fig. 13.3).

The real equilibrium schedule *IS* has been drawn after finding the equilibrium value (r_E) of the exchange rate through the mathematical solution of the model. This schedule is the graphic counterpart to Eq. (13.35) for r given at r_E. In fact, given r_E, p (and so π) is given; besides, as W is given, the argument W/p in the absorption function is given. Since G_R, N and R are also given, Eq. (13.39) determines the taxation level T. It follows that, in Eq. (13.36), disposable income varies only as q varies and so, since the marginal propensity to absorb is smaller than one, absorption increases by less than q. It should be noted that, since this propensity is defined with respect to y_d, this result is based on the implicit assumption that y_d varies approximately by the same amount as q, which in turn implies that the ratio p_h/p in (13.36) does not greatly differ from one. This will always be the case if we set $r = p_h = p = 1$ in the initial equilibrium situation, which involves no loss of generality.[18]

---

[17]From this point of view the model can also be seen as a model for the determination of the exchange rate (a similar observation can be made in relation to the successive analysis of long-run equilibrium and to the case of rational expectations). This, it should be stressed, holds in all models where the exchange rate is flexible and is determined as an endogenous variable. We shall deal with the problem of the equilibrium exchange rate in Sect. 15.6.

[18]When the equilibrium is displaced and a variation in r occurs (see below, in the text), the term p_h/p—which equals $\pi^{1-\alpha}$ by way of (13.26) and (13.27)—will become greater or smaller according as r appreciates or depreciates. In the latter case the result is strengthened, since when q increases y_d will increase by less; in the former, y_d will increase by more than q and so one cannot exclude the possibility that a may increase by more than q. Even in this case, however, since an increase in a causes a decrease in the x balance, it is likely that the sum $(a + x)$ will increase by less than q (a result which is necessarily true when a increases by less than q). In formal terms, by differentiating $(a + x)$ with respect to q, the condition for this sum to increase by less than q turns

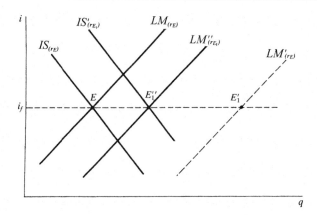

Fig. 13.3 Static expectations: short-run equilibrium and economic policy

The increase in absorption due to the increase in q also causes a deterioration in the x balance (but by less than the increase in a); therefore, the sum $(a + x)$ increases by less than q. Thus a decrease in i is required for a to increase sufficiently and absorb the increase in q entirely (when a increases x decreases but by less, since the marginal propensity to import is smaller than one). This shows that the IS schedule is downward sloping.

As regards the LM schedule, it is the graphic counterpart of Eq. (13.37): since, as we explained above, p and W are given, it has the usual positive slope.

It should be noted that the IS and LM schedules cross in correspondence to the foreign interest rate i_f from which—due to the assumptions of perfect capital mobility and static expectations—the domestic interest rate cannot deviate. For this same reason the broken line parallel to the q axis at a distance i_f can be interpreted (see Chap. 11, Fig. 11.5), as the BB schedule.

Let us now consider the effects of monetary and fiscal policy.

An expansionary monetary policy, namely—in this context—an increase in the stock of money, for example through a purchase of national bonds by the public sector from the private sector,[19] shifts LM to the right, for example to LM'. The excess supply of money puts a downward pressure on the domestic interest rate, which is however brought back to i_f by the incipient capital outflow. Through its effect on the balance of payments, this incipient outflow causes the exchange rate to depreciate, for example from r_E to r_{E1}; this causes a rightward shift in the IS schedule (not shown). Now, in the traditional analysis (see Chap. 11, Sect. 11.4), this shift would cause the IS schedule to cross LM' at E_1'. However, in the present

---

out to be $\pi^{1-\alpha}a_y(1 + x_a) < 1$, where $0 < a_y < 1$ and $-1 < x_a < 0$ by assumption. This condition is related to the term Ω_3 in Eq. (13.88) in the Appendix.

[19]We have assumed above that the public sector does not perform continuative open market operations. This does not prevent it from performing a once-and-for-all operation of this kind.

model, the depreciation in the exchange rate causes an *increase in the price level and so a decrease in the real stock of money*, so that the monetary-equilibrium schedule shifts back from LM' to the left, though not by so much as to return to the initial position. There is a smaller shift in the real-equilibrium schedule than that envisaged by the traditional analysis, because part of the effect of the exchange-rate depreciation is removed by the increase in p.

The new equilibrium will thus be found at an intermediate point between E and E_1', for example at E_1'': domestic production has increased, but by less than was believed by the traditional analysis.

Another result of the traditional analysis is the absolute *ineffectiveness of fiscal policy* under flexible exchange rates, when perfect capital mobility obtains. This was shown in Sect. 11.4; we can now see that it is *no longer valid* if the link between the general price level and the exchange rate is accounted for.

An increase in government expenditure G_R causes a rightward shift in the IS schedule at unchanged exchange rate, and so an increase in income which, as the supply of money is given, brings about an excess demand for money and so an upward pressure on the domestic interest rate. But this rate is kept at i_f by the incipient capital inflow; by giving rise to a balance-of-payments surplus, this inflow also causes an exchange-rate appreciation which makes the IS schedule shift to the left. According to the traditional theory this shift continues until IS returns to the initial position: this is inevitable because LM has stayed put. In the present model, on the contrary, the exchange-rate appreciation, by causing a *decrease* in the price level, causes the real money stock to *increase* so that the LM schedule shifts to the right. To put it differently, the decrease in the price level has generated money in real terms, thus enabling the unchanged nominal stock to satisfy at least in part the increased real demand for money (due to the increase in output). Fiscal policy, as we can see, regains its effectiveness.

Let us now come to *long-run equilibrium*.

Since, as we have already seen, the stock of money, the stock of national bonds and the stock of international reserves are constant by assumption, the public sector's stock equilibrium is ensured. The private sector's stock equilibrium, which implies a constant stock of wealth, is ensured if the stock of privately held net foreign assets is constant; this, in turn, means—account being taken of Eq. (13.33)—that the capital movements balance is zero:

$$K = 0. \tag{13.43}$$

It follows that, since the overall balance of payments is in equilibrium, the current account balance must also be in equilibrium:

$$CA = 0. \tag{13.44}$$

Long run equilibrium thus implies a *full* equilibrium in the balance of payments.

The constancy of all stocks enables us to determine—as we did for the short run equilibrium—the equilibrium values of the exchange rate and domestic production (which are thus constant) and, hence, all the remaining variables.

The analysis of the effects of monetary and fiscal policy in the long run requires a laborious mathematical analysis, for which we refer the reader to the Appendix.

13.4.4 Rational Expectations and Overshooting

The analysis of the model with rational expectations is more complicated, owing to the presence of the terms $\widetilde{\dot{r}}/r$ and $\widetilde{\dot{p}}/p$ in the various functions. As an example, we examine the *short-run effects* of an expansionary monetary policy.

In Fig. 13.4 we have reproduced Fig. 13.3 so as to have a standard of comparison: as we know from Sect. 13.4.3, a monetary expansion ultimately causes a depreciation in the exchange rate and a shift of the *LM* and *IS* schedules to the position *LM″* and *IS′* respectively, so that the equilibrium point shifts from *E* to E_1''. Now, it is possible to show (see below, Fig. 13.6) that, when expectations are rational, the exchange-rate depreciation in the *new* equilibrium will be *less* than that occurring under static expectations, so that ultimately the exchange rate will show an *appreciation* with respect to the case of static expectations. Since this appreciation is exactly anticipated, the domestic interest rate will equal $i_f + \dot{r}/r$, where $\dot{r} < 0$ given the appreciation with respect to the value of r underlying point E_1''. Thus the *LM* and *IS* schedules will cross along a straight line parallel to the horizontal axis at a height $i_f + \dot{r}/r$. Since, as we said, the exchange rate shows an appreciation with respect to that obtaining under static expectations (even if it does not go all the way back to its initial value r_E), the real-equilibrium schedule shifts from *IS′* to the left, for example to *IS″*, and the monetary-equilibrium schedule shifts from *LM″* to the right, for example to *LM‴*; the symbol r_{E2} denotes the new equilibrium exchange rate, lower than r_{E1}, but higher than r_E. The intersection between the two schedules will in any case occur at a point included between A_1, and A_2, but we do not know whether to the left or to the right of E_1'': in Fig. 13.4 we have determined the new equilibrium point E_2 to the left of E_1'' (so that domestic product q increases by less

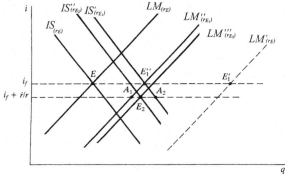

Fig. 13.4 Rational expectations: short-run equilibrium and economic policy

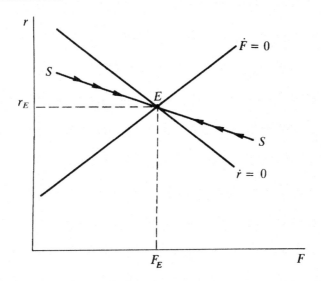

Fig. 13.5 Rational expectations and long-run equilibrium

than under static expectations), but we could have equally well determined it to the right of E_1''.

Long run equilibrium implies a constant value of the stocks and so of the exchange rate: in fact, only with a constant value of the exchange rate (and hence of the price level), will the stocks in real terms also remain constant. This equilibrium can be described in terms of Fig. 13.5, where the schedule $\dot{r} = 0$ denotes all the combinations of r and F such that the exchange rate is constant; similarly, the $\dot{F} = 0$ schedule denotes all the combinations of r and F such that the stock of net foreign assets is constant (for the slope of the two schedules see the Appendix). The long-run equilibrium point is E, where the two schedules cross.

The dynamics of the system is very complex and here we limit ourselves to observing that there exists only one trajectory, denoted by the line SS, along which the movement of the system converges to the equilibrium point (which is a *saddle point*: see the Appendix); any point outside SS will move farther and farther away from E.

Rational expectations are however such as to bring any (non-equilibrium) initial point not lying on SS onto this line with a discrete jump, as described in Fig. 13.6. Unanticipated policy changes which cause the schedules to shift and determine a new point of equilibrium also cause the exchange rate to jump immediately onto the stable trajectory in discrete fashion (and then follow it and smoothly move toward the new long-run equilibrium point). In fact, rational expectations imply that agents know the model, hence they know that the only way to reach the new equilibrium point is to jump on the stable trajectory, a jump that they can bring about since in an efficient foreign exchange market r is free to make discontinuous jumps as a consequence of unanticipated events or *news* (this is why r is called a *jump*

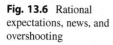

Fig. 13.6 Rational
expectations, news, and
overshooting

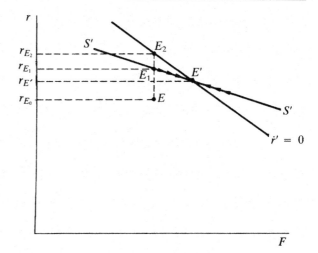

variable). The question might arise, why don't rational agents immediately cause
a jump onto the new equilibrium point, instead of reaching it in such a roundabout
way. The answer lies in the fact that not all variables are jump variables: in fact,
F is a *predetermined* variable and can move only gradually through time. The
distinction between jump and predetermined variables lies, in fact, at the heart of
rational expectations models (see Gandolfo 2009, Sect. 28.4).

In Fig. 13.6 we have considered the case of an unanticipated monetary expansion
which, in the long-run equilibrium, causes an increase in both r and F (see the
Appendix); the two schedules shift and determine the new equilibrium point E'.
For simplicity's sake, we have drawn only the new schedule along which the
exchange rate is constant ($\dot{r}' = 0$) and the new convergent trajectory ($S'S'$). The
system was initially at the previous equilibrium point E, which is now no longer
an equilibrium point, owing to the shifts in the schedules. Given, as we said, the
stock F, the exchange rate jumps to point E_1, on the $S'S'$ line: as can be seen,
at E_1, the exchange rate is *higher* than that corresponding to the new long-run
equilibrium; after the jump, r smoothly *appreciates* towards $r_{E'}$, a value higher
than r_{E0} but lower than r_{E1}. This is the phenomenon known as exchange-rate
overshooting, since the exchange rate initially depreciates by more than required,
and then appreciates. The overshooting phenomenon is typical of the dynamics of
rational expectations models (see Gandolfo 2009, Sect. 28.4), and was introduced in
the exchange rate by Dornbusch (1976), whose model we shall examine in Chap. 15,
Sect. 15.3.2.

13.5 Appendix

13.5.1 Partial-Equilibrium Asset Adjustment

On account of the balance constraint (13.1), the demand functions (13.2) are related by the constraint

$$h(i_h, i_f, y) + g(i_h, i_f, y) + f(i_h, i_f, y) = 1, \tag{13.45}$$

from which it follows that the sum of the partial derivatives of the three functions with respect to each variable must be zero, that is,

$$\underset{(-)}{\frac{\partial h}{\partial i_h}} + \underset{(+)}{\frac{\partial g}{\partial i_h}} + \underset{(-)}{\frac{\partial f}{\partial i_h}} = 0, \quad \underset{(-)}{\frac{\partial h}{\partial i_f}} + \underset{(-)}{\frac{\partial g}{\partial i_f}} + \underset{(+)}{\frac{\partial f}{\partial i_f}} = 0, \quad \underset{(+)}{\frac{\partial h}{\partial y}} + \underbrace{\frac{\partial g}{\partial y} + \frac{\partial f}{\partial y}}_{(-)} = 0. \tag{13.46}$$

Thus, once two partial derivatives are known the third is determined. Under the various partial derivatives we have placed the signs that derive from hypotheses made in the text. The partial derivatives $\partial g/\partial y$ and $\partial f/\partial y$ individually have uncertain signs, but their sum must necessarily be negative.

Given the hypothesized signs of the above-mentioned derivatives, the slopes of the schedules LL, NN, FF, are immediately evident.

Let us now consider the system of equilibrium conditions (13.3) of which, as we know, only two are independent. By choosing the second and third (so as to be consistent with the graphic representation in the text) we have the system

$$N^S - g(i_h, i_f, y)W = 0,$$
$$F - f(i_h, i_f, y)W = 0. \tag{13.47}$$

On the basis of the implicit function theorem, it is possible to express the two variables F, i_h as differentiable function of all other variables (N^S, i_f, y) if the Jacobian of (13.47) with respect to F, i_h is different from zero, that is

$$\begin{vmatrix} 0 & -\dfrac{\partial g}{\partial i_h}W \\ 1 & -\dfrac{\partial f}{\partial i_h}W \end{vmatrix} = \frac{\partial g}{\partial i_h}W \neq 0, \tag{13.48}$$

a condition which certainly occurs. We can thus calculate the partial derivatives: $\partial F/\partial N^S, \partial i_h/\partial N^S$, which give us the effect of a variation of N^S on the equilibrium values of F and i_h. By differentiating (13.47) we have

$$1 - \frac{\partial g}{\partial i_h} \frac{\partial i_h}{\partial N^S} W = 0,$$

$$\frac{\partial F}{\partial N^S} - \frac{\partial f}{\partial i_h} \frac{\partial i_h}{\partial N^S} W = 0, \tag{13.49}$$

from which

$$\frac{\partial F}{\partial N^S} = \frac{\partial f / \partial i_h}{\partial g / \partial i_h} < 0, \quad \frac{\partial i_h}{\partial N^S} = \frac{1}{(\partial g / \partial i_h) W} > 0. \tag{13.50}$$

As can be seen, the signs are unequivocally determined thanks to the assumptions on the signs of various partial derivatives, for which see (13.46).

Lastly, we come to dynamic stability. The hypothesis that i_h moves in response to the excess supply of domestic securities gives rise to the following differential equation:

$$\dot{i_h} = v \left[N^S - g(i_h, i_f, y) W \right], v > 0. \tag{13.51}$$

As far as F is concerned, alternative hypotheses are possible. The simplest one, is that the quantity of foreign securities held is always instantly equal to the desired quantity, that is, to the quantity derived from the demand function. This means that the relationship $F = f \left(i_h, i_f, y \right) W$ always occurs at every instant of time. In this case, (13.51) determines the path of i_h over time and, by substituting it in the demand function for foreign securities, the path of F is obtained.

The second, more general, hypothesis is that, on account of delays, frictions, etc. of various types, the quantity of foreign securities held cannot be instantly equal to the desired quantity, but moves towards it according to a partial adjustment mechanism of the type

$$\dot{F} = \alpha \left[f \left(i_h, i_f, y \right) W - F \right], \tag{13.52}$$

where $\alpha > 0$ denotes the adjustment speed: the reciprocal of α represents the mean time-lag i.e. the time necessary for about 63 % of the discrepancy between the current and the desired value to be eliminated by movements of F. Equations (13.51) and (13.52) constitute a system of two differential equations which, after linearising about the equilibrium point, becomes

$$\bar{\dot{i_h}} = -v \frac{\partial g}{\partial i_h} W \bar{i_h}, \quad \bar{\dot{F}} = \alpha \frac{\partial f}{\partial i_h} W \bar{i_h} - \alpha \bar{F}, \tag{13.53}$$

where a dash above the variable as usual indicates the deviations from equilibrium. The characteristic equation of this system is

$$\lambda^2 + \left(v\frac{\partial g}{\partial i_h} + \alpha\right)\lambda + \alpha v\frac{\partial g}{\partial i_h} = 0. \tag{13.54}$$

Since the necessary and sufficient stability conditions

$$v\frac{\partial g}{\partial i_h} + \alpha > 0, \quad \alpha v\frac{\partial g}{\partial i_h} > 0, \tag{13.55}$$

certainly occur, because $\alpha, v, \partial g/\partial i_h$ are all positive, the equilibrium is stable.

We conclude by observing that, once the path of F has been determined, the path of M will be determined by the constraint (13.4), according to which

$$\dot{M} + \dot{F} = 0. \tag{13.56}$$

13.5.2 Portfolio and Macroeconomic Equilibrium Under Fixed Exchange Rates

We shall now pass to the examination of the model (see O'Connell 1984) for the integration between portfolio and macroeconomic equilibrium, which for convenience is given here

$$
\begin{aligned}
&Y - d\left[(1-u)\left(Y + iN_p + iF_p\right), W\right] - x_0 - (Z - kiN^g) = 0, \\
&\quad \text{where } 0 < d_y < 1, d_W > 0, \\
&M - L\left(Y, W\right) = 0, \text{ where } L_y > 0, 0 < L_W < 1, \\
&W - (M + N_p + F_p) = 0, \\
&\dot{M} = sB + hg, \\
&N^g = (1-s)B + (1-h)g, \\
&g = Z + (1-k)iN^g - u(Y + iN_p + iF_p), \\
&CA = x_o - m\left[(1-u)\left(Y + iN_p + iF_p\right), W\right] - i(N^g - N_p) + iF_p, \\
&\quad \text{where } 0 < m_y < 1, m_y < d_y, \\
&B = CA + \dot{N}^g - \left(\dot{N}_p + \dot{F}_p\right).
\end{aligned}
\tag{13.57}
$$

Note that the hypothesis $0 < L_W < 1$ means that only a part of the increase in wealth goes to increase the demand for money, in line with the theory of portfolio choice, and that the hypothesis $m_y < d_y$ indicates that, given an increase in disposable income, residents increase the demand for domestic goods to a greater degree than the demand for imported goods. The other partial derivatives can be interpreted immediately.

The first three equations make it possible to determine the momentary equilibrium and to express the endogenous variables $Y, (N_p + F_p), W$ as functions of the other variables considered as exogenous (x_0, Z, i, M, N^g) provided always that (implicit-function theorem) the Jacobian of these equations is different from zero with respect to the endogenous variables. This Jacobian is

$$J \equiv \begin{vmatrix} 1 - d_y(1-u) & -d_y(1-u)i & -d_W \\ -L_y & 0 & -L_W \\ 0 & -1 & 1 \end{vmatrix}$$

$$= -\left[1 - d_y(1-u)\right]L_W - L_y d_W - L_y d_y(1-u)i, \qquad (13.58)$$

which is negative, given the hypotheses made regarding the various partial derivatives.

The following differentiable functions thus exist

$$\begin{aligned} Y &= Y(x_0, Z, i, M, N^g), \\ N_p + F_p &= V(x_0, Z, i, M, N^g), \\ W &= W(x_0, Z, i, M, N^g). \end{aligned} \qquad (13.59)$$

13.5.2.1 The Dynamics of the Long-Run Equilibrium

The examination of the dynamic stability of long-run equilibrium is more complex: this requires the combination of equations which define the mechanisms of the movement of stock variables (the fourth and fifth equations) with the remaining equations in the model. If we suppose that the momentary equilibrium equations are valid at any instant,[20] we can differentiate the functions (13.59) with respect to time; what interests us is the derivative of the second function, which expresses the stock of (national and foreign) securities held by the private sector. Therefore, supposing the intrinsic exogenous variables x_0, Z and i to be constants, we have

$$\dot{N}_p + \dot{F}_p = \frac{\partial V}{\partial M}\dot{M} + \frac{\partial V}{\partial N^g}\dot{N}^g, \qquad (13.60)$$

where the partial derivatives $\partial V/\partial M$ and $\partial V/\partial N^g$ can be calculated by using the implicit function rule. If we differentiate the first three equations in (13.57) with

---

[20]This implies the hypothesis that the system moves in the long-run through a succession of momentary equilibria. More generally, it would be possible to analyze a situation in which there is also short-term disequilibrium, by assuming appropriate dynamic adjustment mechanisms for the variables $Y, (N_p + F_p), W$. However, this would give rise to a system of five differential equations, the formal analysis of which would be excessively complex.

respect to M, taking account of (13.59), we get

$$\left[1 - d_y(1-u)\right]\frac{\partial Y}{\partial M} - d_y(1-u)i\frac{\partial V}{\partial M} - d_W\frac{\partial W}{\partial M} = 0,$$

$$-L_y\frac{\partial Y}{\partial M} - L_W\frac{\partial W}{\partial M} = -1, \quad (13.61)$$

$$-\frac{\partial V}{\partial M} + \frac{\partial W}{\partial M} = 1,$$

from which:

$$\begin{aligned}
\partial Y/\partial M &= \left[d_y(1-u)i(L_W - 1) - d_W\right]/J, \\
\partial V/\partial M &= \left\{\left[1 - d_y(1-u)\right](L_W - 1) + L_y d_W\right\}/J, \quad (13.62) \\
\partial W/\partial M &= \left\{-\left[1 - d_y(1-u)\right] - L_y d_y(1-u)i\right\}/J,
\end{aligned}$$

where J is given by (13.58). It is easy to see that $\partial Y/\partial M > 0, \partial W/\partial M > 0$, while $\partial V/\partial M$ has an uncertain sign.

Similarly,

$$\begin{aligned}
\partial Y/\partial N^g &= kiL_W/J, \\
\partial V/\partial N^g &= -L_y ki/J, \quad (13.63) \\
\partial W/\partial N^g &= -L_y ki/J,
\end{aligned}$$

where $\partial Y/\partial N^g < 0, \partial V/\partial N^g = \partial W/\partial N^g > 0$, as it is easy to establish. It should be noted that (13.62) and (13.63) can be considered, from the short-run point of view, as *comparative statics* results.

Now, by substituting (13.60) into the balance of payments equation, we have

$$B = CA + \dot{N}^g - \frac{\partial V}{\partial M}\dot{M} - \frac{\partial V}{\partial N^g}\dot{N}^g, \quad (13.64)$$

which when substituted in the equations that determine the behaviour of the money supply and domestic securities, that is, in the fourth and fifth equations of (13.57), gives

$$\begin{aligned}
\dot{M} &= sCA + s\dot{N}^g - s\frac{\partial V}{\partial M}\dot{M} - s\frac{\partial V}{\partial N^g}\dot{N}^g + hg, \\
\dot{N}^g &= (1-s)CA + (1-s)\dot{N}^g - (1-s)\frac{\partial V}{\partial M}\dot{M} \quad (13.65) \\
&\quad -(1-s)\frac{\partial V}{\partial N^g}\dot{N}^g + (1-h)g.
\end{aligned}$$

By collecting and reordering the terms, we have

$$(1 + s\frac{\partial V}{\partial M})\dot{M} - s(1 - \frac{\partial V}{\partial N^g})\dot{N^g} = sCA - hg,$$

$$(1 - s)\frac{\partial V}{\partial M}\dot{M} + \left[s + (1 - s)\frac{\partial V}{\partial N^g}\right]\dot{N^g} = (1 - s)CA + (1 - h)g.$$

(13.66)

This is a first-order system of differential equations, which can be transformed into the normal form through well-known procedures (Gandolfo 2009; Sect. 18.3.1), that is, expressing $\dot{M}$ and $\dot{N^g}$ in terms of CA and g. The system obtained is

$$\dot{M} = \frac{1}{D}\left\{sCA + \left[s + \frac{\partial V}{\partial N^g}(h - s)\right]g\right\},$$

$$\dot{N^g} = \left\{(1 - s)CA + \left[(1 - h) + (s - h)\frac{\partial V}{\partial M}\right]g\right\},$$

(13.67)

where

$$D \equiv \frac{\partial V}{\partial N^g}(1 - s) + s(1 + \frac{\partial V}{\partial M}).$$

(13.68)

It can be established from (13.62), account being taken of (13.58), that $1 + \partial V/\partial M$ is a positive quantity, so that $D > 0$.

13.5.2.2 The Stability Conditions

In order to examine the local stability of system (13.67) a linearisation is carried out around the point of long-run equilibrium. In order to determine this linearisation correctly, it must be born in mind that CA and g are ultimately functions of M and N^g, account being taken of (13.59). Therefore, by indicating the deviation from equilibrium as usual with a dash over the variable we have

$$\overline{\dot{M}} = \frac{1}{D}\left\{s\frac{\partial CA}{\partial M} + \left[s + \frac{\partial V}{\partial N^g}(h - s)\right]\frac{\partial g}{\partial M}\right\}\overline{M}$$

$$+ \frac{1}{D}\left\{s\frac{\partial CA}{\partial N^g} + \left[s + \frac{\partial V}{\partial N^g}(h - s)\right]\frac{\partial g}{\partial N^g}\right\}\overline{N^g},$$

$$\overline{\dot{N^g}} = \frac{1}{D}\left\{(1 - s)\frac{\partial CA}{\partial M} + \left[(1 - h) + (s - h)\frac{\partial V}{\partial M}\right]\frac{\partial g}{\partial M}\right\}\overline{M}$$

$$+ \frac{1}{D}\left\{(1 - s)\frac{\partial CA}{\partial N^g} + \left[(1 - h) + (s - h)\frac{\partial V}{\partial M}\right]\frac{\partial g}{\partial N^g}\right\}\overline{N^g},$$

(13.69)

where the derivatives $\partial CA/\partial M$ etc., are understood to be calculated at the equilibrium point.

The characteristic equation of this system turns out to be

$$
\lambda^2 - \frac{1}{D}\left[(h-s)\left(\frac{\partial V}{\partial N^g}\frac{\partial g}{\partial M} - \frac{\partial V}{\partial M}\frac{\partial g}{\partial N^g}\right) + s\left(\frac{\partial CA}{\partial M} + \frac{\partial g}{\partial M}\right)\right.
$$
$$
\left. + (1-s)\frac{\partial CA}{\partial N^g} + (1-h)\frac{\partial g}{\partial N^g}\right]\lambda + \frac{1}{D^2}\left\{\left[s\frac{\partial CA}{\partial M}\right.\right.
$$
$$
+ \left(s + \frac{\partial V}{\partial N^g}(h-s)\right)\frac{\partial g}{\partial M}\right]\left[(1-s)\frac{\partial CA}{\partial N^g}\right.
$$
$$
+ \left((1-h) + (s-h)\frac{\partial V}{\partial M}\right)\frac{\partial g}{\partial N^g}\right] - \left[(1-s)\frac{\partial CA}{\partial M}\right.
$$
$$
+ \left((1-h) + (s-h)\frac{\partial V}{\partial M}\right)\frac{\partial g}{\partial M}\right]\left[s\frac{\partial CA}{\partial N^g}\right.
$$
$$
+ \left(s + \frac{\partial V}{\partial N^g}(h-s)\right)\frac{\partial g}{\partial N^g}\right]\right\} = 0.
$$

(13.70)

The necessary and sufficient stability conditions are

$$
(h-s)\left(\frac{\partial V}{\partial N^g}\frac{\partial g}{\partial M} - \frac{\partial V}{\partial M}\frac{\partial g}{\partial N^g}\right) + s(\frac{\partial CA}{\partial M} + \frac{\partial g}{\partial M})
$$
$$
+ (1-s)\frac{\partial CA}{\partial N^g} + (1-h)\frac{\partial g}{\partial N^g} < 0,
$$

(13.71)

$$
\left\{s\frac{\partial CA}{\partial M} + \left[s + \frac{\partial V}{\partial N^g}(h-s)\right]\frac{\partial g}{\partial M}\right\}\left\{(1-s)\frac{\partial CA}{\partial N^g}\right.
$$
$$
+ \left[(1-h) + (s-h)\frac{\partial V}{\partial M}\right]\frac{\partial g}{\partial N^g}\right\} - \left\{(1-s)\frac{\partial CA}{\partial M}\right.
$$
$$
+ \left[(1-h) + (s-h)\frac{\partial V}{\partial M}\right]\frac{\partial g}{\partial M}\right\}\left\{s\frac{\partial CA}{\partial N^g}\right.
$$
$$
+ \left[s + \frac{\partial V}{\partial N^g}(h-s)\right]\frac{\partial g}{\partial N^g}\right\} > 0.
$$

(13.72)

In order to check whether these conditions are satisfied, a preliminary step is to calculate the derivatives $\partial CA/\partial M$ etc. By differentiating the expressions which give CA and g, account taken of Eq. (13.59), we have

$$
\frac{\partial CA}{\partial M} = -m_y(1-u)\frac{\partial Y}{\partial M} - m_y(1-u)i\frac{\partial V}{\partial M} - m_W\frac{\partial W}{\partial M} + i\frac{\partial V}{\partial M},
$$
$$
\frac{\partial CA}{\partial N^g} = -m_y(1-u)\frac{\partial Y}{\partial N^g} - m_y(1-u)i\frac{\partial V}{\partial N^g} - m_W\frac{\partial W}{\partial N^g} - i + i\frac{\partial V}{\partial N^g},
$$

(13.73)

$$\frac{\partial g}{\partial M} = -u\frac{\partial Y}{\partial M} - ui\frac{\partial V}{\partial M},$$

$$\frac{\partial g}{\partial N^g} = (1-k)i - u\frac{\partial Y}{\partial N^g} - ui\frac{\partial V}{\partial N^g},$$

where $\partial Y/\partial M$ etc., and $\partial Y/\partial N^g$ etc., are given by (13.62) and (13.63). Thus by substituting from these into Eq. (13.73), we have

$$\begin{aligned}
\frac{\partial CA}{\partial M} &= -m_y(1-u)\frac{\left[d_y(1-u)i(L_W-1) - d_W\right]}{J} \\
&\quad - m_y(1-u)\frac{i\left[1 - d_y(1-u)\right](L_W-1) + iL_y d_W}{J} \\
&\quad + \frac{i\left[1 - d_y(1-u)\right](L_W-1) + iL_y d_W}{J} \\
&\quad + \frac{m_W\left[1 - d_y(1-u)\right] + m_W L_y d_y(1-u)i}{J} \\
&= J^{-1}\{m_y(1-u)d_W - m_y(1-u)i(L_W-1) \\
&\quad - m_y(1-u)iL_y d_W + i(L_W-1) - d_y(1-u)i(L_W-1) \\
&\quad + iL_y d_W + m_W - m_W d_y(1-u) + m_W L_y d_y(1-u)i\} \\
&= J^{-1}\{i(L_W-1)\left[1 - (1-u)(m_y + d_y)\right] \\
&\quad + d_W iL_y\left[1 - m_y(1-u)\right] + d_W m_y(1-u) \\
&\quad + m_W\left[1 - d_y(1-u) + L_y d_y(1-u)i\right]\},
\end{aligned} \tag{13.74}$$

$$\begin{aligned}
\frac{\partial CA}{\partial N^g} &= -m_y(1-u)\frac{kiL_W}{J} + m_y(1-u)i\frac{L_y ki}{J} + m_W\frac{L_y ki}{J} - i - i\frac{L_y ki}{J} \\
&= m_y(1-u)ki\frac{iL_y - L_W}{J} + (m_W - i)\frac{L_y ki}{J} - i,
\end{aligned} \tag{13.75}$$

$$\begin{aligned}
\frac{\partial g}{\partial M} &= -u\frac{d_y(1-u)i(L_W-1) - d_W}{J} - ui\frac{\left[1 - d_y(1-u)\right](L_W-1) + L_y d_W}{J} \\
&= \frac{u\left[d_W(1 - iL_y) - i(L_W-1)\right]}{J},
\end{aligned} \tag{13.76}$$

$$\frac{\partial g}{\partial N^g} = (1-k)i - u\frac{kiL_W}{J} + ui\frac{L_y ki}{J} = (1-k)i - \frac{u(1-i)ki}{J}(L_W - L_y). \tag{13.77}$$

Let us now begin to examine the extreme cases, which occur when $s = h = 1$ (there is no sterilization of the deficits or surpluses in the balance of payments and the whole of the budget deficit is financed by printing money) and when $s = h = 0$ (complete sterilization of the disequilibrium in the balance of payments, and financing of the whole of the budget deficit through the issue of securities).

As can be seen from (13.67), when $s = h = 1$ the system is reduced to a single differential equation relating to $\dot{M}$, and the stability condition becomes

$$\frac{\partial CA}{\partial M} + \frac{\partial g}{\partial M} < 0. \tag{13.78}$$

By substituting from (13.74) and (13.76), and simplifying, we have

$$J^{-1} \{ i(1 - L_W) \left[(1 - u)(m_y + d_y) \right]$$
$$+ u d_W (1 - i L_y) + d_W i L_y \left[1 - m_y (1 - u) \right] + d_W m_y (1 - u)$$
$$+ m_y \left[1 - d_y (1 - u) + L_y d_y (1 - u)i \right] \}$$
$$< 0. \tag{13.79}$$

Given the assumptions made on the signs of the various partial derivatives, the expression in braces is positive and therefore, as $J < 0$, the stability condition is satisfied. Note that the system is stable irrespective of any assumptions regarding parameter k, which in fact does not appear in (13.79).

When $s = h = 0$, the system is reduced to a single differential equation relating to $\dot{N}^g$ and the stability condition becomes

$$\frac{\partial CA}{\partial N^g} + \frac{\partial g}{\partial N^g} < 0. \tag{13.80}$$

By substituting from (13.74) and rearranging terms we have

$$J^{-1} k i \{ (i L_y - L_W) \left[m_y (1 - u) + u \right] + L_y (m_W - i) \} - k i < 0. \tag{13.81}$$

It will be seen at once that this condition cannot be satisfied when $k = 0$. If, on the other hand, $k > 0$, for (13.81) to occur, it is sufficient (even if not necessary) that

$$i L_y > L_W, \quad m_W > i. \tag{13.82}$$

We conclude with the observation that in all intermediate cases between the two extremes examined, it will be necessary to refer to the general conditions (13.71) and (13.72), after substituting into them from (13.62), (13.63) and (13.74); the stability will also depend on the relative magnitude of h and s, besides the various partial derivatives.

13.5.3 Portfolio and Macroeconomic Equilibrium Under Flexible Exchange Rates

13.5.3.1 The Basic Model

The model explained in the text is reproduced here for convenience of the reader:

$$p = p_h^\alpha r^{1-\alpha},$$

$$\pi = p_h/r,$$

$$\widetilde{p}/p = \begin{cases} 0 \\ \dot{p}/p \end{cases},$$

$$\widetilde{r}/r = \begin{cases} 0 \\ \dot{r}/r \end{cases},$$

$$i = i_f + \widetilde{r}/r,$$

$$G = p_h G_R + iN,$$

$$g = p_h G_R + iN - pT - r i_f R,$$

$$CA = p_h x(\pi, a) + r i_f (F + R),$$

$$x_\pi < 0, -1 < x_a < 0,$$

$$r\dot{R} = CA - r\dot{F},$$

$$q = a\left(y_d, i - \frac{\widetilde{p}}{p}, \frac{W}{p}\right) + x(\pi, a) + G_R,$$

$$0 < a_y < 1, a_i < 0, a_W > 0,$$

$$y_d = \frac{p_h}{p}q + \frac{iN}{p} + \frac{r i_f F}{p} - T,$$

$$\frac{M}{p} = L\left(i, q, \frac{W}{p}\right),$$

$$L_i < 0, L_q > 0, 0 < L_w < 1,$$

$$W = M + N = rF,$$

$$g = 0,$$

$$\dot{N} = 0, N > 0,$$

$$\dot{R} = 0, R = 0,$$

$$\dot{M} = 0.$$

$$(13.83)$$

Before going on to its analysis a few observations are in order. We first note that the fifth equation is the well-known UIP condition discussed in Chap. 4, Sect. 4.2, where the symbology $(\check{r} - r)/r$ has been replaced by the symbology $\dot{r}/r$. A second observation concerns the equations that come last but two and last but one: the constancy of N goes together with a preexisting amount of national bonds ($N > 0$), whilst the perfect exchange-rate flexibility (and so the instantaneous equilibrium in the balance of payments) makes the existence of a stock of international reserves unnecessary, so that this is assumed to be zero.

Finally, we show, in relation to net exports x, that $x_\pi > 0$ means that the critical elasticities condition occurs. If we denote exports in physical terms by $ex(\pi)$ and physical imports by $m(\pi, a)$, in monetary terms (in domestic currency) we get

$$p_h ex(\pi) - rm(\pi, a), \quad ex_\pi < 0, m_\pi > 0, 0 < m_a < 1,$$

and so, in real terms (that is, in terms of domestic output)

$$x(\pi, a) = ex(\pi) - \frac{rm(\pi, a)}{p_h} = ex(\pi) - \frac{1}{\pi}m(\pi, a).$$

Therefore

$$x_\pi = ex_\pi + \frac{1}{\pi^2}m - \frac{1}{\pi}m_\pi = \frac{m}{\pi^2}\left(\frac{ex}{\frac{1}{\pi}-m}\frac{\pi}{ex}ex_\pi + 1 - \frac{\pi}{m}m_\pi\right)$$

$$= \frac{m}{\pi^2}\left(-\frac{ex}{\frac{1}{\pi}-m}\eta_{ex} + 1 - \eta_m\right),$$

where $\eta_{ex} \equiv -(\pi/ex)\,ex_\pi, \eta_m \equiv (\pi/m)\,m_\pi$. If the critical elasticities condition holds, that is if

$$\frac{ex}{\frac{1}{\pi}-m}\eta_{ex} + \eta_m > 1,$$

then $x_\pi < 0$, and vice versa.

By way of suitable substitutions, model (13.83) can be boiled down to the three equations of monetary, real and balance-of-payments equilibrium:

$$\frac{M}{p_h^\alpha r^{1-\alpha}} - L\left(i_f + \frac{\tilde{r}}{r}, q, \frac{M + N + rF}{p_h^\alpha r^{1-\alpha}}\right) = 0, \tag{13.84}$$

$$a\left[\left(\frac{p_h}{r}\right)^{1-\alpha}(q - G_R) + \frac{r i_f F}{p_h^\alpha r^{1-\alpha}}, i_f + \alpha\left(\frac{\tilde{r}}{r} - \frac{\tilde{p}_h}{p_h}\right), \frac{M + N + rF}{p_h^\alpha r^{1-\alpha}}\right]$$
$$+ x\left(\frac{p_h}{r}, a[\ldots]\right) + G_R - q = 0, \tag{13.85}$$

$$r\dot{F} = p_h x\left(\frac{p_h}{r}, a[\ldots]\right) + r i_f F. \tag{13.86}$$

Almost all substitutions are self-evident; thus we only make the following observations. Given the assumptions $g = 0$ and $R = 0$, the expression $(iN/p) - T$ which appears in the definition of y_d can be written, by using the definition of g and the assumptions, as $-(p_h/p)G_R$. If we then consider the definition of p, we have $(p_h/p) = (p_h/r)^{1-\alpha}$; from all this expression which replaces y_d in a follows. As regards the real interest rates, we observe that

$$\frac{\dot{p}}{p} = \alpha\frac{\dot{p}_h}{p_h} + (1-\alpha)\frac{\dot{r}}{r},$$

and so

$$\frac{\tilde{p}}{p} = \alpha\frac{\tilde{p}_h}{p_h} + (1-\alpha)\frac{\tilde{r}}{r},$$

whence, as $i = i_f + \tilde{r}/r$, we have

$$i - \frac{\tilde{p}}{p} = i_f + \alpha\left(\frac{\tilde{r}}{r} - \frac{\tilde{p}_h}{p_h}\right),$$

where the term $\tilde{p}_h/p_h$ will ultimately disappear owing to the assumption of a constant p_h (if, on the contrary, one wishes to extend the model to the consideration of a variable p_h, this term must be kept. For the case of p_h variable, not considered here, see Branson and Buiter 1983 , Sect. 9.5).

We now analyse the two cases of static and rational expectations.

13.5.3.2 Static Expectations

Short-Run Equilibrium

In the case of static expectations, $\tilde{r} = 0$. Let us begin by examining short-run equilibrium, in which all the stocks are considered exogenously given. This amounts to considering the subset consisting in Eqs. (13.84) and (13.85), which form a system of two equations in the two unknowns r and q. According to the implicit-function theorem we can express r and q as differentiable functions of M, N, F, G_R, and so perform exercises in comparative statics, provided that the Jacobian of those two equations with respect to r and q is different from zero. We have

$$J = \begin{vmatrix} -\Omega_1 & -L_q \\ -\Omega_2 & -\Omega_3 \end{vmatrix} \tag{13.87}$$

where

$$
\begin{aligned}
\Omega_1 &\equiv (1-\alpha)\frac{M}{pr} - L_w\frac{M+N-\alpha W}{pr} > 0, \\
\Omega_2 &\equiv (1+x_a)\left\{a_y\left[(q-G_R)(1-\alpha)\right]\frac{\pi^{1-\alpha}}{r} - \frac{\alpha i_f F}{p}\right. \\
&\quad \left. + \frac{a_w}{pr}(M+N-\alpha W)\right\} + \frac{\pi}{r}x_\pi < 0, \\
\Omega_3 &\equiv 1 - (1+x_a)a_y\pi^{1-\alpha} > 0,
\end{aligned}
\tag{13.88}
$$

where the assumptions made on the signs reflect the following hypotheses.

The positivity of Ω_1 reflects the hypothesis that an exchange-rate depreciation, by raising the general price level, causes a decrease in the real money stock greater in absolute value than the possible decrease in the demand for money in real terms. We have said "possible" because this latter decrease only occurs when the stock of real wealth W/p decreases, which in turn only comes about when the country is a net debtor to the rest of the world, i.e. when the stock of private net foreign assets is negative $(F < 0)$. In this case the exchange-rate depreciation, by raising the domestic-currency value of this debt, causes a decrease in nominal wealth W, which, together with the increase in the price level, definitely brings about a decrease in real wealth W/p. If, on the contrary, the country is a net creditor $(F > 0)$, the exchange rate depreciation causes an increase in nominal wealth, which in principle might more than offset the increase in the price level, thus causing real wealth to increase. In this second case real money demand increases, and Ω_1 is certainly positive. But it may be positive also in the previous case if L_w is fairly small, that is, if money is "dominated" by other financial assets in the wealth holders' portfolio equilibrium (many writers believe that because of this dominance, L_w is practically nil).

The negativity of Ω_2 reflects the hypothesis that an exchange-rate depreciation increases the total (domestic plus foreign) demand for domestic output. An exchange-rate depreciation, on the one hand, has a favourable effect on net exports x and hence on total demand if, as was assumed above—see Sect. 13.5.3.1—the critical elasticities condition holds; this effect is captured by the term $(\pi/r)x_\pi$. On the other hand, the depreciation has depressive effects on absorption. The first of these occurs when the depreciation decreases real wealth (the conditions for this decrease have been discussed above): this effect is captured by the term $(a_w/rp)(M + N - \alpha W)$. The second depressive effect is due to the fact that the worsening in the terms of trade (because of the depreciation) reduces the real income corresponding to any given level of domestic production: this is captured by the term $a_y(q - G_R)(1 - \alpha)(\pi^{1-\alpha}/r)$. This effect is however contrasted by the increase in income due to the increased domestic-currency value of interest income on the stock of net foreign assets (if this stock is positive; in the contrary case the increase in interest payments will enhance the depressive effect); this is captured by the term $-a_y(\alpha i_f F/p)$. In conclusion, $\Omega_2 < 0$ if we assume that the expansionary effects via elasticities prevail over the depressive effects.

Finally, Ω_3 is positive owing to the assumption that an increase in domestic output causes an increase in the total demand for it by less than the increase in output.

This said, we can write the Jacobian (13.87) as

$$J = \Omega_1\Omega_3 - L_q\Omega_2 > 0, \tag{13.89}$$

where the positivity is ensured by the assumption made above on the various Ω's. Thus in a neighbourhood of the equilibrium point there exist the differentiable functions

$$r = r(M, N, F, G_R), \quad q = q(M, N, F, G_R). \tag{13.90}$$

The effect on equilibrium of changes in the exogenous variables can be ascertained by way of the suitable partial derivatives, which can be calculated by differentiating Eqs. (13.84)–(13.85), account being taken of (13.90). If we consider, for example, a change in government expenditure, we get

$$\begin{aligned}
-\Omega_1 \frac{\partial r}{\partial G_R} - L_q \frac{\partial q}{\partial G_R} &= 0, \\
-\Omega_2 \frac{\partial r}{\partial G_R} - \Omega_3 \frac{\partial q}{\partial G_R} + \Omega_3 &= 0,
\end{aligned} \tag{13.91}$$

whence

$$\begin{aligned}
\frac{\partial r}{\partial G_R} &= -\Omega_3 L_q / J < 0, \\
\frac{\partial q}{\partial G_R} &= \Omega_1 \Omega_3 / J > 0.
\end{aligned} \tag{13.92}$$

The signs—which derive from the assumptions made above—lend support to the statement made in Sect. 13.4.3 as regards the effectiveness of fiscal policy.

In a similar way we obtain the effects of a change in the stock of money by way of an open market operation:

$$\begin{aligned}
\partial r / \partial M - \partial r / \partial N &= \frac{1}{p} \Omega_3 / J > 0, \\
\partial q / \partial M - \partial q / \partial N &= \frac{1}{p} \Omega_2 / J > 0,
\end{aligned} \tag{13.93}$$

which confirm the statements in the text.

The same procedure will give the effects of a change in the stock of net foreign assets:

$$\begin{aligned}
\partial r / \partial F &= -\left[L_w \frac{r}{p} \Omega_3 + (1 + x_a) \frac{r}{p} (a_w + a_y i_f) L_q \right] / J < 0, \\
\partial q / \partial F &= \left[\Omega_1 (1 + x_a) \frac{r}{p} (a_w + a_y i_f) + L_w \frac{r}{p} \Omega_2 \right] / J > 0,
\end{aligned} \tag{13.94}$$

where the sign of $\partial q / \partial F$ is based on the additional hypothesis that L_w is sufficiently small. An increase in F shifts the RR schedule to the right (see Fig. 13.3) through the wealth effect on absorption. Domestic production increases and the exchange rate appreciates.

Long-Run Equilibrium

As we said in Sect. 13.4.3, in the long-run equilibrium F is constant: this amounts to setting $\dot{F} = 0$ in (13.86) and considering F as an endogenous variable. The system to be examined is now made up of the three equations (13.84), (13.85), (13.86)—where $\tilde{r} = 0$ and $\dot{F} = 0$—whose Jacobian with respect to the three variables r, q, F turns out to be

$$J = \begin{vmatrix} -\Omega_1 & -L_q & -L_w\dfrac{r}{p} \\[2mm] -\Omega_2 & -\Omega_3 & (1+x_a)\dfrac{r}{p}\left(a_w + a_y i_f\right) \\[2mm] -\Omega_4 & -p_h x_a a_y \pi^{1-\alpha} & -\Omega_5 \end{vmatrix}, \tag{13.95}$$

where $\Omega_1, \Omega_2, \Omega_3$ are as defined above in (13.88), and

$$\Omega_4 \equiv x_\pi \pi^2 - i_f F + p_h x_a \left\{ a_y \left[(q - G_R)(1-\alpha)\frac{\pi^{1-\alpha}}{r} - \frac{\alpha i_f F}{p} \right] \right.$$
$$\left. + \frac{a_w}{rp}(M + N - \alpha W) \right\} < 0, \tag{13.96}$$
$$\Omega_5 \equiv r\left(p_h x_a \frac{a_y i_f a_w}{p} + i_f \right) < 0.$$

The negativity of Ω_4 is ensured when a depreciation improves the balance of payments. The elasticity effect (which goes in the right direction because we have assumed that the critical elasticities condition occurs) is strengthened (if we assume $F > 0$) by the increased domestic-currency value of interest receipts from abroad and by the decrease in absorption due to the terms-of-trade effect and the wealth effect.

The negativity of Ω_5 reflects the assumption that the improvement in the interest-income account due to the increase in F is more than offset by the deterioration in the next exports account due to the higher absorption determined by the wealth effect.

If we expand the determinant (13.95), we get

$$J_1 = -\Omega_1 \left[\Omega_3 \Omega_5 + p_h x_a a_y \pi^{1-\alpha} \right] (1 + x_a) \frac{r}{p}\left(a_w + a_y i_f\right)$$
$$+ L_q \left[\Omega_2 \Omega_5 - \Omega_4 (1 + x_a) \frac{r}{p}\left(a_w + a_y i_f\right) \right] \tag{13.97}$$
$$- L_w \frac{r}{p} \left[\Omega_3 p_h x_a a_y \pi^{1-\alpha} + \Omega_3 \Omega_4 \right] > 0,$$

where the positivity of J_1 depends on the assumption made on the various Ω's.

Thus in a neighbourhood of the equilibrium point there exist the following differentiable functions

$$r = r(M, N, G_R), \quad q = q(M, N, G_R), \quad F = F(M, N, G_R), \tag{13.98}$$

which allow us to determine the effects of monetary and fiscal policy on the long-run equilibrium values. To simplify this determination, we shall assume, following Branson and Buiter (1983; p. 267) that the wealth effect in the demand for money (which has already been discussed above) is negligible, so that $L_w = 0$.

To begin with, we consider a change in government expenditure and get

$$
\begin{aligned}
-\Omega_1 \frac{\partial r}{\partial G_R} - L_q \frac{\partial q}{\partial G_R} &= 0, \\
-\Omega_2 \frac{\partial r}{\partial G_R} - \Omega_3 \frac{\partial q}{\partial G_R} + (1 + x_a) \frac{r}{p} \left(a_w + a_y i_f \right) \frac{\partial F}{\partial G_R} + \Omega_3 &= 0, \\
-\Omega_4 \frac{\partial r}{\partial G_R} - p_h x_a a_y \pi^{1-\alpha} \frac{\partial q}{\partial G_R} - \Omega_5 \frac{\partial F}{\partial G_R} + p_h x_a a_y \pi^{1-\alpha} &= 0,
\end{aligned}
\tag{13.99}
$$

whence, by solving and indicating for brevity only the signs of the elements of the various determinants (when this indication is sufficient to determine the sign of the determinant), we obtain

$$
\frac{\partial r}{\partial G_R} = \frac{\begin{vmatrix} 0 & - & 0 \\ - & - & + \\ + & + & + \end{vmatrix}}{J_1} = \frac{-}{+} < 0, \qquad
\frac{\partial q}{\partial G_R} = \frac{\begin{vmatrix} - & 0 & 0 \\ + & - & + \\ - & + & + \end{vmatrix}}{J_1} = \frac{+}{+} > 0,
$$

$$
\begin{aligned}
\frac{\partial F}{\partial G_R} &= \frac{\begin{vmatrix} -\Omega_1 & -L_q & 0 \\ -\Omega_2 & -\Omega_3 & -\Omega_3 \\ \Omega_4 & -p_h x_a a_y \pi^{1-\alpha} & -p_h x_a a_y \pi^{1-\alpha} \end{vmatrix}}{J_1} \\
&= \frac{L_q \left[\Omega_2 p_h x_a a_y \pi^{1-\alpha} + \Omega_3 \Omega_4 \right]}{J_1} \\
&= \frac{L_q \left[\pi^2 x_\pi \left(1 - a_y \pi^{1-\alpha} \right) - i_f F \Omega_3 + p_h x_a \Omega_6 \right]}{J_1} = \frac{-}{+} < 0,
\end{aligned}
\tag{13.100}
$$

where

$$
\Omega_6 \equiv a_y \left[(q - G_R)(1 - \alpha) \frac{\pi^{1-\alpha}}{r} - \frac{\alpha i_f F}{p} \right] + \frac{a_w}{rp} (M + B - \alpha W) > 0. \tag{13.101}
$$

In a similar way we calculate

$$\frac{\partial r}{\partial M} - \frac{\partial r}{\partial N} = \frac{\begin{vmatrix} - & - & 0 \\ 0 & - & + \\ 0 & + & + \end{vmatrix}}{J_1} = \frac{+}{+} > 0,$$

$$\frac{\partial q}{\partial M} - \frac{\partial q}{\partial N} = \frac{\begin{vmatrix} - & - & 0 \\ + & 0 & + \\ - & 0 & + \end{vmatrix}}{J_1} = \frac{+}{+} > 0,$$

$$\frac{\partial F}{\partial M} - \frac{\partial F}{\partial N} = -\frac{1}{p}\frac{\left[\pi^2 x_\pi \left(1 - a_y \pi^{1-\alpha}\right) - i_f F\Omega_3 + p_h x_a \Omega_6\right]}{J_1} = \frac{+}{+} > 0.$$

$$(13.102)$$

Let us now briefly comment on the results. An expansionary fiscal policy causes an increase in domestic production and an appreciation in the exchange rate (which in the short run, on the contrary, depreciated); the stock of net foreign assets decreases. The lower F causes the RR schedule to shift to the left relative to the new short-run equilibrium, in relation to which the exchange rate depreciates and output decreases: with respect to the initial equilibrium, however, the exchange rate appreciates and output increases. Thus the long-run equilibrium will be placed at an intermediate point between the initial equilibrium and the short-run equilibrium determined by the impact effect of the increase in G_R.

An expansionary monetary policy, on the contrary, raises output above its new short-run equilibrium level, and causes the exchange rate to appreciate with respect to its new short-run equilibrium value. This appreciation, however, is not so great as to bring it back to its initial value, with respect to which it shows a depreciation. The stock of net foreign assets increases.

After examining the comparative statics, let us examine the stability of the long-run equilibrium, by way of the dynamic equation (13.86). The idea is that the system *moves in time through a succession of short-run equilibria*, and we must ascertain whether this succession converges towards the long-run equilibrium point. It would be theoretically possible alto to examine the *disequilibrium dynamics* of the system, that is, allowing for the possibility that r and q are *outside* of their short-run equilibrium, but this would greatly complicate the analysis.

To study the *equilibrium dynamics* of the system, we substitute the short-run equilibrium values of r and q—given by the solution of Eqs. (13.84) and (13.85), i.e. the functions (13.88)—in Eq. (13.86) and we obtain a first-order differential equation in F, given the exogenous variables. this equation is

$$\dot{F} = \frac{p_h}{r} x\left(\frac{p_h}{r}, a[\ldots]\right) + i_f F, \qquad (13.103)$$

where r and q are functions of F by way of (13.90). If we linearise Eq. (13.103) at the equilibrium point, we get

$$\bar{F} = \left[\pi x_a a_y \pi^{1-\alpha} \frac{\partial q}{\partial F} - \pi x_a a_y \left(q - G_R \right) (1 - \alpha) \frac{\pi^{1-\alpha}}{r} \frac{\partial r}{\partial F} \right.$$

$$+ \pi x_a a_y \frac{\alpha i_f F \pi^{-\alpha}}{r} \frac{\partial r}{\partial F} + \pi x_a a_y \pi^{-\alpha} + \pi x_a a_w \frac{r}{p}$$

$$\left. - \pi x_a a_w \frac{M + F - \alpha W}{pr} \frac{\partial r}{\partial F} - \frac{\pi}{r} (x + \pi x_\pi) \frac{\partial r}{\partial F} + i_f \right] \bar{F}, \quad (13.104)$$

where, as usual, a bar over the variable denotes the deviations from equilibrium.

The necessary and sufficient stability condition is that the expression in square brackets which multiplies $\bar{F}$ is negative. This condition is what Branson and Buiter (1983; p. 269) call the *"super Marshall-Lerner condition"*.

The stability condition implies that if, for example, F is higher than its long-run equilibrium value, the current account must show a deficit, whence a capital inflow to equilibrate the overall balance and so a decrease in F (a capital inflow means a decrease in foreign assets or an increase in foreign liabilities: in both cases the stock of net foreign assets decreases).

The critical elasticities condition does, of course, operate in favour of the satisfaction of the stability condition because a greater value of F causes, in the short-run, the exchange rate to appreciate $[\partial r / \partial F < 0$ from Eq. (13.94)] and this brings about a deterioration in the trade balance,[21] if the critical elasticities condition holds. This effect is captured by the term[22]

$$- \frac{\pi}{r} (x + \pi x_\pi) \frac{\partial r}{\partial F} < 0.$$

On the other hand a greater value of F implies, *ceteris paribus,* an improvement in the investment income account (term i_f), which tends to offset the previous effect. This improvement does, however, raise disposable income and so absorption, which makes the trade balance deteriorate; this effect is captured by the term

$$\pi x_a a_y i_f \pi^{-\alpha} < 0.$$

In the short-run equilibrium, a greater F brings about a higher output $[\partial r / \partial F > 0$: see Eq. (13.94)], which increases absorption and thus causes the trade balance to

---

[21] In what follows "trade balance" is used in the sense of "net exports" as defined in Sect. (13.5.3.1).

[22] Since we are using foreign-currency-denominated magnitudes (F), we must consider net exports in terms of foreign currency, which—according to the definition explained in Sect. (13.5.3.1)—are x. Thus, $\partial (\pi x) / \partial r = (x + \pi x_\pi) \frac{d\pi}{dr} = - \frac{\pi}{r} (x + \pi x_\pi)$.

deteriorate, as is shown by the term

$$\pi x_a a_y \pi^{1-\alpha} \frac{\partial q}{\partial F} < 0.$$

The exchange-rate appreciation due to the higher F ($\partial r/\partial F < 0$, as said above) has two other effects on income. Firstly, it improves the terms of trade, thus raising real income and hence absorption, with a consequent deterioration in the trade balance, as shown by the term

$$-\pi x_a a_y (q - G_R)(1 - \alpha) \frac{\pi^{1-\alpha}}{r} \frac{\partial r}{\partial F} < 0;$$

secondly, it reduces the domestic-currency value of interest payments from abroad (we are assuming $F > 0$), which works in the opposite direction, as shown by the term

$$\pi x_a a_y \frac{i_f F \alpha \pi^{-\alpha}}{r} \frac{\partial r}{\partial F} > 0.$$

A greater F also means—*ceteris paribus*—greater wealth and hence a greater absorption, which causes the trade balance to deteriorate; this effect is captured by the term

$$\pi x_a a_w \frac{r}{p} < 0.$$

Finally, the exchange-rate appreciation caused by higher F reduces the price level and so raises real wealth,[23] which, via an increase in absorption, causes the balance of trade to deteriorate, as shown by the term

$$-\pi x_a a_w \frac{M + N - \alpha W}{rp} \frac{\partial r}{\partial F} < 0.$$

As we see, almost all the additional effects (additional with respect to the effect due to the elasticities) on the balance of trade work in the right direction, so that this balance is likely to deteriorate as a consequence of the greater F. In opposition to this, there is the improvement in the investment income account (term i_f already commented on), so that it is not possible to establish on a *a priori* grounds whether the current account deteriorates (in which case the stability condition is fulfilled) or improves.

---

[23]This is under the assumption that the capital loss (which comes about when $F > 0$, whose domestic-currency value decreases as a consequence of the exchange-rate appreciation) does not offset the decrease in the price level completely. If $F < 0$ the problem does not arise, as there will also be a capital gain.

13.5.3.3 Rational Expectations

In the case of rational expectations, $\tilde{r} = \dot{r}$ and so system (13.84)–(13.86) must be examined from the dynamic point of view also in the short run. It should be emphasized that the long-run equilibrium is the same as in the case of static expectations, since even with rational expectations, long-run equilibrium implies $\dot{r} = 0, \dot{F} = 0$. From this a very important result follows, namely that the comparative-static analysis carried out on the long-run equilibrium in the static expectations case—see Eqs. (13.95) through (13.102)—holds without modifications in the rational expectations case as well.

On the contrary, the analysis of the dynamic is different, due to the presence of the term $\tilde{r}/r$ in the various equations. If we consider system (13.84)–(13.86), where $\tilde{r} = \dot{r}$, and perform a linear approximation at the long-run equilibrium, we get

$$
\begin{aligned}
\dot{\tilde{r}} &= H_{11}\overline{r} + H_{12}\overline{F}, \\
\dot{\overline{F}} &= H_{21}\overline{r} + H_{22}\overline{F},
\end{aligned}
\tag{13.105}
$$

where (all the following expression are, of course, evaluated at the equilibrium point)

$$
\begin{aligned}
H_{11} &\equiv \left(-\Omega_3\Omega_1 + L_q\Omega_2\right)\Omega_7 > 0, \\
H_{12} &\equiv -[\Omega_3 L_w \frac{r}{p} + L_q\left(1 + x_a\right)\left(a_w + a_y i_f\right)\frac{r}{p}]\Omega_7 > 0, \\
H_{21} &\equiv -\frac{\pi x_a a_i \alpha}{r}\Omega_1\Omega_7 \\
&\quad + \left(-\frac{L_i}{r}x_a a_y \pi^{2-\alpha} + \frac{\pi x_a a_i \alpha}{r}L_q\right)\Omega_2\Omega_7 - \frac{1}{r}\Omega_4 \gtrless 0, \\
H_{22} &\equiv -\frac{\pi x_a a_i \alpha}{r}\frac{L_w r}{p}\Omega_7 \\
&\quad - \left(-\frac{L_i}{r}x_a a_y \pi^{2-\alpha} + \frac{\pi x_a a_i \alpha}{r}L_q\right)\left(1 + x_a\right)\left(a_w + a_y i_f\right)\frac{1}{r}\Omega_5 \gtrless 0.
\end{aligned}
\tag{13.106}
$$

The Ω's subscripted from 1 through 6 are as previously defined, whilst Ω_7 is defined as

$$
\Omega_7 \equiv \left[\Omega_3\frac{L_i}{r} + \frac{\left(1 + x_a\right)a_i}{r}\alpha L_q\right]^{-1} < 0.
\tag{13.107}
$$

The signs of the H's derive from the assumptions on the signs of the Ω's and the various partial derivatives.

The differential equation system (13.105) determines the time path of the exchange rate and of the stock of net foreign assets; the time path of domestic output is determined by the equation, always derived from the linearisation of system (13.84)–(13.86),

$$\bar{q} = \left[-\frac{(1 + x_a) a_i \alpha}{r} \Omega_1 - \frac{L_i}{r} \Omega_2 \right] \Omega_7 \bar{r}$$

$$+ \left[-\frac{(1 + x_a) a_i}{r} \alpha L_w \frac{r}{p} \right]$$

$$+ \frac{L_i}{r} (1 + x_a) \left(a_w + a_{yif} \right) \frac{r}{p} \right] \Omega_7 \bar{F}. \tag{13.108}$$

If we assume that an increase in F causes the trade balance to deteriorate by more than it improves the investment income account, the sign of H_{22} will be positive.[24] Thus the pattern of signs in the coefficient matrix of system (13.105) is

$$\begin{bmatrix} + & + \\ ? & - \end{bmatrix}.$$

This means that, in the phase plane (r, F), the $\dot{r} = 0$ schedule slopes negatively whilst the $\dot{F} = 0$ schedule can be either positively (if $H_{21} < 0$) or negatively (if $H_{21} > 0$) sloping. We have, in fact

$$\left(\frac{dr}{dF} \right)_{\dot{r}=0} = -\frac{H_{12}}{H_{11}} < 0, \quad \left(\frac{dr}{dF} \right)_{\dot{F}=0} = -\frac{H_{22}}{H_{21}} \gtrless 0. \tag{13.109}$$

In the former case (positive slope of $\dot{F} = 0$) the determinant of the matrix is certainly negative and the equilibrium point is certainly a *saddle point* (see Gandolfo 2009, table 21.1), so that a unique convergent trajectory (coinciding with one of the two asymptotes of the saddle point) will exist, as described by SS in Fig. 13.5; any point outside SS will move away from equilibrium.

In the latter case (negative slope of $\dot{F} = 0$), for the determinant of the matrix to be negative the $\dot{F} = 0$ schedule must be steeper than the $\dot{r} = 0$ one, namely $(H_{22}/H_{21}) > (H_{12}/H_{11})$, whence $H_{11}H_{22} - H_{21}H_{12} < 0$ and the equilibrium point will again be a saddle point. When this condition does not occur, the long-run equilibrium point will be singular point of a different nature (a *node* or a *focus*), and the results of the analysis will change.

We conclude by noting that in the case of static expectations the system by definition always moves along the $\dot{r} = 0$ schedule, and—if the system is stable— will reach the same long-run equilibrium which, as we have already noted, is identical with both types of expectations.

---

[24]The coefficient H_{22} can, in fact, be interpreted in a similar way—*mutatis mutandis*—as the multiplicative coefficient of $\bar{F}$ in Eq. (13.104).

References

Branson, W. H. (1974). Stocks and flows in international monetary analysis, in A. Ando, R. Herring, & R. Marston (Eds.), *International aspects of stabilization policies*. Federal Reserve Bank of Boston Conference Series No. 12, 27–50.

Branson, W. H. (1979). Exchange rate dynamics and monetary policy, in A. Lindbeck (Ed.). *Inflation and employment in open economies*. Amsterdam: North-Holland.

Branson, W. H., & Buiter, W. H. (1983). Monetary and fiscal policy and flexible exchange rates, in J.S. Bhandari & B.H. Putnam (Eds.). *Economic interdependence and flexible exchange rates*. Cambridge (Mass.): MIT Press, Chap. 9.

Branson, W. H. & Henderson, D. W. (1985). The specification and influence of asset markets, in R. W. Jones & P. B. Kenen (Eds.). *Handbook of international economics* (Vol. II). Amsterdam: North-Holland, Chap. 15.

Christ, C. F. (1979). On fiscal and monetary policies and the government budget restraint. *American Economic Review, 69*, 526–538.

De Grauwe, P. (1983). *Macroeconomic theory for the open economy*. Hampshire: Gower Publishing Co.

Dornbusch, R. (1976). Expectations and exchange rate dynamics. *Journal of Political Economy, 84*, 1161–1176.

Gandolfo, G. (2009). *Economic dynamics* (4th ed.). Berlin, Heidelberg, New York: Springer.

Markowitz, H. (1952). Portfolio selection. *The Journal of Finance, 12*, 77–91.

McKinnon, R. I. (1969). Portfolio balance and international payments adjustment, in R. A. Mundell & A. K. Swoboda (Eds.), *Monetary problems of the international economy* (pp. 199–234). Chicago: Chicago University Press.

McKinnon, R. I., & Oates, W. (1966). *The implications of international economic integration for monetary, fiscal and exchange rate policy*. Princeton Studies in International Finance No. 16, International Finance Section, Princeton University.

O'Connell, J. (1984). Stock adjustment and the balance of payments. *Economic Notes, 1*, 136–144.

Tobin, J. (1958). Liquidity preference as behavior towards risk. *Review of Economic Studies, 25*, 65–86.

Tobin, J. (1980). *Asset accumulation and economic activity*. Oxford: Blackwell.

Growth in an Open Economy

14

The relations between growth and international economics can be examined both in the context of the theory of international trade (for which see Gandolfo 2014) and in the context of open-economy macroeconomics, as we shall briefly do in this chapter.

The macroeconomic theory of growth, after the pioneering contributions of Harrod (1939), Domar (1946), Solow (1956), Swan (1956), and Kaldor (1957), gave rise in the late 1950s and in the 1960s to an immense literature, which we cannot even mention here (the reader is referred, for example, to the textbooks by Burmeister and Dobell 1970, and Wan 1971). Most of this literature was, however, referred to a closed economy.

After a period of neglect, interest in economic growth as a proper subject of macroeconomic theorizing has been resurgent in the 1990s, with special attention given to endogenous growth in a context of intertemporally optimizing agents (for surveys of the "new" growth theory see Grossman and Helpman 1991, Barro and Sala-i-Martin 2004, Aghion and Howitt 1998).

It would not be possible to give an adequate treatment of the models that attempt an extension to an open economy of the various growth models (both traditional and new), originally conceived for a closed economy. We shall therefore only treat—and at that very briefly—a few topics that are more closely tied with the openness of the economy.

In the context of the traditional approach, three questions in our opinion deserve a closer examination:

(a) whether it is possible to talk of a mechanism of *export-led* growth;
(b) whether, given an economic system which is growing for internal causes, this growth has favourable or unfavourable effects on the system's balance of payments;
(c) what are the conditions for enjoying growth and balance-of-payments equilibrium ("balance-of-payments constrained growth").

© Springer-Verlag Berlin Heidelberg 2016

G. Gandolfo, *International Finance and Open-Economy Macroeconomics*,
Springer Texts in Business and Economics, DOI 10.1007/978-3-662-49862-0_14

This chapter is located in Part III despite the fact that it does not only consider models of the stock-flow type but also models of the pure flow type. The reason is that a unified treatment of the topics listed above is preferable.

The questions concerning growth in an open economy in the context of the intertemporal approach will be examined in Chap. 19.

14.1   Export-Led Growth

The basic idea is that the increase in exports exerts a favourable effect on investment and productivity, so that a country with rapidly expanding exports will see its growth enhanced; an underemployment situation is, of course, assumed. The two best-known and pioneering models in this vein are the Lamfalussy (1963) and the Beckerman (1962) models, which can be considered as complementary.

14.1.1 The Lamfalussy Model

This model (which we propose to treat following the revisions of Caves (1970) and Stern (1973)), starts from the assumption that the investment ratio (to income) also depends on the ratio of exports to income, i.e.

$$\frac{I}{y} = h'\frac{x}{y} + h, \tag{14.1}$$

where the symbols have the usual meanings and the variables are expressed in real terms, in a context of rigid prices and fixed exchange rates. The parameter h can be interpreted as the usual marginal propensity to invest: in fact, if we multiply through by y, we get

$$I = h'x + hy. \tag{14.2}$$

The parameter h' reflects the assumption that a rate of expansion of exports higher than that of the other components of national income causes an increase in the desired investment rate, for example because the sectors producing exportables are the most innovating sectors as they are exposed to international competition.

Saving is a function not only of income (as in the traditional Keynesian model) but also of the change in income, to account for the fact that growth may have a positive influence on the average propensity to save (for example because the distribution of income will change in favour of categories with a higher propensity to save). Thus the proportional saving function is

$$\frac{S}{y} = s'\frac{\Delta y}{y} + s, \tag{14.3}$$

where s is the marginal propensity to save (as can be verified by multiplying through by y) and s' represents the effect of growth. The import equation also embodies a marginal effect (μ') of income growth:

$$\frac{m}{y} = \mu' \frac{\Delta y}{y} + \mu, \tag{14.4}$$

where μ is the marginal propensity to import.

The equation for the determination of income in an open economy

$$S + m = I + x, \tag{14.5}$$

closes the model. If we divide both sides of Eq. (14.5) by y, we get

$$\frac{S}{y} + \frac{m}{y} = \frac{I}{y} + \frac{x}{y}, \tag{14.6}$$

whence, by substituting the previous equations into it and solving for the rate of growth of income, we get

$$\frac{\Delta y}{y} = \frac{1 + h'}{s' + \mu'} \frac{x}{y} + \frac{h - \mu - s}{s' + \mu'}. \tag{14.7}$$

From this equation we immediately see that an increase in x/y has a favourable influence on the rate of growth of income.

14.1.2 The Beckerman Model

The Beckerman model differs from Lamfalussy's model by more directly focussing on productivity, prices, and wages. The basic equations of the model are

$$\frac{\Delta x}{x} = \alpha + \beta \left(1 - \frac{p}{p_f} \right), \quad \alpha > 0, \beta > 0, \tag{14.8}$$

$$\frac{\Delta Q}{Q} = \gamma + \delta \frac{\Delta x}{x}, \quad \gamma > 0, \delta > 0, \tag{14.9}$$

$$\frac{\Delta w}{w} = \theta + \lambda \frac{\Delta Q}{Q}, \quad \theta > 0, 0 < \lambda < 1, \tag{14.10}$$

$$\frac{\Delta p}{p} = \frac{\Delta w}{w} - \frac{\Delta Q}{Q}. \tag{14.11}$$

The export equation relates the rate of growth of exports to the terms of trade, that is, as the exchange rate is rigid by assumption, to the ratio of domestic (p) to foreign (p_f) prices: if this ratio is less than one this means that domestic prices are lower than foreign prices and so exports are enhanced, with a favourable effect on the rate of growth of exports themselves; conversely in the opposite case. The parameter α represents the average rate of growth of world trade (exports), so that Eq. (14.8) states that when domestic goods compete favourably with foreign goods ($p < p_f$), the country's exports will grow at a higher proportional rate than the world average, and conversely in the opposite case.

The second equation considers productivity (Q) and directly relates its rate of growth to the rate of growth of exports. The reason behind this link is similar to that already commented on in Lamfalussy's model: since the sectors which are producing exportables are generally the most innovative and those in which the productivity increases are the largest, a more rapid growth in these sectors has beneficial effects on the productivity of the whole economic system, which are added to the rate of increase of productivity (γ) due to other factors (technical progress, etc.), which are considered exogenous.

The third equation incorporates a precise assumption on the rate of growth of the money wage rate, which states that it grows both for exogenous factors (wage negotiations, etc.), represented by the rate θ, and for the increase in productivity, but in such a way that any increase in the rate of growth of productivity raises the rate of growth of the wage rate by a *smaller* amount, as reflected in the assumption that the coefficient λ is strictly included between zero and one.

The fourth equation expresses the rate of growth of prices as the difference between the rate of growth of the wage rate and the rate of growth of productivity. This can be interpreted either along neoclassical lines or by assuming a price-formation equation based on a markup on unit labour costs.

In the former case, the condition of equality between the value of the (marginal) productivity of a factor and its reward, when applied to labour, gives $pQ = w$. If we consider the changes and neglects the second order of smalls, we get $Q \triangle p + p \triangle Q = \triangle w$; dividing through by $pQ = w$ and rearranging terms we obtain Eq. (14.11).

In the latter case the markup equation is $p = g(w/Q)$, where $g > 1$ is the markup coefficient and Q is interpreted as the average productivity. From the formal point of view the equation can be written as $pQ = gw$. Since g is a constant, the equation for the changes is $Q \triangle p + p \triangle Q = \triangle w$; dividing through by $pQ = gw$ and rearranging terms Eq. (14.11) is immediately obtained.

If we substitute the third equation into the fourth we get

$$\frac{\triangle p}{p} = \theta + (\lambda - 1)\frac{\triangle Q}{Q}, \qquad (14.12)$$

from which we see that, thanks to the assumption $0 < \lambda < 1$, an increase in productivity has a favourable effect on prices, as it tends to curb their rate of increase.

Let us now assume that at a certain moment the country has a competitive advantage in trade, that is, $p < p_f$. From Eqs. (14.8) and (14.9), we see that the rate of growth of productivity is enhanced; consequently—Eq. (14.12)—the rate of increase in prices is depressed. It follows that, as the rate of increase of foreign prices is exogenously given, the initial price disparity is accentuated, hence the rate of growth of productivity is enhanced, and so forth. Thus a kind of "virtuous circle" exports-productivity (and hence growth in income) starts off.

It goes without saying that this circle may also be vicious: if the country has a competitive disadvantage ($p > p_f$) the rate of growth of exports (and hence of productivity) is depressed; this raises the rate of growth of prices and so further increases the disadvantage, etc.

Both the Beckerman and the Lamfalussy models have been criticized especially as regards the basic assumptions, which have been regarded as unwarranted or oversimplified (for example, the inflationary mechanism is more complicated than that embodied in Eq. (14.11), etc.). This criticism is sensible, but it must not lead us to reject the idea than an export-led growth mechanism exists. This idea, however, can receive adequate treatment only in much more complicated models, which we cannot deal with here (for an example, see Gandolfo and Padoan 1990). For a revival of the theory of export-led growth, see McCombie and Thirlwall (1994), Rampa et al. (1998; Chaps. 6, 7), and Frankel and Romer (1999), Federici and Marconi (2002).

14.2 Growth and the Balance of Payments

Let us now consider the second problem. In the context of a simple Keynesian model with rigid prices and fixed exchange rates, and given *exogenously* the time path of exports, the obvious conclusion is that growth for internal causes has an unfavourable effect on the balance of payments. In fact, the more income grows, the more imports grow, so that, given the path of exports, the balance of payments (in the sense of balance on goods and services) is affected unfavourably. It follows that growth finds a limitation in the balance-of-payments restraint or, more precisely, in the international reserves restraint (which enable the country to finance its balance-of-payments deficits, but only up to the point below which reserves cannot be allowed to fall).

But some writers (see, for example, Mundell 1968) pointed to the experience of countries that have grown rapidly in the post-World War II period (Germany, France, Italy, etc.) and enjoyed a surplus, as evidence to the contrary. This would however be easy to rebut by showing that this experience is not in itself sufficient evidence, as these countries might well have enjoyed export-led growth, in which case it is not surprising that growth *and* balance-of-payments surplus have gone hand in hand. In fact if we again take up, for example, the Lamfalussy model, and express the balance of payments as a function of the rate of growth of income, we get

$$\frac{x - m}{y} = \frac{s' - h'\mu'}{1 + h'} \frac{\Delta y}{y} + \frac{s - h - h'\mu}{1 + h'}, \tag{14.13}$$

from which we see that when income growth is led by exports, the balance of payments will be favourably or unfavourably affected according as

$$s' - h'\mu' \gtreqless 0. \tag{14.14}$$

This inequality can in principle occur with either sign, and so, if it occurs with the > sign, income growth will favourably affect the balance of payments.

To avoid confusion, it is therefore necessary to use models in which the rate of growth of income is *not* causally related to exports through an export-led mechanism: only in this way, in fact, the question under consideration can be answered unambiguously. We shall therefore assume that income growth is due to internal factors and is not of the export-led type.

In this framework the answer will depend on the type of model that we use. From the point of view of the traditional Keynesian model the answer given at the beginning of this section remains valid. On the contrary, the answer is exactly the opposite if one uses the monetary approach to the balance of payments (MABP) explained in Chap.12, Sect. 12.3.

In fact, if one assumes that income is growing (always remaining at the full employment level) and that prices and the exchange rate are rigid, it follows from the MABP that the balance of payments will improve. This is so because an increase in income raises the demand for money and so, at unchanged stock of money, an excess demand for money will come about which will generate a balance-of-payments surplus through the mechanism described in Sect. 12.3. It is true that, by increasing the stock of money, this surplus will reduce the excess demand for money and so will tend to automatically eliminate itself; but the continuing income growth will again give rise to an excess demand for money and so the balance-of-payments surplus will be recreated by a continual knock-on effect.

It goes without saying that this phenomenon is valid insofar as the basic model is valid, so that those who do not accept the MABP will not accept the thesis that growth favourably affects the balance of payments. Leaving aside this debate, it is necessary to point out a somewhat paradoxical consequence of the thesis in question, which arises as soon as we abandon the small country framework. In fact, if we look at the problem from a *world* point of view and bear in mind that it is not possible for all countries to have a balance-of-payments surplus at one and the same time, it follows that countries with positive income growth and balance-of-payments surpluses will have to be matched by countries with negative income growth and balance-of-payments deficits. There would then seem to be a kind of exploitation of the latter group by the former, a situation which cannot be maintained in the long run, unless places are swapped every now and then.

We have so far assumed fixed exchange rates. Under flexible exchange rates the balance-of-payments constraint can be eased by appropriate variations in the exchange rate. For an examination of these problems see McCombie and Thirlwall (1994).

The relations between growth and the balance of payments constraint are examined more deeply in the next section.

14.3 Balance-of-Payments Constrained Growth (Thirlwall)

All the major growth paradigms (neoclassical, endogenous growth theory) are basically supply-oriented. The balance-of-payments constrained growth model presented by Thirlwall (1979), and generalizations of it, is a demand-oriented model based on the assumption that factor supplies and technical progress are largely endogenous to an economic system dependent on the growth of output itself. Any explanation of the process of economic growth that simply relies on technology, physical capital and human capital differences across countries is, at some level, incomplete. Thirlwall developed a conceptual framework that emphasizes the notion that growth is demand-led and demand is trade-led, where growth is consistent with external balance.

Should we care about balance of payment equilibrium? The answer is undoubtedly yes for Thirlwall (1979, 1997, 2011). The Thirlwall model is remarkable in its simplicity. The major proposition is that no country can grow faster than its balance-of-payments equilibrium rate for very long, as it is the only way to avoid unsustainable foreign debt.

Let start with the balance of payments equilibrium condition,

$$p_d X = p_f M E, \tag{14.15}$$

where X is exports; M is imports; p_d is the domestic price of exports; p_f is the foreign price of imports, and E is the exchange rate measured as the domestic price of foreign currency.

Export and import demand functions are specified as multiplicative with constant elasticities

$$X = a \left(\frac{p_d}{p_f E} \right)^\eta Z^\varepsilon, \eta < 0, \varepsilon > 0, \tag{14.16}$$

$$M = b \left(\frac{p_f E}{p_d} \right)^\psi Y^\pi, \psi < 0, \pi > 0, \tag{14.17}$$

where η is the price elasticity of demand for exports; ε is the income elasticity of demand for exports; ψ is the price elasticity of demand for imports; π is the income elasticity of demand for imports; Z is the world income, and Y is the domestic income.

Taking logarithms of equations (14.16) and (14.17), and differentiating with respect to time, we obtain

$$x_t = \eta(p_{dt} - p_{ft} - e_t) + \varepsilon z_t, \tag{14.18}$$

$$m_t = \psi(p_{ft} + e_t - p_{dt}) + \pi y_t, \tag{14.19}$$

where lower-case letters indicate the (proportional) growth rates of variables ($x_t = \dfrac{d \ln X_t}{dt} = \dfrac{dX_t}{dt}\dfrac{1}{X_t}$, etc.). Equations (14.18) and (14.19) represent respectively the export and import demand functions expressed in growth terms.

Logarithmic differentiation of Eq. (14.15) gives the balance of payments equilibrium condition in growth terms

$$m_t + p_{ft} + e_t = p_{dt} + x_t. \tag{14.20}$$

Solving the system of equations (14.18)–(14.20) we get the rate of income growth consistent with the balance of payment equilibrium over time:

$$y_B = \left[(1 + \eta + \psi)(p_{dt} - p_{ft} - e_t) + \varepsilon z_t\right]/\pi, \tag{14.21}$$

where y_B denotes the balance-of-payments-constrained equilibrium growth rate of income. From Eq. (14.21) it follows that y_B is the equilibrium rate of growth compatible with perpetual trade balance (Eq. 14.15), and the rate of growth towards which the economy will tend, given that countries are unwilling or unable to attract permanent net inflows of capital, because of the resulting accumulation of foreign indebtedness and consequent debt servicing commitments.

Thirlwall then makes rather simplifying assumptions, that relative prices are constant ($p_{dt} - p_{ft} - e_t = 0$),[1] the rate of growth of world income z_t is exogenously given, and a deficit cannot be financed by capital flows, so Eq. (14.21) reduces to

$$y_{B1} = \varepsilon z_t/\pi \tag{14.22}$$

or similarly,

$$y_{B2} = x_t/\pi, \tag{14.23}$$

which are known in the literature as Thirlwall's laws of growth. These expressions show that the growth rate compatible with the long-term external balance of an economy is determined by the ratio between the growth rate of its exports and the income elasticity of its demand for imports (the Harrod foreign trade multiplier, $1/\pi$). The primary theoretical reference of Thirlwall's laws is the Harrod foreign trade multiplier (Harrod 1933), or more generally, the Hicks super-multiplier (McCombie 1985).

From Eq. (14.21) it follows that:

---

[1]Thirlwall assumes that Purchasing Power Parity (PPP) holds in the long run. Thus, variations in the real exchange rate are considered to be irrelevant to long-term growth. Balance of payments disequilibria are not automatically eliminated over time through changes in relative prices, but the fundamental adjustment variable is the growth of income and exports.

(a) an improvement of terms of trade (or real exchange rate), $p_{dt} - p_{ft} - e_t > 0$, will increase the growth of income consistent with balance of payments equilibrium;
(b) a depreciation of the exchange rate, $e_t > 0$, will improve the growth rate if $(\eta + \psi) > -1$, that is if the Marshall-Lerner condition holds;
(c) a country's growth rate consistent with balance of payments equilibrium is inversely related to the propensity to import;
(d) a country's growth rate is dependent on other countries' growth rates (z_t), but the income elasticity of demand for exports, ε, affects its rate of growth. Furthermore, if the presence of a price effect $(p_{dt} \neq p_{ft} + e_t)$ or of international capital flows $(p_d X \neq p_f ME)$ will affect growth only in the short term, in the long run given condition (14.15) and the assumptions underlying Thirlwall' laws, the rate of growth will tend to condition (14.22); any attempt to boost growth by policy design to stimulate domestic demand will improve growth only in the short run, as imports will increase and condition (14.20) will be violated. The main channels to enhance growth should be export-oriented trade policies (increasing the income elasticity of exports, ε); a sort of "mercantilism" (decreasing the income import propensity, π) or through "global Keynesianism" (increasing world income growth, z) (Setterfield 2011).

Thirlwall (1979) produced convincing empirical evidence of the simple rules. He applied Eq. (14.23) to a selection of developed countries over the time periods 1951–1973 and 1953–1976, and found a remarkable correspondence between the observed growth rate of countries and the growth rate predicted from the balance of payments constrained growth model.

14.3.1 Extensions

The Thirlwall approach relies on stringent assumptions and since his pioneering contribution, the literature has provided many extensions of the original model. One of the major assumption questioned is the absence of capital flows. It is possible to argue that such an assumption is too restrictive as some economies are able to attract net international capital flows that cause long-term deviations from Eq. (14.15). Thirlwall and Hussain (1982; p. 500–1) point out: "...It must be recognized, though, that developing countries are often able to build up ever-growing current account deficits financed by capital inflows (which are then written off!) which allow these countries to grow permanently faster than otherwise would be the case". Indeed, this aspect becomes relevant once we consider developing and emerging countries.

Thirlwall and Hussain (1982) incorporated the capital flows in the original context.

The balance-of-payments equilibrium condition (14.15) (in domestic currency units) at time t can be re-expressed as

$$p_{dt}X_t + C_t = p_f M_t E_t, \tag{14.24}$$

where C_t is the value of foreign capital flows, $C_t > 0$ measures capital inflows and $C_t < 0$ measures capital outflows.

The equation shows that external balance requires that imports of goods and services have to be paid for by the revenues of exports and/or by foreign financial capital.

Expressing Eq. (14.24) in terms of rates of change gives

$$\theta(p_{dt} + x_t) + (1 - \theta)(c_t) = (p_{ft} + m_t + e_t), \tag{14.25}$$

where, as usual, lower case letters indicate rates of growth of the variables. θ and $(1 - \theta)$ are respectively the shares of export and capital flow receipts on total revenues. Let assume the above functional forms for the export and import demand,

$$X = a(\frac{p_d}{p_f E})^\eta Z^\varepsilon, \eta < 0, \varepsilon > 0$$

$$M = b(\frac{p_f E}{p_d})^\psi Y^\pi, \psi < 0, \pi > 0.$$

Taking logarithms and differentiating with respect to time it follows

$$x_t = \eta(p_{dt} - p_{ft} - e_t) + \varepsilon z_t \tag{14.26}$$

$$m_t = \psi(p_{ft} + e_t - p_{dt}) + \pi y_t. \tag{14.27}$$

To characterize the results in this case, let us combine Eqs. (14.26) with (14.27) and (14.25). The balance of payments constrained growth rate is then

$$y_B = [(1 + \theta\eta + \psi)(p_{dt} - p_{ft} - e_t) + \theta(\varepsilon z_t) + (1 - \theta)(c_t - p_{dt})]/\pi. \tag{14.28}$$

On the right-hand side of the equation are respectively, the volume effect of relative price changes on balance-of-payments-constrained real income growth; the terms of trade effect; the effect of exogenous changes in foreign income growth, and the effect of the growth rate of real capital flows.

Keeping the assumption of a constant real exchange rate, Thirlwall and Hussain thus give a modified version of Thirlwall's law in presence of capital flows,

$$y_{B1} = \theta(\varepsilon z_t) + (1 - \theta)(c_t - p_{dt})/\pi, \tag{14.29}$$

which in turn shows that the growth of the economy consistent with the balance-of-payments equilibrium is defined by the ratio between a weighted sum of the growth of exports and of capital inflows (in real terms) and the income elasticity of the demand for imports. This implies that the equilibrium growth rate is what guarantees that the intertemporal sum of the balance-of-payments debits and credits equals zero. Such a result suggests that the access to international capital may enable a long-term economic growth superior to that originally predicted by Thirlwall's model (1979).

The above model misses something important, that is borrowing internationally cannot last indefinitely as it will lead to a growing accumulation of foreign debt, which must be remunerated and repaid, and its dynamics over time can be detrimental to the long-term growth of an economy. This would require imposing a constraint on the level of current account deficits financed by capital inflows.

To deal with this aspect, Moreno-Brid (1998a) proposed to redefine the long-term balance-of-payments equilibrium to further clarify the conditions for a constant ratio of the current account deficit relative to domestic income.

Equation (14.15) is expressed now as follows

$$P_{dt}X + FP_{dt} = P_{ft}ME,\qquad(14.30)$$

where F is the current deficit in real terms and FP_d denotes the nominal capital flows (C) to finance the deficit. Taking rates of change of Eq. (14.30) gives

$$\theta(p_{dt} + x_t) + (1-\theta)(f_t + p_{dt}) = (p_{ft} + m_t + e_t).\qquad(14.31)$$

Exactly the same analysis as above, with the assumption of $f = y$ (i.e. a constant ratio of the current account deficit to GDP), implies

$$y_D = \theta(\varepsilon z_t) + (\theta\eta + \varepsilon + 1)(p_{dt} - p_{ft} - e_t)/\pi - (1-\theta).\qquad(14.32)$$

It is straightforward to verify that if the real exchange rate (or terms of trade) is constant, the constrained growth rate consistent with a fixed deficit/GDP ratio is

$$y_{D1} = \frac{\theta}{\pi - (1-\theta)}.\qquad(14.33)$$

According to Moreno-Brid, the hypothesis of a growth of capital inflows equal to the income growth rate is sufficient for guaranteeing that the foreign debt level of the economy remains asymptotically constant. In addition to capital flows, Elliot and Rhodd (1999), Moreno-Brid (2003), Vera (2006), Alleyne and Francis (2008) also include interest payments on debt to the model. Even if the growth of interest payments is quite high and the debt service ratio is also high, it still makes little difference to the predicted growth rate. In other words, export growth dominates (Thirlwall 2011). Barbosa-Filho (2001) suggests to consider targeted macroeconomic policy to assure that the trajectories of the exchange rate and of income are consistent with the trade balance necessary for maintaining the stability of foreign indebtedness at the level allowed by international credit markets. Meyrelles Filho et al. (2013) developed an alternative analysis to deal with the relationship between economic growth and foreign debt, highlighting the possible contribution of the control of capital inflows.

Araujo and Lima (2007) consider an extension of the Pasinetti (1993, 1981) model of structural change to develop a disaggregated multi-sector version of the Thirlwall law, stressing that key sources of difference in economic performance across countries include the structural aspects of the economy.

Consequently, Eq. (14.21) becomes

$$y_B = \frac{\sum\limits_{i=1}^{n} \omega_{xi} \varepsilon_i(z)}{\sum\limits_{j=1}^{n} \omega_{mj} \pi_j}, \qquad (14.34)$$

where ω_{xi}, ω_{mj} are the sectoral composition of exports and imports and ε_i, π_j the sectoral income elasticity demand for exports and imports. What the multi-sectoral model highlights is that even if sectoral elasticities are constant and there is no change in world income growth, a country can grow faster by structural change that shifts resources to sectors with higher income elasticities of demand for exports and away from sectors with a high income elasticity of demand for imports. This is what import substitution and export promotion policies are meant to achieve (Thirlwall 2011). Cimoli et al. (2010) show that the developing countries that succeeded in reducing the income gap between themselves and developed countries were those that transformed their economic structure towards sectors with a higher income elasticity of demand for exports relative to imports.

The single-country model has been theoretically extended to a multi-country framework. Nell (2003) disaggregates the world income growth variable (z), and considers the different income elasticities of demand for exports and imports to and from each trading partner (p). Thus, Eq. (14.21) is modified as follows

$$y = \frac{(1 + \eta + \psi)(p_{dt} - p_{ft} - e_t) + \sum\limits_{p=1}^{n} \omega_{xp} \varepsilon_p(y_p)}{\sum\limits_{p=1}^{n} \omega_{mp} \pi_p(y_p)}, \qquad (14.35)$$

where y_p is the growth rate of the trading partner $(p = 1, \ldots, n)$; ω_{xp} is the share of exports to country p on total exports; ε_p is the income elasticity of demand for exports to each destination (p); π_p is the income elasticity of demand for imports from each trading partner (p), and ω_{mp} is the share of imports from each trading partner on total imports. Nell (2003) applied the above "generalized" version of Thirlwall's balance-of-payments constrained growth model to analyse the long-run relationships between the output growth rates of OECD countries and two neighbouring regions, South Africa and the rest of the Southern African Development Community. The empirical results find strong support for the multi-country version of Thirlwall's growth model. The picture that emerges reflects the mutual interdependence of the world economy where a country's growth rate depends on the growth rates of the other countries.

For a synthetic presentation of all these developments see McCombie and Thirlwall (2004) and Thirlwall (2011). For a review of the long debate on Thirlwall's law see McCombie (2011). Several empirical papers have applied Thirlwall's model and its extensions to individual countries or to groups of countries. The majority of the studies support the predictions of the balance-of-payments-constrained growth hypothesis with the acknowledgment that Thirwall's approach is not only of theoretical interest, but also sheds light on a range of important empirical patterns in country growth-rates differences (see Thirlwall 2011, for a review and a discussion of the empirical findings).

14.4 Appendix

14.4.1 Exports, Growth, and the Balance of Payments

The thesis that growth can have favourable effects on the balance of payments in the context of the MABP was originally enunciated by Mundell (1968), but was based on a model containing errors, as shown by Reid (1973) and Wein (1974). It is however possible to show the validity of the thesis under consideration, by correctly formalizing it.

For this purpose we shall again take up the model explained in Sect. 12.5.2 and consider the system of differential equations

$$\dot{M} = \alpha(kpy - M),$$
$$\dot{y} = \lambda y, \tag{14.36}$$

where, for simplicity of notation, we have omitted the bar over the symbol y (which, it will be remembered, denotes full employment income).

The first equation is already known, the second expresses the hypothesis that full employment income grows in time at the rate λ. This system can be solved explicitly (see, for example, Gandolfo 2009, Chap. 18); the solution is

$$y(t) = y_0 e^{\lambda t},$$
$$M(t) = \left(M_0 - \frac{\alpha k p y_0}{\alpha + \lambda} \right) e^{-\alpha t} + \frac{\alpha k p y_0}{\alpha + \lambda} e^{\lambda t}, \tag{14.37}$$

where M_0 and y_0 are the initial values of y and M. Since we have, as we know

$$B(t) = \dot{R} = \dot{M}, \tag{14.38}$$

by differentiating the second equation in (14.37) with respect to time, we obtain

$$B(t) = -\alpha \left(M_0 - \frac{\alpha k p y_0}{\alpha + \lambda} \right) e^{-\alpha t} + \lambda \frac{\alpha k p y_0}{\alpha + \lambda} e^{\lambda t}. \tag{14.39}$$

From this equation we see that, since the first term tends to zero asymptotically whilst the second is positive and increasing, in the long run the balance of payments will definitely show an increasing surplus. To ascertain what happens in the short-run, we must know the initial situation in the balance of payments. If, as assumed in the text, the system is initially in equilibrium (that is, the stock of money is equal to the demand for money, hence absorption equals income and the balance of payments is in equilibrium), then

$$M_0 = kpy_0,$$

(14.40)

so that Eq. (14.39) becomes

$$
\begin{aligned}
B(t) &= \frac{-\alpha\lambda kpy_0}{\alpha+\lambda}e^{-\alpha t} + \frac{\alpha\lambda kpy_0}{\alpha+\lambda}e^{\lambda t} \\
&= \frac{\alpha\lambda kpy_0}{\alpha+\lambda}e^{\lambda t}\left(1 - e^{-(\alpha+\lambda)t}\right).
\end{aligned}
$$

(14.41)

Since the term $e^{-(\alpha+\lambda)t}$ is smaller than one for all $t > 0$ and decreases as time goes on, the balance of payments will show a surplus from the instant that immediately follows time zero.

But it is also possible that the system is *not* initially in equilibrium, for example because there is an excessive money stock (in which case the balance of payments will start from an initial deficit) or insufficient (in which case the balance of payments will start from an initial surplus). To examine these cases let us introduce a constant a which measures the excess ($a > 0$) or deficiency ($a < 0$) of money supply at the initial time, that is

$$M_0 = kpy_0 + a.$$

(14.42)

Equation (14.39) now becomes

$$
\begin{aligned}
B(t) &= \left(-\alpha a - \frac{\alpha\lambda kpy_0}{\alpha+\lambda}e^{-\alpha t}\right) + \frac{\alpha\lambda kpy_0}{\alpha+\lambda}e^{\lambda t} \\
&= \frac{\alpha\lambda kpy_0}{\alpha+\lambda}e^{\lambda t}\left(1 - e^{-(\alpha+\lambda)t}\right) - \alpha a e^{-\alpha t}.
\end{aligned}
$$

(14.43)

From this last equation we see that, when $a < 0$ (initial balance-of-payments surplus), $B(t)$ will be *a fortiori* positive for all $t > 0$. If, on the contrary, $a > 0$ (initial deficit), then the balance of payments will remain in deficit in the short-run; this deficit, however, will gradually disappear and become an ever-increasing surplus. The critical value t^* of t—that is, the point in time at which the balance of payments from a situation of deficit becomes zero (equilibrium) to show a surplus immediately after—can be easily calculated from (14.43) by setting $B(t) = 0$ there.

Simple passages yield the exponential equation

$$e^{(\alpha+\lambda)t} = \frac{(\alpha+\lambda)\,a + \lambda k p y_0}{\lambda k p y_0},$$

(14.44)

which, as the right-hand side is positive, admits of a unique real and positive solution, given by

$$t^* = \frac{1}{\alpha+\lambda} \ln \frac{(\alpha+\lambda)\,a + \lambda k p y_0}{\lambda k p y_0},$$

(14.45)

from which we see that the greater is a, i.e. the greater the initial deficit, the greater will be t^*. This is intuitive, since the greater is the initial deficit, the more time will be needed—*ceteris paribus*—to eliminate it.

14.4.2 Dynamic Stability of Thirlwall's Model

Thirlwall's model fails to account for the adjustment to equilibrium. More specifically, Thirlwall provides a steady-state solution, where all variables grow at the same constant rate, and the adjustment mechanism is assumed but not demonstrated.

To address the stability shortcomings of the basic Thirlwall model, Moreno-Brid (1998a) and Barbosa-Filho (2001) extend the model to consider the case of unbalanced trade. We will follow their framework.

Let us recall from the text the original model represented by the export and import demands:

$$X = a\left(\frac{P_d}{P_f E}\right)^{\eta} Z^{\varepsilon}, \eta < 0, \varepsilon > 0,$$

(14.46)

$$M = b\left(\frac{P_f E}{P_d}\right)^{\psi} Y^{\pi}, \psi < 0, \pi > 0.$$

(14.47)

The balance of payment constraint implies current account balance, $X = M$. Let us instead define the balance of payments equilibrium as implying that the current account balance (surplus or deficit) should be equal to a constant and sustainable ratio to national income.

We further normalize net exports, $NX = X - M$, by the nominal home income Y, so that $NX' = X' - M'$, where $NX' = NX/P_d Y$, $X' = X/Y$, $M' = RM/Y$ and $R = (p_f + E - P_d)$. With a constant ratio of the current account balance to Y, the rates of growth of X' and M' should be equal, namely $dX'/dt = dM'/dt$.

By using Eqs. (14.46)–(14.47) and solving for y, we obtain a general expression for the balance of payments constrained growth rate representing the unbalanced trade home growth rate:

$$y = \left(\frac{B\varepsilon}{\pi - 1 + B}\right) z - \left(\frac{1 - \psi - B\eta}{\pi - 1 + B}\right) r,$$

(14.48)

where $B = X'/M'$ is the export-import ratio of home country. Let us note that the more traditional assumption of balanced trade (i.e., a current account balance equal to zero) in the long run is a special case in which $B = 1$.

From Eq. (14.48) we observe that the home country export-import ratio is itself function of the home growth rate. In fact, since

$$\frac{dB}{dt} = d\left(\frac{X'}{M'}\right)/dt = d\left(\frac{X}{M}\right)/dt = B\left(\frac{d\ln X}{dt} - \frac{d\ln M}{dt}\right).$$

Substituting X and M from (14.46) and (14.47) we have

$$\frac{dB}{dt} = B\left[\varepsilon z - \pi y - (1 - \psi - \eta)r\right]. \tag{14.49}$$

Using Eq. (14.48), we get

$$\frac{dB}{dt} = B\left[\frac{(\pi - 1)(1 - B)\varepsilon}{\pi - 1 + B}z - \frac{(1 - \psi - \eta + \pi\eta)(B - 1)}{\pi - 1 + B}r\right], \tag{14.50}$$

where B is not necessarily stationary ($dB/dt = 0$) unless trade is initially balanced. Even if $r = 0$, B is still not necessarily stationary unless trade is initially balanced or the home income elasticity of imports is equal to one ($\pi = 1$).

Following Moreno-Brid (1998b), we assume $r = 0$. To model the dynamics of the rate of home income growth we assume a partial adjustment equation of the type

$$\frac{dy}{dt} = \chi\left[\left(\frac{B\varepsilon}{\pi - 1 + B}\right)z - y\right] \tag{14.51}$$

where χ is a speed of adjustment parameter, which measures how fast the home income growth rate converges to the balance of payments constraint given by (14.48). From Eq. (14.49) we get

$$\frac{dB}{dt} = B(\varepsilon z - \pi y). \tag{14.52}$$

Equations (14.51) and (14.52) constitute a 2×2 non-linear dynamical system for the home growth rate and export-import ratio:

$$\begin{aligned}\dot{y} &= \chi\left[\left(\frac{B\varepsilon}{\pi - 1 + B}\right)z - y\right] \\ \dot{B} &= B(\varepsilon z - \pi y)\end{aligned} \tag{14.53}$$

To study the local stability around the singular point $(y^*, B^*) = (\varepsilon z/\pi, 1)$, let us consider the Jacobian matrix of the system evaluated at the steady state solution (y^*, B^*) :

$$
J = \begin{bmatrix} -\chi & z\chi\varepsilon\,(\pi - 1)\,\pi^{-2} \\ -\pi & 0 \end{bmatrix}.
$$

(14.54)

Local stability depends on the characteristic roots ϱ of J, namely

$$
\begin{vmatrix} -\chi - \varrho & z\chi\varepsilon\,(\pi - 1)\,\pi^{-2} \\ -\pi & 0 - \varrho \end{vmatrix} = 0,
$$

(14.55)

that is

$$
\varrho^2 + \chi\varrho + z\chi\varepsilon\,(\pi - 1)\,\pi^{-1} = 0.
$$

(14.56)

Since $\chi > 0$ by assumption, the stability conditions (Gandolfo 2009; Chap. 18) are satisfied if and only if $\pi > 1$. This means that, given a constant real exchange rate, the home country tends to its balance-of-payments-constrained growth rate with balanced trade when the income elasticity of its imports is greater than one. If, on the contrary, $\pi < 1$, the characteristic equation (14.56) will have two opposite real roots, namely the singular point will be a saddle point. This means that there will be convergence to the singular point only on the stable arm of the saddle. In the particular case in which $\pi = 1$ the characteristic equation will have a zero real root and a negative real root, which means that the system will converge to an arbitrary constant.

Further cases are examined in Barbosa-Filho (2001) and Pugno (1998).

References

Aghion, P., & Howitt, P. (1998). *Endogenous growth theory.* Cambridge (Mass): MIT Press.

Alleyne, D., & Francis, A. A. (2008). Balance of payments constrained growth in developing countries: A theoretical perspective. *Metroeconomica, 59,* 189–202.

Araujo, R. A., & Lima, G. T. (2007). A structural economic dynamics approach to balance-of-payments-constrained growth. *Cambridge Journal of Economics, 31,* 755–774.

Barro, R. J., & Sala-i-Martin, X. (2004). *Economic growth,* 2nd edition. New York: Mc Graw-Hill.

Barbosa-Filho, N. (2001). The balance of payments constraint: from balanced trade to sustainable debt. *Banca Nazionale del Lavoro Quarterly Review, 219,* 381–400.

Beckerman, W. H. (1962). Projecting Europe's growth. *Economic Journal, 72,* 912–925.

Burmeister, E. & Dobell, A. R. (1970). *Mathematical Theories of Economic Growth.* London: Macmillan.

Caves, R. (1970). Export-led growth: The post-war industrial setting, in W. A. Eltis, M. F. G. Scott & J. N. Wolfe (Eds.). *Induction, growth and trade: Essays in honour of Sir Roy Harrod* (pp. 234–255). Oxford: Oxford University Press.

Cimoli, M., Porcile, G., & Rovira, S. (2010). Structural change and the balance-of-payments constraint: Why did Latin America fail to converge? *Cambridge Journal of Economics, 34*, 389–411.

Domar, E. D. (1946). Capital expansion, rate of growth and employment. *Econometrica, 14*, 137–147.

Elliot, D. & Rhodd, R. (1999). Explaining growth rate differences in the highly indebted countries: an extension to Thirlwall and Hussain. *Applied Economics*, 1145–1148.

Federici, D., & Marconi, D. (2002). On exports and economic growth: The case of Italy. *Journal of International Trade and Economic Development, 11*, 323–340.

Frankel, J. A., & Romer, D. (1999). Does trade cause growth? *American Economic Review, 89*, 379–399.

Gandolfo, G. (2009). *Economic Dynamics* (4th ed.). Berlin, Heidelberg, New York: Springer.

Gandolfo, G. (2014). *International trade theory and policy* (2nd ed.). Berlin, Heidelberg, New York: Springer.

Gandolfo, G., & Padoan, P. C. (1990). The Italian continuous time model: Theory and empirical results. *Economic Modelling, 7*, 91–132.

Grossman, G. M., & Helpman, E. (1991). *Innovation and growth in the global economy*. Cambridge (Mass.): MIT Press.

Harrod, R. F. (1933). *International economics*. Cambridge (UK): Cambridge University Press.

Harrod, R. F. (1939). An essay in dynamic theory. *Economic Journal, 49*, 14–33.

Kaldor, N. (1957). A model of economic growth. *Economic Journal, 67*, 591–624.

Lamfalussy, A. (1963). Contribution à une théorie de la croissance en économie ouverte. *Recherches Économiques de Louvain, 29*, 715–734.

McCombie, J. S. L. (1985). Economic growth, the Harrod foreign trade multiplier and the Hicks super-multiplier. *Applied Economic, 17*, 55–72; reprinted in McCombie & Thirlwall (2004).

McCombie, J. S. L. (2011). Criticisms and defences of the balance-of-payments constrained growth model: Some old, some new. *PSL Quarterly Review, 64*, 353–392.

McCombie, J. S. L., & Thirlwall, A. P. (1994). *Economic growth and the balance of payments constraint*. London: Macmillan.

McCombie, J. S. L., & Thirlwall, A. P. (2004). *Essays on balance of payments constrained growth. Theory and evidence*. London: Routledge.

Moreno-Brid, J. C. (1998). On capital flows and the balance-of-payments-constrained growth model. *Journal of Post Keynesian Economics, 21*, 283–298.

Mundell, R. A. (1968). *International economics*. New York: Macmillan, Chap. 9.

Meyrelles Filho, S. F., Jayme, F. G., & Libnio, G. (2013). Balance-of-payments constrained growth: A post Keynesian model with capital inflows. *Journal of Post Keynesian Economics, 35*, 373–397.

Moreno-Brid, J. C. (1998). On capital flows and the balance-of-payments-constrained growth model. *Journal of Post Keynesian Economics, 21*, 283–298.

Moreno-Brid, J. C. (2003). Capital flows, interest payments, and the balance-of-payments constrained growth model: A theoretical and empirical analysis. *Metroeconomica, 54*, 346–365.

Nell, K. (2003). A generalized version of the balance-of-payments growth model: An application to neighbouring regions. *International Review of Applied Economics, 17*, 249–267; reprinted in McCombie & Thirlwall (2004).

Pasinetti, L. (1981). *Structural change and economic growth – A theoretical essay on the dynamics of the wealth of the nations*. Cambridge (UK): Cambridge University Press.

Pasinetti, L. (1993). *Structural economic dynamics – A theory of the economic consequences of human learning*. Cambridge (UK): Cambridge University Press.

Pugno, M. (1998). The stability of Thirlwall's model of economic growth and the balance-of-payments constraint. *Journal of Post Keynesian Economics, 20*, 559–581.

Rampa, G., Stella, L., & Thirlwall, A. P. (Eds.) (1998). *Economic dynamics, trade and growth: Essays on Harrodian themes*. London: Macmillan.

Reid, F. J. (1973). Mundell on growth and the balance of payments: A note. *Canadian Journal of Economics, 6*, 592–595.

Setterfield, M. (2011). The remarkable durability of Thirlwall's law. *PSL Quarterly Review, 64*, 393–427.

Solow, R. M. (1956). A contribution to the theory of economic growth. *Quarterly Journal of Economics, 70,* 65–94.

Stern, R. M. (1973). *The balance of payments: theory and economic policy.* Chicago: Aldine.

Swan, T. (1956). Economic growth and capital accumulation. *Economic Record, 32,* 334–361.

Thirlwall, A. P. (1979). The balance of payments constraint as an explanation of international growth rates differences. *Banca Nazionale del Lavoro Quarterly Review, 128,* 45–53.

Thirlwall, A. P. (1997). Reflections on the concept of balance-of payments-constrained growth. *Journal of Post Keynesian Economics, 19,* 377–386.

Thirlwall, A. P. (2011). Balance-of-payments constrained growth models: History and overview. *PSL Quarterly Review, 64,* 307–351.

Thirlwall, A. P., & Hussain, M. N. (1982). The balance of payments constraint, capital flows and growth rate differences between developing countries. *Oxford Economic Papers, 34,* 498–510.

Vera, L. A. (2006), The balance of payments constrained growth model: a north-south approach. *Journal of Post Keynesian Economics, 29,* 67–92.

Wan, H. Y., Jr. (1971). *Economic growth.* New York: Harcourt Brace Jovanovich.

Wein, J. (1974). Growth and the balance of payments: A note on Mundell. *Economic Journal, 84,* 621–623.

Part V

The Exchange Rate

Exchange-Rate Determination

15

The problem of the forces that determine the exchange rate and, in particular, its equilibrium value, is self-evident under flexible exchange rates, but is also important under limited flexibility and even under fixed exchange rates (if the fixed rate is not an equilibrium one, the market will put continuing pressure on it and compel the monetary authorities to intervene in the exchange market, etc.). This problem has already been mentioned in several places (see, e.g., Sects. 3.2.1, 9.1, 13.4); in this chapter we gather up the threads and try to give a general treatment. This is not easy, since various competing theories exist, of which we shall try to give a balanced view; for general surveys of the problems treated in this chapter see, for example, MacDonald and Taylor (1992), Isard (1995), De Jong (1997), MacDonald and Stein eds. (1999), MacDonald (2007).

15.1 The Purchasing-Power-Parity Theory

The oldest theory of exchange-rate determination is probably the purchasing power-parity (henceforth PPP) theory, commonly attributed to Cassel (1918) even though—as usual—precursors in earlier times are not lacking (on this point see Officer 1982). Two versions of the PPP are distinguished, the *absolute* and the *relative* one.

According to the *absolute* version, the exchange rate between two currencies equals the ratio between the values, expressed in the two currencies considered, of the same typical basket containing the same amounts of the same commodities. If, for example, such a basket is worth US $11,000 in the United States and *euro*10,000 in the European Union, the $/*euro* exchange rate will be 1.1 (1.1 dollars per euro).

According to the relative version, the percentage *variations* in the exchange rate equal the percentage variations in the ratio of the price levels of the two countries (the percentage variations in this ratio are approximately equal to the difference between the percentage variations in the two price levels, or *inflation differential*).

© Springer-Verlag Berlin Heidelberg 2016
G. Gandolfo, *International Finance and Open-Economy Macroeconomics*,
Springer Texts in Business and Economics, DOI 10.1007/978-3-662-49862-0_15

In both versions the PPP theory is put forward as a *long-run* theory of the equilibrium exchange rate, in the sense that in the short-run there may be marked deviations from PPP which, however, set into motion forces capable of bringing the exchange rate back to its PPP value in the long term. The problems arise when one wants better to specify this theory, which implies both a precise singling out of the price indexes to be used and the determination of the forces acting to restore the PPP: the two questions are, in fact, strictly related.

Those who suggest using a price-index based on internationally traded commodities only, believe that PPP is restored by international commodity arbitrage which arises as soon as the internal price of a traded good deviates from that prevailing on international markets, when both prices are expressed in a common unit (the *law of one price*).

On the contrary, those who maintain that a general price-index should be used, think that people appraise the various currencies essentially for what these can buy, so that—in free markets—people exchange them in proportion to the respective purchasing power.

Others suggest using cost-of-production indexes, in the belief that international competition and the degree of internationalization of industries are the main forces which produce PPP.

A fourth proposal suggests the use of the domestic inflation rates starting from various assumptions:

(a) the real interest rates are equalized among countries;
(b) in any country the nominal interest rate equals the sum of the real interest rate and the rate of inflation (the Fisher equation);
(c) the differential between the nominal interest rates of any two countries is equal (if one assumes risk-neutral agents) to the expected percentage variation in the exchange rate (for this relation, see Sects. 4.2 and 4.4).

From these assumptions it follows that there is equality between the expected percentage variations (which, with the further assumption of perfect foresight, will coincide with the actual ones) in the exchange rate and the inflation differential.

None of these proposals is without its drawbacks and each has been subjected to serious criticism which we cannot go into here, so that we shall only mention a few points. For example, the commodity-arbitrage idea was criticized on the grounds that it presupposes free mobility of goods (absence of tariffs and other restrictions to trade) and a constant ratio, within each country, between the prices of traded and non-traded goods: the inexistence, even in the long run, of these conditions, is a well-known fact. Besides, the law of one price presupposes that traded goods are highly homogeneous, another assumption often contradicted by fact and by the new theories of international trade (see Gandolfo 2014, Chap. 7), which stress the role of product differentiation.

The same idea of free markets, of both goods and capital, lies at the basis of the other proposals, which run into trouble because this freedom does not actually exist. Cassel himself, it should be noted, had already singled out these problems and stated that they were responsible for the deviations of the exchange-rate from the PPP. For a survey of the PPP debate see Taylor and Taylor (2004) .

15.1.1 The Harrod-Balassa-Samuelson Model

A good explanation of why PPP does not necessarily hold in terms of aggregate price indices was separately given by Balassa (1964) and Samuelson (1964), and, earlier, by Harrod (1933). Briefly speaking, the Harrod-Balassa-Samuelson effect is the tendency for the consumer price index (CPI) in rich, fast-growing countries to be high (after adjusting for exchange rates) relative to the CPI in poor, slow-growing countries. To understand this, the distinction between tradables (produced by a traded goods sector) and nontradables (produced by a non-traded goods sector) is essential.

Given this distinction, the model starts from the historical observations that labour productivity in rich countries is higher than in poor countries, and that this productivity differential occurs predominantly in the traded goods sector (perhaps because this sector is exposed to international competition and biased towards high-innovation goods). The result is that CPI levels tend to be higher in rich countries. The explanation is the following.

First, assume that wages are the same in both sectors. Second, assume that prices are directly related to wages and inversely to productivity. Now, a rise in productivity in the traded goods sector will cause an increase in wages in the entire economy; firms in the non-traded goods sector will survive only if the wage rise, unmatched by a productivity rise, is offset by a rise in the relative price of non-traded goods. Given the presence of both tradables and nontradables in the CPI, and assuming that PPP holds for tradables, it follows that—after adjusting for the exchange rate—the CPI in rich countries (where productivity grows in the tradables sector) is higher than the CPI in poor countries (with no, or much lower, productivity growth).

A formal proof will be given in the Appendix; here we note that the Harrod-Balassa-Samuelson effect does help explain why PPP may not hold for aggregate price indices like CPI, but by definition it does not help explain the deviations from PPP as regards traded goods!

These deviations, which make the PPP theory useless to explain the behaviour of exchange rates in the short-run, were one of the reasons which induced most economists to abandon it in favour of other approaches. It should however be pointed out that the PPP theory has been taken up again by the monetary approach (which will be dealt with in Sect. 15.3.1), and used as an indicator of the long-run trend in the exchange rate. However, the empirical evidence in favour of long-run PPP is rather poor (Froot and Rogoff 1996; O'Connell 1998; for a contrary view see Klaassen 1999). In addition, in the cases in which the exchange rate shows a tendency to revert to its long-run PPP value after a shock, it has been found that this reversion is not monotonic (Cheung and Lai 2000) .

15.2   The Traditional Flow Approach

This approach, also called the balance-of-payments view or the exchange-market approach, starts from the observation that the exchange rate is actually determined in the foreign exchange market by the demand for and supply of foreign exchange, and that it moves (if free to do so) to bring these demands and supplies into equality and hence (if no intervention is assumed) to restore equilibrium in the balance of payments.

That the exchange rate is determined in the foreign exchange market by the demand for and supply of foreign exchange is an irrefutable fact (we do, of course, exclude economies with administrative exchange controls), but it is precisely in determining these demands and supplies that most problems arise. The traditional flow approach sees these as pure flows, deriving—in the older version—from imports and exports of goods, which in turn depend on the exchange rate and—after the Keynesian-type models—also on national income. This approach has been widely described before, in particular in Chaps. 7 and 9, to which we refer the reader. For further details on this approach see, for example, Gandolfo (1979). The introduction of capital movements as a further component of the demand for and supply of foreign exchange does not alter this view insofar as these movements are also seen as pure flows: this is, for example, the case of the model described in Sect. 10.3.

This approach can be criticized for several shortcomings, amongst which the fact that it neglects stock adjustments, a point that we have already extensively treated (see, for example, Sects. 10.2.2.1 and 11.2.2). It should however be stressed that, if on the one hand the criticism must induce us to consider the traditional approach inadequate in its specification of the determinants of the demands for and supplies of foreign exchange, on the other, it does not affect the fact that it is the interaction between these demands and supplies which actually determines the exchange rate.

15.3   The Modern Approach: Money and Assets
in Exchange-Rate Determination

The modern approach (also called the *asset-market approach*) takes the exchange rate as the relative price of monies (the monetary approach) or as the relative price of bonds (the portfolio approach). The two views differ as regards the assumptions made on the substitutability between domestic and foreign bonds, given however the common hypothesis of perfect capital mobility.

The monetary approach assumes perfect substitutability between domestic and foreign bonds, so that asset holders are indifferent as to which they hold, and bond supplies become irrelevant. Conversely, in the portfolio approach domestic and foreign bonds are imperfect substitutes, and their supplies become relevant.

To avoid confusion it is as well to recall the distinction between (perfect) capital mobility and (perfect) substitutability (see Sect. 4.6). Perfect capital mobility means that the actual portfolio composition instantaneously adjusts to the desired one. This, in turn, implies that—if we assume no risk of default or future capital controls, etc.—*covered interest parity* must hold (see Sect. 4.1). Perfect substitutability is a stronger assumption, as it means that asset holders are indifferent as to the composition of their bond portfolios (provided of course that both domestic and foreign bonds have the same expected rate of return expressed in a common numéraire). This, in turn, implies that *uncovered interest parity* must hold (see Sect. 4.2).

It is important to note that according to some writers (see, for example, Helliwell and Boothe 1983), the condition of covered interest parity itself becomes a theory of exchange-rate determination (the *interest parity model*, where interest parity may be expressed either in nominal or real terms), if one assumes that the forward exchange rate is an accurate and unbiased predictor of the future spot rate (see Sect. 4.5): it would in fact suffice, in this case, to find the determinants of the expected future spot exchange rate to be able to determine, given the interest rates, the current spot rate.

Since classifications are largely a matter of convenience (and perhaps of personal taste) we have chosen to follow the dichotomy based on the perfect or imperfect substitutability between domestic and foreign bonds within the common assumption of perfect capital mobility.

15.3.1 The Monetary Approach

The monetary approach to the balance of payments has been treated in Sect. 12.3; we only recall that it assumes the validity of PPP as a long-run theory. Now, if one considers fixed exchange rates, as in Chap. 12, then the monetary approach determines the effects of (changes in) the stock of money (which is an endogenous variable) on the balance of payments and vice versa; if one assumes flexible exchange rates the same approach (with the money stock exogenous) becomes a theory of exchange-rate determination. In fact, if we again take up the model explained in Sect. 12.3 and for brevity only consider the equilibrium relations (i.e., we neglect the adjustment process set into motion by discrepancies between money demand and supply) we get

$$M = kpy, \; M_f = k_f p_f y_f, \; p = r p_f. \tag{15.1}$$

The first two equations express monetary equilibrium in the two countries (for notational simplicity we have eliminated the bar over y and y_f it being understood that these are exogenously given), whilst the third is the PPP equation. For simplicity's sake, we assume the rest-of-the-world variables as exogenously given. Then, under fixed exchange rates, the third equation determines the domestic price level p, whilst the first determines the equilibrium domestic money stock M. In the case of a divergence of the current from the equilibrium value, the adjustment described in Sect. 12.3 will set into motion and automatically restore equilibrium.

In the case of flexible exchange rates, the domestic stock of money becomes exogenous, so that the first equation determines p and the third determines the exchange rate. After simple manipulations we get

$$r = \frac{M}{M_f} \cdot \frac{k_f y_f}{ky}, \tag{15.2}$$

from which we see that, as stated above, the exchange rate is the relative price of two monies, i.e. of the two money-stocks M and M_f (the other variables are given exogenously).

In this model the interest rate is not explicitly present, but can be introduced by assuming that the demand for money in real terms is also a function of i, that is

$$M = pL(y, i), \; M_f = p_f L_f(y_f, i_f), \; p = rp_f, \tag{15.3}$$

whence

$$r = \frac{M}{M_f} \cdot \frac{L_f(y_f, i_f)}{L(y, i)}. \tag{15.4}$$

From this equation it can clearly be seen that—*ceteris paribus*—an increase in the domestic money-stock brings about a depreciation in the exchange rate, whilst an increase in national income causes an appreciation, and an increase in the domestic interest rate causes a depreciation. These conclusions, especially the last two, are in sharp contrast with the traditional approach, where an increase in income, by raising imports, tends to make the exchange rate depreciate, whilst an increase in the interest rate, by raising capital inflows (or reducing capital outflows), brings about an appreciation in the exchange rate.

These (surprisingly) different conclusions are however perfectly consistent with the vision of the MABP, described in Sect. 12.3. For example an increase in income raises the demand for money; given the money stock and the price level, the public will try to get the desired additional liquidity by reducing absorption, which causes a balance-of-payments surplus, hence the appreciation. This appreciation, by simultaneously reducing the domestic price level p (given that PPP holds), raises the value of the real money-stock (M/p increases), and so restores monetary equilibrium. Similar reasoning explains the depreciation in the case of an increase in i (the demand for money falls, etc.).

The monetary approach to the exchange rate can be made more sophisticated by introducing additional elements, such as sticky prices which do not immediately reflect PPP, arbitrage relations between i and i_f, the possibility that domestic agents hold foreign money, etc.: see the next section.

15.3.2 Sticky Prices, Rational Expectations, and Overshooting of the Exchange-Rate

Let us consider, following Dornbusch (1976), a small open economy under flexible exchange rates with perfect capital mobility and flexible prices but with given full-employment output. We note at the outset that—in the terminology of rational expectations—the exchange rate is a typical *jump* variable, since it is free to make jumps in response to 'news' (see Chap. 13, Sect. 13.4). On the contrary, commodity prices are assumed to be predetermined variables, namely they are assumed to adjust slowly to their long-run equilibrium value, hence the denomination of *sticky-price monetary model* of exchange rate determination.

Perfect capital mobility coupled with perfect asset substitutability implies that the domestic (i) and foreign (i_f) nominal interest rates are related by *uncovered interest rate parity* (UIP) (see Sect. 4.3), namely

$$i = i_f + x, \tag{15.5}$$

where $x \equiv \widetilde{r}/r$ is the expected rate of change of the exchange rate.

Since we are in a small open economy, the foreign interest rate is taken as given. As regards expectations, we observe that in our deterministic context rational expectations imply perfect foresight, so that the expected and actual rate of depreciation of the exchange rate coincide. Letting e denote the logarithm of the spot exchange rate r, we have

$$\dot{e} = d \ln r / dt = \dot{r}/r, \tag{15.6}$$

hence $\dot{e}$ is the *actual* rate of depreciation of the nominal exchange rate r. By setting $x = \dot{e}$ in Eq. (15.5) we have

$$i = i_f + \dot{e}. \tag{15.7}$$

Let us now consider money market equilibrium,

$$\frac{M}{P} = e^{-\lambda i} Y^\phi,$$

where M is the money supply, P the price level, λ the semi-elasticity of money demand with respect to the interest rate, and ϕ the elasticity of money demand with respect to income Y. Taking logs (lower-case letters denote the logs of the corresponding upper-case-letter variables) and rearranging terms we have

$$p - m = \lambda i - \phi y. \tag{15.8}$$

Combining Eq. (15.8) with the UIP condition incorporating rational expectations
—Eq. (15.7)—we obtain the condition for *asset market equilibrium*

$$p - m = -\phi \bar{y} + \lambda i_f + \lambda \dot{e} , \tag{15.9}$$

where output has been taken as given at its full-employment level $\bar{y}$.

In long-run equilibrium with a stationary money supply, we have

$$\bar{p} - m = -\phi \bar{y} + \lambda i_f, \tag{15.10}$$

since actual and expected depreciation are assumed to be zero in long-run equilibrium. Equation (15.10) determines the long-run equilibrium price level.

Subtracting Eq. (15.10) from Eq. (15.9) we obtain:

$$p - \bar{p} = \lambda \dot{e},$$

or

$$\dot{e} = \frac{1}{\lambda}(p - \bar{p}). \tag{15.11}$$

This is one of the key equations of the model, as it expresses the dynamics of the
current spot exchange rate in terms of the deviations of the current price level from
its long-run equilibrium level.

We now turn to the goods market, that—given the assumption of a constant level
of output—will serve to determine the price level.

Since output is given, excess demand for goods will cause an increase in prices.
We first assume that aggregate demand for domestic output depends on the relative
price of domestic goods with respect to foreign goods, rP_f/P (namely $[(e + p_f) - p]$
in log terms), the interest rate, and real income. Thus we have

$$\ln D \equiv d = u + \delta(e - p) + \gamma y - \sigma i, \tag{15.12}$$

where u is a shift parameter and the given foreign price level P_f has been normalised
to unity (hence $p_f = \ln P_f = 0$). The usual dynamic (Walrasian) assumption
that prices move in response to excess demand yields, account being taken of the
logarithmic setting,

$$\dot{p} = \pi(d - y), \quad \pi > 0, \tag{15.13}$$

from which

$$\dot{p} = \pi[u + \delta(e - p) + (\gamma - 1)y - \sigma i]. \tag{15.14}$$

In long-run equilibrium, $\dot{p} = 0$ and $p = \bar{p}$. With these settings in Eq. (15.14) we finally obtain the long-run equilibrium exchange rate

$$\bar{e} = \bar{p} + (1/\delta)[\sigma i_f + (1 - \gamma)y - u], \tag{15.15}$$

where $\bar{p}$ is given by Eq. (15.10) and $i = i_f$ since the long-run equilibrium exchange rate is expected to remain constant.

Let us now turn back to the disequilibrium dynamics of Eq. (15.14). Solving Eq. (15.8) for the interest rate and substituting into Eq. (15.14) we have

$$\dot{p} = \pi[u + \delta(e - p) + \frac{\sigma}{\lambda}(m - p) - \rho\bar{y}], \tag{15.16}$$

$$\rho \equiv \phi\sigma/\lambda + 1 - \gamma.$$

We now subtract from the r.h.s. of Eq. (15.16) its long-run equilibrium value, which is zero, namely $0 = \pi[u + \delta(\bar{e} - \bar{p}) + \frac{\sigma}{\lambda}(m - \bar{p}) - \rho\bar{y}]$. Since $\dot{p} = \mathrm{d}(p - \bar{p})/\mathrm{d}t$ for constant $\bar{p}$, we can thus express price dynamics in terms of deviations from the long-run equilibrium values of the exchange rate and the price level, that is

$$\dot{p} = -\pi(\delta + \frac{\sigma}{\lambda})(p - \bar{p}) + \pi\delta(e - \bar{e}). \tag{15.17}$$

Equations (15.17) and (15.11) govern the dynamics of our system.

The dynamics of the system is rather complex and here we limit ourselves to observing that there exists only one trajectory along which the movement of the system converges to the equilibrium point (which is in the nature of a *saddle point*: a concept that we have already met in Chap. 13, Sect. 13.4). Any point outside this trajectory will move farther and farther away from equilibrium.

Let us now consider the phenomenon of exchange-rate overshooting in response to 'news', i.e. to unanticipated events such as a monetary shock. This can be examined by means of Fig. 15.1. Suppose that the economic system is initially at its long-run equilibrium (point Q). By an appropriate choice of units we can set $\bar{p} = \bar{e}$, so that OR is a 45° line. Suppose now that the nominal money stock permanently increases. Economic agents immediately recognize that the long-run equilibrium price level and exchange rate will increase in the same proportion, as in the long-run money is neutral. In terms of Fig. 15.1 this means that economic agents recognize that the economy will move from Q to Q_1.

The economy, however, cannot instantaneously jump from Q to Q_1 because of the sticky-price assumption. But the exchange rate is a jump variable, hence the economy *can* jump from Q to Q_2 on the (unique) stable trajectory AA. From the economic point of view, the current exchange rate will depreciate because an increase in the money supply, given the stickiness of prices in the short run, will cause an increase in the real money supply and hence a fall in the domestic nominal interest rate. Since the UIP condition is assumed to hold instantaneously, and the nominal foreign interest rate is given, the exchange rate immediately depreciates by

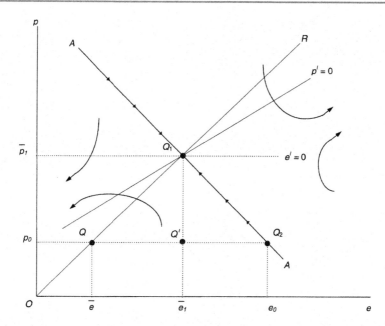

Fig. 15.1   Rational expectations and exchange-rate overshooting

more than the increase in the long-run equilibrium value to create the expectation of an appreciation. This is required from the UIP condition that the interest-rate differential equals the expected rate of appreciation: in fact, given $i = i_f$ in the initial long-run equilibrium, the sudden decrease in i requires $x < 0$—see Eq. (15.5)—namely an anticipated appreciation.

Thus the exchange rate initially overshoots its (new) long-run equilibrium level ($e_0 > \bar{e}_1$), after which it will gradually appreciate alongside with the increase in the price level following the path from Q_2 to Q_1 on the AA line. Overshooting results from the requirement that the system possesses 'saddle-path' stability, which in turn is a typical feature of the dynamics of rational-expectation models, as shown in Gandolfo (2009; Sect. 28.4).

15.3.3 The Portfolio Approach

This approach, in its simplest version, is based on a model of portfolio choice between domestic and foreign assets. As we know from the theory of portfolio selection examined more than once in this volume (see Sects. 13.2, 13.4, 16.1), asset holders will determine the composition of their bond portfolios,[1] i.e. the shares of

---

[1]To simplify to the utmost, we follow Frankel (1983) in considering solely bonds; the introduction of money would not alter the substance of the results (see, for example, Krueger 1983, Sect. 4.3.2).

domestic and foreign bonds on the basis of considerations of (expected) return and risk. If perfect substitutability between domestic and foreign assets existed, then uncovered interest parity should hold, that is

$$i = i_f + \tilde{r}/r \tag{15.18}$$

where $\tilde{r}$ denotes the expected change in the exchange rate over a given time interval; i and i_f are to be taken as referring to the same interval. In the case of imperfect substitutability this relation becomes

$$i = i_f + \tilde{r}/r + \delta. \tag{15.19}$$

Hence, with imperfect substitutability a divergence may exist between i and $(i_f + \tilde{r}/r)$; the extent of this divergence will—*ceteris paribus*—determine the allocation of wealth (W) between national (N) and foreign (F) bonds. For simplicity's sake we make the small-country assumption, i.e. we assume that domestic bonds are held solely by residents, because the country is too small for its assets to be of interest to foreign investors. The model can be extended to consider the general case without substantially altering the results, provided that residents of any country wish to hold a greater proportion of their wealth as domestic bonds (the so-called *preferred local habitat* hypothesis): see Frankel (1983).

Given our simplifying assumption we can write

$$W = N^d + rF^d, \tag{15.20}$$

where the demands are expressed, in accordance with portfolio selection theory, as

$$\begin{aligned} N^d &= g(i - i_f - \tilde{r}/r)W, \\ rF^d &= h(i - i_f - \tilde{r}/r)W, \end{aligned} \tag{15.21}$$

where $g(\ldots) + h(\ldots) = 1$ because of (15.20). If we impose the equilibrium condition that the amounts demanded should be equal to the given quantities existing (supplied), we get

$$N^d = N^s, \quad F^d = F^s, \tag{15.22}$$

and so, by substituting into (15.21) and dividing the second by the first equation there, we obtain

$$\frac{rF^s}{N^s} = \varphi(i - i_f - \tilde{r}/r), \tag{15.23}$$

Alternatively, one could assume that asset holders make a two-stage decision, by first establishing the allocation of their wealth between money and bonds, and then allocating this second part between domestic and foreign bonds.

where $\varphi(\ldots)$ denotes the ratio between the $h(\ldots)$ and $g(\ldots)$ functions. From Eq. (15.23) we can express the exchange rate as a function of the other variables

$$r = \frac{N^s}{F^s}\varphi(i - i_f - \widetilde{r}/r). \qquad (15.24)$$

Equation (15.24) shows that the exchange rate can be considered as the relative price of two stocks of assets, since it is determined—given the interest differential corrected for the expectations of exchange-rate variations—by the relative quantities of N^s and F^s.

The basic idea behind all this is that the exchange rate is the variable which adjusts instantaneously so as to keep the (international) asset markets in equilibrium. Let us assume, for example, that an increase occurs in the supply of foreign bonds from abroad to domestic residents (in exchange for domestic currency) and that (to simplify to the utmost) expectations are static $(\widetilde{r} = 0)$. This increase, *ceteris paribus*, causes an instantaneous appreciation in the exchange rate. To understand this apparently counter-intuitive result, let us begin with the observation that, given the foreign-currency price of foreign bonds, their domestic-currency price will be determined by the exchange rate. Now, residents will be willing to hold (demand) a higher amount of foreign bonds, *ceteris paribus*, only if the domestic price that they have to pay for these bonds (i.e., the exchange rate) is lower. In this way the value of $rF^d = rF^s$ remains unchanged, as it should remain, since all the magnitudes present on the right-hand side of Eqs. (15.21) are unchanged, and the market for foreign assets remains in equilibrium $(F^d = F^s)$ at a higher level of F and a lower level of r.

15.3.3.1 Interaction Between Current and Capital Accounts

The simplified model that we have described is a partial equilibrium model, as it does not consider the determinants of the interest rate(s) and the possible interaction between the current account and the capital account in the determination of the exchange rate. As regards this interaction, we consider a model suggested by Kouri (1983, but the idea was set forth by this author long before that date). As usual we simplify to the utmost by assuming static expectations, absence of domestic bonds in foreign investors' portfolio, an exogenously given interest differential. Thanks to these assumptions, the stock of foreign bonds held in domestic investors portfolios becomes an inverse function of the exchange rate. In fact, from the portfolio equilibrium described in Sect. 15.3.3, we have (in what follows we use F to denote the equilibrium stock of foreign bonds held by residents, $F = F^d = F^s$)

$$rF = h(i - i_f - \widetilde{r}/r)W, \qquad (15.25)$$

so that, given i, i_f, W, and letting $\widetilde{r} = 0$ (static expectations), the right-hand side of Eq. (15.25) becomes a constant, hence the relation of inverse proportionality between r and F.

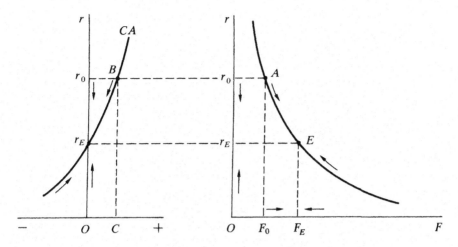

Fig. 15.2 Exchange-rate determination: interaction between the current account and the capital account

As regards the current account, Kouri assumes that its balance is an increasing function of the exchange rate (i.e. the critical elasticities condition occurs). The long-run equilibrium can prevail only when both the current account and the capital account are in equilibrium. In the opposite case, in fact, since any current-account disequilibrium is matched by a capital-account disequilibrium in the opposite sense,[2] the latter will cause a variation in the stock of foreign assets held by residents and so a variation in the exchange rate, which will feed back on the current account until this is brought back to equilibrium.

To illustrate this mechanism we can use Fig. 15.2, where the left-hand panel shows the current account (CA) as a function of the exchange rate (the positive slope means, as we have said, that the critical elasticities condition occurs); the right-hand panel shows the relation between r and F, which is a rectangular hyperbola as its equation is (15.25).

The long-run equilibrium corresponds to the exchange-rate r_E, where the current account (and so the capital account) is in equilibrium. Let us assume, for example, that the exchange rate happens to be r_0, hence a current account surplus OC; the initial stock of foreign assets (which corresponds to r_0) is F_0. The current-account surplus is matched by a capital outflow, i.e. by an increase in foreign assets; thus the residents stock of foreign bonds increases (from F_0 towards F_E, which is the equilibrium stock) and so the exchange rate appreciates (point A moves towards

---

[2]As we know from balance-of-payments accounting, the algebraic sum of the current account and the capital account is necessarily zero (see Sect. 5.1). Here the implicit assumption is that there are no compensatory (see Sect. 5.2) capital movements, i.e. that there is no official intervention (the exchange rate is perfectly and freely flexible). Under this assumption all capital movements are autonomous and originate from private agents.

point E). This appreciation reduces the current-account surplus; the process goes on until equilibrium is reached.

The case of an initial exchange rate lower than r_E is perfectly symmetrical (current-account deficit, decrease in the stock of domestically held foreign assets, exchange rate depreciation, etc.). This—Kouri concludes—explains why the exchange rate of a country with a current account surplus (deficit) tends to appreciate (depreciate); he labels this phenomenon the *acceleration hypothesis*.

Although this is a highly simplified model, it serves well to highlight the features of the interaction between the current account and the capital account. This interaction, it should be emphasized, does not alter the fact that in the short-run the exchange rate is always determined in the asset market(s), even if it tends towards a long-run value (r_E) which corresponds to current-account equilibrium. All the above reasoning, in fact, is based on the assumption that Eq. (15.25) holds instantaneously as an equilibrium relation, so that any change in F *immediately* gives rise to a change in r, which feeds back on the current account, which "follows" the exchange-rate behaviour. The assumption of instantaneous equilibrium in asset markets, or (at any rate) of *a much higher adjustment speed of asset markets compared to that of goods markets*, is thus essential to the approach under consideration. Thanks to this assumption, in fact, the introduction of goods markets and so of the current account does not alter the nature of the relative price of two assets attributed to the exchange rate in the short run.

More sophisticated models, in which the simplifying assumptions adopted here are dropped do not alter the fundamental conclusion stated above.

15.4   The Exchange Rate in Macroeconometric Models

Let us begin with some remarks on the modern approach. As Kouri (1983)—an author who is certainly not against the modern approach—aptly put it, if it is true by definition that the exchange rate is the relative price of two monies, it is vacuous to state that it is determined by the relative supplies of (and demands for) the two monies. To borrow a comparison of his, it is also true that the (money) price of steel is a relative price between steel and money, but nobody would analyze the determination of the price of steel in terms of supply of and demand for money: what one should do is to understand the determinants of the supply of and demand for steel and the mechanism which brings these into equilibrium in the steel market, and so determines the price of steel as well as the quantity exchanged.

Exactly the same considerations apply, according to Kouri, to the exchange rate, which is a price actually determined in the foreign exchange market through the demand for and supply of foreign exchange. This lack of, or insufficient, consideration of the foreign exchange market is, in Kouri's opinion, the main shortcoming of the modern theory. In fact, no theory of exchange-rate determination can be deemed satisfactory if it does not explain how the variables that it considers crucial (whether they are the stocks of money or the stocks of assets or expectations or whatever) actually translate into supply and demand in the foreign exchange market which, together with supplies and demand coming from other sources, determine the exchange rate.

We fully agree with these considerations of Kouri's which, of course, are not to be taken to mean that the modern approach is useless. Indeed, we believe that neither the traditional nor the modern theory is by itself a satisfactory explanation: in this we share Dornbusch's opinion who, after listing the basic views on exchange-rate determination stated "I regard any one of these views as a partial picture of exchange rate determination, although each may be especially important in explaining a particular historical episode" (Dornbusch 1983; p. 45). In fact, as we have already observed, the determinants that we are looking for are both real and financial, derive from both pure flows and stock adjustments, with a network of reciprocal interrelationships in a disequilibrium setting. It follows that only an eclectic approach is capable of tackling the problem satisfactorily, and from this point of view we believe that models like that of Branson and Buiter (1983) explained in Sect. 13.4 are a step forward in the right direction (it should be remembered that the Branson and Buiter model also determines the exchange rate among the other endogenous variables). More complex models seem however called for, because—as soon as one considers the exchange rate as one among the various endogenous variables which are present in the model of an economic system— simple models[3] like those which have been described in this book are no longer sufficient. One should move toward economy-wide macroeconom(etr)ic models. When doing this, however, one should pay much attention to the way in which exchange-rate determination is dealt with.

In order to put this important topic into proper perspective, we first introduce the distinction between models where there is a specific equation for the exchange rate and models where the exchange rate is implicitly determined by the balance-of-payments equation (thus the exchange rate is obtained by solving out this equation). From the mathematical point of view the two approaches are equivalent, as can be seen from the following considerations.

Let us consider the typical aggregate foreign sector of any economy-wide macroeconomic model, and let CA denote the current account, NFA the stock of net foreign assets of the private sector, R the stock of international reserves. Then the balance-of-payments equation (see Chap. 6, Sect. 6.4) simply states that

$$CA - \triangle NFA - \triangle R = 0. \tag{15.26}$$

---

[3]In the sense that, notwithstanding the formal difficulty of many of these, they can account for only a very limited number of variables and aspects of the real world, and must ignore others which may have an essential importance (as, for example, the distinction between consumption and investment, which are usually lumped together in an aggregate expenditure function).

Introduce now the following functional relations:

$$CA = f(r, \ldots), \tag{15.27}$$

$$\Delta NFA = g(r, \ldots), \tag{15.28}$$

$$\Delta R = h(r, \ldots), \tag{15.29}$$

$$r = \varphi(\ldots), \tag{15.30}$$

where r is the exchange rate and the dots indicate all the other explanatory variables, that for the present purposes can be considered as exogenous. These relations do not require particular explanation. We only observe that the reserve-variation equation, Eq. (15.29), represents the (possible) monetary authorities' intervention in the foreign exchange market, also called the monetary authorities' *reaction function*.

System (15.26)–(15.30) contains five equations in four unknowns, but one among equations (15.27)–(15.30) can be eliminated given the constraint (15.26). Which equation we drop is irrelevant from the mathematical point of view but not from the point of view of economic theory. From the economic point of view there are three possibilities:

a) we drop the capital-movement equation (15.28) and use the balance-of-payments equation (15.26) to determine capital movements *residually* (i.e., once the rest of the model has determined the exchange rate, the reserve variation, and the current account).

b) we drop the reserve-variation equation (15.29) and use the balance-of-payments equation to residually determine the change in international reserves.

c) we drop the exchange-rate equation (15.30) and use the balance-of-payments equation as an implicit equation that determines the exchange rate. The exchange-rate, in other words, is determined by solving it out of the implicit equation (15.26). More precisely, if we substitute from Eqs. (15.27)–(15.29) into Eq. (15.26), we get

$$f(r, \ldots) - g(r, \ldots) - h(r, \ldots) = 0, \tag{15.31}$$

which can be considered as an implicit equation in r. This can be in principle solved to determine r, which will of course be a function of all the other variables represented by the dots.

Since the functions f, g and h represent the excess demands for foreign exchange coming from commercial operators (the current account), financial private operators (private capital flows), and monetary authorities (the change in international reserves), what we are doing under approach (c) is simply to determine the exchange *rate through the equilibrium condition in the foreign exchange market (the equality between demand and supply of foreign exchange).*

It should be stressed that if one uses the balance-of-payments definition to determine the exchange rate one is not necessarily adhering to the traditional or "flow" approach to the exchange rate, as was once incorrectly believed. A few words are in order to clarify this point. Under approach (c) one is simply using the fact that the exchange rate is determined in the foreign exchange market, which is reflected in the balance-of-payments equation, under the assumption that this market clears instantaneously. This assumption is actually true, if we include the (possible) monetary authorities' demand or supply of foreign exchange as an item in this market; this item is given by Eq. (15.29), which defines the monetary authorities reaction function. As we have already observed above, in fact, no theory of exchange-rate determination can be deemed satisfactory if it does not explain how the variables that it considers crucial (whether they are the stocks of assets or the flows of goods or expectations or whatever) actually translate into supply and demand in the foreign exchange market which, together with supplies and demands coming from other sources, determine the exchange rate. When all these sources—including the monetary authorities through their reaction function in the foreign exchange market, Eq. (15.29)—are present in the balance-of-payments equation, this equation is no longer an identity (like Eq. (6.18) in Chap. 6), but becomes a *market-clearing condition*. Thus it is perfectly legitimate (and consistent with any theory of exchange-rate determination) to use the balance-of-payments equation to calculate the exchange rate once one has specified behavioural equations for *all* the items included in the balance of payments.

As we have said above, the various cases are equivalent mathematically but not from the economic point of view. In models of type a) and b), in fact, it is in any case necessary to specify an equation for exchange-rate determination, and hence adhere to one or the other theory explained in the previous sections. Besides, these models leave the foreign exchange market (of which the balance-of-payments equation is the mirror) out of the picture. On the contrary, the foreign exchange market is put at the centre of the stage in models of type c). Hence, these models are not sensitive to possible theoretical errors made in the specification of the exchange-rate equation (15.31).

For a survey of exchange-rate determination in actual economy-wide macro-econometric models see Gandolfo et al. (1990a,b).

15.5 Exchange-Rate Determination: Empirical Studies

15.5.1 Introduction

Everybody knows that "explanation" and "prediction" are not necessarily related. Geologists, for example, have very good explanations for earthquakes, but are as yet unable to predict them with any useful degree of accuracy. The effects of putting certain substances together in certain proportions were accurately predicted by alchemists long before the birth of chemistry (prediction without explanation). However, as Blaug (1980; p. 9) noted in general, "...when offered an explanation that does not yield a prediction, is it because we cannot secure all the relevant information about the initial conditions, or is it because (...) we are being handed chaff for wheat?".

Exchange rate determination offers a good example of this dilemma. Surprisingly enough, no rigorous test of the true predictive accuracy of the structural models of exchange rate determination which constitute the modern theory explained in Sect. 15.3 was carried out before the studies of Meese and Rogoff (1983a,b; see also 1988) . The in-sample predictive accuracy, in fact, is not a good test, for it simply tells us that the model fits the data reasonably well. What Meese and Rogoff did was to examine the *out-of-sample* predictive performance of these models. As a benchmark they took the simple random-walk model, according to which the forecast, at time t, of the value of a variable in period $t + 1$, is the current (at time t) value of the variable. To avoid possible confusion, it is as well to stress that to compare the forecasts of a model with those of the random walk does *not* mean that one is assuming that the exchange rate does follow a random walk process (it might or might not, which is irrelevant for our present purposes). It simply means taking as benchmark the simplest type of forecast, which is that of the random walk. This benchmark amounts to assuming a *naïf* agent who has no idea of how the exchange rate will evolve, and feels that increases or decreases are equally likely.

It is well known that the predictive accuracy of the random walk diminishes as the length of the period diminishes, hence it should not be difficult to perform better than this scheme. Instead, what Meese and Rogoff found was that the structural models failed to outperform the random-walk model even when the actual realized values of the explanatory variables were used (*ex post* forecasts[4]). The choice of *ex post* forecasts was made by Meese and Rogoff to prevent a possible defence of the structural models, namely that they do not perform well only because of the forecast errors of the exogenous variables that one has to use to forecast the exchange rate as endogenous variable (Meese and Rogoff 1983a; p. 10) .

We feel that, if one aims at testing the theoretical validity of a model, one should put the model in the most favourable situation for giving predictions, and that these should be true predictions (namely out of sample). If, this notwithstanding, the model fails, then there is certainly something wrong in the model, and "we are being handed chaff for wheat". This methodological vision amounts to following the Meese and Rogoff procedure, by now standard in all empirical studies of the exchange rate.

---

[4]For the student who has no econometric background we recall that by *ex post* forecasts we mean the following procedure. Given a data sample, say from 1970 to 2000, one uses a subsample, say from 1970 to 1995, to estimate the equation(s) of the model. The estimated equations are then used to generate true out-of-sample forecasts over the period left out of the estimation procedure (1996–2000), using however the *actual values of the exogenous variables observed in that period*. In other words, it is as if a hypothetical forecaster acting in 1995 possessed perfect foresight as regards the exogenous variables, having only to forecast the endogenous variable (in our case, the exchange rate), naturally using the equations estimated with the data up to 1995.

15.5.2 The Reactions to Meese and Rogoff, and the Way Out

The results of Meese and Rogoff have been updated and confirmed in several later studies (for example Gandolfo et al. 1993, Flood and Rose 1995, Chinn 1997, De Jong 1997, Cushman 2000, Rogoff and Stavrakeva 2008), and have led to two opposite categories of reactions. On the one hand, there are those who try to improve the performance of the structural models by all means (for example by using more sophisticated techniques, such as error-correction models, time-varying coefficients, cointegration techniques, threshold autoregressive models, etc.), but without appreciable improvement except in sporadic cases (e.g., Choudhry and Lawler 1997, Pippenger and Goering 1998, for a criticism of the alleged improvement obtained in more recent work see Rogoff and Stavrakeva 2008). On the other hand, there are those who take the failure of the structural models as a failure of economic theory and as showing the necessity of moving towards pure "technical analysis", which is a set of procedures for identifying patterns in exchange rates and using them for prediction (prediction without explanation). Forecasters using technical analysis are often called "chartists" in contrast with "fundamentalists", who instead rely on the fundamentals suggested by economic theory for forecasting exchange rates.

We do not agree with either view. In our opinion, the basic fact is that the traditional structural models of exchange-rate determination are of the single-equation, semi-reduced form type, which is inadequate to capture all the complex phenomena underlying the determination of the exchange rate, as we have already observed in Sect. 15.4. To adapt a similitude coined by Edgeworth (1905) for other purposes, there is more than meets the eye in the movement of the exchange rate: this movement, in fact, should be considered as attended with rearrangement of all the main economic variables, just "as the movement of the hand of a clock corresponds to considerable unseen movements of the machinery" (1905, p. 70). In order to explain the movement of the hands of a clock one must have a model of the underlying machinery, and to explain the movement of the exchange rate one must have an economy-wide macroeconometric model. Thus, in front of the failure of the standard structural models, the proper course of action is neither to try to improve their performance by all means nor to abandon the structural models approach. Instead, the proper course of action is to move away from the single-equation, semi-reduced form models, toward suitable economy-wide macroeconometric models capable of capturing all the complex associations between the exchange rate and the other variables (both real and financial, both stocks and flows) of a modern economy.

We wish to add that not every macroeconometric model is inherently suitable for this purpose. In fact, since the exchange rate is just one of the endogenous variables of an economy-wide model, its determination occurs in conjunction with the determination of the other endogenous variables, in a general (dis)equilibrium setting where stocks and flows, real and financial variables, etc., all interact. Thus, economy-wide macroeconometric models that embody a partial view of exchange-rate determination (in the sense that they, like the single-equation models, take account only of some factors), are not suitable.

We have been advocating the systemic approach for many years, starting from non-suspect years, that is when the models now rejected by the data were on the crest of a wave (see, for example, Gandolfo 1979, p. 102). In the next section we briefly summarize the results that we have obtained from our macroeconometric model of the Italian economy.

15.5.3 An Economy-Wide Model Beats the Random Walk

The economy-wide macroeconometric model that we have used is the MARK V version of the Italian continuous time model, for whose details we refer the reader to Gandolfo and Padoan (1990) . Just a few words are in order on the fact that the model is not only specified but also estimated in *continuous time* (instead of assuming and using discrete time-intervals as in conventional econometrics). It has been demonstrated (see, for example, Gandolfo 1981) that one can actually estimate the parameters of a continuous-time model on the basis of the discrete-time observations available. One practical advantage of the continuous time approach is that one can produce forecasts for any time interval and not only for the time unit inherent in the data. This is particularly important in our case. In fact, we had to use quarterly data in estimation (as this is an economy-wide model, most series—in particular national accounting data—were on quarterly basis). However, we wanted to produce out-of-sample forecasts of the exchange rate not only for quarterly intervals but also for other intervals, such one month ahead, one of the intervals considered by Meese and Rogoff. Another important advantage of the continuous time approach that is relevant in the present context is the estimation of adjustment speeds.

In fact, a basic problem in the theoretical debate on exchange-rate determination is the question of the adjustment speeds in the various markets. Now, the continuous time approach enables to determine the adjustment speeds rigorously (whichever the length of the observation interval); therefore, by using the balance-of-payments equation in which all the relevant variables are present and *come from adjustment equations with their specific estimated adjustment speeds*, we do not impose any arbitrary constraint on the data but let them speak for themselves.

Let us now come to the *exchange rate* (as representative exchange rate we have taken the Lira/US \$ rate), which is determined through the balance-of-payments equation. Thus the model follows approach *c* according to our classification. The choice has been made on the grounds that the specification of an exchange rate determination equation would have induced us to accept one or the other theory, whereas the balance-of-payments equation is more "neutral" once properly treated as explained in Sect. 15.4. This of course shifts the problem onto the specification of capital movements and of official intervention.

Capital movements are modelled according to the portfolio approach with non-instantaneous adjustment: the current stock of net foreign assets (as determined by portfolio theory) moves toward the desired stock with an adjustment speed which is estimated. As regards the monetary authorities' intervention, it is modelled as a reaction function, which is rather complicated. It contains, in fact, elements specific to the exchange-rate regime (fixed or floating) as well as elements of permanent nature, such as the leaning-against-the-wind policy and the desired reserves/imports ratio.

Our model has consistently outperformed the random walk (which in turn outperformed the traditional structural models) in out-of-sample forecasts of the exchange rate. The results are shown in detail in Gandolfo et al. (1990a,b) and in Gandolfo et al. (1993) .

A similar exercise was subsequently carried out by Howrey (1994) who, by using the Michigan Quarterly Econometric Model, found that ex post out-of-sample forecasts of the trade-weighted value of the US dollar produced by the model are also superior to forecasts of a random-walk model.

These results confirm the opinion that to outperform the random walk it is necessary to move away from the conventional single-equation, semi-reduced form structural models, toward suitable economy-wide macroeconometric models. It is comforting for economic theory to know that this approach has been able to outperform the random walk.

15.5.4 The Exchange Rate in Experimental Economics

Noussair et al. (1997) have studied exchange-rate determination in an international finance experiment. Experimental economics consists in generating data from appropriately designed "laboratory experiments" mimicking the economics of market decision through the behaviour of appropriately instructed persons participating in the experiment (usually students). These data are then used to test economic theories (for a general introduction to the topic see Smith 1987).

Among their many interesting results we quote the following:

"Result 5. Purchasing power parity is not supported statistically in the data." (p. 845)
"Result 8. The movement of the exchange rate from one market period to the next is influenced by the international demand for and supply of currency [. . .]" (p. 848).

The reason for the failure of PPP is conjectured to be "the differing speeds at which markets adjust in the different countries" (p. 855). We have stressed the importance of the adjustments speeds in *actual* markets in the previous section (p. 352), hence we cannot but agree with this conjecture. Besides, Result 8 lends support to the balance-of-payments approach to exchange rate determination (p. 348), that we have effectively used (see above, p. 352).

15.6   Equilibrium Exchange Rates: BEERs, DEERs, FEERs and All That

The actual exchange rate may not be an "equilibrium" exchange rate, if by such an expression we mean an exchange rate consistent with long-run macroeconomic fundamentals. Hence models for the determination of the actual exchange rate are not necessarily helpful to calculate the equilibrium exchange rate. An exception is PPP (see Sect. 15.1), which is put forward as a *long-run* theory of the equilibrium exchange rate.

The calculation of an equilibrium exchange rate serves to obtain a benchmark to be used to check the possible misalignment of the actual exchange rate and (if required and possible) to take corrective action.

In addition to PPP and its Harrod-Balassa Samuelson variant (see above, Sect. 15.1) several other ways of calculating equilibrium exchange rates have been suggested in the literature (for a survey see MacDonald and Stein eds. 1999).

Nurkse (1945) defined the equilibrium exchange rate as ". . . that rate which, over a certain period of time, keeps the balance of payments in equilibrium".

Williamson (1985, 1994) suggested the notion of "fundamental equilibrium exchange rate" (FEER), defined as that exchange rate 'which is expected to generate a current account surplus or deficit equal to the underlying capital flow over the cycle, given that the country is pursuing internal balance as best it can and not restricting trade for balance of payments reasons' (Williamson 1985; p. 14) . Such a rate should be periodically recalculated to take account of the change in its fundamental determinants (for example the relative inflation rates); thus it must not be confused with a fixed central parity. Bayoumi et al. (1994) specify a variant of the FEER called DEER (Desirable Equilibrium Exchange Rate), which assumes target values for the macroeconomic objectives, such as a targeted current account surplus for each country.

The FEER must not be confused with the BEER (Behavioral Equilibrium Exchange Rate: Clark and MacDonald 1999), the main difference being that the FEER has a normative meaning, while the BEER has a behavioural content. Calculation of the BEER involves the estimation of a reduced-form equation that explains the behaviour of the actual real exchange rate in terms of a set of (long-run and medium-term) economic fundamentals *and* a set of transitory factors affecting the real exchange rate in the short run. The BEER then corresponds to the value obtained from the estimated equation by eliminating the estimated transitory factors. The current misalignment is then defined as the difference between the BEER and the actual real exchange rate.

The *macroeconomic balance framework* for estimating equilibrium exchange rates suggested by Faruquee et al. (1999) can be seen as a specification of the BEER approach, where the macroeconomic fundamentals are derived from the determinants of saving, investment, and current account, given the well known macroeconomic accounting identity $S_N - I_N = CA$. Similarly, the *NATREX* (*NAT*ural

*Real EX*change rate) of Stein (1990, 1995a) can be seen as a generalization of the macroeconomic balance approach.

The ERER (Equilibrium Real Exchange Rate) is a concept set forth by Edwards (1994) and Elbadawi (1994) specifically for developing economies. They start from the definition of real exchange rate (RER) as the relative price of tradable goods (see Chap. 2, Sect. 2.3), and define the ERER as that real exchange rate which is compatible with both internal and external equilibrium for given sustainable values of the economic fundamentals explaining the RER.

Most of these approaches (with the exception of the NATREX) rely on more or less ad hoc assumptions concerning the various fundamentals on which the equilibrium exchange rate is made to depend. In other words (with reference to the distinction we made in Chap. 1 between the "old" or traditional view, and the "new" or modern view), they belong to the traditional view. In our opinion, this is a field in which the intertemporal approach can give us very useful insights. Therefore we refer the reader to Chap. 18, Sect. 18.3.

15.7 Chaos Theory and the Exchange Rate

Chaos theory is a relatively recent and difficult mathematical topic, hence we can do no more than give a few very general ideas (for a more technical introduction see Gandolfo 2009, Chap. 25). For our purpose, the following definition will do: *chaos is apparently stochastic behaviour generated by a dynamic deterministic system.*

By "apparently stochastic" we mean a random path that at first sight cannot be distinguished from the path generated by a stochastic variable. Since it is presumable that readers of this book own (or have access to) a PC, we shall exemplify by a computer pseudo random number generator. The algorithm used by the computer is purely deterministic, but what comes out is a series of numbers that looks random, and that will fool any statistician in the sense that it passes all the standard tests of randomness. As a matter of fact, random numbers generated in this way are usually employed in statistical analysis.

A feature that is often cited as typical of chaotic behaviour of deterministic systems is the *impossibility of predicting the future values of the variable(s) concerned.* This might at first sight seem a contradiction—if we have a dynamic deterministic system, even if we cannot solve it analytically we can simulate it numerically, hence we can compute the value(s) of the variable(s) for any future value of t. This is where another important feature of chaos comes in (as a matter of fact, some take it as defining chaos), that is *sensitive dependence on initial conditions.*

Sensitive dependence on initial conditions (henceforth SDIC) means that even *very small* differences in the initial conditions give rise to *widely* different paths, a phenomenon that occurs for any admissible set of initial conditions. On the contrary, in a "normal" deterministic system, all nearby paths starting very close to one another remain very close in the future, with at most a few exceptions.[5] Hence a sufficiently small measurement error in the initial conditions will not affect our deterministic forecasts. On the contrary, in deterministic systems with SDIC, prediction of the future values of the variable(s) would be possible only if the initial conditions could be measured with infinite precision. This is certainly not the case.

The interest of economists in chaos theory started in the 1980s, more than twenty years after the onset of this theory in physics, which is conventionally dated 1963 (although previous hints can be found in Poincaré and others), when the meteorologist E.N. Lorenz published his paper on what became to be known as the Lorenz attractor (Lorenz 1963). The first to draw the attention of economists to chaos theory was, in fact, Brock (1986), who examined the quarterly US real GNP data 1947–1985 using the Grassberger-Procaccia correlation dimension and Liapunov exponents. Subsequent studies generally found absence of evidence for chaos in macroeconomic variables (GNP, monetary aggregates) while the study of financial variables such as stock-market returns and exchange rates gave mixed evidence.

Studies aimed at detecting chaos in economic variables can be roughly classified into two categories.

(I) On the one hand, there are studies that simply examine the data and apply various tests for chaos (for applications to the exchange rate see Bajo-Rubio et al. 1992; Cuaresma 1998; Guillaume 2000, Chap. 3; Schwartz and Yousefi 2003; Weston 2007). These tests have been originally developed in the physics literature. This approach is not very satisfactory from our point of view, which aims at finding the dynamic model (if any) underlying the data. Besides, in the case of the investigation of individual time series to determine whether they are the result of chaotic or stochastic behaviour, the results could be inconclusive, as shown in the single blind comparative study of Barnett et al. (1997).

(II) On the other hand, structural models are built and analysed. This analysis can in principle be carried out in several ways:

(II.a) showing that plausible economic assumptions give rise to theoretical models having dynamic structures that fall into one of the mathematical forms known to give rise to chaotic motion;

(II.b) building a theoretical model and then

(II.b$_1$) giving plausible values to the parameters, simulating the model, and testing the resulting data series for chaos; or

---

[5]A normal dynamic system with the saddle path property (see Gandolfo 2009, Chap. 28) shows SDIC for initial values that are located in the neighbourhood of the stable asymptote. In fact, even a small deviation from an initial value that falls on such an asymptote will give rise to a completely different path (a divergent path instead of a convergent one), and viceversa. However, SDIC does not exist for other initial conditions, since two divergent paths starting very close to one another will remain very close while diverging from the equilibrium point. Hence we cannot characterize a saddle path dynamic system as showing SDIC.

(II.b$_2$) estimating the parameters econometrically, and then proceeding as in b$_1$.

Existing chaotic exchange rate models (see, e.g., Chen 1999: Da Silva 2000, 2001; De Grauwe and Versanten 1990; De Grauwe and Dewachter 1993a,b; De Grauwe et al. 1993; De Grauwe and Grimaldi 2006a,b; Ellis 1994; Moosa 2000, Chap. 9; Reszat 1992; Szpiro 1994) follow approaches (II.a) or (II.b$_1$). From the theoretical point of view, these models show that with orthodox assumptions (PPP, interest parity, etc.) and introducing nonlinearities in the dynamic equations, it is possible to obtain a dynamic system capable of giving rise to chaotic motion. However, none of these models is estimated, and the conclusions are based on simulations: the empirical validity of these models is not tested.

Exceptions are Federici and Gandolfo (2002), where a continuous time modification of the De Grauwe et al. 1993 model is built and econometrically estimated in continuous time with reference to the Italian lira/\$ exchange rate, and Federici and Gandolfo (2012), where a continuous time model of the Euro/US dollar is built and econometrically estimated in continuous time. In both cases no evidence for chaos has been found. The Federici and Gandolfo (2012) model will be examined in the Appendix.

15.8 Appendix

15.8.1 PPP and the Harrod-Balassa-Samuelson Effect

Consider a two-country, two-sector (tradables-nontradables) model where prices are determined positively by wages and inversely by labour productivity. This is consistent with the condition of equality between the value of the (marginal) productivity of a factor and its reward: this condition, when applied to labour, gives $pQ = w$.

Wages are assumed to be equal in both sectors in each of the two economies $i = 1, 2$:

$$p_i^{NT} = w_i/Q_i^{NT}, \, p_i^T = w_i/Q_i^T. \tag{15.32}$$

Without loss of generality we can assume that country 1 is the "rich" country while country 2 is the "poor" one. We also assume that in the nontradables sector productivities in the rich and poor country are the same. Thus we have

$$Q_1^T > Q_2^T, \, Q_1^{NT} = Q_2^{NT}. \tag{15.33}$$

Finally, PPP is assumed to hold for traded goods:

$$p_1^T = rp_2^T. \tag{15.34}$$

The relative price of non-traded goods in each country is

$$\tau_1 = p_1^{NT}/p_1^T, \quad \tau_2 = p_2^{NT}/p_2^T. \tag{15.35}$$

Substituting from (15.32) into (15.35) we obtain

$$\tau_1 = Q_1^T/Q_1^{NT}, \quad \tau_2 = Q_2^T/Q_2^{NT}, \tag{15.36}$$

so that, using (15.33), we see that

$$\tau_1 > \tau_2, \tag{15.37}$$

namely the relative price of non-traded goods is higher in country 1.

Let us now multiply numerator and denominator of τ_2 in (15.35) by r and use (15.34) and (15.37). We obtain

$$\frac{p_1^{NT}}{p_1^T} > \frac{rp_2^{NT}}{p_1^T},$$

whence

$$p_1^{NT} > rp_2^{NT}, \tag{15.38}$$

which means that PPP does not hold for nontradables. It follows that it doesn't hold either for an aggregate price index. More precisely, if we consider the same basket of tradables and nontradables (q^T, q^{NT}) in both countries and calculate its value, we have

$$p_1^B = p_1^T q^T + p_1^{NT} q^{NT}, \ p_2^B = p_2^T q^T + p_2^{NT} q^{NT}. \tag{15.39}$$

If we multiply p_2^B by r and use (15.34) and (15.38) we obtain

$$rp_2^B = p_1^T q^T + rp_2^{NT} q^{NT} < p_1^B. \tag{15.40}$$

15.8.2 The Dornbusch Overshooting Model

We recall from the text that the dynamics of the model is given by the differential equation system

$$\dot{e} = \frac{1}{\lambda}(p - \bar{p}), \tag{15.41}$$

$$\dot{p} = -\pi(\delta + \frac{\sigma}{\lambda})(p - \bar{p}) + \pi\delta(e - \bar{e}), \tag{15.42}$$

which is already in (log)linear form. The singular point is obtained by letting $\dot{e} = \dot{p} = 0$. The equilibrium point then is $p = \bar{p}, e = \bar{e}$. By an appropriate choice of units we can set $\bar{p} = \bar{e}$, so that in the phase diagram drawn in the text, OR is a 45° line. The $\dot{e} = 0$ locus is a straight line originating from $p = \bar{p}$ and parallel to the e axis. The $\dot{p} = 0$ locus gives rise to the equation

$$-\pi(\delta + \frac{\sigma}{\lambda})(p - \bar{p}) + \pi\delta(e - \bar{e}) = 0,$$

which is a straight line with a slope smaller than unity, since

$$(dp/de)_{\dot{p}=0} = \frac{\pi\delta}{\pi\delta + \pi\sigma/\lambda} < 1. \tag{15.43}$$

The characteristic equation of the dynamic system (15.41)–(15.42) is

$$\begin{vmatrix} -\pi(\delta + \frac{\sigma}{\lambda}) - \mu & \pi\delta \\ \frac{1}{\lambda} & 0 - \mu \end{vmatrix} = \mu^2 + \pi(\delta + \frac{\sigma}{\lambda})\mu - \frac{\pi\delta}{\lambda} = 0, \tag{15.44}$$

where μ denotes the latent roots. Since the succession of the signs of the coefficients is $+ + -$, there will be two real roots, one negative and the other one positive, which means a saddle point (see Gandolfo 2009, Chap. 21, Table 21.1).

The stable arm of the saddle is the straight line AA. It is downward sloping because—as shown in Gandolfo 2009, Chap. 21, Sect. 21.3.2.4, Eq. (21.17)—its equation is

$$(e - \bar{e}) = \frac{\mu_1 + (\pi\delta + \pi\sigma/\lambda)}{\pi\delta}(p - \bar{p}), \tag{15.45}$$

where μ_1 is the stable root of the characteristic equation (15.44). Since the sum of the roots is the trace, we have $\mu_1 + \mu_2 = -(\pi\delta + \pi\sigma/\lambda)$, hence $\mu_1 + (\pi\delta + \pi\sigma/\lambda) = -\mu_2 < 0$ because μ_2 is the positive root.

15.8.3 The Modern Approach to Exchange-Rate Determination

15.8.3.1 The Monetary Approach

Let us consider the simplest form of the monetary approach to the exchange rate, based on the relation derived in the text

$$r = \frac{M}{M_f} \frac{L_f(y_f, i_f)}{L(y, i)}, \tag{15.46}$$

and assume that the money-demand function has the form

$$\frac{L}{p} = L_0 e^{-\varepsilon i} y^{\eta}, \tag{15.47}$$

where ε is the semi-elasticity of money demand with respect to the interest rate and η the real-income elasticity of the same demand. Let us also assume that these functions are internationally identical in our two-country world, so that the parameters $L_{0f}, \varepsilon_f, \eta_f$ are equal to the corresponding parameters L_0, ε, η. Then, if we substitute (15.47) into (15.46) and take the natural logarithms, we get

$$\ln r = \left(\ln M - \ln M_f\right) + \varepsilon \left(i - i_f\right) - \eta \left(\ln y - \ln y_f\right). \tag{15.48}$$

Besides, since, with perfect substitutability between domestic and foreign assets, UIP must hold, we have

$$i - i_f = \widetilde{r}/r. \tag{15.49}$$

In its turn the expected rate of variation in the exchange rate, as *PPP* is assumed to hold also in expectations, equals the difference between the expected rates of inflation

$$\widetilde{r}/r = \widetilde{p}/p - \widetilde{p_f}/p_f, \tag{15.50}$$

so that

$$i - i_f = \widetilde{p}/p - \widetilde{p_f}/p_f, \tag{15.51}$$

and so, by substitution into (15.48), we get

$$\ln r = \left(\ln M - \ln M_f\right) - \eta \left(\ln y - \ln y_f\right) + \varepsilon \left(\widetilde{p}/p - \widetilde{p_f}/p_f\right). \tag{15.52}$$

If we further assume that the time-path of income is exogenously given and that expectations are rational, the expected inflation rate coincides with the proportional rate of change in the money supply, so that

$$\ln r = \left(\ln M - \ln M_f\right) - \eta \left(\ln y - \ln y_f\right) + \varepsilon \left(\dot{M}/M - \dot{M_f}/M_f\right). \tag{15.53}$$

Equations (15.48), (15.52) and (15.53) are alternative formulations of the simple version of the monetary approach. However, there is also a more sophisticated version which, though accepting the validity of *PPP* in the long run, acknowledges that in the short-run the exchange rate may deviate from *PPP* because of price stickiness. According to this version, what happens in the short run is that, for example, an increase in the money stock does not cause an immediate increase in prices owing to their stickiness, but has the effects described in the Mundellian model (see Sect. 11.4): the interest rate tends to decrease, and the incipient capital outflow causes the exchange rate to depreciate. This depreciation, however, is greater than that required by long-run *PPP* (the overshooting phenomenon, see Sect. 15.8.2). To be precise, the amount by which the depreciation is greater than required is exactly such that the expected future appreciation (agents with rational expectations do in fact know that in the long run the exchange rate will have to conform to *PPP*) precisely offsets the interest differential that has come about.

It follows from the above reasoning that Eq. (15.48) and its alternative formulations hold as long-run relations, which we can express by replacing $\ln r$ with $\ln \hat{r}$ into them, where $\ln \hat{r}$ denotes the long-run equilibrium value of the exchange rate satisfying *PPP*. We must now determine exchange-rate expectations: in these, both a short-run and a long-run component are present. In the short-run, agents believe that the exchange rate will tend towards its *PPP* value with a certain speed of adjustment α; and they expect that, when it has reached this value, it will move in accordance with the inflation differential as expressed in (15.50) (which, in this version of the model, is also a long-run relationship). Thus we have

$$\widetilde{r}/r = -\alpha \left(\ln r - \ln \hat{r} \right) + \left(\widetilde{p}/p - \widetilde{p}_f/p_f \right). \tag{15.54}$$

With integrated financial markets and perfect asset substitutability Eq. (15.49) continues to hold, so that, by substituting in Eq. (15.54) and by solving for $(\ln r - \ln \hat{r})$, we have

$$\ln r - \ln \hat{r} = -\left(1/\alpha\right) \left[\left(i - \widetilde{p}/p \right) - \left(i_f - \widetilde{p}_f/p_f \right) \right]. \tag{15.55}$$

The expression in square brackets on the right-hand side is the differential between the *real* interest rates. If we now replace $\ln \hat{r}$ with its PPP value given by (15.52), we finally obtain

$$\ln r = (\ln M - \ln M_f) - \eta(\ln y - \ln y_f)$$
$$-\frac{1}{\alpha} \left(i - i_f \right) + \left(\frac{1}{\alpha} + \varepsilon \right) \left(\widetilde{p}/p - \widetilde{p}_f/p_f \right), \tag{15.56}$$

which is the *sticky-price version* (sometimes called the *overshooting version*) of the monetarist model. It can be readily seen that the simple version (15.48) is a particular case of (15.56), which comes about when the adjustment speed α tends to infinity and Eq. (15.51) holds instantaneously.

15.8.3.2 The Portfolio Approach

The essential feature of the portfolio approach is imperfect substitutability between domestic and foreign assets, so that Eq. (15.49) is no longer valid and must be replaced with

$$\tilde{r}/r = i - i_f - \delta, \tag{15.57}$$

where δ, which denotes the divergence between the interest differential and the expected proportional variation in the exchange rate, is a *risk coefficient* or *risk premium* that asset holders demand on domestic-currency assets relative to foreign currency assets, given the existing stocks of wealth, of assets, and given the expected relative rates of return on the various assets. In other words, δ is the exchange risk premium that must be expected, over and above the interest differential, for asset holders to be indifferent at the margin between uncovered holdings of domestic bonds and foreign bonds.

Instead of pursuing the way outlined in the text, which leads one to introduce the stocks of the various assets, which cannot be easily observed, it is expedient to follow an alternative route, which leads one to express the risk premium in terms of easily observable variables, and then to introduce the result in the equation of the monetary approach. As has been shown by various writers—for example Dooley and Isard (1983) and Hooper and Morton (1982), to whom we refer the reader—the risk premium can be expressed in terms of various factors, amongst which mainly the cumulative imbalances in the current account[6] (or more restrictively, in the trade account) of the two countries,[7] that is

$$\delta = \beta_0 + \beta_1 \sum_{j=0}^{t} x_j + \beta_2 \sum_{j=0}^{t} x_{fj}, \tag{15.58}$$

where x_j is the current account balance of the home country in period j, x_{fj} the current account balance of the rest of the world, and $\beta_0, \beta_1, \beta_2$ are coefficients. It should be noted that the cumulative current accounts could also be taken to represent empirically the role of "news" in exchange-rate determination (the role of news is examined in Sect. 13.4.4 from the theoretical point of view).

---

[6]An elementary explanation (Shafer and Loopesko 1983; p. 30) is the following. As time passes, financial wealth is transferred from the deficit to the surplus countries. If we assume that residents of any country, *ceteris paribus*, have a preference for assets denominated in their own currency (the preferred local habitat assumption, already mentioned in Sect. 15.3.3) this redistribution of wealth alters the relative demands for assets. The currency of a deficit country depreciates to a point from which agents expect it to subsequently appreciates, thus establishing a risk premium.

[7]In a two-country world the two countries' current accounts are mirror-images of each other. In practice one refers to two countries out of n, so that their current accounts are no longer necessarily symmetric.

If we use Eq. (15.57) instead of Eq. (15.49), in the place of Eq. (15.56) we obtain the equation

$$\ln r = \left(\ln M - \ln M_f\right) - \eta \left(\ln y - \ln y_f\right)$$

$$-\frac{1}{\alpha}(i - i_f) + \left(\frac{1}{\alpha} + \varepsilon\right)\left(\widetilde{p}/p - \widetilde{p}_f/p_f\right) + \frac{1}{\alpha}\delta, \qquad (15.59)$$

and so, by substituting (15.58) into (15.59), we get

$$\ln r = \frac{\beta_0}{\alpha} + \left(\ln M - \ln M_f\right) - \eta \left(\ln y - \ln y_f\right)$$

$$-\frac{1}{\alpha}(i - i_f) + \left(\frac{1}{\alpha} + \varepsilon\right)\left(\widetilde{p}/p - \widetilde{p}_f/p_f\right)$$

$$+\frac{\beta_1}{\alpha}\sum_j x_j + \frac{\beta_2}{\alpha}\sum_j x_{fj}, \qquad (15.60)$$

where all the variables are to be taken at time t.

Equation (15.60) is the general form of the asset-market approach which includes, as particular cases, both Eqs. (15.48) and (15.56).

15.8.3.3 Empirical Studies

The models considered by Meese and Rogoff were the flexible-price (Frenkel-Bilson) monetary model, the sticky-price (Dornbusch-Frankel) monetary model, and the sticky-price (Hooper-Morton) asset model. They did not consider the Hooper-Morton model with risk because it had been rejected by Hooper-Morton themselves since the coefficient of the term representing risk has the wrong sign, but we shall include this version for completeness of exposition.

The quasi-reduced forms of the four models can be subsumed under the following general specification:

$$e_t = a_0 + a_1 \left(m - m_f\right)_t + a_2(y - y_f)_t + a_3(i_s - i_{sf})_t + a_4(i_L - i_{Lf})_t$$

$$+a_5(\overline{CA} - \overline{CA}_f)_t + a_6\overline{K}_t + u_t, \qquad (15.61)$$

where the subscript f denotes the foreign country, t is time, and

e = logarithm of the spot exchange rate (price of foreign currency)
m = logarithm of the money supply
y = logarithm of real income
i_s = short-term interest rate
i_L = long-term interest rate
$\overline{CA}$ = cumulated trade balance
$\overline{K}$ = cumulated capital movements balance
u = disturbance term.

The four models are derived as follows:

(FB) Frenkel-Bilson: $a_1 > 0, a_2 < 0, a_3 > 0, a_4 = a_5 = a_6 = 0$;
(DF) Dornbusch-Frankel: $a_1 > 0, a_2 < 0, a_3 < 0, a_4 > 0, a_5 = a_6 = 0$;
(HM) Hooper and Morton: $a_1 > 0, a_2 < 0, a_3 < 0, a_4 > 0, a_5 < 0, a_6 = 0$;
(HMR) Hooper and Morton with risk: $a_1 > 0, a_2 < 0, a_3 < 0, a_4 > 0, a_5 < 0, a_6 > 0$.

The *FB* and *DF* models are monetary models, the difference being that the *FB* model assumes *PPP* in both the short and the long run, while the *DF* model assumes *PPP* only in the long run and allows for sticky prices in the short run. The Hooper-Morton model is a model which draws from both the monetary and the portfolio approach to exchange-rate determination. In the *HM* formulation it follows the *DF* model but introduces the effects of trade-balance surpluses: a persistent domestic (foreign) trade-balance surplus (deficit) indicates an appreciation of the long-run exchange rate. It should be noted that *HM* allowed for different coefficients on the domestic and foreign cumulated trade balances, but it is usual to follow Meese and Rogoff (1983b) in assuming that domestic and foreign trade balance surpluses have an effect on the exchange rate of equal magnitude but opposite sign. Finally, the *HMR* model introduces imperfect asset substitutability hence a risk premium, that we approximate by $\overline{K}$ following Dooley and Isard (1983) .

Subsequent studies by Somanath (1986) suggested that, contrary to the findings of Meese and Rogoff, the introduction of the lagged dependent variable among the explanatory variables improved the forecasting ability of the model, indicating a non-instantaneous adjustment of the actual exchange rate to its equilibrium value as given by the right-hand-side of Eq. (15.61). Thus one should also test the lagged version of the four above models, that is

$$e_t = a_0 + a_1 \left(m - m_f \right)_t + a_2 (y - y_f)_t + a_3 (i_s - i_{sf})_t + a_4 (i_L - i_{Lf})_t$$

$$+ a_5 (\overline{CA} - \overline{CA_f})_t + a_6 \overline{K}_t + a_7 e_{t-1} + u_t. \tag{15.62}$$

It has also been suggested that error correction models (*ECM*) may be better suited for theories that postulate long-run relationships such as, for example, the long-run proportionality between the exchange rate and relative money stocks in the monetary models. In fact, Sheen (1989) found that a modified monetary model (in which *PPP* was dropped and certain restrictive assumptions were made about the generation of the expected future values of the predetermined variables) with an *EC* term outperformed the *RW* in out-of-sample forecasts of the Australian $/US$ exchange rate.

The basic idea of the *ECM* formulation is simply that a certain fraction of the disequilibrium is corrected in the following period. The *ECM* specification of the four models under consideration is

$$\Delta e_t = a_0 + a_1 \left(\Delta m - \Delta m_f\right)_t + a_2 \left(\Delta y - \Delta y_f\right)_t + a_3 \left(\Delta i_s - \Delta i_{sf}\right)_t$$
$$+ a_4 \left(\Delta i_L - \Delta i_{Lf}\right)_t + a_5 \left(\Delta \overline{CA} - \Delta \overline{CA}_f\right)_t + a_6 \Delta \bar{K}_t$$
$$+ a_7 \left(e - m + m_f\right)_{t-1} + a_8 e_{t-1} + u_t. \tag{15.63}$$

The *ECM* specification is equivalent to the cointegration between the relevant variables (Engle and Granger 1987), in our case between the exchange rate and the relative money stocks. A test of $a_7 = 0$ is a test of both the *ECM* specification and the long-run proportionality; the cointegration between the exchange rate and the relative money supplies can be further tested by running the cointegrating regression of Engle and Granger and applying various unit root tests.

As stated in the text, the application of these models (including the *ECM* version) to the *Lira/US$* exchange rate has been a dismal failure: see Gandolfo et al. (1990a,b) and Gandolfo et al. (1993). However, a study based on a panel version of the Engle and Granger ECM procedure finds some support for the monetary model as a long-run phenomenon (Groen 2000; Mark and Sul 2001). Cointegration techniques based on the Johansen procedure have been applied for example to the Canadian/US dollar exchange rate with mixed results (Choudhry and Lawler 1997, found support for the monetary exchange rate model, while Cushman 2000, found no evidence in favour of this model)

Among the various reasons that Meese and Rogoff adduced for the poor performance of the structural exchange rate models, "parameter instability" (Meese and Rogoff 1983a; p. 18) is briefly cited and justified by possible structural breaks due to specific episodes that occurred in the 1970s. Apart from the peculiarity of the sample, a more theoretical justification for parameter instability can be found in the presence of changes of regime, or also as an extension of Lucas' critique. Parameter instability can be dealt with using the time-varying-coefficients (*TVC*) methodology. It should however be noted that the problem in applying *TVC* models is to find a meaningful economic explanation for how the structural coefficients should vary, i.e., why they follow a supposed stochastic process (see Schinasi and Swamy 1989).

In addition, it should be noticed that the use of the rolling regression technique to obtain forecasts (as was done by Meese and Rogoff 1983a) is stochastically equivalent to the use of a *TVC* model where a particular random process—i.e., the multivariate *RW*— is imposed to the coefficients (see Alexander and Thomas 1987):

$$e_t = x_t \beta_t + u_t,$$
$$\beta_t = \beta_{t-1} + \varepsilon_t,$$

where $\varepsilon_t = IIND\ (0; \sum)$ and $E\left[u_t \,|x_t\right] = 0$.

Gandolfo et al. (1993) have used a *TVC* model of the so-called "return-to-normality" type: the coefficients are assumed to follow a generic *ARMA* stationary process around a constant term and are not restricted to the multivariate *RW* process, as in the above model. In other words, this model is equivalent to a restricted Kalman filter where the state variables are deviations from a mean value and follow a stationary process, i.e., the eigenvalues of the transformation matrix, H, are all less than one in absolute value:

$$e_t = x_t \beta_t + u_t,$$
$$\left(\beta_t - \overline{\beta}\right) = H(\beta_{\tau-1} - \overline{\beta}) + \varepsilon_t.$$

Despite the more general framework, the main drawback of this approach still remains: as Schinasi and Swamy (1989) observe in their conclusions, this drawback is the lack of sound and rigorous economic principles to explain the evolution process of the *TVC*'s.

The results of this *TVC* model have however remained poor, and have not outperformed the random walk (Gandolfo et al. 1993). Similarly poor results have been obtained by De Arcangelis (1992) with another type of *TVC* model, based on a Bayesian approach. Thus it does not seem that parameter instability is responsible for the poor performance of the standard structural models of exchange-rate determination.

Pippenger and Goering (1998) suggest the use of the self-exciting threshold autoregressive model (SETAR). They estimate a SETAR model for various monthly US dollar exchange rates and generate forecasts for the estimated models, and find that the SETAR model produces better forecasts than the naive random walk model.

Let us come finally to the results of our continuous-time macroeconometric model of the Italian economy. We first clarify a technical point. In discrete-time models it is usual to make the distinction between (a) single-period (or "static") forecasts and (b) multi-period (or "dynamic") forecasts. The former are those obtained by letting the lagged endogenous variables take on their actual observed values; the latter are those obtained by letting the lagged endogenous variables take on the values forecast by the model for the previous period(s). The equivalent distinction in continuous time models is made according to whether the solution of the differential equation system is (a) re-computed each period, or (b) computed once and for all. In case (a) the differential equation system is re-initialized and solved n times (if one wants forecasts for n periods), each time using the observed values of the endogenous variables in period t as initial values in the solution, which is then employed to obtain forecasts for period $t + 1$. This is equivalent to the single period forecasts in discrete models. In case (b) the observed values of the endogenous variables for a given starting period are used as initial values in the solution of the differential equation system, which is then employed for the whole forecast period. This is equivalent to the dynamic forecasts in discrete models. Although dynamic forecasts are generally less good than static ones because the errors cumulate, we decided to use dynamic forecasts to test the predictive performance of our model, because these are the only ones which can be employed to produce forecasts for a time interval different from that inherent in the data.

The basic random-walk model was used as the benchmark, although some authors suggest that comparing multi-step-ahead predictions of structural models with one-step-ahead predictions of the random-walk model gives the random-walk model an unfair advantage over structural models which do not include a lagged dependent variable; in this case a multi-step-ahead (with or without drift) prediction of the random walk is on a more equal footing with the structural model's predictions (Schinasi and Swamy 1989) . The reason for our choice is that a continuous time model specified as a differential equation system embodies all the relevant dynamics, including the one that discrete-time models try to capture by introducing lagged dependent variables as explanatory variables.

Our results are consistently better than the random walk, as stated in the text. The thesis that a suitable economy-wide model would outperform the random walk has been lent further support by Howrey (1994), who—building on Gandolfo et al. (1990b) —shows that exchange-rate forecasts obtained by the Michigan quarterly econometric model of the US perform better than the random walk.

Finally, a new line of research has been developing, the *microstructural approach* to the exchange market, that aims at explaining the volatility of the exchange rate starting from the study of the high-frequency dynamics of the exchange rate (intradaily data) and of the influence on this dynamics of news and policy announcements. Results are so far mixed (Sarno and Taylor 2001; Evans 2010, 2011). For an attempt at integrating the microstructural with the fundamentals approach see Boubel (2000).

15.8.3.4 Currency Substitution

An interesting theoretical development of the monetary approach to the exchange rate is to be seen in currency-substitution models. In all the models analysed here a common assumption is that residents can hold domestic money but cannot hold foreign currency (they can, of course, hold foreign securities). Currency-substitution models abandon this assumption, so that residents can hold both domestic and foreign currency. Since, by definition, money is the riskless asset and can be assumed to be non-interest bearing,[8] the problem arises of determining how residents will allocate their (real) wealth between domestic and foreign currency. The general criterion is always that of relative expected returns expressed in a common standard, for example domestic currency. The expected real return on domestic is the opposite of the expected rate of domestic inflation, $-\tilde{p}/p$. The expected real return of foreign money equals the difference between the expected proportional rate of change of the exchange rate (the price of foreign currency in terms of domestic currency) minus the expected rate of domestic inflation. The expected differential in returns is thus $\tilde{r}/r$ which in the case of rational expectations

---

[8]The fact that in some countries checking accounts, which are to be considered money for all purposes, bear interest, is an institutional problem that we shall ignore.

(perfect foresight) equals $\dot{r}/r$. While referring the reader to the relevant literature (see, for example, Mizen and Pentecost eds., 1996), we make only two observations. The first is that these models are limited in that the menu offered usually contains only two assets (domestic and foreign money); for a more general approach in which domestic and foreign securities are also considered in a general portfolio balance model, see Branson and Henderson (1985). The second concerns one of the interesting implications of currency substitution models, which is that a high degree of currency substitution may put pressure on the countries, whose currencies are concerned, to move towards a monetary union. According to some writers (see, for instance, Melvin 1985), this may have been a more important reason for the move towards the EMS, than any of the reasons examined in the traditional optimum currency area framework.

15.8.4 Chaos Theory and the Exchange Rate

The Federici and Gandolfo (2012) model mentioned in the text will be examined here.

15.8.4.1 The Model: Formulation in Terms of Excess Demands for Foreign Exchange

Our starting point is that the exchange rate is determined in the foreign exchange market through the demand for and supply of foreign exchange. This is a truism, but it should be complemented by the observation that, when all the sources of demand and supply—including the monetary authorities through their reaction function—are accounted for, that is, once one has specified behavioural equations for *all* the items included in the balance of payments, the exchange rate comes out of the solution of an implicit dynamical equation.

Let us then come to the formulation of the excess demands (demand minus supply) of the various agents. Our classification is functional. It follows that a commercial trader who wants to profit from the leads and lags of trade (namely, is anticipating payments for imports and/or delaying the collection of receipts from exports in the expectation of a depreciation of the domestic currency) is behaving like a speculator.

(1) In the foreign exchange market non-speculators (commercial traders, etc.) are permanently present, whose excess demand only depends on the current exchange rate:

$$E_n(t) = g_n[r(t)],\ g_n' \gtrless 0. \tag{15.64}$$

where $r(t)$ denotes the current spot exchange rate (price quotation system: number of units of domestic currency per unit of foreign currency). Possible transaction costs are subsumed under the non-linear function g_n. On the sign of g'_n see below, Sect. 15.8.4.2.

(2) Let us now introduce speculators, who demand and supply foreign exchange in the expectation of a change in the exchange rate. According to a standard distinction, we consider two categories of speculators, fundamentalists and chartists.[9]

(2a) Fundamentalists hold regressive expectations, namely they think that the current exchange rate will move toward its "equilibrium" value. There are several ways to define such a value[10]; we believe that the most appropriate one is the NATREX (acronym of NATural Real EXchange rate), set forth by Stein (1990, 1995a,b, 2001, 2002, 2006). It is based on a specific theoretical dynamic stock-flow model to derive the equilibrium real exchange rate. The equilibrium concept reflects the behaviour of the fundamental variables behind investment and saving decisions in the absence of cyclical factors, speculative capital movements and movements in international reserves. Two aspects of this approach are particularly worth noting. The first is that the hypotheses of perfect knowledge and perfect foresight are rejected: rational agents who efficiently use all the available information will base their intertemporal decisions upon a *sub-optimal feedback control* (SOFC) rule, which does not require the perfect-knowledge perfect-foresight postulated by the Representative Agent Intertemporally Optimizing Model, but only requires *current* measurements of the variables involved. The second is that expenditure is separated between consumption and investment, which are decided by different agents. The consumption and investment functions are derived according to SOFC, through dynamic optimization techniques with feedback control. Thus the NATREX approach is actually an intertemporal optimizing approach, though based on different optimization rules.

For a treatment of the NATREX, and for an empirical estimation of the \$/€ NATREX, see Belloc et al. (2008), and Belloc and Federici (2010). Let us call N_n the nominal NATREX . Then the excess demand by fundamentalists is given by the function

$$E_{sf}(t) = g_{sf}[N_n(t) - r(t)], \quad \text{sgn}g_{sf}[\ldots] = \text{sgn}[\ldots], \quad g'_{sf} > 0. \qquad (15.65)$$

where N_n is the fundamental exchange rate, that we identify with the nominal NATREX, exogenously given and assumed known by fundamentalists. Transaction costs and the like are subsumed under the non-linear function g_{sf}, which is a sign-preserving function.

(2b) The excess demand by chartists is given by

$$E_{sc}(t) = g_{sc}[ER(t) - r(t)], \quad \text{sgn}g_{sc}[\ldots] = \text{sgn}[\ldots], \quad g'_{sc} > 0, \qquad (15.66)$$

---

[9]For simplicity's sake we neglect the possibility of switch between the two categories.

[10]Typically in the literature the PPP value is used as a measure of the equilibrium exchange rate.

where $ER(t)$ denotes the expected spot exchange rate; the non-linear and sign-preserving function g_{sc} incorporates possible transaction costs. Chartists hold extrapolative expectations:

$$ER(t) = r(t) + h[\dot{r}(t), \ddot{r}(t)], h'_1 > 0, h'_2 > 0, \tag{15.67}$$

where the overdot denotes differentiation with respect to time, and $h[\ldots]$ is a non-linear function. The assumed signs of the time derivatives mean that agents do not only extrapolate the current change ($h'_1 > 0$) but also take account of the acceleration ($h'_2 > 0$). It follows that

$$E_{sc}(t) = g_{sc} \left\{ h[\dot{r}(t), \ddot{r}(t)] \right\}. \tag{15.68}$$

(3) Finally, suppose that the monetary authorities are also operating in the foreign exchange market with the aim of influencing the exchange rate, account being taken of the NATREX, by using an integral policy à la Phillips. The authorities' excess demand $E_G(t)$ can be represented by the following function:

$$E_G(t) = G \left\{ \int_0^t [N_n(t) - r(t)] \, dt \right\}, G' \gtrless 0. \tag{15.69}$$

where $G\{\ldots\}$ is a non-linear function and the integral represents the sum of all the differences that have occurred, from time zero to the current moment, between the NATREX and the actual values of the exchange rate. The sign of G' depends on the policy stance of the monetary authorities. More precisely, if the aim is to stabilize the exchange rate around its NATREX value, then $G' > 0$. In fact, in such a case,

$$sgnG = sgn \left\{ \int_0^t [N_n(t) - r(t)] \, dt \right\}, \tag{15.70}$$

because if the sum of the deviations is positive, this means that the NATREX has been on average greater than the actual exchange rate, so that the latter must increase (depreciate) to move towards the NATREX, hence a positive excess demand for foreign exchange. The opposite holds in the case of a negative sum. Thus the function G passes through zero when moving from negative to positive values, and $G'(0) > 0$.

But the authorities might wish to maintain or generate a situation of competitiveness, which occurs when the actual exchange rate has been on average greater than the NATREX, hence the integral is negative. To maintain or accentuate this situation, the authorities demand foreign exchange, so that

$$sgnG = -sgn \left\{ \int_0^t [N_n(t) - r(t)] \, dt \right\}. \tag{15.71}$$

Thus the function G passes through zero when moving from positive to negative values, and $G'(0) < 0$.

Market equilibrium requires

$$E_n(t) + E_{sf}(t) + E_{sc}(t) + E_G(t) = 0. \tag{15.72}$$

In this way we have only one endogenous variable, $r(t)$, since the fundamentals are subsumed under the NATREX, which is known to both the authorities and the fundamentalists, and is considered exogenous in the present model.

Since the market equilibrium condition (15.72) holds instantaneously (given the practically infinite speed of adjustment of the FOREX market), we can differentiate Eq. (15.72) with respect to time, thus obtaining

$$\dot{E}_n(t) + \dot{E}_{sf}(t) + \dot{E}_{sc}(t) + \dot{E}_G(t) = 0. \tag{15.73}$$

By differentiating Eqs. (15.64), (15.65), (15.68), and (15.69)[11] with respect to time and substituting the result into Eq. (15.73) we obtain

$$g'_n \times \dot{r}(t) + g'_{sf} \times [\dot{N}_n(t) - \dot{r}(t)] + g'_{sc} \times [h'_1 \times \ddot{r}(t) + h'_2 \times \dddot{r}(t)] + G' \times [N_n(t) - r(t)] = 0. \tag{15.74}$$

Collecting terms we get

$$g'_{sc}h'_2 \times \dddot{r}(t) + g'_{sc}h'_1 \times \ddot{r} + (g'_n - g'_{sf}) \times \dot{r}(t) - G' \times r(t) = -G' \times N_n(t) + g'_{sf} \times \dot{N}_n(t), \tag{15.75}$$

whence, dividing through by $g'_{sc}h'_2 \neq 0$,

$$\dddot{r}(t) + \frac{h'_1}{h'_2}\ddot{r} + \frac{(g'_n - g'_{sf})}{g'_{sc}h'_2}\dot{r}(t) - \frac{G'}{g'_{sc}h'_2}r(t) = -\frac{G'}{g'_{sc}h'_2}N_n(t) + \frac{g'_{sf}}{g'_{sc}h'_2}\dot{N}_n(t). \tag{15.76}$$

This equation may seem linear, but it is not so. In fact, the derivative of a function is a function of the same arguments of the function, i.e.

$$g'_n = f_n[r(t)], g'_{sf} = f_{sf}[N_n(t) - r(t)], g'_{sc} = f_{sc}\left\{h[\dot{r}(t), \ddot{r}(t)]\right\}, \quad \text{etc.,} \tag{15.77}$$

so that the coefficients of Eq. (15.76) are to be considered as (non-linear) functions.

The model could be linearised and the resulting linear form analysed, but this would be uninteresting in the present context, since a linear model cannot give rise to chaos. The problem then arises of specifying the non-linearities of our model.

---

[11]Note that $\dot{E}_G(t) = G' \times [\dot{N}_n(t) - \dot{r}(t)]$.

15.8.4.2 The Intrinsic Non-Linearity of the Model

When one abandons linearity (and related functional forms that can be reduced to linearity by a simple transformation of variables, such as log-linear equations), in general it is not clear which non-linear form one should adopt. Further to clarify the matter, let us distinguish between *purely qualitative* non-linearity and *specific* non-linearity.

By *purely qualitative* non-linearity we mean the situation in which we only know that a generic non-linear functional relation exists with certain *qualitative* properties, such as continuous first-order partial derivatives with a given sign and perhaps certain bounds. This is the aspect so far taken by our model, but it is hardly useful for our purposes, because the econometric estimation obviously requires specific functional forms.

By *specific* non-linearity we mean the situation in which we assume a specific non-linear functional relationship. Since in general it is not clear from the theoretical point of view which non-linear form one should adopt, the choice of a form is often arbitrary or made for convenience.

In our case, however, it is possible to introduce a non-linearity on sound economic grounds. This concerns the excess demand of non-speculators. To understand this point, a digression is called for on the derivation of the demand and supply schedules of these agents.

Derivation of the Demand and Supply Schedules of Non-Speculators [12]

The main peculiarity of these demand and supply schedules for foreign exchange is the fact that they are *derived* or *indirect* schedules in the sense that they come from the underlying demand schedules for goods (demand for domestic goods by nonresidents and demand for foreign goods by residents). This has been fully clarified in Chap. 7, Sect. 7.3.1 and Fig. 7.1.

In the case depicted in Fig. 7.1a the function $S(r)$ can be represented by a quadratic, while in the case of Fig. 7.1b a cubic might do. Let us consider the simpler quadratic case, $S(r) = a + br + cr^2, a > 0, b > 0, c < 0$, where a, b, c are constants.[13]

What we propose to do is to introduce the above quadratic non-linearity *while assuming all the other functions to be linear and with constant coefficients*. Thus, assuming that $D(r)$ is linear ($D(r) = d_0 + d_1 r, d_0 > 0, d_1 < 0$, where d_0, d_1 are constants), we can write

$$E_n(t) = D(r) - S(r) = (d_0 + d_1 r) - (a + br + cr^2) = (d_0 - a) + (d_1 - b)r - cr^2$$
$$(15.78)$$

Given this, we have

$$\dot{E}_n(t) = \alpha \dot{r}(t) + \beta r(t) \dot{r}(t), \qquad \text{where } \alpha = (d_1 - b) < 0, \beta = -2c > 0. \quad (15.79)$$

---

[12]For an in-depth treatment of this point see Sect. 7.3.1.
[13]The quadratic function $a + br + cr^2$ as represented in the diagram implies $a > 0, b > 0, c < 0$.

Comparing Eqs. (15.78) and (15.64) we note that

$$g'_n = (d_1 - b) - 2cr(t) = \alpha + \beta r(t). \tag{15.80}$$

As regards the other excess demands, we set

$$E_{sf}(t) = m[N_n(t) - r(t)], m = g'_{sf} > 0$$

$$E_{sc}(t) = n[ER(t) - r(t)], n = g'_{sc} > 0$$

$$ER(t) = h[r(t), \dot{r}(t), \ddot{r}(t)] = r(t) + b_1\dot{r}(t) + b_2\ddot{r}(t), \tag{15.81}$$

$$h'_1 = b_1 > 0, h'_2 = b_2 > 0; \text{ replacing in the previous equation we get}$$

$$E_{sc}(t) = nb_1\dot{r}(t) + nb_2\ddot{r}(t)$$

$$E_G(t) = g\left\{\int_0^t [N_n(t) - r(t)] dt\right\}, g = G' \gtrless 0$$

where m, n, b_1, b_2, g are all constants. Substituting Eq. (15.80), and the parameters defined in (15.81), into Eq. (15.76), and rearranging terms, we obtain

$$\dddot{r}(t) + \frac{b_1}{b_2}\ddot{r}(t) + [\frac{\alpha - m}{nb_2} + \frac{\beta}{nb_2}r]\dot{r}(t) - \frac{g}{nb_2}r(t) = \frac{-g}{nb_2}N_n(t) + \frac{m}{nb_2}\dot{N}_n(t) = 0, \tag{15.82}$$

or

$$\dddot{r}(t) = -\frac{b_1}{b_2}\ddot{r}(t) + [\frac{m - \alpha}{nb_2} - \frac{\beta}{nb_2}r(t)]\dot{r}(t) + \frac{g}{nb_2}r(t) + \varphi(t), \tag{15.83}$$

where

$$\varphi(t) \equiv -\frac{g}{nb_2}N_n(t) + \frac{m}{nb_2}\dot{N}_n(t). \tag{15.84}$$

The homogeneous part of the non-linear third-order differential equation (15.83) is a jerk function,[14] and is known to possibly give rise to chaos for certain values of the

---

[14] A jerk function has the general form

$$x''' = F(x'', x', x).$$

In physical terms, the jerk is the time derivative of the acceleration.

It seems that the denomination "jerk" came to the mind of a physics student traveling in a car of the New York subway some twenty years ago. When standing in a subway car it is easy to balance a slowly changing acceleration. But the subway drivers had a habit of accelerating erratically (possibly induced by the rudimentary controls then in use). The effect of this was to generate an extremely high jerk.

parameters (Sprott 1997; eq. (8)). Besides, since the equation is non-autonomous, the dimension of the state space is increased by one. In fact, Eq. (15.82) can be easily rewritten as a *system* of first-order equations by defining new variables,

$$x_1 \equiv r, \quad x_2 \equiv \dot{r}, \quad x_3 \equiv \ddot{r}. \tag{15.85}$$

The resulting system consists of three first-order equations in the x_i, written as

$$\dot{x}_i = x_{i+1}, \quad i = 1, 2,$$

$$\dot{x}_3 = \frac{g}{nb_2}x_1 + \left[\frac{m-\alpha}{nb_2} - \frac{\beta}{nb_2}x_1\right]x_2 - \frac{b_1}{b_2}x_3 + \varphi(t). \tag{15.86}$$

System (15.86) is obviously non-autonomous, like the original equation. It can be rewritten as an autonomous system at the expense of introducing an additional variable, say

$$x_4 = t. \tag{15.87}$$

In this case x_4 obeys the trivial equation

$$\dot{x}_4 = 1,$$

and system (15.86) becomes an autonomous system of *four* first order equations:

$$\dot{x}_i = x_{i+1}, \quad i = 1, 2,$$

$$\dot{x}_3 = \frac{g}{nb_2}x_1 + \left[\frac{m-\alpha}{nb_2} - \frac{\beta}{nb_2}x_1\right]x_2 - \frac{b_1}{b_2}x_3 + \varphi(x_4), \tag{15.88}$$

$$\dot{x}_4 = 1.$$

In any case, we are not interested in a general numerical analysis of our jerk equation or of its equivalent system, but in its analysis with the *estimated* values of its coefficients.

15.8.4.3 Estimation Results

Estimates of the parameters were found by a Gaussian estimator of the non-linear model subject to all constraints inherent in the model by using Wymer's software for the estimation of continuous time non-linear dynamic models. We use daily observations of the nominal Euro/Dollar exchange rate over the period January 2, 1975 to December 29, 2003 (weekends and holidays are neglected).[15] The derivation

---

[15] Source: EUROSTAT.

of the NATREX series is discussed in detail in Belloc and Federici (2010).[16] The equation estimated is Eq. (15.83), written in the form

$$\dddot{r}(t) = a_1\ddot{r}(t) + [a_2 + a_3 r(t)]\dot{r}(t) + a_4 r(t) - a_4 N_n(t) - a_5 \dot{N}_n(t). \tag{15.89}$$

where

$$
\begin{aligned}
a_1 &\equiv -\frac{b_1}{b_2} < 0, \\
a_2 &\equiv \frac{m-\alpha}{nb_2} > 0, \\
a_3 &\equiv -\frac{\beta}{nb_2} < 0, \\
a_4 &\equiv \frac{g}{nb_2} \gtreqless 0, \\
a_5 &\equiv -\frac{m}{nb_2} < 0.
\end{aligned}
\tag{15.90}
$$

The expected signs of the a_i coefficients reflect our theoretical hypotheses set out in the previous sections. We note that the "original" parameters are seven $(b_1, b_2, m, \alpha, n, \beta, g)$ while we can estimate only five coefficients. Hence it is impossible to obtain the values of the original parameters. What we can do is to check the agreement between the signs listed in (15.90) and the coefficient estimates. The estimates are reported in Table 15.1.

The last column (Ratio) gives the (absolute value of the) ratio of the parameter estimate to the estimate of its asymptotic standard error (ASE). This ratio does not have a Student's t distribution, but has an asymptotic normal distribution. Thus in a sufficiently large sample it is significantly different from zero at the 5 % level if it is greater than 1.96 and significantly different from zero at the 1 % level (i.e., highly significant) if it is greater than 2.58.

The estimation of the model shows a remarkable agreement between estimates and theoretical assumptions. In fact, not only all the coefficients have the expected sign and are highly significant, but, in addition, the observed and the estimated values are very close, as shown by Fig. 15.3 (the correlation coefficient turns out to be 0.9959).

Table 15.1 Estimation results

Coefficient	Estimate	ASE	Ratio
a_1	−12.405	1.538	8.06
a_2	16.976	2.823	6.01
a_3	−27.421	3.545	7.73
a_4	−0.01064	0.003596	2.96
a_5	−1.226	0.184	6.66

Log-likelihood value 0.3287539E+05

---

[16]We have generated daily data over the sample period used in estimation.

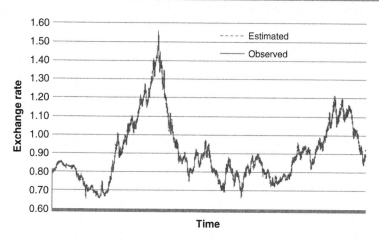

Fig. 15.3 Observed and estimated values

The in-sample root mean square error (*RMSE*) of forecasts[17] of the endogenous variable *r* turns out to be 0.005475, a very good result.

As regards the out-of-sample, ex post forecasts, we simulated the model over the period January 5, 2004 to June 30, 2006 (weekdays only) and obtained a *RMSE* of 0.091338. This value, although higher than the in-sample value (which is a normal occurrence), is satisfactory.

15.8.4.4 Testing for Chaos[18]

Our first step[19] was that of looking for a strange attractor through *phase diagrams*.

Figure 15.4 plots $r'(t)$ against $r(t)$ (these are denoted by $X'(t), X(t)$ in the figure). No discernible structure appears. There does not seem to be a point around which the series evolves, approaching it and going away from it infinite times. On the contrary, the values are very close and no unequivocal closed orbits or periodic motions seem to exist. If we lengthen the time interval for which the phase diagram is built we obtain closed figures, but we cannot clearly classify them as strange attractors because when the data contain such an attractor, this should remain substantially similar as the time interval changes. Such a feature is absent. This test, however, is hardly conclusive, as it relies on impression rather than on quantitative evaluation.

---

[17]To obtain these forecasts, the differential equation is re-initialized and solved *n* times (if one wants forecasts for *n* periods), each time using the observed value of the endogenous variable in period *t* as initial value in the solution, which is then employed to obtain the forecast for period *t* + 1. In other words, the re-initialization is at the same frequency as the sample observations.

[18]All the tests reported here have been carried out using the software by Sprott and Rowlands (1992).

[19]The following tests were carried out using the software Chaos Data Analyzer by Sprott and Rowlands (1992).

Fig. 15.4 Phase diagram

Phase-Space Plots

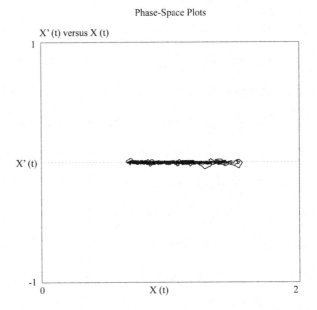

X' (t) versus X (t)

Fig. 15.5 Power spectrum

Power Spectrum

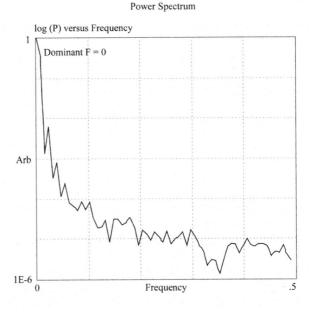

log (P) versus Frequency

We then computed the *power spectrum* (Fig. 15.5). Power spectra that are straight lines on a log-linear scale are thought to be good candidates for chaos. This is clearly not the case.

Quantitative tests are based on the *correlation dimension* and *Liapunov exponents*.

Fig. 15.6 Correlation
dimension

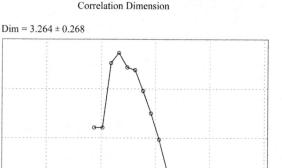

Correlation Dimension

Dim = 3.264 ± 0.268

The Grassberger-Procaccia algorithm for the computation of the *correlation dimension* requires the presence of a flat plateau in the diagram where the log of the dimension is plotted against the log of the radius.

Since no such plateau exists (see Fig. 15.6), the computation of the dimension (which turned out to be 3.264 ± 0.268) is not reliable. In any case, it should be noted that saturation of the correlation dimension estimate is just a necessary, but not sufficient, condition for the existence of a chaotic attractor, since also nonlinear nonchaotic stochastic systems are capable of exhibiting this property (Scheinkman and LeBaron 1989).

Arguably, the only test specific for chaos is provided by *Liapunov exponents*. The Liapunov exponent is a measure of the rate at which nearby trajectories in phase space diverge (SDIC). Chaotic orbits have at least one positive Liapunov exponent.

Inserting the estimated parameters into the original non-linear model and solving the differential equation, we obtained the values (daily data) of the exchange rate generated by the model. Then we applied to this series the Lyapunov exponents test. In this case the greatest Lyapunov exponent is 0.103±0.016. This is evidence for chaos, but the reshuffled (surrogate) data procedure[20] refutes such a result. The basic idea is to produce from the original data a new series with the same distributional properties but with any non-linear dependence removed. The maximum Lyapunov exponent test is then applied to this surrogate series to check whether it gives the

---

[20]See Scheinkman and LeBaron (1989), Theiler (1991) and Rapp et al. (1993) for a discussion about shuffle diagnostics.

same (pro chaos) results as those obtained from the original series. If the results are the same, we should suspect the veracity of our conclusions. We obtained a positive largest Lyapunov exponent of 0.419 ± 0.16. Hence we can conclude that the series generated by the estimated model cannot be considered as chaotic.

The previous results are confirmed by a different procedure, which is the following.

Lyapunov exponents have been calculated from the underlying non-linear model for the estimated parameter values, using the variational matrix equation, and these concentrate information on the nature of the non-linear dynamics. In our case all exponents are negative, and are $-0.218691, -0.620225, -0.620229$.

On the basis of these results the model is stable dynamically (i.e. for a given set of parameters) and structurally stable (i.e. the results did not change in a substantial way even for large changes in the parameter values[21]).

The stability properties of the model suggest its importance not only to the foreign exchange market but, given the events over the past few years, to financial markets more generally. The Lyapunov exponents of the model show that it is stable at the estimated values and apparently in a wide neighbourhood of those values. If parameter a_2 is set to zero, however, which means that fundamentalists are not active in the market, the model is unstable. Moreover, in a fairly wide neighbourhood of the other parameters, the model remains unstable. There is a major change in the dynamic structure depending on whether or not fundamentalists are in the market.

15.8.4.5 Conclusion
Our results have important economic implications.

(I) The implications for the foreign exchange market, and almost certainly other financial markets, is striking. The stabilizing role of fundamentalists is not surprising given their longer horizons, but the need for the presence of fundamentalists to stabilize a market that would otherwise be unstable raises questions about the role of the other players. In recent years, it has been argued that day-traders and other short-term players are important in providing liquidity to the market. If so, they should make the market more stable but they do not. Some (largely anecdotal) evidence suggests that as risk rises these traders disappear from the market. If that is the case their role in providing liquidity is superficial, providing liquidity when it is not needed and not when it is. If that is so, from a macro-economic point of view it is an inefficient use of capital.

---

[21]These changes were not arbitrary. In fact, we are dealing with *estimated* parameters; it follows that the "true" value of the parameter can lie anywhere in the confidence interval, calculated as

point estimate $\pm 1.96\sigma$ (95 %)

or

point estimate $\pm 2.58\sigma$ (99 %)

where σ is the ASE.

On this point see Gandolfo 1992.

(II) The second implications is methodological. As stated in the Introduction, after the failure of the standard structural models of exchange rate determination in out-of-sample ex-post forecasts (the most notable empirical rejection was that by Meese and Rogoff, confirmed by subsequent studies), exchange rate forecasting has come to rely on technical analysis and time series procedures, with no place for economic theory. Economic theory can be reintroduced:
   (a) through a non-linear purely deterministic structural model giving rise to chaos;
   (b) through a non-linear non-chaotic but stochastic structural model.

The fact that our model fits the data well but does not give evidence for chaos means that non-linear (non-chaotic but stochastic) differential equations econometrically estimated in continuous time are the most promising tool for coping with this phenomenon.

References

Alexander, D., & Thomas, L. R. (1987). Monetary/asset models of exchange rate determination: How well have they performed in the 1980s? *International Journal of Forecasting, 3*, 53–64.

Bajo-Rubio, O., Fernandez-Rodriguez, F., & Sosvilla-Rivero, S. (1992). Chaotic behavior in exchange-rate series: First results for the peseta–U.S. dollar case. *Economics Letters, 39*, 207–211.

Balassa, B. (1964). The purchasing power parity doctrine: A reappraisal. *Journal of Political Economy, 72*, 584–596.

Barnett, W. A., Gallant, A. R., Hinich, M. J., Jungeilges, J. A., Kaplan, D. T., & Jensen, M. J. (1997). A single blind controlled competition among tests for non-linearity and chaos. *Journal of Econometrics, 82*, 157–192.

Bayoumi, T., Clark, P. Symansky, S., & Taylor, M. (1994). The robustness of equilibrium exchange rate calculations to alternative assumptions and methodologies. In: J. Williamson (Ed.), *Estimating equilibrium exchange rates* (Chap. 2). Washington (DC): Institute for International Economics.

Belloc, M., Federici, D., & Gandolfo, G. (2008). The euro/dollar equilibrium real exchange rate: a continuous time approach. *Economia Politica: Journal of Analytical and Institutional Economics, 25*, 249–270.

Belloc, M. & Federici, D. (2010). A two-country NATREX model for the euro/dollar. *Journal of International Money and Finance, 29*, 315–335.

Blaug, M, (1980). *The methodology of economic.* Cambridge (UK): Cambridge University Press.

Boubel, A. (2000). Volatilité des taux de change et variables fondamentales. PhD thesis, Université d'Evry-Val d'Issonne, Département d'Économie.

Branson, W. H. & Buiter, W. H. (1983). Monetary and fiscal policy and flexible exchange rates. In: J. S. Bhandari & B. H. Putnam (Eds.), *Economic interdependence and flexible exchange rates* (Chap. 9). Cambridge (Mass.): MIT Press.

Branson, W. H. & Henderson, D. W. (1985). The specification and influence of asset markets. In: R. W. Jones & P. B. Kenen (Eds.), *Handbook of International Economics* (Vol. 2, Chap. 15). Amsterdam: North-Holland.

Brock, A. (1986). Distinguishing random and deterministic systems. *Journal of Economic Theory, 40*, 168–195.

Cassel, G. (1918). Abnormal deviations in international exchanges. *Economic Journal, 28*, 413–415.

Chen, S. (1999). Complex dynamics of the real exchange rate in an open macroeconomic model. *Journal of Macroeconomics, 21*, 493–508.

Cheung, Y. -W., & Lai, K. S. (2000). On the purchasing ower parity puzzle. *Journal of International Economics, 52*, 321–330.

Chinn, M. D. (1997). Paper pushers or paper money? Empirical assessment of fiscal and monetary models of exchange rate determination. *Journal of Policy Modeling, 19*, 51–78.

Choudhry, T., & Lawler, P. (1997). The monetary model of exchange rates: Evidence from the Canadian float of the 1950s. *Journal of Macroeconomics, 19*, 349–362.

Clark, P., & MacDonald, R. (1999). Exchange rates and economic fundamentals: A methodological comparison of BEERs and FEERs. In: R. MacDonald & J. L. Stein (Eds.), *Equilibrium exchange rates* (Chap. 10). Dordrecht: Kluwer Academic Publishers.

Cuaresma, J. L. (1998). Deterministic chaos versus stochastic processes: An empirical study on the Austrian schilling-US dollar exchange rate. Institute for Advanced Studies, Vienna, *Economic Series*, No. 60.

Cushman, D. O. (2000). The failure of the monetary exchange rate model for the Canadian-U.S. dollar. *Canadian Journal of Economics, 33*, 591–603.

Da Silva, S. (2000). The role of foreign exchange intervention in a chaotic Dornbusch model. *Kredit und Kapital, 33*, 309–345.

Da Silva, S. (2001). Chaotic exchange rate dynamics redux. *Open Economies Review, 12*, 281–304.

De Arcangelis, G. (1992). Time-varying parameters and exchange rate forecasting. University of Rome "La Sapienza", CIDEI Working Paper No. 14.

De Grauwe, P., & Versanten, K. (1990). Determistic chaos in the foreign exchange market. Discussion Paper No. 70, Centre for Economic Research.

De Grauwe, P., & Dewachter, H. (1993a). A chaotic model of the exchange rate: The role of fundamentalists and chartists. *Open Economies Review, 4*, 351–379.

De Grauwe, P., & Dewachter, H. (1993b). A chaotic monetary model of the exchange rate. In: H. Frisch, & A. Worgotter (Eds.), *Open economy macroeconomics* (pp. 353–376). London: Macmillan.

De Grauwe, P., Dewachter, H., & Embrechts, M. (1993). *Exchange rate theories. Chaotic models of the foreign exchange markets*. Oxford: Blackwell.

De Grauwe, P., & Grimaldi, M. (2006a). *The exchange rate in a behavioral finance framework*. Princeton: Princeton University Press.

De Grauwe, P., & Grimaldi, M. (2006b). Exchange rate puzzles: A tale of switching attractors. *European Economic Review, 50*, 1–33.

De Jong, E. (1997). Exchange rate determination: Is there a role for fundamentals? *De Economist, 145*, 547–572.

Dooley, M. P., & Isard, P. (1983). The portfolio model of exchange rates and some structural estimates of the risk premium. *IMF Staff Papers, 30*, 683–702.

Dornbusch, R. (1976). Expectations and exchange rate dynamics. *Journal of Political Economy, 84*, 1161–1176.

Dornbusch, R. (1983). Exchange rate economics: Where do we stand? In: J. S. Bhandari & B. H. Putnam (Eds.), *Economic interdependence and flexible exchange rates* (pp. 45–83). Cambridge (Mass.): MIT Press.

Edgeworth, F. Y. (1905). Review of H. Cunynghame's book *A geometrical political economy*. *Economic Journa, 15*, 62–71.

Edwards, S. (1994). Real and monetary determinants of real exchange rate behavior. In: J. Williamson (Ed.), *Estimating equilibrium exchange rates*. Washington (DC): Institute for International Economics.

Elbadawi, I. (1994). Estimating long-run equilibrium real exchange rates. In: J. Williamson (Ed.), *Estimating equilibrium exchange rates*. Washington (DC): Institute for International Economics.

Ellis, J. (1994). Non linearities and chaos in exchange rates. In: J. Creedy, & V. L. Martin (Eds.), *Chaos and non linear models in economics: Theory and applications*. Aldershot: E. Elgar.

Engle, R. P., & Granger, C. W. P. (1987). Co-integration and error correction: representation, estimation and testing. *Econometrica, 55*, 251–276.

Faruquee, H., Isard, P., & Masson, P. R. (1999). A macroeconomic balance framework for estimating equilibrium exchange rates. In: R. MacDonald, & J. L. Stein (Eds.), *Equilibrium exchange rates* (Chap. 4). Dordrecht: Kluwer Academic Publishers.

Evans, M. D. D. (2010). Order flows and the exchange rate disconnect puzzle. *Journal of International Economics, 80*, 58–71.

Evans, M. D. D. (2011). *Exchange rate dynamics*. Princeton: Princeton University Press.

Federici, D., & Gandolfo, G. (2002). Chaos and the exchange rate. *Journal of International Trade and Economic Development, 11*, 111–142.

Federici, D., & Gandolfo, G. (2012). The euro/dollar exchange rate: Chaotic or non-chaotic? A continuous time model with heterogeneous beliefs. *Journal of Economic Dynamics and Control, 36*, 670–681.

Flood, R. P., &. Rose, A. K (1995). Fixing exchange rates: A virtual quest for fundamentals. *Journal of Monetary Economics, 36*, 3–37.

Frankel, J. A. (1983). Monetary and portfolio models of exchange rate determination. In: J. S. Bhandari & B. H. Putnam (Eds.), *Economic interdependence and flexible exchange rates* (pp. 84–115). Cambridge (Mass.): MIT Press.

Froot, K. A., & Rogoff, K. (1996). Perspective on PPP and long-run exchange rates. In: G. Grossman & K. Rogoff (Eds.), *Handbook of international economics* (Vol. III, pp. 1647–1688). Amsterdam: North-Holland.

Gandolfo, G. (1979). The equilibrium exchange rate: Theory and empirical evidence. In: M. Sarnat & G. P. Szego (Eds.), *International finance and trade* (Vol. I, pp. 99–130). Cambridge (Mass.): Ballinger.

Gandolfo, G. (1981). *Qualitative analysis and econometric estimation of continuous time dynamic models*. Amsterdam: North Holland.

Gandolfo, G. (2009). *Economic dynamics* (4th ed.). Berlin, Heidelberg, New York: Springer.

Gandolfo, G. (2014). *International trade theory and policy* (2nd ed.). Berlin, Heidelberg, New York: Springer.

Gandolfo, G., & Padoan, P. C. (1990). The Italian continuous time model: Theory and empirical results. *Economic Modelling, 7*, 91–132.

Gandolfo, G., Padoan, P. C., & Paladino, G. (1990a). Structural models vs random walk: The case of the lira/$ exchange rate. *Eastern Economic Journal, 16*, 101–113.

Gandolfo, G., Padoan, P. C., & Paladino, G. (1990b). Exchange rate determination: Single-equation or economy-wide models? A test against the random walk. *Journal of Banking and Finance, 14*, 965–992.

Gandolfo, G., Padoan, P. C., & De Arcangelis, G. (1993). The theory of exchange rate determination, and exchange rate forecasting. In: H. Frisch & A. Wörgötter (Eds.), *Open economy macroeconomics* (pp. 332–352). London: Macmillan.

Groen, J. J. J. (2000). The monetary exchange rate model as a long-run phenomenon. *Journal of International Economics, 52*, 299–319.

Guillaume, D. M. (2000). *Intradaily exchange rate movements*. Dordrecht: Kluwer.

Harrod, R. F. (1933), *International economics*. Chicago: Chicago University Press.

Helliwell, J. F., & Boothe, P. M. (1983). Macroeconomic implications of alternative exchange rate models. In: P. De Grauwe & T. Peeters (Eds.), *Exchange rate in multicountry econometric models* (pp. 21–53). London: Macmillan.

Hooper, P,. & Morton, J. (1982). Fluctuations in the dollar: A model of nominal and real exchange rate determination. *Journal of International Money and Finance, 1*, 39–56.

Howrey, E. P. (1994). Exchange rate forecasts with the Michigan quarterly econometric model of the US economy. *Journal of Banking and Finance, 18*, 27–41.

Isard, P. (1995). *Exchange rate economics*. Cambridge (UK): Cambridge University Press.

Klaassen, F. (1999). Purchasing power parity: Evidence from a new test. Tilburg University, Center for Economic Research Working Paper No. 9909.

Kouri, P. J. K. (1983). Balance of payments and the foreign exchange market: A dynamic partial equilibrium model. In: J. S. Bhandari & B. H. Putnam (Eds.), *Economic interdependence and flexible exchange rates*. Cambridge (Mass.): MIT Press.

Krueger, A. O. (1983). *Exchange rate determination*. New York: Cambridge University Press.

Lorenz, E. N. (1963). Deterministic non-period flows. *Journal of Atmospheric Sciences, 20*, 130–141.

MacDonald, R. (2007). *Exchange rate economics: Theories and evidence*. New York: Routledge.

MacDonald, R., & Taylor, M. P. (1992). Exchange rate economics: A survey. *IMF Staff Papers, 39*, 1–57.

MacDonald, R., & Stein J. L. (Eds.) (1999). *Equilibrium exchange rates*. Dordrecht: Kluwer Academic Publishers.

Mark, N. C., & Sul, D. (2001). Nominal exchange rates and monetary fundamentals: Evidence from a small post-Bretton Woods panel. *Journal of International Economics, 53*, 29–52.

Meese, R. A., & Rogoff, K. (1983a). Empirical exchange rate models of the seventies: Do they fit out of sample? *Journal of International Economics, 14*, 3–24.

Meese, R. A., & Rogoff, K. (1983b). The out-of-sample failure of empirical exchange rate models: Sampling error or misspecification? In: J. A. Frenkel (Ed.), *Exchange rates and international macroeconomics*. Chicago: Chicago University Press.

Melvin, M. (1985). Currency substitution and Western European monetary unification. *Economica, 52*, 79–91.

Mizen, P., & Pentecost, E. J. (Eds.) (1996). *The macroeconomics of international currencies: Theory, policy and evidence*. Cheltenham (UK): E. Elgar.

Moosa, I. A. (2000). *Exchange rate forecasting: Techniques and applications*. New York: St. Martin's Press.

Noussair, C. N., Plott, C. R., & Riezman, R. G. (1997). The principles of exchange rate determination in an international finance experiment. *Journal of Political Economy, 105*, 822–861.

Nurkse, R. (1945). *Conditions of international monetary equilibrium*. Essays in International Finance No. 4, International Finance Section, Princeton University.

Officer, L. H. (1982). *Purchasing power parity and exchange rates: Theory, evidence, and relevance*. Greenwich (CT): JAI Press.

O'Connell, P. G. J. (1998). The overvaluation of purchasing power parity. *Journal of International Economics, 44*, 1–19.

Pippenger, M. K., & Goering, G. E. (1998). Exchange rate forecasting: Results from a threshold autoregressive model. *Open Economies Review, 9*, 157–170.

Rapp, P. E., Albano, A. M., Schmah, T. I., & Farwell, L. A. (1993). Filtered noise can mimic low-dimensional chaotic attractors. *Physical Review E, 47*, 2289–2297.

Reszat, B. (1992). Chaos and exchange rate. Proceedings of the Pennsylvania Economic Association Seventh Annual Meeting, 4–6 June, 130–138.

Rogoff, K. S., & Stavrakeva, V. (2008). The continuing puzzle of short horizon exchange rate forecasting. NBER Working Paper No. 14071.

Samuelson, P. A. (1964). Theoretical notes on trade problems. *Review of Economics and Statistics, 46*, 145–154.

Sarno, L., & Taylor, M. P. (2001). *The microstructure of the foreign exchange market: A selective survey of the literature*. Princeton University, International Economics Section, Princeton Studies in International Economics No. 89.

Scheinkman, J. A., & LeBaron, B. (1989). Nonlinear dynamics and stock returns. *Journal of Business, 62*, 311–337.

Schinasi, G. J., & Swamy, P. A. V. B. (1989). The out-of-sample forecasting performance of exchange rate models when coefficients are allowed to change. *Journal of International Money and Finance, 8*, 375–390.

Schwartz, B., & Yousefi, S. (2003). On complex behavior and exchange rate dynamics. *Chaos Solitons and Fractals, 18*, 503–523.

Shafer, J. R., & Loopesko, B. E. (1983). Floating exchange rates after ten years. *Brookings Papers on Economic Activity, 1*, 1–70.

Sheen, J. (1989). Modelling the floating australian dollar: Can the random walk be encompassed by a model using a permanent decomposition of money and output? *Journal of International Money and Finance, 8*, 253–276.

Smith, V. L. (1987). Experimental methods in economics. *The new Palgrave: A dictionary of economics* (Vol. 2). London: Macmillan.

Somanath, V. S. (1986). Efficient exchange rate forecasts: Lagged models better than the random walk. *Journal of International Money and Finance, 5*, 195–220.

Sprott, J. C., & Rowlands, G. (1992). *Chaos data analyzer - Professional version*. Physics Academic Software.

Sprott, J. C. (1997). Some simple chaotic jerk functions. *American Journal of Physics, 65*, 537–543.

Stein, J. L. (1990). The real exchange rate. *Journal of Banking and Finance, 14*, 1045–1078.

Stein, J. L. (1995a). The natural real exchange rate of the United States dollar, and determinants of capital flows. In: J. L. Stein, P. Reynolds Allen & Associates (Eds.), *Fundamental determinants of exchange rates* (Chap. 2). Oxford: Oxford University Press.

Stein, J. L. (1995b). The fundamental determinants of the real exchange rate of the US dollar relative to the other G-7 countries. *IMF Working Paper 95/81*.

Stein, J. L. (2001). The equilibrium value of the euro/$ US exchange rate: an evaluation ofrResearch. *CESifo Working Papers*, No. 525.

Stein, J. L. (2002). Enlargement and the value of the euro. *Australian Economic Papers, 41*, 462–479.

Stein, J. L. (2006). *Stochastic optimal control, international finance and debt crisis* (Chap. 4). Oxford: Oxford University Press.

Szpiro, G. (1994). Exchange rate speculation and chaos inducing intervention. *Journal of Economic Behavior and Organization, 24*, 363–368.

Taylor, A. M., & Taylor, M. P. (2004). The purchasing power parity debate. *Journal of Economic Perspectives, 18*, 135–158.

Theiler, J. (1991). Some comments on the correlation dimension of 1/f noise. *Physics Letters A, 155*, 480–493.

Weston, R. (2007). The chaotic structure of the $C/$US exchange rate. *International Business & Economics Research Journal, 6*, 19–28.

Williamson, J. (1985). *The exchange rate system* (revised ed.). Washington (DC): Institute for International Economics.

Williamson, J. (Ed.) (1994). *Estimating equilibrium exchange rates*. Washington (DC): Institute for International Economics.

Capital Movements, Speculation, and Currency Crises

International capital movements have been mentioned several times in previous chapters (see, for example, Sects. 2.2, 2.6, 5.1, 7.4, 10.2, 11.2–11.4, 13.2–13.4). The purpose of the next sections is to bring these together in a unified picture, and to examine the causes and effects of the main types of capital movements in detail; for convenience the traditional distinction between short-term and long-term movements will be maintained. It should be stressed that for obvious reasons our examination will be made from the point of view of international monetary economics, so that we shall ignore—except for a very brief mention—the important problem of the behaviour of multinational corporations in carrying out foreign direct investment (FDI), a problem which more properly belongs to the theory of the firm, as we shall see in the next section.

16.1 Long-Term Capital Movements

The main types of private long-term capital movements are *portfolio investment* and *direct investment* (this is the terminology used in BOP accounting, see Chap. 5; it coincides with the terminology foreign direct investment or FDI). Other types of long-term capital movements are international loans and commercial credits, naturally with a maturity of more than one year. The difference between portfolio and direct investment is that the direct investor seeks to have, on a lasting basis, an effective voice in the management of a nonresident enterprise, whilst portfolio investment is of a purely financial nature. In general, the typical direct investment is in ordinary shares (equities) and the operator is usually a multinational corporation. Portfolio investment covers government bonds, private bonds, bonds issued by international organizations, preference shares, equities (but not so as to gain control over the corporation), various kinds of other securities (certificates of deposit, marketable promissory notes, etc.). At this point the problem arises of determining the percentage of ownership of an enterprise above which one can talk of control.

© Springer-Verlag Berlin Heidelberg 2016
G. Gandolfo, *International Finance and Open-Economy Macroeconomics*,
Springer Texts in Business and Economics, DOI 10.1007/978-3-662-49862-0_16

It is true, of course, that full legal control is achieved by owning just over 50 % of the equities (or other form of ownership of the enterprise), but in the case of big corporations with a widely distributed ownership among numerous small shareholders a much lower percentage is often sufficient to achieve the actual control of the corporation. Thus a conventional accounting solution is inevitable. Most countries, therefore, rely on the percentage of ownership of the voting stock in an enterprise. Some countries use more than one percentage (for example according to whether investment is in foreign enterprises by residents or in domestic enterprises by nonresidents, or depending on the degree of dispersion of foreign ownership among foreign investors, etc.). In general the percentage chosen as providing evidence of direct investment is low, usually ranging from 25 % down to 10 %, with a tendency towards the lower end of the range. Let us now come to a brief consideration of theoretical problems.

Portfolio investment, once assumed to be a function of differential yields and of risk diversification, but without a precise framework to fit in, has received an adequate theoretical placing within the general theory of portfolio selection. This theory—which has already been treated in Sect. 13.2—starts from a given amount of funds (wealth) to be placed in a certain set of admissible domestic and foreign assets, where the rates of return and the direct and cross risk coefficients of the various assets are known. The maximization procedure is then carried out in two stages: first the set of *efficient*[1] portfolios is determined, then the *optimum* portfolio in this set is determined by using the investor's utility function. To put it another way, given the stock of wealth, the optimum stock of each of the various assets included in the portfolio is a function of the rates of return and risk coefficients of all assets as well as of the "tastes" of the wealth holder (at the same rates of return and risk coefficients, the portfolio of a risk-averse investor will be different from that of a risk lover). Now, since portfolio investment is an aggregate of flows, it is self-evident that these arise if, and only if, the currently owned *stocks* of the various assets relevant to the balance of payments (that is, of foreign financial assets owned by residents, and of domestic financial assets owned by nonresidents) are *different* from the respective optimum stocks. As soon as these are reached by way of the adjustment flows, the flows themselves cease. Therefore, according to this theory, the existence of continuous flows of portfolio investment derives *both* from the fact that the elements (yields, risk, tastes) underlying the optimum composition[2] are not constant but change through time (since the optimum composition changes as they change, continuous adjustment flows will be required) and from the fact that

---

[1] A portfolio, that is a given allocation of funds among the various assets, is efficient if a greater return can be achieved only by accepting a greater risk (or a lower risk can be obtained only by accepting a lower return).

[2] It should be remembered that according to the theory of portfolio selection, the composition of both efficient and optimum portfolios is independent of the stock of wealth, which has the role of a scale variable. In other words, one first determines the composition (i.e., the fractions of total wealth allocated to the various assets) and then the stocks of the various assets themselves, by applying these fractions to the stock of wealth.

the stock of wealth is not itself a constant but changes through time, so that even if the composition were constant, the desired stocks of the various relevant assets change just the same and so adjustment flows take place. This is a very plausible picture, as even a casual look at world financial markets will confirm that yields and risks are in a state of continual change in all directions (we neglect tastes, not because they are not important, but because they are not directly observable); the stock of financial wealth is also a magnitude in continual evolution. Thus the theory of portfolio selection is capable of giving a consistent and satisfactory explanation of portfolio investment.

The problem of *direct investment* is much more complicated and does not seem susceptible to a single simple explanation. First of all, it is necessary to dissipate a possible misunderstanding due to the confusion between the *real* and the *financial* aspects of direct investment. Many analyses, in fact, consider direct investment in the context of the theory of international factor mobility treated by the pure theory of international trade: according to this view, direct investment is nothing more nor less than the movement of the productive factor "(real) capital" between countries. If one accepted this point of view, the causes and effects of direct investment would be those already analysed by the pure theory of international trade. As regards the *causes*, these should be seen essentially in the existence of different rewards to capital in different countries (this implies that the factor price equalization theorem does not hold), due both to the presence of tariffs (we remember that by the Stolper-Samuelson theorem there is a precise relationship between tariffs and factor rewards) and to other elements (such as complete specialization, factor-intensity reversals, etc.). As regards the *effects* on the host country, these could be analysed by using first of all the Rybczynski theorem (see Gandolfo 2014, Chap. 5, Sect. 5.4), because the view under consideration implies that direct investment brings about an increase in the capital stock of the country which receives it, and Rybczynski's theorem is the appropriate tool for analyzing the effects of an increase in factor endowments. The problem of the possible repatriation of profits should then be dealt with (if this repatriation were complete, the case might occur in which it entirely absorbs the increase in national income due to the increase in output made possible by the increase in the stock of capital). All these problems can be fully dealt with in the theory of international trade (see, for example, Gandolfo 2014, Chap. 6, Sect. 6.8).

What we wish to stress is that this formulation, if perfectly valid in the context of the pure theory of international trade, is no longer valid with certainty in the context of international monetary economics. A direct investment does *not* necessarily mean an increase in the physical capital stock of the host country. If, for example, the multinational corporation x of country 1 buys the majority of the equities of corporation y in country 2 (previously owned by country 2's residents) the only thing that has happened is an inflow of financial capital (the payment for the equities) into country 2, whose stock of physical capital is *exactly the same* as before. It goes without saying that insofar as the multinational x subsequently transfers entrepreneurship, known-how, etc., to y, there will be "real" effects on country 2, but this is a different story. It has indeed been observed that direct investment is strongly industry-specific: in other words, it is not so much a flow of capital from country 1 to country 2 but rather a flow of capital from industry α of country 1 to industry α of country 2. The typical enterprise which makes direct investment is usually a big corporation which operates in a market with a high product differentiation, and, for this corporation, direct investment is often an alternative to exporting its products, as the ownership of plant in foreign countries facilitates the penetration of foreign markets. From this point of view it is clear that the theory of direct investment belongs to the theory of the firm and, to be precise, to the theory of multinational firms (which has had an enormous development in recent times), rather than to general international economics. Therefore we refer the reader to the relevant literature, amongst which Dunning (1977, 1993), Buckley (1998), UN (1998), Zebregs (1999), Barba Navaretti and Venables (2004). Reference is made to that literature also for the study of the effects of direct investment on the host country, which are the subject of heated debate (see Lipsey 2004). We merely mention the fact that among the pros, the transfer of entrepreneurship and new technology to the host country is pointed out, whilst among the cons the critics point out the exploitation of the host economy (for example when the outflow of repatriated profits becomes higher than the inflow of direct investment), the possible diminution in its sovereignty (the subsidiary responds to the instructions of the parent company rather than to those of the local authorities), the possible checkmating of its economic policies (for example a restrictive monetary policy can be nullified by the subsidiary which has recourse to the financial market of the country of residence of the parent company). On most aspects of this issue there is a wide range of empirical results with little signs of convergence.

An objective balance between the pros and cons is probably impossible, partly because of the political questions that come into play.

16.2   Short-Term Capital Movements and Foreign Exchange Speculation

The economic role of speculation is a moot question also outside international economics. On the one hand, in fact, it is claimed that speculators, by buying when the price is low and reselling when the price is high, help to smooth out and dampen down the fluctuations of the price around its normal value, so that their operations are beneficial (*stabilizing speculation*). On the other hand, the possibility is stressed that speculators buy precisely when the price is rising in order to force a further rise and then profit from the difference (bullish speculation: the case of bearish speculation is perfectly symmetrical), so that their operations *destabilise* the market. It does not therefore seem possible to reach an unambiguous theoretical conclusion, as we shall see below.

This said in general, let us pass to the examination of foreign-exchange specula-tion, in particular of speculation on the *spot* market. The asset concerned is foreign exchange, whose price in terms of domestic currency is the (spot) exchange rate. Therefore, if speculators anticipate a depreciation (i.e. if the expected exchange rate is higher than the current one), they will demand foreign exchange (simultaneously supplying domestic currency) in the expectation of reselling it at a higher price and so earning the difference. It goes without saying that the expected difference will have to be greater than the net costs of the speculative operation.

Conversely they will supply foreign exchange (simultaneously demanding domestic currency) if the expected exchange rate is lower than the current one.

In order better to examine the effects of speculation, we must distinguish a fixed exchange-rate regime of the adjustable peg type (see Sect. 3.2) and a freely flexible exchange-rate regime.

Under an adjustable peg regime (such as the Bretton Woods system), speculation is normally destabilising, for a very simple reason. Since the regime allows once-and-for-all parity changes in the case of fundamental disequilibrium, in a situation of a persistent and serious balance-of-payments disequilibrium it will be apparent to all in which direction the parity change if any will take place, so that speculation is practically risk-free (the so-called *one-way option*). The worst that can happen to speculators, in fact, is that the parity is not changed, in which case they will only lose the cost of transferring funds, the possible interest differential against them for a limited period of time, and the possible difference between the buying and selling prices (which is very small, given the restricted margins of oscillation around parity). It goes without saying, that these speculative transfers of funds make the disequilibrium worse and thus make the parity change more and more necessary: they are *intrinsically destabilising*.

Among the cases of this type of speculation, those that occurred on the occasion of the parity changes of the pound sterling (devaluation of November, 1967), of the French franc (devaluation of August, 1969), and of the Deutschemark (revaluation of October, 1969) are usually pointed out. In fact, in the case of a fundamental disequilibrium of the deficit type, the pressure on the exchange rate is in the sense of a devaluation, and the authorities are compelled, as we know, to sell foreign exchange to defend the given parity. Now speculators demand foreign exchange: this demand has to be added to the demand deriving from the fundamental deficit and, by increasing the pressure on the exchange rate, may cause the monetary authorities defence to collapse (this defence might otherwise have been successful in the absence of speculation). A similar reasoning holds in the case of a fundamental disequilibrium in the surplus direction.

It should be noted that in what we have said, there is an implicit judgement that *destabilising speculation is harmful*. This judgement is generally shared, whether implicitly or explicitly. Friedman (1960) has tried to oppose it, by arguing, for example, that "destabilising" speculation (in an adjustable peg regime) compels the monetary authorities to make the parity adjustment, thus accelerating the attainment of the new equilibrium.

Under a freely flexible exchange-rate system, the situation is different. First of all, the uncertainty about the future path of the exchange rate increases the risk and so tends to put a brake on speculative activity. But the fundamental issue consists in examining the destabilising or stabilizing nature of speculation, to which we now turn.

16.2.1 Flexible Exchange Rates and Speculation

According to one school of thought, speculation under flexible exchange rates is necessarily stabilizing. The basic argument of those supporting this claim is that speculation is profitable insofar as it is stabilizing: consequently, destabilising speculators lose money and must leave the market, where only stabilizing speculators, who make profits, remain. Here is a well-known quotation on the matter from Friedman (1953; p. 175): "People who argue that speculation is generally destabilising seldom realize that this is largely equivalent to saying that speculators lose money, since speculation can be destabilising in general only if speculators sell when the currency is low in price and buy it when it is high".

But the equation destabilization = losses (and so stabilization = profits) does not seem generally valid, as can be easily argued. Assume, for example, that the non-speculative exchange-rate (i.e., the one determined by fundamentals in the absence of speculation) follows a cyclically oscillating path (due for example to normal seasonal factors) around a constant average value. If speculators concentrate their sales of foreign currency immediately after the upper turning point (point *A* in Fig. 16.1) and their purchases immediately after the lower turning point (point *B*), an acceleration of both the downwards and the upwards movement (as shown by the broken lines) follows, with an explosive increase in the amplitude and/or frequency of the oscillations. The effect is destabilising, and it is self-evident that speculators, by selling the foreign currency at a higher price than that at which they purchase it, make profits.

This case shows that profitable destabilising speculation may cause a *speculative bubble,* a term used to describe an episode in which the price of a commodity (for example the Dutch tulips in 1634–1637, which gave rise to the tulipmania bubble) or asset (in our case the foreign exchange) displays an explosive divergence from its fundamental value (on speculative bubbles in general see Flood and Garber 1984).

One can also point out the case, already mentioned above, of bullish or bearish speculation (much like that which takes place in the Stock Exchange). This leaves

Fig. 16.1 An example of profitable destabilizing speculation

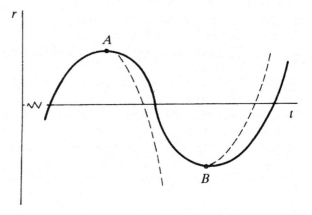

out of consideration any normal or average reference value such as that drawn in Fig. 16.1. Bullish speculators, well aware of the effects that their action will have on the price (in this case the exchange rate), buy foreign exchange with the aim of forcing an increase in its price; their intervention will be followed by other, less sagacious, operators hoping to make a killing by purchasing an asset (in this case, the foreign exchange) which is appreciating. When the exchange rate has depreciated sufficiently, the initial speculators sell the foreign exchange they bought in the first place (which may well give rise to a wave of sales and an abrupt fall in the price): they have certainly made profits and just as certainly destabilised the market.

Naturally in this example someone has to bear the losses and will leave the market, but will be replaced by someone else who wishes to have a go. The idea that professional speculators might on the average make profits while a changing body of "amateurs" regularly loses large sums was already considered by Friedman (1953; p. 175), but he dismissed it as unlikely. On the contrary, other writers (for example Glahe 1966) believe that the existence of professional and non-professional speculators is the norm (into the latter category fall those traders who occasionally speculate by exploiting the leads and lags of trade).

Another interesting example of profitable and destabilising speculation was given by Kemp (1963). For further considerations see Cutilli and Gandolfo (1963, 1972), and Gandolfo (1971); other examples of profitable and destabilising speculation are given by Ljungqvist (1992).

It should be stressed that it would *not* be correct to argue, from what we have said, that under flexible exchange rates speculation is *always* destabilising. It is, in fact, quite possible for speculators to behave as described by Friedman, in which case their stabilizing effect is self-evident. It has even been shown that there are cases in which speculation stabilizes an otherwise unstable flexible-exchange-rate regime (i.e., one which would have been unstable in the absence of speculation): see, for example, McKinnon (1983).

Thus we have seen that, whilst under an adjustable peg regime speculation is generally destabilising, under flexible exchange rates it may have either effect, so that the question we started from has no unambiguous answer. Even this apparently inconclusive answer is of importance, as it denies general validity to both the statements, that speculation is generally stabilizing and that it is generally destabilising.

Neither does the question seem solvable on the basis of the empirical evidence, which has given contradictory results. This is not surprising, if one bears in mind that the nature of speculation is, more than other economic phenomena, strictly related to political, historical and institutional circumstances, so that it becomes difficult if not impossible to obtain general answers, but it is necessary to consider each case separately. It should be stressed that, as the theory refers to freely flexible exchange rates, the empirical evidence derived from the managed float that has prevailed since 1973 is not relevant (we shall come back to the problem of speculation in the managed float in Sect. 17.3).

As regards *forward speculation*, we refer the reader to what we said in Sects. 2.6.1 and 7.4. We simply remind the reader here that forward speculation by itself does *not* give rise, at the moment in which it is undertaken, to any transfer of funds: only when the contract matures will there be a (positive or negative) excess demand for (spot) foreign exchange, deriving from the liquidation of the contract.

16.3 Speculative Attacks, Currency Crises, and Contagion

To improve our understanding of currency crises (collapse of a fixed exchange rate regime) as determined by speculative attacks, several analytical models have been developed in the literature. The pioneers were Krugman, and Flood and Garber (see also Jeanne and Masson 2000; Tavlas 1996). Krugman (1979) drew on the Salant and Henderson (1978) model in which the government uses a stock of an exhaustible resource to stabilize its price. Flood and Garber (1984) made the assumption of linearity and introduced the notion of "shadow floating exchange rate", namely the floating exchange rate that would prevail if international reserves had fallen to the minimum level and the exchange rate were allowed to float freely. Thanks to these assumptions they were able to derive an analytical expression for the collapse time, namely the point in time in which international reserves are exhausted (or drop below the minimum acceptable level), so that the monetary authorities have to abandon the fixed exchange rate.

It is now customary to classify the abundant literature on this topic into three categories: *first generation, second generation,* and *third generation* models (Flood and Marion 1999; Jeanne 2000). Further developments point to (a not yet well defined) fourth generation (see below, Sect. 16.3.4)

The framework of first generation models (also called *"exogenous policy"* models) is quite simple. The country engaged in maintaining a fixed exchange rate is also engaged in domestic expansionary policies, that are financed by expanding domestic credit. With fixed real and nominal money demand, domestic credit expansion brings about international reserve losses. However, money financing of the budget deficit has the higher priority and continues notwithstanding its inconsistency with the fixed exchange rate. Thus international reserves are gradually depleted until they reach a certain *minimum* level, after which they are exhausted in a final speculative attack, that compels the authorities to abandon the fixed exchange rate.

First generation models were applied to currency crises in developing countries (e.g., Mexico 1973–1982, Argentina 1978–1981), where the cause of the crisis could indeed be shown to be an overly expansionary domestic policy.

Second generation models (also called *"endogenous policy"* models or *escape-clause* models) introduce nonlinearities and the reaction of government policies to changes in private behaviour. For example, rather than given targets (the fixed exchange rate, the expansion of the domestic economy) the government faces a trade-off between the various targets (the exchange rate, employment, etc.). More generally, the commitment to the fixed exchange rate is state dependent (hence the name of endogenous policy models) rather than state invariant as in first generation models, so that the government can always exercise an escape clause, that is, devalue, revalue, or float. Nonlinearity is a source of possible multiple equilibria, some of which can be stable, others unstable. Second generation models can explain speculative attacks even when fundamentals are not involved, as they can take into account "bandwagon" effects (if somebody starts selling a currency, others will follow the example, without bothering to look at fundamentals), etc.

Second generation models were applied to currency crises in industrial countries (Europe in the early 1990s) and to the Mexican crisis of 1994, where speculative attacks seemed unrelated to economic fundamentals.

However, neither generation of models seems able to give an explanation of the Asian crisis that broke out in the late 1990s. The budget surpluses (or limited budget deficits) of the Asian economies prior to the crisis are brought as evidence against the fiscal origins of the 1997 crisis, hence against first generation models. The indicators of macroeconomic performance considered by second generation models (output growth, employment, and inflation) were far from weak: GDP growth rates were high, and inflation and unemployment rates low.

The fact that some of the currency crises at the end of the 1990s coincided with turbulences in the financial sector inspired a further development in the currency crises literature, what may be called *third generation* models. An element stressed by this literature is that currency crises cannot be seen in disjunction from banking crises. On the contrary, banking and currency are "twin" crises that should be modelled as interrelated phenomena: the interaction between the exchange rate and the domestic financial sector must be explicitly analyzed. Since the interacting variables turn out to be so many that it is difficult to include them all in an analytical model, the *indicators* approach (which uses a great number of indicators to gain insight in the actual chain of causation) is parallely developing.

A closely related topic is that of *contagious* speculative attacks, which means that speculative attacks tend to spread across currencies. A speculative attack against one currency may accelerate the collapse of a second currency not only when this latter collapse is "warranted", but also when the parity of the second currency is viable.

16.3.1 A First Generation Model

The simplest first generation model is due to Flood and Garber (1984), and combines a monetary equilibrium equation,

$$\frac{M(t)}{P(t)} = a_0 - a_1 i(t), \qquad a_1 > 0, \tag{16.1}$$

with uncovered interest parity

$$i(t) = i_f + \frac{\dot{r}(t)}{r(t)}, \tag{16.2}$$

where the foreign interest rate i_f is assumed constant. Note that the use of the *actual* rather than the *expected* exchange-rate variation in Eq. (16.2) implies perfect foresight.

In Eq. (16.1), M is the stock of monetary base (high-powered money) given by

$$M(t) = R(t) + D(t), \tag{16.3}$$

where R is the domestic currency value of international reserves, and D the domestic credit held by the domestic monetary authority. It is assumed that domestic credit grows at a positive constant rate μ, for example in order to finance increasing government expenditure:

$$\dot{D}(t) = \mu. \tag{16.4}$$

Finally, PPP is assumed, so that

$$P(t) = r(t)P_f, \tag{16.5}$$

where the foreign price level is assumed to be constant.

If we plug Eq. (16.5) into Eq. (16.1) we obtain

$$M(t) = \beta r(t) - \alpha \dot{r}(t), \quad \beta \equiv a_0 P_f - a_1 P_f i_f, \alpha \equiv a_1 P_f, \tag{16.6}$$

where β is assumed to be positive.

If the exchange rate is fixed at $\bar{r}$, from Eq. (16.2) we see that the domestic interest rate must equal the given foreign interest rate. Besides, from Eq. (16.5) we see that the domestic price level is a given constant when the exchange rate is fixed. Monetary equilibrium in Eq. (16.1) must hold in order to keep the interest rate at the value i_f, hence M must remain constant since P is constant. This constant value of M can be calculated using (16.6):

$$M(t) = \beta \bar{r}. \tag{16.7}$$

From (16.7) and ((16.3) we obtain

$$R(t) = \beta \bar{r} - D(t), \tag{16.8}$$

which shows that the only way to keep M constant in the face of an increasing D is to adjust international reserves, namely the economy runs a balance-of-payments deficit and the stock of reserves decreases at the same rate at which domestic credit increases, since from Eq. (16.8) we have

$$\dot{R}(t) = -\dot{D}(t) = -\mu. \tag{16.9}$$

With a finite amount of reserves the fixed exchange rate cannot survive forever, since the stock of reserves will be exhausted in a finite time. When the stock of reserves is exhausted (the result would not change if the monetary authorities set a lower limit below which they do not want their reserves to fall), the fixed exchange rate collapses and the exchange rate is left free to float. From Eqs. (16.9), (16.3), (16.7) we obtain (see the Appendix)

$$
\begin{aligned}
R(t) &= R_0 - \mu t, \\
D(t) &= D_0 + \mu t, \\
M(t) &= R(t) + D(t) = R_0 + D_0 = \beta \bar{r},
\end{aligned}
\tag{16.10}
$$

where R_0 is the given initial stock of reserves. Hence the stock of reserves will fall to zero at time

$$
t = R_0 / \mu.
\tag{16.11}
$$

However, the exchange rate collapses *before* this time, because there will be a final speculative attack that extinguishes any remaining stock of reserves. This is the collapse time.

To find the collapse time, Flood and Garber introduce the notion of *shadow floating exchange rate*, which is the floating exchange rate conditional on a collapse at any arbitrary time z.

If the exchange rate collapses at a time z because of a speculative attack, this means that the monetary authorities will have exhausted the stock of reserves. At the instant immediately following the attack (denoted by z_+), money market equilibrium requires

$$
M(z_+) = \beta r(z_+) - \alpha \dot{r}(z_+),
\tag{16.12}
$$

where $M(z_+) = D(z_+)$ since $R(z_+) = 0$.

It turns out (see the Appendix) that the shadow exchange rate $\tilde{r}(t)$ is given by

$$
\tilde{r}(t) = \frac{\alpha \mu}{\beta^2} + \frac{M(t)}{\beta}, \quad t \geq z,
\tag{16.13}
$$

where (since the reserves are exhausted)

$$
M(t) = D(t) = D_0 + \mu t.
\tag{16.14}
$$

We can now show that, since agents foresee the collapse, at the time of the collapse the shadow exchange rate must be equal to the pre-collapse fixed exchange rate $\bar{r}$. In fact, speculators profit by purchasing foreign exchange from the monetary authorities at the fixed exchange rate $\bar{r}$ immediately prior to the collapse and reselling it at market-determined exchange rate immediately after the collapse, which is the shadow exchange rate. If $\bar{r} > \tilde{r}$ the speculators would not profit by attacking, hence the fixed-exchange-rate regime survives. In the contrary case $(\bar{r} < \tilde{r})$ the speculators who purchase foreign exchange make a profit, but in this perfect foresight example nobody can anticipate the others, and perfect competition ensures that nobody makes a profit. Hence the foreseen attack occurs exactly when there are neither profits nor losses, namely when $\bar{r} = \tilde{r}$. Substituting this into Eq. (16.13), where we also substitute Eq. (16.14), and solving for $t = z$ (the collapse time) we obtain

$$z = \frac{\beta \bar{r} - D_0 - (\alpha \mu / \beta)}{\mu},$$

and finally, since $\beta \bar{r} - D_0 = R_0$ by the third equation in (16.10), we obtain

$$z = \frac{R_0 - (\alpha \mu / \beta)}{\mu}, \qquad (16.15)$$

which is smaller than (16.11).

Further research on first generation models is contained in Flood et al. (1996) and Flood and Marion (2000).

16.3.2 A Second Generation Model

We present here a second generation model (Sachs et al. 1996), where the speculative attack and the consequent currency crisis are not due to excessive money growth or other misaligned fundamentals, but to self-fulfilling panics.

Let us consider a small open economy where the government wants to maximize an objective function. The standard setting of such problems is to minimize a quadratic loss function; in the present model it is specified as

$$L = \frac{1}{2} \left(\alpha \pi_t^2 + x_t^2 \right), \qquad \alpha > 0, \qquad (16.16)$$

where π is the actual rate of exchange-rate devaluation (equal to the inflation rate) and x the flow of net tax revenue (taken as policy determined). The policy maker dislikes both inflation and (interpreting the preferences of the public) taxes. The fact that the rate of inflation and of exchange-rate devaluation coincide is due to the small open economy assumption coupled with the assumption of purchasing power parity. Under fixed exchange rates and assuming no inflation abroad, $\pi = 0$.

The optimization is carried out subject to the government budget constraint

$$Rb_t = x_t + \theta(\pi_t - \pi_t^e), \qquad \theta > 0, \qquad (16.17)$$

where R is the interest rate (equal to the given world interest rate owing to the small open economy assumption coupled with perfect capital mobility), b the inherited stock of net commitments of the consolidated government (including the Central Bank), and π_t^e the exogenously given expected rate of devaluation (inflation). The term $\theta(\pi_t - \pi_t^e)$ can be interpreted as inflation tax revenue (the fact that fully anticipated devaluation yields no revenue is a normalization condition).

It turns out that the optimal value of the loss function is

$$L^d(b_t, \pi_t^e) = \frac{1}{2}\lambda(Rb_t + \theta\pi_t^e)^2, \quad \lambda \equiv \frac{\alpha}{\alpha + \theta^2} < 1, \tag{16.18}$$

where the superscript ^{d} stands for "devaluing".

Let us now consider what happens when the policy maker has precommitted not to devalue, so that $\pi_t = 0$, and the constraint (16.17) becomes $Rb_t = x_t - \theta\pi_t^e$. The problem is now

$$\min L = \frac{1}{2}x_t^2$$

$$\text{sub } Rb_t = x_t - \theta\pi_t^e,$$

that of course admits of no trade off, hence the unique value of $x_t = Rb_t + \theta\pi_t^e$ and of the loss function

$$L^f(b_t, \pi_t^e) = \frac{1}{2}(Rb_t + \theta\pi_t^e)^2, \tag{16.19}$$

where the superscript ^{f} stands for "fixing".

Since $\lambda < 1$, it follows that $L^d < L^f$, hence it might seem that a government committed to a fixed exchange rate could obtain a better outcome by a surprise devaluation. However, things are not exactly like that. In fact, a government that reneges on the promise to maintain a fixed exchange rate incurs costs such as loss of face, voter disapproval, etc. These costs are not necessarily proportional to the devaluation or to macroeconomic variables, hence we shall take them as exogenously given at the amount $c > 0$.

Hence a government pegging the exchange rate will find it optimal to devalue if

$$L^d + c < L^f.$$

Using (16.18) and (16.19), this condition becomes

$$Rb_t + \theta\pi_t^e > k,$$
$$\text{where} \tag{16.20}$$
$$k \equiv (1 - \lambda)^{-1/2}(2c)^{1/2} > 0.$$

This shows that a devaluation will occur in equilibrium when expectations of devaluation are sufficiently high or inherited debt is too great.

We now turn to the private sector, where forward-looking atomistic agents act on the basis of rationally formed devaluation expectations. Starting with a fixed exchange rate, these agents understand the temptation summarized by Eq. (16.20), and act accordingly. Several outcomes of the government-agent interaction are possible, and Sachs et al. (1996; p. 270) summarize them under the following questions:

(1) When will the government *not devalue* regardless of π_t^e?
(2) When will the government *devalue* regardless of π_t^e?
(3) When will the government not devalue if $\pi_t^e = 0$, but devalue if π_t^e is sufficiently high?

To answer these questions we must preliminarily determine π_t^e, which may take on different values.

We first note that, since rational agents know condition (16.20), it will be rational for them to expect a devaluation ($\pi_t^e > 0$) if the accumulated stock of debt is sufficiently high, $Rb_t > k$. In this case, in fact, the condition is satisfied independently of π_t^e, and government will certainly devalue; hence $\pi_t^e = 0$ is not rational.

If the stock of debt is sufficiently low, namely falls short of the critical value, $Rb_t \leq k$, then $\pi_t^e = 0$ is certainly a rational-expectation equilibrium, since setting $\pi_t^e = 0$ in (16.20) this condition is *not* satisfied, hence the government will not devalue and the expectation will be validated. There is, however, another rational-expectation equilibrium implying devaluation. In fact, when $Rb_t \leq k$, condition (16.20) can be satisfied with $\pi_t^e > 0$. To find the relevant range we recall that in a deterministic context rational expectations imply (see the Appendix)

$$\theta \pi_t = \theta \pi_t^e = \frac{1-\lambda}{\lambda} Rb_t. \tag{16.21}$$

Inserting this into Eq. (16.20) yields the result that expectations of devaluation by the amount indicated in Eq. (16.21) will be validated if $Rb_t \geq \lambda k$. In the opposite case, $Rb_t < \lambda k$, condition (16.20) will not be satisfied and the government will not devalue regardless of π_t^e.

These results can be conveniently depicted in Fig. 16.2, adapted from Sachs et al. (1996; p. 271).

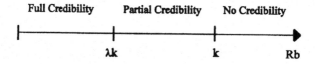

Fig. 16.2 Debt levels, crises, and multiple equilibria

For low levels of debt, such that $Rb_t < \lambda k$, no devaluation will take place regardless of π_t^e. This is the full credibility zone.

For high levels of debt, such that $Rb_t > k$, the government will inevitably devalue regardless of π_t^e. This is the no credibility zone.

For intermediate values of debt, such that $\lambda k \leq Rb_t \leq k$, there are two equally rational equilibria. If agents expect no devaluation, no devaluation will take place. If agents expect a precise devaluation (not any devaluation, but a devaluation of size $\theta \pi_t^e = [(1 - \lambda)/\lambda] Rb_t)$, this will also take place. This is the partial credibility zone, where there are self-fulfilling multiple equilibria. Which one will materialize depends on the "animal spirits" of agents.

This model shows in a simple way the importance of multiple equilibria and self-fulfilling outcomes. The fact that, for the same value of accumulated debt, both a no-devaluation and a devaluation self-fulfilling rational equilibrium can occur, shows that currency crises may arise independently of misaligned fundamentals. It also points out another important feature, that self-fulfilling outcomes cannot occur at any level of debt. Only levels that are sufficiently high, but not too high, can give rise to these outcomes. At too high debt levels a devaluation will inevitably occur, while at sufficiently low levels no devaluation will take place.

16.3.3 Third Generation Models

Although a few formal models of the third generation have been built (see, for example, Chang and Velasco 1998, Corsetti et al. 1999, Mendoza and Velasco eds., 2000), no consensus yet exists on common features. Third-generation models, in fact, emphasize the links between banking crises and currency crises (the *twin* crises), but these links are not clear. The chain of causation might, in fact, run either way. Problems of the financial sector might give rise to the currency crisis and collapse, for example when central banks print money to finance the bailout of domestic financial institutions in trouble (note that, if we abstract from the cause of the excessive money creation, the setting is the same as that of first-generation models). At the opposite side, balance-of-payments problems might be the cause of banking crises, for example when an initial foreign shock (say, an increase in foreign interest rates) in the context of a pegged exchange rate gives rise to a reserve loss. If this loss is not sterilized, the consequence will be a credit squeeze, hence bankruptcies and financial crisis. Finally, there is the possibility that currency and financial crises might have common causes, for example financial liberalization coupled with implicit deposit insurance followed by a boom financed by a surge in bank credit, as banks borrow abroad. When the capital inflows become outflows, both the currency and the banking system collapse.

Anzuini and Gandolfo (2000) suggest a classification based on the three main causes of the crisis set forth in the existing literature:

(A) *Moral Hazard*: the crisis is due to over-investment. Over-investment takes place because domestic firms feel as implicitly insured by the government any investment volume. Corsetti et al. (1999) develop a model utilizing some insights of Díaz-Alejandro (1985) and Krugman (1998). From the 1985 article they take the moral hazard interpretation and its economic consequences. The formalization of moral hazard as the crisis source is the insight of the 1998 article. Domestic firms behave as if their investments were insured by government. In case of need firms expect the government to step in and save them from bankruptcy. Foreign lenders are supposed to share that opinion and continue to lend at the same rate till debt reaches a critical fraction of international reserves.

    This moral hazard interpretation is not extravagant. Any authority announcement of a non-intervention policy is never fully credible ex-ante because agents know that policy intervention will be decided ex-post via a cost-benefits analysis.

(B) *Financial Fragility*: the crisis is due to a liquidity squeeze, caused by panic of foreign or domestic lenders who run on domestic financial intermediaries. Chang and Velasco (1998) assume the liquidity problem to provoke a premature liquidation of intermediaries' assets. Liquidation has real effects because assets prematurely liquidated loose part of their value. This model is inspired by the bank crisis literature pioneered by Diamond and Dybvig (1983) and represents an open economy version of that paper.

(C) *Balance Sheet*: the crisis is due to the firms' foreign debt blowing up following devaluation. The model (Krugman 1999), is developed analyzing the movement of Asian macroeconomic variables and the plan implemented by the International Monetary Fund (IMF). "If there is a statistic that captures the violence of the shock to Asia most dramatically, it is the reversal in the current account" (Krugman 1999; p. 9). For example, Thailand with a 10 % pre-crisis deficit, had to move its current account to an 8 % surplus. This was necessary because of the unexpected and huge capital outflow. An increase in net exports can be obtained reducing imports and\or increasing exports. In the short run this means exchange rate devaluation and\or economic activity reduction. This is exactly what happened in Asia. The exchange rate, few weeks earlier stably anchored to the US dollar, lost in few days almost 50 % of its value. Economic activity fell into a deep recession never experimented by those economies.

The other element highlighted is the IMF plan. During all crises the IMF's main concern has been exchange rate stabilization. That policy was due to the necessity of avoiding the explosion of high foreign debt. The exchange rate defence was to be implemented by rising the interest rate in the short run, and reorganizing (liberalizing) financial structure in the medium term. Had the plan succeeded, the international creditors confidence would have been restored, the interest rate would have been reset to normal level and the Asian economies would have boomed again with a stronger (more similar to the western standard) financial system. But something went wrong. Stabilization policy failed and did not prevent the materialization of a deep recession.

16.3.3.1 A Third Generation Model

We shall consider the Krugman (1999) model that, although less formalized than others, offers useful insights. This model falls into category (c) above, and describes a small open economy that produces a single homogeneous good using capital and labour according to a Cobb-Douglas production function

$$y_t = K_t^\alpha L_t^{1-\alpha}. \tag{16.22}$$

Capital is assumed to last only one period, so that period t's capital is equal to investment carried out in period $t-1$. Residents are divided into two classes: workers who consume all their income and capitalists-entrepreneurs who only save and invest all their income. Commodity y is not a perfect substitute of foreign goods, and there is a unitary elasticity of substitution between home and foreign goods. Hence a constant fraction μ of both consumption and investment is spent on foreign goods (imports), and the remaining fraction $(1 - \mu)$ on domestic goods.

Given the small economy assumption, the domestic economy's exports are exogenously given. More precisely, the value (in terms of foreign goods) X of domestic exports X is exogenously given, and has a value pX in terms of domestic goods, where p is the terms of trade (the relative price of foreign goods or real exchange rate).

In equilibrium, supply (output) of the domestic good equals its demand, that is

$$y_t = (1 - \mu)C_t + (1 - \mu)I_t + p_t X. \tag{16.23}$$

Bearing in mind that workers spend all their income (which is $(1 - \alpha)y_t$ given the Cobb-Douglas production function) we can rewrite (16.23) as

$$y_t = (1 - \mu)I_t + (1 - \mu)(1 - \alpha)y_t + p_t X, \tag{16.24}$$

which gives the real exchange rate as

$$p_t = \frac{[1 - (1 - \mu)(1 - \alpha)]y_t - (1 - \mu)I_t}{X}. \tag{16.25}$$

As regards the determination of investment, Krugman observes that the ability of entrepreneurs to invest may be limited by their borrowing ability and, more specifically, that lenders impose a limit on leverage, so that entrepreneurs can borrow at most θ times their wealth

$$I_t \leq (1 + \theta)W_t, \tag{16.26}$$

where wealth is defined as

$$W_t = \alpha y_t - D_t - p_t F_t. \tag{16.27}$$

In fact, entrepreneurs own all domestic capital, which by the assumptions made above equals their share in domestic output; they may also own claims on, and/or have debt to, foreigners. Such claims/debts are partly denominated in terms of domestic goods (D_t denotes *net debts* of this type, of course $D_t < 0$ means that claims are greater than liabilities), partly in terms of foreign goods (F_t denotes net debts of this type with a value of $p_t F_t$ in terms of domestic goods). Hence Eq. (16.27).

The constraint (16.26) need not be binding: if it is true that entrepreneurs always invest all their wealth, it is not necessarily true that they decide to borrow up to the maximum. In fact, investment decisions are taken by comparing the real return on domestic investment i, that depends on the production function) with the real return on investment abroad (i^*, which is the given foreign real interest rate). A way of doing this is to compare i^* with the return obtained by converting foreign into domestic goods at time t at price p_t, then converting the result obtained in period $t + 1$ back into foreign goods at price p_{t+1}, namely

$$(1 + i)(p_t/p_{t+1}) \geq 1 + i^*, \tag{16.28}$$

which expresses the statement that the return on domestic investment must be at least as large as the return on foreign investment.

Finally, there is the assumption that investment cannot be negative

$$I_t \geq 0. \tag{16.29}$$

The Crisis

According to this model, a decline in capital inflows may cause a crisis because it affects the real exchange rate [Eq. (16.25)] and consequently, the balance sheet of domestic entrepreneurs [Eq. (16.27)]. This reduces the ability of domestic entrepreneurs to borrow and hence to invest [Eq. (16.26)], further reducing capital inflows, and so forth.

More precisely, suppose that the offer of credit depends on what lenders *believe* will be the value of the collateral of borrowers. This value depends on the real exchange rate, because some debt is denominated in foreign goods, and hence on the actual level of borrowing. Thus a rational expectations equilibrium will be a set of self-fulfilling guesses: the *actual* level of investment that will take place given the credit offers will be equal to the *expected* level of investment implicit in those credit offers.

To show this we begin by deriving the wealth-investment relationship Since wealth depends on the real exchange rate [Eq. (16.27)], which in turn depends on I_t

[Eq. (16.25)], it is easy to calculate (for notational simplicity from now on we omit the time subscript)

$$\frac{dW}{dI} = \frac{dW}{dp}\frac{dp}{dI} = \frac{(1-\mu)F}{X}. \tag{16.30}$$

We now define the *financeable* level of investment (I_f) as the level of investment that would occur if the leverage constraint (16.26) were binding, namely

$$I_f = (1+\theta)W. \tag{16.31}$$

From Eqs. (16.30) and (16.31) we immediately obtain

$$\frac{dI_f}{dI} = \frac{dI_f}{dW}\frac{dW}{dI} = \frac{(1+\theta)(1-\mu)F}{X}. \tag{16.32}$$

The magnitude of dI_f/dI is crucial. If it is lower than unity, a highly productive economy may have problems in the adjustment of the capital stock due to financing constraints, but no crisis will occur. On the contrary, when $dI_f/dI > 1$ there may be multiple equilibria and crisis. A possibility is shown in Fig. 16.3, adapted from Krugman (1999). The expected level of investment (on the horizontal axis), via its effect on the real exchange rate and hence on balance sheets, determines how much credit is extended to firms. The resulting level of actual investment is plotted on the vertical axis. At low levels of expected investment firms are bankrupt, and cannot invest at all—the binding constraint is the non-negativity constraint (16.29). At high levels of expected investment the financing constraint (16.26) is not binding, and investment is determined by the rate-of-return constraint (16.28). In the intermediate range, where the schedule is steeper than 45° ($dI_f/dI > 1$), actual investment is constrained by financing.

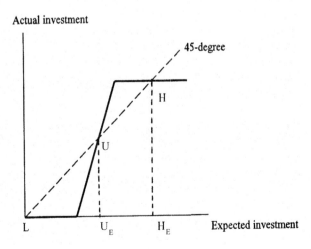

Fig. 16.3 Balance sheets and financial crises

There are three equilibrium points in this model. The two extreme points L and H are stable, while the intermediate point U is unstable, for if lenders become slightly more optimistic or pessimistic (expected investment moves slightly to the right or to the left of U_E), actual investment will start to increase (decrease) along the schedule toward H or L respectively, due to the circular loop from investment to real exchange rate to balance sheets to investment.

At the high-level equilibrium H, investment takes place so as to equalize the domestic and foreign rate of return. At the low level equilibrium L, lenders do not believe that entrepreneurs have any collateral and hence do not offer funds; this implies a depreciated real exchange rate which in turn means that entrepreneurs are actually bankrupt (thus validating the lenders' bad opinion).

The peculiarity of this model is that the stable equilibrium points are *locally stable but globally unstable*. Take for example H : a sufficiently great displacement of expected investment, from H_E to the left of U_E, caused by a wave of pessimism will start a cumulative motion towards L. Thus we have an explanation of the crisis: if, for whatever reason, lenders become suddenly pessimistic, the result is a collapse from H to L. This does not mean that previous investments were unsound, because the true problem is financial fragility.

Equation (16.32) shows the factors at work: high leverage (θ high), large foreign currency debt relative to exports (F/X high), low marginal propensity to import (μ low), are all possible causes of $dI_f/dI > 1$, which is the condition for a financial crisis possibly to occur.

16.3.4 Towards Fourth Generation Models?

Krugman (2001) conjectured about a future fourth-generation crisis model, which need not be a currency crisis model, but might be a more general financial crisis model. In this latter model asset prices other than the exchange rate would play a fundamental role, implying that there might be other elements from which the arrival of an unanticipated and costly currency crisis could be detected. Apart from the three generations mentioned before, there is an increase in additional approaches to the investigation of the causes of a currency crisis that can be named *fourth-generation* or so called *institutional* models (Breuer 2004). They extend the earlier literature by identifying features of the institutional environment and coordination failures to explain currency crises. Actually, weak institutions worsen the problems associated with risk and uncertainty and contribute to a misallocation of resources thereby setting the stage for currency crises. Institutions are informative as they can signal market agents about the future economic fundamentals, and shape market expectations. Furthermore, market failure in international capital markets, and distortion in domestic financial markets, are identified as coordination issues.

Fourth generation models highlight the roles of transparency and supervision over the financial system, rule of law and contract enforcement, protection of shareholder and creditor rights, regulatory frameworks, and the socioeconomic environment. Moreover, the models also considered political variables such as democracy and political instability, and sociological variables such as corruption, trust, culture, and ethnicity. Alesina and Wagner (2006) found that countries with poor institutional quality related to the business environment and the socio-political environment, have difficulty in maintaining an announced peg and are more likely to abandon it. Calvo and Mishkin (2003), in re-assessing the debate about fixed versus floating exchange rates, argues that deeper institutional features related to fiscal, financial, and price stability are crucial to macroeconomic stability and the avoidance of crises. Breuer (2004) noted that models of currency and banking crises share parallel dynamics and stressed that features of the institutional environment affect the stage for the build-up of macroeconomic imbalances, which subsequently give rise to banking problems. Poor institutional factors appear to be the underlying cause for unsustainable policies, excessive borrowing and lending, hyperinflation, among others. Corruption, government instability, weak law and order, as well as a de facto exchange rate regime are some of the strongest results found to affect the probability of crisis with regard to the issue of institutions (Shimpalee and Breuer 2006). Leblang and Shanker (2006) stress the importance of the linkage between political institutions, and the behaviour of currency speculators and speculators' expectations, to currency crises. The authors show how politics can cause crises even if we hold fundamentals constant via the channel of speculators' expectations of one another's behaviour.

Although economic factors also play a role in fourth generation models, the institutional factors set the conditions for economic outcomes. Many databases that quantify institutional factors have become available recently, enabling more research to be carried out.

16.3.5 The Indicators Approach: Can Crises Be Forecast?

In practice it seems possible to use *indicators* to obtain insights in the actual chain of causation of the crisis. The empirical literature concerned with predicting crises has been focusing on developing early warning systems (EWS) which seek to predict future crises. The main body of empirical evidence on forecasting currency crises builds on the seminal works by Kaminsky et al. (1998), Kaminsky and Reinhart (1999) and Frankel and Rose (1996) who monitor a large set of monthly indicators that signal a crisis whenever they cross a certain threshold.

The main macroeconomic indicators proposed are (Kaminsky et al. 1998; Kaminsky and Reinhart 1999):

(1) indicators associated with *financial liberalization* (the M2 multiplier, the ratio of domestic credit to nominal GDP, the real interest rate on deposits, the ratio of lending-to-deposit interest rates);
(2) *other financial indicators* (excess real M1 balances, real commercial-banks deposits, the ratio of M2 to foreign exchange reserves);
(3) indicators related to the *current account* (percent deviation of the real exchange rate from trend, the value of imports and exports, the terms of trade);

(4) indicators associated with the *capital account* (foreign exchange reserves, the domestic-foreign real interest-rate differential);

(5) indicators of the *real sector* (industrial production, an index of equity prices);

(6) *fiscal* variable (the overall budget deficit as a percent of GDP).

Using these indicators and an appropriate statistical methodology, Kaminsky and Reinhart (1999) examine 26 banking crises and 76 currency crises in 20 countries in the period 1970—mid-1995, finding that "problems in the banking sector typically precede a currency crisis-the currency crisis deepens the banking crisis, activating a vicious spiral; financial liberalization often precedes banking crises. The anatomy of these episodes suggests that crises occur as the economy enters a recession, following a prolonged boom in economic activity that was fuelled by credit, capital inflows, and accompanied by an overvalued currency".

An important point made by these authors is that the alleged novelty of the Asian crisis with respect to previous crises (the novelty consisting in the fact that this crisis occurred in the context of immaculate fiscal and economic fundamentals) is not really a novelty, because many of its features and antecedents were common to several previous crisis episodes in Latin America, Europe, and elsewhere: "Consider an economy that had successfully stabilized inflation, enjoyed an economic boom, and was running fiscal surpluses. However, this economy had liberalized its capital account and its domestic financial sector amidst an environment of weak regulation and poor banking supervision. Banking-sector problems emerged and intensified, eventually undermining the ability of the central bank to maintain its exchange-rate commitment. While this profile fits Asia rather well, this was Díaz-Alejandro's description of the antecedents to the fierce Chilean crisis of 1982. At the roots of the melt-down of the Thai baht, Korean won, and Indonesian rupiah lay systematic banking problems. Thus, it would appear that we can only consider these crises as a new breed if we ignore the numerous lessons history offers" (Kaminsky and Reinhart 1999; 494–497). The conclusion is a case for strong banking regulation and supervision to help prevent these crises.

The indicators approach seems to *explain* the Asian crisis well, but the question is, would it have been able to *predict* the crisis? In other words, suppose we had the Kaminsky-(Lizondo)-Reinhart model available in 1997 and applied it: could we have forecast the Asian crisis? Anzuini and Gandolfo (2003) perform this *ex post* forecasting exercise on Thailand's crisis and find that only 27 % of the indicators issued a crisis signal in the 24 months before the crisis, a result that would have induced to assign a small probability to the occurrence of the crisis. They suggest an alternative approach based on speculators' expectations and find that speculators did correctly forecast the crisis.

While currency crises were the subject of investigation in these pioneering studies, the recent literature has tried to include more types of costly crises, including banking crises, debt crises, and financial distress. In addition, recent research has strived to improve early warning models by developing new methodological techniques and employing more extensive data sets. Existing EWS models differ sharply from one another in terms of the empirical procedures adopted. They can be broadly grouped into the following approaches:

(a) linear regression or limited dependent variable probit/logit techniques (e.g. Berg and Pattillo 1999, Demirgüç-Kunt and Detragiache 2000, Bussiere and Fratzscher 2006);
(b) non-parametric, or crisis signal extraction approach (e.g. Kaminsky et al. 1998, Kaminsky and Reinhart 1999);
(c) qualitative and quantitative analysis of the behaviour of various variables around crisis occurrence, splitting countries into a crisis group and non-crisis control group (e.g. Berkmen et al. 2012, Blanchard et al. 2010, Frankel and Saravelos 2012);
(d) use of innovative techniques to identify and explain crisis incidence, encompassing the use of binary recursive trees to determine leading indicator crisis thresholds, artificial neural networks and genetic algorithms (e.g. Lin et al. 2008, Elsinger et al. 2006);
(e) agent-based models to explain the behaviour of the economic systems (e.g. Thurner et al. 2012). Among the numerous contributions to the wide literature on early warning indicators we also mention the following.

Demirgüç-Kunt and Detragiache (2005) reviewed the two basic methodologies adopted in cross-country empirical studies, the signals approach and the multivariate probability model, to study the determinants of banking crises. They found that empirical models have been more useful in identifying factors associated with the occurrence of banking crises than in predicting the occurrence of crises out-of-sample, reflecting the fact that most empirical models were not conceived as forecasting tools.

Drehman and Juselius (2013) evaluated the relative performance of different EWS from the perspective of a macro-prudential policy maker. From this point of view, they suggested the ideal characteristics that an EWS should have and translated these requirements into statistical evaluation criteria. They showed that the credit-to-GDP gap is the best indicator at longer horizons, whereas the debt service dominates at short horizons.

Rose and Spiegel (2012) performed an extensive investigation into over sixty potential variables that could help explain cross-country crisis incidence. They did not find consistently statistically significant variables, pointing out to the difficulty of finding significant leading indicators to explain the cross-country incidence of the 2008–2009 financial crisis.

Berkmen et al. (2012) took into account the change in growth forecasts by professional economists before and after the 2007–2008 global financial crisis. They found that countries with more leveraged domestic financial systems and rapid credit growth tended to suffer larger downward revisions to their growth forecasts, while international reserves did not play a significant role.

Lund-Jensen (2012) showed that the level of systemic risk crucially depends on several risk factors: banking sector leverage, credit-to-GDP growth, changes in banks' lending premium, equity price growth, increasing interconnectedness in the financial sector, and real effective exchange rate appreciation.

Other indicators proposed are, among others: growth and terms of trade as robust leading indicators of banking crises (Davis and Karim 2008); a sharp increase in private indebtedness for banking crises (Reinhart and Rogoff 2011); growth in global credit for costly asset price bubbles (Alessi and Detken 2011); a large real GDP decline for debt crises (Levy Yeyati and Panizza 2011); the level of central bank reserves and real exchange rate appreciation for events such as the recent financial crisis (Frankel and Saravelos 2012), and a combination of several indicators into composite indices for banking crises (Borio and Lowe 2002).

Finally it should be mentioned that a growing number of international financial institutions and central banks are using EWS models in their surveillance activities. Similarly, several investment banks have developed in-house EWS models aimed at providing foreign exchange trading advice to their clients. The IMF (IMF 2002 and 2013) has played a leading role in developing EWS models as part of its wider focus on crisis prevention, to improve its ability to assess the vulnerability of its member countries to currency and financial crisis. These models typically have an empirical structure that attempts to forecast the likelihood of a certain type of "crisis" using factors such as country fundamentals, developments in the global economy and global financial markets, political risks.

16.3.6 Contagion

The fact that speculative attacks on different currencies tend to be temporally correlated has given rise to the "contagion" literature, according to which speculative attacks and the ensuing currency crises are like infectious diseases: they tend to spread contagiously. This is reflected in the very names given to various crises, e.g. the Asian *Flu* in 1997, the Russian *Virus* in 1998, the Brazilian *Sneeze* in 1999. However, diseases do not only spread to disease-prone persons but also to healthy people: can we carry the similitude as far as to state that speculative attacks against a misaligned currency tend to spread not only to other misaligned currencies but also to apparently sound currencies? In other words, contagion is a "disease" but contagion also refers to the "transmission" of a disease. This question has enormous policy implications, because an affirmative answer would warrant the bailout (by international organisations, other governments or groups of governments like the G-7) of any country under speculative attack, so as to prevent contagion to other (sound) countries. It goes without saying that the study of contagion is closely related to the model of currency crisis one has in mind.

Despite the countless papers dedicated to analyse how shocks propagate internationally, there is little agreement on a formal definition of what actually constitutes contagion. The World Bank (http://go.worldbank.org/JIBDRK3YC0) defines contagion in different ways:

Broad Definition: "Contagion is the cross-country transmission of shocks or the general cross-country spillover effects. Contagion can take place both during "good" times and "bad" times. Then, contagion does not need to be related to crises. However, contagion has been emphasized during crisis times".

> *Restrictive Definition*: "Contagion is the transmission of shocks to other countries or the cross-country correlation, beyond any fundamental link among the countries and beyond common shocks. This definition is usually referred as excess co-movement, commonly explained by herding behavior".
>
> *Very Restrictive Definition*: "Contagion occurs when cross-country correlations increase during "crisis times" relative to correlations during "tranquil times."

Some researchers have proposed using the more specific terms "*shift contagion*", when there is a significant increase or "shift" in cross-market linkages after a shock to an individual country (Dornbusch et al. 2000; Forbes and Rigobon 2002), and "*pure contagion*", a residual category. In the latter case any form of contagion is completely unrelated not only to changes in fundamentals but also to the level of fundamentals, be they country-specific or global. Pure contagion may arise from self-fulfilling loss of confidence, from irrational herding behaviour (Chari and Kehoe 2003), or from wealth effects for investors, triggered by capital losses in the country which originated the crisis (Kodres and Pritsker 2002; Kyle and Xiong 2001).

The debate on how to define contagion has important implications for the measurement of contagion and for the evaluation of policy responses. Furthermore, there are many different theories why contagion can occur. This literature can be divided into two broad groups: fundamental causes (including common shocks, trade linkages and financial linkages) and investors' behavior (including liquidity problems, informational asymmetries, market coordination problems, and investor reassessment). Another important issue refers to the channels through which shocks are transmitted across countries. A vast literature models and tests potential sources of contagion (see Claessens et al. 2011; Karolyi 2003; Forbes 2004 and Allen et al. 2009 for reviews of the various definitions and taxonomies of contagion channels and ways to group them). Let us mention the main channel categories: (A) trade, (B) banks, (C) portfolio investors, and (D) wake-up calls, on which more below.

(A) *Trade*

In one of the first empirical studies on contagion, Eichengreen et al. (1996), using thirty years of panel data from twenty industrialized countries, found evidence of contagion; they also found that contagion spreads more easily to countries that are tied by close *international trade linkages* than to countries in similar macroeconomic circumstances.

This appears to be consistent with the formal model of Gerlach and Smets (1995), who use a two-country version of the Flood and Garber model (hence a first generation model: see above, Sect. 16.3.1) and show that, with excessive credit creation in both countries, the collapse of the first currency accelerates the collapse of the second currency. The reason is that the collapse of the first currency implies an appreciation of the (still pegged) second currency, which leads to a fall in income and prices in the second country. As a consequence the demand for money in the second country falls, which causes a loss of foreign exchange reserves that would not otherwise have occurred. The lower foreign exchange reserves reduce the ability to withstand a speculative attack and accelerate the collapse of the second currency. This shows the spreading of the contagion to disease-prone currencies (in both countries there is excessive credit creation).

More interestingly, these authors also show that *contagion may spread to sound currencies*, namely that a pegged exchange rate that is sustainable in the absence of an attack on another currency may be attacked and collapse in the case of a successful attack on that other currency. This is consistent with the evidence of the European Monetary System turmoil in 1992–1993. "The attack on the United Kingdom in September 1992 and sterling's subsequent depreciation are said to have damaged the international competitiveness of the Republic of Ireland, for which the UK is the single most important export market, and to have provoked the attack on the punt at the beginning of 1993. Finland's devaluation in August 1992 was widely regarded as having had negative repercussions for Sweden, not so much because of direct trade between the two countries but because their exporters competed in the same third markets. Attacks on Spain in 1992–1993 and the depreciation of the peseta are said to have damaged the international competitiveness of Portugal, which relies heavily on the Spanish export market, and to have provoked an attack on the escudo despite the virtual absence of imbalances in domestic fundamentals" (Eichengreen et al. 1996, 465). The trade channel has also been emphasized by Glick and Rose (1999), who deemed that Hong Kong, Indonesia, the Philippines and Thailand were affected by the "Mexican crisis" in 1994–1995, while Argentina, Brazil, the Czech Republic, Hungary and South Africa have been among the victims of the Asian Crisis. For empirical evidence on trade channels see also Forbes (2002, 2004), Burstein et al. (2008) and Claessens et al. (2011).

(B) Banks

However, trade linkages are not the only channel of transmission of contagion. It seems difficult, for example, to argue that the so-called *Tequila crisis* (the pressure put on Latin American and East Asian currencies following the collapse of the Mexican peso in 1994) was caused by trade links alone. Argentina and Brazil traded extensively with Mexico, but the same was not true of Hong Kong, Malaysia, Thailand. Similarly, trade links are not sufficient to explain the spread of crisis from Thailand to other East Asian countries in 1997. Furthermore, contagion following the Russian crisis cannot be attributed to trade channels, since Russia is insignificant both as a trade competitor in third nations and as a destination for exports. Hence financial channels must also be taken into account.

A partially overlapping classification of the channels of contagion is due to Masson (1999a,b; 587–88), who suggested the following channels:

(i) *monsoonal effects*, which emanate from the global environment (in particular, from policies in industrial countries), and sweep over all developing countries to a greater or lesser extent;

(ii) *spillover effects,* which explain why a crisis in one country may affect other emerging markets through linkages operating through trade, economic activity, or competitiveness;

(iii) *jumps between multiple equilibria,* which is a residual category: if the first two do not explain the coincidence of crises, it is argued that there is a role for self-fulfilling expectations in which sentiment with respect to a given country changes purely as a result of a crisis in another country.

We have already shown that trade links alone cannot account for contagion in the Tequila, Asian, Russian crises. In addition, the magnitude and timing of developments in industrial countries, such as the tightening of US monetary policy in 1994 and the appreciation of the dollar in 1995–1996, cannot plausibly explain these contagions (Masson 1999b; 588). Thus monsoonal and spillover effects are not sufficient as an explanation, hence the need for effects of the third type. Masson (1999b) builds a simple balance-of-payments model in which all three channels can be taken into account, with particular attention paid to self-fulfilling devaluation or default expectations. The model gives rise to multiple equilibria, and shows that there is a range of fundamentals (international reserves and external indebtedness) where changes in such self-fulfilling expectations—perhaps triggered by a crisis elsewhere—are possible.

The Masson model is clearly related to second generation models of currency crises (see above, Sect. 16.3.2), like the Gerlach-Smets model is related to first generation models. An attempt to empirically study contagion in the context of third generation models has been carried out by Van Rijckeghem and Weder (2001). They focus on spillovers resulting from specific financial linkages, more precisely on spillovers channelled through the so-called *common bank lender effect,* where the common bank lender is identified by the country that lent most to the first country in crisis in each of the major crises. The common creditor in the Mexican crisis was the United States; in the Asian crisis, Japan; and in the Russian crisis, Germany. With large bank exposures, potential losses are large, hence the need to restore capital asset ratios, meet margin calls, or readjust risk exposure, thus accounting for contagion. The authors find that spillovers through this channel are important.

Similar results have been found by Caramazza et al. (2000, 2004), who carry out an extensive empirical investigation of all the factors (external, domestic, and financial weaknesses as well as trade and financial linkages) potentially relevant in inducing financial crises. They find that the indicators of vulnerability to international financial spillovers (the common creditor effect) and of financial fragility (reserve adequacy) are highly significant and appear to explain the regional concentration of these crises.

Allen et al. (2009) provide a survey of the theoretical and empirical literature on financial crises and contagion. The relationship between asset price bubbles, bank, real estate, crises and contagion is discussed at length. The role of banks in causing contagion can be aggravated by their close relationship to the solvency of their sovereign, their high degree of leverage, and their extensive interconnections. Allen et al. (2012) compare the impact of information contagion on systemic risk across asset structures, whereby adverse news about aggregate solvency of the banking system lead to runs on multiple banks. As shown in Greenwood et al, (2015), when a bank experiences a negative shock to its equity, a natural way to return to target leverage is to sell assets, and the sales by one bank's impact on other banks with common exposures. The authors show how this contagion effect adds up across the banking sector, and how it can be empirically estimated using balance sheet data. Van Wincoop (2013) investigates the case of the 2007–2009 financial crises: the crisis originating in the United States witnessed a drop in asset prices and output at least as large in the rest of the world. In their opinion, this could have been the result of transmission through leveraged financial institutions. The paper highlights the various transmission mechanisms associated with balance sheet losses.

However, several developments in the global financial system might have amplified the transmission of the shocks. Financial globalization and securitization led to a complex net of interconnections among financial institutions across economies.

(C) Portfolio Investors

An extensive literature considers *portfolio investors* as one important financial channel for contagion. Models show how increased risk aversion after a negative shock or informational asymmetries could cause investors to sell assets across countries and "overreact". Namely, portfolio rebalancing mechanisms are crucial in explaining contagion patterns, even in the absence of common macroeconomic fundamentals (Masson 1999a,b; Kodres and Pritsker 2002; for a survey see Gelos 2011). Goldstein and Pauzner (2004) showed that a crisis in one country reduces agents' wealth, which makes them more averse to the strategic risk associated with the unknown behavior of other agents in the other country. This increases agents' incentive to withdraw their investments in the latter, that is, the mechanism that triggers contagion originates in the wealth effect (see also Pavlova and Rigobon (2008) for portfolio constraints). One successful speculative attack may lead to rampant collapse of confidence and likely to a series of attacks on other currencies: Taketa (2004), for example, showed how a currency crisis can spread from one country to another even when the countries are unrelated in terms of economic fundamentals. He stated that the propagation mechanism lies in each speculator's private information about his own type and learning behaviour about other speculators' types. Empirical findings in Dasgupta et al. (2011) show that crises tend to spread among economies which are institutionally similar. Speculators believe that a rare crisis in one country may signal systematic weaknesses in other countries that bear institutional similarities to the originating country, because these countries are supposed to operate with the same rules of the game, even though their economic fundamentals may not be correlated. Lee et al. (2007) examine whether the South-East Asian Tsunami of 2004, as an external and unpredictable shock (the paper spotlights the role of rarity: rare crises can shake the public confidence dramatically), influenced the stability of the correlation structure in international stock and foreign exchange markets. Their results indicate that no international stock market suffered contagion, that international foreign exchange markets displayed contagion for one to three months after the Tsunami, and that contagion effects were stronger in developing financial markets than in developed ones. Raddatz and Schmukler (2012), using micro-level data on mutual funds from different financial centers investing in equity and bonds, analysed how investors and managers behave and transmit shocks across countries. Their results show that the volatility of mutual fund investments is quantitatively driven by both the underlying investors and fund managers who respond to country returns and crises and adjust their investments substantially, generating large reallocations during the global financial crisis. There are studies focusing on international equity market contagion. For instance, Tong and Wei (2011) find that the average decline in stock prices, during the 2007–2009 financial crisis, in a sample of 4000 firms in 24 emerging countries was more severe for those firms intrinsically more dependent on external finance (in particular on bank lending and portfolio flows).

(D) Wake-Up Calls

Another channel (closely related to the others) by which contagion can occur is *"wake-up calls"*. The *"wake-up call hypothesis"* states that a crisis initially restricted to one market or country provides new information that may cause investors to reassess the vulnerability of other market or countries, that is, for example, when additional information or a reappraisal of one country's fundamentals leads to a reassessment of the risks in other countries. Under the wake-up call hypothesis, for example, countries without trade or banking linkages to the country where the crisis originates may experience contagion, and the incidence or extent of their exposure depends on the strength of their local fundamentals and institutional factors. However wake-up calls involve a wide range of reassessment—including not only the macroeconomic, financial or political characteristics of the country—but also the functioning of financial markets and the policies of international financial institutions.

The term "wake-up call" originates from Goldstein (1998), who coined this term to capture the sudden awareness of risks in Asian financial systems during the 1997–1998 crisis and, in particular, to explain contagion from Thailand (a relatively small and closed economy) to other Asian countries. He argues that the other countries were affected by the same structural and institutional weaknesses as Thailand, but investors ignored those weaknesses. Such a behaviour is consistent with forms of rational inattention theory (Wiederholt 2010), according to which, given costs in acquiring and processing information, rational agents could optimally choose to ignore some information. Basu (2002) focused on debt markets where the ability to pay the debt depends on the interaction between an imperfectly known risk factor, common across a number of countries, and country-specific economic fundamentals and institutional factors. Investors have a prior about the common risk factor. Default in one country, the wake-up call, prompts investors to revise their priors, not only for the country in question, but for all countries sharing the unobserved common risk factor. Ahnert and Bertsch (2013) study contagion in a global game of speculative currency attacks under incomplete information. Here a successful attack also acts as a wake-up call to investors inducing them to acquire costly information about their exposure to the country attacked. Van Rijckeghem and Weder (2003) find important common bank lender effects during the Mexican and South-East Asian crises, as a channel of contagion, but view the Russian crisis as the outcome of a wake-up call in emerging markets. The wake-up call hypothesis is supported by empirical evidence, such as Bekaert et al. (2014) who, analysing equity markets during the global financial crisis of 2007–2009, identify wake-up calls as the key driver of contagion. Karas et al. (2013) find a wake-up call effect during the Russian banking panic of 2004. Giordano et al. (2013) find empirical evidence for contagion based on the wake-up call of the Greek crisis of 2009–2010.

To sum up, the wake-up call theory of contagion explains how currency crises, bank runs, and debt crises spread across regions without a common investor base, correlated fundamentals or interconnectedness.

16.4 Appendix

16.4.1 A First-Generation Model

We examine the model by Flood and Garber 1984 (see also Flood and Garber 1994, Flood et al. 1996, Flood and Marion 2000).

Let us consider the equation [see the text, Eq. (16.9)]

$$\dot{R}(t) = -\dot{D}(t) = -\mu. \tag{16.33}$$

Integrating both sides we obtain

$$\int \dot{R}(t)dt = A - \int \mu dt,$$

where A is an arbitrary integration constant. Thus we have

$$R(t) = A - \mu t,$$

where A turns out to equal R_0, given that $R(t) = R_0$ for $t = 0$.

As regards the shadow exchange rate, the differential equation

$$M(t) = \beta r(t) - \alpha \dot{r}(t), \quad t \geq z \tag{16.34}$$

is a non-homogeneous first-order linear equation with constant coefficients. Its solution is given by the solution of the corresponding homogeneous part plus a particular solution of the non-homogeneous equation. The solution of the homogeneous part (Gandolfo 2009; Chap. 12) is

$$r(t) = Ae^{(\beta/\alpha)t}, \quad t \geq z, \tag{16.35}$$

where A is an arbitrary constant. To find a particular solution for the non-homogeneous equation we apply the method of undetermined coefficients (Gandolfo 2009; Chap. 12). Since, as we have seen in the text, $\dot{M}(t) = \mu$, it follows that $M(t)$ has the linear form $M(t) = M_0 + \mu t$. Thus as a particular solution $\tilde{r}(t)$ we try a linear function with undetermined coefficients

$$\tilde{r}(t) = \lambda_0 + \lambda_1 t. \tag{16.36}$$

Substituting (16.36) into (16.34) we have

$$M_0 + \mu t = \beta \lambda_0 + \beta \lambda_1 t - \alpha \lambda_1,$$

whence

$$(\mu - \beta \lambda_1)t + (M_0 - \beta \lambda_0 + \alpha \lambda_1) = 0,$$

that will be identically satisfied if and only if

$$\begin{aligned} \mu - \beta \lambda_1 &= 0, \\ M_0 - \beta \lambda_0 + \alpha \lambda_1 &= 0. \end{aligned} \tag{16.37}$$

From Eq. (16.37) we obtain

$$\begin{aligned} \lambda_1 &= \frac{\mu}{\beta}, \\ \lambda_0 &= \frac{M_0}{\beta} + \frac{\alpha \mu}{\beta^2}, \end{aligned} \tag{16.38}$$

so that the particular solution we are looking for is

$$\tilde{r}(t) = \frac{\alpha \mu}{\beta^2} + \frac{M_0}{\beta} + \frac{\mu}{\beta}t = \frac{\alpha \mu}{\beta^2} + \frac{M(t)}{\beta}. \tag{16.39}$$

The particular solution can be interpreted, as usual (see Gandolfo 2009, Chap. 12) as the equilibrium solution of the model, i.e., the shadow exchange rate we are looking for.

16.4.2 A Second-Generation Model

We examine the model by Sachs et al. (1996). The basic problem is

$$\min L = \frac{1}{2}\left(\alpha\pi_t^2 + x_t^2\right), \qquad \alpha > 0, \tag{16.40}$$

subject to

$$Rb_t = x_t + \theta(\pi_t - \pi_t^e), \qquad \theta > 0, \tag{16.41}$$

where the symbols have been defined in the text.

From the Lagrangian

$$\Lambda = \frac{1}{2}\left(\alpha\pi_t^2 + x_t^2\right) + \mu[Rb_t - x_t - \theta(\pi_t - \pi_t^e)], \tag{16.42}$$

where μ is a Lagrange multiplier, we obtain the first-order conditions

$$\begin{aligned}
\frac{\partial\Lambda}{\partial\pi_t} &= \alpha\pi_t - \mu\theta = 0, \\
\frac{\partial\Lambda}{\partial x_t} &= x_t - \mu = 0, \\
\frac{\partial\Lambda}{\partial\mu} &= Rb_t - x_t - \theta(\pi_t - \pi_t^e) = 0.
\end{aligned} \tag{16.43}$$

Given the linear-quadratic setting, the second-order conditions are certainly satisfied. From the two first equations we have

$$x_t = \frac{\alpha}{\theta}\pi_t,$$

whence, introducing the composite parameter

$$\lambda \equiv \frac{\alpha}{\alpha + \theta^2} < 1, \tag{16.44}$$

we obtain

$$x_t = \frac{\lambda}{1 - \lambda}\theta\pi_t. \tag{16.45}$$

Substituting this result into the third equation we get

$$\theta\pi_t = (1 - \lambda)(Rb_t + \theta\pi_t^e). \tag{16.46}$$

Let us note for future reference that, if we impose the deterministic perfect foresight condition $\pi_t^e = \pi_t$, from (16.46) we have

$$\theta \pi_t = \theta \pi_t^e = \frac{1-\lambda}{\lambda} Rb_t. \qquad (16.47)$$

From Eqs. (16.46) and (16.45) we get

$$\begin{aligned} \pi_t &= \theta^{-1}(1-\lambda)(Rb_t + \theta \pi_t^e), \\ x_t &= \lambda(Rb_t + \theta \pi_t^e). \end{aligned} \qquad (16.48)$$

Let us now calculate the optimal value of the loss function by substituting Eqs. (16.48) into Eq. (16.40). We obtain

$$\begin{aligned} L^d(b_t, \theta \pi_t^e) &= \frac{1}{2} \left[\alpha \frac{(1-\lambda)^2}{\theta^2} + \lambda^2 \right] (Rb_t + \theta \pi_t^e)^2 \\ &= \frac{1}{2}\lambda \left[\alpha \frac{\lambda^{-1}(1-\lambda)^2}{\theta^2} + \lambda \right] (Rb_t + \theta \pi_t^e)^2, \end{aligned}$$

where the superscript d stands for "devaluing". Using the definition of λ the expression in the last square brackets turns out to be unity, hence

$$L^d(b_t, \pi_t^e) = \frac{1}{2}\lambda(Rb_t + \theta \pi_t^e)^2. \qquad (16.49)$$

16.4.3 Krugman's Third Generation Model: The Stabilization Dilemma

To show that his model applies to the Asian crisis, Krugman asks, and answers, two questions: Why Asia? why now? The answer to the first question is high leverage: all the Asian economies hit by the crisis had unusually high levels of θ. The answer to the second question—given that high leverage has been a feature of the Asian economies for decades—is that only in the 1990s these economies began to extensively borrow in foreign currencies, adding a second factor of risk of financial collapse.

The standard IMF strategy for coping with these crises has been that of advising the Asian countries to defend their currencies by increasing interest rates. This can be roughly translated in terms of the present model by imagining that the effect of that strategy is to keep the real exchange rate constant in the face of a decline in the willingness of foreign lenders to finance investment. The result is a decline in output, for if we keep p constant and rearrange (16.23) we obtain

$$y = \frac{1}{1 - (1 - \alpha)(1 - \mu)}[pX + (1 - \mu)I], \tag{16.50}$$

that Krugman calls a "quasi-Keynesian" multiplier. Plugging (16.50) into (16.27) we can calculate

$$\frac{dW}{dI} = \frac{\alpha(1 - \mu)}{1 - (1 - \alpha)(1 - \mu)}, \tag{16.51}$$

which shows that a decline in investment will reduce wealth, and once again cause a feedback from actual to financeable investment

$$\frac{dI_f}{dI} = \frac{dI_f}{dW}\frac{dW}{dI} = \frac{(1 + \theta)\alpha(1 - \mu)}{1 - (1 - \alpha)(1 - \mu)}. \tag{16.52}$$

This last equation shows that a high leverage may cause $dI_f/dI > 1$: in fact, the critical value of θ, call it θ_c, turns out to be

$$\theta_c = \frac{\mu}{\alpha(1 - \mu)}, \tag{16.53}$$

so that $dI_f/dI > 1$ for $\theta > \theta_c$.

This illustrates the stabilization dilemma: stabilizing the exchange rate closes one channel for financial collapse but opens another. In fact, "if leverage is high, the economy may stabilize its real exchange rate only at the expense of a self reinforcing decline in output that produces an equivalent decapitation of the entrepreneurial class." In the light of this model, much of the debate on the IMF strategy for dealing with the Asian crisis, vituperated by some (Stiglitz 2000), praised by others (Dornbusch 1999) looks futile. Both answers (defend the exchange rate; leave it go) may be equally bad. Krugman suggests other possible means such as the provision of huge emergency lines of credit or the imposition of a curfew on capital flight.

References

Ahnert, T., & Bertsch, C. (2013). A wake-up call: information contagion and strategic uncertainty, Sveriges Riksbank Working Paper No 282.

Alesina, A., & Wagner, A. F. (2006). Choosing (and reneging on) exchange rate regimes. *Journal of the European Economic Association, 4,* 770–799.

Alessi, L., & Detken, C. (2011). Quasi real time early warning indicators for costly asset price boom/bust cycles: A role for global liquidity. *European Journal of Political Economy, 27,* 520–533.

Allen, F., Babus, A., & Carletti, E. (2009). Financial crises: Theory and evidence. *Annual Review of Financial Economics, 1,* 97–116.

Allen, F., Babus, A., & Carletti, E. (2012). Asset commonality, debt maturity and systemic risk. *Journal of Financial Economics, 104,* 519–534.

Anzuini, A., & Gandolfo, G. (2000). Currency crises and speculative attacks: A comparison between two different forecasting approaches. CIDEI Working Paper No. 60, University of Rome "La Sapienza".

Anzuini, A., & Gandolfo, G. (2003). Can currency crises be forecast? In: G. Gandolfo, & F. Marzano (Eds.), *International economic flows, currency crises, investment and economic development: Essays in memory of Vittorio Marrama* (pp. 61–82). Roma: EUROMA (Publications of the Faculty of Economics of the University of Rome La Sapienza).

Barba Navaretti, B., & Venables, A. J. (2004). *Multinational firms in the world economy.* Princeton: Princeton University Press.

Basu, R. (2002). Financial contagion and investor 'learning': An empirical investigation. International Monetary Fund Working Paper No. 218.

Bekaert, G., Ehrmann, M., Fratzscher, M., & Mehl, A. J. (2014). Global crisis and equity market contagion. *Journal of Finance, 69,* 2597–2649.

Berg, A., & Pattillo, C. (1999). Predicting currency crises: The indicators approach and an alternative. *Journal of International Money and Finance, 18,* 561–586.

Berkmen, S. P., Gelos, G., Rennhack, R., & Walsh, J. (2012). The global financial crisis: Explaining cross-country differences in the output impact. *Journal of International Money and Finance, 31,* 42–59.

Blanchard, O., Mitali, D., & Faruqee, H. (2010). The initial impact of the crisis on emerging market countries. *Brooking Papers on Economic Activity,* 263–323.

Borio, C., & Lowe, P. (2002). Asset prices, financial and monetary stability: Exploring the nexus. BIS Working Papers No. 114.

Breuer, J. B. (2004). An exegesis on currency and banking crises. *Journal of Economic Surveys, 18,* 293–320.

Buckley, P. J. (1998). *International strategic management and government policy.* London: Macmillan.

Burstein, A,, Kurz, C., & Tesar, L. (2008). Trade, production sharing, and the international transmission of business cycles. *Journal of Monetary Economics, 55,* 775–795.

Bussiere, M., & Fratszcher, M. (2006). Towards a new early warning system of financial crises. *Journal of International Money and Finance, 25,* 953–973.

Calvo, G., & Mishkin, F. S. (2003). The mirage of exchange rate regimes for emerging market countries. *Journal of Economic Perspectives, 13,* 43–64.

Caramazza, F., Ricci, L., & Salgado, R. (2000). Trade and financial contagion in currency crises. International Monetary Fund WP/00/55.

Caramazza, F., Ricci L., & Salgado, R. (2004). International financial contagion in currency crises. *Journal of International Money and Finance, 23,* 51–70.

Chari, V. V., & Kehoe, P. J. (2003). Hot money. *Journal of Political Economy, 111,* 1262–1292.

Chang, R., & Velasco, A. (1998). Financial crises in emerging markets: A canonical model. NBER Working Paper No. 6606.

Claessens, S., Tong, H., & Wei, S. J. (2011). From the financial crisis to the real economy: Using firm-level data to identify transmission channels. NBER Working Paper No. 17360.

Corsetti, G., Pesenti, P., & Roubini, N. (1999). Paper tigers? A model of the Asian crisis. *European Economic Review, 43,* 1211–1236.

Cutilli, B., & Gandolfo, G. (1963). The role of commercial banks in foreign exchange speculation. *Banca Nazionale del Lavoro Quarterly Review, 65,* 216–231.

Cutilli, B., & Gandolfo, G. (1972). Wider band and "oscillating exchange rates". *Economic Notes, 1,* 111–124.

Dasgupta, A., Leon-Gonzalez, R., & Shortland, A. (2011). Regionality revisited: An examination of the direction of spread of currency crises. *Journal of International Money and Finance, 30,* 831–848.

Davis, E. P., & Karim, D. (2008). Comparing early warning systems for banking crises. *Journal of Financial Stability, 4,* 89–120.

Demirgüç-Kunt, A., & Detragiache, E. (2000). Monitoring banking sector fragility: a multivariate logit approach. *World Bank Economic Review, 14,* 287–307.

Demirgüç-Kunt, A., & Detragiache, E. (2005). Cross-country empirical studies of systemic bank distress: A survey. *National Institute Economic Review, 192*, 68–83.

Diamond, D. W., & Dybvig, P. H. (1983). Bank runs, deposit insurance and liquidity. *Journal of Political Economy, 91*, 401–419.

Díaz-Alejandro, C. F. (1985). Good-bye financial liberalization, hello financial crash. *Journal of Development Economics, 19*, 1–24.

Dornbusch, R. (1999). The IMF didn't fail. *Far Eastern Economic Review*, p. 28.

Dornbusch, R., Park, Y., & Claessens, S. (2000). Contagion: Understanding how it spreads. *The World Bank Research Observer, 15*, 167–195.

Drehman, M., & Juselius, M. (2013). Evaluating early warning indicators of banking crises: Satisfying policy requirements. *International Journal of Forecasting, 30*, 759–780.

Dunning, J. H. (1977). Trade, location of economic activity and the MNEs. In: B. Ohlin, P. O. Hesselborn & P. M. Wijkman (Eds.), *The international allocation of economic activity*. London: Macmillan.

Dunning, J. H. (1993). Towards an interdisciplinary explanation of international production. In: J. H. Dunning (Ed.), *The theory of transnational corporations* (Vol. 1, pp. 387–412). London: Taylor & Francis.

Eichengreen, B., Rose, A. K., & Wyplosz, C. (1996). Contagious currency crises. *Scandinavian Journal of Economics, 98*, 463–484.

Elsinger, H., Lehar, A., & Summer, M. (2006). Using market information for banking system risk assessment. *International Journal of Central Banking, 2*, 137–165.

Flood, R. P., & Garber, P. M. (1984). Collapsing exchange-rate regimes: Some linear examples. *Journal of International Economics, 17*, 1–13.

Flood, R. P., & Garber, P. M. (1994). *Speculative bubbles, speculative attacks, and policy switching*. Cambridge (Mass): MIT Press.

Flood, R. P., Garber P. M., & Kramer, C. (1996). Collapsing exchange-rate regimes: Another linear example. *Journal of International Economics, 41*, 223–234.

Flood, R., & Marion, N. (1999). Perspectives on the recent currency crisis literature. *International Journal of Finance and Economics, 4*, 1–26.

Flood, R., & Marion, N. (2000). Self-fulfilling risk predictions: An application to speculative attacks. *Journal of International Economics, 50*, 245–268.

Forbes, K. (2002). Are trade linkages important determinants of country vulnerability to crises? In: S. Edwards & J. A. Frankel (Eds.), *Preventing currency crises in emerging markets* (pp. 77–132). Chicago, IL: University of Chicago Press.

Forbes, K. (2004). The Asian flu and Russian virus: The international transmission of crises in firm-level data. *Journal of International Economics, 63*, 59–92.

Forbes, K., & Rigobon, R. (2002). No contagion, only interdependence: Measuring stock market co-movement. *Journal of Finance, 57*, 2223–2261.

Frankel, J., & Rose, A. (1996). Currency crashes in emerging markets: an empirical treatment. *Journal of International Economics, 41*, 351–366.

Frankel, J., & Saravelos, G. (2012). Can leading indicators assess country vulnerability? Evidence from the 2008–09 global financial crisis. *Journal of International Economics, 87*, 216–231.

Friedman, M. (1953). The case for flexible exchange rates. In: M. Friedman (Ed.), *Essays in positive economics* (pp. 157–203). Chicago: University of Chicago Press.

Friedman, M. (1960), In defence of destabilizing speculation. In: R. W. P. Pfouts (Ed.), *Essays in economics and econometrics* (pp. 133–141). Chapel Hill: University of North Carolina Press.

Gandolfo, G. (1971). Tentativi di analisi teorica in tema di cambi flessibili e speculazione. *L'Industria, 1*, 40–60.

Gandolfo, G. (2009). *Economic dynamics* (4th ed.). Berlin, Heidelberg, New York: Springer.

Gandolfo, G. (2014). *International trade theory and policy* (2nd ed.). Berlin, Heidelberg, New York: Springer.

Gelos, G. (2011). International mutual funds, capital flow volatility, and contagion—A Survey. International Monetary Fund Working Paper, No 92.

Gerlach, S., & Smets, F. (1995). Contagious speculative attacks. *European Journal of Political Economy, 11*, 45–63.

Giordano, R., Pericoli, M., & Tommasino, P. (2013). Pure or wake-up-call contagion? Another look at the EMU sovereign debt crisis. *International Finance, 16*, 131–160.

Glahe, F. R. (1966). Professional and non-professional speculation, profitability and stability. *Southern Economic Journal, 23*, 43–48.

Glick, R., & Rose, A. (1999). Contagion and trade: Why are currency crises regional. *Journal of International Money and Finance, 18*, 603–617.

Goldstein, M., (1998), *The Asian financial crisis: Causes, cures, and systematic implications*. Washington D.C.: Institute for International Economics.

Goldstein, I., & Pauzner, A. (2004). Contagion of self-fulfilling currency crises due to the diversification of investment portfolios. *Journal of Economic Theory, 119*, 159–183.

Greenwood, R., Landier A., & Thesmar, D. (2015). Vulnerable banks. *Journal of Financial Economics, 115*, 471–485.

International Monetary Fund. (2002). *Global financial stability report* (Chapter 4). International Monetary Fund.

International Monetary Fund. (2013). *The IMF-FSB early warning exercise: Design and methodological toolkit*. International Monetary Fund Occasional Paper No. 274.

Jeanne, O. (2000). *Currency crises: A perspective on recent theoretical developments*. Special Papers in International Economics No. 20, Princeton University, International Finance Section.

Jeanne, O., & Masson, P. (2000). Currency crises, sunspots, and Markov-switching regimes. *Journal of International Economics, 50*, 327–350.

Kaminsky, G. A., Lizondo, S., & Reinhart, C. M. (1998). The leading indicators of currency crises. *International Monetary Fund Staff Papers, 45*, 1–48.

Kaminsky, G. A., & Reinhart, C. M. (1999). The twin crises: The causes of banking and balance-of-payments problems. *American Economic Review, 89*, 473–500.

Karas, A., Pyle, W., & Schoors, K. (2013). Deposit insurance, banking crises, and market discipline: Evidence from a natural experiment on deposit flows and rates. *Journal of Money, Credit and Banking, 45*, 179–200.

Karolyi, A. G. (2003). Does international finance contagion really exist? *International Finance, 6*, 179–199.

Kemp, M. C. (1963). Profitability and price stability. *The Review of Economics and Statistics, 45*, 185–189.

Kodres, L. E., & Pritsker, M. (2002). A rational expectations model of financial contagion. *Journal of Finance, 57*, 769–799.

Krugman, P. (1979). A model of balance-of-payments crises. *Journal of Money, Credit, and Banking, 11*, 311–325.

Krugman, P. (1998). What happened to Asia? in Paul Krugman's homepage, http://www.mit.edu/people/krugman/index.html#hard.

Krugman, P. (1999). Balance sheets, the transfer problem, and financial crises. In: P. Isard, A. Razin, & A. K. Rose (Eds.), *International finance and financial crises: Essays in honor of Robert. P. Flood, Jr.* Norwell (Mass.): Kluwer.

Krugman, P. (2001). Crises: The Next Generation?, Paper Presented at Conference Honoring Assaf Razin, Tel Aviv.

Kyle, A. S., & Xiong, W. (2001). Contagion as a wealth effect. *Journal of Finance, 56*, 1401–1440.

Leblang, D., & Shanker, S. (2006). Institutions, expectations and currency crises. *International Organization, 60*, 254–262.

Levy Yeyati, E., & Panizza, U. (2011). The elusive costs of sovereign defaults. *Journal of Development Economics, 94*, 95–105.

Lee, H. -Y., Wu, H. -C., & Wang, Y. -J. (2007). Contagion effect in financial markets after the South-East Asia tsunami. *Research in International Business and Finance, 21*, 281–296.

Lin, C. -S., Khan, H. A., Chang, R. -Y., & Wang, Y. -C. (2008). A new approach to modeling early warning systems for currency crises: Can a machine-learning fuzzy expert system predict the currency crises effectively? *Journal of International Money and Finance, 27*, 1098–1121.

Lipsey, R. E. (2004). Home- and host-country effects of foreign direct investments. In: R. E. Baldwin & L. A. Winters (Eds.), *Challenges to globalization*. Chicago: Chicago University Press.

Ljungqvist, L. (1992). Destabilizing exchange rate speculation: A counterexample to Milton Friedman. Seminar Paper 125, Institute for International Economic Studies, Stockholm.

Lund-Jensen, K. (2012). Monitoring systemic risk based on dynamic thresholds. International Monetary Fund Working Paper No. 159.

Masson, P. (1999a). Contagion: monsoonal effects, spillovers, and jumps between multiple equilibria. In: P. R. Agénor, M. Miller, D. Vines, & A. Weber (Eds.), *The Asian financial crisis: Causes, contagion and consequences*. Cambridge (UK): Cambridge University Press.

Masson, P. (1999b). Contagion: macroeconomic models with multiple equilibria. *Journal of International Money and Finance, 18*, 587–602.

McKinnon, R. I. (1983). The J-curve, stabilizing speculation, and capital constraints on foreign exchange dealers. In: D. Bigman & T. Taya (Eds.), *Exchange rate and trade instability: Causes, consequences, and remedies* (pp. 103–127). Cambridge (Mass.): Ballinger.

Mendoza, E. G., & Velasco, A. (Eds.) (2000), Symposium on globalization, capital markets crises and economic reform. *Journal of International Economics, 51*(1) (special issue).

Pavlova, A., & Rigobon, R. (2008). The role of portfolio constraints in the international propagation of shocks. *Review of Economic Studies, 75*, 1215–56.

Raddatz, C., & Schmukler, S. (2012). On the international transmission of shocks: Micro-evidence from mutual fund portfolios. *Journal of International Economics, 88*, 357–374.

Reinhart, C. M., & Rogoff, K. S. (2011). From financial crash to debt crisis. *American Economic Review, 101*, 1676–1706.

Rose, A., & Spiegel, M. M. (2012). Cross-country causes and consequences of the 2008 crisis: Early warning. *Japan and the World Economy, 24*, 1–16.

Sachs, J., Tornell, A., & Velasco, A. (1996). The Mexican peso crisis: Sudden death or death foretold? *Journal of International Economics, 41*, 265–283.

Salant, S., & Henderson, D. (1978). Market anticipation of government policy and the price of gold. *Journal of Political Economy, 86*, 627–648.

Shimpalee, P. L., & Breuer, J. B. (2006). Currency crises and institutions. *Journal of International Money and Finance, 25*, 125–145.

Stiglitz, J. (2000). Insider's account of Asian economic crisis. *The New Republic*, 17 April 2000.

Taketa, K. (2004). A large speculator in currency crises: A single "George Soros" makes countries more vulnerable to crises, but mitigates contagion. IMES (Institute for Monetary and Economic Studies, Bank of Japan), Discussion Paper 2004-E-23.

Tavlas, G. S. (Ed.). (1996). Currency crises. Special issue of *Open Economies Review, 7*(Supplement 1).

Tong, H., & Wei, S. -J. (2011). The composition matters: Capital inflows and liquidity crunch during a global economic crisis. *Review of Financial Studies, 24*, 2023–2052.

Thurner, S., Farmer, J. D., & Geanakoplos, J. (2012). Leverage causes fat tails and clustered volatility. *Quantitative Finance, 12*, 695–707.

UN (1998). *World investment report 1998: Trends and determinants*. United Nations Conference on Trade and Development. New York: United Nations Publications.

Van Rijckeghem, C., & Weder, B. (2001). Survey of contagion: is it finance or trade? *Journal of International Economics, 54*, 293–308.

Van Rijckeghem, C., & Weder, B. (2003). Spillovers through banking centers: a panel data analysis of bank flows. *Journal of International Money and Finance, 22*, 483–509.

Van Wincoop, E. (2013). International contagion through leveraged financial institutions. *American Economic Journal: Macroeconomics, 5*, 152–189.

Wiederholt, M. (2010). Rational inattention, in *The New Palgrave dictionary of economics*. London: Palgrave Macmillan.

Zebregs, H. (1999). Long-term international capital movements and technology: A review. International Monetary Fund WP/99/126.

Fixed Vs Flexible Exchange Rates   17

17.1 The Traditional Arguments

It may seem that the old debate on fixed and flexible exchange rates has been made obsolete by international monetary events, as the international monetary system abandoned the Bretton Woods fixed exchange rate regime (of the adjustable peg type) in the early 1970s, and is now operating under a managed float regime mixed with others (see Sect. 3.3); nor does it seem likely that freely flexible exchange rates will be generally adopted or fixed ones will return. However, a general outline of the traditional arguments (Stockman 1999) is not without its uses, because many of these keep cropping up. The reference to aspects already treated in previous chapters will allow us to streamline the exposition. In examining the main pros and cons of the two systems it should be borne in mind that the arguments for one system often consist of arguments against the other.

The birth of a Keynesian economic policy about 1950 explains one of the main criticisms then directed at the existing system of fixed parities. This system, it was argued, in many instances created dilemma cases between external and internal equilibrium (see Sect. 11.1). Only after the diffusion of external convertibility and interest-sensitive capital flows did the theoretical possibility of solving the dilemma cases (by way of an appropriate mix of fiscal and monetary policy) arise, as shown in Sect. 11.2. However, the numerous criticisms levelled at the policy mix solution (see Sect. 11.2.2) gave good reason to the advocates of flexibility to point out the importance of exchange-rate flexibility for the achievement of external equilibrium, so as to be able to use fiscal and monetary policy to solve internal problems without burdening these tools with external problems.

The critics of flexibility pointed out the serious consequences for international trade and investment that would derive from a situation of uncertainty on the foreign exchange markets (whence higher risks). But the advocates replied that foreign exchange risks can be hedged by way of the forward market (see Sect. 2.5), and pointed out that habitual use of this market would stimulate its development

© Springer-Verlag Berlin Heidelberg 2016

G. Gandolfo, *International Finance and Open-Economy Macroeconomics*,
Springer Texts in Business and Economics, DOI 10.1007/978-3-662-49862-0_17

and efficiency, so that forward cover could be obtained at moderate cost. On the contrary, the lack of development of an efficient and "thick" forward market under the adjustable peg regime,[1] meant that, when parity changes were expected, forward cover could be obtained only at a prohibitive cost. Furthermore, the possibility—which actually became a reality on several occasions—of parity changes, did *not* generate that certainty which the advocates of the adjustable peg claimed against the uncertainty of flexible exchange rates.

Among the further criticisms against fixed exchange rates three more points are worthy of note. One is the observation that this regime has a distortionary effect on markets when the relative competitiveness of two countries varies. If the exchange rate does not reflect this variation because it is fixed, it then means that an additional advantage is created for the country with the lower rate of inflation against the country with the higher one. The former, in fact, sells its commodities to the latter at increasing prices (under the assumption that the exporting country adjusts the price of its exports towards the price of similar goods in the importing country) while maintaining the same rate of conversion, notwithstanding the fact that the latter's currency has been losing purchasing power. It should be added that the country with the lower rate of inflation will probably see a decrease in its exports of capital (capital outflows) and an increase in its imports of capital (capital inflows), which may give rise to a disparity in the growth of the two countries.

The second point is that the maintenance of fixed parities ultimately amounts to subsidizing firms engaged in international trade, as it implies the use of public funds to absorb part of the risks inherent in private international transactions, and so involves a possible misallocation of resources (unless there is a diversion of social costs and benefits from private ones, which, however, has to be demonstrated). As Lanyi (1969; p. 7) aptly put it, "if one should ask an economist whether it is *necessarily* (italics added) desirable to subsidize industry X while not subsidizing industry Y, he would immediately reply in the negative". Therefore the answer to the question "are we *necessarily* (italics added) better off because international commerce is subsidized through the government's bearing the exchange risk, while most types of domestic commerce receive no government assistance in risk-bearing?" must also be in the negative.

---

[1]Under normal conditions, all agents expect parities to remain fixed, so that there is no incentive for them to have recourse to the forward market. It is only in the case of fundamental disequilibrium that agents begin to fear parity changes. But, by the very nature of the adjustable peg, their expectations will be unidirectional, so that there will either be only a demand for forward exchange (if a devaluation is expected) by importers and other agents who have to make future payments abroad, or else only a supply of forward exchange (if a revaluation is expected) by exporters and other agents who are due to receive future payments from abroad. Thus in both cases the other side of the market will be absent, i.e. there will be no supply (if a devaluation expected) or no demand (if a revaluation is expected) to match the demand and supply respectively. This means that banking and other intermediaries will procure the forward cover at cost practically corresponding to the expected devaluation or revaluation.

The third point is the extent to which the fixed or pegged regimes encourage firms to take on excessive foreign currency debt that exposes them to rollover and foreign currency risks. The idea that the exchange rate regime affects firms' incentive to hedge their exposure to currency risk was studied extensively following the Asian financial and currency crises of the 1990s.

Vulnerabilities in the market corporate sector coming from external foreign currency borrowing have been singled out as a key source of financial fragility in emerging market countries. Large exposure to foreign currency debt by the private sector, meaning that private debt is overwhelmingly in foreign currency but revenues are in local currency (currency mismatches) is, therefore, associated with a substantial risk of a crisis. According to the proponents of flexible exchange rate regimes, the authorities' commitment to defend a fixed exchange rate implies an implicit guarantee against major exchange rate changes, which leads to moral hazard problems and excessive foreign currency borrowing. The choice of floating exchange rate regimes would provide incentives for a more cautious management of currency exposure, thereby reducing financial vulnerabilities associated to currency mismatches in the private sector.

Among the advantages of flexible exchange rates the advocates included, in addition to the greater freedom of economic policy to achieve internal equilibrium, already mentioned above, the following:

(a) the possibility of protecting domestic price stability by an appreciating exchange rate with respect to countries with a higher inflation rate. The existence was also claimed of an insulating power against disturbances of a real nature (see, however, Sect. 9.4);

(b) the greater effectiveness of monetary policy: a restrictive monetary policy, for example, by causing a capital inflow, brings about an appreciation in the exchange rate, with depressive effects on aggregate demand which reinforce those already due to the increase in the interest rate; the opposite is true in the case of an expansionary monetary policy (see, however, Sect. 11.3);

(c) the lower need for international reserves, as the elimination of possible deficits is ensured by the exchange-rate flexibility.

Among the disadvantages of flexible exchange rates, in addition to the increased risk in international transactions, already mentioned above, the critics included the following:

(1) the possible non-verification of the critical elasticities condition (see Sects. 7.2, 7.3, 9.1) would make the system unstable (see, however, Sect. 7.3.1);

(2) the possible presence of destabilising speculation; the advocates of flexible exchange rates, however, pointed out the undoubtedly destabilising nature of speculation under the adjustable peg and argued for its stabilizing nature under flexible exchange rates (this debate has been treated in Sect. 16.2);

(3) the resource reallocation costs of flexible rates, which induce resource movements into and out of export and import-competing domestic industries (see, however, Thursby 1981);

(4) the loss of monetary discipline, as the need for restrictive monetary policies in the presence of inflation is reduced if there is no external constraint; the advocates of flexible rates, however, pointed out that an excessive exchange-rate depreciation is as good as an excessive reserve loss as an indicator of the need for monetary restriction;

(5) the alleged inflationary bias due to a ratchet effect of exchange-rate movements on prices: a depreciation, by raising the domestic prices of imported final goods and of imported intermediate goods, raises the domestic general price level, whilst an appreciation does not bring it down at all or not as much. Hence the possibility of a "vicious circle" of depreciation-inflation. This topic will be dealt with at some length in Sect. 17.4.1.

17.2   The Modern View

It is impossible to strike a balance between all the traditional arguments for and against the two regimes. The reason for the impossibility of declaring one regime definitely superior from the theoretical point of view lies in the fact that neither one has inferior costs and superior benefits on all counts. The modern view has tried to overcome this impasse by concentrating on a subset of criteria, namely has tried to assess which regime better stabilizes the economy in the face of shocks of various type. The exchange rate regime that provides the greater stability is considered superior.

To illustrate the effects of shocks let us start from an old acquaintance, the Mundell-Fleming model with perfect capital mobility (see Sect. 11.4). This model does not consider the price level nor the aggregate supply function. Since we take it that the two main goals of society are output and price stability, and we also want to consider the effects of supply shocks, we can introduce a simple aggregate supply function according to which, given the wage rate, supply of output is a positive function of the price at which producers are able to sell their output. We also assume that excess demand for domestic output causes an increase in the supply of output (according to the standard mechanism used in these models) as well as an increase in the price level (according to a Walrasian mechanism). Finally, we assume that the social preference function (or the objective function of the authorities) negatively depends on the deviations of both output and the price level from their respective target levels, where the weights given to the two deviations reflect the relative importance or subjective trade-off between the two targets.

Thus we are equipped for considering various types of shocks. To avoid problems related to expectations, we assume that all shocks are *unanticipated*.

17.2.1 Money Demand Shock

Starting from an equilibrium situation in which both prices and output are at their respective target levels, suppose that there is an unanticipated exogenous rise in money demand. Given the money supply, this tends to cause an increase in the interest rate. Because of the assumed perfect capital mobility, there is an incipient capital inflow.

Under *fixed* exchange rates the incipient capital inflow causes an increase in money supply until it becomes equal to the increased money demand. Output does not change neither does the price level. The fixed exchange rate regime completely stabilizes the economy.

Under *flexible* exchange rates the incipient capital inflow causes an appreciation of the exchange rate which depresses aggregate demand. The (negative) excess demand for output causes both a decrease in output and a decrease in the price level.

It is clear that in these circumstances the flexible exchange rate regime is inferior, independently of the weights attached to output and price stability.

17.2.2 Aggregate Demand Shock

We now consider an unanticipated exogenous decrease in aggregate demand. This tends to cause a decrease in both output and the price level. Hence we have a tendency for the interest rate to decrease, both because money demand decreases and because the real money supply increases owing to the price decrease. Due to perfect capital mobility there is an incipient capital outflow.

Under *fixed* exchange rates the incipient capital outflow causes a decrease in money supply until it is brought in line with the lower money demand. In the final equilibrium both output and the price level will be lower.

Under *flexible* exchange rates the incipient capital outflow causes an exchange-rate depreciation which sustains aggregate demand. Both output and the price level will decline less than in the case of fixed exchange rates, which are clearly inferior.

The same results hold if we consider changes in terms of trade that may affect output growth through their incidence on relative prices and on external demand for domestically produced goods and services: flexible exchange rate regimes help reduce the real impact of terms of trade shocks.

17.2.3 Aggregate Supply Shock

We finally consider an unanticipated exogenous decrease in aggregate supply. The resulting demand fall (though not by the same extent, if we assume a marginal propensity to spend smaller than 1) causes a decrease in money demand and hence an excess supply of money. The excess demand for goods will tend to cause an increase in both output and the price level, but of course output will remain below its previous level. The price rise reduces the real supply of money and the demand rise increases money demand. Depending on the structural parameters of the economy (which determine the extent of the demand and price changes) the initial excess supply of money may remain such or turn into an excess demand for money. The two cases must be considered separately.

(I) In the former case the excess supply for money causes a tendency for the interest rate to decrease and hence an incipient capital outflow.

Under *fixed* exchange rates the incipient capital outflow causes a decrease in money supply until it is brought in line with the lower money demand. In the final equilibrium output will be lower and the price level will be higher than in the initial equilibrium.

Under *flexible* exchange rates the incipient capital outflow causes an exchange-rate depreciation which sustains aggregate demand. Hence in the final equilibrium output will be higher than under fixed exchange rates (though lower than before the shock) and the price level will be higher.

Thus we see that flexible exchange rates are superior as regards output stabilization, inferior as regards price stability. The decision on the better exchange-rate regime will depend on the relative weights given to the two targets.

(II) In the latter case the initial excess supply for money turns into an excess demand for money, which causes a tendency for the interest rate to increase and hence an incipient capital inflow.

Under *fixed* exchange rates the incipient capital inflow causes an increase in the money supply until it is brought in line with the increased demand. In the final equilibrium output will be lower and the price level will be higher than in the initial equilibrium.

Under *flexible* exchange rates the incipient capital inflow causes an exchange-rate appreciation which lowers aggregate demand. In the final equilibrium output will be lower than under fixed exchange rates, and the price increase will also be lower than in the fixed exchange rate regime.

Thus we see that now fixed exchange rates are superior as regards output stabilization, inferior as regards price stability. Also in this case the decision on the better exchange-rate regime will depend on the relative weights given to the two targets. However, given the weights, the decision will be crucially dependent on the structural parameters of the economy. With the same relative weights, in fact, the choice in case I will be exactly the opposite of the choice in case II, and vice versa.

17.2.4 Conclusion

The modern approach has succeeded in reducing the range of uncertainty but has not been able to settle the debate. Notwithstanding its simplicity, the model adopted has clearly shown the reasons for this failure. There is probably no universally 'optimal' regime, the choice between the two regimes does, in fact depends on several factors:

(a) the nature and magnitude of the shock,
(b) the structural characteristics (parameters) of the economy,
(c) the objective function of the authorities.
(d) the associated institutional setup that is needed for the regime to be viewed as 'credible.'

More complicated models (see the Appendix) would not change this conclusion.

Costs and benefits will have to be weighted according to a social preference function, which may vary from country to country (and from period to period in the same country). In addition, the structural parameters of the economy are not given once-and-for-all, but are subject to change, the more so the more dynamic is the economic system. In conclusion, "no single currency regime is right for all countries or at all times" (Frenkel 1999; see also Stockman 1999).

17.3   The Experience of the Managed Float

17.3.1 Introduction

On August 15, 1971, the United States of America announced to the world a series of measures amongst which the imposition of a 10 % additional tax on imports and the *"de jure" inconvertibility of the US dollar into gold.* Although the dollar had long been inconvertible *de facto*,[2] the official declaration of its *de jure* inconvertibility was the beginning of the end of the Bretton Woods era. Subsequent consultations with other countries led to the Smithsonian Agreement in December, 1971, in which new par values or central rates of the main currencies with respect to the dollar were established, and the margins of fluctuation were widened to $\pm 2.25 \%$ of these rates. For their part, the US eliminated the additional tariff and devalued the dollar with respect to gold from \$35 to \$38 per ounce (in February, 1973, this price was further increased to \$42.22, and in 1976 it was abolished: see Sect. 22.7).

The realignment of the various currencies brought into being at the Smithsonian Conference with the aim of giving a certain stability to exchange rates was short-lived. In fact, severe balance-of-payments difficulties soon compelled

---

[2]It should be noticed that the official convertibility of the US dollar into gold put, at least in theory, restraints on the conduct of the US monetary policy (and, more generally, on US economic policy). See, for example, Argy (1981), Chaps. 3 and 6.

various countries to abandon the fixed exchange rate (adjustable peg) regime for a (managed) float: the first to float was the pound sterling (June, 1972), then the Italian lira (January-February 1973).[3] In March, 1973, the EEC (European Economic Community) countries agreed to let their currencies float vis-à-vis the dollar whilst maintaining fixed parities (with predetermined margins) among themselves; this was called a *joint float*. The same countries had previously (in March-April, 1972) agreed to create the *"snake"* in the *"tunnel"*, i.e. to restrict the margins of fluctuation around their partner-country parities to ±1.125 % (snake) whilst maintaining the Smithsonian margins of ±2.25 % (the tunnel) around the parities vis-à-vis non-partner countries. This was clearly an attempt at moving towards a *currency area*, but—as mentioned above—balance-of-payments difficulties led Italy and then France to abandon the snake and let their currencies float (England had been floating since June 1972). This attempt was taken up again more formally in 1979 with the creation of the EMS (European Monetary System), which will be dealt with at some length in Sect. 21.1.

Italy and England were exempted from the March 1973 agreement (together with Ireland, which pegged its currency to the pound sterling); these countries maintained the float vis-à-vis all currencies. In January, 1974, the French franc was compelled to abandon that agreement and float; after returning in July, 1975, France had to abandon it again in March, 1976. In 1973 other countries as well (amongst them Japan and Switzerland) had decided to let their currencies float; Canada had been under a managed float since 1970. With the Kingston (Jamaica) Agreement in January, 1976, the floating exchange-rate regime was legalized within the IMF. This agreement does contemplate the possibility of returning to a system of "stable but adjustable" parities should the Fund ascertain that the presuppositions exist and

---

[3] Some countries, amongst them Italy, adopted a *two-tier* (or dual) exchange rate, that is a regime where there are two distinct markets for foreign exchange: one for commercial (or, more generally, current account) transactions, where a "commercial" exchange rate (usually fixed) exists, and the other for the residual transactions, where a "financial" exchange rate (usually floating, more or less cleanly) is determined. The two markets must of course be completely separated, otherwise no spread could exist between the commercial and the financial exchange rate; the administrative measures required to bring about this segregation need not concern us here. The idea behind a two-tier market is to isolate the current account from disturbances deriving from possible destabilizing capital flows. This implies that the financial exchange rate is left completely free so as to equilibrate the capital account, whilst the commercial exchange rate is fixed or under a heavily managed float. The relative pros and cons of a *perfect* two-tier market are a matter of debate (for a theoretical analysis of dual exchange markets see, for example, Flood 1978, Adams and Greenwood 1985, and references therein); all agree, however, that for an actual two-tier market to approach the ideal theoretical form, the two markets must be effectively segregated. Otherwise, in fact, on the one hand the financial market comes to lose any practical importance and, on the other, clandestine capital movements (which, as we know, take place through current account transactions: see Sect. 5.1.4) are stimulated and a "parallel" market develops. This is what actually happened, for example, in Italy, so that in March, 1974, the dual market—which had lost any practical importance—was abolished. For a detailed description of actual two-tier markets see International Monetary Fund, *Annual Report on Exchange Restrictions* (years 1971 through 1974, and the current year for indication of those countries still maintaining dual exchange markets).

with a majority of 85 %. In the meantime each country is free to adopt the exchange-rate regime it wishes although it must notify its decision to the IMF and accept the surveillance of the Fund, which requires certain very general obligations to be respected.

In March 1979, as mentioned above, the EEC countries constituted the EMS, which involved an adjustable peg regime (with relatively wide margins) among the partner countries, and a float vis-à-vis non-partner countries (see Sect. 21.1). This system evolved into the European Monetary Union, whereby the joining European countries adopt a common currency, the euro (see Sect. 21.2).

17.3.2   New Light on an Old Debate?

The very brief description given in the previous section (further details on the international monetary system will be given in Chap. 22) was meant to provide the reader with the minimum of historical perspective better to appreciate the debate on fixed versus flexible exchange rates. Many, in fact, wonder why the empirical evidence accumulated since 1972–1973 on floating exchange rates is not used to throw new (and possibly conclusive) light on the various arguments examined above. But the evaluation of this evidence requires an important warning: the theoretical regime of flexible exchange rates contemplates *free* flexibility, that is, with no control nor interference on the part of the monetary authorities. This is definitely *not* the case of the managed float, which is what we have been observing. Therefore, it would be wrong simply to extend the results of the managed float experience to flexible exchange rates proper.

This said, the impression that one gets from the observation of almost thirty years of managed floats is that the claims of the advocates of the two extreme regimes in favour of the supported regime and against the other were rather exaggerated. To put it differently, the fluctuation has proved to be neither the panacea that advocates of flexible rates (opponents of fixed rates) claimed, nor the disaster that opponents of flexible rates (advocates of fixed rates) predicted.

An in-depth examination of the empirical evidence would be outside the scope of the present book, so that we shall only give a brief outline (with the exception of the vicious circle problem, on which see Sect. 17.4.1), referring the reader to the relevant literature, e.g. Eichengreen (1996), De Grauwe (1996), Ghosh et al. (2003), Klein and Shambaugh (2010), and Rose (2011).

As regards the problem of *external adjustment*, it does not seem that a substantial equilibrium in the various countries' balances of payments (either on current account or overall) has come about; concomitantly, large stocks of *international reserves* have had to be kept on hand by the various countries. But one should bear in mind that in the 1970s various oil crises (of which the first, at the end of 1973, was particularly serious: see Sect. 22.6) came about. Against these, flexible exchange rates are impotent, because of the short-run price rigidity of the demand for oil and the low short-run price elasticity of imports of oil-exporting countries. One must therefore wonder what would have happened under fixed exchange rates, namely whether the balance-of-payments disequilibria would have been greater than those actually observed. For empirical contributions on this literature see, for example, the following studies. Chinn and Wei (2013) , using data on over 170 countries for the 1971–2005 period, analyze the relationship between the exchange rate regimes and the speed of current account adjustment and do not find any evidence supporting the hypothesis that the current account reversion to its long run equilibrium is faster under flexible exchange rate regimes, even after accounting for the degree of economic development and for trade and capital account openness. Pancaro (2013) provides a systematic analysis of current account reversals across different de facto exchange rate regimes. In particular, the study examines whether current account reversals in industrial economies follow different patterns depending on the exchange rate regime in place and finds that larger deficits and larger output gaps are associated with a higher probability of experiencing reversals across all exchange rate regimes. Lane and Milesi-Ferretti (2012) explore the process of adjustment of external imbalances between 2008 and 2010, considering a large sample of countries and taking into account the countries' de-facto exchange rate regimes. Their results suggest that the external adjustment in deficit countries worked primarily through a contraction of expenditure and output. The real effective exchange rates moved in a destabilizing direction for pegging countries and were weakly tied to the current account in the case of countries with intermediate or floating regimes.

Other studies support the Friedman hypothesis claiming that flexible exchange rates are more suitable for avoiding current account imbalances than fixed exchange rate regimes. Among others, Ghosh et al. (2010) show that current account dynamics differ depending on whether current accounts are in deficit or in surplus, and on whether imbalances are large or small. Flexible exchange rate regimes seem to be associated with faster adjustment of both small deficits and surpluses, and, in particular, of large surpluses. In contrast, flexible exchange rates do not lead to faster adjustment of large deficits: in this case, intermediate regimes exhibit the lowest persistence. Furthermore, Ghosh et al. (2013, 2014) argue that evidence on whether floating exchange rates facilitate external adjustment is contradictory because existing regime classifications do not adequately capture exchange rate flexibility relevant to external adjustment. Using a trade-weighted bilateral exchange rate volatility in order to measure the exchange rate regimes more precisely, the authors show that exchange rate flexibility indeed matters for current account dynamics.

The hoped-for (greater) autonomy of monetary policy has not come about. This is not surprising, if one thinks that monetary independence can be achieved only if exchange-rate flexibility is such as to maintain the balance of payments always in equilibrium, so that there are no effects on the domestic money stock; but this might require exchange-rate variations so ample and frequent as to be undesirable because of their side effects. Leaving aside the problem of inflation which will be dealt with subsequently, the exchange-rate variations influence the current account insofar as resources move into and out of export and import-competing domestic industries, hence causing possible resource reallocation costs (already mentioned under point (3) of the criticism against flexible rates, at the end of Sect. 17.1).

As a matter of fact the European countries have often asked the US to ease its monetary policy so that they could ease theirs and avoid making undesired monetary restrictions at home. It has been rightly pointed out that, whilst in the Bretton Woods era the European countries had to follow US monetary policy to prevent excessive oscillations in their balances of payments, in the floating exchanges era the same countries have again had to follow US monetary policy to prevent excessive oscillations in their exchange rates vis-à-vis the dollar. The reasons are different, but a constraint on monetary policy exists in both regimes. For a review of the trade-off among exchange stability, monetary independence, and capital market openness (trilemma or impossible trinity of open-economy macroeconomic policy,) see Sect. 20.4 and Obstfeld et al. (2005).

International trade does not seem to have been negatively influenced by exchange-rate variability, and has continued growing at sustained rates, but of course it is not easy to ascertain whether it would have grown more (or less) if exchange rates had been fixed. Bacchetta and Wincoop (2000) have shown that, in general, both trade and welfare can be higher under either fixed or flexible exchange rates, depending on preferences and on the monetary-policy rules followed under each system. More recently, the use of refined quantitative methods brings more scepticism about the effects of short-term exchange-rate volatility on international trade (Clark et al. 2004; Tenreyro 2007; Adam and Cobham 2007; Egger 2008). More specifically, the relationship between the two variables seems most likely driven by underlining long-term policy credibility rather than the short-term relation (Klein and Shambaugh 2006; Qureshi and Tsangarides 2010). Broda and Romalis (2010) focus on the reverse causality, that is trade flows help to stabilize real exchange rate fluctuations, thus reducing exchange rate volatility.

The exchange-rate variability has been high, but no evidence seems to exist for destabilising *speculation* of the explosive type described in Sect. 16.2. Speculative episodes have undoubtedly taken place, which—since the regime was a managed float with a more or less heavy intervention of the monetary authorities, sometimes to support a certain (not officially declared) level of the exchange rate—have been more similar to those which occurred under the adjustable peg than to those feared by the opponents of flexible exchange rates. One of the main contributions to the volatility of exchange rates seems to have been given by the huge capital movements induced by interest arbitrage (of course, insofar as these movements were not covered, they also contained a speculative element).

One of the main problems that the individual countries and the international monetary system have had to face since the beginning of the managed float era is *inflation*. It is fair to point out that the inflation concomitant to floating exchange rates had its roots in the previous fixed-exchanges era, but many wonder whether the float, though not the cause of inflation, has been an amplifying agent. This will be dealt with in the next section, but we wish to conclude the present treatment with a very simple observation. The experience of the European Monetary System (which will be dealt with in Sect. 21.1) shows that many parity realignments have taken place among the partner countries. Now, if this has been necessary within a currency area formed by countries already linked by economic integration agreements (the

EEC), it would have been *a fortiori* necessary for other less integrated countries if all the world had been under an adjustable peg in the same period. In other words, it is presumable that a system of fixed exchange rates at world level could not have held out. A return to an international monetary system based on fixed exchange rates does not seem likely in the foreseeable future.

More recent literature emphasizes possible endogeneity of the choice of exchange rate regime and focuses on the determinants of this choice rather than its effects on macroeconomic variables (see for example Levy-Yeyati et al. 2010, Berdiev et al. 2012).

Let us conclude by mentioning a problem related to the managed float, i.e. the criteria for the *authorities' intervention* in the foreign exchange market (see also Sect. 3.2.1). Various criteria are in principle possible: one consists in contrasting excessive movements (if any) of the exchange rate in either direction *(leaning against the wind)*, another in managing the exchange rate towards a desired value or zone *(target approach)*. The latter can in turn take various forms, according to whether the authorities wish to use the exchange rate (insofar as they can) as a *tool* to achieve certain targets (this would be, for example, the case of the new Cambridge school treated in Sect. 12.4) or wish to guide it towards its *equilibrium* value (on the equilibrium exchange rate see Sect. 15.6) thus reducing exchange rate misalignment. Let us mention two other motives behind the authorities' intervention into the foreign exchange market: managing or accumulating foreign currency reserves and ensuring adequate liquidity. After the Asian financial crisis many central banks officially announced that intervention would be conducted for both the purpose of building reserves for precautionary motives and the purpose of ensuring adequate liquidity to counter disorderly markets and avoid financial stress.

In any case the member countries of the IMF have agreed (Second Amendment to the Articles of Agreement of the Fund, which came into force in March, 1978) to adhere to certain general principles in their interventions in the exchange markets, amongst which that of not manipulating exchange rates in order to prevent effective balance of payments adjustment or in order to gain an unfair competitive advantage over other members. The Fund, according to this Amendment, shall exercise firm surveillance over the exchange rate policies of members, which must consult with the Fund in establishing these policies.

For a treatment of these problems both from a theoretical point of view and from the point of view of actual practice, see, for example, Polak 1999. Central bank have often used direct interventions as a tool to stabilize short-run trends or to correct long term misalignments of the exchange rate. The large literature on the impact and the effectiveness of these interventions provides mixed evidence (see Beine et al. 2009, Dominguez 2006, Sarno and Taylor 2001, among others).

17.4   Exchange-Rate Pass-Through

Price responses to exchange rate movements are one of the central topics in international macroeconomics. Why don't the prices of imported goods fully reflect exchange rate movements? Indeed, abundant empirical evidence shows that the exchange rate elasticity of import prices is rather low and documents the phenomenon of incomplete exchange rate pass-through (see the comprehensive literature review by Burstein and Gopinath 2014).

Exchange rate pass-through (ERPT) refers to the degree to which changes in the exchange rate "pass-through" to import prices, and then to the general price level and the corresponding rate of inflation. The original definition referred to the percentage change in import prices in response to a 1 % change in the exchange rate (called Stage 1 pass-through). This effect is believed to occur fairly quickly. Many studies have extended the effect of exchange rate movements on producer or consumer prices (overall pass-through). The effect of a change in import prices on producer or consumer prices is known as Stage 2 pass-through. This second phase is more difficult to estimate, as it depends on a broad range of hard-to-measure variables, including inflation expectations and the credibility of monetary policy, and is believed to be quite slow – with many estimates suggesting roughly 3 to 5 years before most of the adjustment is made. A slow pass-through can give rise to J-curve phenomena (see Sect. 9.2).

The sensitivity to the exchange rate will decline down the price distribution chain, from import prices through producer prices to final consumer prices. ERPT to prices is incomplete if exchange rate changes elicit less than equi-proportionate changes in prices. Although the ERPT concept is simple, many factors can influence this relationship, thus to exactly predict how exchange rate movements affect prices is not straightforward. Much of the existing research focuses on the relationship between movements in nominal exchange rates and import prices. A smaller but equally important strand of the literature concentrates on the macroeconomic exchange rate pass-through to aggregate price indices. In the literature, it is almost universally recognized that at all levels of aggregation, exchange rate pass-through is less than full and much of the price response occurs with a substantial delay. For space reasons we mention only a few important studies that are frequently cited.

Studies conducted for the case of developed countries include Anderton (2003), Campa and Goldberg (2005), Campa and González (2006), Gagnon and Ihrig (2004), Hahn (2003), Faruqee (2006), Ihrig et al. (2006) and McCarthy (2007). There are also many studies applied to emerging market economies, including cross-country comparisons as in Choudhri and Hakura (2006), Frankel et al. (2012) and Mihaljek and Klau (2000). Cunningham and Haldane (2000) show evidence of a reduced pass-through in the United Kingdom, Sweden and Brazil (see Burstein and Gopinath (2014) for a survey of the literature).

A widely cited explanation for incomplete pass-through is the Dornbusch (1987) and Krugman (1987) pricing-to-market model of firms' price-setting behavior in relation to changes in the exchange rate, where incomplete pass-through depends on imperfect competition and market segmentation. Firms adjust their mark-up (and not only prices) to accommodate the local market environment in response to an exchange-rate shock.

Menu costs of price adjustment and long-term contracts are another common explanation for incomplete ERPT. Menu costs include the administrative, technical and informational costs to implement a price change (Ball and Mankiw 1994; Andersen 1994). Burstein et al. (2003) emphasize the role of (non-traded) domestic inputs in the chain of distribution of tradable goods.

Another line of research stresses the role that monetary and fiscal authorities play, by partly offsetting the impact of changes in the exchange rate on prices (Gagnon and Ihrig 2004; see also Mishkin 2008 for a discussion of exchange-rate pass-through and the implications for monetary policy). Devereux and Engel (2002), Bacchetta and van Wincoop (2005), and Devereux et al. (2004) explore the role of local currency pricing and slow nominal price adjustment in reducing the degree of ERPT both at the import price level and at the level of retail prices.

Taylor (2000) analyses the hypothesis that the responsiveness of prices to exchange rate fluctuations depends positively on inflation. He explains that the shift toward more credible monetary policy and thus a low-inflation regime would reduce the transmission of exchange rate changes.

Studies at the level of industries, firms, products or retail goods lend insight into underlying structural price adjustments and the sources of incomplete and changing ERPT (for example Gopinath and Rigobon (2008) , Gopinath et al. (2010), Auer and Schoenle (2015) for the USA). Key findings from the micro-data studies for industrial countries on ERPT are that:

(a) heterogeneous ERPT estimates are typically founded at the sectorial and goods levels, and ERPT is delayed and incomplete for imports, and for both retail and wholesale domestic prices;
(b) goods with frequently adjusting import and export trade prices have a far higher long-run ERPT than low-frequency adjusters. Exporters to the US of homogeneous goods (e.g. raw products) mainly price in dollars and adjust prices more frequently than for product-differentiated goods, for which there is a higher proportion of non-dollar pricers (emphasizing that choice of the invoicing currency is endogenous);
(c) an important source of incomplete ERPT for the destination country's retail and wholesale prices is the combination of non-traded local costs in the destination market and imported inputs into the exporter's good;
(d) mark-up adjustment accounts for the gap between actual and complete ERPT;
(e) the role of nominal rigidities appears small in the longer-run but is important in the delayed response of prices to cost in the short-run.

Burstein et al. (2003) and Goldberg and Campa (2010) suggest that the greater use of imported inputs in traded and non-traded goods across countries and industries is the key contributor to changing ERPT – not changes in distribution margins. Berman et al. (2012), Chatterjee et al. (2013), and Amiti et al. (2014)

find support for the prediction that the extent of pass-through is inversely related to the size of the firm, using large across-industries firm-level data sets from France, Brazil, and Belgium: small non importing firms have a nearly complete pass-through, while large import-intensive exporters have a pass-through around 50 percent, with the marginal cost and markup channels contributing roughly equally. The quality of exported varieties has been investigated as an additional determinant of ERPT heterogeneity (Auer and Chaney 2009; Chen and Juvenal 2014) . Finally, a paper by Gust et al. (2010) suggests that the process of international globalization (that is, lower trade costs) itself may induce a fall in pass-through: lower trade costs increase the exporting firm's relative markup, which in turn allows the firm's prices to be less sensitive to exchange rates.

An important issue that has received attention in the recent literature refers to the sources of the exchange rate fluctuation, i.e. the types of shocks and their persistence.

The most part of theoretical and empirical studies takes the exchange rate fluctuation as given, i.e., as an exogenous shock, although very different types of factors or shocks cause exchange rate dynamics (Shambaugh 2008). Corsetti et al. (2008) stress the importance of controlling for the general-equilibrium effects of the shocks leading to exchange rate movements when measuring pass-through. Bussiere et al. (2015) undertake a broad empirical and theoretical assessment of the relation between appreciation and growth, paying particular attention to underlying factors that triggered the appreciation in the first place. They show that exchange rate appreciations driven by domestic productivity shocks would have different effects on trade prices and growth than movements driven by surges in capital inflows.

Nevertheless, the degree of exchange rate pass-through and the causes of its decline are difficult to pin down with certainty. Whether the behaviour of exchange rate pass-through is attributed to sticky prices or to more structural features of international trade is important. Much of the recent debate deals, mainly, with issues like:

(a) dynamic factors in explaining pass-through. For example, by considering the dynamic menu cost model of price-setting according to which the pricing behavior of firms systematically and substantially differs across domestic and export markets in terms of frequency, timing and size of price changes. By contrast, in static pricing models firms set prices purely according to a fixed schedule or change prices with a fixed probability;

(b) strategic complementarities in pricing generated by different models of consumer demand that can amplify the delays in price adjustment beyond the duration of price rigidity;

(c) firm-productivity heterogeneity, as the availability of extremely large and disaggregated data firm-level customs and balance sheet data representing a wide range of goods and information on both importers and exporters permits to explore the roles of invoice currency, firm characteristics, and market shares in explaining the extent to which firms adjust export prices in response to exchange rate fluctuations;

(d) firms' forward-looking behaviour in price adjustment, that is firms incorporate
    expectations of future exchange rate changes into their current pricing decisions.

17.4.1 The Vicious Circle Depreciation-Inflation

17.4.1.1 Introductory Remarks

The pass-through phenomenon is related to the so-called "depreciation-inflation
vicious circle" (which is only one aspect of the phenomenon of inflation in open
economies), that is a situation in which an exchange-rate depreciation causes a
domestic price increase such as to prevent the hoped-for benefits of the depreciation
(gain of competitiveness and restoration of balance-of-payments equilibrium) from
coming about, thus calling for a new depreciation, and so forth. The opposite
virtuous circle between exchange-rate appreciation and price stability or deflation
also exists, to which one can apply a reasoning symmetrical to that carried out below
in relation to the vicious circle (the symmetry is not perfect, however, given the
downward rigidity of prices).

The danger of the vicious circle, should the existence of this circle be proved,
is obvious. It, in fact, would preclude the use of exchange-rate depreciation as a
means to restore equilibrium in the current account balance and, if the depreciation
were imposed by capital account disequilibria (remember that the depreciationary
pressures on the exchange rate come from the overall payment imbalance), it would
give rise to domestic inflationary effects etc., such as to undermine the economic
stability of the country affected by it.

The problems arising in the study of the vicious circle are many, but, at the
cost of drastic simplification for didactic purposes, they can be condensed into two
questions:

(1) does a depreciation-inflation circle exist?
(2) if so, is this circle really vicious?

In the first question we have purposely omitted the adjective vicious; i.e. we are
only asking whether *causality links* exist between depreciation and inflation which
go from depreciation to inflation and from inflation to depreciation. The presence of
these links is, in fact, a necessary, but not sufficient, condition for the phenomenon
under examination to occur.

For the circle to be really vicious, the inflation induced by the depreciation must
be such as to *prevent* the hoped-for benefits of the depreciation from coming about,
as we said at the beginning.

The next two sections will be dedicated to a brief examination of the two above
questions.

17.4.1.2 The Depreciation-Inflation Circle

The problem of a depreciation-inflation circle, as has been shown by De Cecco
(1983), was already present in seventeenth century writers, and was hotly debated in

the 1920s. We shall, for brevity, refer only to the contemporary debate. The literature on the circle has followed two different approaches: that based on purely statistical tests, and that of *structural macroeconomic models.*

The first approach started with causality tests, that are statistical tests aimed at finding the existence, if any, of a "causality" between two variables (in the case under consideration these are the exchange rate and the price level). It should be noted that the "causality" referred to is to be understood in a statistical sense (to be clarified presently) which does not necessarily coincide with the concept in use in the hard sciences (physics, chemistry, etc.) and in economic theory, but we leave this problem to philosophers of science. Loosely speaking, a variable x is said to "cause" another variable y if, by using the current and past values of x in addition to the past values of y, it is possible to obtain a statistically better prediction of the current value of y than by using the past values of y only. If, besides the fact that x causes y, it is also true that y causes x (in the definition, simply interchange x and y), then two-way or bidirectional causality (also called feedback) is said to exist between x and y.

This brief technical premise (for a detailed treatment of causality and of the relative tests the reader see, for example, Geweke 1984) is necessary to understand that the existence of a depreciation-inflation (or exchange rate-price level) circle is proved, according to this approach, by the existence of causality between the exchange rate and the general price level. It goes without saying that causality must be *bidirectional*, as the presence of mere *unidirectional* causality (from the exchange rate to prices or vice versa) is not enough to prove the existence of the circle.

The early studies which followed this approach did not give unequivocal results. A study by Falchi and Michelangeli (1977) found the circle only as regards Italy, whilst in the UK causality was unidirectional (from prices to the exchange rate, but not vice versa); in the other two countries considered (France and Germany) the result of the test was negative, as it revealed no causality in either direction.

A subsequent study by Kawai (1980) examined ten industrialized countries and found the presence of the circle not only in Italy, but also in Belgium, the Netherlands, Switzerland and Japan (in the last two the circle would be of the virtuous type). Contrary to Falchi and Michelangeli's results, Kawai did not find any causality in the UK.

Causality tests of the depreciation-inflation relationship have subsequently been abandoned in favour of more recent statistical techniques, such as impulse-response functions and variance decompositions derived from VAR (Vector Auto-Regression) analysis. Using this methodology, McCarthy (1999) finds that the pass-through of external factors (exchange-rate depreciation, rise in import prices) to domestic inflation for several industrialised economies is quite modest. A different methodology is employed by Goldfajn and Ribeiro da Costa Werlang (2000), who use panel data analysis on a sample of 71 countries (monthly data, 1980–1998) and find that the pass-through from depreciation to inflation is much lower for OECD countries than for developing countries, but is in any case smaller than 100 %.

The *second approach* consists, as we said, in building structural macroeconomic models, where the relations between the main macroeconomic variables are adequately modelled, so as to explain the circle, find its ultimate economic cause(s) and hence derive policy suggestions to influence it. This is certainly an approach that is more consistent with the way of thinking of an economist, but a problem immediately arises. As we know and have seen in the previous chapters, there are many schools of thought in (open economy) macroeconomics. The vision of the functioning of the economy and so of the transmission mechanisms is different, often drastically different, from one school to the other; different will then be the explanation of the circle and the policy suggestions (Mastropasqua 1984; Spaventa 1983).

There is however a fact which is generally accepted. Suppose that inflation is measured according to a general price index in which the prices of both domestic (p_h) and foreign (rp_f) goods are present, say

$$p = p_h^\alpha (rp_f)^{1-\alpha},$$

where $\alpha, (1 - \alpha)$ are the weights. Suppose that p_f is constant. An exchange rate depreciation (r increases) does cause an increase in p for three reasons. The first is the increase in the (rp_f) term, namely in the domestic-currency price of imported final goods. The second is the possible increase in p_h caused by the increase in the price of imported intermediate inputs (for oil-importing countries this relationship is quite obvious). This is particularly evident in the case of market forms which are not perfectly competitive, in which price setting by firms based on a markup on production costs is often adopted. Flexible wages that adjust to the price level are a possible third cause of increase in the price of domestic goods via markup pricing.

The inflationary bias of an exchange rate depreciation becomes more intense if one accepts the existence of overshooting phenomena (see Sects. 13.4.4 and 15.3.2) coupled with a downward price rigidity (or at least with a lower flexibility downwards than upwards). In such a case, in fact, the huge price increase caused by the exchange-rate overshooting will not be offset (or will only be partially offset) by the subsequent appreciation of the exchange rate.

17.4.1.3 Is the Circle Really Vicious?

To further clarify what we said in the previous section, we point out that the depreciation-inflation circle can either die away spontaneously or perpetuate itself and (even) become explosive: only in the latter case is it correct, in our opinion, to talk of a really vicious circle.

By a circle which dies away spontaneously we mean a situation of the following type: given a $d\%$ exchange-rate depreciation, prices increase *less* than proportionally, say by $p\%$ (where $p < d$: in other words, the *pass-through coefficient* $(p/d)\%$ is smaller than 100%); this increase is followed by a new depreciation which is, however, smaller than the previous one (say $d'\%$, where $d' < d$), hence a new price increase but by a smaller percentage than before, say by $p'\%$, where $p' < p$, and so on and so forth. As can be seen, depreciation and inflation converge to zero, *ceteris paribus*, so that the circle works itself out spontaneously and cannot be said to be vicious. Completely different is the case in which the above percentages are non-decreasing or increasing ($p \geq d, d' \geq d, p' \geq p$, etc.): the circle is really vicious! For a formalization of these aspects by means of a dynamic model see the Appendix.

It is also essential to determine the *adjustment lags*. These are extremely important, as it is obvious that the situation is completely different when the *p%* inflation determined by a *d%* depreciation comes about in one month or in six months or after several years; the same applies to the inflation-to-depreciation causation. The longer the adjustment lags, the less worrying the circle. As mentioned above (page 435), many estimates suggest that it takes several years for most of the adjustment to be made.

17.5 Appendix

17.5.1 The Shock-Insulating Properties of Fixed and Flexible Exchange Rates

There are several models for evaluating the shock-insulating properties of different exchange-rate arrangements. In the text we used a modified form of the Mundell-Fleming model. The model presented here (Pilbeam 1991, 2013) is more elaborated, though based on the same general ideas. In the following exposition all variables except interest rates are expressed in logarithms.

Money demand in real terms is a standard form; u_L denotes a transitory shock term, normally distributed with mean zero:

$$M_d - p = \eta y - \lambda i + + u_L. \tag{17.1}$$

The variable p is an *aggregate price index* where domestic prices and foreign prices enter with weights α and $(1 - \alpha)$ respectively. Since foreign prices p_f are expressed in foreign currency, they are converted into domestic currency by the exchange rate r:

$$p = \alpha p_h + (1 - \alpha)(r + p_f). \tag{17.2}$$

Total aggregate demand for domestic output is positively related to the real exchange rate (on the assumption that the current account positively reacts to a depreciation) and to the 'natural' level of income, while it is negatively related to the real interest rate:

$$y_d = \theta r_R + \pi y_n - \beta i_R + u_{y_d}, \tag{17.3}$$

where u_{y_d} is a transitory aggregate demand shock, normally distributed with mean zero.

The *real exchange rate* is defined according to PPP (see Chap. 2, Sect. 2.3)

$$r_R = r + p_f - p_h. \tag{17.4}$$

The *real interest rate* is defined according to the Fisher equation (see Chap 4), namely as the current nominal interest rate minus the expected inflation rate:

$$i_R = i - (\tilde{p}_h - p_h),$$

where $\tilde{p}_h$ is the (log of the) expected price level; since the variables are in logs, $\tilde{p}_h - p_h$ measures the expected inflation rate.

Perfect capital mobility and perfect asset substitutability are also assumed, so that uncovered interest parity (*UIP*, see Chap. 4) holds

$$i = i_f + (\tilde{r} - r), \tag{17.5}$$

where $\tilde{r}$ is the (log of the) expected nominal exchange rate.

The *supply of domestic output* is inversely related to the real wage rate: if the price at which producers can sell their output rises relative to the (nominal) wage rate, they will increase output, and vice versa:

$$y_s = \sigma(p_h - w) + u_{y_s}, \tag{17.6}$$

where u_{y_s} is a transitory aggregate supply shock with zero mean and normally distributed. *Employment* is related to domestic output through an inverse production function with given capital stock and variable labour input:

$$N = N(y_s), \frac{\partial N}{\partial y_s} > 0, \frac{\partial^2 N}{\partial y_s^2} < 0,$$

where the sign of the second derivative reflects the diminishing marginal productivity of labour.

The setting of the *nominal wage rate* takes place through contracting agreements in each period which establish the wage rate at the level required to generate an expected output y_n which is the 'natural' rate of output and is also the target level of the authorities:

$$w = w^*. \tag{17.7}$$

where w^* is the wage rate required to generate the target output level y_n in the absence of any shocks.

Under fixed exchange rates, perfect capital mobility with perfect asset substitutability implies that the domestic interest rate equals the foreign interest rate and the money supply is endogenously determined. Under flexible exchange rates, the money supply is exogenous and the exchange rate and the domestic interest rate are endogenously determined but tied together through UIP.

Since we are dealing with transitory, self-reversing shocks to equilibrium, the economy is always expected to return to its equilibrium situation, so that we can make the following simple assumption as regards expectations (normal or regressive expectations): if the price or output levels increase above (or decrease below) their current normal equilibrium level due to a shock, they are expected to return to their normal levels next period. Similarly, if the exchange rate depreciates (appreciates) today, it is expected to appreciate (depreciate) back to its normal level tomorrow.

17.5.2 The Effects of Various Shocks

To examine the effects of shocks we start from an equilibrium conditions in which money demand and supply are equal, and output is at its natural level y_n, which is taken as fixed in the short run. Thus we have the equations of change in the neighbourhood of equilibrium:

$$dM_s = \alpha dp_h + (1 - \alpha)d(r + p_f) + \eta dy - \lambda d[i_f + (\tilde{r} - r)] + du_L,$$

$$dy_d = \theta d(r + p_f - p_h) - \beta d\left[i_f + (\tilde{r} - r) - (\tilde{p}_h - p_h)\right] + \pi dy_n + du_{y_d},$$

$$dy_s = \sigma d\left(p_h - w^*\right) + du_{y_s}.$$

Since y_n is constant, $dy_n = 0$, and correspondingly $dw^* = 0$. Starting from an equilibrium situation displaced by a transitory shock, and given the assumption on expectations, it follows that $\tilde{r}$ and $\tilde{p}_h$ are equal to the respective normal equilibrium values, which are constant, hence $d\tilde{r} = d\tilde{p}_h = 0$. Finally, we assume that nothing happens in the rest of the world, so that foreign variables do not change. Thus our equations of change reduce to

$$\begin{aligned}
dM_s - \alpha dp_h - \eta dy - (1 - \alpha - \lambda)dr &= du_L, \\
(\theta + \beta)dp_h + dy - (\theta + \beta)dr &= du_{y_d}, \\
-\sigma dp_h + dy &= du_{y_s},
\end{aligned} \tag{17.8}$$

where $dy = dy_d = dy_s$ is the displacement in the equilibrium value of output. System (17.8) can easily be solved for the effects of the various shocks.

17.5.2.1 Money Demand Shock

We have $du_L \neq 0, du_{y_d} = du_{y_s} = 0$, so that the basic system becomes

$$\frac{dM_s}{du_L} - \alpha\frac{dp_h}{du_L} - \eta\frac{dy}{du_L} - (1 - \alpha - \lambda)\frac{dr}{du_L} = 1,$$
$$(\theta + \beta)\frac{dp_h}{du_L} + \frac{dy}{du_L} - (\theta + \beta)\frac{dr}{du_L} = 0, \qquad (17.9)$$
$$-\sigma\frac{dp_h}{du_L} + \frac{dy}{du_L} = 0.$$

Under *fixed* exchange rates, $dr/du_L = 0$. Thus the system is

$$\frac{dM_s}{du_L} - \alpha\frac{dp_h}{du_L} - \eta\frac{dy}{du_L} = 1,$$
$$(\theta + \beta)\frac{dp_h}{du_L} + \frac{dy}{du_L} = 0, \qquad (17.10)$$
$$-\frac{dp_h}{du_L} + \frac{dy}{du_L} = 0.$$

From the last two equations we see that, since the determinant

$$\begin{vmatrix} (\theta + \beta) & 1 \\ -\sigma & 1 \end{vmatrix} = \sigma + (\theta + \beta)$$

is different from zero, the only solution is $\dfrac{dp_h}{du_L} = \dfrac{dy}{du_L} = 0$. From the first equation we obtain $\dfrac{dM_s}{du_L} = 1$, namely $\dfrac{dM_s}{du_L} = \dfrac{dM_d}{du_L}$. No adjustment to either prices or output is required: all the authorities have to do is to increase the money supply to keep it equal to the increased money demand.

Under *flexible* exchange rates, $\dfrac{dM_s}{du_L} = 0$ (the money supply does not change). Thus we have the system

$$-\alpha\frac{dp_h}{du_L} - \eta\frac{dy}{du_L} - (1 - \alpha + \lambda)\frac{dr}{du_L} = 1,$$
$$(\theta + \beta)\frac{dp_h}{du_L} + \frac{dy}{du_L} - (\theta + \beta)\frac{dr}{du_L} = 0, \qquad (17.11)$$
$$-\sigma\frac{dp_h}{du_L} + \frac{dy}{du_L} = 0.$$

Defining

$$\Delta \equiv \begin{vmatrix} -\alpha & -\eta & -(1-\alpha+\lambda) \\ (\theta+\beta) & 1 & -(\theta+\beta) \\ -\sigma & 1 & 0 \end{vmatrix}$$

$$= -\sigma[\eta(\theta+\beta)+(1-\alpha+\lambda)] - [\alpha(\theta+\beta)+(\theta+\beta)(1-\alpha+\lambda)]$$

$$= -(\theta+\beta)(1+\lambda+\eta\sigma) - \sigma(1-\alpha+\lambda) < 0 \qquad (17.12)$$

and solving, we obtain

$$\frac{dp_h}{du_L} = \frac{\theta+\beta}{\Delta} < 0, \quad \frac{dy}{du_L} = \frac{\sigma(\theta+\beta)}{\Delta} < 0, \quad \frac{dr}{du_L} = \frac{\theta+\beta+\sigma}{\Delta} < 0, \qquad (17.13)$$

hence both prices and output decline (due to the appreciation in the exchange rate). Thus it is clear that fixed exchange rates are superior to flexible exchange rates.

17.5.2.2 Aggregate Demand Shock

We have $du_{y_d} \neq 0, du_L = du_{y_s} = 0$, so that the basic system becomes

$$\frac{dM_s}{du_{y_d}} - \alpha\frac{dp_h}{du_{y_d}} - \eta\frac{dy}{du_{y_d}} - (1-\alpha+\lambda)\frac{dr}{du_{y_d}} = 0,$$

$$(\theta+\beta)\frac{dp_h}{du_{y_d}} + \frac{dy}{du_{y_d}} - (\theta+\beta)\frac{dr}{du_{y_d}} = 1, \qquad (17.14)$$

$$-\sigma\frac{dp_h}{du_{y_d}} + \frac{dy}{du_{y_d}} = 0.$$

Under *fixed* exchange rates we have $\dfrac{dr}{du_{y_d}} = 0$, hence the second and third equations determine $\dfrac{dp_h}{du_{y_d}}, \dfrac{dy}{du_{y_d}}$, which turn out to be

$$\frac{dp_h}{du_{y_d}} = \frac{1}{\sigma+\theta+\beta} > 0, \quad \frac{dy}{du_{y_d}} = \frac{\sigma}{\sigma+\theta+\beta} > 0. \qquad (17.15)$$

The first equation then determines

$$\frac{dM_s}{du_{y_d}} = \frac{\alpha+\eta\sigma}{\sigma+\theta+\beta} > 0. \qquad (17.16)$$

Under *flexible* exchange rates we have $\dfrac{dM_s}{du_{y_d}} = 0$ and the system becomes

$$-\alpha\frac{dp_h}{du_{y_d}} - \eta\frac{dy}{du_{y_d}} - (1 - \alpha + \lambda)\frac{dr}{du_{y_d}} = 0,$$
$$(\theta + \beta)\frac{dp_h}{du_{y_d}} + \frac{dy}{du_{y_d}} - (\theta + \beta)\frac{dr}{du_{y_d}} = 1, \qquad (17.17)$$
$$-\sigma\frac{dp_h}{du_{y_d}} + \frac{dy}{du_{y_d}} = 0.$$

Solving we obtain

$$\frac{dp_h}{du_{y_d}} = -\frac{1 - \alpha + \lambda}{\Delta} > 0, \quad \frac{dy}{du_{y_d}} = \frac{-\sigma(1 - \alpha + \lambda)}{\Delta} > 0, \quad \frac{dr}{du_{y_d}} = \frac{\alpha + \eta\sigma}{\Delta} < 0,$$
$$(17.18)$$

where Δ has been defined above, Eq. (17.12).

Prices and output move in the same direction under both regimes, but under flexible exchange rates the change is smaller in absolute value for both variables. In fact,

$$\left(\frac{dp_h}{du_{y_d}}\right)_{FLEX} < \left(\frac{dp_h}{du_{y_d}}\right)_{FIX},$$

namely, substituting from (17.18) and (17.15),

$$\frac{1 - \alpha + \lambda}{(\theta + \beta)(1 + \lambda + \eta\sigma) + \sigma(1 - \alpha + \lambda)} < \frac{1}{\sigma + \theta + \beta},$$

from which

$$(\sigma + \theta + \beta)(1 - \alpha + \lambda) < (\theta + \beta)(1 + \lambda + \eta\sigma) + \sigma(1 - \alpha + \lambda).$$

Canceling out identical terms we finally obtain

$$-\alpha < \eta\sigma$$

which is certainly true.

Similarly it can be shown that

$$\left(\frac{dy}{du_{y_d}}\right)_{FLEX} < \left(\frac{dy}{du_{y_d}}\right)_{FIX}. \qquad (17.19)$$

Thus flexible exchange rates are unambiguously superior.

17.5.2.3 Aggregate Supply Shock

We have $du_{y_s} \neq 0, du_L = du_{y_d} = 0$, so that the basic system becomes

$$\frac{dM_s}{du_{y_s}} - \alpha \frac{dp_h}{du_{y_s}} - \eta \frac{dy}{du_{y_s}} - (1 - \alpha + \lambda)\frac{dr}{du_{y_s}} = 0,$$

$$(\theta + \beta)\frac{dp_h}{du_{y_s}} + \frac{dy}{du_{y_s}} - (\theta + \beta)\frac{dr}{du_{y_s}} = 0, \qquad (17.20)$$

$$-\sigma \frac{dp_h}{du_{y_s}} + \frac{dy}{du_{y_s}} = -1.$$

Note that we have assumed a *negative* supply shock.

Under fixed exchange rates $\dfrac{dr}{du_{y_s}} = 0$, hence from the last two equations we obtain

$$\frac{dp_h}{du_{y_s}} = \frac{1}{\sigma + \theta + \beta} > 0, \frac{dy}{du_{y_s}} = -\frac{\theta + \beta}{\sigma + \theta + \beta} < 0. \qquad (17.21)$$

From the first equation we then have

$$\frac{dM_s}{du_{y_s}} = \frac{\alpha - \eta(\theta + \beta)}{\sigma + \theta + \beta}. \qquad (17.22)$$

Under flexible exchange rates $\dfrac{dM_s}{du_{y_s}} = 0$, hence solving the system we obtain

$$\frac{dp_h}{du_{y_s}} = -\frac{\eta(\theta + \beta) + (1 - \alpha + \lambda)}{\Delta} > 0, \frac{dy}{du_{y_s}} = \frac{(\theta + \beta)(1 + \lambda)}{\Delta} < 0,$$

$$\frac{dr}{du_{y_s}} = \frac{\alpha - \eta(\theta + \beta)}{\Delta} \gtrless 0, \qquad (17.23)$$

where Δ has been defined above, Eq. (17.12).

The results are qualitatively the same under both regimes. By examining the inequality

$$\left(\frac{dp_h}{du_{y_s}}\right)_{FIX} \gtrless \left(\frac{dp_h}{du_{y_s}}\right)_{FLEX} \qquad (17.24)$$

that is, substituting from (17.21) and (17.23),

$$\frac{1}{\sigma + \theta + \beta} \gtrless \frac{\eta(\theta + \beta) + (1 - \alpha + \lambda)}{(\theta + \beta)(1 + \lambda + \eta\sigma) + \sigma(1 - \alpha + \lambda)}, \qquad (17.25)$$

we obtain that the price increase will be higher (lower) under fixed than flexible exchange rates according as

$$\eta(\theta + \beta) \lessgtr \alpha. \tag{17.26}$$

The same inequality tells us whether

$$\left| \frac{dy}{du_{y_s}} \right|_{FIX} \gtrless \left| \frac{dy}{du_{y_s}} \right|_{FLEX} \tag{17.27}$$

namely whether the output decrease is in absolute value higher (lower) under fixed than flexible exchange rates. Thus when $\eta(\theta + \beta) > \alpha$ fixed exchange rates favour price stability while flexible exchange rates favour income stability; exactly the opposite is true when $\eta(\theta + \beta) < \alpha$. The choice between the two regimes will depend on the structural parameters of the economy as well as on the weights $\omega, (1 - \omega)$ given to output loss and inflation rise in the social preference function:

$$\Omega = \omega(y - y_n)^2 + (1 - \omega)(p_h - p_{h_n})^2, \tag{17.28}$$

$$\Omega_{FIX} = \omega \left[-\frac{\theta + \beta}{\sigma + \theta + \beta} \right]^2 + (1 - \omega) \left[\frac{1}{\sigma + \theta + \beta} \right]^2, \tag{17.29}$$

$$\Omega_{FLEX} = \omega \left[\frac{(\theta + \beta)(1 + \lambda)}{\Delta} \right]^2$$

$$+ (1 - \omega) \left[-\frac{\eta(\theta + \beta) + (1 - \alpha + \lambda)}{\Delta} \right]^2. \tag{17.30}$$

In fact, the inequality

$$\Omega_{FIX} \lessgtr \Omega_{FLEX} \tag{17.31}$$

obviously involves both the weights of the social preference function and the structural parameters of the economy.

17.5.2.4 Conclusion

The results obtained from this model are qualitatively the same as those obtained in the text using the simple Mundell-Fleming model. Different results would however be obtained introducing further complications, for example wage indexation (see Pilbeam 1991), but this—though leading to different choices according to the type of shock—would not eliminate the basic impossibility of declaring one regime definitely superior. On the contrary, it would introduce an additional element of uncertainty: the institutional setting of the economy (such as wage indexation).

It should also be observed that the two regimes have been evaluated in terms of the extent of the output and price changes caused by a shock. A more explicit comparison based on welfare levels could be introduced, in which case additional uncertainty would be introduced, due to the fact that insulation and welfare do not necessarily go hand in hand. For example, Fender (1994) shows that flexible exchange rates may insulate an economy from foreign monetary disturbances, but that such shocks may lead to higher expected welfare under fixed exchange rates.

17.5.3 The Intertemporal Approach

The intertemporal approach (see Chap. 18), being based on the explicit intertemporal optimization of the representative agent's utility function, is well suited to compare the various exchange rate regimes in terms of the optimal levels of this function. If, following a shock, such optimal level is higher under regime X than under regime Y, then regime X is unambiguously better. Obstfeld and Rogoff (2000), using a stochastic extension of their basic model (see Sect. 19.3.1), show that the optimal policy is to allow the exchange rate to fluctuate in response to cross-country differences in productivity shocks. However, as noted by Lane (2001; Sect. 12), welfare results in intertemporal models are highly sensitive to the precise denomination of price stickiness, the specification of preferences and financial structure, etc. This is one reason why the new open-economy macroeconomics is yet of only limited interest in policy circles.

17.5.4 The Simple Dynamics of the Depreciation-Inflation Circle

We present a simple dynamic model of the depreciation-inflation circle based on the pass-through with adjustment lags. Let π and ϱ respectively denote the proportional variations in the price level and in the exchange rate (the exchange rate is defined according to the price quotation system, so that an increase in ϱ means a depreciation). The parameter $k > 0$ measures the degree of pass-through from the exchange rate   proportional variations to the proportional variations in the price level:

$$\pi = k\varrho. \tag{17.32}$$

It is easy to see that $k \lesseqqgtr 1$ means incomplete, complete, overeffected pass-through. Denoting by $\gamma > 0$ the parameter that measures the subsequent proportional change in the exchange rate following the proportional change in the price level, we have

$$\varrho = \gamma\pi, \tag{17.33}$$

where $\gamma \lesseqqgtr 1$ when the feedback from the price level to the exchange rate is incomplete, complete, overeffected.

It should now be noted that the pass-through from exchange rate changes to prices and voice-overs are not instantaneous, but occur after a lag. This can be

introduced by means of partial adjustment equations (Gandolfo 2009, Sect. 12.4),
that is:

$$\frac{d\pi}{dt} = \delta(k\varrho - \pi), \qquad \frac{d\varrho}{dt} = \eta(\gamma\pi - \varrho), \tag{17.34}$$

where the coefficients of adjustment δ and η can be interpreted as the reciprocals of
the *mean time-lags*. Considering for example the first equation in (17.34), the mean
time-lag $1/\delta$ shows the time required for about 63 % of the discrepancy between
$k\varrho$ and π to be eliminated by changes in π following changes in ϱ. Similarly,
$1/\eta$ shows the time required for about 63 % of the discrepancy between $\gamma\pi$ and
ϱ to be eliminated by changes in ϱ following changes in π. A small coefficient of
adjustment implies a long lag, and voice-overs.

The differential equation system (17.34) is a first-order system in normal form,
whose characteristic equation is

$$\begin{vmatrix} -\delta - \lambda & \delta k \\ \eta\gamma & -\eta - \lambda \end{vmatrix} = 0, \tag{17.35}$$

where λ denotes the characteristic roots. Expanding the determinant we obtain

$$\lambda^2 + (\delta + \eta)\lambda + \delta\eta(1 - \gamma k) = 0. \tag{17.36}$$

It can be observed that the roots of Eq. (17.36) are both real: in fact, calculating the
discriminant we obtain

$$(\delta + \eta)^2 - 4\delta\eta(1 - \gamma k) = (\delta - \eta)^2 + 4\delta\eta\gamma k > 0. \tag{17.37}$$

For both roots to be stable all the coefficients of Eq. (17.36) must be positive, hence
the crucial stability condition turns out to be

$$\gamma k < 1. \tag{17.38}$$

This is certainly satisfied if both pass-through coefficients are smaller than 1, but
can be satisfied also when one coefficient is greater than 1, provided that the other is
sufficiently smaller than 1. When the stability condition is satisfied, the circle dies
away spontaneously.

When condition (17.38) is not satisfied, two cases may occur. The first is the
borderline case $\gamma k = 1$, which implies a negative real root and a zero real root, so
that the system converges to a constant (i.e., the circle converges to a constant rate
of inflation and a constant rate of depreciation). The second possible case occurs
when $\gamma k > 1$, so that the succession of the signs of the coefficients in Eq. (17.36)
is $+ + -$, hence the characteristic equation has one negative and one positive real
root. This means that the system possesses conditional stability of the saddle-point
type (Gandolfo 2009; p. 381): along the stable arm of the saddle the circle works
itself out spontaneously, but outside of it the circle is really vicious.

References

Adam, C. & Cobham, D. (2007). Exchange rate regimes and trade. *The Manchester School, 75*, 44–63.

Adams, C. & Greenwood, J. (1985). Dual exchange rate systems and capital controls: An investigation. *Journal of International Economics, 18*, 43–64.

Amiti, M., Itskhoki, O., & Konings, J. (2014). Importers, exporters, and exchange rate disconnect. *American Economic Review, 104*, 1942–1978.

Andersen, T. (1994). *Price rigidity: Causes and macroeconomic implications.* Oxford: Oxford University Press.

Anderton, B. (2003). Extra-Euro area manufacturing import prices and exchange rate pass-through. ECB Working Paper No. 219.

Argy, V. (1981). *The postwar international money crisis: An analysis.* London: Allen&Unwin.

Auer, R., & Chaney, T. (2009). Exchange rate pass-through in a competitive model of pricing-to-market. *Journal of Money, Credit and Banking, 41*, 151–175.

Auer, R., & Schoenle, R. (2015). Market structure and exchange rate pass-through. CEPR Discussion Paper No. 10585.

Bacchetta, P., & van Wincoop, E. (2000). Does exchange-rate stability increase trade and welfare? *American Economic Review, 90*, 1093–1109.

Bacchetta, P., & van Wincoop, E. (2005). A theory of the currency denomination of international trade. *Journal of International Economics, 67*, 295–319.

Ball, L., & Mankiw, N. G. (1994). Asymmetric price adjustment and economic fluctuations. *The Economic Journal, 104*, 247–261.

Beine M., De Grauwe, P., & Grimaldi, M. (2009). The impact of FX central bank intervention in a noise trading framework. *Journal of Banking and Finance, 33*, 1187–1195.

Berdiev, A., Kim, Y., & Chang, C. P. (2012). The political economy of exchange rate regimes in developed and developing countries. *European Journal of Political Economy, 28*, 38–53.

Berman, N., Martin, P., & Mayer T. (2012), How do different firms react to exchange rate variations? *Quarterly Journal of Economics, 127*, 437–492.

Broda, C., & Romalis, J. (2010). Identifying the relationship between trade and exchange rate volatility, NBER Chapters in *Commodity Prices and Markets*, East Asia Seminar on Economics 20,79–110, National Bureau of Economic Research, Inc.

Burstein, A. T., Neves, J. C., & Rebelo, S. (2003). Distribution costs and real exchange rate dynamics during exchange-rate-based stabilizations. *Journal of Monetary Economics, 50*, 1189–1214.

Burstein, A., & Gopinath, G. (2014). International prices and exchange rates. In G. Gopinath, E. Helpman, & K. Rogoff (Eds.), *Handbook of International Economics* (Vol. 4, pp. 391–451). Amsterdam: Elsevier.

Bussiere, M., Lopez C., & Tille, C. (2015). Do large real exchange rate appreciations matter for growth? *Economic Policy, 30*(81), 7–45.

Campa, J., & Goldberg, L. (2005). Exchange rate pass-through into import prices. *The Review of Economics and Statistics, 87*(4), 679–690.

Campa, J. M., & González, J. M. (2006). Difference in exchange rate pass-through in the Euro area. *European Economic Review, 50*, 121–145.

Chatterjee, A., Dix-Carneiro, R., & Vichyanond, J. (2013). Multi-product firms and exchange rate fluctuations. *American Economic Journal: Economic Policy, 5*, 77–110.

Chen, N., & Juvenal, L. (2014). Quality, trade and exchange rate pass-through. IMF Working Papers No 42.

Chinn, M., & Wei, S. -J. (2013). A faith-based initiative meets the evidence: Does a flexible exchange rate regime really facilitate current account adjustment? *Review of Economics and Statistics, 95*, 168–184.

Choudhri, E. U., & Hakura, D. S. (2006). Exchange rate pass-through to domestic prices: Does the inflationary environment matter? *Journal of International Money and Finance, 25*, 614–639.

Clark, P., Tamirisa, N., & Wei, S. -J. (2004). Exchange rate volatility and trade flows: Some new evidence, Occasional Paper No. 235, International Monetary Fund.

Corsetti, G., Dedola, L., & Leduc, S. (2008), High exchange-rate volatility and low pass- ihrough. *Journal of Monetary Economics, 55*, 1113–1128.

Cunningham, A., & Haldane A. G. 2000. The monetary transmission mechanism in the United Kingdom: Pass-through & policy rule. Central Bank of Chile Working Paper No. 83.

De Cecco, M. (1983). The vicious/virtuous circle debate in the twenties and in the seventies. *Banca Nazionale del Lavoro Quarterly Review, 146*, 285–303.

De Grauwe, P. (1996). *International Money: Postwar Trends and Theories* (2nd ed.). Oxford: Oxford University Press.

Devereux, M., & Engel, C. (2002). Exchange rate pass-through, exchange rate volatility, and exchange rate disconnect. *Journal of Monetary Economics, 49*, 913–940.

Devereux, M. B., Engel, C., & Storgaard, P. E. (2004). Endogenous exchange rate pass-through when nominal prices are set in advance. *Journal of International Economics, 63*, 263–291.

Dominguez, K. M. E. (2006). When do central bank interventions influence intra-daily and longer-term exchange rate movements? *Journal of International Money and Finance, 25*, 1051–1071.

Dornbusch, R. (1987). Exchange rates and prices. *American Economic Review, 77*, 93–106.

Egger, P. (2008). De facto exchange rate arrangement tightness and bilateral trade flows. *Economics Letters, 99*, 228–232.

Eichengreen, B. (1996). *Globalizing capital: A history of the international monetary system.* Princeton: Princeton University Press.

Falchi, G., & Michelangeli, M. (1977). Interazione fra tasso di cambio e inflazione: una verifica empirica della tesi del circolo vizioso, in Banca d'Italia. *Contributi alla ricerca economica* (vol. 7, pp. 51–74). Roma: Banca d'Italia.

Faruqee, H. (2006). Exchange rate pass-through in the Euro area. *IMF Staff Papers, 53*, 63–88.

Fender, J. (1994). On the desirability of insulation: A counterexample. *Journal of International Money and Finance, 13*, 232–38.

Flood, R. P. (1978). Exchange rate expectations in dual exchange markets. *Journal of International Economics, 8*, 65–77.

Frankel, J., Parsley, D., & Wei, S. -J. (2012). Slow pass-through around the world: A new import for developing countries? *Open Economies Review, 23*, 213–251.

Frenkel, J. A. (1999). *No single currency regime is right for all countries or at all times.* Essays in International Finance No. 215, International Finance Section, Princeton University.

Gagnon, J., & Ihrig, J. (2004). Monetary policy and exchange rate pass-through. *International Journal of Finance and Economics, 9*, 315–338.

Gandolfo, G. (2009). *Economic dynamics* (4th ed.). Berlin, Heidelberg, New York: Springer.

Geweke, J. (1984). Inference and causality in economic time series models. In: Z. Griliches & M. D. Intriligator (Eds.), *Handbook of econometrics* (Vol. 2, pp. 1101–44). Amsterdam: North-Holland.

Ghosh, A., Gulde, A., & Wolf, H. (2003). *Exchange rate regimes: Choices and consequences.* Cambridge, MA: MIT Press.

Ghosh, A., Terrones, M., & Zettelmeyer, J. (2010). Exchange rate regimes and external adjustment: New answers to an old debate. In: C. Wyplosz (Ed.), *The new international monetary system: Essays in honor of Alexander Swoboda.* London: Routledge.

Ghosh, A., Qureshi, M., & Tsangarides, C. (2013). Is exchange rate regime really irrelevant for external adjustment? *Economic Letters, 118*, 104–109.

Ghosh, A., Qureshi, M., & Tsangarides, C. (2014). Friedman redux: External adjustment and exchange rate flexibility. IMF Working Papers No. 146, International Monetary Fund.

Goldberg, L., & Campa J. (2010). The sensitivity of the CPI to exchange rates: Distribution margins, imported inputs, and trade exposure. *The Review of Economics and Statistics, 92*, 392–407.

Goldfajn, I., & Ribeiro da Costa Werlang, S. (2000). The pass-through from depreciation to inflation: A panel study. Banco Central do Brasil Working Paper No. 5.

Gopinath, G., Itskhoki O., & Rigobon, R. (2010). Currency choice and exchange rate pass-through. *American Economic Review, 100*, 304–36.

Gopinath, G., & Rigobon R. (2008). Sticky borders. *The Quarterly Journal of Economics, 123*, 531–575.

Gust, C., Leduc,S., & Vigfusson, R. (2010). Trade integration, competition and the decline in exchange-rate pass-through. *Journal of Monetary Economics, 57*, 309–324.

Hahn, E. (2003). Pass-through of external shocks to euro area inflation. European Central Bank Working Paper No. 243.

Ihrig, J. E., Marazzi, M., & Rothenberg, A. (2006). Exchange rate pass-through in the G7 economies.Board of Governors of the Federal Reserve System International Finance Discussion Paper No. 851.

Kawai, M. (1980). Exchange rate-price causality in the recent floating period. In: D. Bigman & T. Taya (Eds.), *The functioning of floating exchange rates: theory, evidence, and policy implications* (pp. 197–219). Cambridge (Mass.): Ballinger.

Klein, M., & Shambaugh J. C. (2006). Fixed exchange rates and trade. *Journal of International Economics, 70*, 359–383.

Klein, M., & Shambaugh, J. (2010). *Exchange rate regimes in the modern era*. Cambridge (MA): MIT Press.

Krugman, P. (1987), Pricing to market when the exchange rate changes. In: S. Arndt & J. D. Richardson (Eds.), *Real-financial linkages among open economies* (pp. 49–70). Cambridge (MA): MIT Press.

Lane, P. R. (2001). The new open economy macroeconomics: A survey. *Journal of International Economics, 54*, 235–266.

Lane, P., & Milesi-Ferretti, G. M. (2012). External adjustment and the global crisis. *Journal of International Economics, 88*, 252–265.

Lanyi, A. (1969). *The case for floating exchange rates reconsidered*. Essays in International Finance No. 72, International Finance Section, Princeton University.

Levy-Yeyati, E., Sturzenegger, F., & Reggio, I. (2010). On the endogeneity of exchange rate regimes. *European Economic Review, 54*, 659–677.

Mastropasqua, C. (1984). Il circolo vizioso inflazione-svalutazione: teoria e evidenza empirica. *Quaderni Sardi di Economia, 14*, 3–37.

McCarthy, J. (1999). Pass-through of exchange rates and import prices to domestic inflation in some industrialised economies. BIS Working Paper No. 79, Basle: Bank for International Settlements.

McCarthy, J. (2007). Pass-through of exchange rates and import prices to domestic inflation in some industrialized economies. *Eastern Economic Journal, 33*, 511–537.

Mihaljek, D., & Klau, M. (2000). A Note on the pass-through from exchange rate and foreign price changes to inflation in selected emerging market economies. BIS Papers No. 8, 69–81.

Mishkin, F. S. (2008). Exchange rate pass-through and monetary policy. NBER Working Papers No. 13889.

Obstfeld, M. & Rogoff, K. (2000). New directions for stochastic open economy models. *Journal of International Economics, 50*, 117–54.

Obstfeld M., Shambaugh, J. C., & Taylor, A. M. (2005). The trilemma in history: Tradeoffs among exchange rates, monetary policies, and capital mobility. *Review of Economics and Statistics, 87*, 423–438.

Pancaro, C. (2013). Current account reversal in industrial countries. Does the exchange rate regime matter? European Central Bank WP No. 1547.

Pilbeam, K. (1991). *Exchange rate management: Theory and evidence*. London: Macmillan.

Pilbeam, K. (2013). *International finance* (4th ed.). London: Palgrave Macmillan.

Polak, J. J. (1999). *Streamlining the financial structure of the International Monetary Fund*. Essays in International Finance No. 216, Princeton University, International Finance Section.

Qureshi, M. S., & Tsangarides, C. G. (2010). The empirics of exchange rate regimes and trade: Words vs. deeds. IMF Working Paper No. 48, International Monetary Fund.

Rose, A. (2011). Exchange rate regimes in the modern era: Fixed, floating, and flaky. *Journal of Economic Literature, 49,* 652–672.

Sarno, L. & Taylor, M. P. (2001). Official intervention in the foreign exchange market: Is it effective and, if so, how does it work? *Journal of Economic Literature, 39,* 839–868.

Shambaugh, J. (2008). A new look at pass-through. *Journal of International Money and Finance, 27,* 560–591.

Spaventa, L. (1983). Feedbacks between exchange rate movements and domestic inflation: Vicious and not so virtuous circles, old and new. *International Social Science Journal, 97,* 517–34.

Stockman, A. C. (1999). Choosing an exchange-rate system. *Journal of Banking and Finance, 23,* 1483–1498.

Taylor, J. (2000). Low inflation, pass-through and the pricing power of firms. *European Economic Review, 44,* 1389–1408.

Tenreyro, S. (2007). On the trade impact of nominal exchange rate volatility. *Journal of Development Economics, 82,* 485–508.

Thursby, M. C. (1981). The resource reallocation costs of fixed and flexible exchange rates: A multi-country extension. *Journal of International Economics, 11,* 487–493.

Part VI

The Intertemporal Approach

The Intertemporal Approach to the Balance of Payments, and the Real Exchange Rate

18

18.1 Introduction: The Absorption Approach

Forward-looking behaviour, namely current behaviour (for example *current* saving and investment decisions) determined by calculations based on *expectations* of future variables (for example future income, productivity growth, real interest rates, etc.) is one of the hallmarks of modern macroeconomics, and forms the basis of the new open-economy macroeconomics (NOEM) (for a survey see Lane 2001). Forward-looking calculations have already been illustrated as regards financial variables in Chap. 4; in this chapter we apply them to the current account.

The starting point is the observation that the current account is also national income less absorption, or national saving less investment. This follows from the accounting identities illustrated in Chap. 6, and is by no means a new idea, since it was set forth as far back as the 1950s by the *absorption approach* (Alexander 1952, 1959). To put the question into proper perspective a brief digression on the absorption approach is called for.

If we denote national income (product) by y, total aggregate demand (for both consumption and investment) or *absorption* by A, the balance of payments (current account) by B, we have

$$y = A + B, \tag{18.1}$$

whence, considering the variations and rearranging terms,

$$\Delta B = \Delta y - \Delta A. \tag{18.2}$$

Equation (18.2) shows that for a devaluation to improve the balance of payments it must either cause a *decrease* in absorption at unchanged income, or an *increase* in income at unchanged absorption or (better still) both effects, or suitable combinations of changes in the two variables (for instance, both income and absorption may

© Springer-Verlag Berlin Heidelberg 2016
G. Gandolfo, *International Finance and Open-Economy Macroeconomics*,
Springer Texts in Business and Economics, DOI 10.1007/978-3-662-49862-0_18

increase, provided that the latter increases by less, etc.). No elasticities are actually involved.

Equation (18.2) is of course an accounting identity, and, to give it a causal interpretation, we must answer three questions:

(i) how does the devaluation affect income;
(ii) how does a change in income affect absorption;
(iii) how does the devaluation directly (i.e., at any given level of income) affect absorption.

For this purpose we first recall from Chap. 8 that consumption and investment are functions of income, so that we can write the functional relation

$$\Delta A = c\Delta y - d, \tag{18.3}$$

where c is the sum of the marginal propensity to consume and the marginal propensity to invest, and d denotes the direct effect of the devaluation on absorption. By obvious substitutions we get

$$\Delta B = (1 - c)\Delta y + d. \tag{18.4}$$

Question (i) bears on Δy, question (ii) on the magnitude of c, question (iii) on d.

Table 18.1, taken from Machlup (1955), summarizes the various effects, a synthetic exposition of which is:

Idle-resources effect: if there are unemployed resources, the increase in exports following the devaluation brings about an increase in income via the foreign multiplier.

Terms-of trade effect: the devaluation causes a deterioration in the terms of trade and hence a reduction in the country's real income.

Thus the two effects upon income are in opposite directions, so that Δy may have either sign and the answer to question (i) is ambiguous.

As regards question (ii), Alexander was inclined to believe that c is greater than one, hence $(1 - c)$ is negative, so that as regards effects (i) and (ii) a devaluation will improve the balance of payments if its net effect on income is negative. This is, however, an empirical problem whose answer may be different through space (different countries) and time (changing habits).

Table 18.1 Effects of a devaluation according to the absorption approach

Effects upon and via income	Direct effects on absorption
$(1 - c)\Delta y$	d
Idle-resources effect	Cash-balance effect
Terms-of-trade effect	Income-redistribution effect
	Money-illusion effect
	Three other minor effects

Let us now turn to the direct effects on absorption.

Cash-balance effect: the devaluation causes an increase in the domestic price of imports and hence in the general price level. This brings about a decrease in the real value of wealth held in monetary form (cash balances): the public will try to build up their cash balances (to restore the real value of these) both by reducing absorption and by selling bonds. The sale of bonds causes a decrease in their price, i.e. an increase in the interest rate, which further reduces absorption.

Income-redistribution effect: the increase in prices caused by the devaluation may bring about a redistribution of income (for example from fixed-income recipients to the rest of the economy), and this influences absorption provided that the different groups of income recipients have different marginal propensities to spend.

Money-illusion effect: assuming that prices and money income increase in the same proportion, real income does not change, but if people do not realize this because they are subject to money illusion, they will change their absorption (the direction of change depends on the type of money illusion).

The three other minor direct effects concern the expectation of further price increases (so that people may buy goods in advance to avoid paying higher prices in the future); the discouragement to investment caused by the increased price of imported investment goods; the discouragement to expenditure on foreign goods in general, caused by their increased price.

The absorption vs elasticity approach gave rise to a heated debate in the 1950s, that culminated in an unsatisfactory synthesis which simply amounted to considering the effect of the devaluation as an initial exogenous change in the balance of payments to which the standard multiplier was applied to obtain the final result (for details see Tsiang 1961). At the time nobody, not even its author, seemed to understand the truly innovative content of the absorption approach: the pioneering idea that *it is the set of macroeconomic factors underlying absorption (i.e., saving and investment decisions) that ultimately determine the current account and hence international borrowing or lending patterns* (see the last column of the matrix illustrated in Chap. 6, Table 6.1).

The modern intertemporal approach to the current account can be seen as an extension of the absorption approach, since it starts from the same idea but supplements it with the recognition that private saving and investment decisions are *forward looking*. Foreign borrowing and lending can, in turn, be viewed as *intertemporal trade*, namely as the exchange of goods available on different dates. Insofar as the intertemporal approach to the current account also considers relative prices as determinants of saving and investment decisions, it can be viewed as offering a modern synthesis of the absorption and elasticity approaches.

18.2   Intertemporal Decisions, the Current Account, and Capital Flows

To simplify at the utmost, we start with a pure exchange economy in which no production or investment take place, and consider two time periods. The economy is endowed with a fixed amount of a consumption good (say, corn) in each of the two periods (these endowments can be considered as the economy's incomes in the two periods), and the preferences between current consumption and future consumption can be represented by social indifference curves (or by the indifference curves of a representative individual), according to a well-known diagram introduced by Fisher (1930; Chap. X).

The slope of the indifference curve reflects the marginal rate of time preference, namely the rate at which the consumer is willing to give up a small amount of current consumption to obtain more consumption in the future or, alternatively, the rate at which the consumer is willing to forego future consumption to obtain a marginal increase in current consumption. Note that the marginal rate of time preference is not a constant, but depends on the marginal utility of current and future consumption (see the Appendix).

We have said "reflects", because the slope of the indifference curve is not equal to the rate of time preference ρ, but equals $1 + \rho$ in absolute value. In fact, an increment dC_1 in future consumption is discounted to the current period by the variable subjective discount factor $(1 + \rho)^{-1}$, hence its value in the current period is $dC_1/(1 + \rho)$. The consumer is indifferent between a decrement dC_0 in current consumption and the present value (subjectively discounted) of dC_1 when

$$dC_0 + \frac{1}{1 + \rho} dC_1 = 0,$$

whence

$$\frac{dC_1}{dC_0} = -(1 + \rho), \tag{18.5}$$

which is the slope of the indifference curve.

In market equilibrium the marginal rate of time preference must be equal to the interest rate i, since the latter reflects the market relative price of C_1 in terms of C_0: if someone forgoes a unit of consumption today and lends it to someone else, the lender will get back $1 + i$ tomorrow.

Let us first consider a closed economy endowed with $y_0 \equiv C_{0A}, y_1 \equiv C_{1A}$ respectively in period 0 and 1. Since there is no trade, the consumption of the economy in each period must be equal to the endowment. The equilibrium point E_A in Fig. 18.1 shows the tangency between the indifference curve I_A and the line whose slope in absolute value is $\tan\alpha = 1 + i_A$, where i_A is the autarky interest rate.

Fig. 18.1 Intertemporal trade: pure consumption

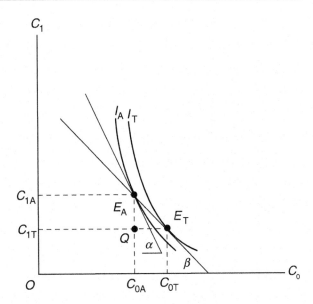

Suppose now that the economy can trade at a world interest rate $i^* \neq i_A$, where without loss of generality we can take $i^* < i_A$. Utility maximization takes place at point E_T, where the (absolute value of) slope of the indifference curve I_T equals $\tan\beta = 1 + i^*$, hence $i^* = \rho$.

The attainment of point E_T implies the borrowing from abroad of an amount $C_{0T}C_{0A}$ in the current period (which is like an import) and the repayment of $C_{1A}C_{1T}$ in the next period (this is like an export). It is easy to see that $E_A Q = E_T Q \cdot \tan\beta$, hence $C_{1A}C_{1T} = C_{0T}C_{0A} \cdot (1 + i^*)$. In general, denoting by CA the current account balance,

$$CA_0 + \frac{CA_1}{1 + i^*} = 0, \qquad (S_0 - I_0) + \frac{S_1 - I_1}{1 * i^*} = 0, \qquad (18.6)$$

where each current account balance is taken with its sign. The second relation in (18.6) follows from the first considering the accounting identity $(I - S) + CA = 0$ (see Sect. 6.2: here we neglect the public sector or, alternatively, we define S and I asnational saving and national investment). Equation (18.6) is the *intertemporal balance constraint*.

The source of intertemporal trade is a difference between i_A and i^*, just as in the static model trade is determined by a difference between the autarkic and world commodity prices.

Unlike the static model, where commodities are exchanged for commodities (barter trade), in the intertemporal model trade involves the *exchange of commodities for assets*, which are claims on future production. Thus in period 0 the importing country will give some sort of bond to the exporting country; this bond states the obligation to repay in period 1 the amount of the commodity borrowed plus interest. Using the terminology of the balance of payments (see Chap. 5), in period 0 our economy runs a *current account deficit* matched by a *capital account surplus* (the sale of the bond to the foreign country). This confirms what we stated at the beginning, that the current account is determined by the saving-investment decisions of the economy. These decisions also determine capital flows as the necessary counterpart to commodity flows. Hence the overall balance of payments is necessarily in equilibrium, and its structure is completely determined.

We have seen that in moving from E_A to E_T the economy attains a higher welfare level. This, however, implies that the agent who borrows is the same as the one who repays, and may be misleading when this is not the case. If, for example, the actual length of the period is very long, so that the agents living in period 1 are different from those who lived in period 0, the outcome is that agents living in period 0 are better off at the expense of those living in period 1, who must repay the debt without having previously enjoyed the benefits of higher consumption. This problem is even more serious when one extends the model to a multi-period setting, and can be dealt with in several ways. One is to assume infinitely-lived agents. Another is to assume *overlapping generations,* namely a multi-period setting in which people live for two periods, so that in any period there are both "old" (those in their second period of life) and "young" (those in their first period of life) agents (see Obstfeld and Rogoff 1996, Chap. 3; Frenkel and Razin 1996). A simpler way out is described here.

In the model so far examined there is no place for investment. Borrowing and lending take place only to smooth consumption between different time periods. Let us now assume that our homogeneous good can be both consumed and invested (used as capital good), so that lower consumption today means higher output and consumption next period (less corn consumed today means more corn planted and hence more corn produced tomorrow). The *intertemporal transformation curve* is drawn in Fig. 18.2; it is concave to the origin due to diminishing marginal productivity of capital.

Let us first consider autarky. The economy is endowed with OQ_{0A}, OQ_{1A} respectively in the current and next period. By consuming less than the current endowment and investing the amount $C_{0A}Q_{0A}$ of saved output, the economy can obtain the additional output $Q_{1A}C_{1A}$ in period 1, thus being able to consume OC_{1A} rather than just OQ_{1A}. This is actually the solution that maximizes the country's welfare, as shown by the tangency between the intertemporal transformation curve and the highest indifference curve I_A at E_A.

The opening up of trade (with the same endowments) at the international interest rate i^*, where $\tan \beta = 1 + i^*$, enables the country to attain the optimum point E_T, clearly superior to E_A. But what is striking is that now *consumption is higher in both periods* (hence no intergenerational conflict can arise). In the previous case (Fig. 18.1), higher current consumption could be achieved only by sacrificing future consumption and vice versa. This is no longer the case when foreign borrowing can also be used to finance productive investment, as illustrated in Fig. 18.2.

More precisely, the production point Q_T means that the economy, by investing $Q_{0A}Q_{0T}$ today, will obtain the additional output $Q_{1A}Q_{1T}$ next period. Part of this

Fig. 18.2 Intertemporal trade: production and investment

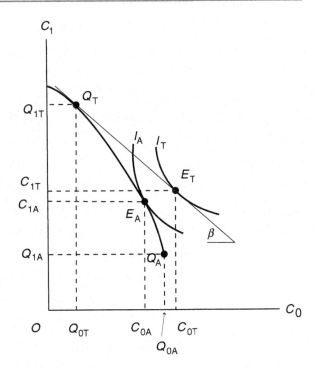

additional output, and precisely $Q_{1A}C_{1T}$, will be consumed together with next period's endowment so as to raise consumption in period 1 to OC_{1T}, higher than the autarky level OC_{1A}. The remaining part $(Q_{1T}C_{1T})$ will be used to repay the loan contracted in period 0 $(Q_{0T}C_{0T})$ that enables the economy to raise period 0's consumption to OC_{0T} (higher than the autarky level OC_{0A}).

It might then seem that foreign borrowing has been entirely used to finance current consumption, but it is not so. The interpretation is the following: the economy consumes all of its current endowment OQ_{0A}, *plus part of the loan* $(Q_{0A}C_{0T})$, using the rest of the loan $(Q_{0T}Q_{0A})$ as investment. The obvious policy implication is that borrowing from abroad is not by itself a negative or harmful action: it will be so if it is squandered in consumption, but if it is used for productive investment it will raise the welfare of future generations.

Let us observe, in conclusion, that the country will run, as in the pure consumption case, a current account deficit matched by a capital account surplus, hence the overall balance of payments will be in equilibrium.

18.2.1 The Feldstein-Horioka Puzzle

In a closed economy (aggregating the public and private sector) national saving equals national investment. This is no longer true in an open economy, where a divergence between saving and investment can be accommodated by a corresponding current account imbalance. The intertemporal approach examined in the previous section shows that such a divergence is indeed the normal outcome of the action of optimizing agents. It should however be pointed out that, since the corresponding current account imbalance is only possible through foreign borrowing and lending, free capital mobility is an essential ingredient.

Thus, while in a closed economy saving and investment cannot but move concomitantly, there is no reason why they should do so in an open economy *provided that* capital is internationally mobile. With capital immobility changes in national saving rates ultimately change national investment rates by the same amount. This is the point made in the Feldstein and Horioka (1980) well-known paper. They reported empirical evidence, showing a high and strong correlation between gross domestic I/Y and gross domestic S/Y in a cross-sectional sample of 16 OECD countries over the period 1960–1984, according to the least-squares regression

$$I/Y = \underset{(0.02)}{0.04} + \underset{(0.07)}{0.89 S/Y}, \qquad R^2 = 0.91,$$

where the numbers below the coefficients are the standard errors. The authors took this as evidence for the lack of capital mobility. Subsequent studies by the same and other authors found similar econometric results.

This poses a puzzle, because these results contradict other evidence that capital is quite mobile within developed countries. The main theoretical point is then whether a high correlation between saving and investment indicates low capital mobility.

Various explanations have been set forth to solve the puzzle, and the most recent ones (for a survey of the previous ones see Obstfeld and Rogoff 1996, pp. 162–163) have focused on two points.

The first is mainly econometric, and starts by observing that saving and investment are cointegrated over time, as is implied by the intertemporal budget constraint. The effect of this constraint seems strong enough to explain the high correlation between saving and investment, without any implication on capital immobility (Jansen 1997).

The second evaluates the appropriateness of the saving-investment correlation as a measurement criterion, and argues that international parity conditions are better criteria. It shows that international parity conditions, properly viewed and estimated, provide strong evidence for capital mobility, and hence the puzzle is resolved (Moosa 1997).

18.2.2 The Harberger-Laursen-Metzler Effect Again

As we have seen in Chap 9, Sect. 9.1, this effect consists in the fact that "as import prices fall and the real income corresponding to a given money income increases, the amount spent on goods and services out of a given money income will fall. The argument is applicable in reverse, of course, to a rise of import prices. In short, our basic premise is that, other things being the same, the expenditure schedule of any given country rises when import prices rise and falls when import prices fall". Since the authors did not look at investment, the fall in saving determined by an import price rise implied a worsening of the trade balance.

The argument was more or less implicitly based on Keynesian theorizing, prevailing at the time. According to the simple Keynesian consumption function, the relations between increases in income and consumption are governed by the marginal propensity to consume which is in any case smaller than unity; besides, the average propensity to consume is also smaller than unity and decreases as real income rises (non-proportional short-run consumption function). Given these premises, the Harberger-Laursen-Metzler assumption is quite obvious.

However, the intertemporal approach shows that these premises are no longer necessarily true: even the simple model examined above shows that consumption can be greater than income thanks to borrowing; besides, an increase in real income may have any effect on real consumption according to the rate of time preference. For an examination of these points see Svensson and Razin (1983), Persson and Svensson (1985), Sen (1994), Fang and Lin (2013), and the appendix to the present chapter, Sect. 18.4.2.

18.3   Intertemporal Approaches to the Real Exchange Rate

18.3.1 Introduction

The real exchange rate has been defined in Chap. 2, Sect. 2.3. Now, it might seem that no specific theory is necessary, since the real exchange rate is a derived concept: knowing the nominal exchange rate and the other magnitudes involved in the definition of real exchange rate, the real exchange rate follows. However, nothing prevents us from building a specific theory of the real exchange rate. One reason for doing so could be the poor empirical performance of the nominal exchange rate models (see Sect. 15.5), while a specific theory of the real exchange rate (henceforth RER) might do better. The intertemporal approach seems to be particularly useful for this purpose (for other theories see, for example, MacDonald and Stein eds., 1999).

Two theories of the RER based on the intertemporal approach are the RAIOM (*R*epresentative *A*gent *I*ntertemporal *O*ptimization *M*odel) and the NATREX (*NAT*ural *R*eal *EX*change rate) approach.

18.3.2 The RAIOM Approach

We use the simple pure exchange model described above, Sect. 18.2. What we need here is to find an explicit expression relating consumption of the representative agent to the agent's permanent income. For this purpose we assume, following

Obstfeld and Rogoff (1995a,b, 1996) , that the utility function is time separable (i.e., the utility in each period only depends on that period's consumption); a further simplification can be obtained assuming that the functional form is logarithmic, namely

$$U(C_t, C_{t+1}) = U(C_t) + U(C_{t+1}) = \ln C_t + \ln C_{t+1}. \tag{18.7}$$

This function has to be maximized subject to the intertemporal budget constraint, that is

$$C_{t+1} = y_{t+1} + (1+i)(y_t - C_t), \tag{18.8}$$

where y_t, y_{t+1} are the given endowments (incomes) in the two periods, and i is the interest rate at which the agent can borrow or lend in the international market. The intertemporal budget constraint states that consumption at time $t+1$ equals resources available at time $t+1$, which are income at time $t+1$ plus (minus) the amount that the agent receives (pays) due to lending (borrowing) in period t.

If we discount future income using the given interest rate we obtain the present value of lifetime income:

$$Y_t = y_t + (1+i)^{-1} y_{t+1}, \tag{18.9}$$

and rewriting the budget constraint as

$$C_t + (1+i)^{-1} C_{t+1} = (1+i)^{-1} y_{t+1} + y_t, \tag{18.10}$$

we see that the present value of lifetime consumption must be equal to the present value of lifetime income.

With the separable logarithmic utility function (18.7), the optimal consumption function takes on the particularly simple form (for proof see the Appendix)

$$C_t = \beta Y_t, \quad \beta \equiv \left[1 + \frac{1}{1+\rho} \right]^{-1}, \tag{18.11}$$

namely consumption is proportional to the present value of lifetime income, the proportionality factor depending on the subjective discount rate. One can call βY_t permanent income Y_t^p, so that (18.11) states that consumption equals permanent income.

The current account is income minus absorption (see Chap. 6, Sect. 6.2), that is

$$CA_t = y_t - C_t. \tag{18.12}$$

Using the value of optimal consumption and the definition of permanent income we have

$$CA_t = y_t - Y_t^p. \tag{18.13}$$

Thus there will be current account deficits (surpluses) when current income is smaller (greater) than permanent income.

In the analysis above, C is social consumption, namely $C = C_p + G$, or private consumption C_p plus public consumption G. In practice a distinction between private and public consumption may be desirable. With this distinction the resources available to the private representative agent are $(y - G)$, and private consumption is $(C - G)$, hence the budget constraint becomes

$$(C_{t+1} - G_{t+1}) = (y_{t+1} - G_{t+1}) + (1 + i)\left[(y_t - G_t) - (C_t - G_t)\right], \qquad (18.14)$$

and consequently optimal private consumption (see the Appendix) equals permanent income less permanent government consumption:

$$C_{pt} = Y_t^p - G_t^p. \qquad (18.15)$$

With these results, the current account (18.12) becomes

$$CA_t = y_t - C_{pt} - G_t = y_t - [Y_t^p - G_t^p] - G_t$$
$$= [y_t - Y_t^p] - [G_t - G_t^p]. \qquad (18.16)$$

Equation (18.16) states that the current account depends on the deviation of national income from its permanent level less the deviation of government consumption from its permanent level.

Equations (18.13) and (18.16) constitute the essence of forward-looking RAIOM; there is no such thing as a current account surplus (deficit) being due to an undervalued (overvalued) exchange rate.

In order to include an exchange rate, we must introduce two sectors: a tradable and nontradable goods sector. Thus we can use the definition of real exchange rate as the domestic relative price of traded and nontraded goods (see Chap. 2, Sect. 2.3), that does not involve the nominal exchange rate:

$$r_{R_t} = \frac{p_t^{NT}}{p_t^T}. \qquad (18.17)$$

If we now assume that the agent's utility function (it now includes both traded and nontraded goods) is logarithmic Cobb-Douglas, it follows (see Appendix) that the optimal pattern of consumption is such that the ratio of expenditures on consumption of tradables $p_t^T C_t^T$ to nontradables $p_t^{NT} C_t^{NT}$ is a constant $1/\gamma$ depending on the utility function. From this and (18.17) we get

$$r_{R_t} = \gamma \frac{C_t^T}{C_t^{NT}}. \qquad (18.18)$$

By definition, nontradables are entirely consumed at home, hence private consumption of them equals their production Q_t minus government consumption G_t (government consumption is assumed to consist entirely of nontradables). The intertemporal budget constraint of the private sector continues to hold as regards tradables, hence private consumption of tradables follows the same rules as above, and equals permanent income involving tradables, continue to call it Y_t^p. It follows that the real exchange rate is given by

$$r_R = \frac{\gamma Y_t^p}{Q_t - G_t}. \tag{18.19}$$

Equation (18.19) is the basic exchange rate equation of the model, and has very important implications.

In fact, since r_R is derived from an optimization procedure, there is no such thing as an overvalued real exchange rate that causes an unsustainable current account deficit and growth of the foreign debt. In any case the foreign debt is not a problem because it is the result of an intertemporal optimization with an intertemporal budget constraint.

For a survey of empirical studies on the RAIOM approach see Obstfeld and Rogoff (1995a; Sect. 4). The results of these studies are actually not very favourable to the theory, as pointed out by Stein (1996; Sects. 2 and 3).

18.3.3 The NATREX Approach: An Overview

The NATREX (*NAT*ural *R*eal *EX*change rate) approach of Stein (1990, 1995a,b, 1996, 1999), though based on intertemporal optimization and micro-unit behaviour, presents two important departures from the standard approach (RAIOM).

The *first* is that the hypotheses of perfect knowledge and perfect foresight are rejected. Rather, rational agents that efficiently use all the available information will base their intertemporal decisions upon a *sub-optimal feedback control* (SOFC) rule (Infante and Stein 1973; Stein 1995a). Basically, SOFC starts from the observation that the optimal solution derived from standard optimization techniques in perfect-knowledge perfect-foresight models has the saddle-path stability property (for this concept see above, Chap. 13, Sect. 13.4), hence the slightest error in implementing the stable arm of the saddle will put the system on a trajectory that will diverge from the optimal steady state. Actual optimizing agents know that they do not possess the perfect knowledge required to implement the stable arm of the saddle without error, hence it is rational for them to adopt SOFC, which is a closed loop control that only requires *current* measurements of the variable(s) involved, not perfect foresight, and will put the economy on a trajectory which is asymptotic to the unknown perfect-foresight stable arm of the saddle.

The *second* is that expenditure is separated between consumption and investment, which are decided by different agents. The consumption and investment functions are derived according to SOFC, through dynamic optimization techniques with feedback control.

Thus the NATREX approach is actually an intertemporal optimizing approach, though based on different optimization rules.

The NATREX is the intercyclical equilibrium real exchange rate that ensures balance-of-payments equilibrium in the absence of cyclical factors, speculative capital movements and movements in international reserves. In other words, the NATREX is the equilibrium real exchange rate that would prevail if the above-mentioned factors could be removed and the GNP were at capacity. Since it is an equilibrium concept, the NATREX guarantees both the internal and the external equilibrium, the focus being on the long run.

The long-run internal equilibrium is achieved when the economy is at capacity output, that is when the GNP is at its potential level. The long-run external equilibrium is achieved when the long-term accounts of the balance of payments are in equilibrium. Short term (speculative) capital movements and movements in official reserves are bound to be short term transactions, since they are unsustainable in the long run. In the long-run equilibrium they must average out at zero, and the current account balance (given by the trade balance plus interest payments on the stock of foreign debt) must be in equilibrium, which means that the NATREX generates a trade balance surplus just enough to pay interest on accumulated foreign debt.

Under these conditions, and abstracting from growth for simplicity, the real market long-run equilibrium condition and the long-term external equilibrium condition coincide:

$$S - I = CA = 0, \qquad (18.20)$$

where CA is the balance of payments' current account, the private and public sectors having been aggregated into a single one. Relation (18.20) is intended in *real terms*: the model assumes neutrality of money and that monetary policy keeps inflation at a level compatible with internal equilibrium (at least in the long run). Therefore, the focus being on the real part of the economy, there is no need to model the money market. More precisely, the NATREX does not ignore monetary-price effects, but separates them from real factors: the nominal exchange rate (r) corresponding to the NATREX (r_R) is $r = r_R(p_f/p_h)$, hence differential rates of inflation produce offsetting changes in the nominal exchange rate. Finally, perfect international capital mobility is assumed: the real interest rate is driven by the portfolio equilibrium condition or real interest parity condition, possibly with a risk premium.

The system is assumed to be self-equilibrating (hence the adjective *natural* in the acronym NATREX). Take for example an initial position of full equilibrium ($S - I = CA = 0$) and suppose an exogenous shock leads to a situation where $S - I < 0$. Given the perfect international capital mobility, the interest rate cannot play the role of the adjustment variable; rather, the difference between national investment and national saving originates a corresponding inflow of long-term capital. The RER appreciates accordingly, leading to a deterioration in the current account. The capital inflow also causes an increase in the stock of foreign debt, which in turn determines a decrease in consumption and hence an increase in saving, until equilibrium is restored. In conclusion, the RER is the adjustment variable in Eq. (18.20).

The model can be solved for its medium run and long run (steady state) solutions. Any perturbation on the real fundamentals of the system pushes the equilibrium RER on a new medium-to-long-run trajectory. Since cyclical, transitory and speculative factors are considered noise, averaging out at zero in the long run, the actual RER converges to the equilibrium trajectory. The PPP theory turns out to be only a special case of the NATREX approach: "the issue is not whether or not the real exchange rate is stationary over an arbitrary period, but whether it reflects the [real] fundamentals." (Stein 1995a; p. 43).

To avoid possible misunderstanding we stress that the NATREX does *not* aim at tracking the *actual* real exchange rate, but is, on the contrary, a measure of the long-run *equilibrium* real exchange rate, the benchmark against which we can measure the misalignment of the actual real exchange rate. Thus expressions like "the domestic currency is weak", "the domestic currency is strong", "the domestic currency is undervalued", "the domestic currency is overvalued", etc., which are often used in a vague sense, can be given a precise meaning. Let us remember that the real exchange rate (and hence the NATREX) is defined in such a way that an *increase* means a (real) *appreciation* of the domestic currency. Hence when the actual real exchange rate is *lower* (*higher*) than the NATREX, it follows that the domestic currency is *undervalued* (*overvalued*).

The NATREX approach has been able to satisfactorily explain the medium-to-long run dynamics of the RER in several industrial countries: USA (Stein 1995a,b, 1999), Australia (Lim and Stein 1995), Germany (Stein and Sauernheimer 1996), France (Stein and Paladino 1998), Italy (Stein and Paladino 1998; Gandolfo and Felettigh 1998; Federici and Gandolfo 2002), Belgium (Verrue and Colpaert 1998). Belloc and Federici (2010) extend this framework to a two-country (US and EU) model and obtain similarly good results.

18.3.4   A More Technical Presentation

We start from the basic behavioural equations, following Stein (1995a).

The *investment function* is derived in the context of an intertemporal optimization problem in which agents apply a suboptimal feedback control (SOFC) rule. The standard optimal control problem gives rise to the rule according to which the marginal productivity of capital must be equal to the sum of the growth rate and discount rate. In the neighbourhood of the steady state the optimal control is that the rate of investment must be proportional to the gap between actual and steady-state capital intensity. The implementation of this control requires, amongst other, the knowledge of the steady-state capital intensity (which also enters into the coefficient of proportionality). The slightest mistake would cause the system to diverge.

It can be shown (see the Appendix) that a SOFC rule which

(i) requires only current measurements of the marginal product of capital,
(ii) is guaranteed to drive the system to the unknown steady-state capital intensity, and
(iii) is robust to perturbations,

is that (sub)optimal investment is positively related to the marginal productivity of capital less the discount rate. Now, "Let the real long-term interest rate substitute for the discount rate. The SOFC law states that one should focus upon the current marginal product of capital less the real long term rate of interest." (Stein 1995a; p. 53). Thus we are led to the function

$$I = I(f_k - i_R; Z), \quad I_{f_k - i_R} > 0, I_{i_R} < 0, \tag{18.21}$$

where f_k denotes the marginal productivity of capital, i_R the real long-term interest rate, and Z represents exogenous fundamentals, such as an exogenous (positive) shock to productivity. Alternatively we could write the investment function as

$$I = I(k, i_R; Z), \quad I_k < 0, I_{i_R} < 0, \tag{18.22}$$

where k is capital per unit of labour (for simplicity's sake we assume that the labour force is stationary). $I_k < 0$ means that an increase in capital reduces the marginal productivity of capital and hence investment; $I_{i_R} < 0$ means that future profit flows are discounted at a higher rate, which lowers current investment.

As regards *consumption*, the NATREX approach assumes that saving (hence consumption) decisions are made independently of investment decisions. This is equivalent to assuming that optimizing agents are functionally separated into two categories, those who take investment decisions (firms) and those who take consumption-saving decisions (consumers). It can be shown (see Appendix) that an appropriate intertemporal optimization process by the representative consumer will give rise to a function

$$C = C(k, F; Z), \quad C_k > 0, C_F < 0. \tag{18.23}$$

The intuition behind the formulation of C is the following. From intertemporal optimization under uncertainty (Merton 1992), consumption is proportional to wealth, here defined as capital k less the foreign debt F. The presence of the term F with a negative effect on C is also due to a feedback control on the part of the government (recall that C is private + public consumption). Part of the government's debt is foreign held. When the government realizes that its foreign debt is increasing, it changes its policy by decreasing current expenditure. Exogenous fundamentals such as the rate of time preference are represented by Z.

Social saving is the difference between GNP net of interest payments to foreigners $y(k; u) - i_R F$, and social consumption C, hence

$$S = y(k; Z) - i_R F - C(k, F; Z) = S(k, F, i_R; Z), \quad S_F > 0. \tag{18.24}$$

An increase in the stock of foreign debt has a two-fold effect on saving. On the one hand, it tends to depress saving because of the higher interest payments, on the other it tends to enhance saving via the reduction in consumption. The NATREX approach requires that savings are an increasing function of foreign debt in order to ensure that the stock of debt does not explode. This is the only condition required, there is no intertemporal budget constraint in which the initial and final debt should be equal. The model allows the economy to pass from a net debtor position to a net creditor position and viceversa.

After explaining how the investment and saving function are derived from intertemporal optimization, we now turn to the other structural equations.

The *trade balance* is negatively related to the real exchange rate (given its definition, an increase in r_R means an appreciation) on the assumption that the critical elasticities condition is satisfied. It is also negatively related to the capital stock, because an increase in k increases wealth and hence desired imports. On the contrary, it is positively related to the foreign capital stock k^*, an increase in which increases foreign wealth and demand for exports. It is also positively related to the stock of foreign debt because an increase in F redistributes wealth from the home to the foreign country, lowering domestic demand for imports and raising foreign demand for exports. Thus we have

$$TB = TB(r_R, k, F; k^*, Z), \quad TB_{r_R} < 0, TB_k < 0, TB_F > 0, TB_{k^*} > 0, \tag{18.25}$$

where Z denotes exogenous fundamentals such as the rate of time preference. The current account balance CA is the trade balance minus interest payments on foreign debt, and in turn equals saving minus investment:

$$CA = TB - i_R F = S - I. \tag{18.26}$$

Since we are considering an equilibrium in which short-term capital flows and reserve movements are absent, the sum of the current account and long-run capital movements balance (which in turn equals the rate of change of the stock of foreign debt) is zero:

$$CA + \frac{dF}{dt} = 0. \tag{18.27}$$

The domestic *goods market equilibrium* condition is

$$y(k; Z) = C(k, F; Z) + I(f_k - i_R; Z) + TB(r_R, k, F; k^*, Z), \tag{18.28}$$

where output is at capacity as determined by the production function $y = f(k; Z)$. The corresponding equation for the rest of the world (denoted by a superscript f) is

$$GDP^f = C^f + I^f - TB(r_R, k, F; k^*, Z). \tag{18.29}$$

The *rate of change of foreign debt* (capital flows) has already been defined in (18.27), hence

$$\frac{\mathrm{d}F}{\mathrm{d}t} = -CA = I - S. \tag{18.30}$$

In the long-run equilibrium (steady state) CA and hence $\mathrm{d}F/\mathrm{d}t$ are zero, which implies a constant value of the stock of (net) foreign debt F, say F^* (a negative value means that the country is a net creditor). This in turn implies, via Eq. (18.26) that the steady-state trade balance surplus (deficit) equals interest payments on foreign debt (interest received on foreign assets). Note that in the short-to-medium run countries are allowed to change their net debtor position into a net creditor position ($F < 0$), and viceversa; in any case the present value of the steady-state stock of net foreign debt goes to zero as time goes to infinity. In fact,

$$\lim_{t \to \infty} \frac{F^*}{(1 + i_R)^t} = 0, \tag{18.31}$$

since F^* is a constant.

We finally have the *portfolio balance equation* which states that the domestic and foreign real interest rates become equal in the long run, but may be different in shorter times:

$$\frac{\mathrm{d}\left(i_R - i_R^f\right)}{\mathrm{d}t} = -a\left(i_R - i_R^f\right), a > 0, \tag{18.32}$$

which represents the convergence of the real interest rates according to a velocity a.

18.3.4.1 Solution of the Model

For ease of reference we summarize the model in tabular form.

Domestic goods market $y(k; Z) = C(k, F; Z) + \mathrm{d}k/\mathrm{d}t + TB(r_F, k, F, k^f; Z)$,
Foreign goods market $\quad GDP^f = C^f + I^f - TB(r_F, k, F, k^f; Z)$,
Investment function $\quad \mathrm{d}k/\mathrm{d}t = I = I(f_k - i_R; Z)$,
Saving function $\quad S = S(k, F, i_R; Z)$,
Rate of change of debt $\mathrm{d}F/\mathrm{d}t = I - S = -CA = -(TB - i_R F)$,
Portfolio equation $\quad \mathrm{d}i_R/\mathrm{d}t = -a(i_R - i_R^f)$.

$$\tag{18.33}$$

In solving the model we must distinguish between the medium run and the long run.

The Medium Run

The medium run is defined as the period in which the stocks of capital and debt are given and the interest differential is zero. If we consider the goods-market equilibrium expressed by the first and second equation in (18.33), we see that the two variables allowed to change are the real exchange rate r_R and the real interest rate $i_R = i_R^f$. Thus in Fig. 18.3 we can represent the goods-market equilibrium as the schedules IS, IS^f respectively for the domestic and foreign economy.

The IS is negatively sloped because a rise in the real interest rate, lowering investment, reduces aggregate demand below capacity. Hence the real exchange rate has to decrease (depreciate) to improve competitiveness and increase aggregate demand via an improvement in the trade balance. The foreign IS^f curve is positively sloped because a depreciation of the domestic currency is an appreciation of the foreign currency.

The intersection of the two schedules determines the medium-run equilibrium at H, where the real exchange rate is R_0 and the real interest rate is $i_{R_0} = i_{R_0}^f$. Suppose now that there is a shift in exogenous fundamentals, for example a productivity shock that increases investment demand such that the IS curve shifts from IS_0 to IS_1. At point H there is excess demand for domestic goods, and to maintain goods-market equilibrium the real interest rate has to rise. Given the real exchange rate R_0 the domestic real interest rate will rise to a, corresponding to point H_1.

Since the foreign real interest rate has remained at $i_{R_0}^f = i_{R_0}$, the interest-rate differential $H - H_1$ will induce agents to purchase domestic securities in exchange for foreign ones. Both the excess demand for goods and the capital inflow will appreciate the real exchange rate, while the redirection of portfolio investment to the home country will lead to a convergence in real interest rates, from H_1 to H_2 at home and from H to H_2 abroad. Thus the new equilibrium point H_2 is reached.

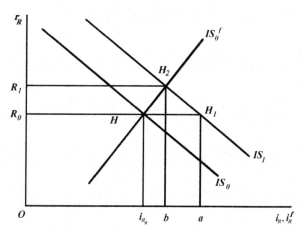

Fig. 18.3 The NATREX model: medium-run dynamics

The Long Run

In the long run, capital and debt change, hence points like H or H_2 move through time. The endogenous movements of k and F in the long run are by the third and fifth equation in (18.33). They can be described by the phase diagram in Fig. 18.4, on the assumption that there is interest-rate convergence (that is, we are considering a succession of point like H or H_2).

The steady-state long run equilibrium (k^*, F^*) is represented as point E in the diagram. F^* has been normalized to zero, but we know that it may have any (constant) value.

The locus of points where $dk/dt = 0$ is graphed as the vertical straight line originating from E. It is independent of F because the foreign debt does not affect the world real interest rate. The rate of change of capital will be zero at $k = k^*$, when the marginal product of capital equals the real interest rate. Since the marginal productivity of capital is a decreasing function of capital, it follows that for lower (higher) values of capital, $k \lessgtr k^*$, the marginal productivity of capital will be greater (smaller) than the real interest rate, which implies a positive (negative) investment. Thus the capital stock increases (decreases) to the left (right) of k^*. The horizontal arrows describe the movement.

The locus of points where dF/dt is graphed as the $L = 0$ curve. Along it there are no capital flows, the stock of debt is constant, saving equals investment, and the current account is zero. It is downward sloping because a rise in capital raises saving relative to investment and reduces the debt. A rise in debt raises saving— see Eq. (18.24)—and hence saving less investment. This means that *above (below)* the $L = 0$ curve saving exceeds (is less than) investment, hence there are current account surpluses (deficits), and the debt declines (rises) towards $L = 0$. This movement is described by the vertical vectors.

Fig. 18.4 The NATREX model: long-run dynamics

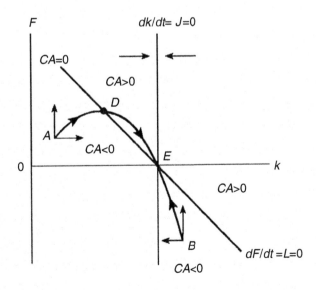

The vectors point toward equilibrium, which is therefore stable (the mathematical proof is given in the appendix). A case is described as the path ADE. Suppose that the system was initially in equilibrium at A, and that an exogenous productivity shock occurs. The new equilibrium point is E, because the productivity shock stimulates investment, so that the new equilibrium capital stock is higher. The higher capital stock raises output, and hence saving less investment, so that the new steady-state level of debt is lower. Let us now consider the dynamics from A to E.

Initially, the higher investment causes $I - S > 0$, namely $CA < 0$ and hence a capital inflow. The debt rises together with capital along AD. At D the current account is in equilibrium, the debt stops moving, but the capital stock continues to increase. Hence the system is pushed to the right, above the $L = 0$ curve, where saving exceeds investment (see above), so that the debt declines while the capital stock continues to increase, and the system moves along the DE curve towards E.

18.4 Appendix

18.4.1 The Two-Period Case

The analytical model corresponding to the simple two-period problem explained in the text is the following standard maximization problem (Obstfeld and Rogoff 1996)

$$\max_{C_0, C_1} U(C_t, C_{t+1}), \tag{18.34}$$

subject to

$$C_t + (1 + i)^{-1} C_{t+1} = Y \equiv y_t + (1 + i)^{-1} y_{t+1}, \tag{18.35}$$

where U is a well-behaved (i.e., increasing and strictly concave in its arguments) function evaluated at time t, Y is the given lifetime income (evaluated at time t) of the representative consumer, and the other symbols have already been defined in the text. Equation (18.35) gives the intertemporal budget constraint, whose slope in the (C_t, C_{t+1}) plane is $dC_{t+1}/dC_t = -(1 + i)^{-1}$.

Let us now observe that along any intertemporal indifference curve we have $U(C_t, C_{t+1}) = \kappa$, where κ is any positive constant, hence using the implicit function differentiation rule we obtain the slope of the intertemporal indifference curve

$$\frac{dC_{t+1}}{dC_t} = -\frac{\partial U/\partial C_t}{\partial U/\partial C_{t+1}} \equiv -(1 + \rho), \tag{18.36}$$

where $(1 + \rho)^{-1} \equiv \dfrac{\partial U/\partial C_{t+1}}{\partial U/\partial C_t}$ is the subjective discount factor representing the agent's time preference. To avoid confusion, it is as well to remind that sometimes the slope of the indifference curve is written in a different way, which depends on the assumption of a *time-separable* utility function, namely

$$U(C_t, C_{t+1}) = U(C_t) + (1 + \delta)^{-1} U(C_{t+1}), \tag{18.37}$$

where $(1 + \delta)^{-1}$ is the constant subjective discount factor that the agent uses to discount future utilities. By the usual procedure we obtain

$$\frac{dC_{t+1}}{dC_t} = -\frac{(1 + \delta) dU/dC_t}{dU/dC_{t+1}}. \tag{18.38}$$

The reason for the difference between (18.36) and (18.38) is that in the general utility function the ratio of the marginal utilities of current and future consumption, evaluated at time t, already embodies the subjective discount rate. This is due to the fact that a partial derivative depends on the same arguments of the function being differentiated, namely $\partial U/\partial C_t = f(C_t, C_{t+1})$, $\partial U/\partial C_{t+1} = g(C_t, C_{t+1})$, which of course is the difference between a general and a time-separable utility function. Formulation (18.36) is more general than (18.38), but in most applications (18.38) is used, due to its analytical manageability especially when we have to deal with longer time-horizons.

The Lagrangian of our optimization problem is

$$L = U(C_t, C_{t+1}) + \lambda \left[C_t + (1 + i)^{-1} C_{t+1} - Y \right], \tag{18.39}$$

that gives the first order conditions

$$\partial U/\partial C_t = -\lambda,$$
$$\partial U/\partial C_{t+1} = -\lambda (1 + i)^{-1},$$

whence

$$\frac{\partial U/\partial C_{t+1}}{\partial U/\partial C_t} = (1 + i)^{-1},$$

and so

$$(1 + \rho)^{-1} = (1 + i)^{-1}, \tag{18.40}$$

that implies the tangency between the intertemporal indifference curve and the intertemporal budget constraint. Given the assumptions on U, the second-order conditions for a maximum are certainly satisfied.

In a similar way we can treat the model with trade examined in the text. We only observe that another way of looking at the representative agent's budget constraint in an open economy is to consider each period's constraint, which is

$$
\begin{aligned}
C_t &= y_t - B_t, \\
C_{t+1} &= y_{t+1} + (1+i)B_t,
\end{aligned}
\tag{18.41}
$$

where B_t is the agent's net stock of internationally traded bonds, a negative value meaning indebtedness. $B_t \lessgtr 0$ means that if the agent wants to consume more (less) than current income in period t, then he issues (buys) bonds on the international market. In our two-period model these bonds will have to be repaid (will be cashed in) together with accrued interest at time $t+1$. Solving for B_t in the first equation and plugging it into the second we obtain

$$
C_{t+1} = y_{t+1} + (1+i)y_t - (1+i)C_t.
\tag{18.42}
$$

If we divide through by $(1+i)$ and rearrange terms we get (18.35). We know that the current account is simply income minus absorption, $CA_t = y_t - C_t$, and taking account of the first equation in (18.41) we have

$$
CA_t = y_t - C_t = B_t,
$$

hence $B_t \lessgtr 0$ means that the country runs a current account deficit (surplus) matched by a capital inflow (outflow), so that the overall balance of payments will be in equilibrium. Of course the current account deficit (surplus) will be matched by a current account surplus (deficit) in period $t+1$. This shows the role of the foreign sector in the intertemporal transfer of wealth so as to smooth consumption through time according to the agent's intertemporal preferences.

18.4.2 The Harberger-Laursen-Metzler Effect

To examine this effect in the context of the intertemporal approach it is convenient to use the simplified formulation of the utility function given in (18.37). Thus the Lagrangian of our optimisation problem becomes

$$
L = U(C_t) + (1+\delta)^{-1}U(C_{t+1}) + \lambda\left[C_t + (1+i)^{-1}C_{t+1} - Y\right],
\tag{18.43}
$$

whence the first-order conditions

$$
\begin{aligned}
\frac{\mathrm{d}U}{\mathrm{d}C_t} + \lambda &= 0, \\
(1+\delta)^{-1}\frac{\mathrm{d}U}{\mathrm{d}C_{t+1}} + \lambda(1+i)^{-1} &= 0, \\
C_t + (1+i)^{-1}C_{t+1} - Y &= 0.
\end{aligned}
\tag{18.44}
$$

The second-order condition requires the appropriate bordered Hessian to be positive, namely

$$H \equiv \begin{vmatrix} \dfrac{\mathrm{d}^2 U}{\mathrm{d}C_t^2} & 0 & 1 \\[1em] 0 & (1+\delta)^{-1}\dfrac{\mathrm{d}^2 U}{\mathrm{d}C_{t+1}^2} & (1+i)^{-1} \\[1em] 1 & (1+i)^{-1} & 0 \end{vmatrix} > 0. \tag{18.45}$$

This gives

$$-\frac{\mathrm{d}^2 U}{\mathrm{d}C_t^2}(1+i)^{-2} - (1+\delta)^{-1}\frac{\mathrm{d}^2 U}{\mathrm{d}C_{t+1}^2} > 0, \tag{18.46}$$

which is certainly satisfied, since $\dfrac{\mathrm{d}^2 U}{\mathrm{d}C_t^2}$ and $\dfrac{\mathrm{d}^2 U}{\mathrm{d}C_{t+1}^2}$ are both negative.

We now apply the method of comparative statics (Gandolfo 2009; Chap. 20). The set of first-order conditions can be considered as a set of implicit functions in the three endogenous variables C_t, C_{t+1}, λ and the three parameters i, δ, Y. The Jacobian of such a set with respect to C_t, C_{t+1}, λ coincides with the Hessian, hence by the implicit function theorem it is possible to express the endogenous variables as differentiable functions of the parameters

$$\begin{aligned} C_t &= C_t(i, \delta, Y), \\ C_{t+1} &= C_{t+1}(i, \delta, Y), \\ \lambda &= \lambda(i, \delta, Y), \end{aligned} \tag{18.47}$$

and to compute their partial derivatives. We are interested in the effect of a change in real income (Y). Total differentiation of conditions (18.44) with respect to Y, account being taken of (18.47), yields

$$\frac{\mathrm{d}^2 U}{\mathrm{d}C_t^2}\frac{\partial C_t}{\partial Y} + \frac{\partial \lambda}{\partial Y} = 0,$$

$$(1+\delta)^{-1}\frac{\mathrm{d}^2 U}{\mathrm{d}C_{t+1}^2}\frac{\partial C_{t+1}}{\partial Y} + (1+i)^{-1}\frac{\partial \lambda}{\partial Y} = 0,$$

$$\frac{\partial C_t}{\partial Y} + (1+i)^{-1}\frac{\partial C_{t+1}}{\partial Y} = 1,$$

whose solution is

$$\frac{\partial C_t}{\partial Y} = \frac{-(1+\delta)^{-1}\dfrac{\mathrm{d}^2 U}{\mathrm{d}C_{t+1}^2}}{H} > 0, \qquad \frac{\partial C_{t+1}}{\partial Y} = \frac{-(1+i)^{-1}\dfrac{\mathrm{d}^2 U}{\mathrm{d}C_t^2}}{H} > 0,$$

$$\frac{\partial \lambda}{\partial Y} = \frac{\dfrac{\mathrm{d}^2 U}{\mathrm{d}C_t^2}(1+\delta)^{-1}\dfrac{\mathrm{d}^2 U}{\mathrm{d}C_{t+1}^2}}{H} < 0.$$

(18.48)

Since both C_t and C_{t+1} increase when Y increases, we conclude that an increase in real income (Y) causes an increase in aggregate expenditure, contrary to the Harberger-Laursen-Metzler hypothesis.

18.4.3 An Infinite Horizon Model

The generalization to the infinite-horizon case is fairly simple. We use the continuous time formulation with a time-separable utility function (ρ is the subjective discount factor)

$$\max_{\{C(t)\}} \int_0^\infty U[C(t)]e^{-\rho t}\mathrm{d}t,$$

(18.49)

subject to the flow budget constraint

$$\dot{a}(t) = [Y(t) + ia(t)] - C(t),$$

(18.50)

and the transversality condition

$$\lim_{t\to\infty} a(t)e^{-it} = 0.$$

(18.51)

The flow budget constraint states that the difference between any instant's current revenue (that includes interest income on the stock of wealth a) and expenditure equals the change in the stock of wealth. Given the transversality condition, the flow budget constraint is equivalent to the intertemporal budget constraint

$$\int_0^\infty C(t)e^{-it}\mathrm{d}t = \int_0^\infty Y(t)e^{-it}\mathrm{d}t + a(0),$$

(18.52)

which states that the present value of consumption expenditure must equal the present value of consumer's income plus initial wealth.

To show the equivalence between the two forms of budget constraint let us multiply through (18.50) by e^{-it} and integrate from $t = 0$ to $t = \tau$:

$$\int_0^\tau \dot{a}(t)e^{-it}dt = \int_0^\tau Y(t)e^{-it}dt + \int_0^\tau ia(t)e^{-it}dt - \int_0^\tau C(t)e^{-it}dt. \qquad (18.53)$$

Integrating by parts the left-hand side we obtain

$$\int_0^\tau \dot{a}(t)e^{-it}dt = \left[a(t)e^{-it}\right]_0^\tau - \int_0^\tau -ia(t)e^{-it}dt = a(\tau)e^{-i\tau} - a(0) + \int_0^\tau ia(t)e^{-it}dt, \qquad (18.54)$$

where $a(0)$ is the given wealth at time zero (inherited from the past). Substituting (18.54) into (18.53) we obtain

$$a(\tau)e^{-i\tau} - a(0) + \int_0^\tau ia(t)e^{-it}dt = \int_0^\tau Y(t)e^{-it}dt + \int_0^\tau ia(t)e^{-it}dt - \int_0^\tau C(t)e^{-it}dt,$$

namely

$$a(\tau)e^{-i\tau} + \int_0^\tau C(t)e^{-it}dt = \int_0^\tau Y(t)e^{-it}dt + a(0). \qquad (18.55)$$

If we now let $\tau \to \infty$ and use the transversality condition (observe that τ and t refer to the same variable) we obtain (18.52).

Since the flow form is more frequently used than the integral form, we shall adopt it. Thus our maximization problem is given by Eq. (18.49) with the constraint (18.50), and can be solved by Pontryagin's maximum principle (see, for example, Gandolfo 2009, Chap. 27, Sect. 27.3). We introduce a costate variable $\mu(t)$ and form the Hamiltonian

$$H = U[C(t)]e^{-\rho t} + \mu(t)[Y(t) + ia(t) - C(t)],$$

that can be rewritten, by introducing the new costate variable $\lambda(t) = \mu(t)e^{\rho t}$, as

$$H = e^{-\rho t}\{U[C(t)] + \lambda(t)[Y(t) + ia(t) - C(t)]\}. \qquad (18.56)$$

The reader should note that this change of variable is a mathematically convenient way of dealing with problems involving discounting.

We now look for the possible existence of an interior maximum to the Hamiltonian by setting

$$\frac{\partial H}{\partial C} = e^{-it}\{U_C - \lambda(t)\} = 0,$$

from which

$$U_C = \lambda(t). \tag{18.57}$$

Then there are the equations of motion of the state and costate variable. The former is given by Eq. (18.50), the latter is given by the condition

$$\dot{\mu}(t) = -\frac{\partial H}{\partial a} = -\mu(t)i$$

or, in terms of the costate variable λ (t),

$$\dot{\lambda}(t) = \lambda(t)(\rho - i). \tag{18.58}$$

We finally have the transversality condition on the costate variable

$$\lim_{t \to \infty} \lambda(t)e^{-\rho t} = 0. \tag{18.59}$$

Let us now rewrite the optimality conditions in the following way. First, logarithmic differentiation of (18.57) with respect to time yields

$$\frac{\dot{\lambda}(t)}{\lambda(t)} = -\sigma(C)\frac{\dot{C}(t)}{C(t)}, \tag{18.60}$$

where $\sigma(C) \equiv -C(U_{CC}/U_C)$ is the elasticity of marginal utility. Substitution of (18.60) into (18.58) yields

$$\dot{C}(t) = \frac{1}{\sigma(C)}[i - \rho]C(t). \tag{18.61}$$

Thus we have reduced the differential equation for the costate variable to a differential equation for the control variable, and the dynamic system that solves the dynamic optimization problem is

$$\begin{aligned}\dot{a}(t) &= Y(t) + ia(t) - C(t), \\ \dot{C}(t) &= \frac{1}{\sigma(C)}[i - \rho]C(t).\end{aligned} \tag{18.62}$$

To determine the existence of a steady state equilibrium we set $\dot{a}(t) = 0, \dot{C}(t) = 0$. The first condition gives $Y(t) + ia(t) = C(t)$, namely revenue and expenditure are equal in the steady state equilibrium, an obvious requirement.

The second condition requires $i - \rho = 0$, namely the market rate of interest and the rate of time preference must be equal. A problem might arise when ρ is constant, because no steady state can exist unless ρ by chance equals i, and if this happens a zero-root problem will arise in the dynamic system (18.62). It should however be noted that $i = \rho$ is the same condition that we found in the simple two-period case treated in the previous section, where the equality was ensured by the variability of ρ. There is no reason why ρ should now be a constant, hence we can plausibly expect the condition to be satisfied.

Further discussion of continuous-time infinite-horizon problems can be found in Sen (1994).

18.4.4 The RAIOM Approach to the Real Exchange Rate

Consider the following two-period maximization problem

$$\max_{C_t, C_{t+1}} U(C_t, C_{t+1}) = \max_{C_t, C_{t+1}} (\ln C_t + \frac{1}{1+\rho} \ln C_{t+1}) \tag{18.63}$$

sub

$$(1+i)^{-1} C_{t+1} + C_t = (1+i)^{-1} y_{t+1} + y_t, \tag{18.64}$$

where y_t, y_{t+1} are the given endowments (incomes) in the two periods, and i is the interest rate at which the agent can borrow or lend in the international market.

Forming the Lagrangian

$$L = (\ln C_t + \frac{1}{1+\rho} \ln C_{t+1}) + \lambda \left[(1+i)^{-1} C_{t+1} + C_t - (1+i)^{-1} y_{t+1} - y_t \right], \tag{18.65}$$

we have the first-order conditions

$$
\begin{aligned}
\partial L / \partial C_t &= \frac{1}{C_t} + \lambda = 0, \\
\partial L / \partial C_{t+1} &= [(1+\rho) C_{t+1}]^{-1} + \lambda (1+i)^{-1} = 0, \\
\partial L / \partial \lambda &= (1+i)^{-1} C_{t+1} + C_t - (1+i)^{-1} y_{t+1} - y_t = 0.
\end{aligned}
\tag{18.66}
$$

From the first and second equation we obtain

$$C_{t+1} = \frac{(1+i)}{(1+\rho)} C_t. \tag{18.67}$$

Substituting C_{t+1} from (18.67) into the intertemporal budget constraint (18.64) we get

$$\frac{1}{1+\rho} C_t = y_t + (1+i)^{-1} y_{t+1} - C_t, \tag{18.68}$$

whence

$$C_t = \beta Y_t, \quad \beta \equiv \left(1 + \frac{1}{1+\rho}\right)^{-1}, \tag{18.69}$$

where

$$Y_t = y_t + (1+i)^{-1} y_{t+1} \tag{18.70}$$

is the present value of lifetime income. The quantity βY_t is called permanent income Y_t^p.

When we introduce the distinction between private and public consumption the problem becomes

$$\max_{C_t, C_{t+1}} U(C_t, C_{t+1}) = \max_{C_t, C_{t+1}} \left(\ln C_t + \frac{1}{1+\rho} \ln C_{t+1}\right) \tag{18.71}$$

sub

$$(1+i)^{-1}(C_{t+1} - G_{t+1}) + (C_t - G_t) = (1+i)^{-1}(y_{t+1} - G_{t+1}) + (y_t - G_t). \tag{18.72}$$

The problem is the same as before, with $(C_{t+1} - G_{t+1})$ replacing C_{t+1}, $(y_{t+1} - G_{t+1})$ replacing y_{t+1}, etc.; consequently the optimality condition (18.68) becomes

$$\frac{1}{1+\rho}(C_t - G_t) = (y_t - G_t) + (1+i)^{-1}(y_{t+1} - G_{t+1}) - (C_t - G_t), \tag{18.73}$$

from which

$$(C_t - G_t) \equiv C_{pt} = \beta Y_t - \beta \Gamma_t = Y_t^p - G_t^p, \tag{18.74}$$

where

$$Y_t = y_t + (1+i)^{-1} y_{t+1}, \quad \Gamma_t = G_t + (1+i)^{-1} G_{t+1}. \tag{18.75}$$

Let us now consider the tradables-nontradables model and assume the logarithmic Cobb-Douglas utility function

$$U(C_t^T, C_t^{NT}, C_{t+1}^T, C_{t+1}^{NT})$$

$$= (1-\alpha) \ln C_t^T + \alpha \ln C_t^{NT} + \left(\frac{1-\alpha}{1+\rho}\right) \ln C_{t+1}^T + \left(\frac{\alpha}{1+\rho}\right) \ln C_{t+1}^{NT}, \tag{18.76}$$

that the representative agent maximizes subject to the following budget constraints

$$(1 + i)^{-1} p_{t+1}^T C_{t+1}^T + p_t^T C_t^T = (1 + i)^{-1} p_{t+1}^T y_{t+1}^T + p_t^T y_t^T, \qquad (18.77)$$

$$p_t^{NT} C_t^{NT} = p_t^{NT}(Q_t - G_t), \qquad (18.78)$$

$$p_{t+1}^{NT} C_{t+1}^{NT} = p_{t+1}^{NT}(Q_{t+1} - G_{t+1}). \qquad (18.79)$$

Since intertemporal consumption smoothing is only allowed through the international market, it only involves tradables, hence the intertemporal constraint (18.77) has the usual meaning, that the present value of lifetime consumption of tradables equals the present value of lifetime income of tradables. Constraints (18.78) and (18.79) state that consumption of nontradables must equal the amount available in each period, which is the period's output Q minus government consumption in the same period G. All quantities are multiplied by the relevant prices since this is a two-sector model.

We can add (18.77)–(18.79) to obtain the overall budget constraint

$$\{[(1 + i)^{-1} p_{t+1}^T C_{t+1}^T + p_t^T C_t^T - (1 + i)^{-1} p_{t+1}^T y_{t+1}^T - p_t^T y_t^T]$$
$$+ p_{t+1}^{NT} [C_{t+1}^{NT} - (Q_{t+1} - G_{t+1})] + p_t^{NT} [C_t^{NT} - (Q_t - G_t)]\}$$
$$= 0. \qquad (18.80)$$

Forming the Lagrangian

$$L = (1 - \alpha) \ln C_t^T + \alpha \ln C_t^{NT} + \left(\frac{1 - \alpha}{1 + \rho}\right) \ln C_{t+1}^T + \left(\frac{\alpha}{1 + \rho}\right) \ln C_{t+1}^{NT}$$
$$+ \lambda \{[(1 + i)^{-1} p_{t+1}^T C_{t+1}^T + p_t^T C_t^T - (1 + i)^{-1} p_{t+1}^T y_{t+1}^T - p_t^T y_t^T]$$
$$+ p_{t+1}^{NT} [C_{t+1}^{NT} - (Q_{t+1} - G_{t+1})] + p_t^{NT} [C_t^{NT} - (Q_t - G_t)]\} \qquad (18.81)$$

we obtain the first-order conditions

$$\partial L/\partial C_t^T = (1 - \alpha)/C_t^T + \lambda p_t^T = 0,$$
$$\partial L/\partial C_{t+1}^T = (\tfrac{1-\alpha}{1+\rho})/C_{t+1}^T + \lambda(1 + i)^{-1} p_{t+1}^T = 0,$$
$$\partial L/\partial C_t^{NT} = \alpha/C_t^{NT} + \lambda p_t^{NT} = 0, \qquad (18.82)$$
$$\partial L/\partial C_{t+1}^{NT} = (\tfrac{\alpha}{1+\rho})/C_{t+1}^{NT} + \lambda p_{t+1}^{NT} = 0,$$
$$\partial L/\partial \lambda = 0 = \text{ overall budget constraint.}$$

The first four conditions can be rewritten as

$$
\begin{aligned}
1 - \alpha &= -\lambda p_t^T C_t^T, \\
\frac{1 - \alpha}{1 + \rho} &= -\lambda (1 + i)^{-1} p_{t+1}^T C_{t+1}^T, \\
\alpha &= -\lambda p_t^{NT} C_t^{NT}, \\
\frac{\alpha}{1 + \rho} &= -\lambda p_{t+1}^{NT} C_{t+1}^{NT}.
\end{aligned}
\tag{18.83}
$$

From the first and second condition we get

$$
(1 + i)^{-1} p_{t+1}^T C_{t+1}^T = \frac{1}{1 + \rho} p_t^T C_t^T.
\tag{18.84}
$$

Substituting from (18.84) into the intertemporal budget constraint (18.77) we get

$$
\frac{1}{1 + \rho} p_t^T C_t^T + p_t^T C_t^T = (1 + i)^{-1} p_{t+1}^T y_{t+1}^T + p_t^T y_t^T,
\tag{18.85}
$$

from which

$$
p_t^T C_t^T = \beta Y_t^T, \quad \beta \equiv \left(1 + \frac{1}{1 + \rho} \right)^{-1},
\tag{18.86}
$$

where $Y_t^T \equiv (1 + i)^{-1} p_{t+1}^T y_{t+1}^T + p_t^T y_t^T$ is the present value of lifetime income involving tradables. Note that Eq. (18.86) is essentially similar to Eq. (18.69).

The share of tradables and nontradable goods in total consumption is constant, a normal occurrence with a Cobb-Douglas logarithmic utility function. To check this, observe that from the first and third equation in (18.83) we have:

$$
\frac{p_t^{NT} C_t^{NT}}{p_t^T C_t^T} = \frac{\alpha}{1 - \alpha}.
\tag{18.87}
$$

Since the real exchange rate is the relative price of the nontradable good, from (18.87) we finally have

$$
r_{R_t} = \frac{p_t^{NT}}{p_t^T} = \gamma \frac{C_t^T}{C_t^{NT}}; \quad \gamma \equiv \frac{\alpha}{1 - \alpha}.
\tag{18.88}
$$

18.4.5 The NATREX Approach

18.4.5.1 The SOFC Rule and the Investment Function

We shall start from a simple dynamic optimization problem of the standard type. Consider a closed economy with an infinitely-lived representative consumer (or a social planner) solving the dynamic optimization problem

$$\max_{\{c\}} \Omega = \int_0^\infty U(c)e^{-\rho t}dt, \quad U_c > 0, U_{cc} < 0, \qquad (18.89)$$

sub

$$\frac{dk}{dt} = f(k) - \lambda k - c, \quad f_k > 0, f_{kk} < 0,$$

where U is the utility function, c indicates consumption per worker and ρ is the discount rate measuring the time preference of consumers. The constraint is the overall goods market equilibrium, where the rate of investment equals output $f(k)$ minus consumption (c) and a term involving the growth rate of labour (n) and the depreciation rate (δ) of capital goods, λk ($\lambda = n + \delta$). Per capita production is a function of capital intensity (k). Both the utility function and the production function are concave, reflecting positive but diminishing marginal utility (productivity). The utility and production functions are further assumed to have the properties

$$\lim_{c \to 0} U_c(c) = \infty, \quad \lim_{c \to \infty} U_c(c) = 0, \qquad (18.90)$$

$$\lim_{k \to 0} f_k(k) = \infty, \quad \lim_{k \to \infty} f_k(k) = 0. \qquad (18.91)$$

To solve the problem let us form the Hamiltonian

$$H = e^{-\rho t}U(c) + \mu[f(k) - \lambda k - c],$$

where μ is a Lagrange multiplier or costate variable. Introducing the new costate variable

$$q = \mu e^{\rho t}, \qquad (18.92)$$

we obtain

$$H = e^{-\rho t}\{U(c) + q[f(k) - \lambda k - c]\}. \qquad (18.93)$$

To check for an interior maximum to the Hamiltonian we set

$$\frac{\partial H}{\partial c} = e^{-\rho t}\{U_c(c) - q\} = 0,$$

from which

$$q = U_c(c). \qquad (18.94)$$

Since the utility function is concave (i.e. $U_{cc}(c) < 0$), $U_c(c)$ is monotonically decreasing from ∞ to zero by properties (18.90). Hence a unique interior solution always exists. Moreover, since $U_{cc}(c) < 0$, this solution is indeed a maximum. Equation (18.94) states that the shadow price of per capita capital accumulation along the optimal path equals the marginal utility of per capita consumption.

Let us now come to the canonical equation for the (original) costate variable, which is

$$\frac{d\mu}{dt} = -\frac{\partial H}{\partial k} = -\mu[f_k(k) - \lambda]. \tag{18.95}$$

To transform (18.95) into a differential equation for the costate variable q (which also represents a shadow price) observe that $\mu = qe^{-\rho t}$, hence $d\mu/dt = (dq/dt)e^{-\rho t} - \rho q e^{-\rho t}$. By substituting into Eq. (18.95) and collecting terms we have

$$\frac{dq}{dt} = -[f_k(k) - \lambda - \rho]q, \tag{18.96}$$

from which

$$f_k(k) + \frac{dq/dt}{q} - \delta - n - \rho = 0. \tag{18.97}$$

We can interpret this equation as the condition of *zero net profit rate*, where the net profit rate is defined as the gross profit rate (the marginal productivity of capital) plus capital gains $[(dq/dt)/q]$ minus the losses due to depreciation (δ) minus the "dilution" due to population growth (n) minus the intertemporal cost of waiting, represented by the subjective discount rate (ρ).

Note that Eqs. (18.94) and (18.97) already give us a lot of information on the optimal growth path even if we do not yet know it.

Let us now observe that by logarithmic differentiation of the optimality condition (18.94) with respect to time we obtain

$$\frac{dq/dt}{q} = \frac{U_{cc}(c)}{U_c(c)}\frac{dc}{dt} = -\sigma(c)\frac{dc/dt}{c},$$

which can be substituted in the canonical equation (18.97) to obtain

$$\frac{dc}{dt} = \frac{1}{\sigma(c)}[f_k(k) - (\lambda + \rho)]c,$$

where

$$\sigma(c) = -c\frac{U_{cc}}{U_c}, \tag{18.98}$$

is the elasticity of marginal utility.

In this way we have reduced the differential equation for the costate variable to a differential equation for the control variable. Thus the canonical equations of our optimal control problem are

$$\frac{dc}{dt} = \frac{1}{\sigma(c)}[f_k(k) - (\lambda + \rho)]c,$$
$$\frac{dk}{dt} = f(k) - \lambda k - c,$$

(18.99)

which are a system of two *autonomous* non-linear differential equations. The singular point of this system gives the steady state values c^*, k^* of c, k :

$$f_k(k) = \lambda + \rho, \qquad \text{which determines } k^*,$$
$$c^* = f(k^*) - \lambda k^*, \text{ which determines } c^*.$$

(18.100)

Note that the first equation always has a unique positive solution given assumption (18.91). The transitional dynamics of system (18.99) is represented by a saddle path in the (c, k) plane (see Gandolfo 2009, Sect. 28.2.1), but we are interested in deriving an *optimal feedback rule* in the $(k, dk/dt)$ plane.

The Optimal Feedback Control Rule

For this purpose let us differentiate the second equation in (18.99) with respect to k, obtaining

$$\frac{\partial(dk/dt)}{\partial k} = f_k(k) - \lambda - \frac{dc}{dk}.$$

(18.101)

We now observe that

$$\frac{dc}{dk} = \frac{dc}{dt} \Big/ \frac{dk}{dt},$$

(18.102)

and substituting from (18.99) into (18.102) and then into (18.101) we obtain

$$\frac{\partial(dk/dt)}{\partial k} = f_k(k) - \lambda - \frac{U_c}{U_{cc}} \frac{[\lambda + \rho - f_k(k)]}{dk/dt}.$$

(18.103)

For notational simplicity we set $V(k) \equiv dk/dt$, $V'(k) \equiv \partial(dk/dt)/\partial k$ and rewrite Eq. (18.103) as

$$V'(k) = f_k(k) - \lambda - \frac{U_c}{U_{cc}} \frac{[\lambda + \rho - f_k(k)]}{V(k)} = F[k, V(k)].$$

(18.104)

Let us now consider a neighbourhood of the steady state, where

$$V'(k)\big|_{k=k^*} = V'(k^*) = f_k(k^*) - \lambda - \frac{U_c}{U_{cc}} \frac{[\lambda + \rho - f_k(k^*)]}{V(k^*)} = \rho - \frac{U_c(c^*)}{U_{cc}(c^*)} \frac{0}{0},$$
(18.105)

where we have used the steady-state values given in (18.100). The indeterminacy can be removed using l'Hopital's rule:

$$V'(k^*) = \rho + \frac{U_c(c^*)}{U_{cc}(c^*)} \frac{f_{kk}(k^*)}{V'(k^*)}.$$
(18.106)

Multiplying through by $V'(k^*)$ and rearranging terms we obtain

$$[V'(k^*)]^2 - \rho[V'(k^*)] - \frac{U_c(c^*)}{U_{cc}(c^*)} f_{kk}(k^*) = 0.$$
(18.107)

This can be considered as a second-degree polynomial equation to determine the unknown $V'(k^*)$. Given the signs of the various derivatives, the succession of the coefficient signs is $+ - -$, hence there will be two real roots, one positive and the other negative. Local stability of the feedback rule requires $V'(k^*)$ to be negative, hence we only consider the negative root of (18.107), that is

$$V'(k^*) = -A(k^*) < 0,$$
(18.108)

where

$$A(k^*) \equiv \frac{\rho}{2} \left[\left(1 + 4 \frac{U_c(c^*)}{U_{cc}(c^*)} \frac{f_{kk}(k^*)}{\rho^2}\right)^{\frac{1}{2}} - 1 \right] > 0.$$
(18.109)

If we integrate (18.108) between k and k^* in the neighbourhood of the steady state we have

$$\int_k^{k^*} V'(\phi)\,d\phi = -\int_k^{k^*} A(k^*)\,d\phi,$$
(18.110)

where

$$\int_k^{k^*} V'(\phi)\,d\phi = [V(\phi)]_k^{k^*} = V(k^*) - V(k) = -V(k),$$
(18.111)

since $V(k^*) = 0$ by definition, and

$$-\int_k^{k^*} A(k^*)\,d\phi = -A(k^*)\int_k^{k^*} d\phi = -A(k^*)(k - k^*),$$
(18.112)

since $A(k^*)$ is a constant. From (18.110)–(18.112) we finally have

$$\frac{dk}{dt} = -A(k^*)(k - k^*), \tag{18.113}$$

that is the optimal control law in feedback form, which relates the change in the state to the deviations of the state from the steady-state equilibrium point. The differential equation (18.113) obviously drives the system to equilibrium.

The problem with this rule is that perfect knowledge of the steady-state values is required. Moreover, there is no reason why the production function should be constant over time. Finally, given the perfect foresight assumption, an even slightly wrong perception will lead the system outside the stable arm of the saddle and hence cause it to diverge from the steady-state.

The Sub-Optimal Feedback Control (SOFC) Rule
The idea is to derive a feedback control rule that only requires observation of current values and that, though being sub-optimal, asymptotically tends to the optimal control law (Infante and Stein 1973; Stein 1996) . For this purpose we expand $f_k(k^*)$ in Taylor's series:

$$f_k(k^*) = f_k(k) + f_{kk}(k)(k^* - k). \tag{18.114}$$

In the steady state we have

$$f_k(k) = \rho + \lambda + f_{kk}(k)(k - k^*), \tag{18.115}$$

whence

$$k - k^* = \frac{f_k(k) - (\rho + \lambda)}{f_{kk}(k)}, \tag{18.116}$$

and so the suboptimal form of the optimal control law is

$$\left(\frac{dk}{dt}\right)_{SOFC} \equiv V_1(k) = \frac{-A(k)}{f_{kk}(k)} [f_k(k) - (\rho + \lambda)] \tag{18.117}$$

where

$$A(k) \equiv \frac{\rho}{2} \left[\left(1 + 4\frac{U_c(c)}{U_{cc}(c)} \frac{f_{kk}(k)}{\rho^2} \right)^{\frac{1}{2}} - 1 \right] > 0, \tag{18.118}$$

so that *only current observed values* are needed.

In the neighbourhood of the origin, the slope of $V_1(k)$ equals that of $V(k)$:

$$\frac{dV_1(k^*)}{dk} = -A(k^*) = V_1'(k^*) = V'(k^*), \tag{18.119}$$

so that

$$\lim_{t \to \infty} [V(k) - V_1(k)] = 0, \qquad (18.120)$$

hence the sub-optimal feedback control tends to the optimal one.

The SOFC rule (18.117) states that investment is an increasing function of the difference between the current marginal productivity of capital and the subjective discount rate.

18.4.5.2 Analysis of the NATREX Equilibrium

The Medium Run
We first recapitulate the equations of the model explained in the text.

Domestic goods market $y(k;Z) = C(k, F;Z) + dk/dt + TB(r_F, k, F, k^f; Z)$,
Foreign goods market $GDP^f = C^f + I^f - TB(r_F, k, F, k^f; Z)$,
Investment function $dk/dt = I = I(f_k - i_R; Z)$,
Saving function $S = S(k, F, i_R; Z)$,
Rate of change of debt $dF/dt = I - S = -CA$,
Portfolio equation $di_R/dt = -a(i_R - i^f_R)$.

$$\qquad (18.121)$$

The medium-run subsystem is

$$
\begin{aligned}
r_R(t) &= H[i_R(t), k(t), F(t); Z], \\
r_R(t) &= H^f[i^f_R(t), k^f(t), F(t); Z], \\
di_R/dt &= -a(i_R - i^f_R),
\end{aligned}
\qquad (18.122)
$$

where the first and second equation are the equations for goods-market equilibrium, derived from the first and second equation in (18.121). They represent the *IS* and *IS^f* curves respectively. In the medium run, capital and debt are taken as predetermined variables, while the interest rate adjusts to satisfy portfolio balance, which is the third equation in (18.122). We have

$$
\begin{aligned}
\left. \frac{\partial r_R(t)}{\partial i_R(t)} \right|_{IS} &= -\frac{I_{i_R}}{TB_{r_R}} < 0, \\
\left. \frac{\partial r_R(t)}{\partial i_R(t)} \right|_{IS^f} &= -\frac{I^f_{i^f_R}}{-TB_{r_R}} > 0,
\end{aligned}
\qquad (18.123)
$$

which are the slopes of the *IS* and *IS^f* schedules.

Medium-run portfolio equilibrium occurs when real interest-rate convergence has taken place, $i_R(t) = i^f_R(t)$. With this condition, system (18.122) can be solved for $r_R(t), i_R(t)$, obtaining

$$r_R(t) = R[k(t), F(t); k^f(t), Z], \tag{18.124}$$

$$i_R(t) = i_R^f(t) = i[k(t); k^f(t), Z]. \tag{18.125}$$

A medium-run equilibrium point does not remain fixed, but moves through time, because in the long run capital and debt change. Thus the dynamics of the real exchange rate is obtained from Eq. (18.122) by differentiation with respect to time:

$$\frac{dr_R}{dt} = b_1 \frac{dk}{dt} + b_2 \frac{dk^f}{dt} - b_3 \frac{dF}{dt} + b_4(i_R - i_R^f) + b_5 \frac{dZ}{dt},$$

where the $b's$ are coefficients related to the partial derivatives of the functions in (18.122).

The Long Run
The dynamics of the long-run system can be reduced to a system of two differential equations. First, substitute the real interest rate equation (18.125) into the equation for investment (third equation in 18.121) to obtain

$$\frac{dk}{dt} = I[f_k(k; Z) - i(k; k^f, Z)] \equiv J(k; k^f, Z), J_k = f_{kk} - i_k < 0. \tag{18.126}$$

Second, substitute the real interest rate equation (18.125) into the equation for saving (fourth equation in 18.121) to obtain

$$S = S(k, F; k^f, Z), S_k > 0, S_F > 0. \tag{18.127}$$

Third, substitute (18.126) and (18.127) into the equation for capital inflow (fifth equation in 18.121) to obtain

$$\frac{dF}{dt} = I - S = J(k; k^f, Z) - S(k, F; k^f, Z) \equiv L(k, F; k^f, Z),$$

$$L_k = J_k - S_k < 0, L_F = -S_F < 0. \tag{18.128}$$

The singular point of the differential equation system

$$\frac{dk}{dt} = J(k; k^f, Z), \frac{dF}{dt} = L(k, F; k^f, Z), \tag{18.129}$$

implies

$$f_k(k^*; Z) - i_R(k^*; k^f, Z) = 0, I - S = CA = 0, \tag{18.130}$$

namely in the long-run equilibrium (steady-state) the capital intensity does not change because the marginal productivity of capital equals the real interest rate, and the stock of foreign debt does not change because the current account is in equilibrium. *This latter implies that the real exchange rate (the NATREX) is such as to generate a trade balance surplus just enough to pay interest on foreign debt.*

The local stability (for simplicity's sake we assume that the foreign country is already in equilibrium, so that k^f does not change) of the differential equation system (18.129) depends on the roots θ of the characteristic equation

$$|\mathbf{A} - \theta \mathbf{I}| = \begin{bmatrix} J_k - \theta & 0 \\ L_k & L_F - \theta \end{bmatrix} = \theta^2 - (J_k + L_F)\theta + J_k L_F = 0. \qquad (18.131)$$

The stability conditions are

$$J_k + L_F < 0, \quad J_k L_F > 0. \qquad (18.132)$$

These conditions are certainly satisfied, given the signs shown above, hence both roots are stable. Note the crucial importance of the positivity of S_F. In the contrary case, in fact, we would have $L_F > 0$, and the second stability condition would not be satisfied. In such a case we would have saddle-path stability. In fact, since the sign of $J_k + L_F$ would be uncertain, the succession of the signs of the coefficients of Eq. (18.131) would be either $+ - -$ or $+ + -$. In both cases the roots are real and, by Descartes' theorem, one is positive and the other negative.

Once determined the dynamics of k, F around the steady state, the dynamics of the real exchange rate is obtained differentiating Eq. (18.124) with respect to time:

$$\frac{dr_R}{dt} = R_k \frac{dk}{dt} + R_F \frac{dF}{dt}.$$

References

Alexander, S. S. (1952). Effects of a devaluation on a trade balance. *International Monetary Fund Staff Papers, 2,* 263–278.

Alexander, S. S. (1959). Effects of a devaluation: A simplified synthesis of elasticities and absorption approaches. *American Economic Review, 49,* 22–42.

Belloc, M., & Federici, D. (2010). A two-country NATREX model for the euro/dollar. *Journal of International Money and Finance, 29,* 315–335.

Fang, C., & Lin, P. -S. (2013). Traded bond denominations, shock persistence and current account dynamics: Another look at the Harberger–Laursen–Metzler Effect. *Pacific Economic Review, 18,* 502–529.

Federici, D., & Gandolfo, G. (2002). Endogenous growth in an open economy and the real exchange rate. *Australian Economic Papers, 41,* 499–518.

Feldstein, M., & Horioka, C. (1980). Domestic saving and international capital flows. *Economic Journal, 90,* 314–329.

Fisher, I. (1930). *The theory of interest.* New York: Macmillan.

Frenkel, J. A., & Razin, A., with the collaboration of C. -W. Yuen (1996). *Fiscal policies and growth in the world economy* (3rd ed.). Cambridge (Mass): MIT Press.

Gandolfo, G. (2009). *Economic dynamics*. Berlin, Heidelberg, New York: Springer.

Gandolfo, G., & Felettigh, A. (1998), The NATREX: An Alternative Approach. CIDEI Working Paper 52, University of Rome "La Sapienza".

Infante, E. F., & Stein, J. L. (1973). Optimal growth with robust feedback control. *Review of Economic Studies, XL*, 47–60.

Jansen, W. J. (1997). Can the intertemporal budget constraint explain the Feldstein-Horioka puzzle? *Economics Letters, 56*, 77–83.

Lane, P. R. (2001). The new open economy macroeconomics: A survey. *Journal of International Economics, 54*, 235–266.

Lim, G. C., & Stein, J. L. (1995). The dynamics of the real exchange rate and current account in a small open economy: Australia, in J. L. Stein, P. Reynolds Allen et al. (1995). *Fundamental determinants of exchange rates* (pp. 85–125). Oxford: Oxford University Press.

MacDonald, R., & Stein, J. L. (Eds.) (1999). *Equilibrium exchange rates*. Dordrecht: Kluwer Academic Publishers.

Machlup, F. (1955). Relative Pprices and aggregate spending in the analysis of devaluation. *American Economic Review, 45*, 255–278.

Merton, R. C. (1992). *Continuous time finance* (revised ed.). Oxford: Basil Blackwell.

Moosa, I. A. (1997). Resolving the Feldstein-Horioka puzzle. *Economia Internazionale, 50*, 437–458.

Obstfeld, M., & Rogoff, K. (1995a). The intertemporal approach to the currentaAccount. Chap. 34 in G. Grossman & K. Rogoff (Eds.). *Handbook of international economics, Vol. III*. Amsterdam: North-Holland.

Obstfeld, M., & Rogoff, K. (1995b). Exchange rate dynamics redux. *Journal of Political Economy, 103*, 624–660.

Obstfeld, M., & Rogoff, K. (1996). *Foundations of international macroeconomics*. Cambridge (Mass; MIT Press.

Persson, T., & Svensson, L. E. O. (1985). Current account dynamics and the terms of trade: Harberger-Laursen-Metzler two generations later. *Journal of Political Economy, 93*, 43–65.

Sen, P. (1994). Savings, investment, and the current account, in F. Van der Ploeg (Ed.), *Handbook of international macroeconomics* (pp. 506–534). Oxford: Basil Blackwell.

Stein, J. L. (1990). The real exchange rate. *Journal of Banking and Finance, 14*(special issue), 1045–1078.

Stein, J. L. (1995a). The natural real exchange rate of the United States dollar, and determination of capital flows, in J. L. Stein, P. Reynolds Allen et al. (1995). *Fundamental determinants of exchange rates* (pp. 38–84). Oxford: Oxford University Press.

Stein, J.L. (1995b). The fundamental determinants of real exchange rate of the US dollar relative to the other G-7 countries. IMF Working Paper 95/81.

Stein, J.L. (1996). Real exchange rates and current accounts: The implications of economic science for policy decisions. *Economie Appliquée, XLIX*, 49–94.

Stein, J.L. (1999). The evolution of the real value of the US dollar relative to the G7 currencies, in: MacDonald, R., & Stein, J.L. (Eds.) (1999). *Equilibrium exchange rates*. Dordrecht: Kluwer Academic Publishers, 67–101.

Stein, J.L., & Paladino, G. (1998). Exchange rate misalignments and crises. Brown University Working Paper 98-7.

Stein, J.L., & Sauernheimer, K.-H. (1996). The equilibrium real exchange rate of Germany. *Economic Systems, 20*, 97–131.

Svensson, L.E.O., & Razin, A. (1983). The terms of trade and the current account: The Harberger-Laursen-Metzler effect. *Journal of Political Economy, 91*, 97–125.

Tsiang, S.C. (1961). The role of money in trade-balance stability: Synthesis of the elasticity and absorption approaches. *American Economic Review, 51*, 912–936.

Verrue, J.L., & Colpaert, J. (1998). A dynamic model of the real Belgian franc. University of Rome "La Sapienza", CIDEI Working Paper 47.

Other Applications

19

19.1 Introduction

In the previous chapter we have explained the basic ideas of the intertemporal approach, and shown how it can shed new light on old problems, such as the causes of current account imbalances, and exchange-rate determination. These results are by now standard. In the present chapter we shall briefly deal with two additional areas of research in the intertemporal approach: endogenous growth in an open economy, and nominal rigidities. Although we have left the more difficult manipulations for the Appendix, the present chapter involves much more mathematics than the average chapter of this book.

19.2 An Intertemporal Model with Endogenous Growth in an Open Economy

The theory of economic growth, after being a very fashionable topic in the late 1950s and in the 1960s, dropped out of fashion for almost two decades. In the 1990s there has been a resurgence of interest, especially thanks to the new theory of endogenous growth (Barro and Sala-i-Martin 2004; Aghion and Howitt 1998). Models of the new growth theory in open economies are relatively few, and those which exist mostly make the assumption that UIP holds, which implies that countries are free to borrow or lend as much as they may wish at the given world interest rate.

In the context of the intertemporal approach we shall consider the consequences of dropping the assumption of UIP. The case of a *net borrower* economy faced with an upward-sloping supply curve of debt is particularly relevant for developing countries.

© Springer-Verlag Berlin Heidelberg 2016

G. Gandolfo, *International Finance and Open-Economy Macroeconomics*,
Springer Texts in Business and Economics, DOI 10.1007/978-3-662-49862-0_19

In this section we consider an endogenous-growth one-country model due to Turnovsky (1997; Chap. 5) . The country produces a homogeneous traded good (that can be both consumed and invested) and is populated by an intertemporally optimizing representative agent.

One of the points made by the endogenous growth literature is that the assumption of decreasing returns to factors (decreasing marginal productivity) should be dropped. The simplest production function with these properties is

$$Y = \alpha K, \tag{19.1}$$

where α is a positive constant that reflects the technological level. For obvious reasons this function has come to be known in the recent literature as the *"AK"* production function (Barro and Sala-i-Martin 2004) , but its use in growth theory dates back at least to Harrod (1939) and Domar (1946).

The labour supply is fixed inelastically.

In consuming the commodity at rate C_j, the representative agent (denoted with the subscript j) derives utility over an infinite time horizon, which is represented by an isoelastic intertemporal utility function

$$\Omega_j \equiv \int_0^\infty \frac{1}{\gamma} C_j^\gamma e^{-\beta t} dt, \quad -\infty < \gamma < 1. \tag{19.2}$$

The exponent β in this function denotes the agent's subjective discount rate; his or her intertemporal elasticity of substitution, s, is given by $s = 1/(1-\gamma)$.

The agent accumulates physical capital via expenditures on increases in the capital stock, I_j, which entail costs of adjustment. These costs are represented by a quadratic function, so that the expenditure required to carry out an investment I_j is

$$\Phi(I_j, K_j) = I_j + h \frac{I_j^2}{2K_j} = I_j \left(1 + \frac{h}{2} \frac{I_j}{K_j}\right). \tag{19.3}$$

Sustenance of a steady-state equilibrium of ongoing growth requires this function to be linearly homogeneous.

The agent also accumulates net traded bonds, b_j (which pay an exogenously given world interest rate r), pays taxes on income from physical capital, bond income, and consumption expenditures at rates τ_k, τ_b, and τ_c, and receives transfer payments T_j from the government. The agent's instantaneous budget constraint is then given by

$$\dot{b}_j = [(1-\tau_k)\alpha K_j + (1-\tau_b)rb_j] - \left[(1+\tau_c)C_j + I_j \left(1 + \frac{h}{2} \frac{I_j}{K_j}\right)\right] + T_j, \tag{19.4}$$

in which the cost function $\Phi(I_j, K_j)$ has been replaced by expression (19.3). The government is assumed to maintain a continuously balanced budget: all tax revenues are rebated back to the private sector in the form of a lump-sum transfer payment

$$T = \tau_k \alpha K + \tau_b rb + \tau_c C. \tag{19.5}$$

Aggregating (19.4) and combining it with (19.5), which is already an aggregate relation, we obtain

$$\dot{b} = rb + \alpha K - C - I\left(1 + \frac{h}{2}\frac{I}{K}\right), \tag{19.6}$$

which expresses the net rate of accumulation of traded bonds by the private sector, which coincides with the current account. Here again we see that the balance of payments depends on the intertemporal saving-investment decisions of the private sector (see Chap. 18).

We now go back to the representative agent's optimization problem, from now on omitting the subscript j for simplicity's sake.

Assuming non-depreciation of capital, the agent faces a physical constraint on capital accumulation

$$\dot{K} = I. \tag{19.7}$$

To maximize the intertemporal utility function (19.2), the agent must choose the level of consumption, C, rate of investment, I, and rate of asset accumulation, $\dot{b}$, subject to the equations constraining financial and physical accumulation, Eqs. (19.4) and (19.7). Skipping the mathematical details (see the Appendix), the following results are obtained.

(a) The optimality conditions for C and I are

$$C^{\gamma-1} = \lambda(1 + \tau_c), \quad 1 + h\frac{I}{K} = q, \tag{19.8}$$

where λ is the shadow value of wealth and q the market value of capital. These conditions imply that in macrodynamic equilibrium the marginal utility of consumption equals the tax-adjusted shadow value of wealth and that the marginal cost of purchasing and installing an additional unit of capital equals the market value of capital. Rewriting the second of these conditions as

$$\frac{I}{K} = \frac{\dot{K}}{K} = \frac{q-1}{h} \equiv \phi \tag{19.9}$$

gives the optimal rate of capital accumulation.

(b) The optimality conditions with respect to b and K imply the arbitrage relationships

$$\beta - \frac{\dot{\lambda}}{\lambda} = r(1 - \tau_b), \quad \frac{\alpha}{q}(1 - \tau_k) + \frac{\dot{q}}{q} + \frac{(q-1)^2}{2hq} = r(1 - \tau_b). \qquad (19.10)$$

The first (the Keynes-Ramsey consumption rule) equates the marginal return on consumption to the after-tax rate of return on holding a foreign bond. Since β, r, τ_b are given constants, this rule implies a constant rate of growth of λ and hence of the marginal utility of consumption. In fact, from the first equation in (19.8) we get $\dot{\lambda}/\lambda = $ dln $C^{\gamma-1}/dt$. This result is worthy of note, because in the standard stationary models of optimal growth, marginal utility of consumption has to be constant in order to ensure a finite steady-state equilibrium.

The second condition in (19.10) equates the after-tax rate of return on domestic capital to the after-tax bond yield. The former consists of three elements. The first is the after tax output per unit of capital (valued at the relative price q). The second is the rate of capital gain. The third is more complicated, but its meaning can be understood by rewriting it—making use of Eq. (19.3)—as $(qI - \Phi)/qK$. This expression measures the rate of return arising from the difference between the value of new capital, qI, and the value of the resources required to obtain it, per unit of installed capital. This component also shows that an additional advantage of higher capital stock is to reduce the installation costs (which depend on I/K) associated with new investment.

To ensure that the agent's intertemporal budget constraint is met, the following transversality conditions are also imposed on the solution:

$$\lim_{t\to\infty} \lambda b e^{-\beta t} = 0; \quad \lim_{t\to\infty} q' K e^{-\beta t} = 0. \qquad (19.11)$$

Turnovsky (1997; Chap. 5) shows that the equilibrium growth rate of domestic consumption varies inversely with the tax on foreign bond income, but is independent of all other tax rates. He also shows, as inspection of (19.9) and (19.10) bears out, that the critical determinant of the growth rate of capital is the relative price of installed capacity, q, whose path is determined by the second arbitrage condition in (19.10).

With the interest rate exogenously determined, this model gives rise to a system without transitional dynamics. In fact, the only solution for q that is consistent with the transversality conditions is that q always be at the (unstable) steady-state solution. In response to any shock, q immediately jumps to its new equilibrium value (Turnovsky 1997; Proposition 5.1).

Thus endogenization of the interest rate becomes essential for both formal and economic reasons. From the formal point of view, endogenizing the interest rate will give rise to a dynamic system with transitional dynamics. From the economic point of view, we must account for the fact that economies are not free to borrow or lend as much as they may wish at the given world interest rate. Following Turnovsky, we shall considers the case of a net borrower economy faced with an upward-sloping supply curve of debt.

19.2.1 The Net Borrower Economy

In the case of a *net borrower* economy, the small open economy faces an upward sloping supply curve of debt that embodies the risk premium associated with lending to a sovereign borrower. A standard formulation of this debt supply function is

$$r(Z) = r^* + v(Z), v' > 0,$$

where r^* is the given interest rate prevailing internationally, and $v(Z)$ is the country-specific risk premium, an increasing function of the stock of debt $Z(\equiv -b)$ issued by the country. It has however been observed that an indebted developing country, by adopting growth-oriented policies, can shift the debt supply function so that a lower risk premium is charged at each debt level. The simplest way of taking this into account is to assume that the risk premium depends upon the stock of debt *relative* to some measure of earning capacity (and therefore of debt-servicing capacity) such as output or, in the context of our AK production function, capital stock. Thus we write the debt supply function as

$$r(Z/K) = r^* + v(Z/K), v' > 0. \tag{19.12}$$

The flow budget constraint (19.4) must then be written as

$$\dot{Z} = \left[(1 + \tau_c)C + I \left(1 + \frac{h}{2} \frac{I}{K} \right) + \tau_k \alpha K + (1 - \tau_b) r \left(\frac{Z}{K} \right) Z \right] - [\alpha K + T], \tag{19.13}$$

which simply means that the agent will increase the stock of (foreign) debt to the extent that his overall expenditure for consumption, investment, tax obligations and outstanding interest payments exceeds his overall revenue from output and transfers. Note that, since τ_b is applied to debt, it is actually a subsidy. The assumption that the government rebates back to the private sector all tax revenues net of subsidies gives

$$T = \tau_k \alpha K + \tau_c C - \tau_b r Z, \tag{19.14}$$

so that the country's net accumulation of foreign debt (capital inflow), the current account deficit, is

$$\dot{Z} = C + I \left(1 + \frac{h}{2} \frac{I}{K} \right) + r \left(\frac{Z}{K} \right) Z - \alpha K. \tag{19.15}$$

It should be noted that the representative agent takes the interest rate as given in performing the optimization. The usual atomistic agent assumption entails that the representative agent assumes to be unable to influence the interest rate facing the debtor nation, which is a function of the economy's *aggregate* debt.

With r no longer a constant, but varying as a function of Z/K, the system yields a transitional dynamics around the steady-state solution in terms of three key variables: consumption, the stock of debt, the price of capital, where consumption and the stock of debt are expressed per unit of capital. This dynamics is of the usual saddle-path type. We have already met such a dynamics in the case of rational expectations (see Chap. 13, Sect. 13.4, and Chap. 15, Sect. 15.3.2); another typical case of saddle-path dynamics comes out of dynamic optimization problems such as the present one (see Gandolfo 2009, Sect. 28.4). In the present case the "jump" on the stable arm of the saddle will require two jump variables, that Turnovsky takes to be consumption and the price of capital, while the stock of debt is constrained to adjust gradually as it is a predetermined variable.

Thus the economy will converge to its steady state, where $C/K, Z/K, q$ are constant. This means that C, K (and hence Y), Z will grow at the same rate. From the point of view of international economics this is an interesting result: in the long-run equilibrium it is *not* necessary for the current account to be in equilibrium. On the contrary, the indebted economy can indefinitely run a current account deficit (thus acquiring a net amount of commodities from the rest of the world) matched by a capital inflow (thus increasing its foreign debt). The reason is that the country, by using its resources (including those coming from abroad) not only to consume, but also to appropriately increase its stock of physical capital, can indefinitely increase output so as to be able to regularly service the debt and not undergo any aggravation in the conditions of the debt itself. This result extends what we have already seen in Chap. 14, Sect. 14.2.

The role of the basic assumption (19.1) will now be clear: only if there are no decreasing returns to capital will it be possible to indefinitely increase output by indefinitely increasing the stock of capital.

While behaving as described, the country will maximize its intertemporal utility function (19.2).

19.3 Nominal Rigidities

An essential element of the traditional open-economy macroeconomics is the presence of nominal rigidities, such as sticky prices and rigid wages. On the contrary, the basic intertemporal approach assumes full flexibility of prices and other nominal variables. A noteworthy line of research in the field of the intertemporal approach is the introduction of nominal rigidities while keeping the other features of the model (e.g. Obstfeld and Rogoff 1995, 1996; Rankin 1998; Obstfeld and Rogoff 2000; Hau 2000). This introduction can be obtained by assuming a monopolistic supply sector, which permits one to justify rigorously the assumption of Keynesian models that output is demand determined in the short run when prices are rigid.

In fact, under monopoly prices are set above marginal cost. If prices are rigid in the short run, a shock that causes an increase in demand will be met by increasing output, as this increases the monopolist's profit.

In this section we shall present a simplified version of a model due to Obstfeld and Rogoff (1995, 1996, Chap. 10, Sect. 10.2). The simplification (a two-period instead of an infinite time horizon) does not affect the results.

The basic model is the tradables-nontradables model already used in the previous chapter, Sect. 18.3.2, enriched by the following elements:

(1) the traded goods sector is perfectly competitive, and the law of one price holds, so that

$$p^T = rp^*, \tag{19.16}$$

namely the domestic price of tradables equals the given international price p^* multiplied by the exchange rate. Thus the nominal exchange rate and the domestic price of tradables change in the same proportion.

(2) the source of price rigidities is the nontradables sector, where the price is independent of international elements. This sector is assumed to be monopolistically competitive, and fixes the price of nontradables a period in advance, so that—while prices in the competitive sector are fully flexible—prices in the monopolistic sector are rigid in the short run, which means that given a shock in period t they will adjust only in period $t + 1$ (the economy is assumed to fully adjust in period $t + 1$). This is admittedly a rough form of price rigidity, introduced for analytical convenience.

(3) there are two financial variables, money (in the form of an exogenously given money supply M), and bonds B, which are denominated in tradables and carry a constant world interest rate in tradables i^*.

(4) the representative agent's utility function does not only depend on consumption, but also on real money balances and on labour disutility. The convenient assumption of a time-separable utility function is maintained, so that we have

$$U = \sum_{s=t}^{t+1} \beta^{s-t} \left[\alpha \ln C_s^T + (1 - \alpha) \ln C_s^{NT} + \frac{\chi}{1 - \varepsilon} \left(\frac{M_s}{P_s} \right)^{1-\varepsilon} - \frac{\kappa}{2} \left(y_s^{NT} \right)^2 \right], \tag{19.17}$$

where $\beta \equiv (1 + \rho)^{-1}$ is the discount factor, and P is a consumption-based aggregate price index,

$$P \equiv \frac{\left(p^T \right)^\alpha \left(p^{NT} \right)^{1-\alpha}}{\alpha^\alpha (1 - \alpha)^{1-\alpha}}. \tag{19.18}$$

The real money supply does not enter the utility function in a simple log form like consumption, but enters according to a general isoelastic function (which tends to the log form when $\varepsilon \to 1$). The reason for this modification is that $\varepsilon \neq 1$ is essential for the occurrence of overshooting.

Finally, the term $-\frac{\kappa}{2}\left(y_s^{NT}\right)^2$ captures the disutility of effort required to produce output. Suppose that the production function is $y = A\ell^\beta$ where ℓ is labour (or effort), while the disutility from effort is given by $-\phi\ell$. Inverting the production function and assuming that $\beta = 1/2, \kappa = 2\phi/A^{1/\beta}$, we obtain the term under examination. It is important to clarify that only nontradables are produced at home; as regards tradables, the assumption is that each representative home citizen is endowed with a constant quantity of the traded good each period (Obstfeld and Rogoff 1996; p. 690) . This explains why the disutility of effort only concerns nontradables.

(5) it is convenient to assume that there is no government spending, so that the government budget constraint (assuming that Ricardian equivalence holds) is

$$0 = \tau_t + \frac{M_t - M_{t-1}}{p_t^T}, \tag{19.19}$$

where lump-sum taxes τ_t are denominated in tradables.

The representative agent maximizes the intertemporal utility function (19.17) subject to the period budget constraint

$$p_t^T (B_t - B_{t-1}) + (M_t - M_{t-1})$$
$$= \left[p_t^T i^* B_{t-1} + p_t^{NT} y_t^{NT} + p_t^T \bar{y}_t^T\right] - \left(p_t^{NT} C_t^{NT} + p_t^T C_t^T + p_t^T \tau_t\right). \tag{19.20}$$

The right-hand side is the difference between current income (which includes interest on bonds and the value of the output of nontradables and tradables, the latter being constant) and current expenditure (which includes consumption of both nontradables and tradables, and taxes). The difference gives the net accumulation of financial assets, which includes the value of the change in the stock of bonds, and the change in the stock of money. To avoid confusion with time subscripts, it is as well to note that stocks are defined end of period, so that M_{t-1}, B_{t-1} are the agent's holdings of nominal money balances and bonds entering period t.

Among the several results of the optimization procedure (see the Appendix) we note the following:

(i) the economy has a balanced current account regardless of shocks to money or to the production of nontradables, because at the optimum

$$C_t^T = \bar{y}^T, \quad \forall t, \tag{19.21}$$

where $\bar{y}^T$ is the constant output of tradables.

(ii) money demand can be expressed as

$$\frac{M_t}{P_t} = \left\{ \frac{\chi}{\alpha} \left[\frac{p_t^T C_t^T / P_t}{1 - \left(\beta p_t^T / p_{t+1}^T\right)} \right] \right\}^{1/\varepsilon}. \tag{19.22}$$

Let us now consider the effects of an unanticipated permanent money shock that occurs in period 1. Prices in the monopolistic nontradables sector are set a period in advance, hence they adjust to the shock in period 2 (the economy is assumed to reach its new long-run equilibrium in exactly one period, namely by period 2). Since there are no current-account effects, money is neutral in the long run, which means that only nominal variables change in the new equilibrium.

In the short run (period 1) the rigidity of nontradables prices means that nontradables output is demand determined, i.e., determined by nontradables consumption.

With a number of manipulations (see the Appendix) we can calculate the short-run (i.e., in period 1) proportional deviations of the price of traded goods, which turns out to be

$$\frac{\left(p_1^T - \bar{p}_0^T\right)}{\bar{p}_0^T} = \frac{\beta + (1 - \beta)\varepsilon}{\beta + (1 - \beta)(1 - \alpha + \alpha\varepsilon)} \frac{(M_1 - \bar{M}_0)}{\bar{M}_0}, \tag{19.23}$$

where $(M_1 - \bar{M}_0)/\bar{M}_0$, $\left(p_1^T - \bar{p}_0^T\right)/\bar{p}_0^T$ are the short-run proportional deviations from the initial steady state (denoted by a subscript zero and an overbar).

We also know that, due to neutrality of money and the fact that the shock is permanent,

$$\frac{\left(\bar{p}^T - \bar{p}_0^T\right)}{\bar{p}_0^T} = \frac{(\bar{M} - \bar{M}_0)}{\bar{M}_0} = \frac{(M_1 - \bar{M}_0)}{\bar{M}_0}, \tag{19.24}$$

where $\left(\bar{p}^T - \bar{p}_0^T\right)/\bar{p}_0^T$, $(\bar{M} - \bar{M}_0)/\bar{M}_0$ are the long-run (i.e., in period 2) deviations of the new steady state with respect to the initial steady state.

Note that, due to (19.16), the exchange rate and the price of traded goods change in the same proportion, hence Eq. (19.23) also gives the proportional change in the exchange rate. The inequality

$$\frac{\beta + (1 - \beta)\varepsilon}{\beta + (1 - \beta)(1 - \alpha + \alpha\varepsilon)} \gtreqless 1 \tag{19.25}$$

can be reduced to

$$\varepsilon \gtreqless 1. \tag{19.26}$$

It follows that, if $\varepsilon > 1$, the nominal exchange rate overshoots its long-run level, namely

$$\frac{\left(p_1^T - \bar{p}_0^T\right)}{\bar{p}_0^T} = \frac{(r_1 - \bar{r}_0)}{\bar{r}_0} > \frac{\left(M_1 - \bar{M}_0\right)}{\bar{M}_0} = \frac{\left(\bar{M} - \bar{M}_0\right)}{\bar{M}_0} = \frac{(\bar{r} - \bar{r}_0)}{\bar{r}_0}. \qquad (19.27)$$

Thus we have reached a result which is the same as that obtained in the context of the traditional approach (see Chap. 15, Sect. 15.3.2).

The crucial role played by ε is obvious. From Eq. (19.22) we see that $1/\varepsilon$ is the consumption elasticity of money demand. Suppose for a moment that following the shock, the price of traded goods rose in proportion to the money supply, say by $x\%$. This would imply a rise in the supply of real balances by only $(1 - \alpha)x\%$, because the price of nontradables is fixed.

As regards the demand for real balances, let us begin by observing that in the short run a rise in the price of tradables causes an equiproportionate increase in the consumption of tradables. Since these have a weight $(1 - \alpha)$ in real consumption, it follows that *real consumption* rises by only $(1 - \alpha)x\%$, since the consumption of nontradables is fixed. Hence the demand for real balances would rise by $(1/\varepsilon)\left[(1 - \alpha)x\%\right]$ if the price of traded goods rose in proportion to the money supply. With $\varepsilon > 1$, this means that the demand for real balances would rise by less than the supply. The conclusion is that the price of tradables has to increase by more than $x\%$, and, by (19.16), the exchange rate must overshoot.

Note however that the overshooting phenomenon would *not* occur if $\varepsilon = 1$ (logarithmic form of the real balances term in the utility function). A still more striking conclusion is obtained if $\varepsilon < 1$: Eq. (19.23) shows that in this case the exchange rate would *undershoot*.

Hence we must conclude that the intertemporal approach may give exactly the opposite result as the traditional theory: a positive shock to the money supply causes the nominal exchange rate to undershoot rather than to overshoot, a counterintuitive result.

19.3.1 Extensions

A more general model has been presented by Obstfeld and Rogoff (2000) in a stochastic context with traded and nontraded goods, where the crucial assumption is that workers set next period's nominal wages (in their domestic currency) in advance of production and consumption. Monopolistic firms set prices by a constant markup over wage costs, hence short-run price rigidity.

Assuming that nominal exchange rates play a key role in the short run in shifting world demand between countries (an assumption which also plays a central role in the traditional Mundell-Fleming model), the authors study monetary policy rules in open economies, and exchange rate regimes. One of their findings is that a constrained optimal monetary policy is procyclical with respect to productivity shocks. They also calculate the welfare costs of keeping the exchange rate fixed in response to asymmetric shocks.

The assumption of the expenditure-switching effects of nominal exchange rate changes is dropped in PTM (Pricing to Market, a name coined by Krugman 1987) models. According to this approach, international markets for manufactures are segmented, so that producers can adapt the prices they charge in different national markets to the specific local demand conditions. Thus, for example, a US firm exporting to Europe may find it optimal to lower its markup on the goods exported in the face of a depreciation of the euro against the dollar. In this case the euro price of the goods would rise proportionally less than the amount of the depreciation: the pass through to the euro price would be less than 100 %. This is a well known phenomenon, that we have already met in dealing with the *J* curve (see Sect. 9.2). The PTM hypothesis is often coupled with an assumption on currency invoicing, according to which exporters' prices are quoted in the buyers' currencies and are temporarily rigid in terms of these currencies (Devereux 1997, calls this the Local-Currency Pricing or LCP assumption). Amiti et al. (2014) show that firms with high import shares and high market shares have low exchange rate pass-through.

In the extreme case of a zero pass through (exporters keep their foreign-currency prices constant absorbing the full extent of the devaluation by an adjustment in their foreign markups and foreign margins), a nominal exchange rate depreciation has no expenditure-switching effect at all in the short run.

For a model using the PTM-LCP assumption see Betts and Devereux (2000) , who show that the absence of expenditure-switching effects of nominal exchange-rate changes can give rise to far higher exchange-rate volatility. They also show that in this environment, monetary policy is a beggar-thy-neighbour policy, because a surprise depreciation of the exchange rate *improves* the country's terms of trade (contrary to the costumary presumption) and raises domestic welfare at the expense of foreign welfare.

The validity of these models is largely a matter of empirical evidence that, according to Obstfeld and Rogoff (2000; Sect. 2) is against the PTM-LCP assumption. We would also like to point out that in practice the reduction of the markup, if too big, would give rise to the suspect of dumping and hence to possible anti-dumping contingent protection (see Gandolfo 2014, Sect. 12.6.1).

19.4 Appendix

19.4.1 The Dynamic Optimization Problem

The basic equations of Turnovsky's model (Turnovsky 1997) have been described in the text. In formal terms we have to solve the continuous-time dynamic optimization problem

$$\max_{C,I}\Omega \equiv \int_0^\infty \frac{1}{\gamma}C^\gamma e^{-\beta t}dt,$$

sub   (19.28)

$$\dot{b} = (1 - \tau_k)\alpha K + (1 - \tau_b)rb - (1 + \tau_c)C - I\left(1 + \frac{h}{2}\frac{I}{K}\right) + T,$$

$$\dot{K} = I,$$

where C, I are the control variables. For this we apply Pontryagin's maximum principle (see Gandolfo 2009, Sect. 27.3) and start from the Hamiltonian

$$H(C, I, b, K, \lambda_1, \lambda_2) = \frac{1}{\gamma}C^\gamma e^{-\beta t} + \lambda_1[(1 - \tau_k)\alpha K + (1 - \tau_b)rb$$

$$-(1 + \tau_c)C - I\left(1 + \frac{h}{2}\frac{I}{K}\right) + T - \dot{b}\Big]$$

$$+\lambda_2\left[I - \dot{K}\right],$$ (19.29)

where λ_1, λ_2 are costate variables or dynamic Lagrange multipliers. A mathematically convenient way of dealing with problems involving discounting is to introduce new costate variables

$$\lambda = \lambda_1 e^{\beta t},$$
$$q' = \lambda_2 e^{\beta t},$$ (19.30)

so that the Hamiltonian becomes

$$H(C, I, b, K, \lambda_1, \lambda_2) = \frac{1}{\gamma}C^\gamma e^{-\beta t} + \lambda e^{-\beta t}\left[(1 - \tau_k)\alpha K + (1 - \tau_b)rb\right.$$

$$-(1 + \tau_c)C - I\left(1 + \frac{h}{2}\frac{I}{K}\right) + T - \dot{b}\Big]$$

$$+ q'e^{-\beta t}\left[I - \dot{K}\right],$$ (19.31)

where λ is the shadow value (marginal utility) of wealth in the form of internationally traded bonds, and q' is the shadow value of the representative agent's capital stock. The analysis of the model is facilitated if we use the shadow value of wealth as numéraire. Consequently we define

$$q \equiv q'/\lambda$$ (19.32)

as the market value of capital in terms of the (unitary) price of foreign bonds.

The optimality conditions are

$$\begin{aligned}
\partial H/\partial C &= C^{\gamma-1}e^{-\beta t} - \lambda e^{-\beta t}(1 + \tau_c) = 0, \\
\partial H/\partial I &= -\lambda e^{-\beta t}\left(1 + h\tfrac{I}{K}\right) + q'e^{-\beta t} = 0,
\end{aligned} \tag{19.33}$$

from which

$$\begin{aligned}
C^{\gamma-1} &= \lambda(1 + \tau_c), \\
1 + h\frac{I}{K} &= q,
\end{aligned} \tag{19.34}$$

that have been commented on in the text. The second equation can be rewritten as

$$\frac{I}{K} \equiv \frac{\dot{K}}{K} = \frac{q-1}{h} \equiv \phi. \tag{19.35}$$

Let us note for future reference that, if we take the logarithmic derivative of the first equation, we get the time path of consumption in terms of the costate variable λ (to be determined later), namely

$$(1 - \gamma)\frac{\dot{C}}{C} = -\frac{\dot{\lambda}}{\lambda}. \tag{19.36}$$

The canonical equations for the (original) costate variables are

$$\begin{aligned}
\dot{\lambda}_1 &= -\frac{\partial H}{\partial b} = -\lambda_1(1 - \tau_b)r, \\
\dot{\lambda}_2 &= -\frac{\partial H}{\partial K} = -\lambda_1\left[(1 - \tau_k)\alpha + \frac{h}{2}\frac{I^2}{K^2}\right].
\end{aligned} \tag{19.37}$$

To transform these into differential equations for λ, q we observe that, from Eqs. (19.30), (19.32) we have

$$\begin{aligned}
\lambda_1 &= \lambda e^{-\beta t}, \\
\lambda_2 &= q'e^{-\beta t}, \\
q &= q'/\lambda = \lambda_2/\lambda_1,
\end{aligned} \tag{19.38}$$

so that

$$\begin{aligned}
\dot{\lambda}_1 &= \dot{\lambda}e^{-\beta t} - \beta\lambda e^{-\beta t}, \\
\dot{\lambda}_2 &= \dot{q}'e^{-\beta t} - \beta q'e^{-\beta t}, \\
\frac{\dot{q}'}{q'} &= \frac{\dot{q}}{q} + \frac{\dot{\lambda}}{\lambda}.
\end{aligned} \tag{19.39}$$

Substitution in (19.37), account being taken of the second equation in (19.34), yields

$$\dot{\lambda}e^{-\beta t} - \beta \lambda e^{-\beta t} = -\lambda e^{-\beta t}(1 - \tau_b)r,$$
$$\dot{q}'e^{-\beta t} - \beta q'e^{-\beta t} = -\lambda e^{-\beta t}\left[(1 - \tau_k)\alpha + \frac{(q - 1)^2}{2h}\right], \tag{19.40}$$

from which

$$\beta - \frac{\dot{\lambda}}{\lambda} = (1 - \tau_b)r, \tag{19.41}$$

$$\frac{\dot{q}'}{q'} - \beta = -\frac{\lambda}{q'}\left[(1 - \tau_k)\alpha + \frac{(q - 1)^2}{2h}\right]. \tag{19.42}$$

Equation (19.42), account being taken of (19.32), (19.41), and the third equation in (19.39), can be written as

$$\frac{\dot{q}}{q} + \beta - (1 - \tau_b)r - \beta = -\frac{(1 - \tau_k)\alpha}{q} - \frac{(q - 1)^2}{2hq},$$

so that we finally have

$$(1 - \tau_k)\frac{\alpha}{q} + \frac{\dot{q}}{q} + \frac{(q - 1)^2}{2hq} = (1 - \tau_b)r. \tag{19.43}$$

The dynamic equations (19.41) and (19.43) constitute the arbitrage relations examined in the text.

To conclude the exposition of the optimality conditions, since this is a problem with infinite terminal time we must add the transversality conditions

$$\lim_{t \to \infty} \lambda b e^{-\beta t} = 0; \quad \lim_{t \to \infty} q'Ke^{-\beta t} = 0. \tag{19.44}$$

These can also be interpreted as ensuring national intertemporal solvency. To ascertain the parameter restrictions implied by these conditions we observe that the first limit requires the sum of the growth rate of λ and b to be smaller than β. Using Eq. (19.41) we have

$$\frac{\dot{b}}{b} - (1 - \tau_b)r < 0.$$

Since in steady-state equilibrium all variables grow at the same rate, $\dot{b}/b = \dot{K}/K$. Given that $\dot{K}/K = I/K = \phi$ by (19.35), we have

$$(1 - \tau_b)r - \phi > 0. \tag{19.45}$$

This inequality will be helpful in the study of the transitional dynamics.

The model with r constant gives rise to a degenerate system (one differential equation) with no transitional dynamics, which is uninteresting from our point of view. Hence we skip it and go on to the case in which r is endogenized.

19.4.2 The Net Borrower Nation

We recall from the text that in the case of a net borrower nation the flow budget constraint of the private sector has to rewritten as

$$\dot{Z} = C + I\left(1 + \frac{h}{2}\frac{I}{K}\right) + r\left(\frac{Z}{K}\right)Z - \alpha K, \tag{19.46}$$

where $Z \equiv -b$ denotes the stock of debt. We also recall that the interest rate is assumed to be an increasing function of Z/K

$$r(Z/K) = r^* + v(Z/K), v' > 0. \tag{19.47}$$

To analyse the existence and stability of a steady-state equilibrium it is convenient to reduce the dynamic system to a set of three differential equations in terms of the variables c, z, q, where

$$c \equiv \frac{C}{K}, \quad z \equiv \frac{Z}{K}. \tag{19.48}$$

Since in steady state the variables C, K, Z should grow at the same rate, the variables c, z will be stationary. The variable q does not need any transformation, as it will be stationary when K grows at a constant rate.

Differentiating the definitions (19.48) logarithmically with respect to time we have

$$\frac{\dot{c}}{c} = \frac{\dot{C}}{C} - \frac{\dot{K}}{K}, \quad \frac{\dot{z}}{z} = \frac{\dot{Z}}{Z} - \frac{\dot{K}}{K}. \tag{19.49}$$

Using (19.36) and combining it with (19.35),(19.41), and (19.48) we obtain

$$\dot{c} = c\left\{\frac{1}{1-\gamma}[(1-\tau_b)r(z) - \beta] - \frac{q-1}{h}\right\}. \tag{19.50}$$

Let us now consider the flow budget constraint (19.46). Dividing through by Z we have

$$\frac{\dot{Z}}{Z} = \frac{C}{Z} + \frac{I}{Z}\left(1 + \frac{h}{2}\frac{I}{K}\right) + r(z) - \alpha\frac{K}{Z}$$

$$= \frac{c}{z} + \frac{I/K}{z}\left(1 + \frac{h}{2}\frac{I}{K}\right) + r(z) - \frac{\alpha}{z},$$

and using (19.49)

$$\frac{\dot{z}}{z} = -\frac{I}{K} + \frac{I/K}{z}\left(1 + \frac{h}{2}\frac{I}{K}\right) + r(z) - \frac{\alpha}{z}.$$

Finally, using (19.35), multiplying through by z, and rearranging terms we get

$$\dot{z} = c - \frac{q-1}{h}z + \frac{q^2 - 1}{2h} + zr(z) - \alpha. \tag{19.51}$$

The third dynamic equation is obtained rewriting (19.43) in the form

$$\dot{q} = (1 - \tau_b)r(z)q - (1 - \tau_k)\alpha - \frac{(q-1)^2}{2h}. \tag{19.52}$$

The dynamic system to analyse consists of the three nonlinear differential equations (19.50), (19.51), (19.52).

19.4.2.1 Steady-State Stability and Comparative Dynamics

The singular point of this system, namely the steady-state growth path, is obtained by setting $\dot{c} = \dot{z} = \dot{q} = 0$. Thus the steady-state values of c, z, q, denoted by tildes, are determined by the system

$$\begin{aligned}
\frac{1}{1-\gamma}[(1 - \tau_b)r(\tilde{z}) - \beta] - \frac{\tilde{q} - 1}{h} &= 0,\\
\tilde{c} - \frac{\tilde{q} - 1}{h}\tilde{z} + \frac{\tilde{q}^2 - 1}{2h} + \tilde{z}r(\tilde{z}) - \alpha &= 0,\\
(1 - \tau_b)r(\tilde{z})\tilde{q} - \frac{(\tilde{q} - 1)^2}{2h} - (1 - \tau_k)\alpha &= 0.
\end{aligned} \tag{19.53}$$

Let us now consider the transitional dynamics. Linearising the system (19.50), (19.51), (19.52) around the steady state solution yields the following characterization of the local dynamics

$$\begin{bmatrix} \dot{\bar{c}} \\ \dot{\bar{z}} \\ \dot{\bar{q}} \end{bmatrix} = \begin{bmatrix} 0 & \dfrac{\tilde{c}}{1-\gamma}r'(\tilde{z})(1 - \tau_b) & -\dfrac{\tilde{c}}{h} \\ 1 & [r(\tilde{z}) + \tilde{z}r'(\tilde{z})] - \dfrac{\tilde{q} - 1}{h} & \dfrac{\tilde{q} - \tilde{z}}{h} \\ 0 & r'(\tilde{z})(1 - \tau_b)\tilde{q} & (1 - \tau_b)r(\tilde{z}) - \dfrac{\tilde{q} - 1}{h} \end{bmatrix} \begin{bmatrix} \bar{c} \\ \bar{z} \\ \bar{q} \end{bmatrix}, \tag{19.54}$$

where an overbar denotes the deviations from the steady-state values ($\bar{c} = c - \tilde{c}$, etc.).

The characteristic roots of the coefficient matrix will be given by the solution of the cubic equation

$$f(\theta) = \theta^3 - \varrho\theta^2 + \sigma\theta - \Delta = 0, \tag{19.55}$$

where ρ is the trace, σ the sum of all second-order principal minors, and Δ the determinant of the coefficient matrix.

The trace is

$$\varrho = \left[r(\tilde{z}) - \frac{\tilde{q}-1}{h} \right] + \tilde{z}r'(\tilde{z}) + \left[(1 - \tau_b)r(\tilde{z}) - \frac{\tilde{q}-1}{h} \right]. \tag{19.56}$$

Since $(1 - \tau_b)r(\tilde{z}) - (\tilde{q} - 1)/h > 0$ by (19.45), it will a fortiori be true that $r(\tilde{z}) - (\tilde{q} - 1)/h > 0$. Hence the trace is certainly positive.

Expanding by the first column we find that the determinant of the matrix is

$$\Delta = - \begin{vmatrix} \dfrac{\tilde{c}}{1-\gamma}r'(\tilde{z})(1-\tau_b) & -\dfrac{\tilde{c}}{h} \\ r'(\tilde{z})(1-\tau_b)\tilde{q} & (1-\tau_b)r(\tilde{z}) - \dfrac{\tilde{q}-1}{h} \end{vmatrix}$$

$$= - \left[(1-\tau_b)r(\tilde{z}) - \frac{\tilde{q}-1}{h} \right] \left[\frac{\tilde{c}}{1-\gamma}r'(\tilde{z})(1-\tau_b) \right]$$

$$\quad - \frac{\tilde{c}}{h} \left[r'(\tilde{z})(1-\tau_b)\tilde{q} \right]. \tag{19.57}$$

Given (19.45) and $r'(\tilde{z}) > 0$, we immediately see that $\Delta < 0$.

Thus the succession of the signs of the coefficients of the characteristic equation (19.55) will be

$$+ \ - \ ? \ + \tag{19.58}$$

Whichever sign the ? represents, there will in any case be one continuation and two changes in the succession of the signs of the coefficients, which by Descartes' theorem implies two positive and one negative real root if the roots are all real.

A cubic may also have one real root and two complex conjugate roots. In this case the real root will be negative and the two complex roots will have a positive real part. In fact, if we consider $f(\theta)$ we see that

$$f(0) = -\Delta > 0, \tag{19.59}$$

$$\lim_{\theta \to -\infty} f(\theta) = -\infty,$$

so that $f(\theta)$ will cross the θ axis at a negative value, say $\theta_1 < 0$. Let $\theta_{2,3} = \delta \pm i\omega$ be the pair of complex conjugate roots. Using the relations between the roots and the coefficients (see Gandolfo 2009, Sects. 16.4 and 18.2.1) we have

$$\rho = \theta_1 + \theta_2 + \theta_3 = \theta_1 + 2\delta > 0. \tag{19.60}$$

Since $\theta_1 < 0$, it follows that $\delta > 0$.

Although we don't know whether the characteristic equation has three real roots or one real and two complex conjugate roots (the conditions for distinguishing between the two cases exist, but do not have any economic meaning), we can conclude that there will in any case be one stable real root and two unstable roots, characterizing the usual saddle-path dynamic behaviour.

Thus we need two "jump" variables, assumed to be c and q. This may cast some doubt on the plausibility of the model, since it is difficult to believe that in real life c would be a jump variable, as most empirical studies have shown aggregate consumption to be, if anything, "excessively" smooth and its adjustment sluggish (Quah 1990). On the other hand, this assumption is perfectly consistent with the intertemporally-optimizing representative-agent framework, where consumption is a control variable.

Since the equilibrium is (saddle-path) stable, we can perform exercises in *comparative dynamics* (see Gandolfo 2009, Sect. 20.6). By the implicit function theorem, if we consider system (19.53) in the neighbourhood of the equilibrium point we can express the equilibrium values $\tilde{c}, \tilde{z}, \tilde{q}$ as continuously differentiable functions of the parameters (tax rates etc.) provided that the Jacobian determinant of (19.53) with respect to $\tilde{c}, \tilde{z}, \tilde{q}$ is different from zero in the neighbourhood of the equilibrium point. This Jacobian turns out to be [see Eqs. (19.54) and (19.55)]

$$|\mathbf{J}| = \begin{vmatrix} 0 & \dfrac{1}{1-\gamma}r'(\tilde{z})(1-\tau_b) & -\dfrac{1}{h} \\ 1 & [r(\tilde{z})+\tilde{z}r'(\tilde{z})]-\dfrac{\tilde{q}-1}{h} & \dfrac{\tilde{q}-\tilde{z}}{h} \\ 0 & r'(\tilde{z})(1-\tau_b)\tilde{q} & (1-\tau_b)r(\tilde{z})-\dfrac{\tilde{q}-1}{h} \end{vmatrix} = \dfrac{1}{\tilde{c}}\Delta, \tag{19.61}$$

where the last equality follows from the fact that by multiplying a row (in this case the first row) of a determinant by a constant the determinant is multiplied by that constant. Thus we can carry out our comparative dynamics exercises and calculate the partial derivatives of the equilibrium values with respect to the parameters. To calculate for example the partial derivative $\partial\tilde{q}/\partial\tau_k$ we differentiate (19.53) with respect to τ_k and obtain

$$\begin{aligned} \left[\frac{1}{1-\gamma}r'(\tilde{z})(1-\tau_b)\right]\frac{\partial\tilde{z}}{\partial\tau_k} - \frac{1}{h}\frac{\partial\tilde{q}}{\partial\tau_k} &= 0, \\ \frac{\partial\tilde{c}}{\partial\tau_k} + \left\{[r(\tilde{z})+\tilde{z}r'(\tilde{z})]-\frac{\tilde{q}-1}{h}\right\}\frac{\partial\tilde{z}}{\partial\tau_k} + \frac{\tilde{q}-\tilde{z}}{h}\frac{\partial\tilde{q}}{\partial\tau_k} &= 0, \\ [r'(\tilde{z})(1-\tau_b)\tilde{q}]\frac{\partial\tilde{z}}{\partial\tau_k} + \left[(1-\tau_b)r(\tilde{z})-\frac{\tilde{q}-1}{h}\right]\frac{\partial\tilde{q}}{\partial\tau_k} &= -\alpha, \end{aligned} \tag{19.62}$$

from which

$$\frac{\partial \tilde{q}}{\partial \tau_k} = \frac{\alpha \left[\frac{1}{1-\gamma} r'(\tilde{z})(1 - \tau_b) \right]}{|\mathbf{J}|} < 0. \tag{19.63}$$

An increase in the tax rate on capital income lowers the price of capital, inducing a decrease in the growth rate of capital and hence of the economy: in fact, by (19.34), a decrease in $\tilde{q}$ implies a decrease in $\tilde{I}/\tilde{K}$. The reader can perform other comparative dynamics exercises following the same lines.

19.4.3 Nominal Rigidities

19.4.3.1 The Consumption-Based Price Index

Let us first calculate the consumption-based aggregate price index, defined as the minimum expenditure P required to purchase one unit of composite real consumption $\left(C^T \right)^\alpha \left(C^{NT} \right)^{1-\alpha}$. The problem is

$$\begin{aligned} P = \min \left(p^T C^T + p^{NT} C^{NT} \right) \\ \text{sub} \\ \left(C^T \right)^\alpha \left(C^{NT} \right)^{1-\alpha} = 1. \end{aligned} \tag{19.64}$$

From the Lagrangian

$$L = p^T C^T + p^{NT} C^{NT} + \lambda \left[1 - \left(C^T \right)^\alpha \left(C^{NT} \right)^{1-\alpha} \right] \tag{19.65}$$

we obtain the first-order conditions

$$\begin{aligned} \frac{\partial L}{\partial C^T} &= p^T - \lambda \alpha \left(C^T \right)^{\alpha-1} \left(C^{NT} \right)^{1-\alpha} = 0, \\ \frac{\partial L}{\partial C^{NT}} &= p^{NT} - \lambda (1 - \alpha) \left(C^T \right)^\alpha \left(C^{NT} \right)^{-\alpha} = 0, \\ \frac{\partial L}{\partial \lambda} &= 1 - \left(C^T \right)^\alpha \left(C^T \right)^{1-\alpha} = 0, \end{aligned} \tag{19.66}$$

that is

$$\begin{aligned} \lambda \alpha \left(C^T \right)^\alpha \left(C^{NT} \right)^{1-\alpha} - p^T C^T &= 0, \\ \lambda (1 - \alpha) \left(C^T \right)^\alpha \left(C^{NT} \right)^{1-\alpha} - p^{NT} C^{NT} &= 0, \\ \left(C^T \right)^\alpha \left(C^T \right)^{1-\alpha} &= 1. \end{aligned} \tag{19.67}$$

Summing the first and second equation, account being taken of the third, we have

$$\lambda = p^T C^T + p^{NT} C^{NT} = P. \tag{19.68}$$

Substituting for λ in the first two equations of (19.67), account being taken of the third, we obtain

$$
\begin{aligned}
C^T &= \alpha P/p^T, \\
C^{NT} &= (1-\alpha)P/p^{NT},
\end{aligned}
\tag{19.69}
$$

and substituting in real consumption

$$
\left(C^T\right)^\alpha \left(C^{NT}\right)^{1-\alpha} = \left(\alpha P/p^T\right)^\alpha \left[(1-\alpha)P/p^{NT}\right]^{1-\alpha} = 1,
\tag{19.70}
$$

hence

$$
P = \frac{\left(p^T\right)^\alpha \left(p^{NT}\right)^{1-\alpha}}{\alpha^\alpha (1-\alpha)^{1-\alpha}}.
\tag{19.71}
$$

19.4.3.2 The Composite Nontraded Good, and Its Demand Function

The nontraded good, which is produced under conditions of monopolistic competition, is actually a composite good. It is convenient to assume that there is a continuum of differentiated goods indexed by $z \in [0, 1]$. We then define the index of the real aggregate consumption of nontraded goods as

$$
C^{NT} = \left[\int_0^1 c(z)^{\frac{\theta-1}{\theta}}\, \mathrm{d}z\right]^{\frac{\theta}{\theta-1}}, \theta > 1,
\tag{19.72}
$$

where $c(z)$ is the consumption of commodity z. Equation (19.72) is the natural generalization of the standard two-commodity CES (Constant Elasticity of Substitution) function. The maximization of real consumption subject to the budget constraint gives us the demand function. The problem is

$$
\begin{aligned}
\max_{c(z)} C^{NT} &= \left[\int_0^1 c(z)^{\frac{\theta-1}{\theta}}\, \mathrm{d}z\right]^{\frac{\theta}{\theta-1}} \\
\mathrm{sub} & \\
Z &= \int_0^1 p(z)c(z)\mathrm{d}z,
\end{aligned}
\tag{19.73}
$$

where Z is any fixed nominal expenditure on nontradables. Since we have to maximize an integral with respect to an unknown function appearing under the integral sign, subject to an integral constraint, we cannot use ordinary calculus, but the appropriate tool is the calculus of variations (see Gandolfo 2009, Sect. 27.2), which leads to the Euler equation

$$
\frac{\partial}{\partial c(z)}\left\{c(z)^{\frac{\theta-1}{\theta}} - \lambda p(z)c(z)\right\} = \frac{\theta-1}{\theta}c(z)^{-\frac{1}{\theta}} - \lambda p(z) = 0,
\tag{19.74}
$$

where λ is a multiplier. Since the optimum condition must hold for all goods, considering any two goods z and z' we have

$$\frac{\theta-1}{\theta}c(z)^{-\frac{1}{\theta}} = \lambda p(z),$$
$$\frac{\theta-1}{\theta}c(z')^{-\frac{1}{\theta}} = \lambda p(z'),$$

(19.75)

which gives

$$\frac{c(z)^{-\frac{1}{\theta}}}{c(z')^{-\frac{1}{\theta}}} = \frac{p(z)}{p(z')},$$

whence

$$c(z) = c(z')\left[\frac{p(z)}{p(z')}\right]^{-\theta} = \frac{c(z')}{p(z')^{-\theta}}p(z)^{-\theta}.$$

(19.76)

Let us now consider the aggregate price index p^{NT} corresponding to the real consumption index (19.72). This price index is defined as the minimum expenditure required to purchase one unit of real aggregate consumption $C^{NT} = 1$:

$$p^{NT} = \min_{c(z)} Z = \min_{c(z)} \int_0^1 p(z)c(z)dz$$

$$\text{sub}$$

(19.77)

$$C^{NT} = \left[\int_0^1 c(z)^{\frac{\theta-1}{\theta}}dz\right]^{\frac{\theta}{\theta-1}} = 1.$$

Applying the calculus of variations we have the Euler equation

$$\frac{\partial}{\partial c(z)}\left\{p(z)c(z) - \lambda c(z)^{\frac{\theta-1}{\theta}}\right\} = p(z) - \lambda\frac{\theta-1}{\theta}c(z)^{-\frac{1}{\theta}} = 0,$$

(19.78)

from which we again obtain condition (19.76), an obvious result since problems (19.77) and (19.73) are dual of each other.

Thus we can plug (19.76) in both the budget constraint in (19.73) and the consumption constraint in (19.77), and obtain

$$Z = \int_0^1 p(z)c(z)dz = \int_0^1 p(z)c(z')\left[\frac{p(z)}{p(z')}\right]^{-\theta}dz$$

$$= \int_0^1 \frac{c(z')}{p(z')^{-\theta}}p(z)^{1-\theta}dz,$$

(19.79)

$$\int_0^1 c(z)^{\frac{\theta-1}{\theta}}dz = \int_0^1 c(z')^{\frac{\theta-1}{\theta}}\left[\frac{p(z)}{p(z')}\right]^{1-\theta}dz = \int_0^1 \frac{c(z')^{\frac{\theta-1}{\theta}}}{p(z')^{1-\theta}}p(z)^{1-\theta}dz$$

$$= 1$$

(19.80)

where in Eq. (19.80) we have used the fact that $\left[\int_0^1 c(z)^{\frac{\theta-1}{\theta}} dz\right]^{\frac{\theta}{\theta-1}} = 1$ implies $\int_0^1 c(z)^{\frac{\theta-1}{\theta}} dz = 1$. Since $c(z'), p(z')$ are constants with respect to the integration variable z, we have

$$
\begin{aligned}
Z &= \frac{c(z')}{p(z')^{-\theta}} \int_0^1 p(z)^{1-\theta} dz, \\
1 &= \frac{c(z')^{\frac{\theta-1}{\theta}}}{p(z')^{1-\theta}} \int_0^1 p(z)^{1-\theta} dz.
\end{aligned}
\tag{19.81}
$$

Dividing the first equation by the second gives

$$
Z = \frac{c(z')}{p(z')^{-\theta}} \left[\frac{c(z')}{p(z')^{-\theta}}\right]^{\frac{1-\theta}{\theta}} = \left[\frac{c(z')}{p(z')^{-\theta}}\right]^{\frac{1}{\theta}},
\tag{19.82}
$$

hence, by taking the θ-th power,

$$
Z^\theta = \frac{c(z')}{p(z')^{-\theta}}.
\tag{19.83}
$$

Substituting (19.83) in the first equation of (19.81) we obtain

$$
Z = Z^\theta \int_0^1 p(z)^{1-\theta} dz \Rightarrow Z^{1-\theta} = \int_0^1 p(z)^{1-\theta} dz,
\tag{19.84}
$$

from which, remembering that $Z = p^{NT} C^{NT}$ and that in the optimization problem C^{NT} has been set to 1,

$$
p^{NT} = \left[\int_0^1 p(z)^{1-\theta} dz\right]^{\frac{1}{1-\theta}},
\tag{19.85}
$$

which is the aggregate price index we are looking for.

We now proceed to finding the functional form of the demand for commodity z. From the first equation in (19.81) we obtain

$$
Z^{\frac{1}{1-\theta}} = \left[\frac{c(z')}{p(z')^{-\theta}}\right]^{\frac{1}{1-\theta}} \left[\int_0^1 p(z)^{1-\theta} dz\right]^{\frac{1}{1-\theta}},
$$

hence, using (19.85),

$$
Z^{\frac{1}{1-\theta}} = \left[\frac{c(z')}{p(z')^{-\theta}}\right]^{\frac{1}{1-\theta}} p^{NT},
$$

and solving for $c(z')/p(z')^{-\theta}$ we obtain

$$\frac{c(z')}{p(z')^{-\theta}} = Z \left(p^{NT}\right)^{\theta-1}. \tag{19.86}$$

Let us now substitute (19.86) into (19.76), obtaining :

$$c(z) = Z \left(p^{NT}\right)^{\theta-1} p(z)^{-\theta} = \left[\frac{p(z)}{p^{NT}}\right]^{-\theta} \frac{Z}{p^{NT}} = \left[\frac{p(z)}{p^{NT}}\right]^{-\theta} C^{NT}, \tag{19.87}$$

where the last equality uses the fact, already noted above, that $Z = p^{NT} C^{NT}$, where p^{NT} is the (minimum) money cost of one unit of composite consumption.

Thus the demand for commodity z takes the constant-elasticity-of-substitution form

$$y_d^{NT}(z) = \left[\frac{p(z)}{p^{NT}}\right]^{-\theta} C_A^{NT}, \tag{19.88}$$

where C_A^{NT} is aggregate home consumption of nontraded goods. Monopolistic producers of nontradables take C_A^{NT} as given. We assume that each differentiated good is produced by a single individual.

19.4.3.3 The Intertemporal Optimization Problem

Let us consider the optimization problem of the representative agent

$$\max U = \alpha \ln C_t^T + (1 - \alpha) \ln C_t^{NT} + \beta \alpha C_{t+1}^T + \beta(1 - \alpha) \ln C_{t+1}^{NT}$$

$$+ \frac{\chi}{1 - \varepsilon} \left(\frac{M_t}{P_t}\right)^{1-\varepsilon} + \beta \frac{\chi}{1 - \varepsilon} \left(\frac{M_{t+1}}{P_{t+1}}\right)^{1-\varepsilon}$$

$$- \frac{\kappa}{2} \left[y_t^{NT}(z)\right]^2 - \beta \frac{\kappa}{2} \left[y_{t+1}^{NT}(z)\right]^2. \tag{19.89}$$

Before going on to solve the constrained optimization problem, we must recall that each representative agent is endowed with a constant quantity of the traded good each period, and has a monopoly over the production of one of the nontraded goods $z \in [0, 1]$. The agent does of course consume an aggregate of the nontraded goods as represented by the real index calculated in Sect. 19.4.3.2. Thus on the side of receipts we must count $p_t^{NT}(z) y_t^{NT}(z)$ while on the side of expenditure we must count $p_t^{NT} C_t^{NT}$, where p_t^{NT}, C_t^{NT} are the indexes calculated in Sect. 19.4.3.2.

With these specifications, the budget constraint in period t is

$$p_t^T (B_t - B_{t-1}) + (M_t - M_{t-1})$$

$$= \left[p_t^T i^* B_{t-1} + p_t^{NT}(z) y_t^{NT}(z) + p_t^T \bar{y}_t^T\right]$$

$$- \left(p_t^{NT} C_t^{NT} + p_t^T C_t^T + p_t^T \tau_t\right),$$

that we can rewrite, rearranging terms and dividing by p_t^T, as

$$B_t + \frac{M_t}{p_t^T}$$

$$= (1 + i^*)B_{t-1} + \frac{M_{t-1}}{p_t^T} + \frac{p_t^{NT}(z)}{p_t^T} y_t^{NT}(z) + \bar{y}_t^T - \frac{p_t^{NT}}{p_t^T} C_t^{NT} - C_t^T - \tau_t.$$

$$(19.90)$$

To obtain the intertemporal budget constraint we first iterate (19.90) to $t + 1$ and bring the result back to t by discounting:

$$\beta B_{t+1} + \beta \frac{M_{t+1}}{p_{t+1}^T}$$

$$= \beta(1 + i^*)B_t + \beta \frac{M_t}{p_{t+1}^T} + \beta \frac{p_{t+1}^{NT}(z)}{p_{t+1}^T} y_{t+1}^{NT}(z) + \beta \bar{y}_{t+1}^T$$

$$- \beta \frac{p_{t+1}^{NT}}{p_{t+1}^T} C_{t+1}^{NT} - \beta C_{t+1}^T - \beta \tau_{t+1}.$$

$$(19.91)$$

We make now the crucial assumption (Obstfeld and Rogoff 1995, p. 656; 1996, p. 690) that $(1 + i^*)\beta = 1$, namely that the subjective discount rate equals the constant world interest rate in tradables i^*. Thus Eq. (19.91), after adding and subtracting M_t/p_t^T and rearranging terms becomes

$$\beta B_{t+1} + \beta \frac{M_{t+1}}{p_{t+1}^T} + \frac{M_t}{p_t^T}\left[1 - \beta \frac{p_t^T}{p_{t+1}^T}\right] - \beta \frac{p_{t+1}^{NT}(z)}{p_{t+1}^T} y_{t+1}^{NT}(z) - \beta \bar{y}_{t+1}^T$$

$$+ \beta \frac{p_{t+1}^{NT}}{p_{t+1}^T} C_{t+1}^{NT} + \beta C_{t+1}^T + \beta \tau_{t+1}.$$

$$= B_t + \frac{M_t}{p_t^T}.$$

$$(19.92)$$

We can now substitute the value of $B_t + M_t/p_t^T$ from Eq. (19.92) into Eq. (19.90), obtaining

$$(1 + i^*)B_{t-1} + \frac{M_{t-1}}{p_t^T} + \frac{p_t^{NT}(z)}{p_t^T} y_t^{NT}(z) + \bar{y}_t^T - \frac{p_t^{NT}}{p_t^T} C_t^{NT} - C_t^T - \tau_t$$

$$= \beta B_{t+1} + \beta \frac{M_{t+1}}{p_{t+1}^T} + \frac{M_t}{p_t^T}\left[1 - \beta \frac{p_t^T}{p_{t+1}^T}\right] - \beta \frac{p_{t+1}^{NT}(z)}{p_{t+1}^T} y_{t+1}^{NT}(z) - \beta \bar{y}_{t+1}^T$$

$$+ \beta \frac{p_{t+1}^{NT}}{p_{t+1}^T} C_{t+1}^{NT} + \beta C_{t+1}^T + \beta \tau_{t+1},$$

$$(19.93)$$

which is the intertemporal budget constraint.

We must now recall that the value of the output of the nontradable (for simplicity of notation we omit the time subscript) is

$$p^{NT}(z)y^{NT}(z) = p(z)y_d^{NT}(z), \tag{19.94}$$

since in equilibrium supply equals demand. From Eq. (19.88) we have

$$p(z) = y_d^{NT}(z)^{-1/\theta}p^{NT}\left(C_A^{NT}\right)^{1/\theta} \tag{19.95}$$

and replacing $p(z)$ in (19.94) we obtain, recalling that $y_d^{NT}(z) = y^{NT}(z)$,

$$p^{NT}(z)y^{NT}(z) = y^{NT}(z)^{\frac{\theta-1}{\theta}}p^{NT}\left(C_A^{NT}\right)^{1/\theta}. \tag{19.96}$$

To solve the optimization problem we form the Lagrangian

$$L = \alpha \ln C_t^T + (1-\alpha)\ln C_t^{NT} + \beta\alpha \ln C_{t+1}^T + \beta(1-\alpha)\ln C_{t+1}^{NT} + \frac{\chi}{1-\varepsilon}\left(\frac{M_t}{P_t}\right)^{1-\varepsilon}$$

$$+\beta\frac{\chi}{1-\varepsilon}\left(\frac{M_{t+1}}{P_{t+1}}\right)^{1-\varepsilon} - \frac{\kappa}{2}\left(y_t^{NT}\right)^2 - \beta\frac{\kappa}{2}\left(y_{t+1}^{NT}\right)^2$$

$$+\lambda\left\{(1+i^*)B_{t-1} + \frac{M_{t-1}}{p_t^T} + \frac{p_t^{NT}(z)}{p_t^T}y_t^{NT}(z) + \bar{y}_t^T - \frac{p_t^{NT}}{p_t^T}C_t^{NT} - C_t^T - \tau_t\right.$$

$$-\beta B_{t+1} - \beta\frac{M_{t+1}}{p_{t+1}^T} - \frac{M_t}{p_t^T}\left[1 - \beta\frac{p_t^T}{p_{t+1}^T}\right] + \beta\frac{p_{t+1}^{NT}(z)}{p_{t+1}^T}y_{t+1}^{NT}(z) + \beta\bar{y}_{t+1}^T$$

$$\left.-\beta\frac{p_{t+1}^{NT}}{p_{t+1}^T}C_{t+1}^{NT} - \beta C_{t+1}^T - \beta\tau_{t+1.}\right\}, \tag{19.97}$$

where $p^{NT}(z)y^{NT}(z)$ are to be replaced by (19.96).

The relevant first-order conditions are

$$\frac{\partial L}{\partial C_t^T} = \alpha/C_t^T - \lambda = 0, \tag{19.98}$$

$$\frac{\partial L}{\partial C_{t+1}^T} = \beta\alpha/C_{t+1}^T - \lambda\beta = 0, \tag{19.99}$$

$$\frac{\partial L}{\partial C_t^{NT}} = (1-\alpha)/C_t^{NT} - \lambda p_t^{NT}/p_t^T = 0, \tag{19.100}$$

$$\frac{\partial L}{\partial C_{t+1}^{NT}} = \beta(1-\alpha)/C_{t+1}^{NT} - \lambda\beta p_{t+1}^{NT}/p_{t+1}^T = 0, \tag{19.101}$$

$$\frac{\partial L}{\partial M_t} = \frac{\chi}{P_t} \left(\frac{M_t}{P_t}\right)^{-\varepsilon} - \lambda \frac{1}{p_t^T} \left[1 - \beta \frac{p_t^T}{p_{t+1}^T}\right]$$

$$= \frac{\chi}{P_t} \left(\frac{M_t}{P_t}\right)^{-\varepsilon} - \lambda \frac{1}{p_t^T} + \frac{\lambda \beta}{p_{t+1}^T} = 0, \tag{19.102}$$

$$\frac{\partial L}{\partial y_t^{NT}} = -\kappa y_t^{NT} + \lambda \frac{\theta - 1}{\theta} y_t^{NT}(z)^{-\frac{1}{\theta}} \left(C_A^{NT}\right)_t^{1/\theta} = 0. \tag{19.103}$$

from which

$$C_{t+1}^T = C_t^T, \tag{19.104}$$

$$C_t^{NT} = \frac{1 - \alpha}{\alpha} \left(\frac{p_t^T}{p_t^{NT}}\right) C_t^T, \tag{19.105}$$

$$\frac{\alpha}{C_t^T} = \chi \frac{p_t^T}{P_t} \left(\frac{M_t}{P_t}\right)^{-\varepsilon} + \beta \frac{p_t^T}{p_{t+1}^T} \frac{\alpha}{C_{t+1}^T}, \tag{19.106}$$

$$\left(y_t^{NT}\right)^{\frac{\theta+1}{\theta}} = \frac{(\theta - 1)(1 - \alpha)}{\kappa \theta} \left(C_A^{NT}\right)_t^{1/\theta} \frac{1}{C_t^{NT}}. \tag{19.107}$$

Here is a sketch of the derivations.

Condition (19.104) immediately derives from (19.98) and (19.99). Note that this condition implies that agents smooth consumption of traded goods independently of nontraded goods, which is a consequence of the additivity and time separability of the utility function. If we assume that the initial stock of net foreign assets is zero, it follows that the consumption of traded goods equals their (constant) production

$$C_t^T = \bar{y}_t^T. \tag{19.108}$$

This condition has an important implication: since consumption of traded goods equals their production, the economy has a balanced current account, independently of nontraded goods production or shocks to money.

Condition (19.105) immediately derives from (19.98) and (19.100).

Condition (19.106) is obtained by using condition (19.98) to replace λ with α/C_t^T, and condition (19.99) to replace $\lambda\beta$ with $\beta\alpha/C_{t+1}^T$.

Condition (19.107) is obtained from condition (19.103) using condition (19.100) to substitute for λ.

We now observe that condition (19.106) can be solved for M_t/P_t to obtain money demand

$$\frac{M_t}{P_t} = \left\{\frac{\chi}{\alpha}\left[\frac{p_t^T C_t^T/P_t}{1 - \left(\beta p_t^T/p_{t+1}^T\right)}\right]\right\}^{1/\varepsilon}, \tag{19.109}$$

where we have used condition (19.104) to replace C_{t+1}^T with C_t^T.

Steady-State Equilibrium

We consider the case in which all prices are fully flexible and all exogenous variables (including the money supply) are constant. In this equilibrium, $C_t^{NT} = y_t^{NT}(z) = (C_A^{NT})_t$ for all z. Thus Eq. (19.107) implies the following value for the steady-state output of nontradables:

$$\bar{y}^{NT} = \bar{C}^{NT} = \left[\frac{(\theta - 1)(1 - \alpha)}{\kappa\theta}\right]^{\frac{1}{2}}. \tag{19.110}$$

In the steady-state equilibrium prices of traded goods must also be constant, assuming no speculative bubbles, so that

$$p_{t+1}^T = p_t^T. \tag{19.111}$$

Since money shocks have no effects on wealth, the only long-run consequence of a money shock is to raise the prices of tradables and nontradables in the same proportion. We are however interested in finding the short-run effects of a monetary shock.

Short-Run Effects of an Unanticipated Money Shock

Let us now consider the effects of an unanticipated permanent money shock that occurs in period 1. Prices in the monopolistic nontradables sector are set a period in advance, hence they adjust to the shock in period 2 (the economy is assumed to reach its new long-run equilibrium in exactly one period, namely by period 2). Since there are no current-account effects, money is neutral in the long run, which means that only nominal variables change in the new equilibrium.

In the short run (period 1) the rigidity of nontradables prices means that nontradables output is demand determined, i.e., determined by nontradables consumption (a variable without a time subscript is a short-run variable):

$$y_d^{NT} = C^{NT}. \tag{19.112}$$

We now combine Eqs. (19.112), (19.108), (19.105) and obtain

$$y^{NT} = C^{NT} = \frac{1 - \alpha}{\alpha}\left(\frac{p^T}{\bar{p}^{NT}}\right)\bar{y}^T, \tag{19.113}$$

where $\bar{p}^{NT}$ is temporarily fixed. Equation (19.113) shows that in the short run y^{NT} and C^{NT} are functions of p^T. To solve for the (fully flexible) tradables prices we first log-linearise Eq. (19.109) in the neighbourhood of the initial steady state. This amounts to taking the logarithms and then considering the differentials, which (evaluated at the initial equilibrium point) give us the proportional deviations: $d \ln M_1 = (M_1 - \bar{M}_0)/\bar{M}_0$, and so on. Taking the logarithms we have

$$\varepsilon \left(\ln M_t - \ln P_t\right) = \ln \frac{\chi}{\alpha} + \ln \left(p_t^T C_t^T / P_t\right) - \ln \left(1 - \beta p_t^T / p_{t+1}^T\right), \qquad (19.114)$$

and computing the differentials we obtain

$$\varepsilon \left(d \ln M_t - d \ln P_t\right) = d \ln \left(p_t^T C_t^T / P_t\right) - d \ln \left(1 - \beta p_t^T / p_{t+1}^T\right)$$

$$= d \ln p_t^T - d \ln P_t - \frac{d \left(1 - \beta p_t^T / p_{t+1}^T\right)}{1 - \beta p_t^T / p_{t+1}^T} \qquad (19.115)$$

since C_t^T is constant by (19.108). We now compute

$$\frac{d \left(1 - \beta p_t^T / p_{t+1}^T\right)}{1 - \beta p_t^T / p_{t+1}^T} = \frac{-\beta}{1 - \beta p_t^T / p_{t+1}^T} \frac{p_{t+1}^T d p_t^T - p_t^T d p_{t+1}^T}{\left(p_{t+1}^T\right)^2}$$

$$= \frac{-\beta}{1 - \beta p_t^T / p_{t+1}^T} \left(\frac{p_t^T}{p_{t+1}^T} \frac{d p_t^T}{p_t^T} - \frac{p_t^T}{p_{t+1}^T} \frac{d p_{t+1}^T}{p_{t+1}^T}\right)$$

$$= -\frac{\beta}{1 - \beta p_t^T / p_{t+1}^T} \left(\frac{p_t^T}{p_{t+1}^T} d \ln p_t^T - \frac{p_t^T}{p_{t+1}^T} d \ln p_{t+1}^T\right)$$

$$= -\frac{\beta}{1 - \beta} \left(d \ln p_t^T - d \ln p_{t+1}^T\right), \qquad (19.116)$$

where in the last equality we have used (19.111). Substituting (19.116) into (19.115) and considering the short-run (period 1) we have

$$\varepsilon \left(d \ln M_1 - d \ln P_1\right) = d \ln p_1^T - d \ln P_1 + \frac{\beta}{1 - \beta} \left(d \ln p_1^T - d \ln p_2^T\right), \qquad (19.117)$$

where $d \ln p_2^T = (\bar{p} - \bar{p}_0^T)/\bar{p}_0^T$ are the long-run (i.e., in period 2, since by assumption all adjustments are completed by period 2) deviations of the new steady state with respect to the initial steady state.

Log differentiating the price index (19.71), account being taken that p^{NT} is fixed in the short run, we have

$$d \ln P_1 = \alpha d \ln p_1^T. \qquad (19.118)$$

In this model, the long-run neutrality of money implies that

$$d \ln p_2^T = d \ln M_2 = d \ln M_1. \qquad (19.119)$$

Substituting (19.118) and (19.119) into (19.117) and solving for $d \ln p_1^T$, we finally obtain

$$d \ln p_1^T = \frac{\beta + (1 - \beta)\varepsilon}{\beta + (1 - \beta)(1 - \alpha + \alpha\varepsilon)} d \ln M_1, \qquad (19.120)$$

which is the relation used in the text to show exchange-rate overshooting.

References

Aghion P., & Howitt, P. (1998). *Endogenous growth theory*. Cambridge (Mass.): MIT Press.

Amiti, M., Itskhoki, O., & Konings, J. (2014). Importers, exporters, and exchange rate disconnect. *American Economic Review, 104*, 1942–1978.

Barro, R. J., & Sala-i-Martin, X. (2004). *Economic growth* (2nd ed.). New York: Mc Graw-Hill.

Betts, C., & Devereux, M. B. (2000). Exchange rate dynamics in a model of pricing to market. *Journal of International Economics, 50*, 215–244.

Devereux, M. B. (1997), Real exchange rates and macroeconomics: Evidence and theory. *Canadian Journal of Economics, 30*, 773–808.

Domar, E. D. (1946). Capital expansion, rate of growth and employment. *Econometrica, 14*, 137–147.

Gandolfo, G. (2009). *Economic dynamics* (Chap. 28). Berlin, Heidelberg, New York: Springer.

Gandolfo, G. (2014). *International trade theory and policy* (2nd ed., Sect. 12.6.1). Berlin, Heidelberg, New York: Springer.

Harrod, R. F. (1939). An essay in dynamic theory. *Economic Journal, 49*, 14–33.

Hau, H. (2000). Exchange rate determination: The role of factor price rigidities and nontradeables. *Journal of International Economics, 50*, 421–447.

Krugman, P. (1987). Pricing to market when the exchange rate changes, in S. W. Arndt & J. D. Richardson (Eds.). *Real-financial linkages among open economies* (pp. 49–70). Cambridge (Mass.): MIT Press.

Obstfeld, M., & Rogoff, K. (1995). Exchange rate dynamics redux. *Journal of Political Economy, 103*, 624–660.

Obstfeld, M., & Rogoff, K. (1996). *Foundations of international macroeconomics*. Cambridge (Mass.): MIT Press.

Obstfeld, M., & Rogoff, K. (2000). New directions for stochastic open economy models. *Journal of International Economics, 50*, 117–154.

Quah, D. (1990). Permanent and transitory movements in labor income: An explanation for "excess smoothness" in consumption. *Journal of Political Economy, 98*, 449–475.

Rankin, N. (1998). Nominal rigidity and monetary uncertainty in a small open economy. *Journal of Economic Dynamics and Control, 22*, 679–702.

Turnovsky, S. J. (1997). *International macroeconomic dynamics*. Cambridge (Mass.): MIT Press.

Part VII

International Monetary Integration

International Monetary Integration: Optimum Currency Areas and Monetary Unions

20

20.1 Introduction

As in the case of commercial integration, also in the case of monetary integration there are various degrees of integration, from the simple currency area to the full monetary union (with a single currency). However, while in the case of commercial integration the various degrees of integration can be precisely classified (see Gandolfo 2014, Sect. 11.7), the same is not true in the case of monetary integration, where a certain amount of terminological confusion exists. Thus a preliminary conceptual and terminological clarification is called for. A good starting point is the definition given in a report to the Council and Commission of the European Economic Community commonly known as the Werner Report (1970). It identifies a first set of conditions (called "necessary conditions" by the subsequent Delors Report 1989) to define a monetary union:

(1) within the area of a monetary union, currencies must be fully and irreversibly convertible into one another;
(2) par values must be irrevocably fixed;
(3) fluctuation margins around these parities must be eliminated;
(4) capital movements must be completely free.

The second set of conditions identified in the Werner Report concerns the *centralization of monetary policy*. In particular, this centralization should involve all decisions concerning liquidity, interest rates, intervention on the exchange markets, management of reserves, and the fixing of currency parities vis-à-vis the rest of the world.

Finally, in the Werner Report the adoption of a single currency, though not indispensable for the creation of a monetary union, is considered preferable to maintaining the various national currencies. This is so for psychological and

© Springer-Verlag Berlin Heidelberg 2016

G. Gandolfo, *International Finance and Open-Economy Macroeconomics*,
Springer Texts in Business and Economics, DOI 10.1007/978-3-662-49862-0_20

political factors, as the adoption of a single currency would demonstrate the irreversible nature of the undertaking.

Some authors (for example Ingram 1973) take the first set of elements listed in the Werner Report and call it *monetary integration*. In practical usage this latter definition is then simplified to fixed exchange rates and freedom of capital movements. Other authors, on the contrary, take the view that monetary integration must imply something more. For example, Corden (1972) states that monetary integration consists of two elements:

(1) "complete exchange-rate union", *i.e.* irrevocably fixed exchange rates and centralization of exchange-rate policy towards the rest of the world and of part of monetary policy, by a supranational body;
(2) "convertibility", namely complete elimination of any control on international (within the area) transactions on both current and capital accounts.

Other authors, however, use "monetary integration" as the analogous, in international monetary theory, of the term "commercial integration" used as a generic term in trade theory. Monetary integration, in other words, is taken as the generic term that contains various categories (including the process of transition from a simple currency union to a full monetary union).

Be it as it may, from the point of view of economic theory the starting point of any analysis of monetary integration (we shall take this term in the generic meaning) is the theory of optimum currency areas. These will be examine in the next section. We shall next analyse the common monetary policy prerequisite and the problem of the single currency in a monetary union. The European monetary union will be examined in the next chapter.

20.2   The Theory of Optimum Currency Areas

The notion of *optimum currency area* or OCA (introduced by Mundell 1961) is an evolution of the concept of currency area.

A currency area is a group of countries which have a common currency (in which case full monetary integration prevails) or which, though maintaining different national currencies, have permanently and rigidly fixed exchange rates among themselves and full convertibility of the respective currencies into one another; instead, the exchange rates vis-à-vis non-partner countries are flexible. Theoretically defined, currency areas do not necessarily correspond to national frontiers, as they might include part of a nation only. But as it would not be viable, we shall not consider this case.

The problem consists in determining the appropriate domain of a currency area (hence the adjective *optimum*) and, specifically, whether the adhesion of a country to a currency area (to be set up or already existing) or its remaining in one is beneficial. Optimality can be judged in various manners, for example on the basis of the capability of maintaining external equilibrium without unemployment at home and with price stability. It is clear that the question of (optimum) currency areas exists insofar as the debate on fixed versus flexible exchange rates has proven to be inconclusive (see Chap. 17). In fact, if either the fixed exchange rate regime or the flexible one could be shown to be definitely superior, then there would be no need for a theory of (optimum) currency areas: the optimum currency area would coincide with the world (if fixed exchange rates were superior) or would not exist (in the case of superiority of flexible exchange rates).

Three approaches can be distinguished in the theory of optimum currency areas. The first is the traditional approach, which tries to single out a crucial criterion to delimit the appropriate domain. The second is the cost-benefit approach, which believes that the participation in a currency area has both benefits and costs, so that optimality has to be evaluated by a cost-benefit analysis. The third is the "new" approach.

20.2.1 The Traditional (or "Criteria") Approach

Several single criteria to delimit the domain of an optimum currency area can be found in the literature, since different authors have singled out different criteria as crucial. Although Mundell (1961), McKinnon (1963), and Kenen (1969) can be considered as the founding fathers of the criteria approach, many other authors have contributed to it. The reader wishing to trace the origins of the different criteria that we are to examine can consult Ishiyama (1975), Tower and Willet (1976), Allen and Kenen (1980; Chap. 14).

(a) One criterion is that of *international factor mobility*: countries between which this mobility is high can profitably participate in a currency area, whilst exchange rates should be flexible between countries with a low factor mobility between them. With a high factor mobility, in fact, international adjustment would resemble the adjustment between different regions of the same country (interregional adjustment), between which, obviously, no balance-of-payments problem exists. Let us assume, for example, that there is a decline in the exports of a region to the rest of the country because of a fall in the demand by the rest of the country for the output of an industry located in that region. The region's income and consumption decrease and, to ease the transition to a situation of lower real income, it is necessary for the region to get outside financing to be able to consume more than the value of output (high mobility of capital, possibly stimulated by policy interventions). Furthermore, the unemployed workers can move to other regions and find a job there (high mobility of labour). A similar process would take place at the international level. It is also clear that, in the absence of the postulated factor mobility, the elimination of the imbalances described above would require exchange-rate variations.

(b) A second criterion is that of the *degree of openness* of the economy, as measured by the relative importance of the sectors producing internationally traded goods or tradables (both exportables and importables) and the sectors producing non-traded goods. A country where traded goods are a high proportion of total domestic output can profitably participate in a currency area, whilst it had better adopt flexible exchange rates in the opposite case. Let us assume that a highly open economy incurs a balance-of-payments deficit: if this is cured by an exchange-rate depreciation, the change in relative prices will cause resources to move from the non-traded goods sector to the traded goods one, so as to meet the increased (foreign) demand for exports and the higher (domestic) demand for import substitutes. This implies huge disturbances (amongst which possible inflationary effects) in the non-traded goods sector because of its relative smallness. This is, in another form, the resource-reallocation-cost argument against flexible exchange rates already mentioned in Sect. 17.1.

In this situation it would be more effective to adopt fixed exchange rates and expenditure-reducing policies which reduce imports and free for exportation a sufficient amount of exportables previously consumed domestically.

(c) A third criterion is that of *product diversification*. A country with a high productive diversification will also export a wide range of different products. Now, if we exclude macroeconomic events which influence the whole range of exports (for example a generalized inflation which causes the prices of all domestically produced goods to increase), in the normal course of events commodities with a fine or brilliant export performance will exist beside commodities with a poor export performance. It is self-evident that these offsetting effects will be very feeble or will not occur at all when exports are concentrated in a very limited number of commodities. On the average, therefore, the total exports of a country with a high product diversification will be more stable than those of a country with a low one. Since the variations in exports influence the balance of payments and so—*ceteris paribus*—give rise to pressures on the exchange rate, it follows that a country with high product diversification will have less need for exchange-rate changes and so can tolerate fixed exchange rates, whilst the contrary holds for a country having low product diversification.

(d) A fourth criterion is that of the degree of *financial integration*. It partially overlaps with criterion (a), but it is especially concerned with capital flows as an equilibrating element of payments imbalances. If there is a high degree of international financial integration, no need will exist for exchange-rate changes in order to restore external equilibrium, because slight changes (in the appropriate direction) in interest rates will give rise to sufficient equilibrating capital flows; in this situation it is possible to maintain fixed exchange rates within the area where financial integration exists. It goes without saying that a condition for financial integration is the elimination of all kinds of restrictions on international capital movements.

(e) A fifth criterion is that of the *similarity in rates of inflation*. Very different inflation rates do, in fact, cause appreciable variations in the terms of trade and so, insofar as these influence the flows of goods, give rise to current-account disequilibria, which may require offsetting exchange-rate variations. When, on the contrary, the rates of inflation are identical or very similar, there will be no effect on the terms of trade and so—*ceteris paribus*—an equilibrated flow of current-account transactions will take place (with fixed exchange rates) within the currency area.

(f) A sixth criterion is that of the *degree of policy integration*. Policy integration can go from the simple coordination of economic policies among the various partner countries to a situation in which these surrender their monetary and fiscal sovereignty to a single supranational monetary authority (necessary for consistently managing the international reserves of the area and the exchange-rates of the partner countries vis-à-vis the rest of the world, for achieving an appropriate distribution of the money supply within the area, etc.) and a single supranational fiscal authority (necessary to coordinate taxation, transfer payments and other measures—for example in favour of those workers who remain unemployed notwithstanding full labour mobility, etc.). It is clear that this ideal situation presupposes complete economic integration which, in turn, cannot be achieved without some form of political integration.

All the above-listed criteria—with the exception of the last one, which is almost a truism, as it amounts to saying that when there is full economic and political integration there also is monetary integration—have been criticized as incomplete and partial. As a matter of fact, the reader who has followed us through the previous chapters will readily see that all these criteria stress only one or the other element present in the adjustment processes of the balance of payments under the various exchange-rate regimes, and that these elements can be subjected to the same criticism examined in the previous chapters. Just as an example, the criterion of financial integration and so of the equilibrating influence of capital flows is susceptible to the same criticisms examined in Sect. 10.2.2.1 (burden of interest payments, stocks and flows, etc.).

20.2.2 The Cost-Benefit Approach

Participation in a currency area involves benefits but also costs, so that to take the best course of action a careful determination of both is necessary by weighting these costs and benefits through some kind of social preference function. It is self-evident that the final decision will depend on the set of weights chosen; as these may vary from country to country (and from period to period in the same country), no general rule can be given. What we shall do is to describe the benefits and costs (Ishiyama 1975; Tower and Willet 1976 ; Allen and Kenen 1980 , Chap. 14; Denton ed., 1974; Robson 1998), with some additional considerations.

The main benefits include the following:

(1) A permanently fixed exchange rate eliminates speculative capital flows between the partner countries. This, of course, depends on the confidence in the fixity of the exchange rates within the area, as in the opposite case, destabilising speculation (of the type which affected the adjustable peg system: see Sect. 16.2)

would inevitably come about. This problem cannot obviously arise in the case of a common currency.

(2) The saving on exchange reserves. The members no longer need international reserves for transactions within the area, exactly as in the case of regions within a country. This, of course, will occur when the credibility of the fixed exchange rates is complete, whilst in the initial stages it may be necessary to hold the same amount of pre-union reserves to ensure the agreed exchange-rate rigidity, i.e. to enforce the fixed parities established within the area.

(3) Monetary integration can stimulate the integration of economic policies and even economic integration. The idea is that participation in a currency area, and so the obligation to maintain fixed exchange rates vis-à-vis the other members, compels all members to make their economic policies uniform (in particular anti-inflationary policies) with those of the most virtuous member, at the same time making more credible domestically the statements of a firm intention to pursue a strong policy against inflation. This is essentially the same argument of monetary discipline already set forth in general in the debate on fixed versus flexible rates (see Sect. 17.1), strengthened by the fact that the commitment to maintain fixed exchange rates within the area would be felt more strongly than the commitment to defend a certain parity vis-à-vis the rest of the world as under the Bretton Woods regime. Another argument, however, suggests that monetary agreements might give rise to more inflation. This might happen (Rogoff 1985) when the national policy authority is involved in a policy game with other institutional agents (trade unions, etc.).

The argument of uniform inflation rates has been the subject of much debate. Firstly, it has been noted that it curiously turns the position of the traditional approach upside down: we have in fact seen in Sect. 20.2.1, criterion (e), that according to the traditional approach the similarity in the rates of inflation is a *precondition* for taking part in an (optimal) currency area, whilst it now becomes a (beneficial) *effect* of taking part. Secondly, many nonmonetarist writers deny that the achievement of a common rate of inflation is ontologically a benefit, and observe that different countries may have different propensities to inflate: some are more or less inflation-shy, others more or less inflation-prone, and this reflects a different structure of their national preference function.

In effect, some confusion seems to exist in relation to this argument. Benefits (1) and (2) are fairly objective ones, in the sense that they do not presuppose the adhesion to a particular school of thought or to a particular preference function (practically everybody agrees that allocative efficiency, elimination of destabilising speculation, saving on international reserves, are benefits). On the contrary, the equalization of the inflation rates is not considered universally desirable. From this point of view the traditional approach seems more neutral, because it only states that if there is similarity in inflation rates, *then* ground exists for participating in a currency area.

Behind the idea that monetary integration is conducive to economic integration there is, if one looks carefully, a psychological expedient of the following type. Assuming that a certain number of countries wish to effect an *economic union*, and assuming also that some of these are not able to implement domestically and as an expression of their autonomy the policies which bring about the characteristics necessary to an economic union (one of these characteristics is the uniformity in inflation rates), then *monetary integration* may well be a useful *instrument* to enable them to implement those policies. This reasoning is based on the belief (or hope) that society is affected by a kind of "economic policy illusion", in the sense that it is not willing to accept certain economic policies as an expression of its own autonomy, but it is willing to accept them if these are presented as deriving from external conditioning, i.e. required by the participation in a currency area. How justified is this belief, is an empirical matter; in general one can only observe that a common currency can certainly be brought into being and manifest all its advantages when it is established as the final step of a process of economic integration, whilst one can doubt its effectiveness as a tool to compel the refractory countries to realize the conditions necessary for economic integration.

(4) Besides the advantages listed above there may also be advantages of a political type, in the sense that a currency area (and, more generally, an economic union) carries more weight than the single countries in negotiating as a whole with outside parties. This of course requires that, although the exchange rates vis-à-vis non members are flexible, the currency area adopts a common exchange policy towards outside currencies. This is in the nature of things when the area adopts a common currency in the strict sense, whilst it not so easily and automatically realizable when the individual members maintain their respective national currencies. In fact, the currency of any member may turn out to be stronger or weaker than those of other members with respect to outside currencies which are key currencies in international transactions (as for example the US dollar). This may give rise to tensions within the currencies of the area (remember what was said in Sect. 2.2 about cross rates), unless there is a coordination of the interventions on the foreign exchange rates; as an extreme case one can envisage a pool of all the international reserves of the partner countries with respect to outside countries and a unified management of this pool.

Let us now come to the *costs*, amongst which we list the following:

(1) loss of autonomy in monetary and exchange policy of the individual members. The financial integration and the related perfect capital mobility makes monetary policy impotent (see Sect. 11.4); in the case of full integration the central banks of the members will merge into one supranational central bank. The disappearance of a possible policy tool such as the managed variations in the exchange rate may give rise to serious problems if wage rates, productivity, and prices, have different trends in different member countries. These problems may become particularly severe in the case of shocks coming from outside the area.

(2) Constraints on national fiscal policy. It is true that fiscal policy—for the same reasons for which monetary policy is ineffective—is fully effective under fixed

exchange rates (see Sect. 11.4), but this is true for an isolated country. In the case of a country belonging to a currency area, its fiscal policy may be constrained by the targets of the area as a whole (for example to maintain a certain equilibrium in the areas balance of payment vis-à-vis the rest of the world). And since the joint management of the single members' fiscal policies is carried out in the interest of the majority, it may happen that some member is harmed (unless a vetoing power is given to each member, in which case, however, there is the risk of a complete paralysis).

(3) Possible increase in unemployment. Assuming that the area includes a country with low inflation and an external surplus, this country will probably become dominant and compel the other members (with greater inflation and an external deficit) to adjust, because—as there are no means to compel the former country to inflate—the deficit countries will have to take restrictive measures which will lead to a decrease in employment. The writers of the monetarist school claim that in the long-run every country will be better off thanks to the lower rate of inflation, but, even allowing this to be true, the problem remains of determining how long is the long-run, as it is clear that in the short-run there are costs to be borne.

4) Possible deterioration of previous regional disequilibria. "Regional" is here used in the strict sense, i.e. referring to single regions within a member country. Since the international mobility of capital (in the absence of controls) is higher than the international mobility of labour, the greater possibilities of finding better-rewarded uses of capital in other countries of the area, together with the relatively low international labour mobility, may aggravate the development problems of the underdeveloped regions of a country. It should be noted that this negative effect occurs insofar as what was listed as the first criterion in the traditional approach (the high international mobility of *all* factors) does not occur, but it seems in any case likely that international labour mobility is lower than that of capital.

Having thus listed the benefits and costs, we conclude by pointing out that the already mentioned problem of weighing them cannot be given a generally valid answer, as it depends on the social welfare functions of the different countries, that may be quite different.

20.2.3 The New Theory

The new theory of currency areas is not really new in its approach, as recent studies continue both to examine various criteria and to enumerate benefits and costs. The novelty resides in the fact that new theoretical results are being applied to old issues (Tavlas 1993). For example, the cost associated with the loss of monetary autonomy was based on the idea that flexible exchange rates would allow a country to choose an optimum point along its Phillips curve. But the idea of a permanent trade-off between inflation and unemployment has been undermined by development in both theory (the displacement of the Phillips curve first by the natural rate of unemployment and then by the NAIRU) and reality (the stagflation problem in the 1970s and early 1980s). Thus it seems that, under this respect, the main benefit of flexible exchange rates is only the ability of choosing a different rate of inflation from other countries.

Recent work on the theory of optimum currency areas and monetary integration, partly stimulated by prospect and then the inception of the European monetary union, concentrates on two issues: the *effects of shocks* and *reputational considerations*.

As regards the effects of shocks, we have already seen in the new debate on fixed versus flexible exchange rates (Sect. 17.2) that the modern approach has succeeded in reducing the range of uncertainty but hasn't been able to settle the debate. There are various reasons for this failure: the shock-absorption capability of fixed and flexible exchange rates depends on several factors, such as the type of shock, the structural parameters of the economy, and the objective function of the authorities. Empirical studies have as usual given mixed evidence (Tavlas 1994).

Reputational considerations start from the observation (see above) that, since policy makers can do little more than choose an optimal rate of inflation, they should aim at a zero (or at least very low) inflation rate. This is so because inflation distorts relative prices through which information is usually transmitted, thus creating uncertainty and inefficient allocation of resources. The more credible the anti-inflation commitment, the lower the costs associated with a given decrease in the inflation rate, and the easier the implementation of an anti-inflation plan. Credibility requires time consistency, namely that the government will pursue the policy in the future because it has no incentive to change it, and the public is convinced that the government will not change it.

Given this, the view has been set forth that a high-inflation country increases its credibility by fixing its exchange rate with respect to the currency of a low-inflation country. But why the exchange rate rather than monetary policy? The answer is twofold. First, the exchange rate is an easily observable and still more easily understood variable by the public, while the money supply is not (or is less) and, to the extent that it is observable, is difficult to control. Second (and related to the first), is the discipline argument: by pegging the exchange rate the monetary authorities tie their hands much more strictly than by a money supply commitment, hence gain in credibility.

Empirical studies have again given mixed evidence (Tavlas 1994), showing that pegging not accompanied by other policy measures is not sufficient for gaining credibility.

Thus we must conclude that recent theoretical and empirical work has produced ambiguous results. But, after all, the move to international integration is more an expression of political will than the outcome of purely economical calculations. The commercial integration of the countries belonging to the European Union (begun in 1958 with the Rome Treaty and now accomplished) has shown this (for a survey of both ex ante and ex post economic calculations see Gandolfo 2014, Sect. 11.7.3) and the monetary integration of the same countries now under way is also showing this (see the next chapter).

20.2.4 Optimum for Whom?

Optimum currency areas are by definition optimum for the participating countries. But what about the rest of the world? This is the *third-country* problem, which means that the agreement to create a currency area may have a negative effect on non-participating countries. Bayoumi (1994) has presented a general equilibrium model with regionally differentiated goods which shows that the formation of a currency area unambiguously lowers welfare for non-participating regions. The reason is very simple: the model shows that the formation of a union will lower the output (because of the interaction between the common exchange rate and lower wage flexibility). This will not only be a cost for the union's members, but also for the rest of the world whose trade will be negatively affected. On the contrary, the benefits from the union are limited to the union's members.

Although these results need not be valid with different models, the possibility of a welfare-lowering effect (for the rest of the world) due to the creation of a currency union should be taken into consideration.

20.3   The Common Monetary Unit and the Basket Currency

The problem of the common monetary unit necessarily arises only in what one could define as the *maximal* form of currency area (i.e., a full monetary integration or monetary union proper), because in the *minimal* form (i.e., that which only establishes fixed exchange rates between the members) all that is needed is the declaration of the bilateral parities between the currencies of the members, namely the so-called *parity grid*. Given n members the parity grid will contain $n(n - 1)$ bilateral exchange rates, since each currency has $(n-1)$ bilateral exchange rates with all the others; it should however be remembered from Sect. 2.2, that it is sufficient to know $(n - 1)$ bilateral exchange rates to determine all. However, also in the minimal form it is possible to define a currency unit of the area which performs certain functions, although the members retain their own currencies.

From the theoretical point of view there are three possible ways to define the common currency unit. The first is to use a unit external to the area, namely not coinciding with any one of the members' currencies or combination of these, for example gold. The second consists in defining a unit internal to the area, which may be either the currency of a member or a combination of their currencies to be duly defined. The third is to create a new currency.

The first way is not considered advisable by many writers because it does not permit a suitable regulation of international liquidity (see Sect. 22.7: actually, gold—once the basis of the international monetary system—has been demonetized, as we shall see in that section). The second way, as we have said, presents two options: the internationalization of the currency of a member or the definition of a *composite unit*, consisting in a bundle or combination of the various currencies belonging to the area. The former option implies that one of the members of the area is the dominant country, with all the inherent problems, as was the case of the Bretton Woods system, which might be considered as a currency area at the world level with the US dollar as the dominant currency. The latter option implies the definition of a "basket" of currencies (hence the name of *currency basket* or *basket-currency*), which contains predetermined and fixed amounts of the single currencies belonging to the area. These amounts are established by a common agreement in accordance with some criterion which usually refers to the relative economic importance of the various members. It should be noted that, in general, a basket-currency is not necessarily linked to a currency area, as it can also be defined and used independently of the existence of a currency area. This is the case, for example, of the International Monetary Fund's Special Drawing Right, on which see Sect. 22.4.

In formal terms, if $q_1, q_2, \ldots, q_n$ are the amounts of the various currencies which make up one unit of the basket-currency; this unit (denoted by N) is defined by the set of numbers

$$(q_1, q_2, \ldots, q_n) = N. \tag{20.1}$$

An alternative way of expressing the same notion is

$$\begin{array}{c} q_1 \text{ units of currency } 1 + q_2 \text{ units of currency } 2 + \ldots + q_n \\ \text{units of currency } n \rightarrow N, \end{array} \tag{20.2}$$

where the arrow instead of the $=$ sign indicates that the left-hand side is not really an arithmetic sum (it is not, in fact, possible to add heterogeneous magnitudes) but an operation which defines the contents of N. Naturally, given the bilateral (fixed) exchange rates r_{ks}, it is possible to define the value of N in terms of any one of the component currencies, that is the exchange rate of the jth currency with respect to N, as

$$R_j = \sum_{k=1}^{n} q_k r_{kj}, \qquad j = 1, 2, \ldots, n, \tag{20.3}$$

where $r_{jj} = 1$. It is also possible to define a weight (b_j) of each currency in the basket, given by the ratio between the amount of the currency in the basket and the value of the basket in terms of the same currency, that is

$$b_j = \frac{q_j}{\sum\limits_{k=1}^{n} q_k r_{kj}}, \qquad j = 1, 2, \ldots, n \tag{20.4}$$

where it must, of course, be

$$\sum_{j=1}^{n} b_j = 1, \tag{20.5}$$

i.e., the sum of the weights equals unity. As can be seen from (20.4), the weights—given the quantities q_j—change as the r_{kj} change. Thus the definition of a currency basket in terms of (fixed) weights is equivalent to that in terms of fixed quantities only if the bilateral exchange rates are irrevocably fixed.

The option of the basket-currency was used by the European countries belonging to the EMS (European Monetary System, see the next chapter).

Finally, there is the option of creating a *new* currency which should be entirely fiduciary (with no relation either with elements external to the system or with the members currencies, issued and regulated by a supranational Central Bank) and would circulate throughout the area, exactly as the US dollar is the currency circulating in all the States of the USA. This is the solution that has been adopted by the countries belonging to the European Union (see the next chapter).

20.4   The Common Monetary Policy Prerequisite, the Inconsistent Triad, and Fiscal Policy

We examine whether a common monetary policy is only desirable or also necessary for monetary integration to be viable. To analyse this question let us begin by recalling a few well-known interest-rate-parity conditions (see Chap. 4). Given perfect capital mobility, we must distinguish the cases of perfect and imperfect asset substitutability (see Sect. 4.6). With perfect asset substitutability, uncovered interest parity must hold, *i.e.*

$$i_h = i_f + \frac{\tilde{r} - r}{r}. \tag{20.6}$$

With imperfect asset substitutability, condition (20.6) has to be modified by the introduction of a risk premium δ, namely

$$i_h = i_f + \frac{\tilde{r} - r}{r} + \delta. \tag{20.7}$$

Let us begin by considering Eq. (20.6). In a currency area (and in higher degrees of monetary integration), exchange rates are irrevocably fixed, hence $(\tilde{r} - r)/r = 0$ and, consequently,

$$i_h = i_f. \tag{20.8}$$

With perfect asset substitutability only the money stock matters (see Sect. 15.3), so that we can consider only money-market equilibrium (demand for money = supply of money) at home and abroad. Thus we have

$$P_h L_h(Y_h, i_h) = M_h, \tag{20.9}$$

$$P_f L_f(Y_f, i_f) = M_f. \tag{20.10}$$

Even if we take the price levels and the outputs as given, the three equations (20.8)–(20.10) form an *undetermined* system, which is unable to determine the four unknowns (the two money supplies and the two interest rates). Thus, there is a fundamental indeterminacy of the money supply and the interest rate in this two-country system. It follows that the two countries will have to (implicitly or explicitly) agree on the conduct of monetary policy. We must stress that the apparent absence of agreement can simply be due to the presence of an implicit agreement. The typical implicit agreement is asymmetric, in the sense that it is based on the dominant role of one country. This means that one country sets its money supply according to its own criteria and the other country adapts its money stock.

Suppose, for example, that the foreign country is the dominant country and fixes M_f. Then Eq. (20.10) determines i_f, which sets i_h by Eq. (20.8). Finally, Eq. (20.9) determines M_h. Thus country 1 must set its money supply at this level.

If it does otherwise, the currency area will break down. In fact, suppose for example that the home country tries to fix a higher money supply. This will depress i_h below i_f. As a consequence, immediate and disrupting capital flows from home to abroad will take place, unless controls on capital flows are introduced (but such controls are hardly compatible with monetary integration). These flows will lead to expectations of a future exchange rate adjustment and to the breakdown of the exchange rate commitment.

Explicit agreements on the conduct of the overall monetary policy are, on the contrary, of the cooperative type. This requires that countries agree to cooperate in setting their money stocks (or interest rates). This presents the well known free-riding problem (Hamada 1985), the same kind of problem that has been treated at length in the theory of international price cartels. In general, this means that once a cooperative agreement has been reached, there are usually incentives for one partner to do something else than was agreed upon. This follows from the fact that, by so doing, the partner will be better off, if the other partners do not retaliate (for example by reneging on the agreement). Thus some institutional mechanism has to be devised to avoid the free-riding problem.

These considerations confirm the essential importance of viable agreements on the conduct of monetary policy within the currency area. We have illustrated this proposition by the simplest model possible, but the results do not change substantially with more complicated models, such as those based on Eq. (20.7) instead of Eq. (20.8). This would require the introduction of the stocks of assets (money and other financial assets) and the determination of the portfolio-balance equilibrium.

Since it is impossible for a country to simultaneously peg its exchange rate and allow unfettered movement of international capital, while retaining any autonomy over its monetary policy, the set of fixed exchange rates, perfect capital mobility, and monetary independence has been called the *trilemma* or the *inconsistent triad* (if we also include perfect commodity mobility, namely free trade, this becomes an *inconsistent quartet*: Padoa Schioppa 1988, p. 373); others call it "the holy trinity" (Rose 1996). An in-depth examination of the trilemma is contained in Klein and Shambaugh (2015).

20.4.1 Fiscal Policy Coordination

The conclusion that a common monetary policy is a prerequisite for a viable monetary union is generally accepted. But what about *fiscal policy*? Levin (1983) found a beggar-my-neighbour effect of fiscal policy in a three-country model of the Mundell-Fleming type (a two-country currency area with a floating exchange rate vis-à-vis the rest of the world considered as country 3). Namely, a fiscal expansion in one country of the area causes a contraction in the other country's national income. If this were generally true the necessity of a common fiscal policy would be obvious—no country in a union would be willing to accept such contractionary effects. It was however shown by Sauernheimer (1984), with a model that is a generalization of Levin's, that Levin's finding depends on the assumption of price rigidity. If this assumption is relaxed, the outcome becomes indeterminate, but a taxonomy can be made. The same results were later rediscovered by Moutos and Scarth (1988), with the same kind of model as Sauernheimer's. Carlberg (1999) shows that the results of fiscal policy also depend on the size of the union, and gives a complete taxonomy of all possible cases. Federici (1996) introduces rational expectations as regards the union's floating exchange rate and shows that each policy maker is in a strategic relationship with other policy makers, and time inconsistency problems emerge.

All these models, as already noted, are descriptive models of the traditional Mundell-Fleming type. Other studies have been carried out in the context of an optimizing approach, in which the authorities maximize a social welfare function (or minimize a loss function). The first model is Sibert's (1992): the framework employed is a two-country, overlapping generations model in which each government chooses tax rate and public goods provision to maximize the utility of its residents. Sibert concluded that lack of coordination leads to a too high level of tax rate and public expenditure. Many other authors used models of the same kind as Sibert's, considering the matter from various angles. Bryson (1994) considered the possibility of stochastic disturbances such as productivity and demand shocks. Andersen and Sorensen (1995) used a model of a unionised economy developed by Dixon (1991) to explore the externality through wages, while Levine and Brociner (1994) considered negative externalities through the capital market. Dixon and Santoni (1997) paid particular attention to the mechanism of money flows within members of a monetary union. In their opinion this kind of mechanism produces positive spillover effects of fiscal policy. Levine and Perlman (2001) examined the effects of fiscal coordination on the stability of a monetary union.

All these authors, with the sole exception of Levine and Brociner (1994), concluded that the establishment of a monetary union should be accompanied by fiscal cooperation among member countries.

The social welfare (or loss) function considered is typically unique, so that there is a trade-off between inflation and other targets (employment, public expenditure, etc.). Consider however the case in which the union's independent central bank (UCB), in charge of monetary policy, has the sole task of guaranteeing price stability, while national governments, in charge of fiscal policy, minimize a loss function depending on inflation, output and public spending, where the weights may be different across countries (Beetsma and Lans Bovenberg 1998). With an optimally designed central bank, fiscal coordination may be counterproductive, in the sense that it decreases welfare.

In fact, without coordination, individual changes in the tax instrument have little effect on the UCB's monetary policy, and this dissuades governments from using such an instrument strategically vis-à-vis the union's central bank in order to induce it to change the inflation rate in the direction preferred by the individual fiscal players. With fiscal coordination public spending and tax rates are chosen by a supranational authority who minimizes an equally weighted sum of the single governments' loss functions. This encourages the (single) fiscal player to use the tax instrument to induce the UCB to change the inflation rate in the direction preferred by the fiscal player. Assuming that the individual governments and hence the common fiscal authority attach a much lower weight to inflation than the UCB, the outcome is that inflation, taxes, and public expenditure all increase.

Thus fiscal coordination weakens discipline and lowers welfare under the assumption of small low money holdings and hence small social benefits from seigniorage (Beetsma and Lans Bovenberg 1998, 254–55).

Van Aarle and Huart (1999), in the context of a two-country optimizing model, consider various cases. A case that they consider is that in which they assume that both countries are symmetric except that the fiscal authorities in country 2 attach a higher weight to government expenditure stabilization than the fiscal authorities in country 1. Moreover, they assume that the common fiscal authority gives the same weight to government spending as the fiscal authority of country 2. The result is that both countries will have less seigniorage revenues available to cover government spending than before under monetary union with national fiscal autonomy.

Contradicting results are obtained in the context of microfounded models of a monetary union. Beetsma and Jensen (2005) find that there are non-trivial gains from fiscal policy coordination, while Okano (2014) finds that fiscal policy cooperation has no benefits. The difference in policy implications for fiscal policy cooperation between Beetsma and Jensen 2005 and Okano (2014) stems from the choice of the households' utility function (see the Appendix).

The need for a common (centralised) fiscal policy to match the common (centralised) monetary policy was already made by one of the founding fathers of the traditional approach (Kenen 1969, p. 45 ff). The basic question that we must now tackle is whether the long-run survival of a monetary union requires a concomitant fiscal union (also called fiscal federalism). Godley (1992) and Goodhart (1998) maintain that it is the presence of a federal budget that makes a monetary union stable. They argue that the link between money creation and sovereignty is crucial. Gabrisch (2013) argues that each currency needs a sovereign for stabilizing financial markets and the real economy. Failing this, a currency union would sooner or later decay. Bordo et al. (2013), using the political and fiscal history of five federal states (the United States, Canada, Germany, Argentina, and Brazil), conclude that a fiscal union is a necessary condition for the euro to succeed.

A related question is whether the instability of a monetary union may be caused by insufficient labour mobility. Labour mobility, as we have seen above (Sect. 20.2.1) is the first and paramount requirement of the traditional approach to OCA. Is perfect labour mobility an essential component of an optimal currency area? Could perfect labour mobility guarantee the stability of a monetary union? Baglioni et al. (2015) thoroughly examine the matter in the context of an intertemporal model, and conclude that "contrary to the conventional vision, a federal budget appears to be essential to the efficient working of a currency area, in particular when labor mobility is efficiency enhancing. Labor mobility and federal budget are not substitutes but complement each other in a optimal currency union" (p. 28).

20.5   The Single-Currency Problem

This problem is best examined in the context of cost-benefit analysis. In fact, the demonstration that the adoption of a single currency is required for the elimination of the inefficiencies linked to the coexistence of national currencies, is not a proof of the necessity of a single currency. What it shows is that a single currency is somehow better than the fixity of exchange rates, not that it is essential for a monetary union. Thus the proper framework is that of cost-benefit analysis.

In general, the full advantages of a monetary union can be obtained only through the perfect substitutability of all the union members currencies in the three basic functions of money: unit of account, means of payment, and store of value. Once the credibility of the irrevocable fixity of exchange rates is firmly established, perfect substitutability in the unit-of-account and store-of-value functions does not present particular problems. Problems are present, on the contrary, in the means-of-payment function. Transaction costs create a wedge between buying and selling rates, since foreign exchange operators charge a cost for their service. These costs could be eliminated for private agents if the authorities subsidize the conversion between the various national currencies. This, however, would simply shift the costs onto the unions budget. In practice bid-ask spreads are such that, by simply converting one currency into another, one after each other, and finally reverting to the initial currency in, say, a fifteen-member union (such as the European Monetary Union), without actually spending a penny one might well end up with less than 40 % of the initial amount! In addition, these spreads are likely to vary from country to country in the union, thus altering the degree of substitutability among currencies.

Let us now come to the examination of the main benefits and costs of a single currency (for a complete treatment see Emerson et al. 1992; De Grauwe 2014, Part I). The benefits are due to the fact that a single currency *by definition* eliminates a number of problems and shortcomings inherent in the use of several national currencies. These are:

(1) The elimination of imperfections in the substitutability of currencies, as detailed above.
(2) The elimination of any possibility, even if remote, of changes in par values. The expression irrevocably fixed exchange rate has no practical significance. Although the international community tries to observe the rule *pacta sunt servanda* (pacts must be observed), history is full of examples of irrevocable commitments to fixed exchange rates that have broken down. The reason is simple: assuming that national governments behave rationally, they will evaluate the costs and benefits of the fixed exchange rate union, as shown in Sect. 20.2.2. If the costs become overwhelming with respect to the benefits, the government concerned may be tempted to change the parity, even if this means breaking an international agreement. This is by no means an impossible occurrence. The evaluation of the costs and benefits, in fact, may vary over time, for example in relation to economic conditions and/or to preference functions of different governments. Thus a fixed exchange rate system does not eliminate the risk that a temporary change in this evaluation might lead a member country to alter the parity. Rational economic agents know this, hence the possibility of speculative capital flows and of an uncertain climate for businesses. Actually, models of currency crises have been built based on the possibility of the government to renege the commitment to fixed exchange rates (an escape clause: see Sect. 16.3).
(3) The elimination of destabilising speculative capital flows within the union, due to expectations of parity changes as detailed under (2).
(4) The elimination of the need for intra-union international reserves, required to make the commitment credible and to offset possible speculative capital flows.
(5) We have shown in Sect. 20.4 that a common monetary policy is necessary for a monetary union. A single currency would greatly facilitate the conduct of this overall monetary policy, and would eliminate free-riding problems.
(6) A single currency would carry more international weight and enable the union to reap the benefits of seigniorage. In addition, the interventions in the foreign exchange market vis-à-vis other currencies would be greatly facilitated and would require a smaller amount of international reserves vis-à-vis the rest of the world.

The main costs that have been stressed are the following:

(1) Costs for the transformation of the system of payments. These include the costs of changing existing monetary values into the new currency, the costs of changing coin machines, etc.

(2) The psychological cost to the public of introducing the new currency and their getting used to it. It is not sufficient to declare a currency legal tender for this to be used willingly in a country in the place of the existing national currency. A new currency cannot be merely imposed by legislative act, but must gain social consensus and be accepted by the market. Thus the authorities will have to ensure that the new currency performs the functions of money at least as efficiently as the existing national currencies. This process of convincing the public is not without costs.

(3) This point is usually presented as an advantage of the fixed exchange rate system, rather than as a cost of the single-currency system. It is called the currency competition argument. Several currencies in competition, so the argument goes, stimulate each national monetary authority in the group to pursue a lower rate of inflation and, more generally, a stable value of its respective currency. This does not seem a theoretically well-founded argument. A system based on competition between monetary policies will result either in the breakdown of the fixed exchange rate commitment or in the dominance of one currency, as shown in Sect. 20.4.

To conclude, we observe that the benefits of a single currency seem much greater than the costs. From this, of course, it does not follow that a single currency is necessary for a monetary union. It is preferable, but not indispensable. There is, however, a further consideration that points to the necessity of a single currency. We have already mentioned above, under entry (2) of the list of benefits, the problems related to maintaining fixed parities. In the absence of capital controls (whose elimination necessarily accompanies the formation of a monetary union), permanently fixed exchange rate is an oxymoron, as Portes (1993; p. 2) aptly put it. No matter what governments say, speculators know that exchange rates between distinct national currencies exist only to be changed. And—confronted with one-way bets thanks to fixed exchange rates, and having practically unlimited resources thanks to free capital mobility—they can indeed compel the authorities to change the parities, as the crises of the EMS (see Sect. 21.1) have shown. Hence there is a strong suspect that a monetary union maintaining distinct national currencies with permanently fixed exchange rates would not be viable.

The same considerations can throw light on the issue of the *process of transition* to a monetary union with a single currency. Two approaches are possible: the gradual approach and the "shock therapy", the latter meaning the sudden introduction of the single currency. With the gradual approach there is the cost-of-credibility question. A major problem in the road to a monetary union is to convince the private sector that the commitment of the national authorities to monetary union is credible. An obvious way of achieving credibility would be to introduce a common currency immediately. A gradual approach may not convince the private sector, because the commitment to a fixed exchange rate is not sufficiently credible, unless this commitment is accompanied by other measures, as has been wisely done in the gradual approach adopted by the European Union (on which see the next chapter, Sect. 21.2).

The single currency issue is not disjunct from the central bank issue. Once a single currency has been decided upon, the necessity arises of the centralization of monetary policy in a single supranational body. This is the union's central bank, which could be of the federal type to take advantage of the long experience that national central banks have built up over the years.

20.6 Appendix

20.6.1 Fiscal Policy in a Monetary Union

We consider here Sauernheimer's model (1984), with slight changes in notation to conform with ours.

The area consists of two countries with a permanently fixed exchange rate normalised at unity (hence we could as well apply the model to a monetary union with a single currency), with a floating exchange rate vis-à-vis the rest of the world. The rest of the world (ROW) is considered exogenous. The basic equations of the model are

$$S_1(y_1) = I_1(\underset{-}{i_1}) + \overline{G}_1 + B_{12}(\underset{-}{y_1}, \underset{+}{y_2}, \underset{+}{p_2/p_1}) + B_{13}(\underset{-}{y_1}, \underset{+}{y_3}, \underset{+}{r\overline{p}_3/p_1}), \tag{20.11}$$

$$S_2(y_2) = I_2(\underset{-}{i_2}) + \overline{G}_2 + B_{21}(\underset{+}{y_1}, \underset{-}{y_2}, \underset{-}{p_2/p_1}) + B_{23}(\underset{-}{y_2}, \underset{+}{y_3}, \underset{+}{r\overline{p}_3/p_2}), \tag{20.12}$$

$$\overline{M_1 + M_2} = p_1 L_1(\underset{+}{y_1}, \underset{-}{i_1}) + p_2 L_2(\underset{+}{y_2}, \underset{-}{i_2}), \tag{20.13}$$

$$i_1 = i_2 = \overline{i}_3, \tag{20.14}$$

where the subscripts 1,2 refer to the countries forming the currency area while the subscript 3 refers to ROW. A bar over a variable denotes that the variable is exogenous. The signs below the arguments in the functions represent the signs of the partial derivatives. The symbols have the usual meaning: $S=$ private saving, $y=$ national income (output), $I=$ private investment, $i=$ nominal interest rate, $B_{ij}=$ trade balance or net exports of country i with respect to country j, $p_i=$ price of country i's domestically produced goods, $r=$ exchange rate of the area vis-à-vis the rest of the world, $M_i=$ country i's money supply. All expenditure variables are measured in real terms, namely in units of the country's goods. Note that to assume the partial derivative of each member country's trade balance (vis-à-vis the ROW) with respect to terms of trade to be positive, implies the assumption that the critical elasticities condition is satisfied.

The first and second equation define the aggregate demand sector of the model (instead of $S(y) = I(i) + \ldots$ one could also write $y = C(y) + I(i) + \ldots$). The third equation is the sum of the two countries' monetary equilibrium conditions, $M_i = p_i L_i(y_i, i_i)$, and reflects the assumption that the *total* money supply of the union is exogenous while that in the single member countries is endogenous. The fourth equation expresses perfect capital mobility *à la Mundell* (on this point see Sects. 4.6 and 11.4) not only within the currency area but also between the area and the ROW. Hence the area's interest rate is exogenously given by the ROW's interest rate.

In Levin's model, "the domestic price of each country's domestically produced good is taken to be constant and normalized at unity on the assumptions of fixed money wages and constant returns in production" (Levin 1983, pp. 330–31). Hence Eqs. (20.11)–(20.13)—if we replace both i_1 and i_2 with $\bar{i}_3$—form a system in the three endogenous variables y_1, y_2, r, from which we can obtain Levin's result by standard comparative statics methods. The economic rationale of this result is easy to see, if we consider for a moment the union as a whole vis-à-vis the rest of the world. We know from Mundell's analysis (see Sect. 11.4) that fiscal policy under perfect capital mobility, flexible exchange rates and price rigidity is completely ineffective. Thus the union's overall income will not change as a consequence of a fiscal expansion. Hence, if now look at the two regions (countries) in the union, it follows that an expansion in one country must cause a depression in the other, since the total is unchanged.

The Sauernheimer model modifies Levin's model by introducing variable prices according to mark-up pricing over a money wage rate that varies through indexation effects. Thus we have

$$p_i = (1 + g_i)(L/y)_i w_i, \tag{20.15}$$

where g_i is the fixed mark-up coefficient, $(L/y)_i$ is the (fixed) labour-input coefficient and w_i is the money wage rate in country $i = 1, 2$. The money wage rate, in turn, depends on the general (consumer) price index in the country (for example through an indexation mechanism)

$$w_i = \omega_i(I_i), \qquad 0 \le dw_i/dI_i \le 1, \tag{20.16}$$

where

$$I_i = \alpha_{i1} p_1 + \alpha_{i2} p_2 + \alpha_{i3} r \bar{p}_3, \qquad \alpha_{i1} + \alpha_{i2} + \alpha_{i3} = 1, \tag{20.17}$$

is a weighted average of the prices of the three goods available to consumers. Note that the case $\omega_{i,I} \equiv dw_i/dI_i = 0$ means a rigid wage rate, while $\omega_{i,I} = 1$ means complete indexation

Equations (20.15)–(20.17) form a subset that allows us to express p_1, p_2 in terms of the exchange rate r and the data ($\overline{p}_3$, etc.). Thus we have

$$p_i = \varphi_i(r, \ldots), \quad i = 1, 2. \tag{20.18}$$

For future reference let us compute the derivatives of the functions φ_i with respect to r. If we substitute I_i from Eq. (20.17) into (20.16) and then the result into Eq. (20.15), we get

$$p_i - (1 + g_i)(L/y)_i \omega_i(\alpha_{i1} p_1 + \alpha_{i2} p_2 + \alpha_{i3} r \overline{p}_3) = 0, \tag{20.19}$$

whose solution gives the functions φ_i. To compute the derivatives $p_{i,r} \equiv \partial p_i / \partial r$, we apply the implicit function theorem. With no loss of generality we can normalize $\overline{p}_3$ at one and choose units such that $p_i = w_i = r$ in the initial situation (this implies choosing output units such that $(1 + g_i)(L/y)_i = 1$). Thus we have

$$(1 - \omega_{1,I}\alpha_{11})p_{1,r} - \omega_{1,I}\alpha_{12}p_{2,r} = \omega_{1,I}\alpha_{13},$$

$$-\omega_{2,I}\alpha_{21}p_{1,r} + (1 - \omega_{2,I}\alpha_{22})p_{2,r} = \omega_{2,I}\alpha_{23}. \tag{20.20}$$

Solving this system we get

$$p_{1,r} = \frac{\omega_{1,I}[\alpha_{13}(1 - \omega_{2,I}\alpha_{22}) + \omega_{2,I}\alpha_{12}\alpha_{2,3}]}{(1 - \omega_{1,I}\alpha_{11})(1 - \omega_{2,I}\alpha_{22}) - \omega_{1,I}\omega_{2,I}\alpha_{12}\alpha_{21}},$$

$$p_{2,r} = \frac{\omega_{2,I}[\alpha_{23}(1 - \omega_{1,I}\alpha_{11}) + \omega_{1,I}\alpha_{21}\alpha_{13}]}{(1 - \omega_{1,I}\alpha_{11})(1 - \omega_{2,I}\alpha_{22}) - \omega_{1,I}\omega_{2,I}\alpha_{12}\alpha_{21}}, \tag{20.21}$$

where both the numerator and the denominator are positive given the definitions of the α's and ω's, except when $\omega_{i,I}$ is zero, in which case $p_{i,r}$ is zero. Also note that $p_{i,r} = 1$ when $\omega_{i,I} = 1$. In general, the weights α and the functions ω will be different in the two countries of the currency area, hence $p_{1,r} \neq p_{2,r}$ (except in the two extreme cases).

If we now substitute Eqs. (20.18) into Eqs. (20.11)–(20.13), where we have also replaced both i_1 and i_2 with $\overline{i}_3$, we can determine y_1, y_2, r in the variable-price case under consideration and perform our comparative statics exercises. In fact, provided that the Jacobian of this system with respect to the endogenous variables y_1, y_2, r is different from zero at the equilibrium point, we can express these variables as differentiable functions of the exogenous variables and compute the relevant partial derivatives by the implicit function theorem. We are interested in $\partial y_i / \partial G_j$, so as to ascertain whether and under which conditions the beggar-my-neighbour effect can materialize.

We compute $\partial y_i / \partial G_1$ (the case $\partial y_i / \partial G_2$ is symmetric) by differentiating system (20.11)–(20.13) with respect to G_1. For simplicity of notation we define

$\mu_{ij} = -\partial B_{ij}/\partial y_i$ $(i = 1, 2; j = 1, 2, 3; i \neq j)$, i.e. country i's partial marginal propensity to import from country j; $\mu_i = \sum_{j \neq i} \mu_{ij}$, i.e. country i's overall marginal propensity to import. We also define $p = p_2/p_1$, the area's internal relative price. Note that, by definition, in the initial situation we have $B_{12} + B_{21} \equiv 0$, from which $\partial B_{12}/\partial y_1 + \partial B_{21}/\partial y_1 = 0$, $\partial B_{12}/\partial y_2 + \partial B_{21}/\partial y_2 = 0$, $\partial B_{12}/\partial p + \partial B_{21}/\partial p = 0$. Finally, we define $S_{iy} = \partial S_i/\partial y_i$, $y_{iG_1} = \partial y_i/\partial G_1$, $B_{ij,p} = \partial B_{ij}/\partial p$, $B_{i3,r} = \partial B_{i3}/\partial r$, $r_{G_1} = \partial r/\partial G_1$, $L_{iy} = \partial L_i/\partial y_i$.

The differentiation of system (20.11)–(20.13) with respect to G_1 yields

$$(S_{1y} + \mu_1)y_{1G_1} - \mu_{21}y_{2G_1} - [B_{12,p}(p_{2,r} - p_{1,r}) + B_{13,r}(1 - p_{1,r})]r_{G_1} = 1,$$

$$-\mu_{12}y_{1G_1} + (S_{2y} + \mu_2)y_{2G_1} + [B_{12,p}(p_{2,r} - p_{1,r}) - B_{23,r}(1 - p_{2,r})]r_{G_1} = 0,$$

$$L_{1y}y_{1G_1} + L_{2y}y_{2G_1} + [L_1p_{1,r} + L_2p_{2,r}]r_{G_1} = 0. \tag{20.22}$$

Solving this linear system we obtain

$$y_{1G_1} = \frac{(S_{2y} + \mu_2)[L_1p_{1,r} + L_2p_{2,r}] - L_{2y}[B_{12,p}(p_{2,r} - p_{1,r}) - B_{23,r}(1 - p_{2,r})]}{\Delta},$$

$$y_{2G_1} = \frac{\mu_{12}[L_1p_{1,r} + L_2p_{2,r}] + L_{1y}[B_{12,p}(p_{2,r} - p_{1,r}) - B_{23,r}(1 - p_{2,r})]}{\Delta},$$

$$r_{G_1} = \frac{-\mu_{12}L_{2y} - L_{1y}(S_{2y} + \mu_2)}{\Delta}, \tag{20.23}$$

where Δ, the determinant of system (20.22), is positive when $p_{1,r} = p_{2,r}$; when $p_{1,r} \neq p_{2,r}$ this determinant will be positive if $B_{12,p}$ is not too large. At any rate, Δ is assumed to be positive (Sauernheimer 1984).

From inspection of Eqs. (20.23) it can be seen that r_{G_1} is negative, while the signs of y_{1G_1}, y_{2G_1} are indeterminate. The appreciation in the union's exchange rate is not a surprise: it is a standard effect of fiscal policy under perfect capital mobility and flexible exchange rates. When the union is considered as a whole we can, in fact, apply the well-known Mundell results (see Chap. 11, Sect. 11.4). The indeterminacy of the signs of y_{1G_1}, y_{2G_1} shows that the beggar-my-neighbour effect is just a possibility, whose materialization depends on the presence of price rigidity (Levin's assumption). In fact, the indeterminacy of the signs of the numerators of the fractions giving y_{1G_1}, y_{2G_1} depends on the presence of the terms $p_{i,r}$. Following Sauernheimer, we examine four main cases.

(1) $p_{1,r} = p_{2,r} = 0$.

This is the fix-price case. The numerator of y_{1G_1} becomes $L_{2y}B_{23,r} > 0$, while that of y_{2G_1} is $-L_{1y}B_{23,r} < 0$. Thus the beggar-my-neighbour effect is present, unless

$B_{23,r} = 0$, namely when there is no relative price effect on country 2s net exports to the rest of the world, a case that we can rule out. Also note that the overall income of the union increases, remains constant or decreases according as $L_{2y} \gtrless L_{1y}$. When $L_{2y} = L_{1y}$, the two countries of the union are so similar that they can be considered as a single country, and we are back in the traditional Mundell setting (see Chap. 11, Sect. 11.4). Fiscal policy under perfect capital mobility and rigid prices is ineffective; the result of an exchange rate appreciation is also standard. Now, if the union's total income is unchanged, then it is obvious that an expansion in one member country causes a depression in the other, as we have already noted above.

(2) $p_{1,r} = p_{2,r} = 1$.

This is the case of complete wage indexation. Levin's result is no longer true. In fact, not only the numerator of y_{1G_1} but also the numerator of y_{2G_1} is positive (if we exclude the abnormal case $\mu_{12} = 0$, which means that country 1 does not import anything from country 2). Hence both countries' income increases, and, of course, the union's income increases. Again, if we consider the union as a whole, we know that the effectiveness of fiscal policy under perfect capital mobility and flexible exchange rates is restored owing to the effects on the real money supply of the price changes induced by the exchange-rate variations (see Sect. 13.4). Note however that, when indexation is not complete, i.e. $0 < p_{1,r} = p_{2,r} < 1$, prices fall less than proportionally to the exchange rate appreciation, and the effect on country 2s income is again indeterminate, while y_{1G_1} is certainly positive. As regards the union as a whole, if $L_{1y} \cong L_{2y}$ the union's income increases.

(3) $1 = p_{1,r} > p_{2,r} = 0$.

In this case the union's internal relative price (p_2/p_1) changes. We now have $y_{1G_1} > 0$, but the sign of y_{2G_1} is indeterminate. This indeterminacy is due to relative price effects in the union, that were not present in the previous cases. Actually, there is an expansionary effect on the aggregate demand for country 2s output due to the tendency of this country's exports to country 1 to increase (term μ_{12} in the numerator of y_{2G_1}). But this effect is counteracted by the diversion of country 2s aggregate demand toward goods produced in countries 1 and 3, due to relative price effects. First, the exchange-rate appreciation diverts demand toward country 3s goods. Second, this appreciation works on the union's internal prices asymmetrically (due to a lack of harmonisation in wage policies), because it causes a decrease in country 1s prices but has no effect on country 2s prices. From this it follows a diversion of country 2s aggregate demand toward country 1s goods induced by the change in the intra-union relative price. Thus the lack of harmonisation of wage policies aggravates country 2s employment problems deriving from the lack of coordination of aggregate demand policies. It should however be noted, as Sauernheimer points out, that changes in relative prices cannot be avoided even with fully coordinated policies when the countries in the union have very different requirements of imported inputs, so that different price-effects of exchange-rate changes cannot be avoided.

(4) $1 \geq p_{2,r} > p_{1,r} = 0$.

This is a particularly interesting case, because country 1—notwithstanding the fiscal expansion and the increase in the union's real quantity of money (due to the price decline induced by the appreciation in the exchange rate)—may suffer an income contraction ($y_{1G_1} < 0$). This, again, is due to a relative price effect: while country 1s prices are rigid, country 2s prices are flexible and decline due to the exchange-rate appreciation. The ensuing relative price effect diverts country 1s aggregate demand toward country 2s goods. Besides, there is the diversion of country 1s demand towards country 3s goods due to the exchange rate appreciation. If the sum of these effects is sufficiently strong, country 1 will suffer a depression.

20.6.2 Fiscal Coordination

In the previous section we have shown that beggar-my-neighbour effects within the currency area are a serious possibility, though not a certainty. When this possibility materializes, fiscal coordination is advisable to avoid these unpleasant effects. However, it is not always the case that fiscal coordination improves welfare. In addition to the papers already examined in the text (Sect. 20.4.1), the issue has been addressed by a number of other papers, based on different models and assumptions. We just give a few examples, other contributions are contained in Bayoumi (1994), Carlberg (1999), Gambacorta (1999), Beetsma and Giuliodori (2010).

Van Aarle and Huart (1999) in the context of a two-country optimizing model show that when moving from a monetary union with national fiscal autonomy to a monetary union that also features a fiscal union, country 1s output declines while for country 2 the effect is most likely to be the opposite. Huizinga and Nielsen (1998) in the context of a two-period optimizing model show that the case for international fiscal coordination depends on the set of tax instruments available in all countries of the union. More precisely, with both investment and saving taxes available, there is no need to coordinate either tax or spending policies. On the contrary, fiscal policies should be coordinated when no saving tax is available. Rodrick and van Ypersele (1999) consider the issue of fiscal coordination in the context of international capital mobility. Starting from the observation that there is no guarantee that free capital mobility makes everyone better off (just as free trade in goods does not necessarily improve everyone's welfare), they show that when agents are highly risk averse and when productivity shocks are negatively correlated, a coordinated tax regime is Pareto superior.

But the most striking results come from a comparison of two similar microfounded models: Beetsma and Jensen (2005), and Okano (2014), already mentioned in the text (Sect. 20.4.1). According to Beetsma and Jensen, in period t, the utility function of the representative household j living in country i is given by

$$U_t^j = E_t \sum_{s=t}^{\infty} \beta^{s-t} \left[U(C_s^j, \epsilon_s^j) + V(G_s^j) - v(y_s^j, z_s^j) \right], \qquad 0 < \beta < 1, \qquad (20.24)$$

where C_s^j is consumption, G_s^j is per-capita public spending, and y_s^j is the amount of goods produced by household j. U and V are strictly increasing and strictly concave functions, and the function v is increasing and strictly convex in y_s^j. This means that households receive utility from consumption and public spending, and experience disutility from their work effort. ϵ_s^j is a shock affecting the demand for consumption goods, and z_s^i is a shock affecting the disutility of work. Finally, β is the discount factor, and E the expectation operator.

Let us now consider the utility function postulated by Okano, which is

$$\cup \equiv E_0 \sum_{t=0}^{\infty} \delta^t U_t, \qquad U_t \equiv \ln C_t - \frac{1}{1 + \varphi} N_t^{1+\varphi}, \qquad (20.25)$$

where C_t denotes consumption, N_t hours of work, φ the inverse of the labour supply elasticity, $0 < \delta < 1$ the discount factor, and E the expectations operator. Thus households receive utility from consumption and disutility from their work effort. Apart from the specification of the functional forms in Okano, the main difference between (20.24) and (20.25) is the presence of public expenditure in (20.24).

Dozens of pages would be required to treat these models in detail, and space limitations do not allow such a treatment, whose conclusion is that, as already noted in the text, in the Beetsma and Jensen (2005) model there are non-trivial gains from fiscal policy coordination, while Okano (2014) finds that fiscal policy cooperation has no benefits. The difference in policy implications for fiscal policy cooperation between Beetsma and Jensen (2005) and Okano (2014) stems from the choice of the households' utility function. In Okano's words (Okano 2014; p. 285), "as long as there is no explicit evidence for which household utility function is suitable or plausible for analyzing the Euro economy or for describing people in EMU countries, we can neither definitely deny Beetsma and Jensen's (2005) policy implication nor strongly insist on our policy implication".

References

Allen Reynolds, P., & Kenen, P. B. (1980). *Asset markets, exchange rates, and economic integration: A synthesis.* Cambridge (UK): Cambridge University Press.

Andersen, T. M., & Sorensen, J. R. (1995). Unemployment and fiscal policy in an economic and monetary union. *European Journal of Political Economy, 11,* 27–43.

Baglioni, A., Boitani, A., & Bordignon, M. (2015). Labor mobility and fiscal policy in a currency union. CESifo Working Paper No. 5159.

Bayoumi, T. (1994). A formal model of optimum currency areas. *IMF Staff Papers, 41,* 537–454.

Beetsma, R. M. W. J., & Lans Bovenberg, A. (1998). Monetary union without fiscal coordination may discipline policymakers. *Journal of International Economics, 45,* 239–258.

Beetsma, R. M. W. J., & Jensen, H. (2005). Monetary and fiscal policy interactions in a micro-founded model of a monetary union. *Journal of International Economics 67,* 320–352.

Beetsma, R. & Giuliodori, M. (2010). The macroeconomic costs and benefits of the EMU and other monetary unions: An overview of recent research. *Journal of Economic Literature, 48,* 603–641.

Bordo, M. D., Jonung, L., & Markiewicz, A. (2013). A fiscal union for the euro: Some lessons from history. *CESifo Economic Studies, 59,* 449–488.

Bryson, J. H. (1994). Fiscal policy coordination and flexibility under European Monetary Union: Implications for macroeconomic stabilization. *Journal of Policy Modeling, 16,* 541–557.

Carlberg, M. (1999). *European monetary union: Theory, evidence, and policy.* Berlin, Heidelberg: Physica-Verlag.

Corden, W. M. (1972). *Monetary integration.* Essays in International Finance No. 93, International Finance Section, Princeton University.

Delors Report (1989). *Report on economic and monetary union in the European community.* Brussels: Commission of the European Communities, 12 April.

De Grauwe, P. (2014). *The economics of monetary integration* (10th ed.). Oxford: Oxford University Press.

Denton, G. (ed.) (1974). *Economic and Monetary Union in Europe.* London: Croom Helm.

Dixon, H. D. (1991). Macroeconomic equilibrium and policy in a large unionised economy. *European Economic Review, 35,* 1427–1448.

Dixon, H. D., & Santoni, M. (1997). Fiscal policy coordination with demand spillovers and unionised labour markets. *Economic Journal, 107,* 405–417.

Emerson, M., Gros, D., Italianer, A., Pisani-Ferry, J., & Reichenbach, H. (1992). *One market, one money.* Oxford: Oxford University Press.

Federici, D. (1996). Fiscal policy in a currency area. CIDEI Working Paper 38, University of Rome "La Sapienza".

Gabrisch, H. (2013). Currency without a sovereign: On the causes of the euro crisis and its overcoming, available at SSRN: http://ssrn.com/abstract=2359047.

Gambacorta, L. (1999). What is the optimal institutional arrangement for a monetary union? Bank of Italy, Temi di discussione del Servizio Studi No. 356.

Gandolfo, G. (2014). *International trade theory and policy* (2nd ed.). Berlin, Heidelberg, New York: Springer.

Godley, W. (1992). Maastricht and all that. *London Review of Books, 14,* No. 19 - 8 October 1992, 3–4.

Goodhart, C. A. E. (1998). The two concepts of money: Implications for the analysis of optimal currency areas. *European Journal of Political Economy, 14,* 407–432.

Hamada, K. (1985). *The political economy of international monetary interdependence.* Cambridge (Mass.): MIT Press.

Huizinga, H., & Nielsen, S. B. (1998). Is coordination of fiscal deficits necessary? Tilburg University, Center for Economic Research Discussion Paper No. 9861.

Ingram, J. C. (1973). *The case for European monetary union.* Essays in International Finance No. 98, International Finance Section, Princeton University.

Ishiyama, Y. (1975). The theory of optimum currency areas: A survey. *International Monetary Fund Staff Papers, 25,* 344–383.

Kenen, P. B. (1969). The theory of optimum currency areas: An eclectic view, in R. A. Mundell & A. K. Swoboda (Eds.). *Monetary problems of the international economy* (pp. 41–60). Chicago: University of Chicago Press.

Klein, M. W., & Shambaugh, J. C. (2015). Rounding the corners of the policy trilemma: Sources of monetary policy autonomy. *American Economic Journal: Macroeconomics, 7,* 33–66.

Levin, J. H. (1983). A model of stabilization policy in a jointly floating currency area, in J. S. Bhandari & B. H. Putnam (Eds.). *Economic interdependence and flexible exchange rates* (pp. 329–349). Cambridge (Mass.): MIT Press.

Levine, P., & Brociner, A. (1994). Fiscal policy coordination and EMU: A dynamic game approach. *Journal of Economic Dynamics and Control, 18*, 699–729.

Levine, P., & Perlman, J. (2001). Monetary union: The ins and outs of strategic delegation. *Manchester School, 69*, 285–309.

McKinnon, R. I. (1963). Optimum currency areas. *American Economic Review, 52*, 717–725.

Moutos, T., & Scarth, W. (1988). Stabilization policy within a currency area. *Scottish Journal of Political Economy, 35*, 387–397.

Mundell, R. A. (1961). A theory of optimum currency areas. *American Economic Review, 51*, 509–517.

Okano, E. (2014). How important is fiscal policy cooperation in a currency union? *Journal of Economic Dynamics and Control, 38*, 266–286.

Padoa Schioppa, T. (1988). The European monetary system: A long-term view, in F. Giavazzi, S. Micossi, & M. Miller (Eds.). *The European monetary system.* Cambridge (UK): Cambridge University Press.

Portes, R. (1993). EMS and EMU after the fall. *World Economy, 16*, 1–15.

Robson, P. (1998). *The economics of international integration* (4th ed.). London: Routledge.

Rodrick, D., & van Ypersele, T. (1999). When does international capital mobility require tax coordination? Tilburg University, Center for Economic Research Discussion Paper No. 9927.

Rogoff, K. (1985). Can international monetary policy cooperation be counterproductive. *Journal of International Economics, 18*, 199–217.

Rose, A. K. (1996). Explaining exchange rate volatility: An empirical analysis of 'The Holy Trinity' of monetary independence, fixed exchange rates, and capital mobility. *Journal of International Money and Finance, 15*, 925–945.

Sauernheimer, K. -H. (1984). 'Fiscal Policy' in einer wechselkursunion. *Finanzarchiv, 42*, 143–157.

Sibert, A. (1992). Government finance in a common currency area. *Journal of International Money and Finance, 11*, 567–578.

Tavlas, G. S. (1993). The 'new' theory of optimum currency areas. *World Economy, 16*, 663–685.

Tavlas, G. S. (1994). The theory of monetary integration. *Open Economies Review, 5*, 211–230.

Tower, E., & Willet, T. D. (1976). *The theory of optimum currency areas and exchange rate flexibility.* Special Papers in International Economics No. 11, International Finance Section, Princeton University.

Van Aarle, B., & Huart, F. (1999). Monetary and fiscal unification in the EU: A stylized analysis. *Journal of Economics and Business, 51*, 49–66.

Werner Report (1970). *Report to the council and the commission regarding the step-by-step establishment of the community's economic and monetary Union.* Brussels: Commission of the European Communities, 8 October.

The European Monetary Union

21

The European Monetary Union (EMU, an acronym that also stands for Economic and Monetary Union, which in turn means the EU or European Union, which is the official denomination), besides its obvious interest for European students, also has a general interest: it is, in fact, the first large-scale experiment of setting up a monetary union among industrialized countries. The precursor of the European Monetary Union was the European Monetary System (henceforth EMS), of which we shall give an overview in the next section, before passing on to treat the European Monetary Union.

21.1   The European Monetary System

On 13th March 1979 the EEC countries (with the exception of Britain), in application of the Bremen Agreement of 7th July 1978, gave birth to a currency area called the European Monetary System (henceforth EMS) based on a unit of account called the European Currency Unit (ECU; it should be noted that this acronym considered as a word, is in French the name of an ancient French coin).

Since the EMS has been superseded by the European Monetary Union, we shall just give a brief overview, especially to check whether it was an optimum currency area.

The EMS was based on three elements: the exchange rate mechanism (ERM), the European Currency Unit (ECU), the credit mechanisms. As regards the ERM, the member countries declared their bilateral parities (giving rise to a so called "parity grid"), around which the actual exchange rates could oscillate within predefined margins. These margins were originally $\pm 2.25\%$ ($\pm 6\%$ in exceptional cases),

© Springer-Verlag Berlin Heidelberg 2016
G. Gandolfo, *International Finance and Open-Economy Macroeconomics*,
Springer Texts in Business and Economics, DOI 10.1007/978-3-662-49862-0_21

and were widened to $\pm 15\%$ from 2 August 1993.[1] The first commitment of the ERM was that the participating countries were obliged to intervene in the foreign exchange market when the market exchange rate hit one of the fluctuation margins. The second commitment of the ERM required central rates to be modified only by collective agreement, with no unilateral action on the part of any partner.

As far as the ERM is concerned, the EMS (with reference to the classification in Sect. 3.1) belonged to the category of limited-flexibility exchange systems, and was a combination of the adjustable peg and the wider band, i.e. an adjustable band system.

The EMS was therefore a currency area *sui generis*, as it officially contemplated the possibility of parity changes. What did then differentiate the EMS from the old snake (mentioned in Sect. 17.3.1)? One of the distinguishing features, in the original intentions, was the ECU and its role in the EMS. In fact, as we have seen in Sect. 20.3, it is when a currency area intends to move toward monetary union that the need for the definition of a common currency unit arises.

The introduction of the ECU was in fact a manifestation of "political will". Indeed, in the negotiations which gave rise to the EMS, the greatest controversies took place in relation to the definition of the operational rules which were to guide the central banks' interventions to maintain their currencies at the given parities with one another. One of the mechanisms considered was simply that of defining a parity grid with an obligation on the part of each country to intervene on the foreign exchange market when its currency had reached one of the margins of fluctuation with respect to any other currency; in practice, nothing more nor less than the old snake. Another mechanism considered was that of defining a basket-currency, to wit the ECU, in terms of which to determine the central rates (parities) of the individual currencies and the respective margins of fluctuation, and so also the obligation to take corrective action in case of divergence.

This was by no means a useless debate, as the mechanisms imply quite different obligations to corrective action. If a single currency, for example the Deutschemark, should begin to deviate too much from its central rate, the second mechanism would require corrective action solely by the Bundesbank (Germany's Central Bank), whilst the first mechanism would require correcting action by *all* central banks. In other words, the ECU mechanism allows to single out the currency which, by diverging too much from the weighted average of the ECU basket, can be considered responsible for the disequilibrium and so has to take corrective action alone; on the contrary, the parity grid mechanism puts the burden of intervention on all currencies, including those not responsible for the disequilibrium. It is, in fact, a simple mathematical property of bilateral exchange rates (see Sect. 2.2) that $r_{ks}r_{sk} = 1$, so that if a currency approaches a margin with respect to all the others, all the others will approach the opposite margin.

---

[1] Actually the margins were asymmetric, namely $+2.275\%$ and -2.225% in the case of the band having total width 4.5%; $+6.18\%$ and -5.82% in the case of the 12% band; $+16.11\%$ and -13.881% in the case of the 30% band.

It was not necessary to be prophets to understand that anticipations of a strong Deutschemark suggested the ECU mechanism to the countries with weaker currencies (amongst them Italy), whilst Germany preferred the parity-grid mechanism. The compromise that was reached was to maintain the parity-grid mechanism as the basis for the *obligation* to take corrective action, and to use the ECU (a) as the unit of reference to define the central rates (parities) of the grid, and (b) as the basis for defining an "indicator of divergence", i.e. an indicator of a currency's divergence from its central ECU price, with the proviso that when this indicator exceeds a certain threshold (the "threshold of divergence"), this results in a *presumption* that the authorities concerned will correct the situation by adequate measures (in the form of exchange market interventions or of internal policy measures). To put it another way, the crossing of the ECU-defined threshold of divergence was intended as a kind of alarm-bell to warn that a currency was deviating too much (though not having yet reached the maximum bilateral margin against any other currency), and so it was presumed that *this* currency would take corrective action (but it was not obliged to: the obligation came into force only when a bilateral margin was reached).

The third element of the EMS was a set of measures of monetary cooperation and of monetary help to currencies under pressure.

21.1.1 The EMS and the Theory of Optimum Currency Areas

By and large, the functioning of the EMS can be divided into three periods: the first goes from its inception to January 1987; the second from January 1987 to August 1992; the third from September 1992 to the inception of the European Monetary Union.

The first period was characterized by frequent realignments, that enabled the system to work like an improved Bretton Woods system at the EEC level. In the second period there was a substantial exchange rate stability: the last true realignment took place in January 1987 (the realignment on the occasion of the definition of the new ECU basket in September 1989, and the realignment on the occasion of the entry of the lira in the narrow band in 1991 had a purely technical nature). In this period the EMS became a fairly stable currency area. In 1991 the Italian lira, that from the inception of the EMS had opted for the wider ($\pm 6\%$) margins, decided to enter the narrow ($\pm 2.25\%$) band. The British pound, that had remained out of the exchange-rate agreement from the beginning of the EMS, decided to enter it.

Several reasons are set forth to explain this stability. The most often quoted are the monetary discipline effect and the wish to pave the way for the European Monetary Union. The monetary discipline argument has already been treated in Sect. 20.2.2, point 3. As regards the second motive, it is clear that the movement towards a monetary union (see below, Sect. 21.2) requires that exchange rates be maintained fixed.

In the third period the system fell into a deep crisis. In September 1992 the lira first depreciated by 7 % and then had to abandon the exchange rate agreement to float. Also the pound left the exchange rate agreement to float. The contingent motives that are invoked to explain this crisis are various, and start from the negative result of the Danish referendum to ratify the Maastricht Treaty (on which see below, Sect. 21.2) and from the consequent confusion on the juridical effects that this refusal would have had on the process of European monetary unification (a new referendum in 1993 approved the Treaty). The uncertainty was increased by the announcement that also in France the Treaty would have been subjected to a referendum in September 1992, whose results were dubious (in fact, the result of this referendum gave only slightly more that 51 % in favour of the Maastricht Treaty). Another element that is often indicated is the refusal (due to internal motives of fight against inflation) of the German central bank to lower interest rates so as to discourage the speculative inflow of capital into Germany. In these conditions, international speculators attacked the structurally weaker currencies of the EMS, namely the lira and the pound.

The same motive (i.e., the refusal of the Bundesbank to lower interest rates) is set forth, together with the breakdown of the so called "Franco-German axis", to explain the still more serious crisis of 31st July–1st August 1993, when—after massive speculative attacks against the French franc and other currencies—it was decided to increase the margins of oscillation around bilateral parities from $\pm 2.25\%$ to $\pm 15\%$ (except for the Dutch Guilder-German DM rate; these two currencies maintained the old margin between themselves). It is clear that the notion itself of fixed exchange rates is devoid of meaning when applied to rates that can move within a 30 % band, so that it would be more correct to speak of a target zone (on which see Sect. 24.5.3). In fact, as a target zone with soft bands the EMS did remarkably well after 1993, and the actual exchange rates stayed well within the band except a few cases (Bartolini and Prati 1999).

Be it as it may, the above-mentioned motives are contingent motives, that only serve to cause the underlying disequilibria to explode. The crises of September 1992 and July 1993 clearly show the impossibility of maintaining fixed exchange rates among countries with divergent economic fundamentals, divergent monetary policies, and perfect capital mobility. An impossibility which is well known since the Bretton Woods era. From the more theoretical point of view, we refer the reader to what we have said in general in Sects. 11.4 (on perfect capital mobility under fixed exchange rates), in Chap. 16 (on speculation under fixed exchange rates and foreign exchange crises), and in Sect. 20.4 (on the common monetary policy prerequisite).

The crisis in the EMS brings us to the next question, namely whether the EMS was an optimum currency area.

If one adopts the traditional approach (see Sect. 20.2.1), then one can say that not all of the criteria for considering the EMS an optimum currency area are fulfilled: factor mobility was, at least in principle, present (it is provided for in the EEC statutes), and the criteria of openness and of product diversification are also satisfied. Financial integration was also present, thanks to the full liberalization of capital movements. Similarity in inflation rates did not exist, and equally absent was the integration of economic policies both within the area and with respect to the rest of the world. Within the area there was the problem of the *hierarchic* relations between strong and weak currencies; with respect to the rest of the world there was the problem of the lack of a common policy vis-à-vis outside currencies, especially the US dollar. Those who believe that the last criteria are the most important, hold the opinion that the EMS was not an optimum currency area, as shown by the crises mentioned above (for a contrary view see Bini Smaghi and Vori 1993).

The situation is more complicated if one adopts the cost-benefit approach. The costs listed in Sect. 20.2.2 did more or less occur in most member countries. The benefits were much less observed, partly because a true fixity of exchange rates was not realised (see what we said on the ECU in the previous sections), so that advantages (1) and (2) of the list in Sect. 20.2.2 did not seem to have come about. In point of fact the only true advantage, which was greatly emphasized by the supporters of the EMS, was number (3) of that list, and, to a lesser extent, number (4).

21.2   The Maastricht Treaty and the Gradual Approach to EMU

The European Council (composed of the Heads of State or Government of the countries forming the European Community), held in the Dutch town of Maastricht on 9–10 December 1991, approved a Treaty containing important modifications to the 1958 Treaty of Rome (which gave rise to the European Economic Community). The final version of the Maastricht Treaty (1992) was signed on 7th February 1992 in Maastricht. We have already touched upon the Danish and French referendum for the approval of this Treaty. Here we shall deal with the main innovations introduced by the Treaty as regards the European Union. The Treaty, adopting the strategy suggested by a report of the Committee for the Study of Economic and Monetary Union (commonly known as the Delors Report 1989), envisaged the movement to EMU in three stages the first of which, already begun in 1990, was further strengthened in the Treaty, that then went on to lay out the second and third stages in detail.

It is evident that the Maastricht Treaty adopted the gradualist approach to monetary union, as opposed to the so-called "shock therapy" approach, which consists of the sudden (or at least very quick) introduction of a *complete* monetary union, i.e., with a common currency.

Let us now consider the various stages in detail.

(I) The *first stage* (1990–1993) consisted of the following main measures:

    (I.1)   abolition of any restriction to capital movements, both within the EC and with respect to third countries. The latter movements may be subjected to restrictions but only if they threaten the functioning of the Union, and in any case cannot be imposed for more than a 6-month period.

    (I.2)   prohibition of financing the public deficit through the central bank.

(I.3)   adoption of programmes of long-run convergence, in particular as regards price stability and public finance issues.

(I.4)   adoption of the narrow band by all countries; avoidance of frequent realignments; prohibition of any modification of the composition of the ECU basket until its transformation in the single European currency.

(II) The *second stage* (1994–6/8) was aimed at securing the convergence of the economies of the EEC countries and to pave the way for the third stage. It contemplated the following main measures:

(II.1)   control of the public deficit and debt, with the aim of reducing the former to 3 % of GDP and the latter to 60 % of GDP.

(II.2)   constitution of the European Monetary Institute (EMI), with the task—amongst others—to coordinate the monetary policies and to pave the way for the European System of Central Banks (to come into being in the stage III). The EMI will dissolve on the starting day of stage three.

(II.3)   obligation on the part of the member countries to conform domestic legislative provisions concerning their central banks to the principles of the Union.

(II.4)   elimination of any automatic solidarity commitment to aid member countries faced with problems.

(III) The *third stage* was to begin on 1st January 1997 or 1st January 1999 at the latest. More precisely, the proviso was that at the end of 1996 the European Council would meet and decide whether the majority of member countries satisfied certain *convergence criteria*: in the affirmative case, the third stage would begin on 1st January 1997. In the negative case, the third stage would be postponed but not later than 1st January 1999, when it would in any case begin with the participation of those countries that met the convergence criteria. The other countries would obtain a temporary derogation and enter when they will satisfy the criteria.

The third stage has actually begun on 1st January 1999.

Let us now examine the convergence criteria, which are the following:

(a) an inflation rate (as measured by the rate of increase of the consumer price index) that does not exceed by more than 1.5 % points the rate of inflation of the three best performing countries (i.e., those having the three lowest inflation rates);

(b) a long-term nominal interest rate (measured on the basis of long-term government bonds) that does not exceed by more than 2 % points the average of those same three countries;

(c) an exchange rate that has respected the normal fluctuation margins in the last 2 years;

(d) a public deficit and debt that satisfies the criteria detailed under (II.1) above.

The measures contemplated in the third stage, that gave rise to the European Union proper, are the following:

(III.1)   creation of the European System of Central Banks (ESCB), which consists of the national central banks plus the European Central Bank (ECB). The ESCB has the task of taking all decisions concerning monetary policy, including the control of the money supply, with the *primary* objective of maintaining price stability and the *subordinate* (i.e., without prejudice to the objective of price stability) objective of supporting the general economic policies in the Union.

(III.2)   the bilateral exchange rates are irrevocably fixed, as well as those vis-à-vis the ECU, that will become a currency by full right.

(III.3)   the ECU will replace the single national currencies at the earliest possible date.

(III.4)   the Community will be entitled to apply appropriate sanctions against the countries which infringe the EC financial regulations after joining stage three.

(III.5)   the position of those countries that were granted a derogation (and were therefore temporarily left out of the Union: see above) will be reconsidered every 2 years.

These measures were subsequently integrated by

(a) the Madrid meeting of the European Council (December 1995), where it was established that the common European currency should be called *euro* rather than ECU, to emphasize that it was a brand new currency and not a basket currency like the ECU (for further details see below, Sect. 21.6);

(b) the Dublin summit in December 1996, where the so-called *stability and growth pact* was adopted on a proposal of the Commission and the Council of the economic and financial Ministers of the European Union (Ecofin), that in turn acted at the behest of Germany. Under this pact the 3 % deficit/GDP ratio is taken as an upper limit, since in normal circumstances the *ins* (i.e., the countries that have been admitted to stage III) should pursue a medium-term balanced budget or even a surplus. Several institutional mechanisms are introduced to ensure the respect of the 3 % upper limit, in particular the possibility of Ecofin to issue fines (up to 0.5 % of GDP in any one year) to members that, after having been admitted to the third phase, do not respect the 3 % deficit/GDP ratio.

21.2.1 Further Developments

The experiences during the first decade of EMU and the euro area crisis led to major changes to the original 1997 framework, including the 2005 reform, the 2011 Six Pack (five regulations and one directive), and the 2013 Two Pack (two regulations), as well as the Treaty on Stability, Coordination, and Governance of 2012 (TCSG, with the relevant articles referred to as the Fiscal Compact).

21.2.1.1 The 2005 Reform

The 2005 reform of the SGP aimed at making the rules more flexible by introducing country-specific medium-term objectives set in structural terms. The reform left the structure of the SGP in place, and did not alter the fundamental elements of the EU fiscal framework enshrined in the Treaty, such as the 3 % deficit and 60 % debt reference values. Within this framework, however, the reform introduced significant changes in the strengthening of budgetary surveillance and coordination of economic policies (the "preventive arm" of the Pact) and on the excessive deficit procedure (the "corrective arm").

Under the preventive arm, refinements include:

– The definition of the medium-term budgetary objective. Rather than being required to target "close to balance or in surplus" budgetary positions, each Member State now presents its own country-specific medium-term objective (MTO) in its stability or convergence programme, which is then assessed by the Council. These country-specific MTOs are differentiated and may diverge from a position of close to balance or in surplus. Members should provide a safety margin with respect to the 3 % of GDP reference value, ensuring rapid progress towards sustainability. For euro area and ERM II Member States, a range for country-specific MTOs, in cyclically adjusted terms and net of one-off and temporary measures, has been set between −1 % of GDP and "in balance or surplus".
– The adjustment path to the medium-term objective. Member States that have not achieved their MTOs are expected to take steps to do so over the cycle. To this end, euro area and ERM II Member States should, as a benchmark, pursue an annual adjustment in cyclically adjusted terms, net of one-off and temporary measures, of 0.5 % of GDP.

As for the corrective arm, the changes introduced more flexibility, in particular by relaxing, adding specificity to or clarifying the availability of escape clauses. Changes include:

– The definition of a "severe economic downturn". The benchmark for a severe economic downturn is now a negative annual real GDP growth rate or an accumulated loss of output during a protracted period relative to potential growth.
– Specification of "other relevant factors". Neither the Treaty nor the original SGP indicated what these other relevant factors might be. The reformed SGP now more explicitly spells out the relevant factors that should be taken into account. Regarding the medium-term economic position, these include, in particular, potential growth, the prevailing cyclical conditions, the implementation of the Lisbon Agenda, and policies to foster research and development and innovation.
– Extension of procedural deadlines. For example, for the correction of excessive deficits. The standard deadline for correcting an excessive deficit remains the "year following its identification unless there are special circumstances".

However, the consideration of whether there are special circumstances justifying an extension by 1 year should be taken into account.

- Unexpected adverse events and repeated recommendations or notices. The SGP reform clarified such matters by explicitly stating that if effective action has been taken in compliance with a recommendation under Article 104(7) or a notice under Article 104(9), and if "unexpected adverse economic events with major unfavorable consequences for government finances" occur after the adoption of the recommendation or notice, the Council may decide to issue a revised recommendation or notice, which may also extend the deadline for the correction of the excessive deficit by 1 year.
- Increasing the focus on debt and sustainability. The ECOFIN Council report of March 2005 also called for a strengthening of debt surveillance. The 2005 SGP reform did not introduce major changes in the area of governance.

21.2.1.2 The Six Pack 2011 Reform

The Six Pack reform of 2011 was designed to improve enforcement. The build-up of severe macroeconomic, financial and fiscal imbalances within the euro area, and the following sovereign debt crisis in several euro area countries called for a strong reinforcement of the EU economic governance framework. The 2011 reform of the SGP is known as the "six pack". These six pieces of law were designed to formalize and strengthen the EU's fiscal surveillance regime—the regime that failed to prevent the Eurozone countries from falling into a deep recession that threatened the monetary union itself. For the SGP's preventative arm major changes include: (a) establishing an annual cycle of economic monitoring called the European Semester. It empowers the Commission and Council to formulate guidelines for economic and employment policy, monitor their implementation, and conduct surveillance to prevent and correct broadly defined 'macroeconomic imbalances'; (b) reforming the SGP's corrective arm, which consists largely of the Excessive Deficit Procedure (EDP). It expands the EDP to focus not only on the Maastricht criterion for an excessive deficit, but also that for an excessive debt. The Commission decides if a Member State has broken or is at risk of breaking either or both rules, with the Council deciding if an excessive deficit or debt then exists. Together, they make recommendations to the Member States, which come with the risk of penalties—including changes in European Investment Bank lending policy, non-interest bearing deposits, and fines; (c) enforcing both arms of the SGP, specifying lodgements and fines for non-compliance with both the corrective and the preventative arms and penalties for falsifying statistics; (d) defining the preventative and corrective arms of a new Macroeconomic Imbalance Procedure (MIP), that allows the Commission and Council conduct much broader fiscal surveillance that is not just limited to the Maastricht criteria. Imbalances are defined as 'any trend giving rise to macroeconomic developments which are adversely affecting, or have the potential adversely to affect, the proper functioning of the economy of a Member State or of the economic and monetary union, or of the Union as a whole'. Like the SGP, the Commission monitors and formulates recommendations under the preventative arm, creating a 'scoreboard' of economic indicators. Finally, the Council requires

EU Member States (except the UK) to adopt national fiscal rules that support compliance with the Maastricht reference values. These national rules must not only specify the target, but also outline the procedure for monitoring compliance and the consequences of failing to comply.

21.2.1.3 The 2012 Treaty on Stability, Coordination and Governance in the Economic and Monetary Union (TSCG)

On 2 March 2012, the Heads of State or Government of all EU Member States, with the exception of the United Kingdom and the Czech Republic, signed the Treaty on Stability, Coordination and Governance in the Economic and Monetary Union (TSCG), which includes the fiscal compact, a fostering of economic policy coordination and convergence as well as measures related to euro area governance. Key elements of the fiscal compact are: a balanced budget rule including an automatic correction mechanism, a strengthening of the excessive deficit procedure, and an ex ante reporting on public debt issuance plans. The contracting parties commit to implementing in their national legislation a fiscal rule which requires that general government budgets are in balance or in surplus. This fiscal rule is deemed to be respected if the annual structural balance is in line with the country-specific MTO—i.e. the MTO as defined in the preventive arm of the SGP—with a lower limit of a structural deficit of 0.5 % of GDP. A higher structural deficit of at most 1 % is only allowed if the government debt-to-GDP ratio is significantly below 60 % and risks to long-term fiscal sustainability are low.

The improvement of governance in the euro area is an important element of the TSCG. It mainly provides for a strengthening of Euro Summits as a forum for regular coordination as well as a strengthening of the role for the European and national parliaments.

21.2.1.4 The 2013 Two Pack

To consolidate the six pack reforms, the TSCG, and ad hoc arrangements created to deal with States facing worse economic situations, two more regulations were approved in 2013, known as the 'two pack'. One regulation applies to Eurozone states not in distress—those not operating under Economic Adjustment Programmes. It adds a common timeline to the European Semester, strengthening the cycle of monitoring and policy recommendations carried out by the Commission. The other half of the two-pack, applies to Eurozone states in distress—those receiving financial aid. It essentially formalizes and regularizes the procedure for the future.

Taken together, these legislative changes and policy decisions represent the most comprehensive set of governance reforms at the European level since the introduction of the single currency, leading to a substantial reinforcement of the mutual surveillance framework.

In 2015, the SGP was made more flexible to encourage investment and structural reforms, and to deal with the economic cycle. The revised guidance states the margin of interpretation, which is left to the Commission, in line with the rules of the Pact, without modifying existing legislation. In particular, the Commission clarifies how three specific policy dimensions can best be taken into account in applying the rules. These relate to: (a) investment, in particular as regards the establishment of a new European Fund for Strategic Investments as part of the Investment Plan for Europe; (b) structural reforms; and (c) cyclical conditions.

21.2.1.5 Europe 2020

In 2010 was launched the "Europe 2020" 10-year jobs and growth strategy to create the conditions for smart, sustainable and inclusive growth. Five headline targets have been agreed for the EU to achieve by the end of 2020: employment; research and development; climate/energy; education; social inclusion and poverty reduction. Progress towards the Europe 2020 targets is encouraged and monitored throughout the European Semester, the EU's yearly cycle of economic and budgetary coordination. All Member States have committed to achieving the Europe 2020 targets and have translated them into national targets.

21.2.2 Fiscal Governance

Fiscal governance in the European Union relies on rule-based policy coordination. Let discuss how the European Fiscal Governance has changed from the original 1997 Pact to the current reform by stressing changes intervening in the mixture of "hard law" and "soft law" rules. Hard law corresponds to a situation in which hard obligation (a legally binding rule) and hard enforcement (judicial control) are connected. Soft law refers to those norms situated in-between hard law and non-legal norms. In the EU hard law encompasses both treaty provisions and regulations, directives and decisions. Soft law comprises: (1) nonbinding rules, which define specific objectives and foresee specific enforcement mechanisms aimed at framing their implementation, and (2) binding rules that do not fall under the control of the European Court of Justice. The EMU as established by the Maastricht Treaty comprised hard law (monetary policy) as well as the two kinds of soft law (nonbinding rules in the field of economic coordination, and binding rules with a soft control mechanism in the field of fiscal policy).

Furthermore, it is possible distinguish between hard law and soft law following the framework introduced by Abbott et al. (2000). The level of being binding is established by using the continuum with which the degree of obligation, precision, and delegation of a rule is determined. The level of obligation refers to the extent to which the member state is legally bound by a specific rule and whether its behavior is subject to scrutiny. Indicators for obligation are active monitoring by third parties and the ability to impose sanctions. A high level of precision means that a rule or norm is unambiguously defined, refraining from vague terms that offer room for various interpretations. Indicators for precision are whether the Council refers to specific national policy programs and whether precise deadlines indicate the terms of implementation. The extent to which delegation takes place is assessed by viewing whether third parties, in this case the European Commission or the Council, have the authority to implement, interpret and apply the rules and may propose new rules. In its most extreme form, hard law has maximum levels at all three properties.

From the original 1997 Pact to the current reforms, the obligation dimension has been progressively weakened while the other two dimensions have acquired more and more strength. For example, in the 2005 Reform the "obligation" dimension was softened: the budget deficit limit was redefined to focus on longer-term deficits rather than single years and on deficits that occur because of slow growth as well as outright recessions. In the same spirit, a country would be judged more favorably if its debt to GDP ratio is low and if it is making progress in dealing with its long-term reforms (for example, pension system). As for the delegation and the precision, the 2005 Reform seems to have moved towards a strengthening of these two dimensions. This evolution might be regarded as a rational process pointing to give more space to country specific contingencies without jeopardizing the Pact credibility. Setting clearer and more precise rules—even in a context in which the application of sanctions (hard law obligations) is rarely enforced—might help Member States to develop reputation and market discipline.

21.3   The Institutional Aspects

The ESCB consists of the newly established ECB and the existing national central banks (NCBs) of the EU. However, the NCBs of the member countries that do not participate in the euro area are members of the ESCB with a special status, namely they do not take part in the decision-making regarding the single monetary policy for the euro area and the implementation of such decisions.

The ECB is managed by an *Executive Board* consisting of the President, the Vice-President, and four more members, and by a *Governing Council* consisting of the governors of the central banks of the countries that entered the third phase (see the previous section) plus the six components of the Executive Board. There also exists a *General Council*, which comprises the President and Vice-President of the ECB and the governors of all the NCBs. The components of the Executive Board are appointed by common accord of the Heads of state or government for a period of 8 years and cannot be reappointed. This relatively long term in office and the non-renewability of the appointment are directed at insulating monetary policy makers from political pressure. In fact, the principle of independence of the central bank has been fully accepted: the ESCB decides in autonomy, namely without either seeking or taking instructions from national governments or supranational EU authorities. The Community authorities and the national governments agree to respect this principle and not to seek to influence the members of the decision making bodies of the ESCB.

The Governing Council's main responsibilities are:

(a) to adopt the guidelines and make the decisions necessary to ensure the performance of the tasks entrusted to the ESCB;

(b) to formulate the monetary policy of the Community, including, as appropriate, decisions relating to intermediate monetary objectives, key interest rates and the supply of reserves in the ESCB, and to establish the necessary guidelines for their implementation.

The Executive Board has the following main responsibilities:

(i) to implement monetary policy in accordance with the guidelines and decisions laid down by the Governing Council of the ECB and, in doing so, to give the necessary instructions to the NCBs;
(ii) to execute those powers which have been delegated to it by the Governing Council of the ECB. The General Council has some minor responsibilities: it contributes to the collection of statistical information, to the preparation of the ECB's quarterly and annual reports, to the preparations for irrevocably fixing the exchange rates of the currencies of the member countries that were granted a derogation in phase III (see the previous section).

As we said, the ESCB is responsible for all monetary policy decisions, with the primary objective of price stability and the subordinate objective of giving support to the economic policy of the Union. It also has the task of carrying out intervention in foreign exchange markets, holding and managing the official international reserves of the member countries, promoting the orderly functioning of the payment system. Besides, the ESCB shall contribute to the smooth conduct of the prudential supervision activity of the single national authorities over credit institutions.

Can we say that—once adopted the single European currency—the ESCB will function like a true central bank of the Union?

The answer to this question requires the examination of the functions of a central bank. If one believes that, in addition to conducting monetary policy, a central bank should also have the power of surveillance over the banking system, then the answer is clearly in the negative. The closest that one finds in the Treaty is article 105(5): "The ESCB shall contribute to the smooth conduct of policies pursued by the competent authorities relating to the prudential supervision of credit institutions and the stability of the financial system". But of course to "contribute" means that the supervisory power remains with the single national authorities. Actually, under article 105(6) it is possible to confer upon the ECB specific tasks concerning the prudential supervision, but only through an unanimous deliberation of the European Council acting "on a proposal from the Commission and after consulting the ECB and after receiving the assent of the European Parliament".

However, prudential supervision and monetary policy do not necessarily go hand in hand. In fact, a moot question is whether prudential supervision should be assigned to the same institution (the central bank) that is responsible for monetary policy, or to a separate agency (see Goodhart and Schoenmaker 1995; Peek et al. 1999). One reason for separation is that price stability might conflict with prudential supervision. Price stability, in fact, might require high interest rates, which might conflict with the wish of keeping interest rates low so as to help banks' debtors in avoiding default (which could weaken the balance sheets of banks). On the other hand, monetary policy measures would gain additional force if the central bank can influence bank policy through regulatory pressure. In practice, in a sample of 167 countries, bank supervision is conducted by the central bank in over 60 %; however, in the Western hemisphere the percentage is only 50 % (Tuya and Zamalloa 1994; Prati and Schinasi 1999).

Whichever solution is adopted, centralized prudential supervision in a monetary union with complete financial integration appears necessary, as it will be impossible to contain possible banking and financial crises within national boundaries. The ECB does not possess this power, and an EC agency endowed with the prudential supervision power has not been contemplated in the Treaty, with the result that the Union shall have a single currency and a centralized monetary policy, but no centralized prudential supervision. The lack of a centralized agency possessing powers of surveillance and bankingregulation might be a problem (Obstfeld 1998b, Sect. 5; Prati and Schinasi 1999).

21.4   The Maastricht Criteria

The aim of these criteria was clearly to prevent the destabilisation of the Union by the premature admission of countries whose economic fundamentals are not compatible with a permanently fixed exchange rate. Each of these criteria can be (and has been) criticized on economic grounds: see, for example, Begg et al. (1991), Eichengreen (1993), and Deissenberg et al. (1997). There is particularly severe criticism on the criteria concerning public deficit and debt (see, for example, Bean 1992, Buiter et al. 1993), and it has even be said that "The two numerical fiscal criteria of the Maastricht treaty make no sense and should be jettisoned" (Buiter 1997; page 24). However, they have been further emphasized by the EU institutions through the *stability and growth pact* (see above).

Thus there seems to be a contrast between economists and politicians. However, before concluding that politicians ignore (as they often do) the suggestions of economic theory, a closer look at the Maastricht criteria has to be taken, to determine whether they are at least qualitatively consistent with economic theory.

Let us begin with the *inflation* criterion (inflation rate not higher than the inflation rate of the three most virtuous countries plus 1.5 % points). Now, from Chap. 7 we know that terms-of-trade modifications due to inflation differentials are a cause of trade-balance disequilibria; furthermore, taking PPP as valid in the long run, stability of exchange rates requires identical inflation rates (see Sect. 15.1). Similarity in inflation rates is a very reasonable criterion contemplated in the traditional theory of optimum currency areas (see Sect. 20.2.1).

The *interest-rate* criterion (long-run interest rate not exceeding that of the three most virtuous countries plus 2 % points) is also obvious if one recalls that under perfect capital mobility and permanently fixed exchange rates (more so under a common currency) UIP must hold, hence $i_h = i_f + \delta$, where δ is a possible risk premium (see Sect. 4.3). The 2 % points should account for possible risk premia among government bonds issued in different countries of the European Union.

The *deficit* and *debt* criteria are not to be seen independently, as they are related by simple mathematical relations. Let g, D, Y respectively denote the budget deficit, the stock of public debt, and GDP, all in nominal terms. We now seek the conditions under which the debt to GDP ratio (D/Y) is non-increasing (sometimes called the "*sustainability*" condition of public finance). Since a fraction remains constant (decreases) when the numerator changes in the same proportion as (proportionally less than) the denominator, it follows that $\Delta (D/Y) \leq 0$ is equivalent to

$$\frac{\Delta D}{D} \leq \frac{\Delta Y}{Y},$$

hence multiplying through by D/Y,

$$\frac{\Delta D}{Y} \leq \frac{\Delta Y}{Y} \frac{D}{Y}.$$

If we remember that $g = \Delta D$ due to the prohibition of financing the public deficit by issuing money, we finally have

$$\frac{g}{Y} \leq \frac{\Delta Y}{Y} \frac{D}{Y}. \tag{21.1}$$

This inequality determines a zone of sustainability, whose boundary (constant debt/GDP ratio) is defined by

$$\frac{g}{Y} = \frac{\Delta Y}{Y} b, \tag{21.2}$$

where b is the constant value of D/Y.

There are two ways of looking at these relations, one concerning *positive* economics, the other one *normative* economics. The former takes *actual* figures, and checks where the economy is situated. For example, in Italy in 1997 the data were $D/Y = 121.6\%, \Delta Y/Y = 4.1\%, g/Y = 2.7\%$, hence this country was well within the sustainability zone, since 2.7 % is lower than 121.6 % × 4.1 % = 4.98 %. This meant that the actual debt to GDP ratio was decreasing.

The second way fixes some numbers as normative values, which is the line followed in the Maastricht treaty, but of course these numbers should be mutually consistent according to relations (21.1) and (21.2), as an increasing debt-to-GDP ratio is ruled out. Using (21.2) it is easy to see that $g/Y = 3\%, D/Y = 60\%$, and $\Delta Y/Y = 5\%$ constitute a triplet of mutually consistent values on the boundary. But there is an infinite number of such triplets, hence we may wonder whether the numbers chosen have an economic rationale or come out of the blue sky (in this latter case we should agree that the two fiscal criteria make no economic sense).

Let us begin with the rate of growth of nominal GDP, $\Delta Y/Y$, that can be decomposed into

$$\frac{\Delta Y}{Y} = \frac{\Delta p}{p} + \frac{\Delta y}{y}, \tag{21.3}$$

where $\Delta p/p$ is the inflation rate and $\Delta y/y$ is the rate of growth of real GDP.

In a target-instrument policy framework, it is perfectly reasonable that policy makers assign "desired" values to real growth and inflation. For example 3 % and 2 % respectively, are plausible *desired* values, whence $\Delta Y/Y = 5\%$.

As regards b, oral tradition says that $b = 60\%$ simply came out of the fact that 60 % happened to be the EEC average when the Maastricht treaty was drafted. Given these values of b and $\Delta Y/Y$, the value of $g/Y = 3\%$ followed. Another oral tradition says that a public deficit was to be allowed only for public investment expenditures, whose value could be taken as 3 % of GDP, hence $g/Y = 3\%$. Given $\Delta Y/Y = 5\%$, the value of $b = 60\%$ followed.

We can however argue in favour of a less casual explanation. We know that equilibrium on the real market *consistent with current-account equilibrium* requires

$$g + (I - S) = 0, \tag{21.4}$$

which follows from Chap. 6, Eq. (6.1), letting $CA = 0$. Thus we have

$$\frac{g}{Y} = \frac{S - I}{Y}, \tag{21.5}$$

namely the deficit/GDP ratio consistent with current-account equilibrium should equal the ratio of the excess of private saving over private investment to GDP.

Now, in mature industrialized economies the $(S - I)/Y$ ratio is becoming very low (in the United States, for example, it is around zero), and in several European countries it has a similarly downward trend. In EU-15 this ratio was around 3–4 % on average in 1991–1992. Hence $g/Y = 3\%$ was a perfectly reasonable value for European countries, from which $b = 60\%$ follows given the inflation and real-growth targets.

Up to now we have dealt with g as a whole, but we know that the budget deficit can be decomposed into interest payments on the public debt and the rest, which is called the *primary deficit*. Thus we have

$$g = iD + g_P, \tag{21.6}$$

where i is the nominal interest rate and g_P the primary deficit (let us remember that according to our convention, which descends from the definitions of the variables in the accounting matrix of Chap. 6, Sect. 6.1, a negative value of g_P means a primary surplus). Substitution of (21.6) into (21.1) yields

$$\frac{g_P}{Y} \leq \left(\frac{\Delta Y}{Y} - i\right)\frac{D}{Y}. \qquad (21.7)$$

Using (21.3) and the definition of real interest rate $i_R = i - \Delta p/p$, Eq. (21.7) can be rewritten as

$$\frac{g_P}{Y} \leq \left(\frac{\Delta y}{y} - i_R\right)\frac{D}{Y}. \qquad (21.8)$$

Equation (21.7) shows that sustainability requires a *primary surplus* as long as the rate of growth of nominal GDP is smaller than the nominal interest rate, while Eq. (21.8) states the same result in terms of real rates. Only when the nominal (real) rate of growth is higher than the nominal (real) interest rate would sustainability allow a primary deficit.

We have seen in the previous chapter, Sect. 20.4 that a common monetary policy is essential for the viability of a monetary union. This has been achieved by transferring the conduct of monetary policy in the hand of the ESCB and by forbidding monetary financing of the public deficit. We have also seen that a common fiscal policy, though desirable, is not so essential like the common monetary policy. The two fiscal criteria are an attempt to impart a minimum common amount of fiscal discipline.

21.5   The New Theory of Optimum Currency Areas and EMU

There is an ongoing debate on whether the European Monetary Union is an optimum currency area proposing an ex-post assessment in light of the predictions of the theory of OCA. Several weaknesses and limitations emerged over time as many OCA properties are difficult to measure unambiguously and by necessity are backward-looking (for a review see Mongelli 2008 and Pasimeni 2014). Many authors share the opinion of M. Feldstein that "a European monetary union would be an economic liability", "the economic consequences of EMU, if it does come to pass, are likely to be negative" (Feldstein 1997, p. 32 and 41). This opinion is based on the observation that the basic criteria for an optimum currency area, in particular high factor mobility and high flexibility of wages and prices (see Chap. 20, Sect. 20.2.1) are not satisfied.

This is undoubtedly true, but these criteria should be seen in the context of the degree of similarity in economic structure. In general, given that each member of a single-currency area cannot use the exchange-rate tool to cope with asymmetric shocks, the main possibilities are migration from low to high growth countries, wage flexibility, unilateral transfers from high to low income countries (Froehlich 1999). None of these seems to exist in EMU. However, it should be pointed out that predominantly asymmetric shocks do require high factor mobility and high price/wage flexibility for being absorbed *in areas with pronounced regional disparities* (such as the United States). These requirements are however much less important when *similarity in economic structure reduces the likelihood of asymmetric shocks.*

In effect, the analysis of the consequences of exogenous disturbances on the participating countries is one of the main areas of research in the so-called "new" theory of optimum currency areas (see Chap. 20, Sect. 20.2.3).

The empirical question is then whether the EU is indeed more homogeneous than the US. Bini Smaghi and Vori (1993) validate this thesis; Bayoumi and Eichengreen (1994), on the contrary, show that only the countries of the "German bloc" have really similar economic structures, but a further paper by the same authors (Bayoumi and Eichengreen 1997; p. 769) finds support for "the notion that EMU and the Single Market can constitute a virtuous, self-reinforcing circle". Helg et al. (1995) find that only Greece, Portugal and Ireland have productive structures out of line with the other EU countries, and Fatás (1997) finds that the increasing trade and monetary policy coordination within the EMS has reduced dishomogeneity. Trento (2000) examines the sources of cyclical fluctuations in industrial output in EU, and finds that these fluctuations are better explained at the country rather than at the industry level, which seems to imply that shocks hit member countries asymmetrically.

The paramount importance attributed to the symmetry or asymmetry of shocks has been challenged by Collignon (1999), who suggests to treat the choice of joining or leaving EMU as an *investment decision*, where investment is defined as the act of incurring an immediate cost in the expectation of future rewards. The immediate cost (transitional cost of moving from the domestic currency to the euro) is a *sunk cost.* The prospective benefits from sharing the single currency are uncertain, since they may vary over time according to changing policy preferences, economic shocks, union size. However, in the long run the benefits will be high if price stability is maintained in the union. Given this, temporary shocks and policy preferences volatility are no threat to the union's sustainability.

Unfortunately in the context of cost-benefit analysis it becomes difficult to make definitive statements, and we fully agree with Wyplosz when he writes "Assessing the costs and benefits of a monetary union quantitatively is both frustrating and useless. It is frustrating because, frankly, as economists we are unable to compute them with any precision, and we owe it to the profession to admit so in public. Our understanding of monetary and exchange rate policy is regrettably limited, and the lack of a precedent leaves us with more conjectures than certainties. Moreover, quantitative estimates are useless unless they are sized up against the costs and benefits of the relevant alternatives, which is equally beyond our current ability. The best that can be done in this situation is to gain an understanding of where the costs and benefits are likely to reside" (Wyplosz 1997, pp. 18–19). A similar point has been made byBayoumi et al. (1997; p. 85): "...research has advanced the state of knowledge but does not permit one to conclude with confidence that the benefits of monetary unification will exceed the costs".

In this respect, it is interesting to point out a usually neglected cost-benefit, due to the distribution of *seigniorage* in EMU. As noted by Sinn and Feist (1997; p. 666), some countries joining the EMU "will win more than others, because they will receive a better currency than the one they lose. A good currency is highly demanded as a medium of transactions and a store of value and its wide usage creates a substantial seigniorage wealth for the issuing country. With the introduction of the euro, national currencies will disappear and seigniorage wealth will be socialized". Some countries stand to gain, others to lose, depending on different scenarios as regards membership in the euro.

To evaluate the European monetary union in accordance with the cost-benefit approach it is necessary, as we have said in general, to give weights to the various costs and benefits (even the hope of an advantage can be given a weight, though in the probabilistic sense). The obvious consideration, from a European point of view, is an elementary one: nowadays (and still more in the future) no European country belonging to the EU is able to compete *single-handed* with the world economic giants (USA, Japan, China and possibly other countries or groups of countries). Only by merging into a single politic-economic entity may these European countries hope to deal on an equal footing with these giants and avoid becoming provinces of the empire(s). If this argument has a paramount weight in the policy maker's (or the reader's) preference functions, then even a very small probability will make advantage (4) in Sect. 20.2.2 preponderant with respect to the costs, which are by no means negligible.

We would also like to point out the important distinction between *ex ante* and *ex post* evaluations, a distinction widely used in the field of commercial integration but usually overlooked in the evaluation of monetary integration. Ricci (1995, 2008), showed that the degree of trade integration and the symmetry in business cycles cannot be considered separately, since both criteria are inter-related and *endogenous* to the process of monetary integration. This means that entry in EMU, possibly motivated by non-economic reasons, may give ex post results quite different from ex ante evaluations. This idea has been taken up by Frankel and Rose (1997), who find that greater trade integration historically has resulted in more highly synchronized cycles. Since one of the effects of EMU is certainly that of further stimulating trade integration, this may give rise to more highly correlated business cycles. "That is, a country is more likely to satisfy the criteria for entry in a currency union ex post than ex ante!" (Frankel and Rose 1997). Similar results have also been found by Fontagné and Freudenberg (1998). This is also called the "endogeneity of optimum currency area" effect, that is, the euro area may turn into an optimum currency area after the launch of monetary integration even if it wasn't before (i.e. the decision to start a monetary union has a self-fulfilling property). There are different mechanisms that can make the OCA-criteria endogenous: monetary union can effect trade flows and intensify trade integration, thus increasing the benefits of the monetary union; monetary integration leads to more intense financial integration reducing the costs of asymmetric shocks; monetary union affects the functioning of the labour markets and can potentially increase their flexibility, thereby reducing the costs of adjusting to asymmetric shocks in the union.

The Maastricht Treaty introduced the free movement of capital and so, the criterion of capital mobility can therefore be considered as fully satisfied by the

EMU. This growing financial integration is key to understand the development of the Eurozone (Obstfeld 2013). Labour mobility in the Eurozone has not reached the same extent as capital mobility, due to cultural and language barriers, and regulatory constraints. Many studies focused on intra-EMU labour mobility (Eichengreen 1991; Fenge and Weizsäcker 2009; OECD 2012; Kahanec 2013; EPC 2013). Recent analyses (EC 2013; EPC 2013; Dao et al. 2014) show that geographical mobility of workers as an immediate response to shocks in the EU is significantly increasing over time.

Alesina et al. (2003) found that the adoption of the Euro has been associated with an acceleration of the pace of structural reforms in the product market, increasing price flexibility. Furthermore, Verhelst and Van den Poel (2010) in a comparative study between Europe and the US, analyzed the levels of price rigidity in the two areas and showed similar levels of price rigidity in the two areas (see also Dhyne et al. 2006). Wages are, in general, more rigid than prices, but it seems that in the Eurozone they are following the same dynamics. The decline in real wage rigidity in Europe has been showed by Goette et al. (2007). Alesina et al. (2008) argued that the run up to the Euro adoption seems to have been accompanied by wage moderation.

Several studies have focused on the effects of a monetary union on trade. Rose (2000) and Frankel and Rose (1997) using a large international panel data sets, showed that membership in a currency union leads to a multiplication of trade. If trade within the union is mostly intra-industry, integration in the goods market increases the similarities in union members' production structures and their exposure to similar shocks, de facto bringing the currency union closer to being optimal (Frankel and Rose 1998; Alesina et al. 2003; Barro and Tenreyro 2007). The findings from these exercises should not be viewed as conclusive tests of the OCA criteria. It is ex post synchronization of shocks that matters for optimality.

In conclusion it should be observed that the creation of a monetary union is not merely an economic fact, but is also (or mainly?) an expression of political will. Unless (or maybe even if) economic theory and empirical analysis conclusively show that this creation is a huge loss, the union will go on. In the case of the European monetary union neither economic theory nor empirical research have reached unambiguous results. Hence this union will go on if the political will continues, exactly as happened with the European commercial integration (begun with the 1958 Treaty of Rome).

21.6   The Euro and the Dollar

According to the Maastricht treaty the ECU was to become the (future) single European currency. In the Madrid meeting of the European Council (December 1995) it was decided that the future European single currency should be called *euro* instead of the generic term ECU, that from 1st January 1999 the official ECU basket would cease to exist, and the euro would become a non-circulating currency in its own right, that it would begin to circulate together with national currencies as legal tender from 1st January 2002, totally replacing national currencies from 1st July

2002 at the latest. In the case of contracts denominated in ECUs substitution by the Euro has been at the rate one to one.

At the Dublin Summit (December 1996) the Heads of State or Government ascertained that the Maastricht criteria were not satisfied by a majority of the fifteen Member States, hence the third step (see Sect. 21.2) could not begin on 1st January 1997.

On 1st May 1998 the Council of the economic and financial Ministers of the European Union (Ecofin) reviewed the situation of the member states as regards the Maastricht criteria and identified the member states that satisfied the conditions for the introduction of the euro; these countries (called the *"participating countries"*) were, so to speak, the "founding members" of the euro. On 2nd May the recommendation of Ecofin was ratified by the Council of the Heads of State and Government of the EU and by the European Parliament. The eleven founding members turned out to be Belgium, Germany, Spain, France, Ireland, Italy, Luxembourg, the Netherlands, Austria, Portugal and Finland (on 1st January 2001 Greece has been admitted). It is important to note that, as regards the two fiscal criteria, the one concerning the deficit/GDP ratio was applied *strictly*, while the criterion of the debt/GDP ratio was interpreted *"dynamically"*, in the sense that a country to qualify should have shown a consistently decreasing trend towards the 60 % reference value, even if the current debt/GDP ratio was actually higher (this was the case, for example, of Italy, where in the last few years this ratio had been on the decrease, but in 1997 was still 121.6 %).

On 3rd May 1998 Ecofin also established the fixed bilateral exchange rates among the participating countries, valid in the transitional period May-December 1998, when the euro did not yet exist. This gave rise to some fears (see for example De Grauwe and Spaventa 1997; Apel 1998, Chap. 5; Obstfeld 1998a) concerning possible destabilising speculation before the conversion rates of the various currencies into the euro after this transitional period were established, or possible withdrawals by prospective EMU members, but nothing happened.

In addition to the future role of the euro as an international currency (see McCauley 1997; Hartmann 1998; Mundell 1998), a problem that has attracted attention is the behaviour of the euro *vis-à-vis the dollar* and, in particular, the euro/dollar exchange-rate variability. The question is: will the euro/dollar exchange rate be more or less variable than a pre-euro comparator basket of the currencies of the participating countries? According to some authors, e.g. Kenen (1995), Bergsten (1997), Benassy-Quéré et al. (1997), the creation of the euro will lead the participating countries to attach a lower weight to exchange-rate stability in their policy reaction functions, a kind of "benign neglect" such as the one attributed to the US authorities as regards the dollar during the Bretton Woods era (see Sect. 23.4), and hence to eliminate or reduce one of the EU's main interests in international cooperation in managing exchange rates (on international policy cooperation see Sect. 24.2). In an optimizing framework it is then easy to show that a lower weight attached to exchange-rate variability will result in less intervention to manage the exchange rate and hence in higher exchange-rate variability.

This problem has been examined in depth by Ricci and Isard (1998), who show that optimizing economic policies do not necessarily imply that EMU will lead to greater exchange-rate variability.

The final problem remains of how to determine the dollar/euro (real) equilibrium exchange rate. In fact, expressions like "the dollar is strong", "the euro is weak", or "the euro is over/under-valued" are meaningless without a benchmark. A reliable theoretical framework for the calculation of a fundamental equilibrium exchange rate is of invaluable importance for the EU, not only as regards monetary matters (which are delegated to the ECB), but also as regards real matters such as trade and growth, which are the main responsibility of the Council, which may formulate general orientations for exchange rate policy (article 109(2) of the Maastricht Treaty on European Union). Trade and growth are clearly related to the EU's welfare.

There is no reason to believe that the exchange rate theories examined in Chap. 15 would perform better with the dollar/euro exchange rate than with the dollar/single currencies (where they have shown a poor performance). A better research program might be to concentrate on the real exchange rate in the context of the intertemporal approach according to the NATREX model (see Chap. 18, Sect. 18.3.3).

21.6.1   The International Role of a Currency

Like a domestic currency, an international currency performs the three functions of money—as a medium of exchange, a unit of account, and a store of value. It does so at two distinct levels, private and public transactions, and accordingly plays the following roles. As a medium of exchange, it is used by private agents to settle international economic transactions or by governments as a foreign exchange market intervention currency. As a unit of account, for instance, to invoice international transactions in goods and services while at the public level playing the role of an anchor to which governments peg their currencies. As a store of value, it is used as the currency of denomination of financial assets held by central banks, sovereign wealth funds and international private investors or as a reserve currency. Economies of scope suggest that these functions tend to reinforce each other.

What are the major costs and benefits of issuing an international currency? The main economic benefits consist of international seigniorage, macroeconomic flexibility, reduced currency mismatch problems, efficiency gains in financial intermediation and lowers transaction costs, lower exchange rate risks for firms. Among the main costs are the constraints on domestic monetary policy due to foreign holdings of the currency, which compromise the gain in macroeconomic flexibility through international currency issuance. Seigniorage, i.e. interest-free loans to the issuing central bank from non-residents who hold the international currency, is generated when foreigners hold the domestic currency, or financial claims denominated in it, in exchange for traded goods and services (Cohen 2012). It is also increasingly stressed that the ability to borrow internationally in the domestic currency reduces the problem of currency mismatches, from which many emerging market economies have suffered severely during financial crises including the recent global one (Dobson and Masson 2009; Genberg 2010). The international use of a currency is affected by the scale of the issuing country's transactional network in the world economy since the benefits of using a particular currency increase with the number of others using it due to network externalities. Yet network externalities also give rise to inertia and path dependency in the choice of use of an international currency, thus creating incumbency advantages for the dominant international currency (Chinn and Frankel 2007).

The roles played by the US dollar and the euro in the global monetary and financial system have remained broadly unaltered since 2007–2008, despite signs of a gradual ascent of the Chinese renminbi. This is also true, albeit to a somewhat lesser extent, during the phase of the European sovereign debt crisis in 2011–2012. This stability testifies to the importance of inertia effects.

The ECB publish annual reports on the international role of the euro. Reports present the main findings of the continued monitoring and analysis conducted by the Eurosystem as regards the development, determinants and implications of the use of the euro by non-euro area residents.

References

Abbott, K. W., Keohane, R. O, Moravcsik, A., Slaughter, A. M., & Snidal, D. (2000). The concept of legalization. *International Organization, 54*, 401–419.

Alesina A., Barro R. J., & Tenreyro, S. (2003). Optimal currency areas. In NBER Macroeconomics Annual 2002, 17, 301–356. MIT Press.

Alesina, A., Ardagna S., & Galasso, V. (2008). The Euro and structural reforms. NBER Working Paper No. 14479.

Apel, E. (1998). *European monetary integration: 1958–2002*. London: Routledge.

Barro, R., & Tenreyro, S. (2007). Economic effects of currency union. *Economic Inquiry, 45*, 1–23.

Bartolini, L., & Prati, A. (1999). Soft exchange rate bands and speculative attacks: Theory, and evidence from the ERM since August 1993. *Journal of International Economics, 49*, 1–29.

Bayoumi, T., & Eichengreen, B. (1994). *One money or many? Analyzing the prospects for monetary unification in various parts of the world*. Princeton Studies in International Finance No. 76, International Finance Section, Princeton University.

Bayoumi, T., & Eichengreen, B. (1997). Ever closer to heaven? An optimum-currency-area index for european countries. *European Economic Review, 41*, 761–770.

Bayoumi, T., Eichengreen, B., & Von Hagen, J. (1997). European monetary unification: Implications of research for policy, implications of policy for research. *Open Economies Review, 8*, 71–91.

Bean, C. (1992), Economic and monetary union in Europe. *Journal of Economic Perspectives, 6*, 31–52.

Begg, D. et al. (1991). *European monetary union-The macro issues*. London: CEPR.

Benassy-Quéré, A., Monjon, B., & Pisani-Ferry, J. (1997). The euro and exchange rate stability. In: P. R. Masson, T. H. Krueger, & B. G. Turtelboom (Eds.), *EMU and the international monetary system* (pp. 157–193). Washington (DC): International Monetary Fund.

Bergsten, F. (1997). The impact of the euro on exchange rates and international policy cooperation. In: P. R. Masson, T. H. Krueger, & B. G. Turtelboom (Eds.), *EMU and the international monetary system* (pp. 17–48). Washington (DC): International Monetary Fund.

Bini Smaghi, L., & Vori, S. (1993). Rating the EC as an optimal currency area: Is it worse than the US? In: R. O. O'Brian (Ed.), *Finance and the international economy* (Vol. 6, pp. 78–104) (The AMEX Bank Review prize essays). Oxford: Oxford University Press.

Buiter, W. H. (1997). The economic case for monetary union in the European union. In: C. Deissenberg, R. F. Owen, & D. Ulph (Eds.), *European economic integration*, special supplement to *Review of International Economics, 5*, issue 4, 10–35.

Buiter, W. H., Corsetti, G., & Roubini, N. (1993). Excessive deficits: Sense and nonsense in the treaty of Maastricht. *Economic Policy: A European Forum, 8*, 57–100.

Chinn, M., & Frankel, J. A. (2007). Will the Euro eventually surpass the dollar as leading international reserve currency. In: R. H. Clarida (Ed.), *G7 current account imbalances: sustainability and adjustment* (pp. 285–322). Chicago: The University of Chicago Press.

Cohen, B. J. (2012). The benefits and costs of an international currency: Getting the calculus right. *Open Economies Review, 23*, 13–31.

Collignon, S. (1999). European Monetary Union, Convergence and Sustainability. A Fresh Look at Optimum Currency Area Theory, in Various Authors (1999). *The sustainability report.* Supplement to *Economia Internazionale, LII*(1).

Dao, M., Furceri, D., & Loungani, P. (2014). Regional labor market adjustments in the United States and Europe. IMF Working Paper No. 26.

De Grauwe, P., & Spaventa, L. (1997). Setting conversion rates for the third stage of EMU. *Banca Nazionale del Lavoro Quarterly Review, L*(201), 131–146.

Deissenberg, C., Owen, R. F., & Ulph, D. (Eds.) (1997). *European economic integration*, special supplement to *Review of International Economics, 5*(4).

Delors Report (1989). *Report on economic and monetary union in the European community.* Brussels: Commission of the European Communities, 12 April.

Dhyne, E., Álvarez, L. J., Le Bihan, H., Veronese, G., Dias, D., Hoffmann, J., et al. (2006). Price changes in the Euro Area and the United States: Some stylised facts from individual consumer price data. *Journal of Economic Perspectives, 20*, 171–192.

Dobson, W., & Masson, R. (2009). Will the renminbi become a world currency? *China Economic Review, 20*, 124–135

Eichengreen, B. (1991). Is Europe an optimum currency area? NBER Working Paper No 3579.

Eichengreen, B. (1993). European monetary unification. *Journal of Economic Literature, 31*, 1321–1357.

European Commission. (2013). Employment and social situation in the EU. *Quarterly Review.* DG Employment, Social Affairs and Inclusion.

European Policy Center (EPC). (2013). Making progress towards the completion of the single European labour market. *Issue Paper* No. 75.

Fatás, A. (1997). EMU: countries or regions? Lessons from the EMS experience. *European Economic Review, 41*, 743–751.

Feldstein, M. (1997). The political economy of the European economic and monetary union: Political sources of an economic liability. *Journal of Economic Perspectives, 11*, 23–42.

Fenge, R., & Weizsäcker, J. von (2009). Public pension systems and distortions of intra-EU mobility: The lodge test. *Journal of Pension Economics and Finance, 9*, 262–279.

Fontagné, L., & Freudenberg, M. (1998). Endogenous (a)symmetric shocks in the monetary union. Université de Paris I and CEPII Working Paper.

Frankel, J. A., & Rose, A. K. (1997). Is EMU more justifiable ex post than ex ante? *European Economic Review, 41*, 753–760.

Frankel, J. A., & Rose, A. K. (1998). The endogeneity of the optimum currency area criteria. *The Economic Journal, 108*, 1009–1025.

Froehlich, H. -P. (1999). Wage behavior and the sustainability of EMU. In: Various Authors, *The Sustainability Report*, Supplement to *Economia Internazionale, LII*(1).

Genberg, H. (2010). The calculus of international currency use. *Central Banking, 20*, 63–68.

Goette L. F., Sunde U., & Bauer T. K. (2007). Wage Rigidity: Wage rigidity: Measurement, causes and consequences. *Economic Journal, 117*, 499–507.

Goodhart, C., & Schoenmaker, D. (1995). Should the functions of monetary policy and banking supervision be separated? *Oxford Economic Papers, 47*, 539–560.

Hartmann, P. (1998). *Currency competiton and foreign exchange markets: The dollar, the yen and the euro.* Cambridge (UK): Cambridge University Press.

Helg, R., Manasse, P., Monacelli, T., & Rovelli, R. (1995). How much (a)simmetry in Europe? Evidence from industrial sectors. *European Economic Review, 35,* 1017–1041.

Kahanec, M. (2013). Labor mobility in an enlarged European Union. In: A. F. Constant. & K. F. Zimmermann (Eds.), *International handbook on the economics of migration* (Chap. 7, pp. 137–152). Celtenham: Edward Elgar.

Kenen, P. B. (1995). *Economic and monetary union in Europe: Moving beyond Maastricht.* Cambridge (UK): Cambridge University Press.

Maastricht Treaty. (1992). *Treaty on European union.* Brussels: Office for Official Publications of the European Communities.

McCauley, R. N. (1997). *The euro and the dollar.* Princeton Essays in International Finance No. 205, International Finance Section, Princeton University.

Mongelli, F. P. (2008). European economic and monetary integration and the optimum currency area theory, DG Economic and Monetary Affairs, European Commission, Ecoomic Papers No. 302

Mundell, R. A. (1998). What the euro means for the dollar and the international monetary system. *Atlantic Economic Journal, 26,* 227–237.

Obstfeld, M. (1998a). A strategy for launching the euro. *European Economic Review, 42,* 975–1007.

Obstfeld, M. (1998b). *EMU: Ready or not?* Princeton Essays in International Finance No. 209, International Finance Section, Princeton University.

Obstfeld, M. (2013). Finance at center stage: some lessons of the euro crisis. European Economy: Economic Papers No 493.

OECD (2012). Economic Survey of the European Union. OECD: Paris.

Pasimeni, P. (2014). An optimum currency crisis. *The European Journal of Comparative Economics ,11,* 173–204.

Peek, J., Rosengreen, E., & Tootell, G. (1999). Is bank supervision central to central banks? *Quarterly Journal of Economics, 114,* 629–653.

Prati, A., & Schinasi, G. J. (1999). *Financial stability in European economic and monetary union.* Princeton Studies in International Finance No. 86, International Finance Section, Princeton University.

Ricci, L. (1995). Exchange rate regimes and location. International Economics Working Papers No. 291, University of Konstanz (SFB178), December; revised version circulated as International Monetary Fund Working Paper No. 97/69, June 1997.

Ricci, L. (2008). A model of an optimum currency area. *Economics, 2,* 1–33.

Ricci, L., & Isard, P. (1998). EMU, adjustment, and exchange rate variability. International Monetary Fund Working Paper No. 98/50, April 1998.

Rose, A. K. (2000). One market: Estimating the effect of common currencies on trade. *Economic Policy, 30,* 7–33.

Sinn, H. -W., & Feist, H. (1997). Eurowinners and eurolosers: The distribution of seignorage wealth in EMU. *European Journal of Political Economy, 13,* 665–689.

Trento, S. (2000). European monetary union and fiscal policy coordination. Unpublished dissertation, University of Rome La Sapienza, Faculty of Economics.

Tuya, J., & Zamalloa, L. (1994). Issues on placing bank supervision in the central bank. In: T. Balino & C. Cottarelli (Eds.), *Frameworks for monetary stability.* Washington (DC): IMF.

Verhelst B., & Van den Poel, D. (2010), Price rigidity in Europe and the US: A comparative analysis using scanner data. Ghent University Working Paper, No. 684.

Wyplosz, C. (1997). EMU: Why and how it might happen. *Journal of Economic Perspectives, 11,* 3–22.

Part VIII

Problems of the International Monetary (Non)System

Key Events in the Postwar International Monetary System

22

22.1 Introductory Remarks

Although a detailed description of the international monetary system in the period after the Second World War is outside the scope of the present work, a brief treatment of the key events is necessary to provide the minimum amount of information necessary to set the theoretical problems against the institutional and historical background (for detailed treatment we refer the reader, e.g., to Triffin 1968, Kenen et al. 1994, De Grauwe 1996, Eichengreen 1996).

As we have already mentioned in Sect. 3.2, the postwar international monetary system was reconstructed on the basis of the Bretton Woods agreements signed in 1944, which gave birth to the IMF (International Monetary Fund) and to the IBRD (International Bank for Reconstruction and Development, also called World Bank) and laid the foundations for international monetary cooperation.

The members of the IMF agreed to tie their currencies to gold (or to the US dollar, which was the same thing, as the dollar was then officially convertible in gold at the official price of $ 35 per ounce) by declaring a par value or parity, with the obligation of keeping the actual exchange rate within narrow margins ($\pm 1\,\%$) of this parity. The features of the system qualified it as a gold exchange standard, though a "limping" one (see Sect. 3.1) and mitigated by the possibility of altering the parity, as we shall see. However, for various reasons most countries used the facility for converting their dollar reserves into gold very moderately or not at all, so that the system became a *de facto* dollar standard. The *de facto* inconvertibility of the US dollar became a *de jure* one in 1971 (see below).

The countries participating in the IMF also agreed to change their currencies' parity solely in the case of fundamental disequilibrium, and even then with certain limitations: changes up to 10 % were discretional, whilst for greater changes the country had first to obtain the Fund's assent. For its part the IMF also had a support role for currencies in difficulty, by using its general resources to extend financial assistance.

© Springer-Verlag Berlin Heidelberg 2016
G. Gandolfo, *International Finance and Open-Economy Macroeconomics*,
Springer Texts in Business and Economics, DOI 10.1007/978-3-662-49862-0_22

Another element in the agreement was the undertaking made by each member to eliminate foreign exchange restrictions and restore the convertibility of its currency, after the necessary transitional period (generalized convertibility was restored at the end of the 1950s), so as to contribute to the establishment of a multilateral system of payments.

The obligation to maintain the par value (typical of fixed exchange rates), coupled with the possibility of changing it as described above, caused this regime to be dubbed an *adjustable peg* system. Obviously it was a compromise between (irrevocably) fixed exchange rates and flexible rates, and it is clear that the degree of closeness to one or the other end depended essentially on the interpretation given to the concept of "fundamental disequilibrium" (not defined in the Articles of Agreement of the Fund) and to the frequency of the possible parity changes. The interpretation implicitly adopted in practice was usually restrictive, in the sense that the par value had to be defended at all costs, and changed only when further defence was impossible. This, amongst other consequences, gave rise to the continuing application of restrictive domestic measures in the case of balance-of-payments difficulties and to destabilising speculative capital flows (see Sect. 16.2). It should also be noted that, whilst in the intention of the Bretton Woods agreements the burden of adjusting balance-of-payments disequilibria was to be equally shared by both deficit and surplus countries, in practice this burden fell mostly on the shoulders of deficit countries only. As a matter of fact, the so-called *scarce currency clause*— which gave the Fund the power formally to declare a members currency scarce, with several consequences, amongst which the automatic authorisation for any other member to impose limitations on the freedom of exchange operations in the scarce currency—was practically never applied. Thus the principle of joint responsibility of surplus and deficit countries for disequilibrium became a dead letter.

Among the main events in the working of the postwar international monetary system are the following (in chronological order):

(a) the restoration of the convertibility of the currencies of the major industrialized countries;
(b) the formation of the Eurodollar market;
(c) the creation of the Special Drawing Rights (SDR) within the Fund;
(d) the *de jure* inconvertibility into gold of the US dollar in 1971 and the subsequent abandonment of the par value by all major currencies, which adopted a managed float system (the collapse of the Bretton Woods system);
(e) the acquisition of enormous dollar surpluses by the oil exporting countries and the consequent "recycling problem";
(f) the demonetization of gold and the legalization of the float;
(g) the creation of the European Monetary System and the European Monetary Union;
(h) the international debt crisis;
(i) the Asian crisis.

Any choice of a limited number of events is inevitably arbitrary and may reflect the idiosyncrasies of the writer. While pleading guilty, we believe that most events listed above would be considered by most writers as key events in the international monetary system. A very brief treatment of each of these (or reference to other parts of the book, in which an event has already been previously treated) will now be given. Current events will be treated in Chap. 24.

22.2 Convertibility

The restoration of the convertibility of currencies, which is indispensable for a multilateral system of payments and trade, was one of the purposes of the IMF. "Convertibility" of a currency is not to be taken here as under the gold standard (convertibility into gold) nor as unlimited convertibility (which means that anybody—resident or nonresident—can freely convert domestic currency into foreign exchange) but as *nonresident convertibility* or convertibility of foreign-held balances of domestic currency. This means that a nonresident (as defined in Sect. 5.1), for example a foreign exporter, who has acquired a domestic currency balance, is entitled to freely obtain the conversion of this into foreign exchange, which may be his own currency or any other currency participating in the convertibility agreement. It is self-evident that, in the absence of convertibility, the nonresident is obliged to spend the domestic currency in the purchase of domestic goods, services, etc., or to transfer it to someone else (for example, an importer in his own country) who has to pay for the purchase of domestic goods, etc. This means a *bilateral* settlement of international exchanges which hampers trade; the fostering of world trade requires a *multilateral* system of payments which, in turn, requires nonresident convertibility.

It is important to point out that in the Articles of Agreement of the IMF, convertibility of the currencies into one another was provided for as regards current transactions, whilst controls over capital movements were not prohibited. The articles—in consideration of the abnormal conditions expected to prevail in the postwar period—also provided for the exemption from convertibility when the domestic situation of a country would have made convertibility impossible. As a matter of fact, the necessities of postwar reconstruction induced all countries (except the USA) to avail themselves of this proviso. It was not until December 1958 that the pound sterling and twelve other European currencies shed restrictions, which had the effect of securing convertibility, which was full for the pound and the D-mark, whilst the other currencies maintained a very limited part of the previous bilateral payment agreements. This move was soon followed by many other currencies, so that by 1961 generalized convertibility prevailed. This passage to convertibility and hence to a (limping) gold exchange standard (as we have said the US dollar was convertible into gold for central banks) rendered more evident the problem of international liquidity (see Chap. 23).

22.3   Eurodollars

The Eurodollar market was born in the fifties, when deposits denominated in US dollars were placed with European banks (in the sense of banks resident in Europe according to the residency criteria illustrated in Sect. 5.1), which used them to extend dollar loans. A European bank can also hold deposits and extend loans in currencies other than the dollar (and, of course, other than the currency of the country where the bank is resident), in which case we talk of *Euro-currencies*. Still more generally, the term *Xeno-currency* (and *Xeno-market*) has been suggested by Machlup (1972; page 120) to include deposits and loans denominated in currencies other than that of the country where the bank—whether European or non-European—is resident.

The "mystery story" of the origins of the Eurodollar market was briefly mentioned in Sect. 2.8; for further historical details see, for example, Johnston (1983).

Estimates of the size of the Euromarket are published by various sources; the best known are those of the Bank for International Settlements (BIS) of Basle. It is as well to warn the reader that the estimates are often revised.

The estimates are based on data provided by banks located in major European countries, in Canada, Japan, the US (international banking facilities only), and in the various offshore centres (the Bahamas, Bahrain, Cayman Islands, Hong Kong, Netherlands Antilles, Panama, Singapore, etc.). The greatest part of Xeno-currencies consists of US dollars: around 80 % of the total up to the 1980s; subsequently this percentage has been decreasing in favour of other currencies, mainly yens.

Thus "Euro-currency market" is to be understood in the broad sense, i.e. in the sense of "Xeno-currency market' in Machlup's terminology. Indeed, although European banks still account for the greatest part of the total, the presence and growing importance of the other international banking centres listed above, and the fact that not only dollars and European currencies, but also other currencies (for example yen) are involved, makes the terms Euromarket and Euro-currency somewhat misleading. In fact, one now also hears the terms "Asia-currency" and "Asia-currency market". Therefore the terms Xeno-currency and Xeno-currency market to indicate the whole market seem more to the point. However, it is still common practice to use "Euro-currency" and "Euro-currency market" in the broad sense; in any case, no danger of confusion exists as long as one clearly defines what one means. Finally, the difference between the gross and net size is that the latter excludes interbank liabilities within the market area.

For a theoretical treatment of the Euro-currency market see Sect. 23.6.

22.4   Special Drawing Rights

The creation of SDRs (*Special Drawing Rights*), which was decided—after many years of discussion and formulation of plans—at the 1967 Rio de Janeiro Annual Meeting of the IMF, is best seen in conjunction with the problem of international liquidity. "International liquidity" is usually taken to cover those financial assets available for use by a country's monetary authorities in meeting balance-of-payments needs and in intervening on the foreign exchange market. This concept leaves undefined the range of actual assets to be considered as international liquidity, which thus may be wider or narrower. Some writers suggest the inclusion in international liquidity not only of the reserve of international means of payment but also the borrowing capacity of individual central banks, so that international liquidity would be taken to mean the capacity of any central bank to meet its foreign obligations, and would thus become a still vaguer concept (Machlup 1964a,b).

Although it must be recognized that from the theoretical point of view there may be a difference between "international reserves" and "international liquidity", to avoid confusion we take these as synonymous: in fact, the definition of international liquidity that we have given above coincides with the concept of international reserves.

As regards the range of assets which are taken to be international reserves, the convention usually adopted is to limit this range to the following assets: reserves of gold and foreign exchange, reserve position in the IMF,[1] and—after 1970, the year in which the agreement of 1967 was put into practice—the SDRs; for the countries belonging to the EMS there were ECUs.

---

[1] The reserve position in the Fund is given by what was once called the "gold tranche position", plus the country's lending to the Fund, if any. The gold tranche position, after the demonetization of gold (see below), is now called the reserve tranche position (henceforth r.t.p.). To understand what the r.t.p. is, it should be remembered that any member country has to pay out a quota to the Fund; this payment is usually made partly in the member's currency (75 % of the quota) and partly in SDRs (once in gold, hence the earlier denomination). The amount of the member's currency owned by the Fund increases when the country buys another currency from the Fund in exchange for its own and, vice versa, decreases when another member buys from the Fund the currency of the country under consideration. Now, the r.t.p. of a country equals the total of the country's quota minus the Fund's holdings of this country's currency; thus it will be equal to, smaller or greater than, the SDR percentage of the quota, according as the Fund's holdings of the country's currency are equal to, greater or smaller than, the amount initially paid out by the country to the Fund. In the latter case (i.e. when the amount of a country's currency owned by the Fund is smaller than 75 % of the country's quota because other countries have purchased this currency from the Fund) it was customary to talk of "super gold (now reserve) tranche position". The drawing (i.e. the purchase of foreign currency from the Fund in exchange for domestic currency) is automatic—namely no assent on the part of the Fund is required—within the limits of the r.t.p., which gives it a status identical with that of foreign exchange reserves. When the r.t.p. falls to zero the automaticity disappears, and further drawings are conditional, i.e. require the Fund's assent.

A numerical example may help to clarify the notion of r.t.p.. Let us assume that, in the initial situation, the Fund owns 75 units of currency x and that country's x quota is 100 (measured in the currency of country x). The r.t.p. is $100 - 75 = 25$, i.e. the SDR part of the quota. If another country borrows 35 of x from the Fund, the Fund's holdings of x drop to 40, and country's x r.t.p. is now $100 - 40 = 60$. Conversely, if—starting from the same initial situation—country x sells 15 units of its currency to the Fund to purchase other currencies, the Fund's holdings of x rise to 90 and the r.t.p. of this country is $100 - 90 = 10$. It is thus clear that the r.t.p. of a country can vary between zero and 100 % of its quota (when the Fund's holdings of that country's currency have fallen to zero because other countries have bought it from the Fund).

Now, if we ignore the reserve position in the Fund (which is a small part of the total), and consider gold and foreign exchange, it must be observed that the determination of their amount depends on elements which are in fact arbitrary and unrelated to the needs of the international monetary system. The increase in the stock of gold depends on this metal being produced by only a few countries, and this production cannot easily be made to increase or decrease in accordance with these needs. As regards foreign exchange, most of it consists of US dollars, whose accumulation by every country (other than the USA) is obviously related to a deficit in the US balance of payments. But a situation of continuing deficits in the US balance of payments weakens the dollar. Hence the *Triffin dilemma* (so named because the first to clearly state it was Triffin 1960): to avoid a shortage in international liquidity, the United States would have to run balance-of-payments deficits, and this would undermine confidence in the dollar; on the other hand, the cessation of the US deficits to strengthen the dollar would create a liquidity shortage. The agreement to introduce SDRs reflects, amongst other things, an attempt at creating a new form of international reserve asset unrelated to any particular currency and intended to become a true *fiat money* issued by the IMF.

The value of the SDR was initially fixed in terms of gold (with the same content as the 1970 US dollar, namely 0.888671 grams of fine gold); in July 1974 it was transformed into a basket-currency (see Sect. 20.3). The currencies composing the basket were initially 16, then reduced to 5 in 1981 (US dollar, Deutsche mark, French franc, Japanese yen, pound sterling) and to 4 in 1999 (the weights were 45 % US dollar, 29 % euro, 15 % Japanese yen, 11 % pound sterling). In November 2015 the Executive Board decided the introduction of the Chinese renminbi: the respective weights of the U.S. dollar, euro, Chinese renminbi, Japanese yen, and pound sterling are 41.73 %, 30.93 %, 10.92 %, 8.33 %, and 8.09 %. These weights will be used to determine the amounts of each of the five currencies to be included in the new SDR valuation basket that will take effect on October 1, 2016.

It is important to note that the newly created SDRs are allocated by the IMF to member countries in proportion to their quotas and that they are indeed fiat money, though their use is subject to certain limitations.[2]

In order to make the SDR competitive with reserve currencies, it was decided to attribute an interest on holdings of it, depending on the short-term interest rates of the countries whose currencies compose the basket. Besides, the SDR has been given the role of numéraire of the international monetary system, in the sense that any declaration of a par value by a country to the Fund is to be made with reference

---

[2]Initially the use of SDRs to settle payment imbalances between member countries was subjected to a partial "reconstitution requirement", whereby members were obliged to maintain, over time, a minimum average level of SDRs holdings as a percentage of their net cumulative allocation. This was eliminated in April 1981; other limitations, however, still persist, for example as regards the maximum total amount of SDRs that a country is obliged to accept.

to the SDR and not to gold or dollars. Finally, after the demonetization of gold (see below, Sect. 22.7) the 25 % of all quota increases (and of quota subscriptions by new members) once to be paid in gold, is paid in SDRs (or in the currencies of other countries specified by the Fund, with the concurrence of the issuers, or in any combination of SDRs and these currencies).

The role of SDRs in international liquidity is however quantitatively irrelevant.

On the problem of international liquidity and in particular of the demand for international reserves, see Chap. 23. Here we only mention the plan for the creation of a substitution account within the IMF. This plan aimed at strengthening the role of the SDR and, at the same time, at solving the problem of not weakening the dollar should some countries wish to change the composition of their international reserves by selling dollars in exchange for other reserve currencies. The idea was that official holders of dollars could deposit them with the IMF in exchange for claims denominated in SDRs. But the plan was dropped both because of disagreement on how that should be done and because the dollar was getting stronger: the official holders wanted to keep their dollars, not turn them in for SDRs.

Another proposal is to issue SDRs in favour of countries that have been recently joining the international monetary system while owning negligible amounts of international reserves. For the most part these are Eastern-European countries belonging to the former communist bloc. The group of the most industrialised countries (G-7 or G-10) has acknowledged that these countries' lack of reserves is a problem that might jeopardize the stability of the international monetary system. According to some authors (for example Polak 1994, Sect. 3) a SDR allocation mechanism is much more suitable than other mechanisms.

22.5 Collapse of Bretton Woods

The events that are identified with the collapse of the Bretton Woods system have already been described in Sect. 17.3, to which we refer the reader. We only add that where the Jamaica Agreement of January 1976 contemplated the possibility of returning to a generalized regime of "stable but adjustable" par values, the reference was not to an adjustable peg of the Bretton Woods type, which has proved to be no longer viable, but to a regime which should ensure a better flexibility than the adjustable peg. Thus the problem of the choice of the appropriate exchange-rate regime is far from obsolete, and we refer the reader to Chap. 17 for a discussion of the various possible alternatives.

It is now as well to examine the causes of the collapse of the Bretton Woods regime. As we know, in this system the country at the centre of the system, or dominant country (the United States) guaranteed the convertibility of its currency into gold at a fixed price. The other participating countries, in turn, guaranteed the convertibility of their currencies into dollars at a fixed exchange rate. Hence the system was, in fact, a gold exchange standard of the limping type (since the convertibility of dollars into gold was limited to central banks). It should be noted that generalised convertibility was fully actuated only in 1959, so that the life span of the regime was a dozen years. There is no doubt that the Bretton Woods regime of convertibility and fixed exchange rates was close to a regime with a single world currency. And, in fact, it brought about great benefits, amongst which the impetus to the growth of international trade. We must then ask ourselves why it didn't last longer.

The answer to this question is generally based on the Triffin dilemma (see above, page 590). In a growing world with growing trade, there is a growing world demand for money for transaction purposes. In the system under consideration, this amounts to a growing demand for international reserves, that is for dollars, which can be acquired only by running balance-of-payments surpluses, excluding of course the US. In fact, given the international consistency condition (see Chap. 5, Sect. 5.2)

$$\sum_{i=1}^{n} B_i = 0,$$

the nth country (the US) has to incur a balance-of-payments deficit for the rest-of-the-world to acquire dollars, namely $B_n < 0$ for $\sum_{i=1}^{n-1} B_i > 0$. But this process poses the problem of the convertibility of dollars into gold. Since the gold stock owned by the US cannot grow at the same rate as the growth of dollars held by the rest-of the-world central banks, there is a loss of confidence in the ability of the US to guarantee convertibility. If, on the other hand, the United States brought their balance of payments into equilibrium, the international monetary system would suffer from a liquidity shortage, with the possible collapse of international trade. Hence the dilemma: if the US allow the increase in international liquidity through deficits in their balance of payments, the international monetary system is bound to collapse for a confidence crisis; if, on the other hand, they do not allow such an increase, the world is condemned to deflation.

In addition to the Triffin dilemma, two other concurring causes are brought into play: the rigidity of the system and the "seigniorage" problem. The *rigidity of the system* was due to the fact that the idea of "fixed but adjustable" exchange rates (*adjustable peg*) was interpreted, as observed in Sect. 3.2, in the restrictive sense. That is to say, the parity was to be defended by all means, and was to be changed only when any further defence turned out to be impossible. Hence the external adjustment could not come about through reasonably frequent exchange-rate variations, but through deflationary policies.

The *seigniorage* problem was related to the reserve-currency role of the dollar. This enabled the US to acquire long-term assets to carry out direct investment abroad in exchange for short-term assets (the dollars). These dollars were usually invested by the rest-of-the-world central banks in short-term US Treasury bills, that carried a relatively low interest and whose purchasing power was slowly but steadily eroded by US inflation. This problem was much felt by several countries (especially by France under De Gaulle) and, though not being an essential cause, certainly weakened the will to save the system when it came under pressure.

The Triffin dilemma is the commonly accepted explanation of the collapse of the Bretton Woods system. But there is an alternative explanation, put forth by De Grauwe (1996; Chap. 4). This explanation (hints of which can be found in Niehans 1978) is based on Gresham's law ("bad money drives out good money"). This law, originally stated for a bimetallic standard (gold-silver, etc.), can be applied to any monetary system based on the use of two moneys whose relative price (or conversion rate) is officially fixed by the authorities, who commit themselves to buying and selling the moneys at the official price. If one of the two currencies becomes relatively abundant, its price in the private market will tend to decrease: economic agents will then buy it there and sell it to the authorities at the official price, which is higher. In the same way they obtain the scarce currency from the authorities at a cheaper price than in the private market. This means that the scarce currency will go out of the monetary circuit to be used for non-monetary purposes (hoarding etc.). Only the abundant currency will remain in use in the monetary circuit.

Now, if we apply this law to the gold-dollar system, it can be seen that the increase in dollars was not matched by an increase in the gold stock, and that the official price was losing credibility. In fact, while the purchasing power of dollars (in terms of goods and services) was decreasing because of inflation, it remained fixed in terms of gold. Thus the so-called "gold pool", made up of the central banks of the most industrialised countries (see Sect. 22.7), that acted buying and selling gold in the private market with the aim of stabilizing its price at the official level, was faced with increasing gold demand from the private sector at the given official price ($ 35/oz.). The ensuing gold loss compelled the gold pool to discontinue any intervention and leave the private price free, maintaining the official price for transactions between central banks only. The August 1971 official declaration of inconvertibility of the dollar was just the *de jure* acknowledgment of a *de facto* situation, namely the functioning of the system as a *dollar standard*.

If one accepts this analysis of De Grauwe's, the corollary follows that, even if the Triffin dilemma were hypothetically solved, the working of Gresham's law would not make it advisable a "return to gold" (as some still advocate).

Whichever explanation is accepted for the collapse of Bretton Woods, the following fundamental point should be stressed: *to base the international monetary system on fixed exchange rates is tantamount to assuming that the world as whole is an optimum currency area*. If this assumption is not true (and it does not seem that the Bretton Woods system was an optimum currency area for all participants), then the system is bound to collapse.

22.6   Petrodollars

The repeated increases in oil prices[3] charged since October 1973 by the oil producing and exporting countries united in the OPEC cartel (Organization of Petroleum Exporting Countries), gave rise to serious balance-of-payments problems in the importing countries and to the accumulation of huge dollar balances (also called petrodollars) by these countries, as oil was paid in dollars. In 1974 the (flow) financial surplus of the OPEC countries was about $ 56 billion, which mirrored an equal overall deficit of the oil importing countries vis-à-vis OPEC countries. About two-thirds of this deficit concerned industrial countries, the remaining third the non-oil-producing developing countries (also called the "Fourth World"). The problem of financing the oil deficits was particularly acute given the suddenness and the amount of the price increases, and was solved by means of the so-called *recycling* process, through which the oil surpluses of the OPEC countries were lent back (indirectly) to the deficit countries. The recycling process operated mostly through market mechanisms, both international (the Eurodollar market) and national (the US and UK financial markets), and for the rest through *ad hoc* mechanisms. Since the OPEC countries invested their petrodollar surpluses in the Eurodollar market and in the US and UK financial markets, the deficit countries could borrow the dollars that they needed by applying to the Eurodollar market and to the US and UK financial markets. The *ad hoc* mechanisms concerned both bilateral agreements between an oil importing and an oil exporting country, and agreements brought into being by international organizations. The latter included the IMF's "oil facility" brought into being in 1974 (and discontinued in 1976) and financed by borrowing agreements between the IMF and other countries (mostly oil producing countries) with the aim of granting loans, with certain conditions, to countries facing balance-of-payments disequilibria due to oil deficits. The agreements also included borrowing facilities created within the EEC and OECD organisations.

The recycling process was of course a short-run solution as it did not solve the problem of the elimination of the oil surpluses and deficits. This can be examined in the context of the transfer problem (see Sects. 8.6, 9.1.3, 10.2.2.1), and in the context of the theory of cartels (see Gandolfo 2014, Sect. 10.6.3).

22.7   Demonetization of Gold

The problem of the function of gold in the international monetary system and of its price was finally solved by the Jamaica Agreement of 1976 (referred to above), which ruled the *demonetization* of gold, thus removing the privileged status that

---

[3]To set the problem into proper perspective it should be pointed out that these increases occurred after a long period of low oil prices, which, though stable in nominal terms, had actually been decreasing in real terms. Many people wonder therefore whether a more far sighted policy (on the part of all those concerned), of gradual price increases in the period before 1973, might not have avoided the problems created by the huge and sudden 1973 price increase.

gold had previously enjoyed, with the ultimate aim of making it like any other commodity.

The Bretton Woods Agreement gave gold a central place in the international monetary system, as was clear from the obligation on the part of members to pay out 25 % of their IMF quota in gold and to declare the par value of their currencies in terms of gold or, alternatively, in terms of the US dollar (which was the same thing, since the dollar was convertible into gold at the irrevocably fixed price of $ 35 per ounce). The maintenance of this official price of gold was an easy matter until the end of the fifties. But in 1960 problems began to arise because of the fact that the free market price of gold, quoted in some financial centres (mainly London), began to diverge from the official price, mostly because of speculative hoarding. To counter this, the central banks of eight countries, by the Basle Agreements of 1961 and 1962, constituted the so-called *Gold Pool*, which—by using the gold provided by the central banks themselves—had the task of intervening on the free market to stabilize the gold price, by buying (selling) gold when its price fell (rose) beyond certain limits with respect to the official price. In this way a single price of gold prevailed. But in 1968 the Gold Pool was discontinued, because of huge losses of gold due to increasing speculative pressures on the market, and the two-tier market for gold was established: the official market for the transactions between central banks, at the agreed price of $ 35 per ounce, and the free market, where the price was formed by the free interplay of supply and demand; the central banks agreed not to intervene on the free market.

The increase in the official price of gold to $ 38 and then to $ 42.44 (see Sect. 17.3) did not solve the problem of the gold-reserve freeze, due to the fact that no central bank was willing to meet its international payments by releasing gold at the official price when the market price was much higher. Partial solutions were found within the EEC in 1974, with the settlement in gold of payment imbalances at an official gold price related to the market price. However, the general problem of the function of gold remained, and there were two main schools of thought.

The first aimed at maintaining the monetary function of gold as the paramount international means of payment and suggested the revaluation of its official price to bring it close to the market price. The second aimed (a) at eliminating this function and (b) at replacing gold with an international fiduciary means of payments, as had happened in the individual national economic systems (where the link with gold had long been eliminated and replaced by fiduciary money). The second school prevailed within the IMF, and this is reflected in the Kingston (Jamaica) Agreement of January 1976 [in which, however, only point (a) is fully accepted]. As regards gold, the main provisos of this Agreement are the following:

(1) The elimination of the function of gold as numéraire of the system, i.e. as common denominator of par values of currencies and as the unit of value of the SDR.
(2) The abolition of the official price of gold.

(3) The abolition of any obligation on the part of member countries to make payments to the Fund in gold, in particular as regards quota increases and interest payments.

(4) The authorization for the Fund to return one sixth of its gold (25 million ounces), at the official price of grams 0.888671 per SDR, to the countries which were members as of 31st August 1975, in proportion to their quotas at the same date and in accordance with certain procedures. This return has been carried out.

(5) The authorization for the Fund to sell another sixth of its gold for the benefit of developing countries which were members as of 31st August 1975. A part of the profit (excess of the selling price over the old book value), proportional to these countries' quotas, was to be transferred directly to them, whilst the remaining part was to be used for subsidized loans through a newly created Trust Fund. These gold sales were to be carried out by the IMF, acting as agent of the Trust Fund, through auctions over a period of four years, as has in fact been done.

(6) The possibility for the Fund (the IMF) to release its residual stock of gold (but only after a decision taken each time by a majority of 85 %) through further returns or sales. In the latter case the profits are to be disposed of in prescribed ways.

(7) The obligation for the Fund to avoid stabilizing the gold price (or otherwise regulating the gold market) as a consequence of its sales of gold.

(8) The member countries pledged themselves to cooperate with the Fund and with one another in the management of reserve assets, with the aim of enhancing the role of the SDR as the principal reserve asset (this was the only concession made to point (b) above).

As a consequence of this Agreement, the countries of the Group of Ten stipulated an arrangement (to which other countries adhered) under which they may sell and buy gold at market-related prices, provided this does not involve pegging the price of gold or increasing the total stock of gold then held by the Fund and the Group of Ten.

In parallel to the demonetization[4] of gold the role of the SDR was strengthened (see above, Sect. 22.4).

Although gold has been demonetized, it is still owned by central banks, and included in their international reserves. However, the central banks intervene in the foreign exchange market and meet their country's international payment obligations by using foreign exchange and possibly other reserve assets (SDRs, euros), but not gold as such, which—after the demonetization—has to be "mobilized" before it can be used. The problem then arises of the value that has to be given to gold holdings by central banks. The "official" price of 35 SDRs per ounce is merely conventional, and

---

[4]It should be noted that here "demonetization"—in the sense of the elimination of the function of gold as common denominator of par values, etc., as described above—has a different meaning from the one it has in balance-of-payments accounting (see Sect. 5.1.3).

in practice each country adopts its own criteria, which in general relate the valuation in some way to the market price.

What is clear is that if several countries simultaneously were to realize their gold by selling it on the free market, the price would crash. This is so because the gold market—on which the price is formed in accordance with the law of supply and demand—could not stand the impact of huge gold sales; therefore, the mobilization of gold reserves is by no means an easy matter.

22.8   EMS and EMU

The European Monetary System and the European Monetary Union have been treated in Chap. 21, to which we refer the reader.

22.9   The International Debt Crisis

This term refers to the incapability of the governments of some developing countries (mainly in Latin America) to service their foreign debt, namely to pay the interest and/or repay the principal as scheduled. The beginning of this crisis is usually placed in August 1982, when the Mexican government informed the US Treasury, the IMF, and the creditor foreign banks that it could no longer service its foreign debt. Similar cases had already occurred previously, but the fact that the country concerned was Mexico, an oil producing country and with a debt of more than US $ 80 billion of the time, shook the international financial markets. The example of Mexico was soon followed by several other Latin-American countries.

The reasons why these countries had indebted themselves are clear (mainly the financing of ambitious development programs). The reasons for their difficulties are also clear: the hoped-for huge export increases with which to get the foreign exchange to service the debt did not materialize, partly because of the unfavourable international economic situation. But to create a debt situation there must be two parties, the debtor and the lender who supplies the funds. Hence we must examine the reasons why the main international banks had so easily granted huge amounts of credit to the countries under consideration. A widely shared thesis starts from the oil shocks (see above, Sect. 22.6). These had generated huge financial surpluses in the OPEC countries, that had deposited them with the main international banks in London and in the United States. The enormous funds received by these banks raised the issue of how they could profitably utilise the money. A good outlet was found to be the lending to less developed countries in the course of industrializing their economies (the Newly Industrializing Countries or NICs). The fact that the debt was incurred by governments or guaranteed by governments made the risk of default look fairly low. Paradoxically, it were the increasing prices of the export goods of these countries (hence increasing proceeds in foreign currency) rather than the industrialization to give confidence to the lending banks, so much so that about one half of the stock of debt outstanding in 1992 had been contracted in the previous two years.

The debt problem will be further examined in Sect. 24.3.

22.10 The Asian Crisis

The economies of South Korea, Thailand, Indonesia and Malaysia had for many years been growing at fast rates in real terms, ranging from 6.9 % in Indonesia to 8.4 % in Korea. The high growth from 1986 to 1996 was suddenly followed by a collapse of the real economy. The crisis began in early July 1997, with the Thai baht's devaluation, and spread into a virulent contagion—leaping from Thailand to South Korea, Indonesia, the Philippines, and Malaysia. It led to severe currency depreciations and to an economic recession. Hence the Asian crisis surprised virtually all observers and it is now seen as one of the most significant economic events in recent world history.

In January 1997 a Korean *chaebol* (a chaebol is a large conglomerate company with greatly diversified interests) went bankrupt, and in May 1997 the Thai baht was attacked by speculators; on July 2, 1997, the central bank of Thailand adopted a managed float, which entailed a 15–20 % devaluation of the baht. In the following months the crisis spread to neighbouring regions: on July 11 the Philippine peso was allowed to float against the dollar, and two days later it was the turn of the Malaysian ringgit. In August the Indonesian rupiah plunged and started floating. In a couple of months the crisis reached Taiwan, which also had to devalue. The Korean won collapsed at the beginning of November. In Japan, Yamaichi Securities closed its doors, generating doubts on the soundness of Japan's financial sector.

In August 1998 the partial repudiation by Russia of its public debt spread new troubles over emerging markets and financial centre. In January 1999 the epicenter of the financial earthquake moved to Brazil, which had to devalue the real by almost 30 %, Argentina and Turkey in 2001. In the meantime Thailand, Korea and Indonesia had adopted the IMF recommendations, obtained its help, and avoided complete collapse.

Three elements seem to explain the onset of the Asian crisis: the appreciation of real exchange rates; the rise in nominal and real interest rates; the fragility of the local banking and financial system (for a review of the literature see Dean 2001 and Williamson 2004). Furthermore, scholars contended that international financial liberalization, characterized by free and rapid mobility of short-term capital, played the central role in causing the crisis. Other ingredients that contributed to the crisis included pegged exchange rates that encouraged excessive unhedged foreign borrowing, inadequate reserve levels, and a lack of transparency—particularly about the true level of usable foreign exchange reserves. The capital inflows produced a high ratio of external debt and debt service obligations relative to export earnings. Doubts about the soundness of financial institutions and corporate firms quickly spread across national borders. This set off a vicious circle of capital outflows, plummeting exchange rates, and crippling balance sheet effects, so that the former capital inflows turned to outflows. Private demand collapsed and output in the most affected countries declined sharply. Stiglitz and Yusuf (2001) asserted that the danger associated with capital market liberalization was one of the most important lessons of the Asian crisis. Stein (2006) reviewed the Asian financial crisis from two related perspectives: whether the crisis was caused by a failure of the real exchange rate to be aligned with its fundamental determinants, and/or whether the crisis was precipitated by a divergence of the foreign debt from its optimal path. The results allowed him to assess that there were signs of financial distress before the crisis.

The mixture of these causes has prompted the elaboration of "third generation" models of currency crises and contagion, for which the refer the reader to Sects. 16.3.3 and 16.3.6.

A few additional words are called for, about the role of the IMF in handling the crisis (see Takatoshi 2007, for an overview). This is a topic hotly debated by economists and commentators. On the one hand there are those who argue that the IMF mishandled the crisis through too tough prescriptions (such as stringent monetary policies and financial sector reforms attached to the Fund's loan programs), turning a financial crisis into an economic and social one (Stiglitz 2000). On the other hand there are those who defend the IMF, pointing out that its policies worked (Dornbusch 1999) and helped to dampen the effects of the crisis. Probably the truth lies in between (Corden 1999). See also above, Sect. 16.3.3.1.1.

References

Corden, W. M. (1999). *The Asian crisis: Is there a way out?* Singapore: ISEAS.

Dean, J. (2001). East Asia through a glass darkly: Disparate lenses on the road to Damascus, in G. C. Harcourt, H. Lim, & U. K. Park, (Eds.), *Editing economics: Essays in honour of Mark Perlman*. Oxford: Routledge.

De Grauwe, P. (1996). *International money: Postwar trends and theories* (2nd ed.). Oxford: Oxford University Press.

Dornbusch, R. (1999). The IMF didn't fail. *Far Eastern Economic Review*, 2 December 1999, p. 28.

Eichengreen, B. (1996) *Globalizing capital: A history of the international monetary system*. Princeton: Princeton University Press.

Gandolfo, G. (2014). *International trade theory and policy* (2nd ed.). Berlin, Heidelberg, New York: Springer.

Johnston, R. B. (1983). *The economics of the euro-market: History, theory, and policy*. London: Macmillan.

Kenen, P. B., Papadia, F., & Saccomanni, F. (Eds.) (1994). *The international monetary system*. Cambridge (UK): Cambridge University Press.

Machlup, F. (1964a). *International payments, debt, and gold*. New York: Schribner's Sons.

Machlup, F. (1964b). The fuzzy concepts of liquidity: International and domestic, in F. Machlup (Ed.), *International payments, debt, and gold*. New York: Schribner's Sons.

Machlup, F. (1972). Euro-dollars, once again. *Banca Nazionale del Lavoro Quarterly Review, 25*, 119–137.

Niehans, J. (1978). *The theory of money*. Baltimore: Johns Hopkins University Press.

Polak, J. J. (1994). The international monetary issues of the Bretton Woods era: Are they still relevant? in: Kenen, P. B., Papadia, F., & Saccomanni, F. (Eds.) (1994). *The international monetary system* (pp. 19–34). Cambridge (UK): Cambridge University Press.

Stein, J. L. (2006). *Stochastic optimal control, international finance and debt crises*. Oxford: Oxford University Press.

Stiglitz, J. (2000). Insider's account of Asian economic crisis. *The New Republic*, 17 April 2000.

Stiglitz, J., & Yusuf, S. (2001). *Rethinking the East Asian miracle*. New York: Oxford University Press.

Takatoshi, I., (2007). Asian currency crisis and the International Monetary Fund, 10 years later: Overview. *Asian Economic Policy Review, 2*, 16–49.

Triffin, R. (1960). *Gold and the dollar crisis*. New Haven and London: Yale University Press.

Triffin, R. (1968). *Our international monetary system: Yesterday, today and tomorrow*. New York: Random House.

Williamson, J. (2004). The years of emerging market crises: A review of Feldstein. *Journal of Economic Literature, 37*, 822–837.

International Liquidity, the Demand for International Reserves, and Xeno-Markets

23

23.1 Introductory Remarks

As we have already mentioned above (Sect. 22.4), international liquidity and international reserves are often used as synonyms, although they are not. International reserves are financial assets representing liquid international purchasing power in the hands of the monetary authorities. International liquidity is a broader concept and also includes access to loans as well as the monetary authorities' ability to convert illiquid assets into liquid purchasing power through international asset markets. A further distinction is made between "owned" and "borrowed" reserves. The former, unlike the latter, have no offsetting foreign liability of the monetary authorities. There is a last distinction to be recalled: that between official and private international liquidity. The former only includes liquid international assets held by the monetary authorities, while the latter includes all the foreign liquid assets held by residents of the country.

In any case, the problem remains of what determines the need for, or the adequacy of, international liquidity. To this problem we now turn.

According to an old definition, international liquidity can be considered "adequate" when it allows the countries suffering from balance-of-payments deficits to finance these without having to undertake adjustment policies which are undesirable for the growth of their economies and international trade. This definition was rightly criticized because it inevitably leads one to say that international liquidity is always inadequate, as no reasonable amount of it will be sufficient to allow financing without adjusting balance-of-payments deficits (by way of more or less undesirable policies).

If we consider official international reserves, since the holders and users are the central banks of the various countries, the problem can be seen from the point of view of the central banks' demand for international reserves: international liquidity should then be considered adequate when its actual amount equals the amount which is *desired* in accordance with this demand function. This is quite different from

© Springer-Verlag Berlin Heidelberg 2016
G. Gandolfo, *International Finance and Open-Economy Macroeconomics*,
Springer Texts in Business and Economics, DOI 10.1007/978-3-662-49862-0_23

determining the *need* for international reserves on the basis of allegedly *objective* parameters (such as, for example, a particular ratio of reserves to imports, etc.) as was once customary, because a demand also implies subjective elements and may not be entirely justifiable on the basis of objective ones.

From this point of view it is possible to single out two main approaches to the problem, the *interpretative* or *descriptive approach* and the *optimizing* one. The former aims at finding, by way of theoretical and empirical studies, the determinants which explain the demand for international reserves which *actually* emanates from central banks. The latter approach aims at determining the *optimum* level of international reserves (in accordance with some criterion of optimality to be defined), irrespective of whether this optimum level is what is actually demanded. If we use a distinction widely used in general economics, we may say that the descriptive approach belongs to *positive* economics, the optimizing approach to *normative* economics, since an obvious suggestion deriving from the optimizing approach is that the actual level of international reserves should be brought to the optimum level when the two are different.

23.2 The Descriptive Approach

The basic idea underlying this approach is that central banks demand international reserves for motives analogous to those for which people demand money in accordance with traditional demand-for-money theory.

There is, firstly, the transactions motive: the lack of synchronization between international payments and receipts requires the availability of a stock of means of payment. It should be noted that the international payments imbalances in question are the regular and recurring (hence anticipated) ones such as those due to seasonal factors; some writers also include possible random fluctuations.

There is, secondly, the precautionary motive, according to which countries usually wish to hold balances to meet sudden and unexpected events which may cause a decline in receipts or an increase in payments (crop failures, abnormal changes in capital movements, exceptional drops in foreign demand or increases in foreign prices, political unrest, etc.).

The speculative motive is more difficult to make concrete; it is usually related to the fact that reserve assets—insofar as they give a yield—are an alternative use of national resources.

Once the motives have been determined, the problem arises of singling out the factors to include as independent variables in the demand-for-reserves function. Traditionally the main variable considered was the *period of financial covering of imports*, that is, how long the current flow of imports can be maintained if the existing stock of reserves are used for this purpose. This period can be calculated simply by taking the reserve/imports ratio; thus the demand for reserves is an increasing function of imports. It was however noted that the desired period, and so the desired reserve/imports ratio (hereafter called the reserve ratio), cannot be considered constant, but depends in its turn on a number of variables, amongst which (Flanders 1971; p. 24):

(1) The instability of exports: the higher this instability, the higher the reserve ratio.
(2) The efficiency of the private market in foreign exchange and foreign credit: the more efficient this market, the lower the reserve ratio.
(3) The opportunity cost of holding reserves (measured, for example, by the yield of productive investment forgone): the higher this cost, the lower the reserve ratio.
(4) The rate of return on reserves: the higher this return, the higher the reserve ratio.
(5) The variability of reserves (due to the variability in payment imbalances): the higher this variability, the higher the risk of remaining without reserves and the higher the reserve ratio.
(6) The willingness to change the exchange rate instead of financing payment deficits: insofar as these changes are effective in restoring external equilibrium, a greater willingness to use them implies a lower reserve ratio.
(7) The willingness to accept the cost of adjustment (by tools other than exchange-rate changes, for example restrictive domestic policies) of balance-of-payments deficits: the higher this willingness, the lower the need to finance disequilibria and the lower the reserve ratio.
(8) The existence of inventories of traded goods: if a country has sizeable inventories of traded goods (both exports and imports) the reserve ratio will be lower, as the country can use these goods as substitutes for reserves.
(9) The cost of international borrowing. An alternative to holding reserves is to borrow abroad when the need arises. The higher this cost (where cost is measured not only by interest rate, but also by terms of repayment, ease of obtaining the credit, etc.), the higher the reserve ratio.
(10) Income, which in monetary theory is often considered to be a determinant of the demand for money over and above the effect of the volume of transactions.

In any case—and this is an interesting result that we have already mentioned in Sect. 17.3.2—it does not seem that the passage from fixed to floating exchange rates has significantly reduced the demand for reserves. This result, which some writers have considered as surprising, is—in our opinion—to be considered totally plausible, if only one remembers that the existing regime of floating exchange rates is actually a managed float, where central banks intervene in the foreign exchange markets. In fact, as it seems that the criteria adopted to manage the float have required continuing, heavy interventions, it is by no means surprising that the reserve needs have remained high.

23.3 The Optimizing Approach

This approach aims at determining the optimum level of international reserves by way of a maximizing procedure, and has been developed along two main lines.

The first is that of *cost-benefit-analysis*, which entails beginning by singling out the costs and benefits of holding reserves; the optimum level of these will then be determined by maximizing the difference between benefits and costs.

The second line begins by finding a *social welfare function* (or an objective function of the policy maker), which has the reserve level among its arguments; this function is then maximized subject to all the constraints of the problem.

It is possible to make a further distinction based on the static or dynamic nature of the optimizing process; this distinction can be applied to both lines previously distinguished.

It should be noted that the descriptive approach must not be seen as totally distinct from the optimizing approach. As a matter of fact many explanatory variables considered by the former are again found in the latter, as for example when one singles out costs and benefits.

Let us now begin by briefly examining the cost-benefit line; in general, the optimum will be found by equating marginal costs and benefits. The first example of this kind of analysis applied to reserve holding is Heller's (1966) which, thanks to its simplicity, allows a clear illustration of the procedure.

The cost of holding reserves is an opportunity cost, given by the difference between the yield that the resources held in the form of reserves would have if employed productively at home (this yield is identified with the social yield on capital) and the yield on reserve assets. The latter is assumed to be lower that the former, so that the opportunity cost is positive. It is also assumed that the above yields are constant, so that the net marginal opportunity cost of reserves, denoted by i, is constant.

The benefits consist in the fact that the use of reserves enables the country to finance a possible balance-of-payments deficit instead of having to adjust. In the simple model used by Heller (fixed exchange rates, no capital movements, etc.) the only way of adjusting to an external deficit is to reduce national income to the point where the consequent reduction in imports eliminates the deficit. If we consider a deficit of D, an equal reduction in imports ($\triangle m$) will be required to eliminate it, and since $\triangle m = \mu \triangle y$, income will have to decrease by D/μ. Therefore, if we assume that the marginal propensity to import is constant, the marginal benefit will be $1/\mu$. This benefit, however, is not certain, as it will occur only in the case where a cumulative balance-of-payments deficit of a certain magnitude actually occurs. Thus we must multiply the marginal benefit by prob (R_j), the probability of the occurrence of j consecutive deficits of a size necessitating the use of the jth reserve unit. Thus the optimum condition turns out to be

$$\text{prob}(R_j)\frac{1}{\mu} = i. \tag{23.1}$$

To determine prob(R_j) we need to know the stochastic process underlying the changes in international reserves. If we assume that this process is symmetric, so that the probability of an increase of size h is equal to the probability of a decrease of the same size, we see that the probability of j successive deficits of size h such that

$$jh = R_j, \qquad (23.2)$$

i.e., such that the amount of reserves held by the country is exhausted, is, by the multiplicative theorem for probabilities,

$$\text{prob}(R_j) = (0.5)^j. \qquad (23.3)$$

Substitution in (23.1) yields

$$(0.5)^j = \mu i, \qquad (23.4)$$

from which

$$j = \frac{\log(\mu i)}{\log(0.5)}. \qquad (23.5)$$

Substituting (23.5) into (23.2) we obtain the optimum reserve level

$$\hat{R} = h \frac{\log(\mu i)}{\log(0.5)}.$$

It is self-evident that this is a very much simplified model, which is rendered more complex if one introduces the possibility of adjusting to external deficits not only by income decreases but also by exchange-rate variations, interest-rate changes, and so on; in addition, the monetary authorities can borrow abroad. Other refinements can be introduced by extending the time horizon into the future and going more deeply into the probabilistic aspect. However, the basic idea remains the same as that set forth by Heller; what changes is the determination of costs and benefits.

As regards the *maximization of the policy maker's utility* (or of the social welfare) *function*, the initial step is to single out the arguments of this function. The first studies (Clark 1970; Kelly 1970) assume that this objective function depends positively on the level and negatively on the variability of income. As regards the constraint, one must take into consideration the fact that random disturbances to the balance of payments can be offset either by financing, i.e. using reserves, or by adjusting, i.e. causing income variations (the reasoning is exactly like that made in the cost-benefit analysis above). In the former case there will be a lower income level due to the income sacrificed in holding reserves (i.e., the alternative return which these assets could yield), whilst in the second case there will be a higher income variability. It is thus possible to construct a trade-off between the level of income and its variability. By maximizing the objective function subject to this constraint one obtains the optimum point from which one can derive the optimum level of reserves. This maximization is carried out in a static context; a more satisfactory approach is to consider the problem in an intertemporal maximization framework (see the Appendix to this chapter).

These models also are very simplified: as regards the structure of the economy, the determination of the arguments of the objective function, the implicit assumption of the symmetry of the adjustment policies (on the contrary the pressure on the monetary authorities is quite different according to whether they face a deficit or a surplus, and it also is quite different according as they are indebted to abroad, as noted by Ben-Bassat and Gottlieb 1992). They have, however, the merit of showing the complementarity between reserve policy and the other policy instruments in achieving external and internal equilibrium.

23.4   Is International Liquidity Still a Problem?

A widespread opinion is that, given the current system of floating exchange rates and the huge increase in international asset markets, in particular xeno-markets, the problem of international liquidity, once prominent, is now less important if not irrelevant. With free international asset markets, a creditworthy country (just like a creditworthy firm) can borrow all the liquidity it needs. Although it is now obvious that, under conditions of capital mobility, the external constraint on a country comes from its creditworthiness rather than from its liquidity (Williamson 1994; Crockett 1994) , the opinion under examination cannot be fully shared.

Apart from the fact that not all countries in need of international liquidity are equally creditworthy (think of the East-European countries), a current problem is that of currency crises (see Sects. 16.3 and 22.10). Central banks of creditworthy countries may suddenly lose their creditworthiness due to contagion from outside, and have to face a currency crisis with their own reserves, which are by no means sufficient. Within a country a financial crisis can be coped with by the domestic lender of last resort. At the international level there is no lender of last resort capable of issuing an unlimited amount of international liquidity in an emergency, since the International Monetary Fund has to rely on its resources and cannot create liquidity.

Thus the problem of international liquidity reappears under the form of the problem of an international lender of last resort, on which see Fischer (1999).

23.4.1 Some Facts and Puzzles

Despite the diffusion of greater exchange rate flexibility and international capital markets integration, which were thought to reduce the need for reserves, recent years have witnessed a substantial increase in central banks' foreign exchange holdings. This phenomenon represents a puzzle for the traditional theoretical frameworks laid out above and raises doubts on their ability to account for the rapid accumulation of reserves, particularly in emerging Asia. The starting point of the build-up of international reserves reflects the growing exposure of emerging markets to financial turbulence after the Asian financial crisis. In 2009–2010, the IMF advanced the argument that excessive reserve accumulation was jeopardizing the stability of the international monetary system, and changed its perspective shifting from the longer-standing concerns about the risks from current account global imbalances to the risks posed by excessive reserve accumulation (IEO 2012). In this context, in 2011 the IMF developed a new stream of indicators to assess reserve adequacy. The new metric provided represents an improvement over the simple ratio-of-reserves-to-short-term debt indicator that had been traditionally used. The experience of past balance of payment crises, which were characterized by multiple market pressure channels, suggested the need for a metric encompassing a broad set of risks. The new set of indicators explicitly recognizes that drains on reserves can originate elsewhere than in the obligation to service short-term debt. Specifically, the new metric combines short-term debt, other (medium- and long-term debt and equity) portfolio liabilities, the stock of broad money, and exports in a composite proxy of potential foreign exchange pressure. The relative weights of each of the factors are determined by the size of the drains in past periods of stress in the foreign exchange market. The role played by reserves, and hence a country's reserve needs, differs depending on the structure and level of development of the economy, suggesting that different variables are appropriate to assess reserve adequacy in different types of economies. To capture this heterogeneity, the IMF groups countries on the basis of the depth and liquidity of their markets and flexibility of their economy. Consistently with this, countries are grouped into mature markets, deepening financial markets, and constrained market access, which largely correspond to the commonly used per capita income classification. Within each group, methodologies are developed to help reserve adequacy assessments (for more details see IMF 2014).

Whereas the early literature focused on using international reserves as a buffer stock and on determining their optimal levels, a number of recent studies look for new statistical and analytical paradigms to rationalize why it became increasingly attractive to accumulate reserves, particularly given the tendency to adopt more flexible exchange rate regimes.

What causes such a growth in international reserve holding? The emerging market crisis of the late 1990s and early 2000s and the takeoff of reserve hoarding by China and other countries in the 2000s added new factors to the list of determinants of foreign exchange holdings. Here we offer a short review of the literature on the issue.

Recent literature provides two main arguments, namely *precautionary* and *mercantilist* reasons. In the precautionary view, international reserves are desired for self-insurance against exposure to future sudden stops and capital flow reversals, and currency crises. Rodrik (2006), evaluating the cost-benefits of such a strategy, concludes that emerging markets "have over-invested in the costly strategy of reserve accumulation and under-invested in capital-account management policies to reduce their short-term foreign liabilities." Steiner (2013) explains the accumulation of reserves as a side effect of the liberalization of capital markets and, more specifically, of the integration of emerging and developing economies in the world capital market. According to this hypothesis, central banks do not optimize their reserve levels, but suffer from a "fear of capital mobility", therefore, they might take precautionary measures in the form of foreign exchange hoardings. Aizenman and Lee (2007) focused on international reserves as output stabilizers, arguing that international reserves can reduce the probability of output drops induced by a sudden-stop crisis. Bussière et al. (2015) support this argument finding that countries with more reserves relative to short-term debt fared better during the global financial crisis of 2007–2009.

Another version of self-insurance, views the precautionary hoarding of international reserves as needed to stabilize fiscal expenditure in developing countries (see Aizenman and Marion 2003) . Jeanne and Rancière (2011) and Jeanne (2007) provide strong predictions on the optimal reserve level in the context of self-insurance against sudden stops and currency crises. Obstfeld et al. (2010) and Obstfeld (2013) connect the dynamics of reserve hoarding to several factors such as the "fear of floating", the adoption of active policies to develop the domestic financial intermediation market to complement the deepening of domestic financial intermediation with an increase in the financial integration of the developing country in international financial markets. All factors that increase the exposure of the economy to financial storms. Aizenman and Riera-Crichton (2008) provide evidence that holding "adequate" reserves could lower exchange rate volatility. Gourinchas et al. (2010) explore the implications of recurring financial crises on reserves behaviour (see also Steiner 2014).

In the mercantilist framework, international reserve accumulation is a by-product of export promotion. In this case, reserve accumulation facilitates export growth by preventing currency appreciation, thus protecting export markets (Dooley et al. 2005; Aizenman 2007; Bahmani-Oskooee and Hegerty 2011; Cheung and Qian 2009; Ferguson and Schularick 2007; Pontines and Rajan 2011; Wan and Chee 2009)   .

Chen (2008) argues that the current and capital account surpluses are frequently identifies as causes of the large foreign exchange reserves that lead to the accumulation of excess reserves. For example, China's export-led strategy builds up a large current account surplus, which is interpreted as strong economic fundamental and attracts intensive capital inflows. These twin surpluses build up foreign exchange reserves to high levels (Knight and Wang 2011; Zhang 2009) .

The literature also analyses the optimality of reserve holdings, and finds evidence of excess reserves in the sense that reserves exceed those explained by economic fundamentals (see for example, Jeanne and Rancière 2011; Park and Estrada 2010). The implications of excess reserve hoarding policies, pursued by several Asian economies are potentially economically costly (Rodrik 2006; Bianchi et al. 2013; Filardo and Grenville 2012), especially when assuming that the returns on international reserves remain relatively low.

Common understanding of global liquidity remains elusive and answers to this issue may change over time and may differ across countries. Aizenman et al. (2015) evaluate whether the financial crisis of 2007–2009 and the recent structural changes in the global economic environment are associated with new patterns in the reserves stock hoarding, and find that the optimal level of reserves evolves with the dynamics of the global economy. Since factors that affect reserve holdings may change over time, they conclude stating that "...given the dynamic nature of the forces that shape the hoarding of reserves, there is no reason to expect future stability in the patterns of hoarding international reserves".

23.5   The Composition of International Reserves

In the previous sections we have talked of international reserves as if these were a single entity. Independently of the relevance of the study of the level of international reserves, the fact remains that central banks do own international reserves, which are made up of various assets. Hence the problem of the *composition* of international reserves remains a topic worth studying.

The problem of the composition of international reserves was mainly studied, until the end of the 1960s, by examining the choice between gold and US dollars (see, for example, Kenen 1963; Hageman 1969). This focus was justified by the fact that these were the preponderant components.

After the crisis of the dollar standard which took place in 1971 and the subsequent collapse of the Bretton Woods system and the demonetization of gold (see Sect. 22.7), attention shifted onto the process of diversification within the stock of foreign exchange held by central banks, due to the emergence of new currencies which play the role of *vehicle currencies*, i.e. of international means of payment. We can apply to this problem the theory of portfolio selection, duly enlarged by other considerations (Heller and Knight 1978; Ben-Bassat 1984; Roger 1993. See the Appendix).

We should now ask why particular currencies become an international means of payment. This depends in part on historical accident but also—as Swoboda (1968; p. 10) observed—on economic factors which it is possible to single out.

Firstly, there are the *costs of converting* foreign exchange assets into domestic currency. For instance, traders will prefer to conduct transactions in US dollars rather than in Italian lire if the exchange costs from dollars to domestic currency are lower than those from lire to domestic currency. Asset-exchange costs depend inversely on the size of the market for a particular currency, owing to the presence of economies of scale in financial intermediation. This size partly depends, in turn, on the size of the foreign transactions (volume of foreign trade and balance-of-payments structure) of the country which issues the currency.

Secondly, since asset holders are in general risk-averse, the financial market of a vehicle currency must show *"depth, breadth, and resiliency"*, so that the risk of capital loss (if investors want to sell an asset denominated in that currency) is low.

Thirdly, the *expected behaviour of the currency's exchange rate* plays a role, since, if the currency is expected to fluctuate wildly, it is unlikely to be a good candidate for use as a vehicle currency.

According to Swoboda, it is also interesting to note the self-reinforcing tendency towards the use of a particular currency as vehicle currency. In fact, the more a currency is used as an international means of payments, the lower the asset exchange costs become and the more its use in international payments will expand.

On the other hand, a vehicle currency may cease to be such, if the necessary conditions cease (this was, for example, the case of the pound sterling, replaced by the US dollar), although this loss of status does not usually take place abruptly, but gradually. But the self-reinforcing mechanism noted above may be so strong that a currency may keep its status as vehicle even if one of the basic conditions fails, as is the case of the US dollar, which remains the main vehicle currency notwithstanding its wide swings on exchange markets.

It should also be observed that, on the one hand, the use of a currency as a vehicle can give the issuing country certain advantages, such as, for example, the so-called *seigniorage*, consisting in the fact that the country can use its own currency to settle balance-of-payments deficits. On the other, it can give the country certain disadvantages, such as constraints—dictated by the need to stabilize the exchange rate—on its domestic economic policy, the possible upsetting of its domestic financial markets due to the investment or disinvestment of reserves by foreign central banks, etcetera. These disadvantages are, of course, inversely proportional to the economic, financial, etc. strength of the issuing country, and this partly explains on the one hand the *"benign neglect"*[1] of the US monetary authorities as to the behaviour of the US dollar on foreign exchange markets (this situation, however, has been changing in recent times), on the other the reluctance of some countries (for example Germany before the advent of the euro) to permit its currency to become a vehicle currency.

Let us now go back to the study of the problem of the composition of international reserves. The application of the theory of portfolio selection (whose general principles we have often mentioned: see, e.g., Sects. 13.2, 15.3.2) to this problem starts from the idea that a central bank allocates its wealth—made up of its international reserves—among the various vehicle currencies so as to achieve an efficient portfolio (minimum risk for a given yield or maximum yield for a given risk), much as a private asset-holder allocates his wealth. However, as Heller and Knight (1978) observe, this straightforward extension of the private asset-holders behaviour to a central banks behaviour is not correct for a number of reasons, the main one being that a central bank has much wider tasks and objectives than the mere optimization of its portfolio. Therefore, besides the principles underlying the choice of an efficient portfolio, a central bank will also take into account the prevailing exchange-rate arrangements, the structure of international trade and payments of the country, etc., in determining to what extent its foreign exchange portfolio should be concentrated or diversified.

---

[1]This expression was coined during the dollar standard period of the 1960s, to denote that the US did not take any policy actions to alleviate its balance-of-payments problems (nor should it have done, according to one school of thought: see McKinnon 1979, p. 256).

If, for example, country x pegs its exchange rate to currency y, it is presumable that country x's central bank will allocate a relatively greater proportion of its reserves to holdings of currency y (with respect to the average of the other countries). It is also presumable that the quantity of a particular vehicle currency held by a country will be an increasing function of the amount of trade (and, more generally, of international transactions) of that country with the country which issues the vehicle currency. For a study in this direction and survey of earlier literature see Flamini (1999).

23.6 The Analysis of Euro-Markets

23.6.1 The Fixed-Multiplier Approach

To explain the rapid growth of Euro-markets (in the broad sense of Xeno-markets) two types of models were suggested. The first is the traditional approach, based on multipliers. These multipliers work like the traditional closed-economy monetary-base multipliers in the old view of domestic money creation, account being taken of the peculiarities of Xeno-markets (such as the absence of any legal minimum reserve ratio). Consider, for example, the simple case in which there is an initial deposit of H dollars with the Euro-bank system, which will hold a fraction c of this deposit by way of reserve and will lend the rest (note that, since no legal requirements on the reserve ratio exist as regards Euro-deposits, the reserve ratio c will be purely voluntary). The loan, amounting to $(1-c)H$ dollars, will partly be redeposited with the Euro-bank system and partly go outside, given a leakage coefficient of g. Thus the Euro-bank system receives an additional amount of deposits equal to $(1-g)(1-c)H$. Since an amount $c(1-c)(1-g)H$ will be held as reserve, the new loan will amount to

$$(1-g)(1-c)H - c(1-g)(1-c)H = (1-g)(1-c)^2 H,$$

of which a fraction $(1-g)$, namely an amount $(1-g)^2(1-c)^2 H$ will be redeposited, and so on. Thus the redepositing chain will be given by the infinite geometric series

$$\Delta D = H + (1-g)(1-c)H + (1-g)^2(1-c)^2 H + \ldots + (1-g)^n(1-c)^n H + \ldots$$

whose sum is

$$\Delta D = \frac{1}{1-(1-g)(1-c)}H,$$

which is the familiar multiplier formula.

More complicated multipliers can of course be calculated (see Johnston 1981), but all are based on two assumptions. The first is that the various ratios (the reserve ratio voluntarily held by banks, the various leakages, etcetera) are constant. Hence the name of *fixed multiplier* approach. The second assumption is that the demand for Euro-dollars loans accommodates any increase in lending (irrespective of the conditions) that Euro-banks wish to undertake, so that the technical limits to the expansion of deposits represented by the various multipliers are always reached.

23.6.2 The Portfolio Approach to Euro-Markets

The assumptions of the traditional theory are rejected by the portfolio approach. This approach consists in the application to the Euro-market of the modern view of money and banking theory in a closed economic system; the modern approach—as well as the traditional one—presupposes that the maturity structures of both assets and liabilities match perfectly (on this problem see, for example, Niehans and Hewson 1976).

According to the modern view, there is no substantial distinction between money and other assets and between banks and other non-bank financial intermediaries, which all operate in a market environment where there are demands for and supplies of financial assets depending on yields and risks. These considerations are the more relevant for the Euromarket insofar as it is subjected to no legal minimum reserve ratio. It should be noted that this absence enhances the conclusions of the new view, but is not essential. This approach, in fact, could be applied even if, as is the case for domestic money, there were a legal minimum reserve ratio: it will then be the free reserve ratio that will be sensitive to the interest rate. Thus, for example, a Euro-bank will be willing to reduce this ratio if—*ceteris paribus*—the interest rate that it can earn on loans increases sufficiently, and vice versa. Similar considerations can be made as regards the other coefficients which are present in the various multipliers postulated by the traditional theory.

It follows that the first assumption of the traditional theory—the constancy of the said coefficients—is invalid. The multipliers are not fixed but change as circumstances change (and, in fact, the modern approach is sometimes called the *flexible multiplier* approach), so that any analysis of the Euromarket based on the calculation of multipliers is not very helpful, as is also shown by the disparate estimates that have been obtained (Crockett 1976, footnote 9 to p. 382, reports that the values of the multipliers resulting from the various estimates range from 0.5 to 100!).

The second assumption of the traditional theory is also invalid, because the demand for loans in the Euromarket will be a decreasing function—*ceteris paribus*—of the interest rate.

This said, it is possible to give an idea—in a partial equilibrium context—of how the portfolio approach determines the size of the Euromarket by way of a simple diagram. In Fig. 23.1 we have drawn the demand (*LL*) for loans and the

Fig. 23.1 The portfolio approach to the Euro-market

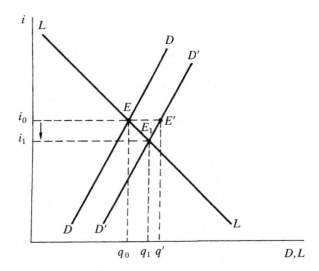

supply (DD) of loan funds (deposits)[2] as functions of the interest rate with the usual normal slopes. Following Hewson and Sakakibara (1974), we make the simplifying assumptions that Euro-banks hold zero reserves and that the margin between the deposit and loan rates is zero.[3] Under these conditions the equilibrium interest rate (i_0) is determined where the supply of loan funds to banks (deposits) equals the demand for funds from banks (loans); concomitantly, the size of the market (q_0) is determined.

Let us now assume that, at the given initial equilibrium interest rate (i_0) there is an exogenous inflow of new deposit funds to the Euro-banking system (the H of the traditional analysis), so that the schedule of the supply of loan funds from the Euro-banks shifts to the right to $D'D'$. This—at an unchanged position of the demand-for-loans schedule and at the given initial interest rate i_0—causes an excess supply of loan funds and hence a decrease in the interest rate. This decrease will, on the one hand, stimulate the demand for loan funds, and on the other depress the supply of these, so that the new equilibrium will not be situated at E' but at E_1. Hence—owing to the decrease in the interest rate from i_0 to i_1—the (final) increase in Euro-deposits will be q_0q_1 (whilst the initial increase is q_0q'). What has happened is that some marginal holders of Euro-currency deposits have shifted their funds out of the Euromarket (because of the decrease in the Euro-interest rate), thus partly offsetting the initial inflow.

If we so wish we can calculate an "implicit" or "ex post" multiplier, defined as q_0q_1/q_0q', i.e. as the ratio of the final to the initial increase in deposits: from the

---

[2]Since the Euro-banks are assumed to operate so as to equate the volume of loans and deposits, it is possible to identify the supply of loan funds with the demand for deposits.

[3]The removal of these simplifying assumptions would not change the substance of the analysis: see Johnston (1981).

diagram it is clear that this multiplier is smaller than one. But it is self-evident that only by knowing the form and the position of the demand and supply schedules (which depend on i and on all the other variables that we have assumed constant under the *ceteris paribus* clause) can we perform that calculation: we are faced with a flexible multiplier.

The above analysis can be formally developed in a general equilibrium context, thus obtaining formulae which express the flexible multiplier as a function of the interest rate and of all the other relevant variables (see the Appendix).

23.7   An Evaluation of the Costs and Benefits of Xeno-Markets

One of the topics in international monetary economics where widely divergent opinions exist is that of the effects of the Euromarket. What we shall do is to try to list the main advantages and disadvantages that have been advanced in the literature, so as to give the basic elements for a kind of cost-benefit analysis. In neither list are items arranged in increasing or decreasing order of importance.

To begin with, we list the disadvantages.

(C1) The market in question is a source which helps to feed speculative capital movements; given its size it follows that if these movements were of the destabilising type, no central bank (nor even any pool of central banks) would be able to oppose them. Thus this market allows speculation to gain an overwhelming victory over the central banks.

(C2) The market increases international capital mobility, and so contributes to the loss of effectiveness of national monetary policies under fixed exchange rates (as we know, given perfect capital mobility, a country's monetary policy becomes ineffective under fixed exchange rates: see Sect. 11.4). The market sets limits to national economic policies also under the managed float. For example, a restrictive monetary policy can be thwarted by the fact that national economic agents borrow funds on the Xeno-market. Besides, the capital inflow related to this borrowing tends to cause the exchange rate to appreciate and, if the monetary authorities do not wish this to occur (for example to avoid a loss of competitiveness of national exporters), they must intervene and buy foreign exchange, which entails an increase in the domestic money supply and hence sterilization problems.

(C3) The growth in the market and so in world liquidity is considered to be one of the causes of the acceleration in world inflation in the early 1970s.

(C4) The market is based on a "paper pyramid", which may collapse at any moment (especially if the Xeno-banks grant medium-term or even long-term loans whilst having mostly short-term liabilities), with catastrophic consequences, partly because the Xeno-banking system has no equivalent of the national banking system's lender of last resort.

(C5) The market's capability for creating international reserves is not only a threat to international liquidity, as already listed under (C3), but also implies that international monetary authorities lose control over the growth of world reserves.

(C6) The possibility of borrowing on the Xeno-market enables the central banks to finance balance-of-payments deficits beyond what is opportune and so to avoid making the necessary adjustments.

Economists worried with these negative aspects believe that it is necessary to move towards a strict control of Xeno-markets by way of international agreements. Let us now list the advantages.

(B1) The market acts as an intermediary between agents wishing to lend and agents wishing to borrow, who belong to different countries. In this manner the situation of imperfect competition prevailing in national credit markets is reduced, and the efficiency of the capital market increased. The contribution to capital mobility is not so great as to give rise to the disadvantages listed under (C2).

(B2) The intermediation function of the market is beneficial not only to private agents but also to central banks, which have an additional source from which they can borrow to finance balance-of-payments deficits.

(B3) The contribution of the market to the recycling of petrodollars has proved invaluable: suffice it to recall that in 1974, the most dramatic year following the first oil price increase, no less than 40 % of the financial surplus of OPEC countries (in absolute terms the amount was about $ 22.5 billion out of a total of about $ 56 billion) was deposited by these countries with the Euromarket and from here recycled to the countries suffering from oil deficits. Should there arise crises in the future (concerning oil or other things) the market will be able to fulfil an analogous task, thus preventing a collapse of the international monetary system.

Those economists who are more sensitive to these positive aspects are against any form of control of a market whose growth, after all, is beneficial.

As in any case of cost-benefit analysis, to strike a balance it is necessary to give a weight to each cost and benefit, a task in which subjective elements are necessarily introduced. Thus we shall only offer a few general considerations which may help the reader in forming a personal opinion:

(a) some controversial points are (implicitly or explicitly) due to different conceptions of the functioning of Xeno-markets. Only if one accepts the traditional theory of the creation of Euro-currencies can one unhesitatingly subscribe to point (C3)—which entails in addition the adhesion to a monetarist view of inflation, not shared by all—and to points (C4), (C5). If, on the contrary, one accepts the modern view, one must acknowledge that the capacity of the Euromarket for creating liquidity is limited, so that the causes of the increase in international liquidity etc., must be looked for elsewhere. Another example of difference in point of view is that of the relations between the

Euromarket and capital mobility: some consider higher capital mobility (due to the relaxation of the restrictions on capital movements and to the expansion of multi-national corporations) to have caused the Euromarket to grow, and not vice versa. Even if one does not accept this view (opposite to that underlying point (C2) above), one must acknowledge that, at least, there is a bidirectional relationship.

(b) The effects which are considered to be positive by some, are considered to be negative by others, and this also reflects different theoretical points of view. The "adjustment or financing" dilemma is reflected in the conflict between (C6) and (B2). It is clear that the supporters of some automatic mechanism which restores equilibrium in international trade and payments (like, for instance, the gold standard or the MABP: see Chap. 12) will stress (C6), whilst the advocates of discretionary interventions will emphasize (B2) as this enables the authorities, if not to eliminate, at least to dilute the painfulness of the adjustment over time.

We would like to conclude with the observation that when international capital flows where to some extent subject to controls (this was the normal situation during the Bretton Woods system, and also after its collapse several countries maintained capital controls), a specific analysis of the costs and benefits of Xeno-markets, by their very nature exempt from national controls (a situation that worried central bankers very much), was very important. But in the early 1990s completely free international mobility of capital became the rule rather than the exception, hence this importance no longer exists.

23.8   Appendix

23.8.1 The Maximization of a Welfare Function

As an example of this second approach to the determination of the optimum reserve level, we shall first consider Kelly's model (1970). The country's welfare depends on the level and variability of income. The reduction in income caused by tying up resources in reserves is

$$y' - y = Ri, \qquad (23.6)$$

where y' is the level of income which could be attained if no reserves were held and y is the level which is obtained when an amount of reserves R is held, at their (net) opportunity cost i. Since we are working within a probabilistic framework, we must consider the expected values (denoted by the symbol E applied to the relevant variable), so that the argument $E(y') - E(y)$ will enter with a negative marginal utility in the welfare function. Also negative will be the marginal utility of income variability as measured by the variance (V) around its expected level when no reserves are held. Kelly considers the quadratic (dis)utility function

$$U = -a\left[E(y') - E(y)\right]^2 - b\left[y - E(y)\right]^2, a > 0, b > 0, \tag{23.7}$$

and then by substituting from (23.6) and taking expected value he gets expected utility as

$$E(U) = -ai^2 E(R)^2 - bV(y). \tag{23.8}$$

Let us now determine the constraint. If we start from an initial equilibrium situation and consider only current transactions, the change in reserves in any period t will be given by the change in exports (considered exogenous) minus the change in imports (considered endogenous) i.e.

$$\Delta R_t = \Delta x_t - \Delta m_t. \tag{23.9}$$

To determine the endogenous change in imports Kelly introduces an import response coefficient f which links Δm to Δx and depends on the willingness of the authorities to allow income to change as exports change, due to external disturbances. This coefficient can be considered as the product of the effect of a change in exports on income (denoted by $g = \Delta y / \Delta x$)[4] and the marginal propensity to import μ, so that $f = \mu g$.

Thus by letting $\Delta m = f \Delta x$, we have

$$\Delta R = \Delta x(1 - f). \tag{23.10}$$

We now compute the variances of reserves and income, $V(R)$ and $V(y)$, which turn out to be

$$V(R) = E(\Delta R^2), V(y) = E(\Delta y^2). \tag{23.11}$$

If we substitute (23.10) and the definition of g in (23.11), we get

$$V(R) = E\left[\Delta x^2 (1 - f)^2\right] = V(x)(1 - f)^2, \tag{23.12}$$

$$V(y) = E(g^2 \Delta x^2) = g^2 V(x). \tag{23.13}$$

Let us assume now that there is some minimum reserve level, R', below which the authorities do not wish the actual reserve level to fall (R' may also be zero, which amounts to saying that the authorities are willing to accept the eventuality of running

---

[4]As a matter of fact g will be a multiplier (see Chap. 8), corrected for possible policy interventions.

out of reserves). Since we are in a stochastic context, the authorities will establish an (arbitrarily) small probability level e such that

$$P\left[R < R'/E(R), V(R)\right] = e. \tag{23.14}$$

Equation (23.14) tells us that the probability that reserves fall below R', given the average level (expected value) and the variance of reserves, equals e. This equation is the constraint in the problem. To be able to solve problem it is necessary to take an explicit probability density function, which Kelly assumes to be

$$e = cV(R)/E(R)^2, \tag{23.15}$$

which has the property that, given e, $dE(R)/dV(R) > 0$; this is a property typical of any regularly behaved probability density function.

If we combine (23.15) and (23.12) we get

$$E(R) = \sqrt{c/e} S(R) = \sqrt{c/e} S(x)(1-f), \tag{23.16}$$

where $S(\ldots) \equiv \sqrt{V(\ldots)}$ is the standard deviation. From (23.13) we obtain $g = S(y)/S(x)$ and, letting $f = \mu g$ and substituting into Eq. (23.16), we obtain the final form of the constraint

$$E(R) = \sqrt{c/e}\left[S(x) - \mu S(y)\right]. \tag{23.17}$$

We must now minimize the loss function (23.8) with respect to $E(R)$ and $S(y)$ with the constraint (23.17). For this purpose we form the Lagrangian

$$L = -ai^2 E(R)^2 - bV(y) + \lambda\left\{E(R) - \sqrt{c/e}\left[S(x) - \mu S(y)\right]\right\}, \tag{23.18}$$

where λ is a Lagrange multiplier. The first order conditions for an extremum are

$$\begin{aligned}
\partial L/\partial E(R) &= -2ai^2 E(R) + \lambda = 0, \\
\partial L/\partial S(y) &= -2bS(y) + \lambda\sqrt{c/e}\mu = 0, \\
\partial L/\partial \lambda &= E(R) - \sqrt{c/e}\left[S(x) - \mu S(y)\right] = 0.
\end{aligned} \tag{23.19}$$

By simple manipulations[5] we get the optimum average level of reserves $\widehat{E(R)}$, which turns out to be

$$\widehat{E(R)} = \frac{S(x)}{\sqrt{e/c} + \sqrt{c/e}\mu^2 i^2(a/b)}. \tag{23.20}$$

---

[5]Solve the first equation for λ and substitute the result in the second, which is then used to express $S(y)$ in terms of $E(R)$. Then substitute the result in the third equation and obtain the result given in Eq. (23.20).

Thus—given the standard deviation of exogenous shocks, $S(x)$, and the various structural parameters—it is possible to determine the optimum level of reserves. This is an increasing function of $S(x)$ and of b (the marginal disutility of income variability) and a decreasing function of a (the marginal disutility of income variations), of i (the opportunity cost of reserves), ad of μ (the marginal propensity to import). Finally, optimum reserves will vary inversely with e (the probability of reserves falling below the minimum specified level) except in particular cases.[6]

23.8.2 Intertemporal Maximization and the Normative Theory of Economic Policy

The approach to the optimum reserve level based on the maximization of the welfare function can be extended in various directions. We examine here the model by Nyberg and Viotti (1976), who consider the problem in an intertemporal maximization framework and in the presence of capital movements. this model will also give us the cue for some considerations on the theory of economic policy in general.

The basic model is the standard Keynesian-type model under fixed exchange rates (see Sect. 10.2), that we rewrite here (with a slightly different symbology)

$$
\begin{aligned}
y &= A(y, i) + x - m(y) + G, \\
L(y, i) &= M, \\
\dot{R} &= x - m(y) + K(y, i).
\end{aligned}
\tag{23.21}
$$

The third equation defines the changes in reserves, which is equal to the surplus or deficit in the balance of payments (a dot over the variables denotes the time derivative). The policy instruments are government expenditure (G) and the money supply (M). In the traditional analysis (see Sect. 11.2) the basic assumption is that the policy maker manages the instruments so as to achieve the prescribed targets. Full employment and external equilibrium (note that the latter amounts to saying that the policy maker wishes to maintain a constant level of international reserves).

Nyberg and Viotti take a different approach, one already suggested by Williamson (1971),[7] which consists in the intertemporal maximization of a welfare function subject to the constraints imposed by the positive economy as represented by model (23.21). They assume that the policy maker's utility (or objective function) U depends positively on income and the stock of reserves, and negatively on the

---

[6]Whilst the previous results are immediately evident by simply inspecting Eq. (23.20), this last result requires the computation of the partial derivative $\partial \widehat{E(R)}/\partial e$, which turns out to be negative unless a is much greater than b.

[7]The idea of intertemporal maximization is of course much older and underlies all the optimal economic growth literature, which dates back at least to the work of Ramsey in 1928.

rate of interest,[8] i.e.

$$U = U(y, i, R). \tag{23.22}$$

It should be pointed out that U denotes the *instantaneous* utility; if one were in the context of a static model one would then—as in Kelly's model treated in Sect. Q.1—maximize U subject to the constraint of model (23.21). In a dynamic context the policy maker acts to maximize utility over time, i.e. the objective is to maximize

$$\Omega = \int_0^\infty e^{-rt} U(y, i, R) dt, \tag{23.23}$$

which represents the present value of total utility over the entire period of time (which goes from zero—the current moment—to the farthest future—infinity—, but it would also be possible to consider a finite time period). The constant r is the social rate of discount or rate of social time preference, which represents the way the policy maker weights the future with respect to the present (the idea is that the more distant the utility enjoyed, the less weight it should be given).

Thus the optimization problem can be expressed as

$$\begin{aligned} &\max \Omega = \int_0^\infty e^{-rt} U(y, i, R) dt \\ &\text{subject to} \\ &\dot{R} = \quad x - m(y) + K(y, i), \end{aligned} \tag{23.24}$$

where y and i are functions of the two instruments G and M through the first two equations in set (23.21); it follows that the only "state variable" is R. Problem (23.24) is a typical *dynamic optimization problem,* which can be solved by applying Pontryagin's *maximum principle* (see, for example, Gandolfo 2009, Chap. 27, Sect. 27.3). For this purpose we form the *Hamiltonian*

$$H = e^{-rt} \{ U(y, i, R) + \lambda [x - m(y) + K(y, i)] \}, \tag{23.25}$$

and set to zero its partial derivatives[9] with respect to the instruments after elimination of the term e^{-rt}; thus we obtain

$$\begin{aligned} \partial H / \partial G &= [U_y - \lambda(m_y - K_y)] y_G + (U_i + \lambda K_i) i_G = 0 \\ \partial H / \partial M &= [U_y - \lambda(m_y - K_y)] y_M + (U_i + \lambda K_i) i_M = 0. \end{aligned} \tag{23.26}$$

---

[8]"Having the rate of interest as a target variable may seem a bit unusual, but is very realistic in many countries (such as Sweden) where high priority is given to keeping down housing costs" (Nyberg and Viotti 1976, p. 127).

[9]For brevity we use subscripts to denote partial derivatives: $U_y \equiv \partial U / \partial y$, etc. We also note that the derivatives y_G etc. are obtained after expressing y and i as functions of G and M by way of the first two equations in set (23.21); the relevant Jacobian is assumed to be different from zero: see below, in the text.

If we now add the canonical equations (with the opportune boundary conditions at terminal time and at initial time)

$$\frac{d}{dt}\left(e^{-rt}\lambda\right) = -\frac{\partial H}{\partial R}, \lim_{t\to\infty} e^{-rt}\lambda = 0,$$

$$\dot{R} = \frac{\partial H}{\partial \lambda}, R(0) = R_0, \tag{23.27}$$

and consider Eqs. (23.26) and (23.27) together, we shall obtain the solution to the problem. We refer the reader to Nyberg and Viotti (1976) for a complete examination of the solution and only consider the optimal path of reserves, which is obtained through the differential equation system (23.27). If we perform the partial differentiation operations on H indicated in that system we get

$$\dot{\lambda} = r\lambda - U_R,$$
$$\dot{R} = x - m(y) + K(y, i). \tag{23.28}$$

By letting $\dot{R} = 0, \dot{\lambda} = 0$, we find the stationary equilibrium point (singular point) of the system, (R^*, λ^*). In the phase plane we get the phase diagram illustrated in Fig. 23.2. The curve $\dot{\lambda} = 0$ is decreasing because of the assumption of decreasing marginal utility of reserves ($U_{RR} < 0$). In fact, from $r\lambda - U_R = 0$ we get

$$\left(\frac{d\lambda}{dR}\right)_{\dot{\lambda}=0} = \frac{U_{RR}}{r} < 0. \tag{23.29}$$

Fig. 23.2 The dynamics of the optimum reserve level

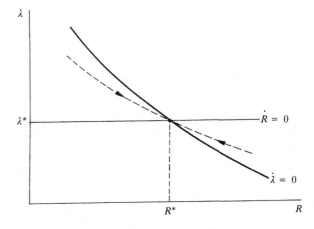

The condition $\dot{R} = 0$ instead gives rise to a straight line parallel to the R axis: in fact, since R does not appear in the equation $x - m(y) - K(y, i) = 0$, we have

$$\left(\frac{d\lambda}{dR}\right)_{\dot{R}=0} = 0. \tag{23.30}$$

We could now analyse the stability of the equilibrium point by way of an arrow diagram, but it is more rigorous to determine the nature of the singular point by way of mathematical analysis. For this purpose we perform a linear approximation of system (23.28) at the singular point. Since y and i are functions of G and M, which in turn depend on λ through the optimum conditions (23.26), we obtain

$$\dot{\bar{\lambda}} = r\bar{\lambda} - U_{RR}\bar{R}, \quad \dot{\bar{R}} = Q\bar{\lambda}, \tag{23.31}$$

where (derivatives are, as usual, taken at the equilibrium point)

$$Q \equiv -\left(m_y - K_y\right)y_\lambda + K_i i_\lambda, y_\lambda = y_M M_\lambda + y_G G_\lambda, i_\lambda = i_M M_\lambda + i_G G_\lambda, \tag{23.32}$$

and the bars over the symbol denote deviations from equilibrium. The sign of Q is essential to determine the nature of the singular point, so that we must carefully examine the expressions which appear in (23.32). From the first two equations in set (23.21) we can, as stated more than once, express y and i as functions of G and M provided that the Jacobian

$$|J_1| = \begin{vmatrix} 1 - A_y + m_y & -A_i \\ L_y & L_i \end{vmatrix} = L_i\left(1 - A_y + m_y\right) + A_i L_y \tag{23.33}$$

is different from zero. The usual assumption on the signs of the partial derivatives ($0 < A_y < 1 + m_y, 0 < m_y < 1$, etc.) allow us to ascertain that $|J_1| > 0$. We can thus calculate the derivatives we are interested in and obtain

$$\begin{aligned} y_G = L_i/\,|J_1| > 0, \quad i_G = -L_y/\,|J_1| > 0, \\ y_M = A_i/\,|J_1| > 0, \quad i_M = \left(1 - A_y + m_y\right)/\,|J_1| > 0. \end{aligned} \tag{23.34}$$

In order to calculate the derivatives M_y and G_y we consider the optimum conditions (23.26). Account being taken of (23.34), Eqs. (23.26) can be satisfied if and only if

$$\begin{aligned} U_y - \lambda\left(m_y - K_y\right) &= 0, \\ U_i + \lambda K_i &= 0. \end{aligned} \tag{23.35}$$

Since y and i are functions of M and G as shown above, we could use set (23.35) to express M and G as functions of λ and then calculate the partial derivatives M_λ and G_λ to be substituted in Eqs. (23.32). Alternatively, and more simply, we can use Eqs. (23.35) to express y and i directly as functions of λ and then compute the derivatives y_λ and i_λ that we are interested in. This is possible provided that the Jacobian of Eqs. (23.35) with respect to y, i is different from zero. This Jacobian is

$$|J_2| = \begin{vmatrix} U_{yy} - \lambda \left(m_{yy} - K_{yy} \right) & U_{yi} - \lambda \left(m_{yi} - K_{yi} \right) \\ U_{iy} + \lambda K_{iy} & U_{ii} + \lambda K_{ii} \end{vmatrix}. \tag{23.36}$$

If we assume, as Nyberg and Viotti do, that the utility function is separable $\left(U_{yi} = U_{iy} = 0 \right)$ and that the balance-of-payments equation is linear (all second-order derivatives of m and K are zero), our Jacobian reduces to

$$|J_2| = U_{yy} U_{ii} > 0, \tag{23.37}$$

where the positivity derives from the assumption of decreasing marginal utilities. Thus we have

$$y_\lambda = \left(m_y - K_y \right) / U_{yy} < 0, i_\lambda = -K_i / U_{ii} > 0, \tag{23.38}$$

where, following Nyberg and Viotti, we have introduced the further assumption that $m_y - K_y > 0$, namely that the Johnson effect (see Sect. 11.2.2, point 4), measured by K_y, is smaller than the marginal propensity to import. Given these results and assumptions, it turns out that Q is positive.

The characteristic equation of system (23.31) is

$$\begin{vmatrix} r - \mu & -U_{RR} \\ Q & -\mu \end{vmatrix} = \mu^2 - r\mu + QU_{RR} = 0, \tag{23.39}$$

so that, since the succession of the signs of the coefficients is $+ - -$, the roots are real, one positive and one negative; hence the singular point is a *saddle point* (Gandolfo 2009, Chap. 28, Sect. 28.1). This means that there will be one and only one optimal trajectory converging towards the stationary equilibrium (which is that drawn with a broken curve in Fig. 23.2) which will determine the time path of the optimal reserve level. Still using the linear approximation (23.31) this path turns out to be

$$R(t) = R^* + (R_0 - R^*)e^{\mu t}, \tag{23.40}$$

where μ is the negative root of the characteristic equation (23.39). As regards the problem we are interested in, we could stop here, but it is important briefly to mention some general considerations on the *normative theory of economic policy* or *optimizing approach to economic policy*.

The traditional theory of the Tinbergen type, which is still widely used, is based, as we have seen in Sect. 11.2.1, on the instrument-target approach. The optimizing approach—instead of starting from prescribed values of the targets—defines an objective function (which can be a social welfare function, a preference function of the policy maker, a loss function, etc.) which is maximized (subject to the constraint consisting of the model which represents the positive economy) in a static or dynamic context. From this optimization process, amongst other results, one obtains the optimum value (or the optimal time paths, if we are in a dynamic context) of the instruments.

For example, in the Nyberg and Viotti model, once the optimal path of R has been obtained as shown above, it is possible to obtain the time paths of λ, of the instruments M and G, and, of course, of the other target variables y and i; all these paths are optimal because they are derived by way of the described optimization procedure. See Nyberg and Viotti (1976) for details.

This approach is undoubtedly interesting and its diffusion is on the increase, but its practical application meets with several difficulties (for example, the determination of the objective function or, more simply, of the weights to be given to the various objectives). Research on this approach is going on, but, as it pertains to the general theory of quantitative economic policy, which lies outside the limits of the present book.

23.8.3 The Composition of International Reserves

The studies of Ben-Bassat (1980, 1984) examined the problem within the framework of the theory of portfolio selection and taking into account the basket of currencies used to pay for imports, since reserves are held, amongst other reasons, to meet import payment needs. As is known, the portfolio selection approach requires one first to determine the set of efficient portfolios (minimum risk for any given return, or, equivalently, maximum return for any given risk).

If we assume that risk can be represented by the variance, the standard mean-variance problem can be formulated as

$$\min \sigma^2 = \sum_{i=1}^{n} a_i^2 \sigma_i^2 + 2 \sum_{i=1}^{n} \sum_{j=1, j>i}^{n} a_i a_j R_{ij} \sigma_i \sigma_j$$

subject to   (23.41)

$$\varrho = \sum_{i=1}^{n} a_i \varrho_i, \quad \sum_i a_i = 1, a_i \geq 0,$$

where a_i is the optimum share (to be found) of currency i in the portfolio, σ_i^2 is the variance of the returns on currency i, R_{ij} is the correlation coefficient between the returns on currencies i and j, and ϱ_i is the return on currency i. Problem (23.41) requires one to minimize the overall variance of the portfolio (σ^2) taken as a measure of risk, for any given overall return ϱ.

It is the definition of the return on currency i (and so of its variance) that the notion of the basket of import currencies (which enter into the basket with weight derived from their actual utilization in import payments) comes into play. The return on currency i is, in fact, defined in terms of the interest rate on funds placed in currency i (denoted by r_i), account being taken of the variation in the exchange rate of currency i with respect to the basket of import currencies (this variation is denoted by E_i). Thus we get

$$1 + \varrho_i = \frac{1 + r_i}{1 + E_i}, \varrho_i = \frac{1 + r_i}{1 + E_i} - 1, \qquad (23.42)$$

from which we see that $\varrho_i = r_i$ if currency i does not change of value with respect to the basket, whilst $\varrho_i \lessgtr r_i$ according to whether currency i depreciates ($E_i > 0$) or appreciates ($E_i < 0$) with respect to the basket. This is an ingenious expedient, as it uses the concept of basket-currency (see Sect. 20.3) defined in terms of the import currencies, to correct the interest rate on each currency.

It is self-evident that, as the import currency basket is different from country to country, the return on the same currency i will be different in different countries, so that the set of efficient portfolios will have to be computed separately for each country.

One the efficient set has been determined, the optimum portfolio will be determined by maximizing the utility function of the agent (in this case the central bank of the country considered) under the constraint of the efficient set. This utility function is assumed to be increasing with respect to return and decreasing with respect to risk.

In the (ϱ, σ^2) plane the problem is then to find the highest-indexed indifference curve compatible with the efficient set. Since the utility function (and hence the indifference map) is not given, Ben-Bassat solves the problem by using Sharpe and Lintner's market price theory (CAPM), according to which the optimum portfolio of risky assets is determined at the point of tangency between the efficiency curve and the straight line which rises from the riskless asset return. It is usual to take the rate of interest on Treasury bills as the riskless asset rate; in the case in question, since the objective function is a currency basket. Ben-Bassat uses the average of the interest rates on Treasury bills in these currencies, expressed in terms of the import basket.

By this procedure the author calculates the optimum reserve composition for 69 countries, then aggregates the results[10] into two groups of countries (the semi-industrialized and developing countries, and the industrialized countries; the latter are further divided into snake countries and floaters) and compares them with the actual reserve composition.

As the author points out, the correspondence between the actual and optimum reserve composition is much closer for the semi-industrialized and developing countries than for the industrialized countries. In the latter group the correspondence is so poor that one is led to believe that in these countries the profit-risk factor

---

[10]This requires that each country's portfolio be given a weight; the weights used by Ben-Bassat are the reserves of the countries.

is of secondary importance. This, we might add, lends support to the argument of those who believe that the optimizing approach, relevant from the point of view of normative economics, is hardly (if at all) useful in explaining the *actual* behaviour of central banks.

For another interesting study of optimal portfolio diversification across currencies see Flamini (1999).

23.8.4 A Portfolio Model of the Euro-Market

In this section we explain the model of Niehans and Hewson (1976; see also Niehans 1984), which, though adopting some simplifying assumptions, lends itself well to the illustration of the portfolio approach to the Eurodollar market.

The model starts from some known definitional and accounting relations and then introduces the appropriate behavioural functions into them. Only liquid assets (denominated in dollars)[11] held by the (nonbank) public are considered (M^*), and it is assumed that they consist of currency (C) and sight deposits, which can be held both with US banks (D) and with Euro-banks (e), so that we have

$$M^* = C + D + e. \tag{23.43}$$

Given a certain amount of monetary base $(\overline{B})$, exogenously determined by the USA, this will be held partly by the public (C) and partly by US banks as bank reserves (R), so that

$$C + R = \overline{B}. \tag{23.44}$$

The balance-sheet identity of US banks is

$$L + R = D + r, \tag{23.45}$$

where the assets are the loans (L) and the reserves (R), whilst the liabilities are the deposits, made up of the public's deposits (D) and the Euro-banks' deposits (r): the latter, in fact, are assumed to hold all their dollar reserves (related to their Eurodollar operation) with US banks. The balance-sheets identity of Euro-banks is

$$l + r = e, \tag{23.46}$$

---

[11]Thus we ignore portfolio choices among other assets (bonds etc.), possibly denominated in different currencies. For more complex models which take this wider spectrum of assets into account see, for example, Hewson and Sakakibara (1976) and Freedman (1977). We also inform the reader that, in order to facilitate comparison with the original source, we shall use the same simbology as Niehans and Hewson.

where the assets are loans (l) and reserves (r), whilst the liabilities are the Eurodollar deposits (e).

Having thus completed the definitional and accounting framework, we now pass to the behavioural relationship. We first have the US banks' demand for reserves as a function of the volume of total deposits

$$R = R(D + r), 0 < R_D < 1. \tag{23.47}$$

More generally, one could consider R as also depending on interest rates, but for simplicity the influence of interest rates is disregarded. Another simplification is that the Euro-banks' reserves are assumed to be independent of Euro-deposits and of any other variables in the model, so that they can be taken as exogenous. As the authors point out, this simplification does not seem excessively unrealistic, and, if anything, it would tend to bias the result in favour of a higher value of the multiplier which will be derived below.

If we denote by I and i, respectively, the US and the Euromarket interest rates and disregard—in each market—the spread between the borrowing and lending rates, we can write the supply of deposits by the public (which depends on the two interest rates and on total liquid assets) as the following functions

$$
\begin{aligned}
D &= D(I, i, M^*), \quad D_I > 0, \ D_i < 0, \ 0 < D_M < 1, \\
e &= e(I, i, M^*), \quad e_I < 0, \ e_i > 0, \ 0 < e_M < 1, \\
L &= L(I, i), \quad\quad L_I < 0, \ L_i > 0, \\
l &= l(I, i), \quad\quad\ l_I > 0, \ l_i < 0,
\end{aligned}
\tag{23.48}
$$

where the signs of the partial derivatives are self-evident.

The equilibrium interest rates are determined by equating the supply of deposits (after due allowance for reserves) and the demand for loans in each market. Thus, if we substitute the various functions (23.48) in the previous equations, and if we also substitute C from (23.44) into (23.43), we get the system

$$
\begin{aligned}
M^* &= \overline{B} - R(D + r) + D(I, i, M^*) + e(I, i, M^*), \\
L(I, i) + R(D + r) &= D(I, i, M^*) + r, \\
l(I, i) + r &= e(I, i, M^*).
\end{aligned}
\tag{23.49}
$$

This system determines the three unknowns, I, i, M^*, whence we can determine the size of the Euromarket, e. To ascertain the effects on e of an exogenous shift of the public's deposits from US banks to Euro-banks, we introduce a shift parameters α to be deducted from D and added to e. Thus we get the following system in implicit form (note that the parameter α cancels out in the first equation)

$$
\begin{aligned}
M^* - \overline{B} + R(D + r) - D(I, i, M^*) - e(I, i, M^*) &= 0, \\
L(I, i) + R(D + r) - D(I, i, M^*) + \alpha - r &= 0, \\
l(I, i) + r - e(I, i, M^*) - \alpha &= 0.
\end{aligned}
\tag{23.50}
$$

By using the implicit function theorem, we can express I, i and M^* as differentiable functions of α if the Jacobian of Eqs. (23.50) with respect to the three variables is different from zero, i.e.

$$
J = \begin{vmatrix}
1 - (1 - R_D)D_M - e_M & -(1 - R_D)D_I - e_I & -(1 - R_D)D_i - e_i \\
-(1 - R_D)D_M & -(1 - R_D)D_I + L_I & -(1 - R_D)D_i + L_i \\
-e_M & l_I - e_I & l_i - e_i
\end{vmatrix} \neq 0.
$$
(23.51)

If this condition occurs, we can use the method of comparative statics to determine the final effect of α on e, which is

$$
de = \left(e_I \frac{dI}{d\alpha} d\alpha + e_i \frac{di}{d\alpha} d\alpha + e_M \frac{dM^*}{d\alpha} d\alpha \right) + d\alpha,
$$
(23.52)

where $d\alpha$ is the initial effect (the initial shift of deposits) whilst the expression in parentheses is the sum of the induced effects, which come from the effects that the initial shift causes on the endogenous variables I, i, M^*. Equations (23.52) can also be written as

$$
\frac{de}{d\alpha} = 1 + \left(e_I \frac{dI}{d\alpha} + e_i \frac{di}{d\alpha} + e_M \frac{dM^*}{d\alpha} \right),
$$
(23.53)

which represents the *flexible* Eurodollar multiplier. Whether this multiplier is greater or smaller than one thus depends on the induced effects, i.e. on the expression in parentheses; to determine it, it is necessary to calculate $dI/d\alpha, di/d\alpha, dM^*/d\alpha$. For this purpose we must check whether J is indeed different from zero and, if possible, determine its sign. The correspondence principle of Samuelson (see, for example, Gandolfo 2009, Chap. 20) can be applied here: it can in fact be shown that $J > 0$ if the equilibrium is stable.

The dynamic system that we are going to set up for the examination of stability is based on standard assumptions, namely that in each market the interest rate varies in relation to the excess demand for loans; we also add the (inessential) assumption that there is an adjustment lag in the equation of the public's total liquid assets, which—dynamically speaking—takes the form of a partial adjustment equation. We thus obtain the following differential equation system

$$
\begin{aligned}
\dot{M}^* &= k_1 \{[B - R(D + r) + D(I, i, M^*) \\
&\quad + e(I, i, M^*)] - M^*\}, & k_1 > 0, \\
\dot{I} &= k_2 \{[L(I, i) + R(D + r)] - [D(I, i, M^*) + r]\}, & k_2 > 0, \\
\dot{i} &= k_3 \{[l(I, i) + r] - e(I, i, M^*)\}, & k_3 > 0,
\end{aligned}
$$
(23.54)

where the k's represent adjustment speeds. If we linearise system (23.54) at the equilibrium point and consider the deviations from equilibrium, we get

$$\dot{\mathbf{x}} = \mathbf{KHx}, \tag{23.55}$$

where $\mathbf{x} = \left\{\overline{M}^*, \overline{I}, \overline{i}\right\}$ is the vector of the deviations from equilibrium (these are denoted by a bar over the variable), $\mathbf{K}$ is the diagonal matrix of the adjustment speeds and $\mathbf{H}$ is the coefficient matrix defined as

$$\mathbf{H} \equiv \begin{bmatrix} e_M + (1 - R_D) D_M - 1 & (1 - R_D) D_I + e_I & (1 - R_D) D_i + e_i \\ -(1 - R_D) D_M & -(1 - R_D) D_I + L_I & -(1 - R_D) D_i + L_i \\ -e_M & l_I - e_I & l_i - e_i \end{bmatrix}. \tag{23.56}$$

The characteristic equation of system (23.55) is

$$-\lambda^3 + c_1 \lambda^2 - c_2 \lambda + k_1 k_2 k_3 |\mathbf{H}| = 0, \tag{23.57}$$

where $|\mathbf{H}|$ denotes the determinant of the matrix $\mathbf{H}$. Among the necessary stability conditions (Gandolfo 2009; Sect. 18.2.2) there is the condition

$$k_1 k_2 k_3 |\mathbf{H}| < 0, \quad \text{i.e. } |\mathbf{H}| < 0. \tag{23.58}$$

It can easily be checked that $|\mathbf{H}|$ coincides with J (with the sign of the elements of the first row changed), so that, by the rules on determinants,

$$-|\mathbf{H}| = J > 0. \tag{23.59}$$

Thus the correspondence principle enables us to ascertain that J is positive. It is then possible to differentiate Eqs. (23.50) with respect to α, account being taken of the fact that M^*, I, i are functions of α. We thus get the system

$$[1 - (1 - R_D)D_M - e_M] \frac{dM^*}{d\alpha} - [(1 - R_D)D_I + e_I] \frac{dI}{d\alpha}$$

$$- [(1 - R_D)D_i + e_i] \frac{di}{d\alpha} = 0,$$

$$-(1 - R_D)D_M \frac{dM^*}{d\alpha} + [-(1 - R_D)D_I + L_I] \frac{dI}{d\alpha} \tag{23.60}$$

$$+ [-(1 - R_D)D_i + L_i] \frac{di}{d\alpha} = -1,$$

$$-e_M \frac{dM^*}{d\alpha} + (l_I - e_I) \frac{dI}{d\alpha} + (l_i - e_i) \frac{di}{d\alpha} = 1.$$

If we solve this system for $dM^*/d\alpha, dI/d\alpha, di/d\alpha$, and insert the resulting expressions into Eq. (23.53), after some simplifications and rearrangement of terms, we get the equation

$$\frac{de}{d\alpha} = 1 + \frac{N}{J},\tag{23.61}$$

where

$$N = (1 - R_D) \{[e_I D_i - e_i D_I] + (1 - D_M)[e_i (L_I + l_I) - e_I (L_i + l_i)]$$
$$+ e_M [D_i(L_I + l_I) - D_I (L_i + l_i)]\}$$
$$+ R_D [e_i(L_I + l_I) - e_I (L_i + l_i)].\tag{23.62}$$

In order to evaluate expression (23.62) and so establish whether the flexible multiplier (23.61) is higher or lower than one, we must introduce a further plausible assumption, namely that each rate of interest has a greater influence on its own market than on the other market. To put the same thing another way, the US interest rate (I) has an influence on L and D which is greater in absolute value than its influence on l and e, whilst the Eurodollar interest rate (i) has a stronger effect on l and e than on L and D. This gives rise to the inequalities

$$|L_I| > |l_I|, |l_i| > |L_i|, |D_I| > |e_I|, |e_i| > |D_i|.\tag{23.63}$$

Thanks to (23.63) it is possible to establish that, of the four expressions in square brackets on the r.h.s. of (23.62), the first, second and fourth are negative, whilst the third is positive. The negative expressions tend to make N smaller and so to push the multiplier below unity, whilst the positive one has the opposite effect. This means that a high marginal propensity to hold Euro-dollars (represented by high values of e_M) tends to raise the multiplier. It is however likely, though not certain, that the multiplier will be below one, as it is presumable that the negative expressions in N prevail over the positive one, so that N is likely to be negative; hence $de/d\alpha < 1$. It should be noted that—whilst in the simplified analysis presented in the text (Fig. 23.1), in which the cross effects of the various interest rates on the various markets have not been considered, the multiplier was sure to be below one—now this results is only likely. In any case, we are not facing a simple fixed-coefficient multiplier as in the traditional analysis, but a *flexible* multiplier in the sense already explained.

References

Aizenman, J. (2007). Large hoarding of international reserves and the emerging global economic architecture. NBER working paper No. 13277.

Aizenman, J., & Marion, N. (2003). The high demand for international reserves in the far east: What's going on? *Journal of the Japanese and International Economies, 17*, 370–400.

Aizenman, J., & Lee, J. (2007). International reserves: Precautionary versus mercantilist views, theory and evidence. *Open Economies Review, 18*, 191–214.

Aizenman, J., & Riera-Crichton, D. (2008). Real exchange rate and international reserves in an era of growing financial and trade integration. *The Review of Economics and Statistics, 90*, 812–815.

Aizenman, J., Cheung Y. -W., & Ito, H. (2015). International reserves before and after the global crisis: Is there no end to hoarding? *Journal of International Money and Finance, 52*, 102–126.

Bahmani-Oskooee, M., & Hegerty, S. W. (2011). How stable is the demand for international reserves? *Applied Economics Letters, 18*, 1387–1392.

Ben-Bassat, A. (1980). The optimal composition of foreign exchange reserves. *Journal of International Economics, 10*, 285–295.

Ben-Bassat, A. (1984). *Reserve-currency diversification and the substitution account.* Princeton Studies in International Finance No. 53, International Finance Section, Princeton University.

Ben-Bassat, A., & Gottlieb, D. (1992). Optimal international reserves and sovereign risk. *Journal of International Economics, 33*, 345–362.

Bianchi, J., Hatchondo, J., & Martinez, L. (2013). International reserves and rollover risks. IMF Working Paper No. 33.

Bussière, M., Cheng, G., Chinn, M. D., & Lisack, N. (2015). For a few dollars more: Reserves and growth in times of crises. *Journal of International Money and Finance, 52*, 127–145.

Chen, Y. (2008). Chinese economy and excess liquidity. *China & World Economy, 16*, 63–82.

Cheung, Y. W,. & Qian, X. (2009). Hoarding of international reserves: Mrs Machlup's wardrobe and the Joneses. *Review of International Economics, 17*, 824–843.

Clark, P. B. (1970). Optimum international reserves and the speed of adjustment. *Journal of Political Economy, 78*, 356–376.

Crockett, A. D. (1976). The Euro-currency market: An attempt to clarify some basic issues. *IMF Staff Papers, 23*, 375–386.

Crockett, A. (1994). The role of market and official channels in the supply of international liquidity. In: P. B. Kenen, F. Papadia & F. Saccomanni (Eds.), *The International Monetary System* (pp. 82–100). Cambridge (UK): Cambridge University Press.

Dooley, M., Folkerts-Landau, D., & Garber, P. (2005). An essay on the revived Bretton Woods system. *Proceedings*, Federal Reserve Bank of San Francisco.

Ferguson, N. & Schularick, M. (2007), 'Chimerica' and the global market asset boom. *International Finance, 10*, 215–239.

Filardo, A. & Grenville, S. (2012). Central bank balance sheets and foreign exchange rate regimes: Understanding the nexus in Asia. Bank for International Settlements, BIS Papers No 66.

Fischer, S. (1999). On the need for an international lender of last resort. *Journal of Economic Perspectives, 13*, 85–104.

Flamini, A. (1999). The management of international reserves and the italian reserve composition at the end of the 1980s. University of Rome "La Sapienza", CIDEI Working Paper No. 58.

Flanders, M. J. (1971). *The Demand for International Reserves.* Princeton Studies in International Finance No. 27, International Finance Section, Princeton University.

Freedman, C. (1977). A model of the Euro-dollar market. *Journal of Monetary Economics, 8*, 139–161.

Gandolfo, G. (2009). *Economic Dynamics* (4th ed.). Berlin, Heidelberg, New York: Springer.

Gourinchas, P. -O., Rey, H., & Govillot, N. (2010). Exorbitant privilege and exorbitant duty. IMES Discussion Paper Series 10-E-20, Institute for Monetary and Economic Studies, Bank of Japan.

Hageman, H. A. (1969). Reserve policies of central banks and their implications for U.S.balance of payments policy. *American Economic Review, 59*, 62–77.

Heller, H. R. (1966). Optimal international reserves. *Economic Journal, 76*, 296–311.

Heller, H. R., & Knight, M. (1978). *Reserve-currency preferences of central banks.* Princeton Essays in International Finance No. 131, International Finance Section, Princeton University.

Hewson, J., & Sakakibara, E. (1974). The Euro-dollar deposit multiplier: A portfolio approach. *IMF Staff Papers, 21*, 307–328.

Hewson, J., & Sakakibara, E. (1976). A general equilibrium approach to the eurodollar market. *Journal of Money, Credit and Banking, 8*, 297–323.

IEO (Independent Evaluation Office). (2012). International reserves: IMF and concerns and country perspectives, Evaluation Report.

International Monetary Fund. (2014). Assessing reserve adequacy - Specific proposal, IMF Staff Report.

Jeanne, O. (2007). International reserves in emerging market countries: Too much of a good thing? *Brookings Papers on Economic Activity, Issue 1* (Spring 2007), 1–79.

Jeanne, O., & Rancière, R. (2011). The optimal level of international reserves for emerging market countries: A new formula and some applications. *Economic Journal, 121*, 905–930.

Johnston, R. B. (1981). Theories of the growth of the Euro-currency market: A review of the Euro-currency deposit multiplier. *BIS Economic Papers*, No. 4, Bank for International Settlements, Basle.

Kelly, M. G. (1970). The demand for international reserves. *American Economic Review, 60*, 655–667.

Kenen, P. B. (1963). *Reserve asset preferences of central banks and stability of the gold exchange standard*. Princeton Essays in International Finance No. 10, International Finance Section, Princeton University.

Knight, J. & Wang, W. (2011). China's macroeconomic imbalances: Causes & consequences. *The World Economy, 34*, 1476–1506.

McKinnon, R. I. (1979). *Money in international exchange*. Oxford: Oxford University Press.

Niehans, J. (1984). *International monetary economics*. Oxford: Philip Allan.

Niehans, J., & Hewson, J. (1976). The Euro-dollar market and monetary theory. *Journal of Money, Credit and Banking, 8*, 1–27.

Nyberg, L., & Viotti, S. (1976). Optimal reserves and adjustment policies. In: E. M. Claasen & P. Salin (Eds.), *Recent issues in international monetary economics* (pp. 124–145). Amsterdam: North-Holland.

Obstfeld, M. (2013). The international monetary system: Living with asymmetry, NBER Chapters. In: *Globalization in an age of crisis: Multilateral economic cooperation in the twenty-first century*, pp. 301–336. National Bureau of Economic Research, Inc.

Obstfeld, M., Shambaugh, J. C. & Taylor, A. M. (2010). Financial stability, the trilemma, and international reserves. *American Economic Journal: Macroeconomics, 2*, 57–94.

Park, D., & Estrada, G. E. B. (2010). Does developing Asia have too much foreign exchange reserves? An empirical examination. *The Journal of The Korean Economy, 11*, 103–128.

Pontines, V., & Rajan, R. S. (2011). Foreign exchange market intervention and reserve accumulation in emerging Asia: Is there evidence of fear of appreciation? *Economics Letters, 111*, 252–255.

Rodrik, D. (2006). The social cost of foreign exchange reserves. NBER Working Paper, No. 11952.

Roger, S. (1993). The management of foreign exchange reserves. BIS Economic Papers No. 38, Basle: Bank for International Settlements.

Steiner, A. (2013). The accumulation of foreign exchange reserves by central banks: Fear of capital mobility? *Journal of Macroeconomics,*textit38, 409–427.

Steiner, A. (2014). Reserve accumulation and financial crises: From individual protection to systemic risk. *European Economic Review, 70*, 126–144.

Swoboda, A. K. (1968). *The Euro-dollar market: An interpretation*. Princeton Essays in International Finance No. 64, International Finance Section, Princeton University.

Wan, L. F., & Chee, Y. L. (2009). Macroeconomic considerations in regional reserve pooling. *Applied Financial Economics, 19*, 1143–1157.

Williamson, J. H. (1971). On the normative theory of balance-of-payments adjustment. In: G. Clayton, J. C. Gilbert, & R. Sedgwick (Eds.), *Monetary theory and monetary policy in the 1970s* (pp. 235–256). Oxford: Oxford University Press.

Williamson, J. (1994). The rise and fall of the concept of international liquidity. In: P. B. Kenen, F. Papadia, & F. Saccomanni (Eds.), *The international monetary system* (pp. 53–64). Cambridge (UK): Cambridge University Press.

Zhang, C. (2009). Excess liquidity, inflation and the yuan appreciation: What can China learn from recent history? *The World Economy, 32*, 998–1018.

Current Problems

24

24.1 Introduction

In this final chapter we examine some current problems of the international monetary system. Any choice is inevitably arbitrary, and what is a problem today might cease to be such tomorrow. However, the problems that we are going to consider are likely to present a theoretical interest to students of international finance and open-economy macroeconomics also in the future. In any case we owe an explanation for the omission of the standard topic called "plans for reform of the international monetary system". We have already mentioned the current international monetary "non system" (see Chap. 3, Sect. 3.3). Proposals for the creation of a new international monetary system have been made from time to time (see Fischer 1999, Fratianni et al. 1999, Rogoff 1999, Sneddon Little and Olivei 1999), but the amount of international agreement that this creation would require (see the section on international policy cooperation, below) makes it unlikely. Hence we prefer to concentrate on a much more modest but viable alternative, the international management of exchange rates.

Thus the topics we are going to deal with are:

(1) international policy coordination;
(2) the debt problem;
(3) the subprime crisis;
(4) the management of exchange rates.

© Springer-Verlag Berlin Heidelberg 2016
G. Gandolfo, *International Finance and Open-Economy Macroeconomics*,
Springer Texts in Business and Economics, DOI 10.1007/978-3-662-49862-0_24

24.2   International Policy Coordination

24.2.1   Policy Optimization, Game Theory, and International Coordination

The international policy coordination problem arises from the observation that—as a consequence of the ever increasing economic interdependence of the various countries—the economic policies of a country influence (and are influenced by) the economic situation and the economic policy stance in other countries. Hence if all countries pursue totally independent policies, undesirable outcomes may result for the world as a whole.

From the terminological point of view, "coordination" and "cooperation" are usually considered as synonyms. Some authors, however, take coordination as implying a significant modification of domestic policies in recognition of international interdependence, namely as something more than cooperation (which in turn is something more than simple "consultation"). We shall use cooperation and coordination interchangeably.

As any case of strategic interdependence, international cooperation can be interpreted in terms of game theory. The players are the governments of the various countries, that adopt various 'strategies' (the policy actions) to pursue their objectives (employment, balance of payments, etc.). Each combination of strategies will give rise to a precise result in terms of each country's objectives. Table 24.1 illustrates the simple case of two countries (United States and Europe), two targets (employment and current account balance) and two strategies (expansionary and restrictive economic policy) under fixed exchange rates.

If both countries expand, both will be better off in terms of employment and there will also be current account balance. If both adopt a contractionary policy, both will suffer a recession though reaching current account balance. If one expands and the other contracts, there will be no employment change in either country, but the expanding country will suffer a deterioration in the current account (the current account of the contracting country will of course improve). In terms of game theory we are in the so-called 'prisoner's dilemma' situation. If one country expands and the other does not, the latter will gain (in terms of current account balance) at the expense of the former. Without coordination, it is impossible to reach the optimal situation in which both countries expand.

Table 24.1 Payoff matrix of the international policy game

Europe	USA	
	Contraction	Expansion
Contraction	Recession in both countries	No employment change
	$B = 0$	B favourable to Europe
Expansion	No employment change	Boom in both countries
	B favourable to US	$B = 0$

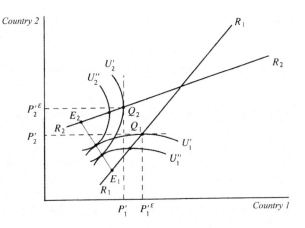

Fig. 24.1 The Hamada diagram

In reality no economic policy consists of two alternatives only (expansion/contraction in our example), but can vary in more or less continuous manner from restriction to expansion. Furthermore, even a partial fulfilment of a target (maybe at the expense of another target) has a value. This means that the government of a country does not carry out its economic policy in a fixed-target context (in our example this would be a given level of employment and equilibrium in the current account balance), but in a flexible-target context. The flexible-target approach means that there is a trade-off among the various targets, in the sense that a higher fulfilment of a target can compensate for a lower fulfilment of another, given a social welfare function. The policy maker aims at maximizing the social welfare function, which depends on the degree of fulfilment of the objectives, given the constraints (represented by the economic system and the other countries' actions). This more general framework can be represented by a diagram due to Hamada (1974, 1985).

If we assume that the policy configuration of each country can be represented by a synthetic and continuous variable, we can show the policy configurations of the two countries on the axes of a diagram like Fig. 24.1. The policy configuration of country 1 is shown on the horizontal axis, while the policy configuration of country 2 is shown on the vertical axis. A point in the diagram thus represents a combination of the two countries policy configurations, which will give rise to a well-defined result in terms of the two countries targets. This result will of course depend on the underlying model representing international interdependence, on which more below (Sect. 24.2.2). The welfare level corresponding to the result will depend on each country's social welfare function. Such a function has a maximum corresponding to the best possible result for the country concerned (the bliss point). Let us assume that E_1, E_2, are the points which give rise to the maximum welfare for country 1 and country 2 respectively. We can then draw around these points the welfare indifference curves of the two countries. A welfare indifference curve of a country is the locus of all points (combinations of the two countries' policy configurations) that give rise to results (in terms of the country's targets) which are considered equivalent by the country under consideration. Let us consider country 1: since we have assumed that the bliss point corresponds to E_1, any other point in the diagram represents a lower welfare. The closer the indifference curve is to point E_1, the better off country 1 is: any policy combination that puts it on indifference curve U_1'' is preferred to any policy combination that puts it on U_1'. The welfare indifference curves are closed curves around the bliss point, but for graphical simplicity we have drawn them only partly. In a similar way we can draw country 2s welfare indifference curves around point E_2. Let us note in passing that in the special case of no interdependence (total independence) the indifference curves of country 1 would be vertical straight lines and those of country 2 horizontal straight lines. In this case each country could achieve its optimal welfare independently of the policy pursued by the other country. Hence the case for coordination is based on the presence of international interdependence (see Martinez Oliva 1991, for the proof that interdependence is necessary, though not sufficient, for coordination to be welfare improving).

In the case of interdependence, the first step of the constrained-optimum problem that each country has to solve is to determine its welfare-maximizing policy configuration for any given policy configuration of the other country. Let us consider, for example, country 1, and let us assume that country 2 adopts the policy configuration represented by point P_2'. If we draw from this point a straight line parallel to country 1s axis, we see that the highest indifference curve that country 1 can reach is U_1', tangent to the aforesaid straight line. U_1', in fact, is the indifference curve nearest to E_1 compatibly with the given policy choice of country 2 (the constraint). Hence, given P_2', the optimal choice for country 1 is policy $P_1'^E$. Going on in like manner, we obtain a set of points that give rise to the $R_1 R_1$ curve, called the *policy reaction function* of country 1 (for graphical simplicity we have drawn it linear).

In a similar way we obtain country 2s policy reaction curve. Given for example country 1s policy configuration P_1', the indifference curve of country 2 which is nearest to the bliss point E_2 is the curve tangent to the straight line originating from P_1' and parallel to country 2s axis.

In the diagram we also have drawn a segment joining the two ideal points E_1, E_2, which is the locus of all points where the two countries' indifference curves are tangential to one another. This locus is a Pareto-optimal or contract curve, since in each of these points the property holds that it is not possible to increase the welfare of one country without decreasing the other country's welfare.

We can now use the Hamada diagram to illustrate what happens without coordination, and the advantages of coordination. In Fig. 24.2 we have drawn the two policy reaction curves obtained as explained above, and we now want to know what will be the behaviour of the two countries.

Let us begin by considering a non-cooperative behaviour, that can take on various forms. The most commonly used are the Cournot-Nash and Stackelberg scenarios.

In the former, each country maximizes its welfare by choosing its own optimal policy taking as given the policy configuration of the other country, on the assumption that this configuration is beyond its influence. Given for example country 1s

Fig. 24.2 The international policy game: Cournot-Nash, Stackelberg, and cooperative solution

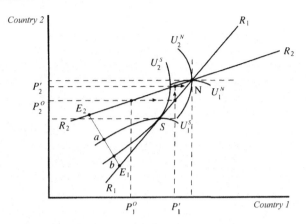

policy P_1^0, country 2 will choose P_2^0 on its own reaction function. In its turn country 1, taking as given country 2s P_2^0, will change its policy to P_1' on its own reaction curve; country 2 will then react to P_1' by changing its policy to P_2' and so forth, until point N is reached. This is the Cournot-Nash equilibrium, where the welfare achieved is U_1^N for country 1 and U_2^N for country 2.

The Stackelberg or leader-follower solution is obtained when one country (for example, country 2) is dominant (the *leader*) and takes account of the fact that its actions influence the other country's decisions, while the *follower* country behaves like in the previous case. The leader knows the reaction curve of the follower and hence country 2 knows that country 1 will react to the leader's policy choices by choosing a policy along the $R_1 R_1$ curve. Thus country 2 maximizes its welfare function taking account of this curve as a constraint. This means that country 2 will choose the highest indifference curve compatible with the constraint. This curve is U_2^S which is tangential to $R_1 R_1$ at point S. In fact, U_2^S is country 2s indifference curve that is nearest to E_2 compatibly with $R_1 R_1$.

It can be seen that in the Stackelberg equilibrium point S, country 2 (the leader) obtains a welfare level clearly higher than in the Cournot-Nash equilibrium, while the follower may or may not be better off. Both equilibria are however inefficient, since they do not lie on the contract curve $E_1 E_2$. Both countries could be better off if they agreed to cooperate: if they coordinate their policies they can reach a point on the segment at of the contract curve. Any such point is clearly superior to both the Cournot-Nash and the Stackelberg equilibrium. The precise point where the two countries will end up will of course depend on the relative bargaining power.

24.2.2 The Problem of the Reference Model and the Obstacles to Coordination

If coordination is favourable to all, why is it not universally adopted, and why many talk of obstacles to coordination? Let us first note that forms of "weak" international cooperation (or "consultation", according to the terminology introduced in

Sect. 24.2) are very frequent. All countries hold routine consultations in the context
of various international economic organizations such as the IMF. The industrialised
countries routinely consult with one another in the context of the OECD; the seven
and the five most industrialised countries hold the G7 and G5 summits respectively,
and there are several other subsets of countries holding routine consultations. But
here we are dealing with policy coordination proper, and we focus on economic
rather than political obstacles.

The first and foremost impediment can be illustrated by an illuminating analogy
due to Cooper (1989). It took over seventy years (from 1834 to 1907)—he
observes—for the various countries to reach an agreement on the best way to
prevent the spread of virulent diseases such as cholera. The reason is that in the 19th
century there were two completely different theories or models on the transmission
of such diseases: the "miasmatic" and the "contagionist". Miasmatists held that
infectious diseases were not transmitted by diseased persons but originated in
environmental "miasms". Contagionists supported the view that such diseases were
transmitted by contact with diseased persons. Now, "epidemiology in the nineteenth
century was much like economics in the twentieth century: a subject of intense
public interest and concern, in which theories abounded but where the scope for
controlled experiment was limited" (Cooper). Hence both views had the support
of the scientific community. It is clear that the age-old technique of quarantining
ships infected (or suspected of being infected) was a decisive measure according
to the contagionist theory, but a pointless measure for the miasmatists. Quarantine
represented a severe burden on trade and shipping, the cost of which would have
been unequally shared, as the greater part of merchandise and passenger trade was
carried out by Britain (a country that, not surprisingly, vigorously supported the
miasmatist theory). Hence an agreement on the means to prevent the spread of
virulent diseases could not be reached until the validity of the contagionist model
was demonstrated. Thus, as Cooper observes, the world was in a situation in which
all agreed on the objectives (the prevention of the spread of virulent diseases),
but sharp disagreement existed on the instruments and on the related cost sharing,
because of the conflicting views on the theory that could explain the facts.

In economics all agree on the objectives: high employment, and growth without
inflation are universally considered as desirable. The attached weights may be different in different countries, but this is normal given the different welfare functions
of the various countries, and does not cause particular problems. It is the *lack of
agreement on the model* that explains the transmission of the effects of economic
policies (both domestically and from one country to another) that determines the
impossibility of international policy coordination to fight unemployment, low (or
negative) growth, inflation as the case may be. In Fig. 24.1 we have taken for
granted the existence of a model—no matter which—accepted by all countries. In
the contrary case there is no ground at all for carrying out the analysis.

A further problem is *model uncertainty*. Assuming that all countries agree on a
model, it might happen that this model turns out to be wrong. Cooperation based on
an incorrect model might be negative rather than positive.

In addition to these theoretical problems, there are practical problems, the most important of which are the free-rider problem and the third-country problem.

The *free-rider* problem is a typical problem of all cooperative equilibria (including all kinds of agreements) in games of the prisoner's dilemma type. Even if the various countries happen to be in a situation in which cooperation is beneficial to all, the problem remains of how to ensure that all countries participating in the cooperative equilibrium respect the agreement. In terms of Table 24.1, each country—taking the other country's expansionary policy for granted—has the incentive not to respect the agreement and adopt a restrictive policy, leaving the burden of expansion on the other country. Each country tries to move from the last cell in the payoff matrix to a cell outside the main diagonal. Such a behaviour causes retaliation from the other country, and we are back in the non-cooperative inferior situation. Hence institutional mechanisms have to be devised and introduced to enforce the cooperative agreement.

The *third-country* problem is related to the fact that the agreement to coordinate usually includes only a subset of the countries of the world. It may then happen that the coordinated policies undertaken by the participating countries have a negative effect on non-participating countries. These third countries could react by carrying out policies that negatively affect the cooperating countries, which might then ultimately be worse off. For example, suppose that country 1 and 2 cooperate and decide to deflate their economies. The adverse effects on country 3s exports may lead this country to deflate as well. This aggravates the recession in country 1 and 2 above what they had anticipated in deciding their coordinated policy, making them worse off than if they had not jointly deflated.

These problems have given rise to a copious literature (see Bryant 1995; Daniels and Vanhoose 1998; Frankel and Rockett 1988), that has amongst others tackled the question of the consequences of uncertainty on the "true" model. The conclusion has been that the case of internationally coordinated policies based on an invalid model can give rise to a lower welfare than the case of no coordination. Ghosh and Masson (1991) have introduced the assumption that policy makers have a learning ability, namely the ability of adjusting the model towards the true model by modifying it through the observation of the policy results. In this case international policy coordination is better than no coordination.

Attempts at empirically estimating the benefits of coordination have been carried out using linked macroeconometric models (e.g. Oudiz and Sachs 1985; Frankel 1988; McKibbin and Sachs 1991) and have shown that the potential gains from coordination are not very high (generally around 0.5% of GDP) even under favourable circumstances. These potential gains will have to be weighed against the potential losses deriving for example from the use of a wrong model.

These results may help explain why, while (as noted by McKibbin 1997) during the 1970s and 1980s the coordination of macroeconomic policies among major industrialized economies was an important issue (especially during periods of crisis), in the 1990s there has been a dramatic fall in both the perception of any need for macroeconomic policy coordination, and (apart from some brief periods) there has apparently been a decline in the practice of coordination at the global level.

In contrast, at the regional level there is increased interest in the coordination of macroeconomic policies, as the debate on European Monetary Union (see Chap. 21) highlights.

The coordination problem can also be studied in the context of the new open economy macroeconomics (the intertemporal approach examined in Part V), but only few and contradictory results exist (Lane 2001, Sect. 7).

24.3 The Debt Problem

The international debt crisis (see Chap. 22, Sect. 22.9) has given rise to serious problems, not only for the creditor banks but also for the governments of the countries to which these banks belong, and for the international monetary system. The possible bankruptcy of these banks, in fact, would create serious dangers for both the national monetary system and the international monetary system. Furthermore, banking crises most often either precede or coincide with sovereign debt crises, as the government takes on massive debts from the private banks, thus undermining its own solvency. Reinhart (2000, 2011) argue that government debts typically rise about 86 % in the three years following a systemic financial crisis, and that banking crises (even those of a purely private origin) increase the likelihood of a sovereign default.

Although several general proposals and plans have been put forward (the Baker plan, the Brady plan, and so on: see Pilbeam 2013, Chap. 15), in practice the international debt problem has been tackled on a case by case approach, through a combination of measures that can be summarized into three categories:

(a) *debt restructuring*. One can generally distinguish two main elements in a debt restructuring procedure: *debt rescheduling*, defined as a lengthening of maturities of the old debt, possibly involving lower interest rates or postponing the payment of interest and/or principal to some date in the future; and *debt reduction*, defined as a reduction in the face (nominal) value of the old instruments. Both types of debt operations involve a loss in the present value of creditor claims. Another measure in this category consists of *debt-equity swaps*, namely part of the debt is exchanged, through third parties, for equities of corporate firms of debtor countries.
(b) *economic reforms* in debtor countries, which are usually summed up in the *d*-triad: devaluation, deflation, deregulation. These are the macroeconomic measures usually suggested by the IMF to debtor countries. The IMF focus attention on the magnitude of vulnerabilities from the rapid growth of cross-border securitized lending, from the heavy dependence of many emerging market countries on foreign capital, and from the speed and interconnectedness of changes in capital market sentiments. Their aim is to improve the economic situation of debtor countries and hence their capability of servicing the debt. In the short run these measures may however have strong negative effects on employment and so turn out to be politically destabilising for debtor countries.

(c) *debt forgiveness*. This can take place in various ways (in addition to simply writing off part of the debt): for example, by allowing the debtor to buy back its debt at a huge discount, or by drastically reducing the interest rate.

What causes international debt crises? The debt crisis of the early 1980s prompted a surge of empirical work to identify the factors contributing to debt. Most defaults and restructuring episodes were triggered by one or more of the following factors: worsening of the terms of trade; dynamics of the country's foreign asset and liability positions; increase in international borrowing costs; poor macroeconomic policies, leading to a build up of vulnerabilities; crisis in a large country that causes contagion across goods and financial markets; problems in cross-border creditor-debtor contracts. Debt crises and restructuring can be self-fulfilling and caused by contagion. The structure of the debt portfolio also impact on the likelihood and timing of default and debt negotiation. Factors that determine the debt profile (e.g., currency composition, fixed vs. floating interest rate, maturity, and creditor composition) may have implications for liquidity, as well as solvency conditions and, therefore, the decision to restructure.

Given the very different nature of the situation and prospects of debtor countries, it is difficult if not impossible to make generalizations. Hence the case-by-case approach seems indeed the most suitable.

24.3.1 NATREX Model of External Debt and Real Exchange Rate

The study of foreign debt dynamics has been based for decades on the "solvency-sustainability" framework. Debt sustainability aims at answering a simple question: when is a country's debt becoming so big that it will not be fully serviced? The IMF's own definition of sustainability is: a debt "is sustainable if it satisfies the solvency condition without a major correction [...] given the costs of financing" (IMF 2002; p. 5). Debt solvency will be achieved when future primary surpluses are large enough to pay back the debt, principal and interest. Solvency, and therefore sustainability as it builds upon solvency, is entirely forward looking.

A non-increasing foreign debt to GDP ratio is seen as a "practical" sufficient condition for sustainability: a country is likely to remain solvent as long as the ratio is not growing. In a country where the debt to GDP ratio is growing, the trade-balance gap, that is the difference between the current trade balance and the trade surplus required to stabilize the debt to GDP ratio, is growing. Such a required trade surplus will be larger the bigger are the debt to GDP ratio and the differential between the real interest rate and the growth rate of the economy. Hence, a country is solvent if the external liabilities/GDP ratio will remain bounded and the debt service payment/GDP will not explode. The debt burden, indicator of "vulnerability", is the trade surplus required to keep the debt/GDP ratio constant at its current level. The higher is the debt burden the more burdensome is the debt and the greater is the likelihood of default. Let us also note that the dynamics of the real exchange rate and of a terms of trade shock importantly affect the debt dynamics. For example, a real depreciation of the currency leads to an increase in the foreign debt to GDP ratio, as it increases the value of foreign currency denominated liabilities of a country, and will worsen its debt sustainability: i.e., a larger trade surplus will be required to stabilize the debt to GDP ratio. Various authors suggested alternative criteria such as external debt to GDP, external debt to exports, debt service to GDP, debt service to exports.

The issue of sustainability of the current account critically depends on projecting into the future the current behaviour of the private sector and the current policy of the government. Because debt sustainability is a forward-looking concept, it cannot be assessed with certainty. While the "practical criterion" for external debt sustainability provides an useful benchmark, it does not directly provide a tool to assess whether a certain stock of debt is sustainable or not, given the uncertainty about future external and domestic shocks, about growth enhancing policy changes and about political and institutional changes that may or may not occur. What it turns out to be crucial for sustainability is the path of the foreign debt that has to be consistent with "intertemporal solvency", remembering that the current account deficit is the sum of consumption, investment, the servicing of the long-term debt or equivalently investment less saving, as shown in Sect. 18.3.4.

Critics (see, for example, Stein 2006) of this approach to the debt issues pointed out that it is impossible to know the future value of the debt because the future growth rates and the interest rates are unknown. Furthermore, the trade deficit and the growth rate are not independent. Thus, as Stein (2006) has forcefully argued, it is unhelpful and often misleading to look at the debt burden based upon current values to address debt evolution and debt crisis.

Stein (2006) proposed a different approach based on the Natural Real Exchange Rate NATREX model of the equilibrium real exchange rate and external debt. The NATREX model permits to take into account the fact that the external debt ratio is an endogenous variable determined by economic fundamentals in a dynamic manner and, in turn, the fundamentals are determined by the actions of both the public and the private sector.

The NATREX, as described in Chap 18 (Sects. 18.3.3 and 18.3.4), and in Stein (Chap. 4 2006), explains the fundamental determinants of the medium-run equilibrium and the dynamic trajectory of the real exchange rate and the external debt to the long-run equilibrium.

Measuring investment I_t, saving S_t, the trade balance TB_t, and the foreign debt F_t as shares of GDP we have:

$$[(I_t - S_t) + TB_t - i_t F_t] = 0, \qquad (24.1)$$

where r_t is the real interest rate. In both the medium run and the longer run the NATREX equilibrium real exchange rate satisfies Eq. (24.1). There are both *internal balance* (where the rate of capacity utilization is at its longer-term mean), and *external balance* (where the long-term accounts of the balance of payments are in equilibrium in the absence of cyclical factors, speculative capital movements and changes in international reserves). The endogenous current account generates an evolving external debt, which feeds back into the medium-run (Eq. 24.1). A trajectory to longer-run equilibrium is generated. The dynamics of the debt/GDP ratio $F(t)$ is:

$$\frac{dF_t}{dt} = (I_t - S_t) - g_t F_t = (i_t F_t - TB_t) - g_t F_t \qquad (24.2)$$

$$= (i_t - g_t) F_t - TB_t,$$

where g is the growth rate. The real exchange rate affects the trade balance TB_t in Eq. (24.1), and the trade balance affects the evolution of the actual debt ratio in Eq. (24.2). Thus, there is a dynamic interaction between the endogenous real exchange rate and debt ratio. The long-term debt ratio stabilizes at a value that satisfies the following equation:

$$(i_t - g_t) F_t - TB_t = 0. \qquad (24.3)$$

The long-run equilibrium real exchange rate r_t^* and debt/GDP ratio F_t° are endogenous variables that satisfy both Eqs. (24.1) and (24.3); as we know, both depend on economic fundamentals. As the fundamentals change over time, the equilibrium real exchange rate and external debt ratio will change over time.

The NATREX model permits to understand the effects of policies and external disturbances upon the trajectories of the equilibrium real exchange rate r_t and equilibrium external debt ratio F_t, which depend upon the vector of fundamentals Z_t. The fundamentals change as a result of policies, exogenous variables and shocks. To give an idea of the logic of the NATREX model, Stein (2006) assumes different scenarios with different elements in the vector Z_t of the fundamentals, and shows the different effects on the equilibrium trajectories of the real exchange rate NATREX and of the external debt. The effects of changes in Z_t are analysed in Sect. 18.4.5.2.

24.4 The Subprime Crisis

The financial crisis that has been gradually developing since summer 2007, can be divided into two distinct phases. The first phase from August 2007 to August 2008 stemmed from losses in the United States (US) subprime residential mortgage sector. In mid-September 2008, however, the financial crisis entered a far more virulent phase. In rapid succession, the investment bank Lehman Brothers went bankrupt, the insurance firm American International Group (AIG) collapsed, there was a run on the Reserve Primary Fund money market fund, and the highly publicized struggle to pass the Troubled Asset Relief Program (TARP) began.

Between September 20 and October 12, 2008, a large liquidity crisis emerged. The crisis started with a deteriorating quality of US subprime mortgages and then became a liquidity event. The liquidity crisis was generated by a global crisis of the financial system. Its propagation across different asset classes and financial markets is attributable to an amplification mechanism due to asymmetric information resulting from the complexity of the structured mortgage products developed over the years through the securitization process. However, although residential mortgages have been the most important element in the evolution of securitization (during 2000–2007, subprime mortgages grew by 800 % and, by the end of this period, 80 % of these mortgages were being securitized), the growing importance of market-based financial intermediaries was a more general phenomenon that extended to other forms of lending—including consumer loans, commercial real estate or corporate loans. The collapse of the American housing market in 2006 and 2007 had a profound effect on the US and global banking systems. Since many large financial institutions had heavily invested in mortgages, the bursting of the housing bubble led to a steep deterioration in bank balance sheets. Questions about bank solvency affected investor confidence.

The US policy responses to the financial crisis came through policies applied to the financial and banking system: conventional (targeting a lower federal funds interest rate) and unconventional monetary policies (liquidity provision and asset purchases), bank "stress tests," and bailouts of some banks and financial institutions. Fiscal stimulus to increase aggregate demand was another part of the government response to the global financial crisis. However, despite these interventions, the global financial crisis quickly evolved into a sharp recession with a global jobs crisis, as the crisis-induced credit crunch strangled the real economy and trade flows collapsed.

A thorough overview of the events preceding and during the financial crisis is provided in Adrian and Shin (2010), Brunnermeier (2009), Taylor (2009a), Allen and Carletti (2010) .

Thus, the global financial crisis had its roots in the credit over extension developed in the US. While the crisis initially had its origin in the US, it rapidly spread across virtually all economies, both advanced and emerging, as well as across economic sectors. A substantial percentage of the world's wealth was destroyed within 18 months of the subprime crisis, with effects ranging from the collapse of major financial institutions to the near bankruptcy of national economies. The financial crisis also led to growing fears about public debt levels, which contributed to the sovereign debt crises that erupted in Greece and Ireland in 2010.

There is still no full agreement among policymakers and researchers on what caused the build-up of financial imbalances globally. Different factors played a role at different stages of the crisis. Some may be considered root causes while others only aggravating circumstances. There are a myriad of contributing factors and lessons to be learned from the crisis, far too many to review here. Drawing from a comprehensive review of crisis-related studies, the main interrelated factors that can be identified refer to interest rates, global imbalances, perceptions of risks, and regulation of the financial system. Taylor (2009b) stresses that the excessively loose US monetary policy fuelled the credit boom. Loose monetary policy that resulted in a low short-term interest rate, may have reduced the cost of wholesale funding for intermediaries, leading those intermediaries to build-up leverage (Adrian and Shin, 2010); may, more generally, have caused banks to take more risks, including credit and liquidity risks (Borio and Zhu 2012); and may have increased the supply and demand for credit (mortgages), causing asset (house) prices to rise (Hirata et al. 2012). The combination of cheap credit together with the easy availability of funds and financial innovations contributed in generating and propagating the financial and economic crisis facing the global economy. Phillips and Yu (2011) suggest that bubbles emerged in the housing market before the subprime crisis and collapsed with the subprime crisis. The bubble then migrated from the housing market to commodity markets and the bond market after the crisis erupted into public dimension. All these bubbles collapsed as the financial crisis impacted the real economic activity.

In the years leading up to the crisis, a number of researchers argued that the large current deficit in the United States was unsustainable. Based on previous current-account driven crises, many observers felt that the US deficit size was a signal of a potential crisis. Global imbalances are associated with a greater dispersion of current account positions across countries and larger net flows of capital between countries. Milesi-Ferretti and Tille (2011) and Broner et al. (2013) analyze the link between gross capital flows and the emergence of the 2008–2009 crises. As argued by some (e.g., Acharya and Richardson 2009, Obstfeld and Rogoff 2009), it may have been a combination of accommodative monetary policy and growing global imbalances that caused the crises.

Supervision and regulation of the financial system have a key role in preventing crises, by controlling moral hazard and discouraging excessive risk-taking on the part of financial institutions. Inadequate supervision and regulation by the governments and international institutions have played a major role in allowing banks and other financial institutions to capitalize on loop-holes in the regulatory system to increase leverage and returns. Regulatory authorities failed to address the dangers that had been building in the global financial system, and the development of new complex financial products outside the scope of existing rules proved to be one of the major regulatory failure. Among these products the best known are CDS (Credit Default Swaps) and CDO (Collateralized Debt Obligations), that belong to the category of credit derivatives (see Chap. 2, Sect. 2.7.4).

Many studies stressed the importance of shocks to liquidity and to risk in explaining the dynamics of the global crisis. In particular the squeeze of liquidity in 2008, which implied a drying up of liquidity among financial institutions, forced many banks and investors to repatriate capital to finance investment and meet redemption calls, thus severely restricting the capital available to the real side of the economy and triggering a major global recession (Adrian and Shin 2010; Brunnermeier 2009; Tirole 2011).

A growing literature on the 2007–2008 financial crisis focused on the US policy responses and on the question of whether and in which ways these policies have been successful. A number of observers point out that households and firms have seen a higher cost of credit during the recession, from which they conclude that monetary policy has not been effective during the financial crisis (for example, Krugman 2008). Others stress that many government actions were ineffective, while others argue that   government interventions prolonged and worsened the crisis by misdiagnosing the structural causes of the collapse of the bank credit markets, and thereby responding inappropriately by focusing on liquidity rather than risk (for example Taylor 2009b). Mishkin (2009) considers what would be the course of events without the policy interventions and concludes that the government actions helped to prevent a deeper recession.

The crisis was largely unexpected, and, due to its complex roots, it continues to puzzle policymakers, economists, and observers. The search for its underlying causes has revived academic interest in financial crises and their history (see Rajan 2011; Gorton 2012; Schularick and Taylor 2012; Reinhart and Rogoff 2009, among others). Furthermore, the recent crisis has renewed the interest for Early Warning Systems to reduce the risks of future crises, that is, whether a set of useful early warning indicators of financial crises can be identified. There is no consensus on potential warning signals and whether macroeconomic or financial factors play a more important role in predicting financial crises. Frankel and Saravelos (2012) surveyed the literature on early warning indicators to see which leading indicators were the most reliable in explaining the crisis incidence. They find that foreign exchange reserves, the real exchange rate, credit growth, real GDP growth and the current account balance are the most reliable indicators to explain crises.

The global financial crisis of 2007–2009 showed that contagion poses an important systemic risk. Although the crisis is seen as having origin in the overheated housing markets and in the associated mortgage backed securities market, it propagated across firms, markets and countries. Many studies focused on the several potential and alternative channels of contagion in the transmission of the crises (among others, Eichengreen et al. 2012; Acharya and Schnabl 2010; Mendoza and Quadrini 2010).

The global financial crisis of 2007–2009 has also given rise to a renewed impetus to reform the international monetary and financial system (Ocampo 2015).

It is important to note that the 2007–2009 crisis has led to a huge increase in public debt as governments increased public expenditure to recover from the crisis, raising serious concerns about its economic and financial impact on investment and growth (Reinhart and Rogoff 2015) .

24.5   Proposals for the International Management of Exchange Rates

24.5.1 Introduction

A much more modest but viable alternative to a plan for creating a new international monetary system starts from the observation that it would be desirable to prevent excessive (and often disrupting) oscillations in the exchange rates of the major currencies (see Chap. 16, Sect. 16.3 on currency crises). There is however no agreement on the best way to reach this goal. Among the various proposals the best known are McKinnon's global monetary objective, John Williamson's target zones, and the Tobin tax. For a general survey of the proposals for the international management of exchange rates see Clarida (2000).

24.5.2 McKinnon's Global Monetary Objective

This proposal was set forth in the 1970s by Ronald McKinnon, who has later perfectioned it (see McKinnon 1988, 1997). It is based on fixed exchange rates (that he considers superior to flexible rates), with a $\pm 5\%$ band, integrated by a precise intervention rule to be followed by the monetary authorities.

According to McKinnon, the main cause for exchange-rate volatility is currency substitution. In a world practically free from controls on international capital flows, private international economic agents (multinational enterprises, portfolio investors, etc.) wish to hold a basket of various national currencies. McKinnon holds that the overall demand for this currency basket is, like the traditional domestic demand for domestic money, a stable function (the Friedman thesis extended to international economics), but that the desired *composition* of the global basket may be very volatile. This implies that the control of the single national domestic supplies is unsuitable, and that Friedman's monetary rule (according to which the money supply must grow at a constant predefined rate) should be shifted from the national to the international level.

In practice this means that, once the nominal exchange rates (McKinnon suggests a PPP rule) and the rate of growth of the world money supply have been fixed, the national monetary authorities interventions in the foreign exchange markets to maintain the fixed parities should consists of *non sterilized* purchases and sales of foreign currencies. Such interventions cause changes (an increase in the case of a purchase of foreign exchange, a decrease in the case of a sale) in the national money supplies. Thus the currency substitution desires of the international agents, which are the cause of the excess demands and supplies of the various currencies, give rise to changes in the national money supplies while leaving the world money supply unchanged and the exchange rates fixed. Hence currency substitution will have no effect on the national economies.

This proposal has been criticized for various reasons. The first and foremost concerns the foundation itself of the proposal: currency substitution seems to be neither the main cause of exchange-rate volatility nor the main determinant of exchanges rates. Rather, it is *asset* substitution concerning assets denominated in the various currencies that appears to have a much greater role. Besides—the critics continue—by fixing nominal exchange rates no room is left for real exchange-rate adjustments. These adjustments might be required not so much because of differences in inflation rates (these could not occur according to the proposal), but to offset different productivity changes in the various countries. For a clear exposition of the various criticisms see Dornbusch (1988).

24.5.3 John Williamson's Target Zones

The idea of trying to combine the advantages of both fixed and flexible exchange rates while eliminating the disadvantages of both is at the basis of this proposal (antecedents of which can be found in the gliding wider band, see Sect. 3.3). John Williamson (1985, 1993) and his coworkers (see, for example, Edison et al. 1987) have elaborated this idea in much detail, giving rise to the target zone proposal, which is based on two main elements.

The first is the calculation of a fundamental equilibrium exchange rate (FEER), defined as that exchange rate 'which is expected to generate a current account surplus or deficit equal to the underlying capital flow over the cycle, given that the country is pursuing internal balance as best it can and not restricting trade for balance of payments reasons' (Williamson 1985; p. 14). Such a rate should be periodically recalculated to take account of the change in its fundamental determinants (for example the relative inflation rates); thus it must not be confused with a fixed central parity.

The second element is the possibility for the current exchange rate to float within wide margins around the FEER (at least $\pm 10\%$). These margins should be soft margins, namely there would be no obligation for the monetary authorities to intervene when the current exchange rate hits a margin; this is aimed at preventing destabilising speculation of the kind that was present in the Bretton Woods system.

The target zone proposal has been criticized for various reasons (see Frenkel and Goldstein 1986). We shall just point out two of them. The first is the difficulty of calculating the FEER. Even by using the most sophisticated econometric techniques and models, there remains a rather wide error margin. The second concerns the "credibility" of the target zone. A target zone is viable only if economic agents find it credible. The experience of the European Monetary System (see Sect. 21.1), which before August 1993 could be considered as a target zone with narrower margins, shows that the monetary authorities, even when there are monetary cooperation agreements like in the ERM, are helpless when credibility lacks.

The credibility problem has been theoretically studied in several models that can be divided into first-generation and second-generation models. First-generation models are based on simplified assumptions: economic agents are convinced that the exchange rate will not go beyond the margins and that the central parity will not be changed. Second-generation models are based on more general assumptions: economic agents assign non-zero probabilities to both events (the exchange rate going beyond the margins and the central parity being changed). For a survey of these models see Svensson (1992), De Arcangelis (1994), Kempa and Nelles (1999), Portugal Duarte et al. (2011).

24.5.4 The Tobin Tax

Tobin (1974, 1978, 1996) suggested a tax (with a modest rate) on all foreign-exchange transactions[1] as a means of "throwing sand in the wheels" of international speculation, namely of contrasting speculative capital flows without disturbing medium-long term "normal flows". Such a tax should be applied on all foreign-exchange transactions (both inflows and outflows) independently of the nature of the transaction. This is necessary to avoid the practically insurmountable enforcement problem of distinguishing between foreign exchange transactions for "speculative" purposes and for other purposes. Such a tax, in fact, given its modest rate would not be much of a deterrent to anyone engaged in commodity trade or contemplating the purchase of a foreign security for longer-term investing, but might discourage the spot trader who is now accustomed to buying foreign exchange with the intention of selling it a few hours later, and who would have to pay the tax every time he buys or sells foreign exchange. A tax of, say, 0.1 % (Tobin 1974, p. 89 originally suggested 1 % but later—1996, p. xvii—recommended a lower rate, between 0.25 and 0.1 %), namely 0.2 % on a round trip to another currency, would cost 48 % a year if transacted every business day, 10 % if every week, etc., but would be a trivial charge on commodity trade and long-term foreign investment.

The Tobin tax has raised much less discussion than it would deserve: in the words of the author himself, "it did not make much of a ripple. In fact, one may say that it sunk like a rock. The community of professional economists simply ignored it" (Tobin 1996, p. x). Raffer (1998) gives a historical survey of the debate on the Tobin tax as well as reasons (mainly political, in his opinion) why this debate has been so scanty. A book edited by ul Haq et al. (1996) contains several papers both pro and against (see also Frankel 1996; Various Authors 1995).

In the discussion on the Tobin tax two aspects are closely tied: enforceability and effects. Let us for a moment suppose that it is enforceable and examine its effects from the theoretical point of view.

The Tobin tax, when seen from the point of view of the agent engaged in international capital movements, is a tax on the relevant foreign exchange transactions, but can be translated into an equivalent tax on interest income. It is, in fact, equivalent either to a tax on foreign interest income at a rate which is an increasing function of the Tobin tax rate θ, or to a negative tax (i.e., a positive subsidy) on domestic interest income at a rate which is an increasing function of θ (see the Appendix). Hence it acts by suitably modifying the interest-rate differential which enters into the CIP and UIP calculations. Here the opinions become divergent: on the one hand there are those who maintain that its effects would be negligible (for example Davidson 1997; Reinhart 2000), on the other those who hold the opposite view (for example Eichengreen et al. 1995). Both views are, however, based on purely theoretical

---

[1] It should be clarified at the outset that in this chapter the Tobin tax is taken in its original meaning, namely as a tax on all *foreign-exchange transactions,* without considering extensions to all kinds of financial transactions, as is sometimes suggested.

models without any empirical testing. Two exceptions are Gandolfo and Padoan (1992) and Jeanne (1996).

Gandolfo and Padoan (1992) simulate the introduction of a Tobin tax in their estimated continuous-time macrodynamic econometric model of the Italian economy by suitably modifying the interest-rate differential on which capital flows depend. In such a way the effects of this introduction on the macroeconomic system can be examined taking account of all the dynamic interrelations between the relevant variables. The results of the simulations show that the introduction of a Tobin tax provides a crucial contribution to the stabilization of the system with full capital liberalization and speculative capital flows. However there is more to it than that. A Tobin tax allows the system to operate with a lower level of the domestic interest rate as it makes the constraint represented by the foreign interest rate less stringent. This obviously gives more room for domestic financial policy (in terms of e.g. the financing of the domestic public debt).

Jeanne (1996) builds a target zone model in which an optimizing government is faced with a trade-off between its foreign exchange and domestic objectives. The introduction of a Tobin tax would improve the credibility of the peg by relaxing the foreign exchange constraint and reducing the cost for the government of pegging the exchange rate. The author also applies the model to the French franc and shows that the stabilizing effect of a 0.1 % Tobin tax would have been quite sizeable.

Let us now come to the enforceability problem. The main practical argument against such a tax is well known: if not all countries adopt it, then the business would simply go to the financial centres where the tax is not present (tax havens). Minor arguments concern the possibility of loopholes (Garber and Taylor 1995, p. 179), which could however by counteracted as soon as they are discovered.

Thus the real problem is generality of application. However, it has been argued (Kenen 1996) that it would be sufficient if major dealing sites (the European Union, the United States, Japan, Singapore, Switzerland, Hong Kong, Australia, Canada and perhaps some other countries) implemented the tax, *charging punitive tax rates for transactions crossing the border between "Tobin countries" and "tax havens"*. But the amount of international agreement that this implementation would require makes it unlikely in the light of the obstacles to international policy cooperation treated above, Sect. 24.2.2. In fact, the formation of a "Tobin area" would ultimately be a manifestation of political will, just as the formation of the European Monetary Union has been (in spite of all its critics). And in any case a Tobin tax would be a much less traumatic measure than the introduction of capital controls, which are from time to time considered as a means of reducing international financial instability (Edwards 1999).

24.6 Appendix

24.6.1 International Policy Coordination

We consider a very simple two-country model (Frankel 1988), in which each country has three targets (real output, current account balance, inflation rate) and two policy instruments (fiscal and monetary policy). As is usual in this type of problem, the policy maker's welfare function is expressed as a loss function, namely as a function of the deviations of the actual values of the targets from an assumed ideal value or optimum. Furthermore, this loss function is assumed to be quadratic for mathematical convenience. The optimal policy problem consists in minimising the loss function. For simplicity's sake we remain in the static context rather than considering a dynamic optimization framework.

Let us denote by y, x, π the deviations of real output, current account, inflation rate from the respective optimum values in the home country; an asterisk will denote the rest-of-the world variables. Hence we have the two welfare (loss) functions

$$
\begin{aligned}
W &= \tfrac{1}{2}y^2 + \tfrac{1}{2}w_x x^2 + \tfrac{1}{2}w_\pi \pi^2, \\
W^* &= \tfrac{1}{2}y^{*2} + \tfrac{1}{2}w_{x*}^* x^{*2} + \tfrac{1}{2}w_{\pi *}^* \pi^{*2},
\end{aligned}
\tag{24.4}
$$

where the w's represent the weights placed on the current account and on the inflation rate relative to the weight placed on output. By g (government expenditure) and m (money supply) we denote the two instruments.

For our purposes it is irrelevant the kind of model(s) used to represent our two-country world system. It is enough to observe that, whichever the model(s) used, the interdependence of the two economies will give rise to the functions

$$
\begin{aligned}
y &= y(m, g, m^*, g^*), \\
x &= x(m, g, m^*, g^*), \\
\pi &= \pi(m, g, m^*, g^*), \\
y^* &= y^*(m, g, m^*, g^*), \\
x^* &= x^*(m, g, m^*, g^*), \\
\pi^* &= \pi^*(m, g, m^*, g^*),
\end{aligned}
\tag{24.5}
$$

where the nature of the functions and the various partial derivatives with respect to the policy variables (these derivatives are also called *policy multipliers*) depend on the underlying structural model(s). Hence the welfare functions can ultimately be written in terms of the policy instruments, namely

$$
\begin{aligned}
W &= \omega(m, g, m^*, g^*), \\
W^* &= \omega^*(m, g, m^*, g^*).
\end{aligned}
\tag{24.6}
$$

When there is only one policy instrument per country (or each country's policy instruments can be bundled into one), we have the diagrams described in the text (Figs. 24.1 and 24.2). Note, however, that in these diagrams the welfare is taken in the conventional way rather than as a loss function.

The marginal welfare effects of changes in the instruments are given by

$$\frac{\partial W}{\partial m} = y\frac{\partial y}{\partial m} + xw_x\frac{\partial x}{\partial m} + \pi w_\pi\frac{\partial \pi}{\partial m},$$

$$\frac{\partial W}{\partial g} = y\frac{\partial y}{\partial g} + xw_x\frac{\partial x}{\partial g} + \pi w_\pi\frac{\partial \pi}{\partial g},$$

$$\frac{\partial W}{\partial m^*} = y\frac{\partial y}{\partial m^*} + xw_x\frac{\partial x}{\partial m^*} + \pi w_\pi\frac{\partial \pi}{\partial m^*},$$

$$\frac{\partial W}{\partial g^*} = y\frac{\partial y}{\partial g^*} + xw_x\frac{\partial x}{\partial g^*} + \pi w_\pi\frac{\partial \pi}{\partial g^*},$$

$$\frac{\partial W^*}{\partial m} = y^*\frac{\partial y^*}{\partial m} + x^*w_x^*\frac{\partial x^*}{\partial m} + \pi^*w_\pi^*\frac{\partial \pi^*}{\partial m}, \qquad (24.7)$$

$$\frac{\partial W^*}{\partial g} = y^*\frac{\partial y^*}{\partial g} + x^*w_x^*\frac{\partial x^*}{\partial g} + \pi^*w_\pi^*\frac{\partial \pi^*}{\partial g},$$

$$\frac{\partial W^*}{\partial m^*} = y^*\frac{\partial y^*}{\partial m^*} + x^*w_x^*\frac{\partial x^*}{\partial m^*} + \pi^*w_\pi^*\frac{\partial \pi^*}{\partial m^*},$$

$$\frac{\partial W^*}{\partial g^*} = y^*\frac{\partial y^*}{\partial g^*} + x^*w_x^*\frac{\partial x^*}{\partial g^*} + \pi^*w_\pi^*\frac{\partial \pi^*}{\partial g^*},$$

The first-order conditions for an overall optimum require all these derivatives to equal zero ($\partial W/\partial m = 0$, and so on). In the case in which each country ignores the effects of its policy actions on the other country (a Nash noncooperative equilibrium), we need only the two first ($\partial W/\partial m = \partial W/\partial g = 0$) and the two last ($\partial W^*/\partial m^* = \partial W^*/\partial g^* = 0$) equations for the solution. But this solution is clearly suboptimal. With the values of m, g, m^*, g^* thus obtained, in fact, there is no reason why the other optimum conditions ($\partial W/\partial m^* = 0$, etc.) should be satisfied. This shows that the cooperative solution is superior.

The set of optimum conditions also clearly illustrates the obstacles to coordination. As Frankel observes, the actual process of policy coordination can be ideally divided into three stages. At the *first* stage, each country has to decide what policy changes by the other countries would best suit its interests, and—on the other hand—what changes in its policy it would be willing to concede in the other countries' interest. At the *second* stage, the countries must negotiate the distribution of the gains from coordination. At the *third* stage, rules must be set to enforce the agreement and avoid the free rider problem (see Sect. 24.2.2).

It is already at the first stage that three kinds of uncertainty might hamper the very beginning of the coordination process. These are (Frankel 1988):

(a) uncertainty about the initial position of the economy with respect to the ideal values (we all know that macroeconomic data are often revised by substantial amounts). In terms of our optimum equations, the initial values y, x, π are not known with certainty (the same holds, of course, for the rest of the world): the policy maker does not know where it stands exactly.
(b) uncertainty about the correct weights to be put on the various possible target variables. Different values of the w's might give rise to different results.
(c) uncertainty about the effects of a policy change (domestic or foreign) on the relevant variables (both domestic and foreign). This is the most serious kind of uncertainty, because it means that policy makers do not know the policy multipliers. Hence what should they ask the other countries for, and what should they give up in turn?

Disagreement over the "true" model is clearly one of the main reasons for type (c) uncertainty (see Sect. 24.2.2).

These three types of uncertainty are also present in domestic policy-making, but at the international level they are more severe, due to the international spillover effects ($\partial y / \partial m^*, \partial y^* / \partial m$, etc.). These are essential for the determination of the cooperative optimum, and are much more difficult to determine than the domestic effects of domestic policies. This difficulty is further aggravated by disagreement over the "true" model, especially when models embodying different visions of the functioning of an economy yield different results already at the qualitative level (i.e., at the level of the signs of the policy multipliers).

24.6.2 Target Zones

The target zone proposal has given rise to a burgeoning theoretical literature, aimed at evaluating the credibility of a target zone. This literature (for surveys see De Arcangelis 1994; Kempa and Nelles 1999) is very interesting but does not give any practical indication on how the central "equilibrium" exchange rate should be calculated—which, after all, is what the target zone proposal is all about. Such calculations (Williamson 1993; MacDonald and Stein 1999); can be performed using macroeconometric models, in which targets and instruments are clearly specified. Suppose, for example, that the targets are the level of output (y) and the balance of payments (B), and that the policy instruments are fiscal (G) and monetary (M) policy. The model endogenously determines y, B, as well as the exchange rate, r (on exchange-rate determination in macroeconometric models see Sect. 15.4). The reduced form of the model allows to express the endogenous variables in terms of the policy instruments, namely

$$
\begin{aligned}
y &= f(G, M, \mathbf{Z}_1), \\
B &= h(G, M, \mathbf{Z}_2), \\
r &= g(G, M, \mathbf{Z}_3),
\end{aligned}
\tag{24.8}
$$

where $\mathbf{Z}_i, i = 1, 2, 3$, are vectors (not necessarily distinct) of other exogenous variables. From the first and second equation, given the targets y^*, B^*, and assuming that the conditions on the Jacobian are satisfied, we can express G and M in terms of the targets, namely

$$
\begin{aligned}
G^* &= G(y^*, B^*, \mathbf{Z}_1, \mathbf{Z}_2), \\
M^* &= M(y^*, B^*, \mathbf{Z}_1, \mathbf{Z}_2).
\end{aligned}
\tag{24.9}
$$

If we now substitute these values in the third equation of (24.8), we get

$$
r^* = g(G^*, M^*, \mathbf{Z}_3) = g(y^*, B^*, \mathbf{Z}_1, \mathbf{Z}_2, \mathbf{Z}_3),
\tag{24.10}
$$

which is the expression of the FEER in terms of the targets and the set of exogenous variables of the model. This approach, which is based on the traditional Tinbergen target-instrument approach, can give rise to problems when the number of targets and (independent) instruments are unequal. Hence an optimal control approach is preferable.

In general, given a loss function of the policy maker (it is usually expressed in terms of the deviations of the actual paths of the variables from some desired path), the optimal control problem can be formulated as

$$
\begin{aligned}
&\min_{\{\mathbf{u}(t)\}} J = -\int_{t_0}^{t_1} I(\mathbf{x}, \mathbf{u}, t) \mathrm{d}t \\
&\text{sub } \dot{\mathbf{x}} = \mathbf{f}(\mathbf{x}, \mathbf{u}, t), \\
&\quad \mathbf{x}(t_0) = \mathbf{x}_0, \\
&\quad \mathbf{x}(t_1) = \mathbf{x}_1,
\end{aligned}
\tag{24.11}
$$

where I is the loss function, $\mathbf{x}$ the vector of endogenous variables, $\mathbf{u} \in U$ the vector of control variables belonging to a set U, t time (including other exogenous variables), and $\dot{\mathbf{x}} = \mathbf{f}(\mathbf{x}, \mathbf{u}, t)$ is the dynamic econometric model of the economy, whose initial and terminal states are $\mathbf{x}_0$ and $\mathbf{x}_1$. The specification of the econometric model as a set of differential equations is not just for mathematical convenience, since continuous time econometric models can actually be estimated. Note that, if the model is not already a first-order system in normal form (because in the specification there are higher-order time derivatives and in some equations the time derivatives also appear on the right hand side), it will have to be reduced to this form. This can be done by well-known mathematical methods (see, for example, Gandolfo 2009, Sect. 18.3).

The solution to this standard optimal control problem can be obtained through Pontryagin's maximum principle, which implies defining the Hamiltonian

$$H(\boldsymbol{\lambda}, \mathbf{x}, \mathbf{u}, t) = I(\mathbf{x}, \mathbf{u}, t) + \boldsymbol{\lambda} \ \mathbf{f}(\mathbf{x}, \mathbf{u}, t), \tag{24.12}$$

where $\boldsymbol{\lambda}$ is a vector of costate variables. One then applies the following conditions

$$\max_{\{\mathbf{u} \in U\}} H \qquad\qquad t_0 \le t \le t_1,$$

$$\dot{\mathbf{x}} = \ \partial H/\partial \boldsymbol{\lambda}, \quad \mathbf{x}(t_0) = \mathbf{x}_0, \tag{24.13}$$

$$\dot{\boldsymbol{\lambda}} = -\partial H/\partial \mathbf{x}, \quad \boldsymbol{\lambda}(t_1) = \ \mathbf{0}.$$

Note that the equations of motion for the state variables are the model, since $\partial H/\partial \boldsymbol{\lambda} = \mathbf{f}(\mathbf{x}, \mathbf{u}, t)$.

For our purposes, let us *suppose* that the exchange rate is a control variable (Gandolfo and Petit 1987, Sect. 6). This of course is *not* true, but what we are doing is to carry out an exercise to determine the optimal path of the exchange rate. This exercise is relevant for two reasons.

First, it can serve as a rational starting point for the determination of the central rate of the target zone. Instead of using the fixed target approach implicit in the Tinbergen method, the optimizing method allows for flexibility in the targets, which appear in the loss function with their appropriate weight reflecting the policy maker's preferences. Hence the optimal path of the exchange rate contains all the relevant information to constitute a guideline for the monetary authorities in their management of the actual exchange rate.

Second, if all countries participating in the target zone agreement perform the exercise (each using its preferred econometric model and its own welfare function—no need to impose a common model and a common welfare function!), it will be easy to determine the consistency of the cross rates emerging from the exercise. Consequently, it will be easy to determine the necessary adjustments so as to establish a mutually consistent set of "optimal" exchange rates as the basis for the target zone. The use by each national authority of its preferred econometric model and its own welfare function avoids all the problems—mentioned in Sect. 24.6.1—due to the lack of agreement on the reference model. Hence the use of multicountry models—unless accepted by all countries—is not advisable in this context.

24.6.3 The Tobin Tax

Let us first calculate the interest wedge brought about by the Tobin tax. For this purpose, we consider the UIP condition (see Sect. 4.2). The agent who owns an amount x of domestic currency can choose between

(a) investing his funds at home (earning the interest rate i_h), or
(b) converting them into foreign currency (and paying the Tobin tax at the rate θ) at the current spot exchange rate r, placing them abroad (earning the interest rate i_f), and converting them (principal plus interest accrued) back into domestic currency (and again paying the Tobin tax) at the end of the period considered, using the expected spot exchange rate ($\tilde{r}$) to carry out this conversion.

The no-profit condition is

$$x\left(1+i_h\right) = \left\{\left[\frac{x(1-\theta)}{r}\left(1+i_f\right)\right]\tilde{r}\right\}(1-\theta) = x\left(1+i_f\right)(1-\theta)^2\frac{\tilde{r}}{r}, \quad (24.14)$$

where the interest rates and expectations are referred to the same time horizon.

For θ sufficiently small (the values suggested by Tobin do satisfy this criterion) we have

$$(1-\theta)^2 \simeq \frac{1-\theta}{1+\theta}, \quad (24.15)$$

hence expression (24.14) becomes

$$x\left(1+i_h\right) = x\left(1+i_f\right)\frac{1-\theta}{1+\theta}\frac{\tilde{r}}{r}. \quad (24.16)$$

If we divide both members of (24.16) by $x(1+i_f)(1-\theta)/(1+\theta)$, and then subtract 1 from the result, we obtain

$$\frac{(1+\theta)(1+i_h) - (1-\theta)(1+i_f)}{(1-\theta)(1+i_f)} = \frac{\tilde{r}-r}{r},$$

whence

$$\frac{(1+\theta)i_h + 2\theta}{(1-\theta)(1+i_f)} - \frac{i_f}{(1+i_f)} = \frac{\tilde{r}-r}{r}.$$

Neglecting, as is usually done in UIP calculations (see Sect. 4.2), the denominator $(1+i_f)$, we finally have

$$\frac{(1+\theta)i_h + 2\theta}{1-\theta} - i_f = \frac{\tilde{r}-r}{r} \quad (24.17)$$

or

$$i_f = \frac{(1+\theta)i_h + 2\theta}{1-\theta} - \frac{\tilde{r}-r}{r}. \quad (24.18)$$

If we introduce a risk premium δ (see Sect. 4.3), then the condition would be

$$i_f = \frac{(1+\theta)i_h + 2\theta}{1-\theta} - \frac{\tilde{r} - r}{r} - \delta. \tag{24.19}$$

The r.h.s. of Eq. (24.18) or (24.19) gives the value that the foreign interest rate (referred to the appropriate time interval) has to exceed for capital outflows to be profitable.[2]

Finally, it is interesting to show that the Tobin tax on foreign transactions is equivalent to a negative tax (i.e., a *subsidy*) on funds invested at home. In this latter case, in fact, we would have

$$x(1+i_h)(1+\sigma) = \left[\frac{x}{r}(1+i_f)\right]\tilde{r},$$

where σ is the subsidy rate. Hence

$$x(1+i_h) = x\left(1+i_f\right)(1+\sigma)^{-1}\frac{\tilde{r}}{r}. \tag{24.20}$$

If we compare (24.14) and (24.20), and let

$$\frac{1}{1+\sigma} = (1-\theta)^2,$$

we can obtain the subsidy rate equivalent to the Tobin tax, which turns out to be

$$\sigma = \frac{1 - (1-\theta)^2}{(1-\theta)^2}. \tag{24.21}$$

For example, a Tobin tax at the rate 0.2 % would correspond to a subsidy at the rate of about 0.4 %.

24.6.3.1 A Simple Model

Gandolfo and Padoan (1992) used Eq. (24.19) to simulate the introduction of the Tobin tax in a context in which both long-term fundamentalist investors and short-term speculators are present. The model is a set of 24 nonlinear stochastic differential equations, and it cannot even be summarized here. We shall instead treat the very simple model by Frankel (1996), which allows us to show the stabilizing effects of the Tobin tax.

Assume that the spot exchange rate in log form, $s = \ln r$, is determined according to the portfolio view (see Sect. 15.3.3), namely by the ratio of the stock of domestic assets (relative to foreign assets), m in log form, to the relative demand for domestic

---

[2]Frankel (1996; p. 22) gives a slightly different formula, in which the term $(\tilde{r} - r)/r$ is neglected.

assets, d in log form:

$$s = m - d + u, \tag{24.22}$$

where u is a stochastic term. A fraction w of participants in the foreign exchange market are long-term investors, who hold regressive expectations, namely expect the current exchange rate to move towards its long-run equilibrium value $\bar{s}$, so that

$$E_i(\mathrm{depr}) = -\alpha(s - \bar{s}), \alpha > 0, \tag{24.23}$$

where $E_i(\mathrm{depr})$ is the exchange-rate depreciation expected by investors.

The remaining fraction $(1 - w)$ are short-term speculators, who expect the exchange rate to diverge, as along a speculative bubble path, at the rate β

$$E_s(\mathrm{depr}) = \beta(s - \bar{s}), \beta > 0. \tag{24.24}$$

Thus the demand d is given by

$$d = wd_i + (1 - w)d_s,$$

where d_i, d_s are investors' and speculators' demands, respectively.

We now assume that these demands depend on expectations according to each group's elasticity of demand with respect to their expectations. If f_i is the elasticity of investors' demand, we shall have $d_i = f_i\alpha(s - \bar{s})$, because investors will demand foreign (domestic) assets if they expect a depreciation (appreciation). Similar considerations show that $d_s = -f_s\beta(s - \bar{s})$. It follows that total demand for domestic assets is

$$d = wf_i\alpha(s - \bar{s}) - (1 - w)f_s\beta(s - \bar{s}) = [wf_i\alpha - (1 - w)f_s\beta](s - \bar{s}). \tag{24.25}$$

From (24.22) and (24.25) we obtain

$$s = \frac{m + [wf_i\alpha - (1 - w)f_s\beta]\bar{s} + u}{1 + [wf_i\alpha - (1 - w)f_s\beta]}. \tag{24.26}$$

Thus the variability of the exchange rate is

$$\mathrm{Var}(s) = \frac{\mathrm{Var}(m + u)}{\{1 + [wf_i\alpha - (1 - w)f_s\beta]\}^2}. \tag{24.27}$$

From Eq. (24.27) it is easy to see that a Tobin tax, by lowering f_s (the responsiveness of speculators to their expectations) and/or raising w (the number of investors in the market) would contribute to lower the variance of the exchange rate.

Using a different kind of model, Reinhart (2000) shows that a Tobin tax might make the price of domestic equities more variable. The question is still open.

24.6.3.2 A More Sophisticated Model

In Sect. 15.8.4 we have presented the Federici and Gandolfo (2012) model of exchange rate determination in a possibly chaotic framework. This model is well suited to examine the introduction and consequences of the introduction of a Tobin tax. This has been done by Gandolfo (2015), whose results are summarised here.

As shown above, Eq. (24.19), a Tobin tax influences the interest rate differential on which capital flows are based. We do not have the interest rate differential as a relevant variable in the Federici and Gandolfo (2012) model; however, since in this model the non-linear function g_{sc} incorporates possible transaction costs, denoting by θ a generic transaction cost we have that

$$\frac{\partial g'_{sc}}{\partial \theta} < 0.$$

Thus the introduction of a Tobin tax has the effect of decreasing the value of the adjustment speed n. A lower adjustment speed means that, ceteris paribus, speculative capital flows are lower.

It can be further shown that a Tobin tax has the effect of reducing speculators' profit. Profit (in terms of domestic currency) over a certain time interval is given by the change in the speculative stocks over that interval. The *instantaneous* change in speculative stocks is equal to the excess demand for domestic currency which, in turn, is equal to the opposite of the excess demand for foreign exchange multiplied by the exchange rate. Thus, with reference to a time interval $s_1 \mapsto s_2, s_2 > s_1$, profit is given by

$$P = -\int_{s_1}^{s_2} E_{sc}(t) r(t) dt.$$

Equivalently, profit can be defined as the excess of the sums collected by speculators over the sums paid out by them over a given time interval, all measured in domestic currency. The sums collected in each instant are given by the amount of foreign exchange supplied, $S(t)$, multiplied by the exchange rate, and the sums paid out in each instant are given by the amount of foreign exchange demanded, $D(t)$, multiplied by the exchange rate. With reference to a given time interval we have

$$P = \int_{s_1}^{s_2} O_{sc}(t) r(t) - \int_{s_1}^{s_2} D_{sc}(t) r(t) = -\int_{s_1}^{s_2} E_{sc}(t) r(t) dt.$$

Given the definition of $E_{sc}(t)$ in our model, we have[3]

$$P = -\int_{s_1}^{s_2} [nb_1 \dot{r}(t) + nb_2 \dot{r}(t)] r(t) dt.$$

It is easy to see that a decrease in n (caused by a Tobin tax) would determine a decrease in P. In fact,

$$\frac{\partial P}{\partial n} = -\int_{s_1}^{s_2} [b_1 \dot{r}(t) + b_2 \dot{r}(t)] r(t) dt > 0,$$

since we assume that $P > 0$. Thus the qualitative effect of a change in n on P is completely determined, without having to calculate the integrals. Lower profits can discourage speculative activity.

Let us finally analyse the effects of a change in n on the model's parameters. We have

$$\frac{\partial a_2}{\partial n} = \frac{\partial}{\partial n}(\frac{m-\alpha}{nb_2}) = -\frac{m-\alpha}{n^2 b_2} < 0,$$

$$\frac{\partial a_3}{\partial n} = \frac{\partial}{\partial n}(-\frac{\beta}{nb_2}) = \frac{\beta}{n^2 b_2} > 0,$$

$$\frac{\partial a_4}{\partial n} = \frac{\partial}{\partial n}(\frac{g}{nb_2}) = -\frac{g}{n^2 b_2} > 0,$$

$$\frac{\partial a_5}{\partial n} = \frac{\partial}{\partial n}(-\frac{m}{nb_2}) = \frac{m}{n^2 b_2} > 0.$$

Recall that we have been able to pinpoint the sign of g (which was uncertain) through the estimation of a_4 (Federici and Gandolfo 2012).

Given this, we have that a *decrease* in n causes:

- an increase in a_2,
- a decrease in a_3,
- a decrease in a_4,
- a decrease in a_5.

---

[3]To ensure that profit is not merely notional, namely not actually realised, it is necessary to consider an interval over which the speculative stocks of foreign exchange are the same at the beginning and at the end. Thus s_1, s_2 must be such that

$$\int_{s_1}^{s_2} E_{sc}(t) dt = 0.$$

In conclusion, the main results are that the introduction of a Tobin tax:

(a) causes a decrease in speculators' activity (which confirms what Tobin had in mind);
(b) makes the model less liable to chaotic motions. In fact, in our previous paper (Federici and Gandolfo 2012; Sect. 5) we found that when a_2 is set close to zero, the model becomes unstable and chaotic motions arise. Thus the decrease in n due to a Tobin tax, by causing an increase in a_2 as shown above, is a stabilizing factor (which is, in our opinion, an interesting new result).

References

Acharya, V. V., & Richardson, M. (2009). Causes of the financial crisis. *Critical Review: A Journal of Politics and Society, 21*, 195–210.

Acharya, V. V, & Schnabl, P. (2010). Do global banks spread global imbalances? Asset backed commercial paper during the financial crisis of 2007–09. *IMF Economic Review, 58*, 37–73.

Adrian, T., & Shin, H. (2010). Liquidity and leverage. *Journal of Financial Intermediation, 19*, 418–437.

Allen, F., & Carletti, E. (2010). An overview of the crisis: Causes, consequences, and Solutions. *International Review of Finance, 10*, 1–26.

Borio, C., & Zhu, H. (2012). Capital regulation, risk-taking and monetary policy: A missing link in the transmission mechanism? *Journal of Financial Stability, 8*, 236–251.

Broner, F., & Didier, T., & Erce, A., & Schmukler, S.L. (2013). Gross capital flows: Dynamics and crises. *Journal of Monetary Economics, 60*, 113–133.

Brunnermeier, M. K. (2009). Deciphering the 2007–08 liquidity and credit crunch. *Journal of Economic Perspectives, 23*, 77–100.

Bryant, R. C. (1995). *International coordination of national stabilization policies.* Washington (DC): Brookings Institution.

Clarida, R. H. (2000). *G-3 exchange-rate relationships: A review of the record and of proposals for change.* Essays in International Economics No. 219. Princeton University, International Economics Section.

Cooper, R. N. (1989). International cooperation in public health as a prologue to macroeconomic cooperation. In: R. N. Cooper et al. (Eds.), *Can nations agree? Issues in international economic cooperation* (pp. 178–254). Washington (DC): Brookings Institution.

Daniels, J. P., & Vanhoose, D. D. (1998). Two-country models of monetary and fiscal policy: What have we learned? What more can we learn? *Open Economies Review, 9*, 263–282.

Davidson, P. (1997). Are grains of sand in the wheels of international finance sufficient to do the job when boulders are often required? *Economic Journal, 107*, 671–686.

De Arcangelis, G. (1994). Exchange rate target zone modeling: Recent theoretical and empirical contributions. *Economic Notes, 23*, 74–115.

Dornbusch, R. (1988). Doubts about the McKinnon standard. *Journal of Economic Perspectives, 2*, 105–112.

Edison, H. J., Miller, M. H., & Williamson, J. (1987). On evaluating and extending the target zone proposal. *Journal of Policy Modeling, 9*, 199–227.

Edwards, S. (1999). How effective are capital controls? *Journal of Economic Perspectives, 13*, 65–84.

Eichengreen, B., Tobin, J., & Wyplosz, C. (1995). Two cases for sand in the wheels of international finance. *Economic Journal, 105*, 162–172.

Eichengreen, B., Mody, A., Nedeljkovic, M., & Sarno, L. (2012). How the subprime crisis went global: Evidence from bank credit default swap spreads, *Journal of International Money and Finance, 31*, 1299–1318.

Federici, D., & Gandolfo, G. (2012). The euro/dollar exchange rate: Chaotic or non-chaotic? A continuous time model with heterogeneous beliefs. *Journal of Economic Dynamics and Control, 36*, 670–681.

Fischer, S. (1999). On the need for an international lender of last resort. *Journal of Economic Perspectives, 13*, 85–104.

Frankel, J. A. (1988). *Obstacles to international macroeconomic policy coordination.* Princeton Studies in International Finance No. 64, International Finance Section, Princeton University.

Frankel, J. A. (1996). How well do foreign exchange markets function: might a Tobin tax help? NBER Working Paper 5422. In: M. ul Haq, I. Kaul, & I. Grunberg (Eds.), *The Tobin tax: Coping with financial instability* (pp. 41–81). Oxford: Oxford University Press.

Frankel, J. A., & Rockett, K. (1988). International macroeconomic policy coordination when policymakers do not agree on the true model. *American Economic Review, 78*, 318–340.

Frankel, J. A., & Saravelos, G. (2012). Can leading indicators assess country vulnerability? Evidence from the 2008–09 global financial crisis. *Journal of International Economics, 87*, 216–231.

Fratianni, M., Salvatore D., & Savona, P. (Eds.) (1999). *Ideas for the future of the international monetary system.* Dordrecht: Kluwer Academic Publishers (reprinted from *Open Economies Review*, Vol. 9, special supplemental issue).

Frenkel, J. A., & Goldstein, M. (1986). A guide to target zones. *IMF Staff Papers, 33*, 633–673.

Gandolfo, G. (2009). *Economic dynamics* (4th ed.). Berlin, Heidelberg, New York: Springer.

Gandolfo, G. (2015). The Tobin tax in a continuous-time non-linear dynamic model of the exchange rate. *Cambridge Journal of Economics, 39*, 1629–1643.

Gandolfo, G., & Petit, M. L. (1987). Optimization in continuous time and policy design in the Italian economy. *Annales d'Économie et de Statistique, 6/7*, 311–333.

Gandolfo, G., & Padoan, P. C. (1992). Perfect capital mobility and the Italian economy. In: E. Baltensperger & H. -W. Sinn (Eds.) (1996), *Exchange rate regimes and currency unions* (pp. 26–61). London: Macmillan.

Garber, P., & Taylor, M. P. (1995). Sand in the wheels of foreign exchange markets: A sceptical note. *Economic Journal, 105*, 173–180.

Ghosh, A. R., & Masson, P. R. (1991). Model uncertainty, learning, and the gains from coordination. *American Economic Review, 81*, 465–479.

Gorton G. B. (2012). *Misunderstanding financial crises: Why we don't see them coming.* New York: Oxford University Press.

Jeanne, O. (1996). Would a Tobin tax have saved the EMS? *Scandinavian Journal of Economics, 98*, 503–520.

Hamada, K. (1974). Alternative exchange rate systems and the interdependence of monetary policies. In: R. Z. Aliber (Ed.), *National monetary policies and the international financial system.* Chicago: Chicago University Press.

Hamada, K. (1985). *The political economy of international monetary interdependence.* Cambridge (Mass.): MIT Press.

Hirata, H., Kose, M. A., Otrok, C., & Terrones, M. E. (2012). Global house price fluctuations: Synchronization and determinants. In: F. Giavazzi, & K. West (Eds), *NBER International Seminar on Macroeconomics* (2012), National Bureau of Economic Research, 119–166.

IMF (2002). Assessing sustainability. Policy Development and Review Department No. 166. http://www.imf.org/external/np/pdr/sus/2002/eng/052802.htm.

Kempa, B., & Nelles, M. (1999). The theory of exchange rate target zones. *Journal of Economic Surveys, 13*,173–210.

Kenen, P. (1996). The feasibility of taxing foreign exchange transactions. In: M. ul Haq, I. Kaul, & I. Grunberg (Eds.) (1996), *The Tobin tax: Coping with financial instability* (pp. 109–128). Oxford: Oxford University Press.

Krugman, P. (2008). *The return of depression economics and the crisis of 2008*. New York: W. W. Norton & Company.

Lane, P. R. (2001). The new open economy macroeconomics: A survey. *Journal of International Economics, 54*, 235–266.

MacDonald, R., & Stein, J. L. (Eds.) (1999). *Equilibrium exchange rates*. Dordrecht: Kluwer Academic Publishers.

Martinez Oliva, J. C. (1991). One remark on spillover effects and the gains from coordination. *Oxford Economic Papers, 43*, 172–176.

McKibbin, W. (1997). Review of: R.C. Bryant, International coordination of national stabilization policies. *Journal of Economic Literature, 35*, 783–784.

McKibbin, W. J., & Sachs, J. D. (1991). *Global linkages: Macroeconomic interdependence and cooperation in the world economy*. Washington (DC): Brookings Institution.

McKinnon, R. I. (1988). Monetary and exchange rate policies for international financial stability. *Journal of Economic Perspectives, 2*, 83–103.

McKinnon, R. I. (1997). *The rules of the game*. Cambridge (Mass.): MIT Press.

Mendoza, E. G., & Quadrini, V. (2010). Financial globalization, financial crisis and contagion. *Journal of Monetary Economics, 57*, 24–39.

Milesi-Ferretti, G. M., & Tille, C. (2011). The great retrenchment: international capital flows during the global financial crisis. *Economic Policy, 26*, 285–342.

Mishkin, F. S. (2009). Is monetary policy effective during financial crises? *American Economic Review, 99*, 573–77.

Obstfeld, M., & Rogoff, K. (2009). Global imbalances and the financial crisis: products of common causes. Centre for Economic Policy Research Discussion Paper No. 7606.

Ocampo, J. A. (2015). Reforming the international monetary and financial architecture. In: J. A. Alonso & J. A. Ocampo (Eds.), *Global governance and rules for the post-2015 era*. London: Bloomsbury.

Oudiz, G., & Sachs, J. (1985). International policy coordination in dynamic macroeconomic models. In: W. H. Buiter & R. C. Marston (Eds.), *International economic policy coordination* (pp. 274–319). Cambridge (UK): Cambridge University Press.

Phillips, P. C. B., & Yu, J. (2011). Dating the timeline of financial bubbles during the subprime crisis. *Quantitative Economics, 2*, 455–491.

Pilbeam, K. (2013). *International finance* (4th ed.). London: Palgrave Macmillan.

Portugal Duarte, A., Sousa Andrade, J., & Duarte, A. (2011). Exchange rate target zones: A survey of the literature. *Journal of Economic Surveys, 27*, 247–268.

Raffer, K. (1998). The Tobin tax: Reviving a discussion. *World Development, 26*, 529–538.

Rajan, R. G. (2011). *Fault lines: How hidden fractures still threaten the world economy*. New York: Princeton University Press.

Reinhart, V. R. (2000). How the machinery of international finance runs with sand in its wheels. *Review of International Economics, 8*, 74–85.

Reinhart, C. M., & Rogoff, K. (2009). The aftermath of financial crises. *American Economic Review, 99*, 466–72.

Reinhart, C. M., & Rogoff, K. (2011). From financial crash to debt crisis. *American Economic Review, 101*, 1676–1706.

Reinhart, C. M., & Rogoff, K. (2015). Financial and sovereign debt crises: Some lessons learned and those forgotten. *Journal of Banking and Financial Economics, 2*, 5–17.

Rogoff, K. (1999). InternationalInstitutions for reducing global financialInstability. *Journal of Economic Perspectives, 13*, 21–42.

Schularick, M., & Taylor, A. M. (2012). Credit booms gone bust: Monetary policy, leverage cycles, and financial crises, 1870–2008. *American Economic Review, 102*, 1029–61.

Sneddon Little, J., & Olivei, G. P. (Eds.) (1999). *Rethinking the international monetary system*. Boston: Federal Reserve Bank of Boston, Conference Series No. 43.

Stein, J. L. (2006). *Stochastic optimal control, international finance and debt crises*. Oxford: Oxford University Press.

Svensson, L. E. O. (1992). An interpretation of recent research on exchange rate target zone. *Journal of Economic Perspectives, 6*, 119–144.

Taylor, J. B. (2009a). Economic policy and the financial crisis: An empirical analysis of what went wrong. *Critical Review: A Journal of Politics and Society, 21*, 341–364.

Taylor, J. B. (2009b). Getting off track: How government actions and intervention caused, prolonged and worsened the financial crisis. Stanford Hoover Institution Press.

Tirole, J. (2011). Illiquidity and all its friends. *Journal of Economic Literature, 49*, 287–325.

Tobin, J. (1974). *The new economics one decade older.* Princeton: Princeton University Press.

Tobin, J. (1978). A proposal for international monetary reform. *Eastern Economic Journal, 4*, 153–159.

Tobin, J. (1996), Prologue. In: M. ul Haq, I. Kaul, & I. Grunberg (Eds.), *The Tobin tax: Coping with financial instability* (pp. ix-xviii). Oxford: Oxford University Press.

ul Haq, M., Kaul, I., & Grunberg, I. (Eds.) (1996). *The Tobin tax: Coping with financial instability.* Oxford: Oxford University Press.

Various Authors (1995). Policy forum: Sand in the wheels of international finance. *Economic Journal, 105*, 161–192.

Williamson, J. (1985). *The exchange rate system* (revised ed.). Washington (DC): Institute for International Economics.

Williamson, J. (1993). *Equilibrium exchange rates: An update.* Washington (DC): Institute for International Economics.

Index

© Springer-Verlag Berlin Heidelberg 2016

667

G. Gandolfo, *International Finance and Open-Economy Macroeconomics*,
Springer Texts in Business and Economics, DOI 10.1007/978-3-662-49862-0

CPSIA information can be obtained
at www.ICGtesting.com
Printed in the USA
LVOW13*2022141217

559742LV00015B/316/P